Design Computing and Cognition '10

John S. Gero
Editor

Design Computing and Cognition '10

Springer

Editor
John S. Gero
Krasnow Institute for Advanced Study
University Avenue 4400
22030 Fairfax Virginia
USA
E-mail: john@johngero.com

ISBN 978-94-007-0509-8 e-ISBN 978-94-007-0510-4

DOI 10.1007/978-94-007-0510-4

Typesetting: Data supplied by the authors

Cover Design: Scientific Publishing Services Pvt. Ltd., Chennai, India

Printed on acid-free paper

9 8 7 6 5 4 3 2 1

springer.com

Preface

Design research has two strands exemplified by the terms science of design and design as science. Both are commonly referred to as design science. The former studies designing scientifically and the latter treats designing as a science. The ways that designing can be studied scientifically include both computational modeling and cognitive modeling. Many computational models of designing are not founded directly on results of cognitive studies. They are founded on conjectures about designing using concepts from artificial intelligence with its focus on ways of representation and on processes that support simulation and generation. Artificial intelligence continues to provide an environmentally rich paradigm within which design research based on computational constructions can be carried out. Increasingly design cognition research, founded on concepts from cognitive science. It provides tools and methods to study human designers in both laboratory and practice settings. It is beginning to allow us to test the claims being made about designing whether carried out individually or in teams and to study the effects of the introduction of novel technologies into the acts of designing.

Just as design cognition is starting to provide evidence-based support for computational studies, so cognitive neuroscience is starting to provide support for cognitive acts in designing.

Design thinking, the label given to the unique acts of designing, has become as paradigmatic view that has transcended the discipline of design and is now widely used in business and elsewhere. As a consequence there is an increasing interest in design research and government agencies are gradually increasing funding of design research, and increasing numbers of engineering schools are revising their curricula to emphasize design. This is because of the realization that design is part of the wealth creation of a nation and needs to be better understood and taught. The continuing globalization of industry and trade has required nations to re-examine where their core contributions lie if not in production efficiency. Design is a precursor to manufacturing for physical objects and is the precursor to implementation for virtual objects. At the same time, the need for sustainable

J.S. Gero (ed.): *Design Computing and Cognition'10*, pp. v–vi.

development is requiring design of new products and processes, and feeding a movement towards design innovations and inventions.

This conference series aims at providing a bridge between the fields of design computing and design cognition. The confluence of these two fields continues to provide the foundation for further advances in each of them.

The papers in this volume are from the *Fourth International Conference on Design Computing and Cognition (DCC'10)* held at the University of Stuttgart, Germany. They represent the state-of-the-art of research and development in design computing and design cognition. They are of particular interest to researchers, developers and users of advanced computation in design and those who need to gain a better understanding of designing.

In these proceedings the papers are grouped under the following nine headings, describing both advances in theory and application and demonstrating the depth and breadth of design computing and design cognition:

Design Cognition
Framework Models in Design
Design Creativity
Lines, Planes, Shape and Space in Design
Decision-Making Processes in Design
Knowledge and Learning in Design
Using Design Cognition
Collaborative/Collective Design
Design Generation

There were 125 full paper submissions to the conference of which 38 were accepted. Each paper was extensively reviewed by three reviewers drawn from the international panel of 115 active reviwers listed on the next pages. The reviewers' recommendations were then assessed before the final decision on each paper was taken. Thanks go to them, for the quality of these papers depends on their efforts.

Mercedes Paulini worked to turn the variegated submissions into the conference format to produce a unified volume, special thanks go to her.

July 2010

John S. Gero
Krasnow Institute for Advanced Study

Contents

Preface v

List of Reviewers xiii

Part I: Design Cognition

A Comparison of Cognitive Heuristics Use between Engineers and Industrial Designers 3
Seda Yilmaz, Shanna R. Daly, Colleen M. Seifert, Richard Gonzalez

Studying the Unthinkable Designer: Designing in the Absence of Sight 23
Ann Heylighen

Design Heuristics: Cognitive Strategies for Creativity in Idea Generation 35
Seda Yilmaz, Colleen M. Seifert, Richard Gonzalez

An Anthropo-Based Standpoint on Mediating Objects: Evolution and Extension of Industrial Design Practices 55
Catherine Elsen, Françoise Darses, Pierre Leclercq

Part II: Framework Models in Design

Beyond the Design Perspective of Gero's FBS Framework 77
Gaetano Cascini, Luca Del Frate, Gualtiero Fantoni, Francesca Montagna

A Formal Model of Computer-Aided Visual Design 97
Ewa Grabska, Grażyna Ślusarczyk

Design Agents and the Need for High-Dimensional Perception 115
Sean Hanna

A Framework for Constructive Design Rationale 135
Udo Kannengiesser, John S. Gero

Part III: Design Creativity

The Curse of Creativity 157
David C. Brown

Enabling Creativity through Innovation Challenges: The Case of Interactive Lightning 171
Stefania Bandini, Andrea Bonomi, Giuseppe Vizzari, Vito Acconci

Facetwise Study of Modelling Activities in the Algorithm for Inventive Problem Solving ARIZ and Evolutionary Algorithms 189
Céline Conrardy, Roland de Guio, Bruno Zuber

Exploring Multiple Solutions and Multiple Analogies to Support Innovative Design 209
Apeksha Gadwal, Julie Linsey

Creative and Inventive Design Support System: Systematic Approach and Evaluation Using Quality Engineering 229
Hiroshi Hasegawa, Yuki Sonoda, Mika Tsukamoto, Yusuke Sato

Part IV: Line, Plane, Shape, Space in Design

Line and Plane to Solid: Analyzing Their Use in Design Practice through Shape Rules 251
Gareth Paterson, Chris Earl

Interactions between Brand Identity and Shape Rules 269
Rosidah Jaafar, Alison McKay, Alan de Pennington, Hau Hing Chau

Approximate Enclosed Space Using Virtual Agent 285
Aswin Indraprastha, Michihiko Shinozaki

Associative Spatial Networks in Architectural Design: Artificial Cognition of Space Using Neural Networks with Spectral Graph Theory 305
John Harding, Christian Derix

Part V: Decision-Making Processes in Design

Comparing Stochastic Design Decision Belief Models: Pointwise versus Interval Probabilities 327
Peter C. Matthews

A Redefinition of the Paradox of Choice 347
Michal Piasecki, Sean Hanna

Rethinking Automated Layout Design: Developing a Creative Evolutionary Design Method for the Layout Problems in Architecture and Urban Design 367
Sven Schneider, Jan-Ruben Fischer, Reinhard König

Applying Clustering Techniques to Retrieve Housing Units from a Repository 387
Álvaro Sicilia, Leandro Madrazo, Mar González

Part VI: Knowledge and Learning in Design

Different Function Breakdowns for One Existing Product: Experimental Results 405
Thomas Alink, Claudia Eckert, Anne Ruckpaul, Albert Albers

A General Knowledge-Based Framework for Conceptual Design of Multi-disciplinary Systems 425
Yong Chen, Ze-Lin Liu, You-Bai Xie

Learning Concepts and Language for a Baby Designer 445
Madan Mohan Dabbeeru, Amitabha Mukerjee

Organizing a Design Space of Disparate Component Topologies 465
Mukund Kumar, Matthew I. Campbell

Part VII: Using Design Cognition

Imaging the Designing Brain: A Neurocognitive Exploration of Design Thinking 489
Katerina Alexiou, Theodore Zamenopoulos, Sam Gilbert

A Computational Design System with Cognitive Features Based on Multi-objective Evolutionary Search with Fuzzy Information Processing 505
Michael S. Bittermann

Narrative Bridging 525
Katarina Borg Gyllenbäck, Magnus Boman

Generic Non-technical Procedures in Design Problem Solving: Is There Any Benefit to the Clarification of Task Requirements? 545
Constance Winkelmann, Winfried Hacker

Virtual Impression Networks for Capturing Deep Impressions 559
Toshiharu Taura, Eiko Yamamoto, Mohd Yusof Nor Fasiha, Yukari Nagai

Part VIII: Collaborative/Collective Design

Scaling Up: From Individual Design to Collaborative Design to Collective Design 581
Mary Lou Maher, Mercedes Paulini, Paul Murty

Building Better Design Teams: Enhancing Group Affinity to Aid Collaborative Design 601
Michael A. Oren, Stephen B. Gilbert

Measuring Cognitive Design Activity Changes during an Industry Team Brainstorming Session 621
Jeff W.T. Kan, John S. Gero, Hsien-Hui Tang

Part IX: Design Generation

Interactive, Visual 3D Spatial Grammars 643
Frank Hoisl, Kristina Shea

A Graph Grammar Based Scheme for Generating and Evaluating Planar Mechanisms 663
Pradeep Radhakrishnan, Matthew I. Campbell

A Case Study of Script-Based Techniques in Urban Planning 681
Anastasia Koltsova, Gerhard Schmitt, Patrik Schumacher, Tomoyuki Sudo, Shipra Narang, Lin Chen

Complex Product form Generation in Industrial Design: A Bookshelf Based on Voronoi Diagrams 701
Axel Nordin, Damien Motte, Andreas Hopf, Robert Bjärnemo, Claus Christian Eckhardt

A Computational Concept Generation Technique for Biologically-Inspired, Engineering Design 721
Jacquelyn K.S. Nagel, Robert B. Stone

First Author Email Address 741

Author Index 743

List of Reviewers

Henri Achten, Czech Technical University, Czech Republic
Tom Arciszewski, George Mason University, USA
Uday Athavankar, IIT Bombay, India
Petra Badke-Schaub, TU Delft, Netherlands
Stefanie Bandini, University of Milano-Bicocca, Italy
Adeliade Blavier, University of Liege, Belgium
Lucienne Blessing, University of Luxembourg, Luxembourg
Frances Brazier, TU Delft, Netherlands
Dave Brown, Worcester Polytechnic Institute, USA
Jon Cagan, Carnegie Mellon University, USA
Luisa Caldas, Instituto Superior Técnico, Portugal
Hernan Casakin, Ariel University Center of Samaria, Israel
Amaresh Chakrabarti, Indian Institute of Science, India
Scott Chase, Aarlborg University, Denmark
Per Christiansson, Aarlborg University, Denmark
John Clarkson, University of Cambridge, UK
Mark Clayton, Texas A&M University, USA
Graham Coates, Durham University, UK
Nathan Crilly, University of Cambridge, UK
Umberto Cugini, Polytecnico Milan, Italy
Steve Culley, University of Bath, UK
Francoise Darses, CNRS, France
Bharat Dave, University of Melbourne, Australia
Bauke De Vries, TU Eindhoven, Netherlands
Ellen Do, Georgia Institute of Technology, USA
Andy Dong, University of Sydney, Australia
Jose Duarte, Instituto Superior Técnico, Portugal
Alex Duffy, University of Strathclyde, UK
Chris Earl, Open University, UK
Claudia Eckert, Open University, UK
Georges Fadel, Clemson University, USA
Susan Finger, CMU, USA
Gerhard Fischer, University of Colorado, USA
Xavier Fischer, ESTIA, France
Christian Freksa, University of Bremen, Germany
Gerhard Friedrich, University of Klagenfurt, Austria

Renate Fruchter, Stanford University, USA
Haruyuki Fujii, Tokyo Institute of Technology, Japan
Kikuo Fujita, Osaka University, Japan
John Gero, George Mason University, USA
Pablo Gervas, Universidad Complutense de Madrid, Spain
Ashok Goel, Georgia Institute of Technology, USA
Gabirella Goldschmidt, Technion, Israel
Andres Gomez De Silva, ITAM, Mexico
Mark Gross, Carnegie Mellon University, USA
David Gunaratnam, University of Sydney, Australia
Balan Gurumoorthy, Indian Institute of Science, India
Winfried Hacker, TU Dresden, Germany
John Haymaker, Stanford University, USA
Ann Heylighen, KU Leuvan, Belgium
Urs Hirschberg, TU Graz, Austria
Koichi Hori, University of Tokyo, Japan
Walter Hower, Albstadt-Sigmaringen Universit, Germany
Jan Yin, University of Southern California, USA
Leo Joskowicz, Hebrew University of Jerusalem, Israel
Richard Junge, Technical University of Munich, Germany
Julie Jupp, University of Technology Sydney, Australia
Jeff Kan, Taylor's University College, Malaysia
Udo Kannengiesser, NICTA, Australia
Yong Se Kim, Sungkyunkwan University, Korea
Terry Knight, MIT, USA
Branko Kolarevic, University of Calgary, Canada
Maria Kozhevnikov, George Mason University, USA
Ramesh Krishnamurti, Carnegie Mellon University, USA
Bimal Kumar, Glasgow Caledonian University, UK
Pierre Leclercq, University of Liege, Belgium
John Lee, University of Edinburgh, UK
Noel Leon, ITESM, Mexico
Andrew Li, Chinese University of Hong Kong, China
Hod Lipson, Cornell University, USA
Peter Lloyd, Open University, UK
Ardeshir Mahdavi
Mary Lou Maher, University of Sydney, Australia
Bob Martens, Technical University of Vienna, Austria
Janet McDonnell, University of the Arts - London, UK
Alison McKay, University of Leeds, UK
Harald Meerkamm, University Erlangen-Nuremberg, Germany
Anja Meier, Cambridge University, UK
Douglas Noble, University of Southern California, USA
Rivka Oxman, Technion, Israel

Panos Paplambros, University of Michigan, USA
Rafael Perez y Perez, UNAM, Mexico
Rabee Reffat, KFUPM, Saudi Arabia
Yoram Reich, Tel Aviv University, Israel
Duska Rosenberg, RHUL, UK
Stephan Rudolph, University of Stuttgart, Germany
Somwrita Sarkar, University of Sydney, Australia
Gerhard Schmitt, ETH Zurich, Switzerland
Chris Schunn, University of Pittsburgh, USA
Kristi Shea, TU Munich, Germany
Li Shu, University of Toronto, Canada
Greg Smith, CSIRO, Australia
Steve Smith, Texas A&M University, USA
Tim Smithers, Fatronik, Spain
Ricardo Sosa, ITESM, Mexico
Ram Sriram, NIST, USA
Martin Stacey, de Mountford University, UK
Rudi Stouffs, Technical University of Delft, Netherlands
Masaki Suwa, Keio University, Japan
Hsien-Hui Tang, National Taiwai University of Science and Technology, Taiwan
Ming Xi Tang, Hong Kong Polytechnic University, China
Toshiharu Taura, Kobe University, Japan
Jan Treur, Vrije Universiteit Amsterdam, Netherland
Barbara Tversky, Columbia University, USA
Andrew Vande Moere, University of Sydney, Australia
Noe Varga-Hernandez, University of Texas El Paso, USA
Willemien Visser, INRIA, France
Christian Weber, Ilmenau University of Technology, Germany
Rob Woodbury, Simon Fraser University, Canada

DESIGN COGNITION

A comparison of cognitive heuristics use between engineers and industrial designers
Seda Yilmaz, Shanna Daly, Colleen Seifert and Richard Gonzalez

Studying the unthinkable designer
Ann Heylighen

Cognitive heuristics in design: Instructional strategies in idea generation
Seda Yilmaz, Colleen Seifert and Richard Gonzalez

An anthropo-based standpoint on mediating objects: Evolution and extension of industrial design practices
Catherine Elsen, Françoise Darses and Pierre Leclercq

A Comparison of Cognitive Heuristics Use between Engineers and Industrial Designers

Seda Yilmaz, Shanna R. Daly, Colleen M. Seifert, and Richard Gonzalez
University of Michigan, USA

The present study focuses on an exploration and identification of design heuristics used in the ideation process in both industrial designers and engineering designers. Design heuristics are cognitive strategies that help the designer generate novel design concepts. These cognitive heuristics may differ based on the design problem, the context defined, and designers' preferences.

In a think-aloud protocol study, five engineers and five industrial designers were asked to develop product concepts for a novel problem. We analyzed these protocols to document and compare industrial designers' and engineers' concept generation approaches, and the use of design heuristics in their proposed solutions. The results show evidence of heuristics use, and that they are effective in generating diverse, creative, and practical concepts. Some differences were observed between the designers from the two domains in their approaches to the design problem and in the design heuristics used in generating alternatives.

Introduction

How do designers explore design spaces? Does the concept generation phase differ between engineers and industrial designers? Both groups are often called upon to create new products and innovative redesigns; yet, their training in creative techniques differs greatly. In industrial design, training emphasizes repeated experience with design concepts along with a critique process. In engineering, greater emphasis is typically placed on solving technical issues within a design; however, training also includes

J.S. Gero (ed.): *Design Computing and Cognition'10*, pp. 3–22.

creativity techniques, as engineers are often called upon to create novel designs [1].

Past studies have examined general approaches used in ideation [2] and [3], and the importance of design heuristics is well recognized [4]; however, it is still unclear how multiple and varied ideas are generated. What cognitive strategies do designers really use, and how do these strategies differ between the domains of engineering and industrial design?

In previous work, we found evidence for specific design heuristics that supported designers in exploring the space of potential designs, leading to the generation of varied and creative solutions [5], [6]. This was particularly noted for heuristics that connect the design context to specific concept transformations [7]. Design heuristics may guide the designer's exploration of possible solutions by varying overall strategies, product characteristics, or element modifications. An example heuristic is "*Adding on, taking out, or folding away components when not in use*," evident when the designer minimized added components by creating concepts integrated within an existing product. Because design heuristics appear to support the generation of multiple and diverse concepts, it seems likely that explicit training in effective heuristics may support the development of ideation skills for designers.

Design Heuristics

The aim of this research was to explore and identify both the types of design heuristics and the frequency of their use in the ideation process. By including both industrial designers and engineers, we hoped to learn about the generality of design heuristics across these domains. Following Newell and Simon [8], we define design as occurring within a "design space" consisting of all feasible designs. Some of these potential designs are easy to consider because they involve simple combinations of known features, or involve already-known elements. However, a designer may never consider some of the possible solutions within this space because they do not naturally come to mind. An alternative process to assist in this exploration is the application of cognitive strategies, defined as "design heuristics," that help to move the designer into new parts of the design space. The key to innovative solutions, then, is to apply different heuristics to assist in creating novel designs within this potential design space [5] [6].

Research in psychology describes heuristics as simple, efficient rules to explain decision making, judgments, and problem solving, especially when faced with complex problems with vague information [9]. Behavioral research shows that experts can utilize heuristics effectively, and suggests that their use of heuristics is one feature that distinguishes them from novices (e.g., [10]). Design heuristics may vary with regard to where and how they are applied, how they impact a design or trigger moves within the design space as a whole, and the amount of time invested in applying them. The usefulness of a particular heuristic will depend on the problem context, so that by definition, there is no determinate heuristic that will lead to a definitive solution.

We propose design heuristics differ from other approaches used in idea generation. Some existing approaches, such as brainstorming, brainwriting, and checklists, are open-ended to allow naturally occurring ideas to flow, often prompted by criteria, constraints, or other ideas. Other approaches have proposed more directed approaches, which can also be called heuristics; specifically, SCAMPER [11], Synectics [12], and TRIZ [13]. These heuristic approaches have a similar foundation in that they provide specific prompts to support the generation of new concepts. However, the heuristics proposed in SCAMPER and Synectics are quite general (e.g., "amplify a feature"), while the heuristics proposed in TRIZ focus more specifically on mechanical devices and systems and are most applicable in later stages of design. None of these approaches have observed heuristics in studies of designers, nor have they been empirically validated. Thus, the present study aims to examine the heuristics that arise in idea generation.

In previous work [14], we characterized three types of cognitive design heuristics that prompted different types of movements in the design space:

- *Local* heuristics define characteristics and relationships of design elements within a single concept, for example, adjusting functions by moving the product's parts.
- *Transitional* heuristics provide ways to transform an existing concept into a new concept, for example, substituting a form.
- *Process* heuristics prompt a designer's general approach to idea generation; for example, changing the context to give rise to new aspects of the product. They serve as cognitive tools used to initially propose ideas by directing the designer's navigation of the solution space.

These heuristics serve as a base set of hypotheses for the types of heuristic use we expect to see in both engineering and industrial designers as they create novel designs. The questions addressed in this study were: How does heuristic use lead designers to potential solutions in the design space? Does heuristics use differ between the two types of designers? How can evidence of heuristics guide design education across both disciplines?

Experimental Approach and Research Questions

Our design heuristics approach suggests that there are cognitive strategies that can aid in navigating and exploring design spaces. Therefore, for both groups of designers, we hypothesized that the application of design heuristics during the creative process would enhance the variety, quality, and creativity of potential designs generated during the ideation stage. We proposed that specific design heuristics would help designers explore new types of potential designs, leading to the generation of innovative solutions. The design task selected was open-ended and involved creating a new product, with very little information about constraints.

In the study, we compared those with industrial design backgrounds to engineers. We expected participants within industrial design to have learned how to generate concepts for vaguely defined design problems, and so would exhibit more creative and diverse design behavior. On the other hand, we expected engineers who have learned to solve technical problems would exhibit more practical and methodical design behavior. Specifically, we hypothesized that, compared to the industrial designers, engineers will: (1) have more technical and practical, but less creative design concepts, and (2) have less diverse concepts since they may have less experience with open-ended design tasks.

Participants

Participants were recruited from professional conferences and a mid-western university. In this study, we report a set of ten case studies. The list of participants with their age, gender, and experience level is shown in Table 1. These ten cases represent a range in domain experience for both fields, as well as a range in performance through the sessions. Within these case studies, we hope to find some suggestive differences between industrial designers and engineers that may be addressed in future studies.

Method

In a think-aloud protocol study, we documented designers' approaches to generating concepts in a single design task. The problem involved designing "a solar-powered cooking device that was inexpensive, portable, and suitable for family use." The design problem statement also specified some design criteria and constraints, but it was intended to serve as an open-ended problem with many potential solutions. The instructions prompted participants to generate diverse creative ideas for the solutions.

Participants were given thirty minutes for the task. After ten minutes, the experimenter provided a few paragraphs of additional information about transferring solar energy into thermal energy in case participants did not feel they had the technical knowledge to proceed. This information encouraged the designers to move past the need for specific technical information for their solutions. Throughout the session, the experimenter asked the participants to keep talking if they became silent at any point.

Table 1 Participants' age, gender, and design-related experience

Participant	Age	Gender	Design-related Experience
Ind. Designer 1	27	Female	2+ years in industry, 2+ years in design graduate school
Ind. Designer 2	29	Male	1+ years in industry, 5+ years in design graduate school
Ind. Designer 3	21	Female	Senior in design school
Ind. Designer 4	21	Female	Senior in design school
Ind. Designer 5	20	Male	Junior in design school
Engineer 1	53	Male	25+ years in industry, 4 years in design management graduate school
Engineer 2	27	Male	4+ years in engineering graduate school
Engineer 3	25	Male	2+ years in engineering graduate school
Engineer 4	23	Female	1+ years in engineering graduate school
Engineer 5	22	Male	Senior in engineering school

The designers' drawings were captured, along with their verbal comments, using an electronic audio recording pen, which also captured the movements of the pen during sketching. After the task was over, participants were asked to review their drawing, and to verbally describe the concepts they had generated, how they moved from one concept to another, and their approaches to ideation. Finally, they were asked to provide demographic information, and rate their performance.

Verbal data from the experimental sessions were transcribed to supplement the audio and visual sketching data, and all data was analyzed for evidence of heuristic use. Two evaluators, one experienced in industrial design and the other in engineering design, examined all of the protocols. The goal of the analysis was to characterize the various decision patterns evident in participants' performance on the task. Thus, the analysis included identifying each concept generated as a separate idea, categorizing characteristics of the solution concepts generated, determining the number and diversity of the concepts, and determining specific design heuristic evident in the concepts. These features were coded for each concept, between concepts, and over the experimental session. The coders worked independently, and then resolved any disagreement through discussion. Initial interrater agreement was 80% across the protocols.

In majority of the cases, heuristics were not consciously articulated by the participants; however, heuristic use was evident in comments such as, "I'll use both a magnifying glass and a mirror, since I'm not sure if the energy will be enough to cook the food." This was evaluated as indicating the use of a "*Using multiple components to achieve one function*" heuristic. The sketches also provided separate evidence of heuristic use in the specified characteristics of the products, the product contexts drawn, and the relationship of these concepts to other solutions. Thus, both verbal and visual (sketched) data were considered for any evidence of heuristic use.

Additional coding was performed on each concept using two criteria: creativity and practicality. First, questions characterizing creativity and practicality for the given design task were identified by the two evaluators, and then each concept was coded by both raters individually. Some of the questions considered for rating creativity included: "Does it address a design criterion unique from the other designers' concepts? Is it considerably different from an existing well-known product? Does it use unexpected materials?" For practicality, some of the questions included: "Is it easy to use? Is it going to work? Is it portable?" The questions were used as guidelines, and the ratings completed in a subjective manner [15].

Results

The results reported here include a discussion of the types of solutions generated, instances of local, transitional, and process heuristics observed, and the relationship of the heuristics used to the diversity of the concepts generated, along with creativity and practicality. In each of these analyses, emphasis was given to differences between protocols from industrial designers and engineers. Because the sample size is small, comparisons across the two groups are likely to be limited in their generalizability.

Types of Concepts Generated

Major elements and key features of the concepts were identified in terms of functionality, form, and user-interaction, Table 2. This allowed us to see the diversity of concepts generated from within this design space. For example, solutions could direct sunlight using mirrors, maintain heat by creating a closed product with a clear lid (so the sunlight could get in), or include straps so the product could attach to the user. Alternatively, a solution could use a magnifying glass to direct sunlight, an insulated box to maintain heat, or a foldable container for easy transport. These were each coded as distinct concepts.

Table 2 Solution characteristics for the solar-powered cooker problem

Diversity Criteria	Examples	Industrial Designers	Engineers
Way of Directing Sunlight	1. Magnifying glass / Lens	10	11
	2. Reflective surface / Mirror / Aluminum foil	9	14
Method of	1. Closed product	6	11

Maintaining Heat	2. Glass / Plastic lid	3	5
	3. Insulation	1	8
	4. Metal	0	2
Method of Cooking or Warming Food	1. Direct sunlight	20	20
	2. Hot surface	5	1
	3. Incorporating fluids	0	5
	4. Solar panels	4	2
	5. Steam / Smoking / Fire	1	2
Product Materials	1. Flexible material	2	4
	2. Open surface	11	7
	3. Pot	6	7
	4. Tube	0	3
Approach to Compactness or Portability	1. Attachment to user	1	1
	2. Carrying case	0	1
	3. Detachable components	3	7
	4. Foldable components	9	4
	5. Rollable components	1	3
	6. Separate pieces	2	10
	7. Wheels	1	0
Other Features	1. Attached to pre-existing things in the environment	0	2
	2. Adjustable settings	6	8
	3. Stand	2	4
	4. Thermometer	1	1
Total number of concepts generated		**28**	**23**

The number of concepts was defined, in part, through the use of cues from participants as they indicated the beginning and ending to a given concept. New concepts were also evident in drawings when moving to a new illustration of an idea. However, number of concepts generated alone does not necessarily reflect the diversity of the concepts, as similar concepts or evolution of one concept could appear at any point within the session. Thus, we report the number of *different* concepts generated by each participant. Criteria used to classify the content of designs and understand the diversity of the space is presented in Table 2.

A difference in technical knowledge was evident in comparing the engineers' solutions to the industrial designers' solutions. For example, the five engineers used insulation more frequently, while the five industrial designers' solutions did not commonly consider the need to maintain the heat. The engineers also created closed surface products more often, while the industrial designers were more likely to have open surfaces for cooking, which would not allow heat to be maintained as effectively.

Another engineering solution was to use multiple mirrors to collect sunlight, reflecting concern about the function of the product, while only one of the industrial designers included this feature. In most cases, industrial designers selected a hot surface as the method of cooking, with open surface designs. In other concepts, engineers generated solutions incorporating fluids like water or oil for cooking; while none of the industrial designers did so. This may reflect a lack of technical knowledge among industrial design compared to the engineers, which may have resulted in more frequent use of existing products as models.

Another interesting difference was that engineers more often used separate pieces and detachable components, while industrial designers more often created single unit products that folded inside. Because of these dissimilarities, it is possible these two groups of designers could benefit from sharing their different approaches with each other.

Evidence of Heuristic Use

The main focus of this study was to document how subjects moved through the design space; that is, the ways they approached concept generation, developed solutions, and transitioned between design concepts. The coding for the evidence of heuristics began with a base set of heuristics from TRIZ principles [13], and from our previous work [7]. We adapted some of these, and added other heuristics to better describe the changes in concepts apparent in the protocols. Table 3 presents the local and transitional design heuristics coded in the concepts generated by the ten participants. Local and transitional heuristics are listed together because the same heuristic can be used for defining the relationship of the elements within one design concept, or as a transition in moving from one concept to a new one. Whether the heuristic was observed as a local or transitional heuristic, or both, is indicated in Table 3.

Table 3 (continues on next page) Partial list of Local (LH) and Transitional (TH) heuristics identified in the content analysis of concepts generated by engineers and industrial designers

Heuristic	Heuristic Description	LH	TH
Adjust functions by moving parts	By moving the product's parts, the user can achieve a secondary function	X	X
Attach components with different functions	Adding a connection between two parts that function independently	X	
Attach the product to another existing item	Utilizing an existing product as part of the function of the new product	X	
Attach the product to the user	The user becomes part of the product's function	X	
Change the configuration of elements	Performing different functions based on the orientation or the angle of the design elements in the product	X	X
Change the geometrical form	Using different geometrical forms for the same function and criteria		X
Compartmentalize	Separating the product into distinct parts or compartments with different functions	X	
Cover	Overspreading the surface of the product with another component to utilize the inner surface	X	X
Combine into a system	Connecting parts with different functions to develop a multi-stage process to achieve the overall goal	X	
Detach / Attach	Making the individual parts attachable /detachable for additional flexibility	X	X
Elevate	Raising up either the entire product or its parts from a lower place to a higher one	X	
Fold	Creating relative motion between parts by hinging, bending, or creasing to condense the size	X	X
Nest	Placing a component inside another identical component or an existing product, entirely or partially	X	
Offer optional components	Providing additional components that can change the function or adjustability	X	X
Provide sensory feedback to the user	Returning some of the output of a system as input to provide control in the process	X	

Heuristic	Heuristic Description	LH	TH
Repeat	Dividing single continuous parts into two or more elements, or repeating the same design element multiple times, in order to generate modular units	X	X
Replace solid material with flexible	Changing a product's material into a flexible one for creating different structural and surface characteristics	X	X
Roll	Revolving a part or the entire product over on a center point or a supporting surface	X	
Rotate around a pivot point	Changing an object's function by manipulating its geometrical surfaces around an axis	X	
Scale	Changing the size of a feature of the product	X	X
Split	Taking a piece of the previous concept to generate a new concept		X
Substitute	Replacing the material, form, or a design component with another to achieve the same function		X
etc			

Table 4 presents the process heuristics observed. Process heuristics are those applied by the designers to the idea generation process as a whole, and reflect a designer's general approach to ideation within the session. The process heuristics observed do not include all possible heuristics for any design task; however, they represent a set of possible heuristics appropriate for idea generation for this design problem.

The protocols demonstrated evidence of all three types of heuristics (local, transitional, and process heuristics) found in our previous work [14]. In sum, heuristics were identified 259 times (local heuristics=216, transitional heuristics=29, and process heuristics=14). The total number of local heuristics per concept ranged from 1 to 10, and multiple heuristics were observed in most of the concepts (47 of 51). Concepts with only one local heuristic seemed to be either very simple solutions (i.e. a plate capturing sunlight), or were vague and undefined. Concepts not emerging from transitional heuristic use indicated that the designer had abandoned the prior concepts and began a new search for a different concept, either with or without the use of a process heuristic.

Table 4 Process Heuristics (PH) identified in the content analysis of concepts generated by engineers and industrial designers

Process Heuristic	Heuristic Explanation
Brainwriting	Using naturally occurring ideas, without judgment, as starting points for concepts
Constraint Prioritizing	Putting more emphasis on certain criteria than others and using the emphasized criteria to focus and guide concept development
Contextualizing	Changing the context in which the product would be used, and using that context to inspire a concept that satisfied the nature of the context
Elaborating	Building on a foundational concept by increasing the details of the concept
Evaluating	Placing value to a concept and generating additional concepts by building on what is seen as effective or adjusting problems found in the evaluation of the concept
Problem Restructuring	Shifting or redefining what the actual problem is and generating products that satisfy the identified real problem
Redesigning	Re-designing existing products with similar functions
Simplifying	Generating and building on the simplest way to solve the problem
Using a Morphological Approach	Identifying different ways of achieving each function the product needs to perform and combining them in different ways to generate concepts

For both engineers and industrial designers, one of the most commonly applied local heuristics was "*Attaching components that have different functions*". For example, in Figure 1, Engineer 5 attached the handle to the pot and the lens, connecting both, and Industrial Designer 4 attached a continuous mirror inside the pot, wrapping it entirely.

Fig. 1. Examples using "*Attach components that have different functions*"

The other most common local heuristics for both groups were, "Covering", "*Elevating*", "*Folding*", and "*Repeating*". The least frequent local heuristics were "*Stacking*", "*Wrapping*", "*Attaching the product to the user*", and "*Using the environment as part of the product*". These differences appear to arise from the specific functions within the design problem. Thus, the context of the problem seemed to impact heuristic use. Applying the last two heuristics could have had a notable impact on the function of the product; however, we did not observe the designers utilizing these heuristics. The most common transitional heuristic for designers from both domains was "*Changing the configuration*". The designers simply rearranged the orientation of the design elements to structure new concepts.

There was little difference in the total number of heuristics used by each group; however, we did observe differences in the type of heuristic used. Engineers more often used "*Repeating*" (11 vs. 6) as a local heuristic, repeating elements such as mirrors to enhance the function of capturing sunlight. Many engineers mentioned their concerns about the adequacy of the energy produced for cooking food, which may have led them to repetition. "*Combine into a system*" was also used by engineers, but not by industrial designers (5 vs. 0). This might also be related to engineers' common practice of systems design as part of their education and experience. Engineers used the heuristic "*Use multiple sources to achieve one function*" in 8 of the 23 concepts that they generated, while this heuristic was evident in only one of the concepts that an industrial designer created. The reason may be that engineers were concerned about function, and continuously evaluated whether or not their concepts would work.

Industrial designers, on the other hand, used "*Elevate*" more frequently than engineers (11 vs. 6), perhaps because they were considering the interaction between the user and the product, which would lead to adjusting the height of the product for the user. In fact, industrial designers included representations of users in multiple concepts, while no engineers did so. The other heuristic more commonly used by industrial designers was "*Attach the product to another item*". Perhaps some of the industrial designers may not have had the technical knowledge or confidence to feel comfortable generating a novel concept from scratch, and do built from a related product. Local heuristics were evident in greater numbers than transitional ones; so, rather than developing early ideas further, they appeared to generate new ideas from scratch each time.

Finally, process heuristics, used as problem solving strategy for the entire session, were observed for some of the designers, which served to move them throughout the design space. For example, one designer strategically chose to consider different potential foods for heating in the oven, resulting in generating several new designs. Based on this data set, there were no distinctions in types of process heuristics used by designers from both disciplines.

Characterizing Design across Sessions

To understand the results, it is helpful to follow individual designers through their session, and explore how heuristics were applied in their work. The following paragraphs describe a sample of engineers' and industrial designers' protocols, including those who generated many diverse concepts and those who produced just one. We highlight the use of local, transitional, and process heuristics in these examples.

Engineer 1 generated seven diverse concepts, Figure 2. For his first concept, he chose a container that could be transported by users to a larger community gathering. The second concept was a large Fresnel lens, adjustable to the angle of the sun as well as to the best angle for cooking. For his next concept, he extended the previous one by segmenting his original lens into four separate lenses. The fourth concept was a spit cooker, which utilized a lens to focus on a line of heat, rather than a point. The fifth concept was a double boiler, consisting of a system pumping hot water from a boiler into an outer pot. Concept 6 was a synthesis of previous concepts: the design combined a double boiler with a Fresnel lens. The seventh concept was a blanket with reflectors and a drying rack. The reflective blankets are lightweight, allowing them to be transported easily, while serving as a windbreak. The eighth concept proposed a smoking chamber. It also included a Fresnel lens, and had two box-like structures on top of the other. The final concept was a three-stage boiler, comprised of a solar heater to warm up water to be utilized for steaming or boiling food.

To generate these diverse concepts, Engineer 1 used multiple process heuristics. One that he applied was the heuristic "*Contextualizing*". For most of his concepts, he first suggested a type of food, and then generated a concept that could cook that food. For example, he said "Other things to eat. We've got shish-kabobs, jerked meat, the dried herbs, the soups and things; um, let's see." He also emphasized different constraints from the problem as he worked; in concept 3, he focused on "maximizing the intensity of the sunlight", while in concept 7, he emphasized the constraints of being "inexpensive and portable".

A number of local heuristics were also documented in the concepts Engineer 1 generated. For example, in concept 3, he applied "*Adjust functions by moving the product's parts*", as the angles of the lenses on all four sides could be altered to change the amount of sunlight directed onto the food. He also applied "*Repeat*", as he added multiple lenses to direct the sunlight. Engineer 1 also used transitional heuristics; for example, he moved from concept 5 to 6 by using "*Cover*" as the transitional heuristic, where he covered the container with a Fresnel lens.

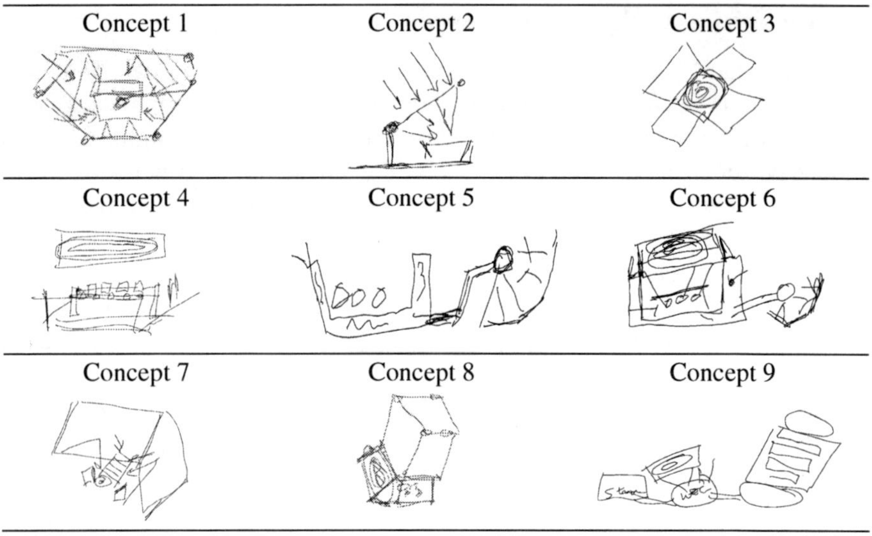

Fig. 2. Sequential concepts generated by Engineer 1

Industrial Designer 2 generated four concepts; all were considered diverse, Figure 3. In the first concept, he described a context in which the user was a hiker, and designed an integrated backpack with a heat pot attached to it. The second concept was a barbeque using solar panels on one side, and a cooking surface on the other. Solar energy was captured when the panels were unfolded fully, and the product was used when it was folded. The next concept used multiple mirrors to direct sunlight onto one part of the product that could be attached to another part for cooking. The location of those components could be switched; the heat unit was on top of the pot for collecting sunlight, and switched below it for providing heat from the bottom when cooking. His final concept was a set of small black cubes that could be utilized to absorb heat, and their orientation could be changed for cooking according to users' needs.

In this ideation process, we observed evidence of the local heuristic '*Change the configuration of elements*" in his third concept, where two components of the product were switched from top to bottom depending on the function to be achieved (cooking or trapping heat). With no evidence of transitional heuristics, Industrial Designer 2 seemed to use an approach of sampling from very different ideas within the problem space. The only consistency among his design ideas was capturing the heat during one time period and using it at another. He also used "*Contextualizing*" as a process heuristic throughout his ideation process. Using this heuristic allowed this designer to compose diverse ideas for very different settings.

Fig. 3. Sequential concepts generated by Industrial Designer 2

We saw a similar approach in Engineer 2's protocol. He seemed to leave each concept behind and started a new one rather than continue to transform a current concept. Each of this engineer's concepts was an expanded idea from an explicit "brainstorming" session he conducted at the beginning of the session.

In contrast, Industrial Designer 3 limited her generation to only one concept; however, she then worked through 7 iterations of that concept, Figure 4. The designer began by attaching two existing components to each other -- a magnifying glass and a griddle -- to create a surface with focused sunlight. In her second concept, she transformed the magnifying glass to a square magnifying glass attached to the tray. In the following concept, she made the lens height adjustable, and, in the forth concept, she added sides to it to maintain the heat more effectively. She then considered portability by adding a rigid handle, which was changed to a flexible handle in concept 6. In addition to all of the features included in the previous versions of the concept, the final concept also included an attachment that held utensils and a spout for draining fluids from the cooking surface.

Industrial Designer 3 applied "*Elaborate*" as a process heuristic, and further developed the first concept in succeeding concepts to explore the design space. She was successful in utilizing transitional heuristics to move about and explore within this concept's range. For example, from concept 2 to concept 3, she used transitional heuristics, "*Adjust functions by moving the product's parts,*" and "*Fold*", and then from concept 5 to concept 6, the transitional heuristic, "*Replace solid material with flexible*", as she changed the material of the handle. Table 5 displays the local heuristics within each concept. The total number of local heuristics increased in each concept while maintaining the changes already introduced. The designer did not leave the heuristics she used in the previous concepts, but instead carried them along, iterating on the concept and adding more to further the design.

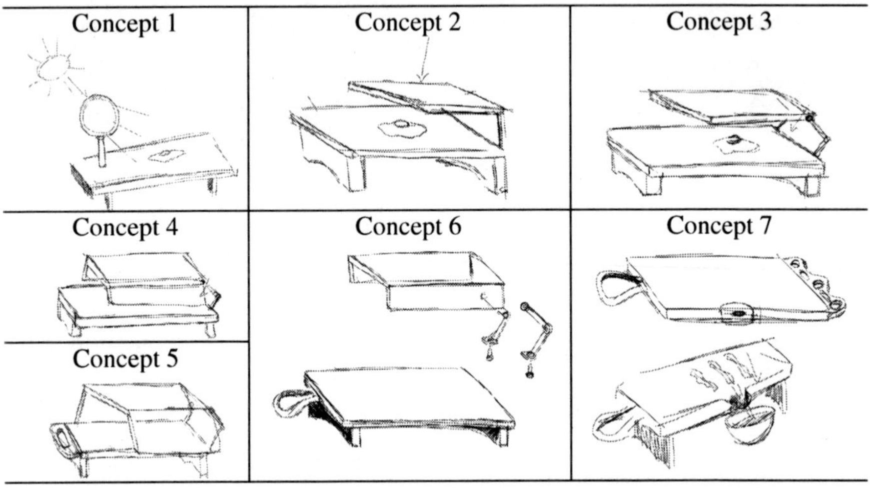

Fig. 4. Sequential concepts generated by Industrial Designer 3

Another example of a designer who generated only a few concepts was Engineer 3, who generated two concepts with no apparent process heuristics. Her first concept was a parabolic reflector in which the shape of the reflector allows the sun to be targeted onto a specific point. The second was a water-heating device in which heat would be stored in water that is heated by the sun.

In this case, two separate ideas are evident, but their generation did not lead to further transformations of concepts, nor to more novel ones. Heuristic use was not evident in these design concepts, suggesting a relationship between the use of design heuristics and the generation of multiple, diverse concepts.

Table 5 Local heuristics observed in Industrial Designer 3's concepts

	C1	C2	C3	C4	C5	C6	C7
Attach components that have different functions	•	•	•	•	•	•	•
Elevate	•	•	•	•	•	•	•
Compartmentalize		•	•	•	•	•	•
Adjust functions by moving the products' parts			•	•	•	•	•
Fold			•	•	•	•	•
Rotate around a pivot point			•	•	•	•	•
Cover				•	•	•	•
Detach or Attach					•	•	•
Replace solid material with flexible						•	•
Offering optional components							•
Repeating							•

Design Heuristics and Concept Diversity, Creativity, and Practicality

We next examined how the use of heuristics throughout the session related to the number and variety of design concepts produced by each individual designer. Figure 5 displays the number of diverse (meaning different in content) concepts for each participant, and characterizes how the use of multiple process heuristics was associated with those concepts. However, as noted above, those with the most diverse concepts were not necessarily the designers who generated creative solutions. There were examples in the case studies that prove both designers with diverse concepts and designers following a single concept through multiple iterations could produce creative outcomes in design.

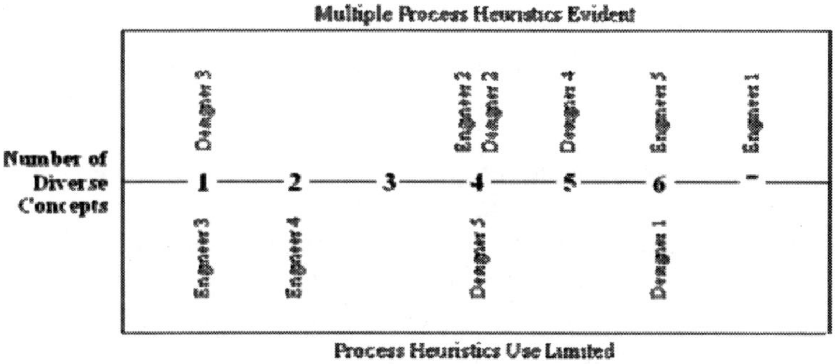

Fig. 5. Number of diverse concepts generated per participant

Comparing the engineers to the industrial designers, the average ratings show there were no mean differences between the engineers and industrial designers on either creativity or practicality (*t*s < 1). This is not surprising because there is relatively little power (five subjects in each group). However, across the whole sample of difference design concepts, the averaged creativity (r=.54) and practicality (r=.53) scores correlate highly with the number of heuristics identified in each ($p<.01$ for both criteria). That is, the designs with more heuristics observed were also rated as more creative and practical.

These correlations are driven almost entirely by the industrial designers' data. This suggests that engineers may have also used other means, such as their technical knowledge, to generate alternative concepts, whereas industrial designers tended to use heuristics to identify different solutions. This result also suggests that the industrial designers were not blocked by their lack of technical knowledge; instead, they may have used design heuristics to compensate for this lack of knowledge.

Discussion

The results provide empirical evidence of heuristic use in design, and show that heuristics are effective in generating diverse concepts. Design heuristics may, at times, be sufficient to stimulate divergent thinking. Furthermore, the study reveals some differences between these two types of designers in how they approached this open-ended, novel design problem. Specifically, we found that engineers produced a more diverse set of designs from among all of the concepts generated. Industrial designers, however, generated more design concepts in the same period. Nevertheless, the number and ways in which they used heuristics was very similar in the two groups, which suggests that design heuristics may be an effective means of ideation in each of the two design domains.

The differences observed in industrial and engineers may arise directly from the type of training emphasized during the educational process, and the types of problems typically experienced during training. Despite lacking technical knowledge, industrial designers generated more concepts. On the other hand, engineers' solutions were more diverse, and they used more diverse criteria (see Table 2). Their concepts were also more detailed, and provided more technical information about their practicality. Industrial designers structured the context and approached the problem from a user perspective, such as families versus individual hikers, the product's use in kitchens versus backyards, and the product as a single entity versus attached to existing products (such as a grill or stove).

It also appeared that engineers did not propose concepts unless they felt it was viable. They constantly evaluated their solutions according to functional principles and practicality in use. On the other hand, industrial designers were not as concerned about carrying the idea to a realistic level. Since this ideation stage was more about generating as many concepts as they could, industrial designers seemed more comfortable proposing concepts with less regard for how they would function.

The results of this empirical study must be considered in context. Certainly, there were differences in experience among the engineers and industrial designers participating in the study, and the number of participants was small. Second, the study was specific to one design task, constraining the generalization of findings across other tasks. Another limitation is that the design task was an isolated, one-time, half an hour session, not a typical work setting for many designers.

However, the success of this heuristic analysis method in characterizing differences among designers may suggest ways to assist designers in adding to their ideation skills. Further, the identification of heuristics may suggest ways for computational tools to assist in design. For example, the frequency of heuristics applied could be analyzed to understand which of the heuristics are most commonly used, what kind of design problems they were frequently applied to, what kind of new concepts they generated, and which heuristics may be relevant given the observable patterns. In particular, this approach may hold promise in instruction for novices as they build their experience with heuristic use and design in general.

Conclusions

Pedagogy for enhancing design creativity is essential because most engineering and industrial design problems demand innovative approaches in the design of products, equipment, and systems. The present study showed that design heuristics can enhance innovation effectively in both engineering and industrial design domains. How can design heuristics be effectively taught?

Exposure to a variety of heuristics and experience in applying them on many different problems may lead to the development of expertise in innovation. For many design students, simply having an arsenal of design heuristics to try might lead to improvement in concepts generated. In fact, one factor may be motivational: it is possible that demonstrating the effectiveness of heuristics for creative tasks may, through feelings of efficacy, motivate creative efforts, just as the outcomes of creative efforts lead to an appreciation of creative work [16].

This study suggests that in design problems, making use of specific design heuristics may lead to more varied, creative, and practical solutions. Future research will continue to study the hypothesis that design heuristics developed by expert, innovative designers may be useful to all practitioners, including novices.

Acknowledgement

This research is supported by The National Science Foundation, Engineering Design and Innovation (EDI) Grant 0927474.

References

1. Court, A.W.: Improving creativity in engineering design education. European Journal of Engineering Education 23, 141–154 (1998)
2. Christiaans, H.H.C.M., Dorst, K.H.: Cognitive models in industrial design engineering: a protocol study. In: 4th International Conference on Design Theory and Methodology, ASME, vol. 42, pp. 131–140 (1992)

3. Park, J.A., Yilmaz, S., Kim, Y.S.: Using visual reasoning model in the analysis of sketching process. In: Workshop Proceedings of 3rd International Conference on Design Computing and Cognition, DCC (2008)
4. Finke, R.A., Ward, T.B., Smith, S.M.: Creative cognition: theory, research, and applications. The MIT Press, Cambridge (1992)
5. Yilmaz, S., Seifert, C.M., Gonzalez, R.: Cognitive heuristics in design: Instructional strategies to increase creativity in idea generation. Journal of Artificial Intelligence in Engineering Design and Manufacturing (AI EDAM), Special Issue: Design Pedagogy (2009)
6. Yilmaz, S., Seifert, C.M.: Cognitive heuristics employed by design experts: A case study. In: 4th International Conference of International Association of Society of Design Research, IASDR, Seoul, Korea
7. Yilmaz, S., Seifert, C.M.: Cognitive heuristics in design ideation. In: 11th International Design Conference, DESIGN 2010, Dubrovnik, Croatia (2010)
8. Newell, A., Simon, H.A.: Human problem solving. Prentice-Hall, Englewood (1972)
9. Nisbett, R.E., Ross, L.: Human inference: Strategies, and shortcomings of social judgment. Prentice-Hall, Englewood Cliffs (1980)
10. Klein, G.: Sources of power: how people make decisions. The MIT Press, Cambridge (1998)
11. Eberle, B.: Scamper, Prufrock, Waco, Texas (1995)
12. Gordon, W.J.J.: Synectics. Harper & Row, New York (1961)
13. Altshuller, G.: Creativity as an exact science. Gordon and Breach, NY (1984)
14. Daly, S.R., Yilmaz, S., Seifert, C.M., Gonzalez, R.: Cognitive heuristic use in engineering design ideation. In: Daly, S.R., Yilmaz, S., Seifert, C.M., Gonzalez, R. (eds.) 117th American Society for Engineering Education Annual Conference. ASEE (2010)
15. Amabile, T.: Social psychology of creativity: a consequensual asessment technique. Journal of Personality and Social Psychology 43, 997–1013 (1982)
16. Basadur, M.S., Graen, G.B., Wakabayashi, M.: Identifying differences in creative problem solving style. In: Parnes, S.J. (ed.) Source book for creative problem-solving, Creative Education Foundation Press, Buffalo (1992)

Studying the Unthinkable Designer: Designing in the Absence of Sight

Ann Heylighen
K.U. Leuven, Belgium

This paper aims at contributing to a more articulate understanding of the nature of design ability. To this end, we study a designer that may seem unthinkable at first glance: an architect who lost his sight and continues designing in this new condition. Based on the analysis of interviews, lectures and documents, we report and illustrate what this "unthinkable" designer and his way of working tell us about the nature of design ability, and the role of visual thinking therein. The results we obtain shed an entirely new light on issues that have been central to the study of design so far, and suggest promising directions for future research.

Aim

In studying the nature of design ability, there are several possible approaches to choose from. Quite a few studies are based on novice designers (*e.g.* students) or designers of relatively modest talents (see for instance [1]). Other studies focus on designers who are considered to have outstanding and exceptional ability, in order to gain an understanding of design at the highest level that it is practiced [2]. Yet other studies observe design tutors [3], as they tend to develop a more articulate view of and discourse about design than most other designers do [4].

The study reported in this paper adopts yet another approach. It focuses on a designer who may seem unthinkable at first glance: a designer who lost his sight and yet continues designing in this new condition. Key to design ability is said to be a characteristic form of cognition which is generally described as "visual thinking": designers are notoriously visually aware and sensitive, and use models and codes that heavily rely on graphic

J.S. Gero (ed.): *Design Computing and Cognition'10*, pp. 23–34.

image. In designing architecture, for instance, the visual seems to be so important that architecture students have been characterized as "the vis kids of architecture" [5]. Even authors who argue that this visual mode of thinking in design is a philosophical construct that can be dispensed with, acknowledge that this does not undermine the significance of the visual dimension [6]. By consequence, it may be hard to imagine how one can design in the absence of sight. However, the fact that the designer we study does so—and with fascinating results at that—offers a unique opportunity to extend our understanding of design ability.

Based on the analysis of interviews, lectures and documents, the paper aims to explore what this "unthinkable" designer and his way of working tell us about the nature of design ability. In doing so, the paper sheds an entirely new light on issues that have been central to (the study of) design so far.

Significance

The core features of design ability have been found to comprise abilities to think and communicate in non-verbal modes [2]. From his observations of the way design tutors work, Donald Schön [3] commented that, through sketches, "*[the designer] shapes the situation, in accordance with his initial appreciation of it; the situation 'talks back', and he responds to the backtalk.*" Even though Schön [3] points out that the language of designing is made up of drawing and talking, and that the non-verbal and verbal dimensions are closely connected, Cross [2] concludes from his observations that "*Design ability therefore relies fundamentally on* non-verbal *media of thought and communication*" (our emphasis).

These abilities to think and communicate in non-verbal modes at the core of design ability are described as a field, including a significant use of mental imagery "in the mind's eye" [2]. However, they are said to be particularly evident in the designer's use of models and "codes" that rely heavily on graphic images—drawings, diagrams and sketches [ibid.]. While these graphic images can obviously serve to communicate ideas and instructions to others, they have been studied especially in their role as aids to internal thinking.

Researchers assert that design is characterized by systematic exchanges between imagery and sketching [5]: designers use imagery to generate tentative solution concepts which they represent through sketching, but they also do the reverse; they sketch to recognize emergent features and properties of the developing concept [2]. This two-way exchange between imagery and sketching has been acknowledged by professional designers, but not always applauded. As Herman Hertzberger formulates it in an interview with Bryan Lawson [7]: "*A very crucial question is whether the pencil works after the brain or before. In fact what should be is that you have an idea, you think and then you score by means of words or drawing what you think. But it could be the other way round, that while drawing, your pencil, your hand is finding something, but I think that's a dangerous way. It's good for an artist, but it's nonsense for an architect.*"

Nonsense or not, numerous studies on the role of sketching all have emphasized its inherent power as design aid. Some have tried to further articulate why sketching is so powerful and essential for crystallizing design ideas, by examining what information architects think of and read off from their own freehand sketches, and how they perceptually interact with and benefit from them, *e.g.* [1]. Overall, the conclusion of these studies is that "*[t]he key 'tool' to assist design cognition remains the traditional sketch. It seems to support and facilitate the uncertain, ambiguous and exploratory nature of conceptual design activity*" [2]. Exceptions to this rule are studies which suggest that sketching is not necessary for certain aspects or stages of design, based on comparisons between "*imagery-alone and externalization conditions*" [8] or on blindfolding designers [9, 10, 11]. However, these studies are rare, as compared with the enormous amount of studies that confirm the key role of sketching in design.

In view of the latter, it may seem rather unlikely that a designer who loses his sight is able to continue designing. In the absence of sight, the key tool to assist design cognition—the traditional sketch—seems to lose its power: while making a sketch may still be possible to some extent, reading off information from it is certainly not. More in general, blindness seems at odds with the visual modes of thinking and communicating that are found to be at the core of design ability. The fact that the designer we study does design in the absence of sight, allows us to further expand our understanding of design ability and, in particular, of the role of visual thinking therein.

The rationale underlying the study of this "unthinkable" designer is just the opposite of why "split-brain" and other studies in the field of neuropsychology have been found relevant to improving our understanding of design. Key to the latter is the finding that damage to the right hemisphere impairs brain functions that relate strongly to design ability [2]; by contrast, our study starts from the observation that loss of sight does *not* to damage design ability (or at least, not necessarily).

Designer and Method

Carlos Mourão Pereira was born in Lisbon, in 1970. He graduated from the Faculdade de Arquitectura of the Universidade Técnica de Lisboa, distinguished with the prize Comendador Joaquim Matias, in 1997. He worked in Lisbon, with Aires Mateus, Carrilho da Graça, Costa Cabral and Gonçalo Byrne; in Zürich, with Toni Geser; and in Genoa, with Renzo Piano. He established his own office in 1998 and had his projects published and presented at seminars and conferences, in museums and at (world) exhibitions. He taught design studio in the Architecture Course of the Universidade de Arquitectura da Beira Interior, in Covilhã, and in the Master in Architecture of the Instituto Superior Técnico in Lisbon, where he currently conducts PhD research. In 2006 he lost his sight and since then maintains his professional activity, in research, teaching as in architectural practice. The work he designed after he lost his sight has been the subject of international exhibitions and publications [12].

The study of Pereira's design strategies and approaches is based on personal conversations and a more formal audio-taped interview. In addition, we attended a lecture where he presented eight projects, all designed after he lost his sight, and a seminar on his PhD research; both lectures were videotaped and notes were made. Furthermore, we analyzed representations of his design projects [13, 14], writings by the designer himself [15] and by other authors [12].

Based on a combination of these data, we first give a flavor of Pereira's design work, and then continue to analyze and illustrate what this "unthinkable" designer and his way of working tell us about the nature of design ability.

Unthinkable Designs

As a child, Carlos Mourão Pereira spent his summer holidays at a beach north of Lisbon, known for its slippery rocks, crashing waves and strong currents. A few days a year, however, the ocean's vigorous and loud waves transformed into a flat and silent sea, temporarily exposing its colorful and texture-rich interior. This cyclical phenomenon allowed walking along aquatic gardens that were most of the time hidden under the waves' foam.

After Pereira lost his sight, he designed a series of bathing facilities that offer the exceptional multi-sensory experience he remembers from his childhood to swimmers and waders of all ages and abilities. His sea bathing facility for Lourinhã (Portugal), for instance, converts the concrete structure of redundant fisheries into a tank for swimming and engaging with sea life (Figure 1). Ramps and handrails tracing the natural incline of the beach enable gradually immersing yourself in sea water baths with varying water levels and temperatures, away from the vigorous waves and currents. Receptacles for various marine species offer a collage of colors, textures and concavities within reach.

Fig. 1. (a) Sea bathing facility, Lourinhã (Portugal) © Carlos Mourão Pereira

Pereira has designed similar facilities for lagoons, lakes and rivers (Figure 2). Other work includes the design of a house, the redesign of a penthouse and 18th century building, and the design of an inclusive party installation. The latter is a bathroom that is accessible and comfortable for all users (including wheelchair users, but also people with low vision), yet without the typically medical look of accessible bathrooms (Figure 3).

Findings

When asked how he works now he has lost his sight, Pereira responds that his way of working is very different, but also not different from before.

Among the things that are not different from before, he mentions the generation of ideas. This may seem surprising at first sight, since "the key 'tool'" to assist design thinking—traditional sketches—are no longer available to him. For Pereira, however, sketching appears to have never been that instrumental to idea generation: "*The moments of creativity before I became blind were, for example, at night, before I go to sleep, during shower ... I was not drawing in these moments*" [14].

Rather than sketching, a more important "tool" to aid his design thinking seems to be his memory. When explaining the concept of the sea bathing facilities, for instance, he refers in detail to experiences from his childhood [15].

Major differences with how Pereira worked before relate to what Schön [16] has called the designer's "*transaction*" or "*conversation with the material design situation*"; in case of architecture, this material design situation includes both the building site and the representations of the designed building.

In the absence of sight, Pereira contends, it is extremely important to visit the building site and to touch everything, to feel the place. In order to have references to work with, his collaborators make small movies that he can listen to afterwards. For sound, he points out, changes a lot: "*a market place at 4 pm is completely different from one at 3 pm*" [14]. For similar reasons, he is looking for ways to register smells, but so far has not find any [14].

What has also changed since he lost his sight, Pereira mentions, is the way he communicates design ideas with other members of the design team [14]. In this respect, he considers himself very lucky to be teaching design, as his students helped him a lot to find alternatives for the traditional sketch. The most important way of communicating for him is gesture: when wanting to describe something to his collaborators, he forms it with his hands. For most complex forms, he uses models in clay; for orthogonal forms Lego turns out to be excellent.

Besides changing Pereira's mode of conversation with the design situation, his blindness also triggered the exploration of new ways of learning about other designers' work. Architects are known to spend considerable time studying what other architects design, by browsing through magazines and books, or visiting buildings. However, when arriving at a city, Pereira can no longer enter a bookshop and read images in books and magazines; therefore he has developed an alternative way to "*meet the architecture of the place, visiting studios and touching models and meeting its ideas directly from working teams*" [15]. In addition, he visits buildings of which he has seen images before, but now he "*sees*" them for the first time with hands, ears and nose. These re-visits result in various memories [14]; auditory: "*Sometimes a space is very noisy*" [14]; olfactory: "*In contemporary spaces with lots of plastic, you smell the plastic*" [14]; tactile: contemporary architects tend to design edges that are more "*cutting*", especially compared to architects from Art Nouveau or the Modern Movement [14]; but also visual: "*because when I touch something I always imagine it*" [15].

Fig. 2. River bathing facility, Schafhousen (Switzerland) © Carlos Mourao Pereira

This attention for sensory qualities can also be found in Pereira's own design work. As the sea bathing facilities illustrate, he approaches his projects with meticulous attention to the different senses [1è]: the feeling of water, wood or concrete; the smell of seaweed and arnica; the sound of sea waves or a waterfall. Even in the choice of the exact location of his projects, sensory qualities play a determining role. A case in point is his design of a river bathing facility in Switzerland (Figure 2). Originally, he had imagined locating the bathing facility very near to the falls. Yet, eventually, he decided to change the place. Listening to tapes recorded near and far from the water falls, he realized that in his first proposal, people would be unable to talk and listen: the sound of the falls would become noise when they are talking. Whence the decision to go a little further so that you can talk [14].

However, Pereira's heightened attention for non-visual sensory qualities does not seem to have damaged his visual awareness and sensitivity. On the contrary. When talking about the use of Lego blocks to communicate his design ideas with others, he hastens to point out that all blocks need to have the same color [14]. Also very telling is that in his design of an inclusive party installation, a major concern for him was to avoid the typical medical "*look*" [14].

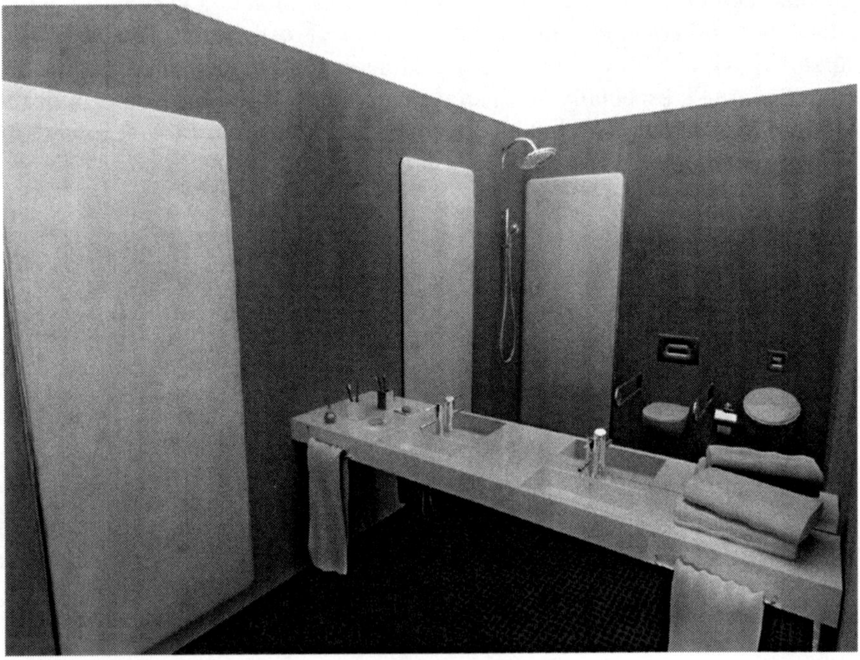

Fig. 3. Inclusive party installation © Carlos Mourão Pereira

Discussion

Our analysis of the "unthinkable" designer and his way of working sheds an entirely new light on issues that have been central to (the study of) design so far.

To start with, it puts the role of **sketching** as key "tool" in design into a entirely different perspective. The fact that his blindness does not allow to perceptually interact with sketches, Pereira contends, does not alter his way of generating design ideas. This is in line with research suggesting that "*externalizing*" may not be necessary for expert designers in the early, conceptual design stages [11].

Striking in Pereira's descriptions of his design work are the explicit and detailed references to memories of experiences in the past. These descriptions strongly suggest that, for him, **mental imagery** of memorable places can present essential information for designing future places [18].

Pereira's reliance on **gesture** to communicate his design ideas seems to be in line with earlier findings on designing in blindfolded conditions [10]. His use of clay and Lego during design, and his visits of other architects' offices, draws attention to the role of **models** in design. Anyone who has visited an architectural office may have noticed the presence of various models. Nevertheless, there are few accounts of their roles in design, as compared with the enormous amount of writing on drawings [19]. Moreover, studies that do look into models, tend to focus primarily on their role in visualization/visual procedures [19], with little attention being paid to their non-visual (*e.g.* haptic [20]) features. The fact that Pereira uses other senses than vision to interact with his models is significant for several reasons. It reminds us that the essence of non-verbal media in design is their ability to "*talk back*" [3] and, at the same time, demonstrates that this "*backtalk*" [3] may occur through any of the senses.

Moreover, the outspoken attention to **sensory qualities** in Pereira's work brings to light the limitations of media that provide only visual talkback, such as sketches and drawings, as design tools in architecture. The way we experience architecture is intrinsically multi-sensory in nature: the quality of space, matter and scale is assessed by a combination of multiple senses [21]. How a space looks is obviously important, but also how it feels, the sound and smell of a place plays a role in how people experience it. Nevertheless, most architecture is currently produced under consideration of mainly one sense: sight. The absence of non-visual features in traditional architectural spatial representations indicates how these are disregarded as important elements in conceiving space [22].

It is exactly these non-visual features that seem to set apart Pereira's design work. That is not to say that visual features have moved entirely to the background. On the contrary, as we have illustrated above, even in the absence of sight the designer's **visual awareness and sensitivity** seems to remain intact.

By studying this "unthinkable" designer, our study adopts a refreshing view on the notion of **disability** (*c.q.* blindness). Disability tends to be associated with "limitations" or "disorders" resulting from the medical condition of an individual. Increasingly, however, rather than a purely medical dysfunction, disability is considered as a complex phenomenon reflecting an interaction between features of a person's body and features of the society in which one lives. This move can be traced, for instance, in the International Classification of Functioning, Disability and Health of the WHO [23], or in the new UN Convention on the Rights of Persons with Disabilities [24]. Our study of the "unthinkable" designer moves a step further, by demonstrating that the perspective and experience of disability (*c.q.* blindness) can become a source of new knowledge and insights. As such, our study can be framed within a cultural model of disability [25], which acknowledges that disability challenges the way the world is organized, but at the same takes advantage of its potential to contribute to that same world.

Conclusions

Given the central role of visual thinking in design, it seems hard to imagine that one can design in the absence of sight. The fact that the designer we study does so, offers a unique opportunity to expand our understanding of design, and the role of different issues therein.

Judging from our analysis, some of these issues remain invariant in the absence of sight: the ways and moments in which ideas are generated, the continuous accumulation of experience of objects designed by others and, perhaps surprisingly, even the outspoken visual awareness and sensitivity.

Other issues seem to change considerably when sight is lost. Major changes relate to the role of non-visual sensory qualities, both in accumulating experience of designed objects and in designing objects, and to the mode of conversation with the materials of the design situation [16]—the site, representations of the design, other members of the design team.

Together, these invariants and changes seem to support the view expressed by Santiago Calatrava [7] that "*the working instrument of the designer is not the hand, but the order, or transmitting a view of something.*" They also suggest promising directions to further expand our understanding of design ability, *e.g.* by calling attention to the role of mental imagery and of models in design.

Limitations of the study include the fact that it is based on the experience of one designer only, and to a large extent his (verbal and written) accounts thereof. Future research will therefore extend our findings with other modes of enquiry (*e.g.* observations in the designer's office), so as to gain a more articulate understanding of his way of working. Questions that need further research are, for instance, how does he interact with models? What information does he "read off" from his own clay models, for instance, and how does he perceptually interact with and benefit from them? In addition, we will compare his way of working with that of another "unthinkable" designer [26].

Awaiting the results of this future research, this study has offered a refreshing twist on disability (c.q. blindness), in that it considers disability as a source of opportunities to explore, rather than as a source of problems to be solved [27].

Acknowledgements

This study has received funding from the European Research Council under the European Community's Seventh Framework Programme (FP7/2007-2013) / ERC grant agreement n° 201673. I would like to thank Stijn Baumers, Jasmien Herssens, Greg Nijs, Megan Strickfaden, and Peter-Willem Vermeersch for their help in the data collection. Last but not least, I would like to thank Carlos Mourão Pereira for his time, enthusiasm, patience and honesty.

References

1. Masaki, S., Barbara, T.: What do architects and students perceive in their design sketches? Design Studies 18(4), 385–403 (1997)
2. Cross, N.: Designerly Ways of Knowing. Springer, London (2006)
3. Schön, D.A.: The Reflective Practitioner. Basic Books, New York (1983)
4. Chen, J., Heylighen, A., Neuckermans, H.: Minding the mind in design tutoring and guiding. In: Zehner, R., Reidsema, C. (eds.) Proceedings of ConnectED 2007. International Conference on Design Education, University of New South Wales (CD-Rom), Sydney (2007)
5. Goldschmidt, G.: On visual design thinking: the vis kids of architecture. Design Studies 15(2), 158–174 (1994)
6. Kathryn, M.: Overlooking the visual. The Journal of Architecture 8, 25–40 (2003)
7. Lawson, B.: Design in Mind. Butterworth-Heinemann, London (1994)
8. Verstijnen, I.M., van Leeuwen, C., Goldschmidt, G., Hamel, R., Hennessey, J.M.: Creative discovery in imagery and perception: Combining is relatively easy, restructuring takes a sketch. Acta Psychologica 99(2), 177–200 (1998)
9. Athavankar, U.: Mental Imagery as a Design Tool. Cybernetics and Systems 28(1), 25–42 (1997)
10. Athavankar, U.: Gestures, Imagery and Spatial Reasoning. In: Gero, J.S., Tversky, B. (eds.) Visual and Spatial Reasoning, pp. 103–128. MIT, Cambridge (1999)
11. Bilda, Z., Gero, J.S., Purcell, T.: To sketch or not to sketch? That is the question. Design Studies 27(5), 587–613 (2006)
12. Lowther, C., Schultz, S. (eds.): Beachlife. In: Architecture and Interior Design at the Seaside, Frame Publishers, Amsterdam (2008)
13. Mourão Pereira C.: Carlos Mourão Pereira, Arquitecto (2010) `http://www.carlosmouraopereira.com/` (accessed: April 14, 2010)
14. Mourão Pereira, C.: Organizing the air and the water. Lecture given (March 19, 2009) (Leuven)
15. Mourão Pereira, C.: Extracts from letters to Juhani Pallasmaa, Lisbon (2007)

16. Schön, D.A.: Designing as reflective conversation with the materials of a design situation. Knowledge-Based Systems 5(1), 3–14 (1992)
17. Vermeersch, P., Heylighen, A.: Blindness and multi-sensoriality in architecture. In: Proceedings of the ARCC/EAAE 2010 International Conference on Architectural Research, ARCC/EAAE (forthcoming) (2010)
18. Downing, F.: Conversations in imagery. Design Studies 13(3), 291–319 (1994)
19. Yaneva, A.: Scaling Up and Down: Extraction Trials in Architectural Design. Social Studies of Science 35(6), 867–894 (2005)
20. Herssens, J., Heylighen, A.: Haptics and Vision in Architecture. In: Lucas, R., Mair, G. (eds.) Sensory Urbanism Proceedings 2008, pp. 102–112. The Flâneur Press (2008)
21. Pallasmaa, J.: The Eyes of the Skin. John Wiley & Sons, Chichester (2005)
22. Dischinger, M.: The Non-Careful Sight. In: Devlieger, P., Renders, F., Froyen, H., Wildiers, K. (eds.) Blindness and the Multi-Sensorial City, pp. 143–176. Garant, Antwerp (2006)
23. World Health Organization International Classification of Functioning, Disability and Health: World Health Organization, ICF (2001)
24. UN Convention on the rights of persons with disabilities (2006), `http://www.un.org/disabilities/convention/conventionfull.shtml` (accessed April 14, 2010)
25. Devlieger, P., Rusch, F.R., Pfeiffer, D. (eds.): Rethinking Disability. The Emergence of New Definitions, Concepts and Communities. The Emergence of New Definitions, Concepts and Communities. Garant, Antwerp (2003)
26. Nijs, G., Vermeersch, P., Devlieger, P., Heylighen, A.: Extending the Dialogue between Design(ers) and Disabled Use(rs). In: Proceedings of International Design Conference - Design 2010, Design Society (in press, 2010)
27. Pullin, G.: Design Meets Disability. MIT Press, Cambridge (2009)

Design Heuristics: Cognitive Strategies for Creativity in Idea Generation

Seda Yilmaz, Colleen M. Seifert, and Richard Gonzalez
University of Michigan, USA

This paper explores the use of heuristics as cognitive strategies invoked during the process of design. We propose new heuristics for design that provide ways to explore the problem space of potential designs, and often lead to the generation of creative solutions. We test whether Design Heuristics can be taught to novices, and whether doing so will improve the creativity of their resulting designs.

In the present empirical study, we evaluate a set of six instructional heuristics, and validate their effectiveness with product concepts generated by novice designers. Six hundred and seventy three drawings were created by 120 first-year college students under four instructional conditions. Drawings were coded according to the use of heuristics, and scored for creativity. The results showed that the most creative concepts emerged from the experimental conditions where heuristics were introduced. Heuristics appeared to help the participants "jump" into a new problem space, resulting in more varied designs, and a greater number of designs judged as more creative.

Our findings suggest that the simple demonstration of design heuristics may, at times, be sufficient to stimulate variation and creativity in design.

Introduction

One of the key educational issues in engineering design is the enhancement of creative abilities. Which cognitive strategies enhance creativity in design? Although the importance of cognitive strategies is well recognized [13], little is known about how designers apply them, and how they affect the quality or creativity of the resulting design. Pedagogy for enhancing design creativity is essential because most engineering problems demand innovative approaches in the design of products,

J.S. Gero (ed.): *Design Computing and Cognition'10*, pp. 35–53.

equipment, and systems. The result of engineering design activity is often expected to be original, adding value to the base of existing designs by solving technical problems in new ways.

This paper presents an empirical approach to the study of cognitive processes in design, and the utility of explicit instruction on strategy use for the creativity of designs. We begin by reviewing previous research on cognitive processes in engineering design, leading to our research questions and experimental approach described in the following sections. Then, we present an empirical study examining the use of heuristics in a conceptual design task.

The long-term objective of this research is the development of cognitive strategies for design generation that will increase the variety, creativity, and quality of designs. Designers are often driven by the need for innovation and a competitive edge in product design, and by problem characteristics where existing designs do not meet requirements and constraints. Our goal is to enhance the effectiveness of idea generation by identifying the cognitive strategies used by experts, and developing corresponding pedagogical approaches to benefit novice designers.

Designers' Cognitive Processes

Design researchers and cognitive scientists have developed a variety of process models to account for creativity in design. These models are often based on observations of design processes in verbal protocols of experts solving design problems. French [14] proposed a model that includes analysis of the problem, conceptual design, embodiment of schemes, and detailing. Cross [8] described a four-stage model of exploration, generation, evaluation, and communication. Finally, Benami and Jin [4] introduced a cognitive model to capture interactions between cognitive processes, design entities, and design operations.

One specific cognitive process identified is "problem framing," where Schon [22] suggested, "In order to formulate a design problem to be solved, the designer must frame a problematic design situation: set its boundaries, select particular things and relations for attention, and impose on the situation a coherence that guides subsequent moves." Csikszentmihalyi and Getzels [10] demonstrated that art students who spent time initially defining the design problem produced more successful designs, and were later judged more successful as professional artists. This "problem finding" process relates to the co-evolution of problem and solution. Dorst and Cross [11] confirmed through a series of protocol studies that creative design involves a period of exploration in which the problem and solution spaces evolve together into a problem–solution pairing.

It is estimated that 70% of a product's cost is defined during conceptual design [20]. Perhaps, as a result, much research has investigated the cognitive processes that occur in the *idea generation phase* of design creation [1], [6], and [11]. There have been some descriptions of the varieties of ways that ideas are generated. Finke et al. [13] divided these creative processes into *generative* (analogical transfer, association, retrieval, and synthesis) and *exploratory* (contextual shifting, functional inference, and hypothesis testing). Christensen and Schunn [5] suggested studying the cues designers use in order to understand what leads to creative outcomes. They propose that, as a cue promotes one type of generative process, it may constrain another exploratory one. Alternatively, a cue might aid the cognitive process within the design domain, while hindering the information processing between domains. Therefore, a more detailed understanding of cognitive processes and their functions is needed.

Design Heuristics

Our approach arises from the intuition that designers appear to generate questions and choose directions from within an internal dialogue, choosing to follow known strategies with or without conscious reflection. Observational studies of designers at various levels of expertise have demonstrated the use of strategies (e.g. Adams & Atman [1]). For example, Park et al. [21] found that expert designers who used generation, transformation, and external representation when performing a sketching task produced more creative alternatives than designers who used perception, maintenance, and internal representation as defined by their visual reasoning model. Other studies have identified some design strategies employed by expert designers in the product design process [9] and [17]. However, many questions remain surrounding the use of strategies. For example, which are the most effective? Does strategy use change depending on level of expertise? How can such strategies be effectively taught in engineering design courses?

Previous theories of heuristics for design, including SCAMPER [12], Synectics [15], and TRIZ [3], have provided suggested heuristics. The TRIZ heuristics were designed to address specific mechanical trade-offs in design based on past patents. The forty heuristics proposed are designed to apply to a very specific set of mechanisms. They appear most useful in solving design tradeoffs within already well-developed concepts as problems are refined. Two other approaches have suggested heuristics for creating new designs during the ideation stage: SCAMPER and Synectics. These heuristics provide very general guidelines to assist in generating concepts. For example, SCAMPER includes *substitute, adapt*, and *modify* as heuristics to apply to a design problem. Synectics heuristics focus more on themes such as adding "animation" to a product. These general strategies may be helpful, but are potentially too abstract to assist a designer in applying them within a given problem.

Most importantly, none of the existing heuristics proposed for design have been examined empirically to determine whether they do assist in idea generation. In the present study, our goal was to define a set of design heuristics, and to determine whether their use leads to more creative designs. In most design problems, the space of possible designs is never fully defined, and may include new features not previously applied to the problem, and not already identified as relevant. Following Newell and Simon [18], we define Design Heuristics as transformative strategies that set up a new space of potential designs, with new features to consider. Design Heuristics are strategies that assist the designer in exploring new parts of this potential design space.

For example, consider the problem of creating a novel design for a desktop accessory, Figure 1) How would a designer approach this problem, and generate a new concept? One heuristic recalled by an expert industrial designer in a pilot study involved a transformation of form by "flipping" the object across an axis, in this case top to bottom. She described seeing a photograph of a flower vase made of circles with overlapping edges, Figure 1(a). By expanding on this form, she created a drawing of circular shapes with one long end hanging from each circle, leading to the "J" shaped object in Figure 1(b). Then, to add interest to the form, she "flipped" the center to go in opposition to the aligned shapes, Figure 1(d). The resulting accessory is striking in the novelty of its design. The "flipping" heuristic created a new form that introduces novel variation because its flipped form adds novelty in the variation introduced.

In this example, the "flipping" heuristic led to an interesting, creative outcome. However, Design Heuristics are not guaranteed to produce either high quality or innovative designs. Instead, heuristics serve as a way to "jump" to a new subspace of possible designs, or to expand the space of possible designs. Design Heuristics allow the designer to move into other types of design concepts, and provide the opportunity to generate variation in candidate design concepts.

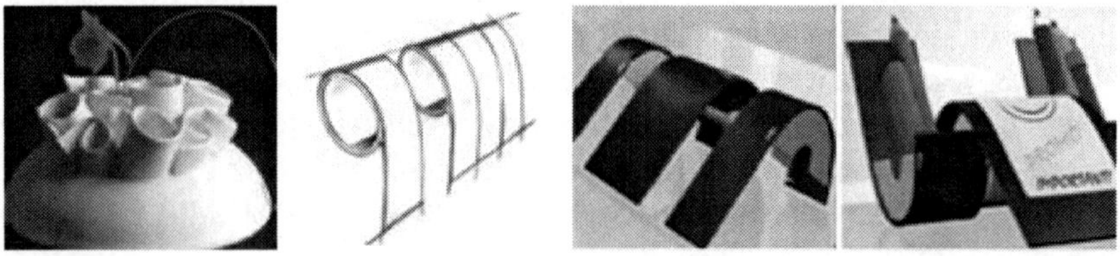

(a) Vase (b) Concept Sketch (c) Prototype (d) 'Flipped' form

Fig. 1. An expert designer's application of the "flip" design heuristic to a form

How is a heuristic applied to a candidate concept? Heuristic application is context-dependent in that there will be more than one way to apply a specific heuristic to a candidate form (for example, flipping upside down or from side to side). An even more challenging question is how the selection of Design Heuristics is accomplished. There appears to be no general prioritized ordering of heuristics; instead, designers may recall and use heuristics based on specific cues or factors within the problem, such as needs of the user, or a specific functional requirement, such as securing closure. Design Heuristics vary in where and how they can be applied, how they impact a resulting design or the design space as a whole, and the amount of time invested in applying them. Their application may be as simple as altering the form, or may set up an entirely different concept that functions in a different way in a different context.

For this study, we defined a set of Design Heuristics that address changes to three dimensional forms. These simpler heuristics lend themselves to the limitations of controlled experiments, time constraints, and our participants' levels of design expertise. Each heuristic requires specific features within the design problem in order to be applied, and produces a changed concept altered in a specific fashion. We specify the nature of each heuristic, the conditions where it can be applied, and the nature of the transformation it provides for a design. A single heuristic can produce alternative versions depending on how it is applied, so that the same heuristic can be applied repeatedly to produce variant designs.

The power of Design Heuristics is that they may result in a more varied set of potential design solutions. Each heuristic can potentially form a possible design that varies significantly from the ones considered so far. For the designer, following these cognitive strategies can prevent lingering in already-considered concepts. Fixation in a type of solution has been shown to be a problem in the creative process [13]. The application of Design Heuristics avoids this problem by providing a means for the designer to consider variations in concepts they may not have otherwise considered. Though Design Heuristics do not guarantee the best solution, they help to reduce ideation time, and assist the designer in generating concrete alternatives.

A main question is whether the use of Design Heuristics may guide the designer toward discovering more creative solutions. Theoretically, the use of heuristics has been assumed to relate to the generation of novel variants in designs, leading to more creative outcomes. But to date, there is no empirical evidence of a relationship between heuristic use and the creativity of the resulting designs.

To test these ideas, the present study begins with a set of heuristics culled from protocols of an expert industrial designer [23] working on a professional engineering project. The project involved designing a bathroom with functions accessible by disabled users while fitting within the standard bathroom spaces in traditional homes. It is this type of expertise – the very effortful process of attempting to create a variety of new design ideas – that is the target for our Design Heuristics approach. By examining the expert's progression of designs over time, we identified a set of six candidate Design Heuristics that provided ways of altering a three-dimensional form: *merging, changing the configuration, substituting, rescaling, repeating*, and *nesting*. The question of interest in this paper is whether novice designers can be instructed to use these same design heuristics successfully, and whether doing so results in designs that are considered more creative.

Experimental Approach and Research Questions

Our hypothesis is that design heuristics offer a means of generating possible designs by guiding designers to consider specific types of variations on concepts. We conducted a study that controlled how participants learned about a series of Design Heuristics, and measured how they used those heuristics when generating multiple product concepts.

In this context of novice designers, we seek to answer the following research questions:

Q1: Can Design Heuristics be taught with simple instructions?

Q2: Does the use of Design Heuristics lead to more creative designs?

Q3: Which Design Heuristics are most effective for creativity?

Participants

To test the utility of instruction on Design Heuristics, we selected a population of novice designers. One hundred and twenty first-year students participated at a mid-western university for course credit. Their age ranged from 18 to 21; 73 (61%) of them were females. Participants were randomly assigned to one of the four instructional conditions.

The choice of novice participants for the study is appropriate for a number of reasons. First, our hypothesis is that expert designers acquire these design heuristics over their lengthy education and experience as designers. By choosing subjects with no training in design or engineering, we avoid the question of the heterogeneity of pedagogies individuals may have been exposed to in the past and potential prior knowledge of design strategies. In addition, because participants had no formal technical training in sketching or drafting, the designs would reflect a similar baseline in drawing ability. Finally, these university participants allowed us to gather a sample of novice designers with a wide range of demographic and educational backgrounds and interests, potentially supporting the effectiveness of this approach in a broader variety of educational programs.

Materials

The experiment used the task of redesigning an everyday object: a pair of salt and pepper shakers. These objects are very familiar in Western cultures, with many prototypical designs available as commercial products. This redesign task was selected based on the time limitation of the study. The redesign task allowed subjects to focus on a product that was easy to grasp in function and use. This allowed time for the instructional conditions that required participants to absorb information on Design Heuristics, and still have time to generate a set of concepts. Because the redesign of every day products is part of design education in engineering and industrial design, this task fits with the types of design problems confronting professional designers.

The Design Task

In all four experimental conditions, a short written description of the design task was provided, along with a picture of simple geometric shapes to use in the generation of design concepts. Simple block shapes were included to encourage thinking in three dimensions, but also to help constrain designs to a manageable set of possible forms. The geometric shapes and instructions provided are presented in Figure 2.

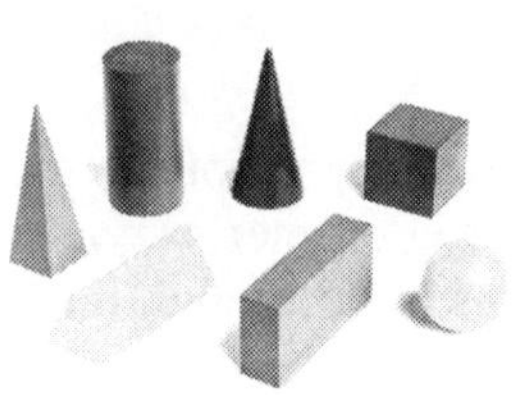

Problem Statement: Imagine that you are working as a product designer in a design consulting firm, and that you are given a rather fun assignment just for today: design salt and pepper shaker sets by utilizing simple geometrical forms and adding as much detail as needed. Think about the functionality of the product, where they will be used, how they will used, how they will be cleaned, how to fill them up, etc.

Fig. 2. Problem statement for the design problem used in the empirical study

Design Heuristics Instructional Materials

Six heuristics were included in the experiment and examples of concepts using each heuristic, Figure 3, were provided to subjects in the Heuristics conditions.

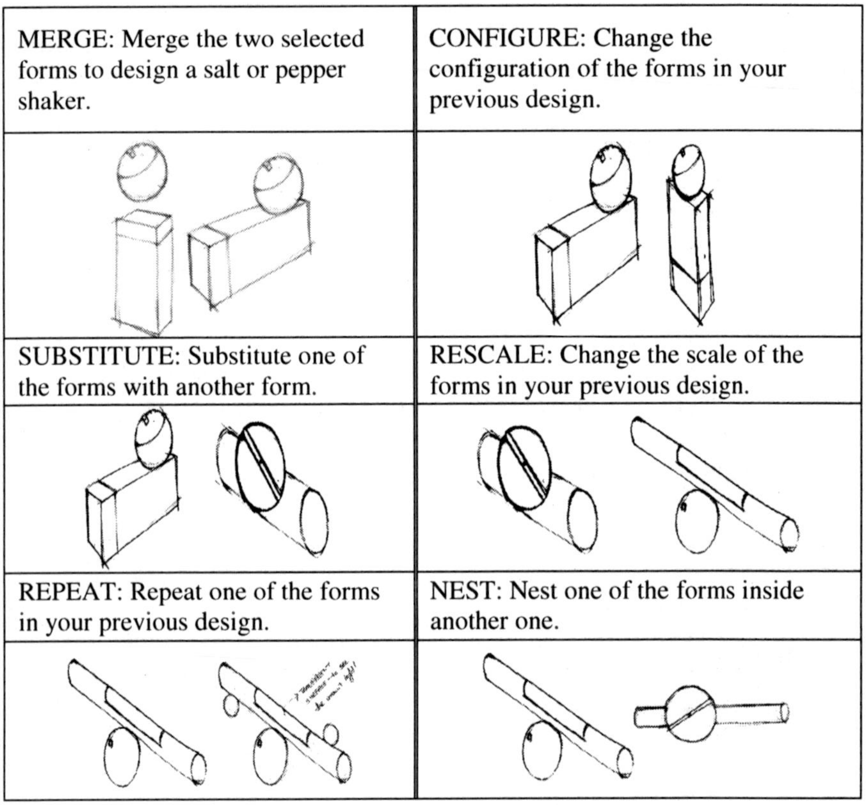

Fig. 3. The six heuristics selected for the study proposing changes to a form to increase variation in the design

These Design Heuristics were selected because they represent the most frequently used strategies throughout this expert designer's concept generation process [23]. This set also appeared to be simple enough for our non-expert participants to understand and use.

The chosen Design Heuristics bear important similarities to the engineering design heuristics proposed in SCAMPER [12], TRIZ [3] and Synectics [15]. For example, the *changing the configuration* heuristic used in this study is also proposed as the "Another Dimension" engineering redesign strategy in TRIZ, as "rearrange/reverse" in SCAMPER, and as the

"Transfer" strategy in Synectics approach. Other Design Heuristics in the study, such as *Nest,* do not appear in the other heuristic approaches.

The training materials provided for each heuristic consisted of a brief written explanation of the heuristic and a visual example of how it can be applied to simple forms, Figure 3. Participants were told that these examples could be used to understand how the heuristic works, but they should not be repeated in their own designs.

Experimental Design

The effects of the Design Heuristics on the creativity of the observed designs were evaluated through four instructional conditions. The design task provided simple geometrical shapes as starting points that allow a wide variety of alternative designs. The participants were free to choose single or multiple shapes to incorporate into their designs. The four experimental conditions in the study are shown in Table 1.

Table 1 Overview of the experiment design

Control Group: No instructions about design heuristics were provided.
Serial Order 1: The six design heuristics were presented one at a time in a single order determined at random, with *merge* as the first heuristic, followed by *change the configuration, substitute, rescale, repeat,* and *nest.*
Serial Order 2: The six heuristics were presented one at a time in a different standard order determined at random, with *merge* as the first heuristic, followed by *repeat, change the configuration, substitute, nest*, and *rescale*.
Heuristic Choice: All six design heuristics were presented together in a list, with *merge* as the first heuristic. Subjects were free to choose which heuristic to attempt next. The order of presentation of the remaining five heuristics was randomized for each subject.

Two different serial orders were tested to determine whether the value of the heuristics used was independent of order of presentation. However, in all three heuristic conditions, the session began with the presentation of *merge* as the first heuristic. The rationale was that this resulted in a concept with at least two shapes in the first candidate concept, making the application of the other heuristics as transformations of an existing concept more feasible. The experimental design employed the four experimental conditions shown in Table 1.

Procedure

Participants were assigned to experimental conditions at random, with 30 participants per group. The sessions were conducted in a classroom in small groups of two to twelve participants. All participants within a testing session were in the same experimental condition.

Participants in all four conditions were given an introduction page summarizing the design task and presenting the task guidelines, Figure 2. Because prior research [16] has shown that creativity test scores are influenced by explicit instructions to "be creative," we included these instructions: "This task involves drawing creatively. That is, please create concepts that are both original (novel, uncommon) and also appropriate (artistically effective)." Participants were given eight letter-size response papers to depict their designs, and were also asked to write labels and notes to clarify their designs.

In all conditions, participants were instructed to complete each design concept on one page, and to label their drawing to indicate specific elements that might help to explain the design. In the two *Serial Order* conditions, the experimenter directed subjects' progress throughout the experiment. Within each group, the heuristics instruction sheets were presented one at a time with a blank response sheet. Subjects were given six minutes to create a design using that particular heuristic, and then the experimenter asked them to turn the page to the next heuristic, and so on. For the *Heuristic Choice* condition, subjects received all of the instructional sheets for the six heuristics together. They were asked to choose the heuristics they wished to use to help them generate designs on the following blank response sheets. In the *Control* condition, subjects were not given any heuristics, and were asked to generate as many concepts as possible on the blank response sheets within the given time for the task.

Results

Three senior undergraduate students with no formal design training rated all 667 designs. The three judges were blind to condition and to the experimental hypothesis. Of course, individuals vary in their ability and comfort with sketching, and as a result may have a more difficult time expressing their creative ideas visually. But because the participants were assigned to experimental condition at random, these individual differences would occur in all of the conditions, and not bias the results towards more creativity in any particular one. To further mitigate differences in drawing skill, we instructed the judges to focus on the creativity of the concepts, and not the skill level shown in the sketches.

Following Amabile's [2] consensual assessment rating technique, the judges were asked to form their own subjective impression of overall creativity, including uniqueness and unexpected solutions. Each of the designs was rated for its creativity on a seven point scale, from "1" meaning "Not at all creative," to "7," meaning "extremely creative." The average score over all judges' rating was 3.35 (SD = 1.1), indicating that the majority of the designs were scored as "low in creativity." The reliability of the judges' scores (computed using Cronbach's Alpha: *Serial Order 1* = .744; *Serial Order 2* = .742; *Heuristic Choice* = .717; *Control* = .855) suggests some inconsistency among the judgments, but overall an acceptable level of agreement.

Because so many designs received consensual rating scores below the midpoint of the seven point creativity scale (4), we decided to select only the designs rated "5" or above by at least one of the three judges for further analysis. The rationale for selecting this smaller set was that we are interested in whether heuristics are evident in creative designs; therefore, only designs rated 5 and over were included. The selected set was comprised of 266 of the 667 original designs, representing approximately 40% of the total. The designs in this selected set were generated by 93 of the 120 subjects (77.5%), and each of these subjects contributed between 2 and 6 designs to the selected. Differences in the frequency of selection by experimental condition are apparent in a generalized linear mixed model with the logit link function and subjects as a random factor: The fewest designs were selected from the *Heuristic Choice* group (24.9%), significantly fewer than the *Serial Order 1* group (51%; z = 4.74, p < .0001), the *Serial Order 2* group (49%; z = 4.57, p < .0001), and the *Control* group (35%; z = 1.98, p = .04). The two *Serial Order* groups did not differ (z = 0.27, p > .05), though both differed from the *Control* group (*Serial Order 1:* z = 2.66, p = .0077; *Serial Order 2*: z = 2.41, p = .015).

Subjects in the two *Serial Order* conditions produced significantly more designs than those in the *Heuristic Choice* condition, which produced significantly more creative concepts than the *Control* condition. This pattern may result from the experimenter-directed procedures in the two serial order conditions, where subjects were instructed when to read about each heuristic and were given six minutes to complete a concept using that heuristic. By contrast, subjects in the *Control* and *Heuristics Choice* conditions were given initial instructions, but then left to work their way through multiple concept pages on their own for the forty minute period. As a result, the *Serial Order* participants may have been kept on task, and paid more attention to the instructions. In addition, the *Heuristic Choice* condition required subjects to spend time in the selection of a heuristic for each concept. Another explanation for the advantage of the *Serial Order* groups in producing creative designs is that, by following the instructions, they may have produced more varied concepts because the procedure required them to use a different heuristic in each.

In sum, the *Serial Order* conditions and the *Heuristic Choice* conditions produced more creative designs (by at least one judge scoring them as "somewhat creative") compared to the *Control* group. However, of these creative designs, which were judged to be the most creative?

Average Creativity Ratings of Selected Designs

Because our focus is on examining whether heuristics lead to better solutions, we hoped to improve the consistency of the rating process by asking the same three judges to compare all of the selected designs within a single rating session. Each of the selected designs was removed from the subjects' booklets, and shuffled into a different randomized order for each judge. This allowed each concept to be considered independently of the sequence of its generation by the subject. A modified rating task was employed where judges, blind to condition, placed each drawing into one of seven piles, each representing a point on the seven-point creativity scale. This "sorting" procedure allowed the judges to shorten the time required to complete the ratings to less than one hour. The result of this rating procedure was higher interrater reliability scores using Cronbach's Alpha, as shown in Table 2.

The average creativity ratings show differences for designs in the four instructional conditions: Designs generated under the *Heuristic Choice* instructions were rated highest in creativity, followed by *Serial Orders 1* and *2*.

Table 2 Inter-rater reliability statistics (Cronbach's alphas) and average creativity ratings for designs in the selected set

Experimental Condition	Cronbach's Alphas	Creativity Means	Standard Deviations
Serial Order 1	.909	3.51	1.348
Serial Order 2	.891	3.30	1.357
Heuristic Choice	.809	3.73	1.205
Control	.900	2.92	1.664

A One-Way ANOVA using a random effects model with designs nested within subjects found that both *Serial Order 1* and the *Heuristic Choice* conditions differed significantly from the *Control* condition ($p < .05$); however no other pairwise comparisons reached significance (neither by the more liberal uncorrected Type I error rate nor the Bonferroni corrections). An *a priori* contrast was conducted comparing all three heuristic instructional groups (*Serial Order 1, Serial Order 2*, and *Heuristic Choice*) against the *Control* group. This pattern was significant, $z = 2.105$, $p = .04$. Further, a contrast testing the prediction that the *Heuristic Choice* were rated highest, followed by both *Serial Order* conditions, followed by the *Control* condition, was also significant ($z = 2.209$, $p = .0339$). This suggests the choice of heuristics produced somewhat higher creativity ratings than the *Serial Order* heuristics instructions, with the *Control* condition designs rating lowest.

As noted above, more of the selected designs came from the *Serial Order* conditions; yet, the *Heuristic Choice* designs were rated higher in creativity than the *Serial Order* conditions. This may appear contradictory; however, the experimenter-driven procedure in the *Serial Order* conditions led participants to produce more designs. However, while producing fewer designs, and fewer creative designs, overall, the *Heuristic Choice* condition produced the highest quality of creative designs.

The higher creativity ratings observed for the three *Heuristics* conditions suggests these instructions resulted in more successful designs compared to the control condition. When compared with the *Control* group, the highly creative concepts in the Heuristics conditions are visually more detailed, have indications (directional arrows) of how they will be used and how contents will come out of the container, have variations in the arrangement of the design elements, and are rarely labeled. These differences suggest the heuristics allowed the participants consider the design form differently, resulting in greater novelty in the resulting design forms.

Heuristic Use

A final analysis involved coding each of the designs for the presence of one or more of the six Design Heuristics. Each of the concepts was examined and scored for the application of each of the six heuristics included in the study. Judges coded for the heuristic by analyzing the relationships of the design elements with each other; for example, whether two forms were merged, or repeated, in the concept. Figure 4 (over two pages) shows examples depicting the use of the six heuristics by participants.

Design Heuristic	Example Design by a Participant
MERGE	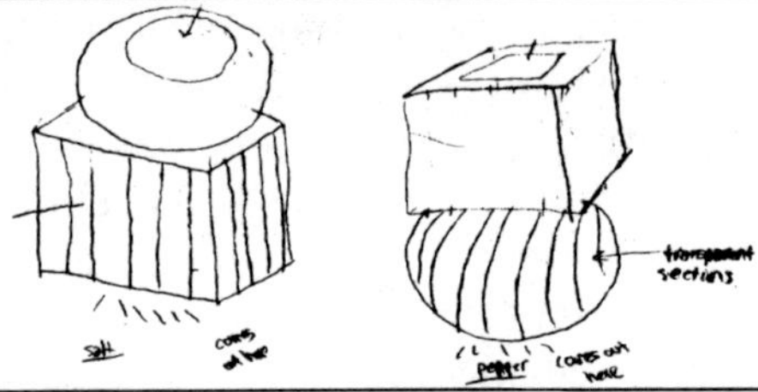
CONFIGURE	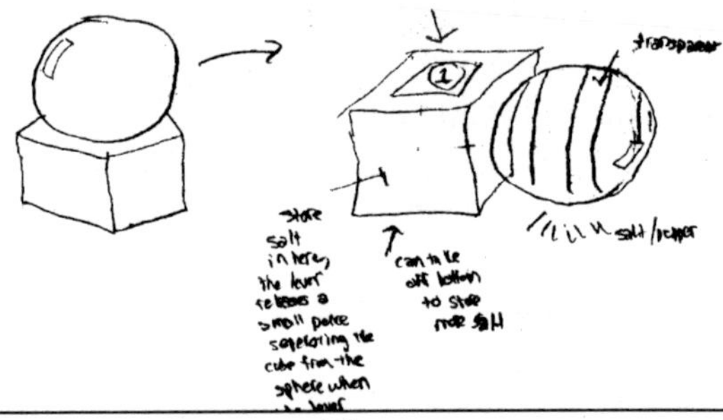
SUBSTITUTE	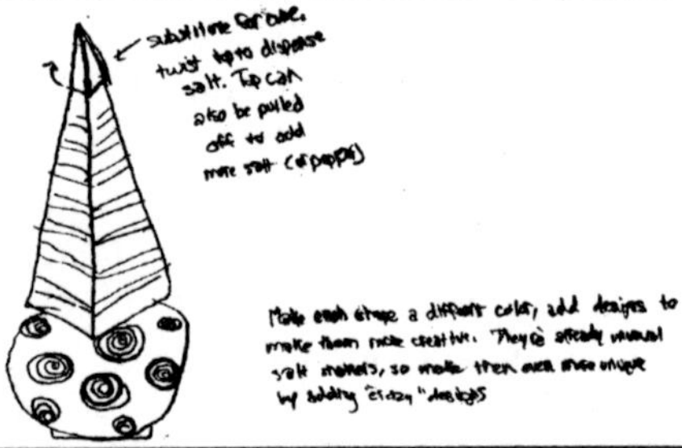

Fig. 4. (continued next page) Example concept drawings showing the presence of specific design heuristics among the final designs by participants

RESCALE	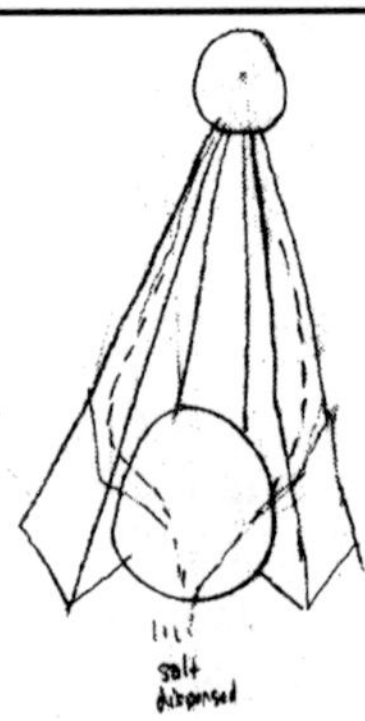

REPEAT	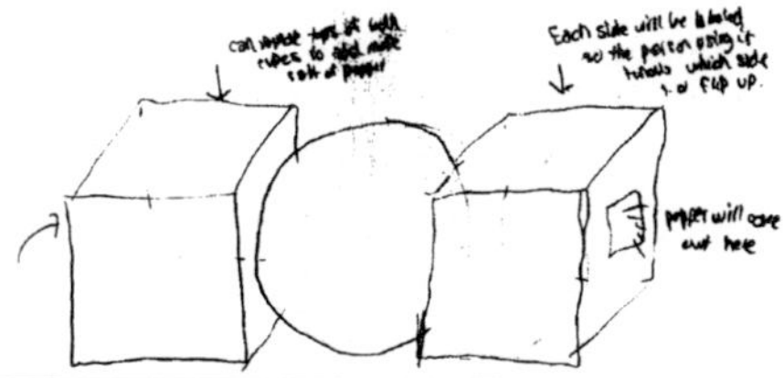
NEST	

Fig. 4. (continued) Example concept drawings showing the presence of specific design heuristics among the final designs by participants

The three instructional groups on average used more than 2 heuristics within each design. Table 3 shows the number of times each heuristic was observed (consensually by all three judges) by instructional condition. In terms of the number of heuristics observed, the two *Serial Order* conditions were coded as showing many more uses of heuristics than the *Heuristic Choice* or *Control* conditions. Again, given the experimenter-driven task procedure discussed above, the Serial Order conditions may have followed instructions to produce a new drawing using each heuristic as directed.

Across all of the concepts, all three heuristic conditions show the greatest use of *merge* and *configure,* used in over 85% of designs in the three heuristics instruction groups, and less than 45 of the control designs. In the *Heuristic Choice* condition alone, these two heuristics appeared in over 85% of the designs. The other four heuristics were selected for use in between 20-40% of the designs, and in the *Serial Order* conditions where subjects were asked to use each heuristic, these four were observed in 20-45% of the designs. *Substitute* and *repeat* were used the least in the *Heuristics* conditions. *Serial Order 1 and 2* appeared to make more use of *nest* and *rescale.*

Table 3 Observed frequency of heuristic use in the selected set designs including scores from all three raters for all four conditions

	Serial Order 1	Serial Order 2	Heuristic Choice	Control	**Total Number of Heuristics**
Merge	71	76	37	26	210
Configure	70	77	36	16	199
Substitute	23	18	10	31	82
Nest	21	34	15	4	74
Repeat	23	18	16	10	67
Rescale	35	21	8	3	67
Total Number of Designs	77	89	43	57	266

Surprisingly, in the *Control* group, where there was no instruction on heuristics, heuristic use averaged more than one for each design, with "substitute" and "merge" used most often. Our results also indicate that more than eighty percent of the participants in the *Control* condition used one or more heuristics without any instruction. The evidence of heuristic use in the *Control* condition may suggest that the heuristics selected were already known or easy to use in the design task, even for these novice designers. Most prominently, *substitute* appeared most frequently in the designs created in the *Control* condition.

In sum, the *Control* condition designs include many with simple forms, and the variations added a new function, detail, or theme. In the three *Heuristic* conditions, the designs show more intentional variation and greater complexity of form (i.e. unexpected attachments, forms that were cut and flipped in various directions, surfaces covered with patterns), presumably through the use of the Design Heuristics provided. This analysis of design content supports the conclusion that heuristic instruction can assist even novice designers in creating more varied visual forms, leading to designs rated as more creative.

Discussion

This empirical study suggests the potential effectiveness of instruction on Design Heuristics. Even for novice designers, a few minutes of text and illustration on six specific heuristics led to designs reliably judged as more creative. In the context of this empirical study of design creation with novice designers, we sought to answer the following research questions:

Q1: Can Design Heuristics Be Taught with Simple Instructions?

This research study shows that heuristic use can be supported with simple written instructions along with visual examples. Another implication is that heuristics are applied frequently once they are learned even when not under instructions to do so. This implies that generating concepts using heuristics may be a natural approach to design, and that providing specific instructions on design heuristics will take further advantage of their utility.

Q2: Does the Use of Design Heuristics Lead to More Creative Designs?

Design Heuristics in the study increased the creative success of concepts. The concepts guided by heuristics appeared more diverse and unusual, concentrated more on visual form, and were judged as more creative. This result has important implications for teaching designers how to think about design creation, and for the kinds of cognitive strategies they may learn through instruction in design.

Q3: Which Design Heuristics Are Most Effective for Creativity?

Six Design Heuristics were compared in the study; of these, merge and *configure,* were used substantially more often by the three heuristics instruction groups, suggesting they were a major factor in the success of these designs. Both heuristics focus attention on the individual forms and their relative composition. This may encourage the consideration of alternative combined forms that are more complex, and therefore more distinctive. Other heuristics (*nest, rescale, repeat*, and *substitute*) may be more appropriate in only some candidate designs.

The results of this empirical study must be considered in context. More specifically, the results here were observed in a study of novice designers without regard to potential design ability, interest, or motivation. Certainly, they were less technically sophisticated than industrial design or engineering design students, and presumably had little exposure to this type of design task. The study also involved a one time, short design task, which may not reflect the typical setting for ideation in product design. Despite these limitations, this study provides evidence for the effectiveness of Design Heuristics in creative ideation. In a simple redesign problem, instruction on specific Design Heuristics successfully led to creative solutions.

Our findings suggest that simple demonstration of Design Heuristics may, at times, be sufficient to stimulate divergent thinking, perhaps because these heuristics are readily learned. Over time, these Design

Heuristics may become internalized, and be applied in design problems where the need to be creative is a driving concern. Indeed, simple exposure to relevant heuristics has proven effective for divergent thinking in other studies [7].

Acknowledgements

This research is supported by National Science Foundation, Engineering Design and Innovation (EDI) Grant 0927474.

References

1. Adams, R.S., Atman, C.J.: Cognitive processes in iterative design behavior. In: Proc. 29th ASEE/IEEE Frontiers in Education Conference, San Juan, Puerto Rico, pp. 11a6-13–11a6-18 (1999)
2. Amabile, T.M.: Creativity in Context. Westview Press, Boulder, Co. (1996)
3. Altshuller, G.: Creativity as an exact science. Gordon and Breach, New York (1984)
4. Benami, O., Jin, Y.: Cognitive stimulation in conceptual design. In: Proc. ASME 2002 Design Engineering Technical Conferences and Computer and Information in Engineering Conference, DETC2002/DTM-34023, Montreal, Canada, pp. 1–13 (2002)
5. Christensen, B.T., Schunn, C.D.: Putting blinkers on a blind man: Providing cognitive support for creative processes with environmental cues. In: Wood, K., Markman, A. (eds.) Tools for Innovation, pp. 48–74 (2008)
6. Christiaans, H.H.C.M., Dorst, K.H.: Cognitive models in industrial design engineering: A protocol study. In: Proc. Fourth International Conference on Design Theory and Methodology, ASME, pp. 131–140 (1992)
7. Clapham, M.M.: Ideation skills training: A key element in creativity training programs. Creativity Research J. 10(1), 33–44 (1997)
8. Cross, N.: Engineering design methods: strategies for product design, 3rd edn., Chichester. John Wiley & Sons Ltd, UK (2000)
9. Cross, N.: Expertise in design: An overview. Design Studies 25(5), 427–441 (2004)
10. Csikszentmihalyi, M., Getzels, J.W.: Discovery-oriented behavior and the originality of artistic products: A study with artists. J. of Personality and Social Psychology 19(1), 47–52 (1971)
11. Dorst, K., Cross, N.: Creativity in the design process: Co-evolution of problem–solution. Design Studies 22(5), 425–437 (2001)
12. Eberle, B.: Scamper. Prufrock, Waco, Texas (1995)
13. Finke, R.A., Ward, T.B., Smith, S.M.: Creative cognition: theory, research, and applications. MIT Press, Cambridge (1992)

14. French, M.J.: Conceptual Design for Engineers. The Design Council/Springer, London, UK (1985)
15. Gordon, W.J.J.: Synectics. Harper & Row, New York (1961)
16. Harrington, D.M.: Effects of explicit instructions to "be creative" on the psychological meaning of divergent thinking test scores. J. of Personality 43(3), 434–454 (1975)
17. Kruger, C., Cross, N.: Solution driven versus problem driven design: Strategies and outcomes. Design Studies 27(5), 527–548 (2006)
18. Newell, A., Simon, H.A.: Human Problem Solving. Prentice Hall, Englewood Cliffs (1972)
19. Nisbett, R.E., Ross, L.: Human Inference: Strategies, and Shortcomings of Social Judgment. Prentice-Hall, Inc., Englewood Cliffs (1980)
20. Pahl, G., Beitz, W.: Engineering Design: A Systematic Approach, 2nd edn. Springer, London (1996)
21. Park, J.A., Yilmaz, S., Kim, Y.S.: Using visual reasoning model in the analysis of sketching process. In: Workshop Proc. Third International Conference on Design Computing and Cognition (DCC 2008), pp. 15–22 (2008)
22. Schon, D.A.: Designing: Rules, types and worlds. Design Studies 9(3), 181–190 (1988)
23. Yilmaz, S., Seifert, C.: Cognitive heuristics employed by design experts: A case study. In: Proc. Third International Design Research Conference (IASDR 2009), Seoul, Korea (2009)

An Anthropo-Based Standpoint on Mediating Objects: Evolution and Extension of Industrial Design Practices

Catherine Elsen[1], Françoise Darses[2], and Pierre Leclercq[1]
[1] *University of Liège, Belgium*
[2] *University of Paris Sud, France*

This paper questions the new uses of design tools and representations in the industrial field. A two months *in situ* observation of real industrial practices shows (i) how strongly CAD (Computer-Aided Design) tools are integrated in work practices, in preliminary design phases as well, and (ii) how design actors sometimes deviate this tool from its initial objectives to use it in complement of sketches' contributions. A multi-layered study built on an anthropo-based approach helps us to deepen the "mediating objects" analysis. It also suggests considering the complementarities of design tools instead of their differences in order to propose another kind of design support tool.

1 Introduction - A Shift in Design Tools' Consideration

Research in the design field deals with numerous topics, among which the support of early stage processes in design, that has gathered a lot of attention in architecture, industrial or mechanical design. Distinct communities emerge: some of them improve CAD (Computer-Aided Design) tools to carry through "quick and dirty" representations; others make SBIM (Sketch Based Interfaces for Modeling) more efficient; or enlarge sketch potentials. The argumentation principle in literature is more or less similar. Most of the authors list sketches' advantages or shortcomings as well as CAD tools' powers or limitations to support ideation (table 1). The core of the comparison lies at the "end of the preliminary design stage", usually defined as the shift from free-hand sketching to Computer-Aided Design detailed drawing [1]. This comparison of the two design tools' benefits and limitations enables the authors to finally confront them (free-hand sketches *versus* Computer Aided Design tools), before presenting one own technical, methodological or theoretical proposition.

J.S. Gero (ed.): *Design Computing and Cognition'10*, pp. 55–74.

Table 1 Free-hand sketch and CAD tool pros and cons.

FREE-HAND SKETCH	
PROS	• is fast, easy, allows an efficient problem/solution exploration through minimal content [2] • makes easier the apprehension of complex and wide problem space • allows unexpected discoveries through its high opportunist aspects [3] and the "see-transform-see" mechanisms [4], keeping the exploration dynamic • allows different levels of abstraction [2] and a certain ambiguity (incoherencies between several representations of a same object are allowed) [5] • enables a "width" strategy (exploration of more alternatives) [6] • constitutes a "paper memory" : deletion is never totally completed • lightens spatial memory load [7]; constitutes a real "external working memory" relieving the internal short-term memory from additional cognitive costs; is a mnemonic help [8] • supports communication and construction of common reference systems [9] • stays a natural, intuitive and traditional "interface"
CONS	• is lacunar, ambiguous, highly personal with complex, implicit content and low level of structuration, stays rigid and static (non-reactive representation) • has a slow production-time (although it can help to mature ideas and get "insights")
COMPUTER AIDED DESIGN TOOL (WITH 3D DYNAMIC MANIPULATIONS)	
PROS	• is a very powerful tool for feasibility studies : allows to calculate, optimize, simulate any kind of reaction to multiple constraints (physical constraints, production constraints, ...) and to reach high levels of complexity • enables a relatively quick access to 3D visualization for evaluation • eases modifications through parametrizing • eases technical communication and data exchanges through formats' unification • sometimes leads to positive premature fixation [10]
CONS	• involves a "depth" strategy during the ideation process: less alternatives are produced [6] • proposes a WIMP interface (Windows, Icons, Menus, Pointing device) that is unnatural and distracts the user from the design task • can cause (in case of altered use): loss of documents, transfer and incompatibility issues, hazardous misinterpretations, ... • requires several months of training for an adequate use • is not well suited for the support of opportunistic creativity • sometimes leads to negative premature fixation [10] • induces frequent deletions or modifications operations that limit the possibility to capture design rationale

Whatever the point of view, both tools present respective particularities that can (in)efficiently equip the design process. Less is said nevertheless about how designers effectively exploit these tools: how do they select them, and according to which characteristics ? is this choice subjected to changes all along the process ? and what are the specificities of these changes ? which factors do "shape" the use of design tools ? On the other hand, free-hand sketches are considered as more "traditional" than CAD tools. How have these "new" design tools impacted the everyday work practices ?

To answer these questions, our paper suggests that once a tool is integrated in work practices - whatever its pros and cons - there is a *reciprocal impact* of, on the one hand, the adaptation of the tool and, on the other hand, the evolution of work practices.

Moreover, a new tool should not be considered as impairing the work but rather as enriching what already exists. In other words, we suggest that it is not worth considering free-hand sketch *against* CAD tools, since these "mediating tools" are useful and complementary in their respective contributions. The paper will show that CAD tools are indeed now fully integrated in designers' work practices while free-hand sketches remain a powerful design tool. This observation also questions the widening of the traditional borders of "the early stage of design" and its "traditional tools".

To better understand these "mediating tools" evolutions and modulations, we examine various factors, such as operating methods, collaborative modalities or cognitive demands all along the design process. The next section will present the theories that structure this examination, while the third section will detail our methodologies. We will then present our main observations and test our previous suggestions. Our hope is that our multidisciplinary approach contributes to a more effective convergence to "augmented design tools" that stay closer to real practices.

2 Rationale of the Study: Understanding the Use of Design Tools through a Three Phases Proposition

Several schools of thought appear in research on design tools:

- The first one holds the situation just as it is: sketches are powerful for preliminary design, CAD tools for detailed design. Mitchell & al [11] share this conservative point of view. They argue that "because creativity is associated with novelty, comprehensive computer tools for creative work will be neither possible nor necessary to develop, any more than it is necessary for a pencil to include all functions for drawing". For this community, CAD tools are not considered as design tools but just as drawing tools, and there are other domains to be explored in design research;
- The second tries to avoid both sketches and CAD tools limitations by proposing parallel techniques, like SBIM (Sketch Based Interfaces for Modeling, for a complete survey, see [12]) or Virtual Reality systems

including a sketch input. These systems deal with "quick and dirty" representations but are not linked to designers' work tools and practices, and being so do not answer the professionals' expectations [12];
- Finally, the third gives up on traditional (and sometimes obsolete) free-hand sketch techniques and focuses on CAD tools, sometimes augmented by haptic or immersive interfaces.

To reach our goal, that is to say to gain insight into design tools evolution and to get closer from current professional realities, we prefer to first put aside such "techno" decisions. Our reasoning is built on 3 main phases: first to take an "anthropo-based" standpoint, then to focus on mediating objects and finally to study the tools' complementarities. Dorst [13] proposes the same type of approach and bases it on 4 main steps: *observe - describe - explain - prescribe.*

2.1 First Phase: Addressing the Question from an "Anthropo-Based" Standpoint

In order to keep the actors of design activity at the core of our research, we adopt a comprehensive ergonomic approach. Ergonomics provide sound methods to conduct empirical *in situ* studies and adopt a multidisciplinary point of view. The aim of these "anthropo-based" methods is to analyze all concerned actors, without focusing only on obvious "end-users". These methods enable us to study the designer's profile, the definition of the real and prescribed tasks, the strategies, the required competences, and so on. Ergonomics particularly fits to the logic of business, reliability, productivity and competition inherent to design environment. This discipline also enables us to take into account two major impacts: the impact of new technologies and the impact of work contexts.

As far as new technologies are concerned, as we underlined before, there is a need to evaluate how designers have been able to adapt their work and competences since CAD tools' introduction. The importance to consider practices' evolution is underlined by Dorst [13]: "likewise, we are surprised that the tools we are developing are not widely used in design practice [...]. The momentous changes in design practice that are taking place at this time do not seem to influence design research at all. But they should […]".

Regarding the impact of the work contexts, as suggested by McGown and Green [14], the linear models of design processes developed in design engineering or psychological studies need to re-introduce the "loops" of actions. There is as well a need to put forward the external constraints of context [13; 1]. We would even emphasize the multiplicity of elements by putting it in the plural: contexts of work, of cooperation with colleagues, of physical environment, of types of project.

2.2 Second Phase: Focusing on "Mediating Objects"

Our interest goes to the evolution of design tools' usages in real practices. As a reference of analysis, we consequently choose to focus on the "mediating tools" of the design activity. We even extend our focus to the "mediating objects". In addition to the physical tools (the pen; the computer, the prototyping machine, ...), the mediating objects include the external representations linked to them (respectively the free-hand sketch; the 3D model or print, the physical model, ...). By considering them this way, we try to avoid a general misunderstanding that can occur between "tool" and "representation". For CAD for instance, a polysemy can occur between (i) the tool itself, with its Human-Machine Interface, its modalities of use and sharing, the techniques of 3D modeling (box modeling; mesh or surface modeling: extrude-edge;...); (ii) the *cognitive* artifact, visual basis of a virtual design (in 2D or 3D) or (iii) the external representation, physical production as 2D prints or 3D prototypes. This polysemy commonly appears during designers' verbalizations and it reveals the multiplicity of significations that an "object" can have.

In order to study these mediating objects, we adopt the instrumental theory as theoretical framework. Developed by Rabardel and Vérillon [in 15] this theory introduces the notion of instrument as the combination of an artifact (material, symbolic, cognitive, or semiotic) and one or more associated schemes. The artifact can be commonly defined as the physical part of a tool. On the other hand, the scheme is the result of "a construction specific to the subject, or through the appropriation of pre-existing social schemes" [16]. The example usually given is the hammering scheme, ordinarily associated with a hammer, that could be adapted to a shifting spanner in case of necessity. Both poles of the instrumental entity (the artifact and its utilization scheme(s)) act together as the mediator between the subject and the "object of his activity" [16], defined here as the "act of designing" (fig.1).

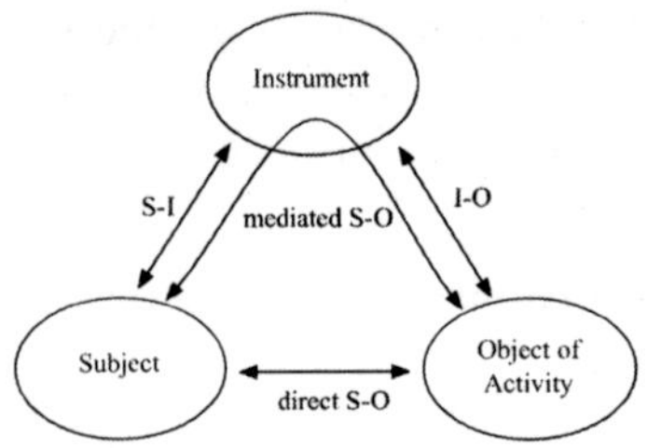

Fig. 1. IAS Model, "Instrumented Activity Situations", by Rabardel & Vérillion, 1995 [15].

Among all the possible approaches of Human-Machine relationships, we adopt this "mediation of the activity through the usage of *objects*". It helps us to put forward the actual characteristics of industrial designers' work through the use, the sequence of use and the modifications of "objects" inputs.

2.3 Third Phase: Undoing the Comparative/Dichotomous Approach to the Benefit of the Study of Complementarities

As we underlined before, new digital design tools and modified contexts of work inevitably affect each other. Some authors argue that the schemes of use of these new tools are in contradiction with the traditional schemes (free-hand sketch schemes), this maladjustment being the cause of a constraining work environment [17]. In contrast, we would suggest not to consider two opposite profiles of designers working in dichotomous worlds and using incompatible schemes (traditional schemes *vs.* CAD tools schemes), but (as figure 2 shows), rather to consider a flexible mid-way profile taking advantages of the objects' diversity and complementarities (in regard to the appearing constraints and the contexts).

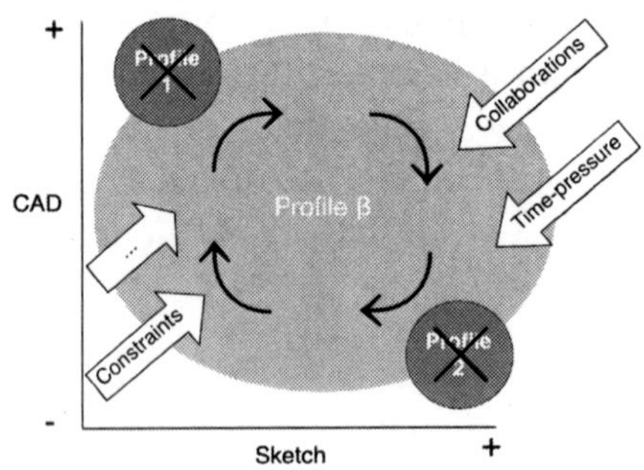

Fig. 2. The undoing of the dichotomous approach to the benefit to complementarities' study.

What could be seen as a paradox - the use of tools that seem inappropriate – will be showed later as the human capacity to adapt to a constraining environment, or to deviate the tools from their original usages.

These three theories (the last one remaining to be proved) structure our study of design tools' evolution as well as our research methods that are presented in the next paragraph.

3 Method

A two-stage method is proposed. Both aim at understanding the reciprocal impacts between the contexts and the mediating objects, as well as analyzing their consequences on work practices' evolution. On top of that, the first stage more particularly aims at (i) listing the designers' work habits and (iii) defining global work profiles. The second detailed stage tests the complementarity thesis and deepens the mediating objects' analysis.

3.1 Twelve Conversations to List the Context Factors: An Exploratory Research

A single research move is not enough to explore all the factors that could exhaustively explain the design tools' evolution, and consequently there is a need to select a few of these factors. This exploratory research tries to embrace the diversity of their origins to better manage this selection.

We organized twelve conversations with designers representing the diversity of the design profession. The representativeness of the sample is exhaustive, since all the designers have different careers (textile designer; light designer; industrial designers; architect/interior designer; furniture designers; teacher in design school; designer specialized in virtual graphic creations; designer of advertising structures and stands).

Among them, 7 can be considered as experts (more than 5 years of business experience in the design field, have been exposed to numerous situations individually or as part of the team); 5 as juniors (less than 5 years of experience). Another type of expertise level can also be underlined: the expertise in CAD tools. Indeed, among the 5 juniors, 4 are considered as experts in CAD tools, and among the experts, only 2 out of 7 are able to use these tools.

Nine of the twelve designers are coming from the same design school. This could be seen as either a limitation of the sample representativeness or the possibility to fix the education variable (that could also explain the expertise level toward CAD tools).

The interview protocol is "semi-directive" and is structured on a retrospective analysis of past projects. The retrospective analysis consists in asking to the designers to choose two projects they consider as representative of their work (achieved or not). They collect all the graphical/digital/physical traces they can find back from these projects. Asking the designers to refer to these real traces helps to found the verbalization on tangible memories and tends to avoid biased speeches. The questions can be classified in 5 themes: (i) general questioning for the sample definition; (ii) presentation of the design process of both selected projects (methods, inspiration sources, collaborations,...); (iii) operative methods of the everyday work; (iv) use of design tools (and representations) and (v) modalities of collaboration.

3.1.1 Data Analysis

The data gained through these 12 interviews is classified in several context factors, each of them presenting a double variable. Five of them are turning to profit in this paper:

- Expertise level in design field: *Junior / Expert*;
- Exploitation of CAD tools: *Him(her)self / Sub-contract*;
- Use of CAD tools: *In production phase only / In preliminary design and production phase*;
- Recourse to free-hand sketching: *Yes / No;*
- Possibility of co-working with a draughtsman: *Yes / No.*

This classification enables us to do a descriptive and quantitative (but preliminary) counting and to classify the 12 subjects in their corresponding variables.

3.1.2 Results of the Interviews

We present a few conclusions that emerge from this exploratory phase. For further contents, please refer to [18]. The interviews' results are summed up in the following matrix (table 2).

Table 2 Each number, situated at the crossing of two diagonals, represents the number of designers positively satisfying to the two variables the diagonals are referring to.

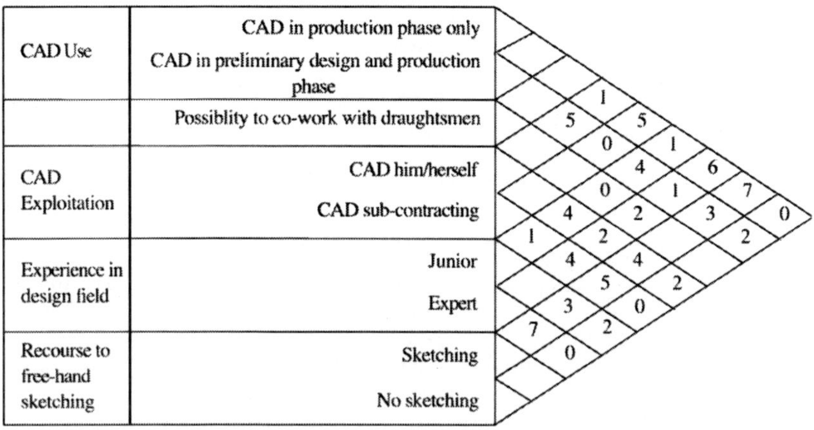

A first difference appears between juniors and experts (in design field), as far as CAD use is concerned. A majority of juniors, educated to CAD tools during their training, does not hesitate to use this design tool as soon as possible. Less interested or trained to CAD tools, experts only use them during the detailed design phase and under time, market and economic pressure. The verbalization makes appear a second difference. The recourse to CAD tools depends on the possibility of co-working with a draughtsman. Juniors usually work individually and do not have access to larger structures introducing draughtsmen. On the other hand, experts have more possibilities of working in such structures, and some of them indeed co-work with them. On top of that, a link between the experience level and the fact of sub-contracting (or not) the CAD detailed phase could exist.

There is no clear link between the use of free-hand sketching and the personal exploitation made of CAD tools. On the other hand, designers that argue not being in need of free-hand sketching never sub-contract the use of the CAD tools. In a similar manner, the link between the CAD tools' use and the recourse to free-hand sketching reveals the remaining importance of both design tools, as well as the impact CAD tools have on work habits and more traditional tools.

From these preliminary results, we propose a first prognostic in terms of three designers' profiles (table 3). This table sums up (i) the recourse to each type of tool in regard to the design phase; (ii) the relation maintained with free-hand sketches and CAD tools and (iii) the relationship with the potential draughtsman.

Table 3 Proposition of three designers' profiles.

Profile number	Supposed relationship with sketches	Supposed relationship with CAD tools	Supposed collaboration with the draughtsman
1 - Sub- contracting CAD phase	•during preliminary design phase principally •iterative loops see-transform-see conversation with the sketch	•minimal •evaluation; checking; communication	• distributed design • negotiation ?
2 - Preliminary iterative design using sketch *and* CAD tools	• during preliminary design *and* production phases • iterative loops • see-transform-see conversations with both representations	•during preliminary design *and* production phases •iterative loops •see-transform-see conversation with CAD representations	• collaboration • co-design
3 - Preliminary iterative design using the CAD tools only	• minimal • reminder sketch • crystallization sketch	•during preliminary design *and* production phases •iterative loops •see-transform-see conversation with CAD representations only	• No information at this stage.

On top of that, this exploratory research enables us to attest some of the impact factors that contribute undoubtedly to the evolution of designers' practices:

- The impact of CAD tools introduction on more traditional tools (here, free-hand sketches), as already underlined by many authors;
- The impact of contexts elements on the use of mediating objects (whatever they are): external constraints, time-pressure, customers expectations, levels of experience in design field and expertise in CAD tool usage;
- The impact of the chosen mediating objects on the design process;
- The impact of a new type of collaboration with the draughtsman.

The second stage, presented next, enables us to go on with the exploration of these factors' impact on our research questions and to refine the profiles definition in regard to the "complementarity" thesis. We decided to take advantage of the anthropo-based approach in order to study industrial designers in their real working context.

Indeed, referencing to a specific domain is more efficient than exploring a wide field of design [12]. Consequently we focus on an industrial design team (i) made up of designers with diverse profiles; (ii) collaborating with draughtsmen, and (iii) working in contexts presenting rich variability.

3.2 Detailed Research

A design team hosted us for a two months *in situ* observation. This team is active in the field of heating devices, and is acknowledged for its high aesthetic and high quality products. The team is composed of 5 designers (all experts in the design field, and among them 3 with high expertise in CAD tools) and 3 draughtsmen (all experts in CAD tools; one expert in the specific design field, 2 juniors).

The observer stayed 8 hours a day inside the open-space office. She was allowed to interview the subjects and capture (recording or filming) every stage of the current designs and all the interactions (between the team, between members of the team and extern members such as the CEO or the prototypists).

This type of *in situ* intervention presents three advantages. First, it avoids the limitations of a non-realistic lab situation by providing the essential contexts elements. Second, it avoids the possible disturbance of a think-aloud protocol. Third, it enables a qualitative approach of the fine-grained details of the design process that would be ignored in a more quantitative study. These details indeed constitute a stumbling block of the whole project rationale but remain very punctual.

On top of the 8 interviews (based on the same semi-directive and retrospective analysis protocol than the exploratory research) we selected 5 different products as a basis of study. These projects were selected for their representativeness. They indeed provide a good range of use of mediating objects, and present diverse states of progression (formal, technical and productive). They provide a relatively complete view of the design process and methods without following a 2 or 3 years complete project.

3.2.1 Data Analysis

Collected Data (interviews based on retrospective analysis as well as *in situ* observations) has been coded [see 18]. This coding aims at gaining information about the tools' and representations' uses in relation to what occurs as "external" factors. Further explanations have been gained (through questioning) in case of uncertainty.

The code applies to distinct units of designing actions. One action is defined as soon as the mediating object changes. This change usually goes with a shift in design process (shift from one support to another, one piece to another, one constraint evaluation to another,...). This coding scheme is exploited to construct the timelines of the projects (fig 3). Timelines aim at reproducing the design process of the 5 selected projects. The X-axis figures the project evolution in time, and represents different time scales since the data proceed from interviews' or observations' coding. The Y-axis sums up the various variables of the coding scheme. These variables are classified according to the use of one specific tool (sketch; CAD tool or prototype). For each tool, variables are again classified in different levels: (i) an "utility level" (or function inside the process) answers the question "what is it useful for ?"; (ii) a "cognitive level" designates the designer's cognitive activity: gathering information or knowledge, generating solutions, evaluating or modifying, searching in iterative loops and (iii) a "productive level" lists the type of representations obtained (in terms of content, spatial representation or underlying model). In parallel on the Y-axis appears the modality of collaboration (with whom, for doing what).

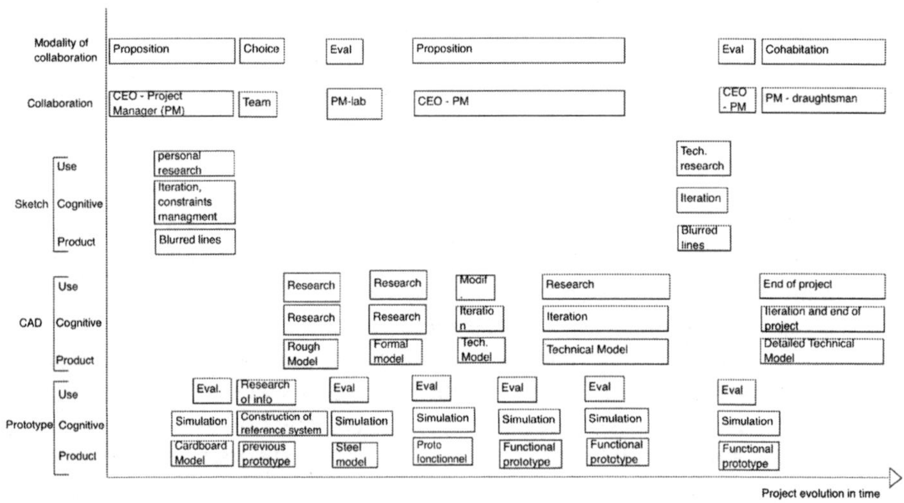

Fig. 3. An example of timeline with some variables (non exhaustive listing).

3.2.2 Intermediate Observations Resulting from Timelines Analysis

The first intermediate results are provided by a comparison of the five timelines. We observed 5 impacts that external context has on project rationale.

First, *the impact of time pressure*. Surprisingly the designers can use CAD tools as a "rough" formal tool and then come back to sketches in order to solve a more technical point for instance. Consequently there is a need to distinguish "rough" sketches and "rough" CAD models or representations (that stay ambiguous and support ideation), from "technical" sketches and "detailed" CAD models (that focus on a more specific sub-problem). Simple 3D primitive forms characterize the "rough"

CAD models or representations. These models are very quickly created without taking care of real dimensions and proportions. As rough sketches, they support the rapid evaluation of more formal or functional ideas.

Second, *the impact of project management*. Some projects indeed suffered from late decisions; tools maladjustments to the design task, necessity to start again detailed 3D models, CEO choices, ...

Third *the impact of tools selection*. Projects are highly structured by several back and forth between different mediating objects (free-hand sketch; CAD tool; prototype). The selection principles depend on the respective properties of both tool and representation. For instance, a sketch on a 2D print will be used to test dimensions or pieces conflicts; 2D hand-drawn perspectives to test a cinematic principle, ...

Then *the impact of collaborations*. The projects present various types of collaborations (complex and laborious co-activities; efficient co-design) that modulate tool selection or task repartition.

And finally, *the impact of a new co-worker*. In parallel with various tasks repartitions, some projects are impacted by the tasks supported by the draughtsmen. They take a great part in the design process, as the following two graphs of actors' activity demonstrate. Based on the activity theory, these graphs give insight into tasks distribution between designers and draughtsmen as well as into the role of mediating objects. The first graph presents the global activity of a designer (fig 4).

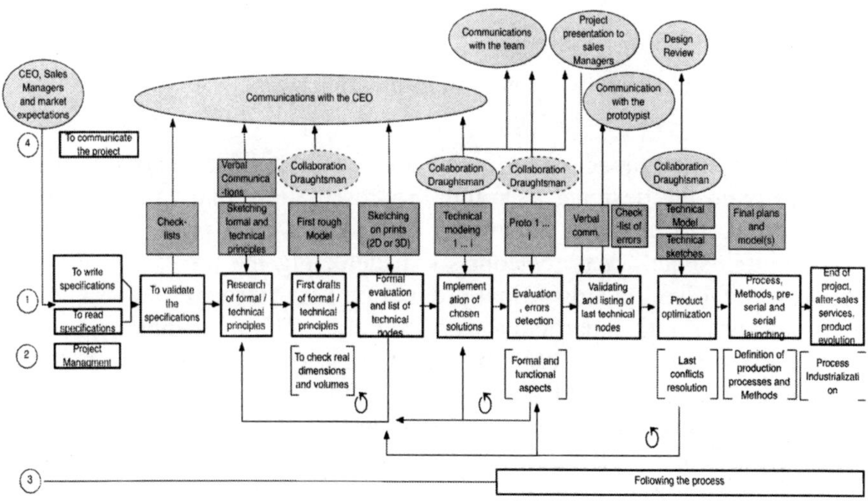

Fig. 4. Designer activity graph, with its multiple layers.

This graph is composed of three linked layers. Bold layer indicates the iterative model of the designer activity. The various tasks of a designer are presented. The circular arrows show the multiple points where iterations might appear. This "task model" has to be considered as a simplification of the whole activity. The dark grey layer represents the several objects (tools or representations) used all along the process, in mediation between the designer and his/her colleagues or with him/herself. The light grey accounts for the occurrence of a collaboration, with specific persons and according to specific modalities of collaboration, and always through the use of a specific mediating representation. The quasi-systematic exchanges appear in continuous lines, while the occasional ones appear in hatching lines.

Similarly, the draughtsman activity is presented in fig.5.

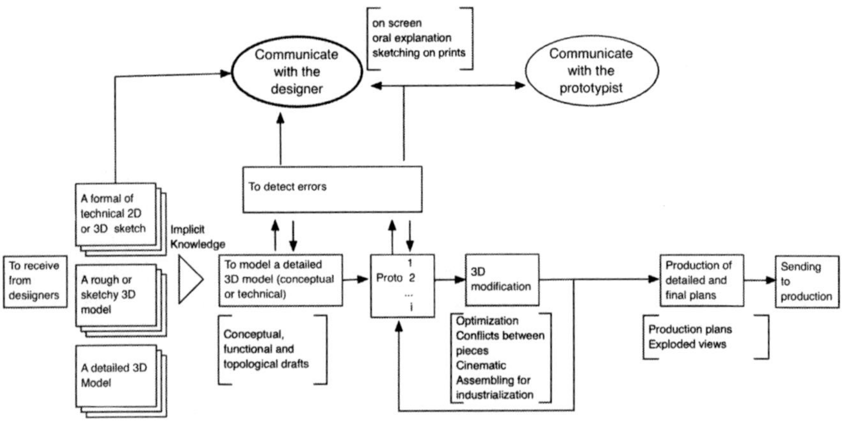

Fig. 5. Draughtsman activity graph, with its multiple layers.

This simplified model underlines 4 observations. First, the draughtsman receives from the designer a "rough" representation, that can either be a free-hand sketch, a rough 3D model or a sketch on a print. Second, the main draughtsman's activity consists in detecting the errors and making the project evolve towards a final production plan (through the production of prototypes in this particular design field). Third, his/her activity is deeply impacted by the type of CAD tool used (Pro-Engineer here). He/She adapts to this tool's possibilities and limitations. Finally, he/she develops in a few years a great expertise in this specific (and very technical) design field and is totally able to co-operate with the designer in a win-win relationship.

4 Results

This section presents our results in terms of mediating objects' evolution and testing of the complementarity thesis.

4.1 Testing the Complementarities

The previous section and the study of the draughtsman's activity graph tends to position the draughtsman not anymore as an executive drawer but as a "designer-draughtsman", which activities are part and parcel of a reassessed design task. The complementarity thesis that we presented above seems even to push further the notion of dichotomy. The next section discusses the corresponding results.

4.1.1 Dichotomy between "Designers That Design" and "Draughtsman That Execute"

As our results tend to prove, the usual dichotomy (or hierarchy) that links designers and draughtsman disappeared with the recurrent use of CAD tools. Required as early as possible in a project (for economic, time or productivity reasons), these tools are being integrated in designers' tasks, and lead to a new type of collaboration between designers and draughtsmen. A shared referential is being constructed between both actors as a function of the expertise and experience levels. This leads to a situation of "co-design" in the highest and more effective situation. This collaboration was already quoted by some authors, but expressed in a different context. For Lebahar, the draughtsmen's mission is beyond a simple verification of representations. They oppose their own vision of the representation and impose, in a certain way, their own models [1]. Marjchzach and al (1997) and Löwstedt (1993) [quoted in 19] argued at the beginnings of the CAD era that [3D models] were "a technology at disposal which implantation deeply and durably transform the organizations and functioning of a company". These affirmations were right at such times where the CAD tools still were an inaccessible technology for designers but should be reconsidered now, since the situation has evolved. We do not talk about "opposition of representations and models" anymore but about "co-design", and we do not consider the CAD models just as "a technology at disposal" but instead as a complementary tool justifying the introduction of a new co-worker in the design field.

We do not argue that both profiles are strictly equal nowadays. There are still differences that make them complementary. For instance, one draughtsman explained that "the question of how to model is more often asked that the question of what to model". The draughtsmen indeed have to develop a specific "way of thinking" to start the 2D or 3D virtual model, that lead them to question the essence of the sketchy representations they receive. Where and what are the "technical nodes" (or difficulties) of the product ? What kind of cinematic behavior will the product have ? How will it be possible for the prototypists to physically put a screw in such a tiny fold ? And last but not least, how will this piece co-exist with the pre-existing environment ? Draughtsmen even talk about a "programming" of the model to think about before starting the modeling. This programming can be defined as an efficient strategy to quickly represent the 3D model in respect with the future potential modifications and with the hierarchical structure imposed by the software (called "referencement tree"). To conclude with the diversities, we can say that (i) mental transitions (from 2D to 3D and vice-versa) are different between designers and draughtsmen, i.e. between the author of the sketchy representation and the interpreter; (ii) these specific draughtsmen develop a "Pro-E" way of thinking that can be or not appropriate to mental representations and tools' utilization schemes. In case of maladjustments, the subjects are able to adapt themselves to the constraining environment.

4.1.2 Dichotomy between "Designers That Design" and "Designer That Model"

Likewise, the dichotomy between "sketch in a preliminary phase" and "CAD in a detailed phase" also have to be revisited. As well as, in extension to what was previously said, the dichotomy between "designers that design" and "designers that model".

The profiles of designers we defined (table 3) have to be extended. All designers, at least in this particular research, sometimes resort to free-hand sketch, and sometimes to CAD tools. It depends on the particular constraint or task they are dealing with or on the current modality of collaboration[1]. Such constant backs and forths between the tools and representations vary from one designer to another and co-exist efficiently in order to reach the design goal. We suggest that these iterations depend on the level of adaptability of the tools and the schemes of utilization. We also underline that there is not anymore one type of free-hand drawing (the "rough" drawing) and one type of detailed CAD model.

As pointed by our observations and by the verbalizations, the content varies from one rough-sketch to a technical sketch, from a rough-model to a detailed model. They can be all used at any time of the design process. A "mediating objects' graph of use" presented in the next section assesses this observation.

[1] For instance, we observed that involved partners always tend to cooperate using the external representation the closer to their shared system of reference (for instance, designers and prototypists cooperate using a physical model; designers and draughtsmen use a 2D print, or designate on screen).

4.2 Going Further in the Analysis of Mediating Objects

The figure 6 is the “mediating objects’ graph of use” that deepens the understanding of the loops that appear in the usage of the design tools. It enables us to identify on which principles designers shift from one object to another and what are the tools’ respective contributions. The X-axis designates the mediating objects appearing in their chronological order, as they appeared in the designers’ activity graph previously presented. The Y-axis presents the three levels of tools “functionality”, as they appeared previously in the timeline. In parallel of the X-axis, the various “drawings registers” of Lebahar [1] test the evolution of abstraction levels. The first drawings register includes the topological representations. The second regroups the projective representations (no account of real measures and angles but organization of the abstract parts in a figural entity). The third register gathers the Euclidian representations (defined by the geometrical invariants and preventing the deformations, unlike the projective register). Again, inside the graph are presented the several variables (coming from our coding scheme) that sum up the global process of the 5 analyzed projects.

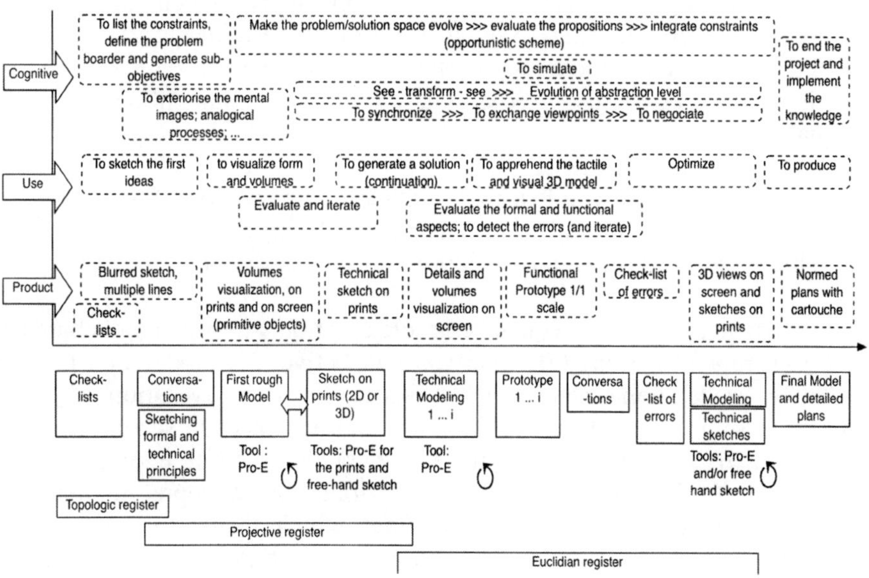

Fig. 6. This graph shows the evolution and extension of mediating objects.

This graph reveals numerous iterative loops. The iterative process is a commonly well accepted concept in design literature, but this graph enables us to enter more deeply into the study of these loops. The study of the objects (or instruments of the mediated activity) shows that in a first loop, relating to a free-hand sketch phase, the sketch stays blurred, dynamic and "open" to creativity. The rough 3D model, when relevant, stays simple, deprived of details and easy to read, as well as also easily modifiable and parametrizable. This loop concludes a first definition of formal concepts. The process then stays relatively linear till the emergence of a more complex model. A new loop can then take place thanks to the emerging constraints (revealed by CAD visual facilities and integration in a pre-existing environment), thanks to interactions with colleagues or consideration of new technical knots. The iteration materializes again through a sketch, but this one presents another type of content. It aims at other objectives: it stays more technical, more focused on the resolution of a specific node and does not consider anymore the global formal aspect. Once the node solved, a model is put together, and this "bottom-up" kind of loops tend towards a more detailed 3D model. Sometimes, a prototype is used to evaluate the project in its real scale and its real mechanisms. This prototype itself reveals new proportions that can be quickly re-evaluated through a formal sketch, and so on.

Other observations can be drawn from this graph:

- the iterative process' loops match the loops of use of mediating objects: rough sketch > technical sketch > 3D model (leading to 2D views) > technical sketch > model > prototype > formal sketch;
- the prototypes are also used during tests phases or simulations of the final product;
- representations' contents evolve in a more continuous way, the abstraction level going towards a more detailed representation [as underlined in 20]. Loops nevertheless remain in the choice of representation type. This is shown through the evolution of Lebahar drawing registers.

The dichotomy is consequently obsolete not only between tools but also between representations (2D plan vs. perspective; 2D model vs. 3D). An iterative model combined with an abstraction level evolution is better suited. To this abstraction evolution we can add that the schemes of use (tools schemes and representations schemes) also seem to evolve inside a single project. The repeated appeal to various tools or representations afford the realization of a redundancy effect, which, thanks to Rabardel, allows the subject to make the better choice and achieve a balance between economic and efficient cognitive objectives [15]. Some "tools" are also used simultaneously: for instance the collaboration on prototypes goes with an enormous amount of gestures, while the giving of a personal sketch is always commented. Such a multi-modality functioning happens very often during each step of the design process, and contributes to our complementarity proposition.

5 Conclusions - Toward Augmented Design Tools Closer to Real Practices

The approach of real design practices through mediating objects enabled us to establish the relevance of the complementarity approach when considering co-workers, tools and representations in a design team. We presented the impact of tools on elements of contexts and vice-versa, as for instance the impact of time pressure on tool selection. We also underlined the need to focus not only on "obvious" end-user actors, but to widen our field studies to all practitioners that impact in a certain way the process. There are not dichotomous profiles but flexible ones, actors adapting their work habits to the contexts. Ergonomics provide researchers sound methods to analyze the profiles, the various contexts and the adaptations in order to dedicate efficient specifications.

The usage of traditional or CAD tools has significantly evolved these past few years, and their respective impacts lead to the extension of what is usually called the "preliminary design". From now on we suggest that CAD tools (in some conditions of use) could be considered as potentially effective also in this part of the process if considered jointly with sketches.

The use of sketches is also expanded since they can help make technical decisions that come out from conceptual design. Both tools offer respective qualities since they are deviated by users, adapting to appearing constraints. A better combination of design tools advantages (in terms of schemes of use, functions and models of representations as well as Human-Machine interfaces) could lead to an interesting design support system.

The presented results deserve to be enriched by complementary observations, in other design teams creating other products (other scale, other relation to the human body) and working with other CAD tools for instance. These futures researches will lead to the definition of more technical specifications for the design of an industrial design support tool that could contribute to free-hand sketches' and CAD tools' facilities by taking advantage of their complementarities.

References

1. Lebahar, J.-C.: La conception en design industriel et en architecture. Désir, pertinence, coopération et cognition, Eds Lavoisier (2007)
2. Cross, N.: Strategies for Product Design, 3rd edn. N. Cross. The open University. Ed. Wiley, Milton Keynes (2000)
3. Visser, W.: The cognitive Artifacts of designing. Ed. L. Erlbaum, London (2006)

4. Schön, D.A., Wiggins, G.: Kinds of Seeing and Their Functions. Designing Design Study 13(2), 135–156 (1992)
5. Goel, V.: Sketches of Thought. Bradford MIT Press, Cambridge (1995)
6. Ullman, D.G., Wood, S., Craig, D.: The importance of drawing in the mechanical design process. In: NSF engineering design research conference (1989)
7. Suwa, M., Purcell, T., Gero, J.: Macroscopic analysis of design processes based on a scheme for coding designers' cognitive actions. Design Studies 19(4), 455–483 (1998)
8. Bilda, Z., Gero, J.: Does sketching off-load visuo-spatial working memory? Studying Designers 2005. In: Gero, J.S., Bonnardel, N. (eds.) Centre of Design Computing and Cognition, Univeristy of Sydney, Australia (2005)
9. McGown, A., Green, G.: Visible ideas: information patterns of concep-tual sketch activity. Design Studies 19(4), 431–453 (1998)
10. Robertson, B.F., Radcliffe, D.F.: Impact of CAD tools on creative problem solving in engineering design-CAD, vol. 41(3), pp. 136–146. Elsevier, Amsterdam (2009)
11. Mitchell, W.J., Inouye, A.S., Blumenthal, M.S. (eds.): Beyond productivity: information technology, innovation and creativity. National Academic Press, London (2003)
12. Olsen, L., Samavati, F.F., Sousa, M., Jorge, J.A.: Sketch-based modeling: A survey. Computers and Graphics 33, 103–856 (2009)
13. Dorst, K.: Viewpoint-Design research: a revolution-waiting-to-happen. Design Studies 29, 4–11 (2008)
14. Howard, T.J., Culley, S.J., Dekoninck, E.: Describing the creative design process by the integration of engineering design and cognitive psychology litterature - Design Studies, vol. 29, pp. 160–180 (2008)
15. Rabardel, P.: Les hommes et les technologies, approche cognitive des instruments contemporains. Armand Colin, Paris (1995)
16. Beguin, P., Rabardel, P.: Designing for instrument-mediated activity - Scandinavian. Journal of Information Systems (2000)
17. Béguin, P.: Le schème impossible, ou l'histoire d'une conception malheureuse. Research innovation revue, Quadrature, Paris, vol. 10, pp. 21–39 (1997)
18. Elsen, C.: Extension & modulation of mediating objects' use in industrial design. Master Th. Work & Society Sc., Ergonomics Research ULg-CNAM, Paris (2009)
19. Béguin, P.: De la complexité du problème à la complexité entre les indi-vidus dans les nouvelles stratégies de conception - Actes du colloque de l'école d'architecture de Marseille-Lunigny (1996)
20. Rasmussen, J.: Mental models and the control of action in complex environments. In: Ackerman, D., Tauber, M.J. (eds.) Mental models and human computer interaction, vol. 1, Elsevier, Holland (1990)

FRAMEWORK MODELS IN DESIGN

Beyond the design perspective of Gero's FBS framework
Caetano Cascini, Luca Del Frate, Gualtiero Fantoni and Francesca Montagna

Formal model of computer-aided visual design
Ewa Grabska and Grażyna Ślusarczyk

Design agents and the need for high-dimensional perception
Sean Hanna

A framework for constructive design rationale
Udo Kannengiesser and John S Gero

Beyond the Design Perspective of Gero's FBS Framework

Gaetano Cascini[1], Luca Del Frate[2], Gualtiero Fantoni[3] , and Francesca Montagna[4]
[1]*Politecnico di Milano, Italy*
[2]*Delft University of Technology, Netherlands*
[3]*Università di Pisa, Italy*
[4]*Politecnico di Torino, Italy*

Among the various model based theories, the Gero's FBS framework is acknowledged as a well-grounded, effective and tested reference for describing both analysis and synthesis design tasks. Despite its design-centric nature, the FBS model can provide a valid support also to represent processes and tasks beyond its original scope. The specific interest of the authors is to extend the FBS application to model also uses and misuses of objects, interpretations of the users, needs and requirements. In fact, as partially addressed also in literature, some issues arise when the classical FBS framework is adopted to model particular aspects such as the user's role, values and needs, as well as to produce an explicit representation of failures and redundant functions. The full paper presents an extended classification of aspects, beyond the design perspective, which currently cannot be represented by the FBS model and some directions for its possible extension. Several examples clarify the scope and the characteristics of the proposed model.

Introduction

Since its first formulation in 1990 [1], Gero's Function-Behavior-Structure (FBS) framework evolved in the last two decades. Gero himself has further developed and integrated his model as in [2]. Many authors have adopted the FBS model as a reference to describe design processes and tasks, while others started a scientific debate about the FBS framework by underlining

J.S. Gero (ed.): *Design Computing and Cognition'10*, pp. 77–96.

some ambiguities (e.g. the absence of a stable definition of function [3]), a few limitations (e.g. in the representation of human-machine interactions [4]) or difficulties in its extension [5, 6]. The Situated FBS is here assumed as the reference starting point [2] for its extension to product use context. Although the FBS framework has been conceived as designer centric model and the aim of Gero and co-authors was to describe and explicit the designers' behavior, the model seems stable enough to allow possible variants and extensions as for example proposed by the authors in [7]. Indeed, in [8] while introducing the concept of value system as a key to interpret innovation, it is highlighted the need to include in the model both producers and adopters, their interactions with the artifact and with each other.

The goal of this paper is to provide a contribution in this area of study, by extending the FBS framework to the representation of product use context through a deeper analysis of Gero's External World and the formalization of some cognitive issues of user-product interaction.

The paper starts with a critical analysis of the FBS model according to the aim of the present work. Then its limits and the reasons of the extension are presented, while in the following chapter the extended model is described with details about the integrated representation framework. A simple, but comprehensive example, related to the design and use of a microwave oven, clarifies the characteristics of the proposed model. The paper ends with some conclusions and foresights for further extensions.

Related Work

The introduction of the product use context requires to manage a series of different entities (actors, interactions and environments). In view of that, more comprehensive models capable to represent product affordances and their user's perception, user's knowledge and its relationships with failures and misuses are required. Therefore, to deal also with such design issues, a wide range of literature works has been used by the authors to ground the extended framework:

1. Actors and relations in the External World. Several researchers have proposed extensions of the FBS model to build a more comprehensive and detailed representation of the External World: the authors in [5] introduced the user needs; in the FEBS (Function-Environment-Behavior-Structure) design model [9] the need of another player, "the working environment" with its boundaries and resources, was highlighted; Brown et al [10] introduced the "rest of the world" (all that is not the device), where the product is used; Norman [11] underlined

the interpretations of artifacts, based on actor's past knowledge and experience, and drawn the attention towards the actor's perception.

2. Product usability and use context [12]. Kuipers [13], Keuneke and Allemang [14] and Chandrasekaran and Josephson [15] proposed the "mode of deployment" for describing the implicit assumption of context for making use of a device. The concept behind guess is that the user of a device can imagine the context the device is intended for, according to "its general usage". The use context is the stage in a product's life cycle when the product performs its functions to satisfy the user's needs. Using a product means instantiating "a goal-directed series of considered actions which includes manipulations of the product" [12].
3. Product affordance [10]. "Affordances are possible actions" and in particular "the affordances A of a device are the set of all potential human behaviors (Operations, Plans, or Intentions) that the device might allow". Affordances can be recognized from experience, can be learned and also inferred by analogy. Perceived affordances (originally introduced in [16]) are context dependent manipulation possibilities from the point of view of a particular actor [10]. The actor is considered to be the entity, human or otherwise, capable of taking action.
4. Failures and their perception. Failures can be observed by several points of view: a device stops working, its performances are reduced, its use in not intuitive etc. A detailed survey of the existing multiple meanings of the notion of failure in engineering is available in [17]. Becattini et al. [18] linked failures to all kinds of FBS model variables: their analysis focused on the loss of ideality of a device in terms of reduced performance, presence of undesired side effects and excessive consumption of resources to make the system work. Brown and Blessing [9], looking at the affordance, claimed that, unlike functions, affordances may or may not be associated with a goal. Thus, when a goal is fixed, affordances may or may not support it, or even in case of "negative affordances" may be undesirable and clashing with the goal.
5. Alternative Uses. Keuneke and Allemang [14] stated that product alternative uses are all the possible uses connected to the context and to the material decomposition of the device. Actually, the detailed material description provides making use of a device for other purposes (e.g. due to its weight a battery can be utilized as a paper holder not only as a voltage source; this functionality can be derived by the theory of physics and the weight descriptions of the components). Thus, the alternative uses are the possible behaviors B (interpreted by the user as possibilities of achieving goals G) of the system coming from its structure, but totally disconnected from the goals the designer interpreted as user needs and the product was designed for. As detailed in the next sections, alternative uses can be described as $G_u \neq G_d$, $Bs_u \neq Bs_d$.

6. Misuses are defined as those conditions in which the user manipulates the product in ways that were not intended by the designer, still keeping the same goal. It is proposed to distinguish between two kinds of misuse. The first case occurs when user's manipulation is based on his/her belief that the product affords A, but A was not intended by the designer. The second case occurs when user and designer agree on the affordances, but the user has erroneous expectations, about product's behavior. Summing up, the misuses are the possible behaviors (interpreted by the user as possibilities of achieving goals) of the system, coming from its structure and linked to the goals the product was designed for. According to the notation proposed in this paper, misuses can be described as: $G_u = G_d$, $Bs_u \neq Bs_d$.

Several research works analyzed in the State of the Art review proposes to extend the domain of the FBS framework. Nevertheless, since their attempts mostly apply the FBS model beyond its intended scope, it is not appropriate to consider the limits they highlight as intrinsic restrictions of the FBS framework. Table 1 summarizes the most relevant issues for the present work through a link between topics and possible approaches, with related references. Other aspects cited in literature, as for example the user interface and the concept of function, are just partially covered in table 1: the user interface and its relationship with product's structure and interacting interface constitute a main contribution of the present work and will be detailed in section 3. Besides, the concept of function and its nuances, even very interesting and not necessarily conflicting each other [3], are out of focus of this paper.

Wrapping up, it appears that FBS potential has not yet been fully exploited for representing design activities related to user's actions and interpretation processes. It could be observed that a user designs how to use an artifact for herself/himself and, consequently, from this point of view the FBS model might be reinterpreted according to this user perspective. Nevertheless, the goal of the authors is to propose a comprehensive representation of the cognitive aspects related to the product use context, in order to strengthen the design process, thus still with a close link with the designer's perspective.

Therefore, the authors have formulated a proposal for an extended FBS model that is simple in principle and effective to represent aspects of design and product development as well as users' behaviors, erroneous uses, misuses and failures. The intention is to represent a wider context for FBS model application and to extend the designer centric perspective to less traditional aspects. For doing that, it is proposed to split the concept of product's structure (S) in two separated parts: the user interacting interface Int and the inaccessible (directly by the user) or not used portion of the structure. The authors aim at demonstrating in the following paragraphs that such introduction (typical for computer science) can bring relevant advantages also for product design.

Table 1 (continued next page) Themes related to product use context not explicitly represented by the classical FBS model [2] and related literature

Theme	Reference	Approach
The behavior of a system is influenced by its environment	FEBS (Function-Environment-Behavior-Structure) design model [9]	"The working environment" with its boundaries and resources supplies both the environmental elements that contribute to the functions of the design and also those that contribute to failures".
Affordances are context dependent	Affordances are context dependent manipulation possibilities from the point of view of a particular actor. The actor is considered to be the entity, human or otherwise, capable of taking actions [10].	The user acts on the basis of beliefs and expectations he has about the product's behavior in a given environment. These beliefs and expectations are part of the user knowledge (Ku).
Affordance as a property	"The term affordance refers to the perceived and actual properties of the thing that determine just how the thing could possibly be used" [11, p.9].	Norman focused on the term "properties". Since he used the adjective "perceived" it is possible to infer that "affordances are information (signals) coming from a device and interpreted (see the next raw) by the user".
Interpretation of reality by the user as a key factor for the design process	"Affordances result from the mental interpretations of things, based on our past knowledge and experience applied to our perception of the things about us." [11, p.219]	User's knowledge acts as a "filter" (interpretation) of the information coming from a product: some information cannot be perceived, understood, etc..
Beyond the design-centric perspective	The introduction of the users' needs in [7] enlarges the number of variables and related relationships modeled by the FBS framework.	When other actors, e.g user and environment, are introduced in the FBS model, the relationships among the actors increase, thus enriching the whole picture.
User's Situatedness	Design as a dynamic process in which the view of the designer changes in time depending on the outcome [2]. The change of the external and the internal world of the designer determine the dynamic "situatedness" throughout the process.	It is interesting to notice that Gero's classification in terms of External world, Interpreted world and Expected world still remains useful also to "situate" new actors interacting with a product in different environments.

Alternative uses	Umeda and Tomiyama [19] focused on two related concepts: "alternative uses" and "redundant functions".	"... a component of a system might perform some functions that can be used in other ways than intended by the designer."
Interface	Interface is defined as the computer-based means by which workers obtain information about, and control the state of, a socio-technical system and it is composed of displays and controls [20].	The role of interface and its distinction from user interface allows to better investigate the reasons and ways by which the user acts on a product.
Misuses	Use of a product, process or service under conditions or for purposes not intended by the supplier, but which can happen, induced by the product, process or service in combination with, or as a result of, common human behavior [21].	The distinction done by Gero about expected and actual becomes even more important within the user perspective. Actually, the user can misunderstand the product behavior despite the achievement of a certain goal or, even, can use consciously a product in a wrong (out-of-design) way.
Failures	Failure. Termination of the ability of an item to perform a required function [22].	It is a common user experience that not all the instances of product's manipulation end up with the successful achievement of the user's goal. The unsuccessful uses are hereafter distinguished from misuses.

Extended Model

In this section the main features of the proposed Extended Model (EM) are introduced. The chapter consists of five subsections. The first one, which explains the basic notions, begins establishing the conceptual connection between the proposed extended model and the FBS Situated framework and continues introducing additional concepts and relations. The second subsection clarifies the notions of "use" and of "Interacting interface". The following subsection examines user's cognitive processes, how user's knowledge is organized and how it shapes user-product relationship. The fifth subsection analyzes how the model deals with misuse and failure phenomena.

Basic Notions

The proposed EM is largely based on Gero and Kannengiesser (2003) Situated FBS framework, with which it shares both the cognitive modeling approach and several key concepts. EM's primitive concepts are:

- Product's structure (S) is the physical constitution of the product, its components and their relationship, i.e. what it is.
- Product's behavior (Bs) is the observable attributes derived from the structure (S), i.e. what it does.
- Product's function (F) is the product's teleology, i.e. what it is for. This last definition is introduced for completeness purposes only since it will not be used in this paper.

A few clarifications are in order. The notion of "observable attributes derived from S" refers to the set of flows of energy, matter and signal (EMS) coming from the product which are potentially observable. Even for small and relatively simple products Bs is a vast set. It includes visible radiation, audible sounds, smells, and variations thereof. It is worth noting that the product is not emitting steadily the entire EMS set and not all subjects are equally exposed to the multiple components which are part of it. This implies that different subjects exposed to different subsets of the entire flow are experiencing different parts of Bs. For example, maintenance personnel are exposed to Bs aspects that are usually inaccessible to final users. For this reason, a distinction within Bs is introduced, namely Bs_u and Bs_d: Bs_u refers to the part of Bs that is observed by the user U; Bs_d is the part of Bs that, according to designer intent, is expected to affect U.

The argument developed in this paper rests on the observation that in many cases Bs_d and Bs_u diverge. The reasons of this divergence are several, but the paper focus does not investigate them. Usually Bs_d is larger than Bs_u, because the designer tries to anticipate all possible user's needs and actions. Sometimes, however, users may be able to identify product's behaviors the designer was not aware of.

The Bs_d and Bs_u divide has also a structural counterpart. Designer's expectations about the features and phenomena that are part of Bs_d are based on his/her knowledge about the product's user interface (UI) that is the part of a product which has been intentionally devised by the designer for hosting the user-product interaction. In this sense, the term applies both to computer's graphical interface as well as to handlers, knobs and other physical features adopted in human-machine interface. As for Bs_d, also UI is designed with the aim of meeting the broadest range possible of users' needs and actions. However, not all users interact in the same way, some are more experienced than others, some are more explorative, and others are very conservative, and so on. Consequently, the extension of S with which users interact is variable. In this paper, the part of S with which U interacts is labeled Int (Interacting Interface). Experienced and explorative users interact extensively and, in their case, Int will tend to coincide with UI. For less experienced users, Int will be smaller than UI. Then, in some cases, users may interact with parts of the product that were unanticipated by the designer, in this way including into Int parts that are not included into UI.

Given the above definitions the following relations hold between structural entities and behavioral phenomena. Bs is the sum total of observable EMS flows from the entire S. Bs_d is the EMS flow from the UI part of S. Bs_d represents the designer expectation about product's behavior during user-product interaction. Finally, Bs_u is the EMS flow from the Int part of S. Consequently, any product's behavior that affects U has to come from Int, and product's behaviors generated outside Int are not perceived by U.

Bs_u has a fundamental role in influencing the way in which U steers his interaction with the product. Bs_u influence is conveyed by two cognitive modules. The first one is Be_u, i.e. user's expectations about product's behavior. The second one is A_u that is user's expected product's affordances or, stated in another way, the possible uses U envisions the product will afford. Considered together, A_u and Be_u constitute user's knowledge (K_u) about products. For instance, a swing chair affords sitting (A_u) and U also expects it will swing (Be_u). A metal door handle affords pulling (A_u) and U also expects the metal will feel cold (Be_u). These two examples show that the product interacting interface (Int), through the observable behavior Bs_u, determines the content of A_u and Be_u. In turn, A_u and Be_u influence U's manipulations with the product. For instance, because of its expected behavior, U will not use a swing chair for reaching a high place. And because a door handle affords pulling, U will not push the door. However, the content of A_u and Be_u is not limited to the inputs conveyed by Int through Bs_u, alternative sources being product's documentation, advertisement, fellow users' opinions, and so on.

Finally, A_u and Be_u dynamical sets and their content may change as long as U is interacting with the product, reading documentation, receiving comments from other users, and so on.

It should be stressed that both affordance and expected behavior are actively shaping users manipulations. Consider for example the case of a user (U) whose goal (G) is to replace a light bulb in his patio. U has bought a brand new light bulb, but needs a lifting device for completing the replacing operation. A basket chair is standing right beneath the old bulb. Affected by the chair's Int through the observable chair features (Bs_u), U believes that the basket chair affords "liftability" A_u, i.e. the capability to hold a standing person. Since U previously used the chair for sitting, she expects the chair will sag to some acceptable extent (Be_u). Then U starts the replacing procedure lifting on the chair. Differently from expectations, the chair sags considerably and nearly collapses before the replacement is completed. At this point U quickly updates his K_u about basket chairs (replacing A_u with A_u' and Be_u with Be_u') and wisely sets out for the shed where a ladder is stored, the ladder providing (according to K_u) the needed affordance and a safer behavior.

Before entering the detailed analysis of the EM, the concepts expressed above are schematically represented in Fig 1, where Bs and S domains of the FBS framework are divided according to the proposed classification into Bs_u-Bs_d and Int-UI respectively. Moreover, the diagram shows the relationships between these entities and the concepts of affordance, expected behavior by the user and user's goal. Eventually, the distinction between External World and Interpreted World is kept and represented.

Figure 1 is a representation, according to the EM presented here, of a single cycle of user-product interaction. The links between the model entities are briefly summarized below, while the following paragraphs provide a more comprehensive description of each item and related concept.

1. Interaction begins because U wishes to achieve the goal G.
2. Among U's knowledge (K_u) there are pieces of information according to which a product having structure S has the appropriate affordances (A_u) and expected behavior (Be_u).
3. A_u is one of the possible sets of true affordances related to the product (At). A_d represents the set of affordances the product should have according to design intent. Designers ambition is that product's affordances included in A_d are also part of At and that A_u falls within A_d. However, it may happen that some elements are not shared between A_u and A_d. Moreover, both of them may include false affordances (¬A).
4. Given G and K_u, U performs a manipulation (M) of the product. M points to a specific part of S that is the Interacting interface (Int).
5. Int is the part of S where the actual user-interaction takes place. Int may differ from the part of structure where the interaction should take place according to design intent, the user interface (UI).
6. Because of user's actions, the product responds with a set of potentially observable behaviors (Bs). Designer expectations are that U will be affected by the subset Bs_d of Bs generated by UI.
7. Since Int and UI may diverge, the actual product's behavior perceived by U (Bs_u) may diverge from Bs_d as well.

8. U interprets the feedback received from the product and compares it with the initial K_u. As a result U updates A_u and Be_u, either confirming or amending them.

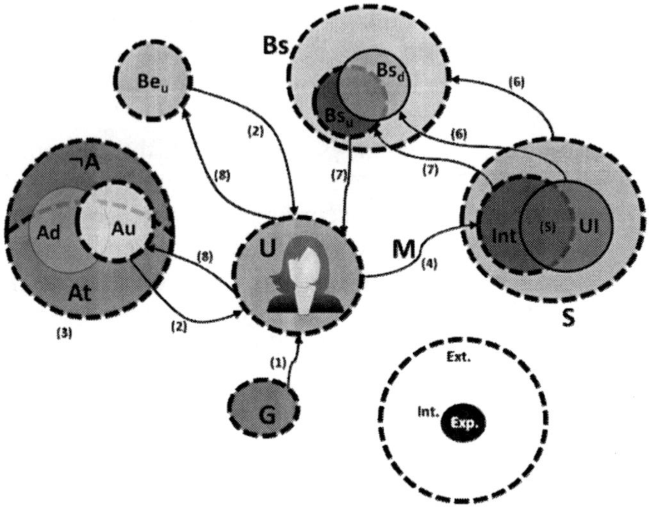

Fig. 1. Schematic representations of the links between the entities of the proposed extension of the FBS framework and relations with the situated model

Using and Interacting: The Interacting Interface

The product use context is the stage in a product's life cycle where the product performs its functions in order to satisfy the user's needs. Therefore, use is the instantiation, by a user, of "a goal-directed series of considered actions which includes manipulations" of product [12].

Two terms appearing in this definition deserve a closer look. Firstly, it should be noticed that the term "manipulation" (M) covers both direct physical manipulation and indirect user's actions pointing at the product. For instance, users control television sets' functions, without direct manipulation, by means of a remote control. By physically manipulating the remote, users are able to use television sets. Secondly, the term "considered actions" means that user's manipulations are based on his/her beliefs and expectations about product's behavior in a given environment. Beliefs and expectations are part of user's knowledge (K_u) as discussed later on.

In order to analyze relevant features of the product use context, it is worth to introduce a distinction between *using a product* and *interacting* with it. It is assumed that when a user U is manipulating a product for the achievement of a goal G, the user is using the entire product, but U is interacting directly only with a part of it. Direct interaction means a bidirectional flow of energy, material and signal between U and D, such that U can perceive it with any of the five senses and the flow is not actively filtered, elaborated or otherwise altered by other products or product's components. For instance, it can be stated that U is interacting directly with the car's brake pedal, even though U is wearing shoes. Nevertheless, U is interacting indirectly with the braking disks via the levers, pipes, pumps, springs and the rest of the braking system. Therefore, it is proposed to introduce the notion of *Interacting Interface* (Int), which is the part of a product's structure that interacts directly with the user.

Let's consider the example of using a car. The user U uses the entire car when driving it on the road from home to work. U is interacting *directly* with the gauges and displays in the cockpit by means of the sense of sight, with the steering wheel and pedals by means of the sense of touch, with the navigation system by means of sight and sound and so on. However, U is using the entire car, including parts of it with which U has only indirect interaction, if any, for example tires and suspensions, lubricant and coolant in the engine, external lights, spare wheel, and so on.

Int should not be confused with the more familiar user interface (UI) concept. In some cases the two interfaces may coincide, but it is not always the case. An important difference between Int and UI is that while, for a given product, UI is fixed by design intent, Int might change and develop in time. Let's consider the remote of a VHS. The user U is not using it for recording movies anymore; nowadays U downloads the movies from the Internet or rents DVDs from a shop nearby. However, U still likes to occasionally watch the footage taken during the holidays. The change of habits is reflected by a corresponding change in the manipulations performed on the VHS remote in order to achieve his/her goals. The remote UI has not changed, of course. The "play" key and the "rec" key are still where they used to be. However, U does not interact anymore with the "rec" key, and it is no longer part of Int.

The VHS example represents a case where Int is contained in UI. However, it is easy to conceive cases where Int is larger than UI, i.e. it includes further elements of the structure. Let's move a few years back in time and consider the case of a teenager who loves assembling his own PC and over-clocking it in order to run his favorite games at full resolution. Since U knows that overheating is a threat, he/she has installed an additional fan. However, the fan is not powerful enough and excessive heat might still damage the machine after prolonged gaming. Luckily, after some tests, U realizes that the additional fan emits a hissing noise when processor temperature is exceedingly high. In this way, he/she knows when it is time to quit the game before severe damage occurs. Of course, the hissing noise has never been intended as a part of the UI. Nevertheless, it is part of the system's Int and helps U to properly interact with it.

Knowledge, Affordances and the Interacting Interface

It has been already anticipated that user's cognitive state, represented by K_u, is a fundamental factor in the proposed model of the use stage. As it will be detail hereafter, K_u is connected to usage via a feedback loop. First, K_u (co)determines manipulation; then receives feedback from manipulation's effects. The feedback is internalized into K_u; and eventually a new manipulation instance occurs. This loop is analogous to the one introduced in FBS for explaining design iterative process [1, 2]. Both models need to distinguish between product's expected and actual behavior. But, in one case the distinction is from the design stage perspective, and in the other is from the use stage perspective. In order to prevent confusion, in this paper the classical FBS terminology is slightly revised and abbreviations are modified in the following way: Be_d replaces Be and represents product's expected behavior according to the designer. Similarly, Bs_d replaces Bs and represents designer's interpretation of product's actual behavior. From the use stage perspective it is introduced Be_u, representing how the user expects the product will perform; and Bs_u representing user's interpretation about product's actual behavior.

It is assumed that M is determined by two main factors, namely: user's knowledge about the product and the operating environment (K_u); and user's goals for using the product (G). Moreover, it is worth distinguishing two domains within K_u. The first is Be_u as already defined above. The second domain (A_u) includes all the affordances that, according to the user, are provided by the product in the operating environment.

Following Maier and Fadel [23], affordances are defined as potential uses of a product. Affordances are product's properties that exist whether or not the user is aware of them [16]. Borrowing an example from Maier and Fadel, a typewriter affords typing behavior to a person, and the corresponding affordance could be dubbed "typeability". However, not all persons have the capacity to perceive the affordance. Having or not the capacity depends on the user's knowledge (Ku), both knowledge from direct interaction with typewriters and similar products, and indirect knowledge such as from seeing others using the product or being told by others. Therefore, it is quite safe to say that all inhabitants of Western countries (small children included) are able to perceive the "typeability" affordance when they see a typewriter. Not all of them have interacted directly with typewriters, nevertheless they have used similar devices; they have seen friends using one; they have seen movies where someone was typing, and so on. On the other hand, it is possible to conceive that the capability to perceive this affordance is absent in traditional communities that have never been exposed to western products (true, this is a very remote possibility nowadays). We may imagine bringing a typewriter to one of these traditional communities. Initially they will manipulate randomly the device. In a short time, from the observable behavior manifested by the device in response of their actions (Bs_u) they will realize, at least, that the keys afford pushing.

Admittedly, the example is rather fictional and simplifies the complexities of human-device relationship. Still, given the paper's aims, it provides an effective way to introduce the relations between user knowledge, affordances and behavior. It is assumed that, ideally, for each product it is possible to define a comprehensive set including all its true affordances (At). This set is potentially very vast. The designer himself could possibly be not aware of all of them, so let's dub A_d the set of product's affordances according to the designer. Similarly, A_u is the set of product's affordances according to the user.

It is important to note that A_d and A_u may include affordances that are not counted in At and vice versa. For example, U might think that a certain window affords to be both opened horizontally and tilted vertically, but it does not. The belief that the window affords to be tilted vertically is included in A_u but not in At, therefore, it can be considered as an instance of false affordance (¬A). This erroneous belief may be disproved simply by trying to tilt the window. The user initially manipulates the window in accordance with A_u expecting it will tilt vertically (Be_u). After few unsuccessful attempts, U will realize that window's observable behavior (Bs_u) is incompatible with Be_u. Thanks to the feedback, U will update K_u and, as a consequence, A_u will change to A_u' and Be_u to Be_u'.

Indeed, U does not need to manipulate directly the device for confirming or disproving believes about product's affordances. U might be relying on information provided by other users he trusts. For example, although a U does not own an iPhone, he/she can be aware that it is possible to interact with an iPhone by orienting it in space. The reason is that U has been told so by a friend who directly interacted with an iPhone.

Despite the fact that in the window example A_u and Be_u overlap to a great extent, it should be stressed that the two concepts are distinct and might diverge. The typewriter provides again a good example. Let's assume that U wants to write a letter. U notices a typewriter on the desk. According to U's Ku, it affords "typeability" (A_u). Moreover, U expects the device will emit a noticeable clicking sound during typing (Be_u). To his/her surprise, the device is very quiet indeed (Bs_u). Hence, after comparing Bs_u with Be_u, U will revise Be_u, but will leave A_u unchanged.

Norman with his book *Psychology of Everyday Things* [11] firstly performed an analysis of product's design by means of the affordance perspective. The analysis was accompanied by a series of instructive examples showing how users perceive product's affordances and how this could be exploited by designers. His message nicely fits with the framework of the present work. Perceptible affordances are those affordances conveyed by Int. As [16] remarks, perceptual information may suggest affordances that do not exist (false affordance), i.e. are not part of At. Similarly, Int may fail to convey to the user a real affordance (hidden affordance). In the next section the notions of perceptible, false and hidden affordance are employed for analyzing failure and misuse phenomena.

Failure and Misuse

First of all it is necessary to emphasize that the above distinctions among perceptible, false and hidden affordance are made from the user's perspective. Looking at a product from the designer's perspective, the following classification applies: true intended affordances (Ai_d), true unintended affordances (Aun_d), false intended affordances ($\neg Ai_d$).

Ai_d are affordances that the designer intentionally implemented into the product. Still these affordances may pass unnoticed by the user and end up as hidden affordances. For instance, in the Netherlands it is very common to find door locks such that, in order to lock the door, the user has to rotate the handler upwards and simultaneously turn the key into the keyhole. If the handler is not turned upwards, the key will not turn. The handler has not distinctive features that could suggest the presence of this affordance, which is therefore hidden from users deprived of the right K_u. In brief, there is at least one affordance about Dutch door locks that is included both in At and Ai_d, but absent in A_u for some users. In this and in analogous situations, the user may be prevented from a successful manipulation of the product.

A different kind of – possibly unsuccessful – interaction may ensue when A_u includes an affordance that is absent from Ai_d. In this case the user may manipulate the product in a way that was unforeseen by the designer, therefore misusing it. Two alternatives may be envisioned. On one hand, the affordance the user wants to take advantage of is a true affordance, even though it is an unintended one, and U is able to achieve G. This happens when screwdrivers are used to open paint cans, for instance. On the other hand, the user is mistaken and the product does not actually have the affordance. U might try to misuse it, unsuccessfully.

Aun_d are affordances that the product truly has, but was not explicitly designed for having them. These are very common and could be considered side effects allowing for alternative uses. The fact that Aun_d are unintended does not say anything about their desirability. An ashtray may be used as a paperweight, but also, because of its "throwability", as a weapon. As for the previous category, also Aun_d may pass unnoticed. For instance, differently from secret agents, the majority of common users are not aware that credit cards can be used to break open locked doors.

$\neg Ai_d$ are affordances that the product should have according to the design intent, but it has not. Two scenarios can be distinguished: on the one hand, there are products missing the intended affordance before entering the use stage, either because of a design error or manufacturing mistake. A dangerous situation may arise when, regardless of the false affordance, Int still conveys to the user the clues usually associated with the true affordance. When it opened on 10 June 2000, thanks to its innovative design (and also because of the charming weather) the London Millennium Bridge attracted crowds of visitors all sharing the belief that it would have afforded a safe and stable crossing. To everyone surprise the bridge started wobbling noticeably and it was closed out of safety concerns. Following extensive investigation it was decided to retrofit the bridge with 37 viscous dampers and since then, it is behaving as expected [24]. On the other hand, there are products that lose the affordance during the use stage. Again, the interacting interface may or may not reliably convey the situation to the user. A pedal, for instance, affords "pushability". Let's compare two kinds of malfunction that may affect a pedal. In the first case the pedal is jammed in the depressed position. The anomaly is clearly evident and the user promptly realizes that the pedal does not afford pushing. On the contrary, a pedal jammed in the lifted position is perceived by the user as still having "pressability". Lastly, a misuse appears where user and designer agree on affordances, but K_u contains a wrong Be_u. For instance, a U buys an energy saving fluorescent lamp with the aim of illuminating a large salon with reduced consumption; the expected affordance coincides with the designer's one, nevertheless, it may happen that the power has been underestimated on the base of K_u, and the resulting amount of visible light is not enough ($Bs \neq Be_u$).

Exemplary Application of the Proposed Model and Discussion

In order to clarify the meaning of the proposed extension of the FBS model, and to provide means to appreciate its potential in terms of describing cognitive aspects of uses, misuses and failures, the present section summarizes all the concepts introduced in the previous chapter through an exemplary model related to a microwave oven.

The basic elements of a microwave oven are the followings:

- a power supply to provide energy to the magnetron with a suitable intensity and for a given duration;
- control and display: I/O devices for the definition of the cooking program/duration;

- magnetron: vacuum tube where electrical energy is converted into electromagnetic waves, characterized by a frequency of 2450 MHz;
- waveguide: rectangular metal tube which delivers the microwaves generated by the magnetron to the cooking cavity; a direct exposure of the magnetron to the cooking cavity is typically avoided, in order to prevent its contamination with food particles;
- a stirrer to homogenize the energy delivered on the food by distributing the microwaves fed by the waveguide in the cooking cavity;
- a turning platform which rotates the food along a vertical axis to produce an uniform exposure to microwaves;
- cooking cavity: volume where the food is heated by exposure to microwaves;
- door: closable opening of the cooking cavity through which food is entered or removed; the door shields the microwaves to prevent unhealthy exposure of people or other objects in the environment.

These elements can be considered as the fundamental components of the Structure of a microwave oven. Through their properties and interactions these elements deliver the function of heating food according to a Behavior essentially based on the capability of microwaves to excite polar molecules and ions inside the food, with consequent increase of temperature due to molecular frictions. Microwaves can pass through materials like glass, paper, plastic and ceramic, and be absorbed by foods and water; but they are reflected by metals. Thus, according to the designer's intention, metal containers should not be used into microwave oven, in order to prevent sparks and possibly fire.

Let's assume that the goal G of a user U, who has never used a microwave oven, is to bake a pizza. Among the ingredients, a few cherry tomatoes and some basil leaves are added to garnish the pizza. Once that the pizza is ready for cooking, U interacts with the oven by opening the door through its handle, entering the pizza in the cooking cavity, closing the door again through its handle, regulating the intensity and the duration of the microwaves by means of the control keyboard and/or knob, getting a feedback about the regulation and the time left through the display and possibly from a beep warning at the end of the cooking process.

According to this "goal-directed series of considered actions" M, driven by the user's knowledge K_u, which is based on general information about microwave cooking and not on previous experiences, the user expects that the oven will produce a crispy and flavored pizza, by heating all the ingredients (Be_u). The manipulation M is accomplished through the Int constituted by the handle (en energy flow is applied to door to open/close it), the cavity opening (a material flow for entering/extracting the pizza), the control (an input signal flow determines microwaves power and duration), the display/beep (an output signal flow provides a feedback to U). In such a situation, Int coincides with the UI; nevertheless, due to the limitations of the K_u, a bad surprise ruins the user's meal: the cherry tomatoes explode, because of the increase of pressure following the evaporation of their water content. According to the extended FBS formalism, this situation can be described as follows:

$G_d = G_u$; Int = UI; Bs = Be_d; Bs ≠ Be_u; misuse due to inadequate K_u.

A similar situation occurs if U makes use of a combined microwave, grill and convection (hot-air); let's assume that a plastic container is adopted to hold the food, as suggested by K_u derived from previous experiences with microwave cooking. The plastic tray will melt because of the heat delivered by the grill and the hot air. Also in this case, the misuse is due to inadequate K_u, with consequent Bs ≠ Be_u despite $G_d = G_u$, Int = UI and Bs = Be_d.

Besides, in the latter situation, assuming that food preparation proceeds in the expected way, it may happen that U keeps the hands close to the side air vents while waiting for the pizza, to get them warmer. This affordance of the oven is clearly unintended by the designer, but still belongs to the possible benefits of the oven Bs. It is worth to notice also that the Int in this case is larger than the UI, since it includes also the air vents (an energy output flow delivers heat to the hands of U). According to the proposed formalism the model is characterized by:

Gd ≠ Gu; Int ⊃ UI; Bs = Bed; Bs = Beu; Aund alternative use.

Besides, Int can be constituted by a subset of the UI elements, in several different scenarios as: U is not aware of all the oven functionalities, thus activates just part of the control buttons; the display is broken, thus the output information about microwave power and duration is not delivered; the oven is in a room with an open window, facing on a highly busy road, therefore the environmental noise is so high that no acoustic signals can be perceived by U at the end of the cooking process. All these situations are characterized by Int ⊂ UI, whatever are the G_u and the obtained result Bs in comparison with the expectations Be_u.

Finally, after an inappropriate maintenance by the user, or even a not sufficiently robust design, it may happen that the stirrer remains blocked due to some food particles obstruction (Bs ≠ Be_d failure); as a consequence, the microwaves power is not equally distributed in the design cavity, with consequent inhomogeneous heating effect. Since U, whatever is his/her K_u about microwave oven usage, has no means to realize the failure, any manipulation M of the device won't produce a satisfactory cooking. The proposed model describes this situation as follows:

$G_d = G_u$; Int = UI; $Bs \neq Be_d$; $Bs \neq Be_u$; $\neg Ai_d$ false intended affordance.

All the above examples reveal the possibility to describe several different product use contexts with a simple formalism, through the proposed extension of the FBS model.

Conclusions

The goal of the present work is to share a proposal of extension of the FBS framework aimed at representing uses, misuses, failures of a product and their mechanisms through a distinction between the designer and the user perspectives. Compared with the original FBS framework, the proposed EM enriches the External World in order to represent the use context of a product. As illustrated through an example related to microwave ovens, the proposed model provides a simple formalism to describe many different situations related to proper/improper uses of a product, its alternative uses and its possible failures.

The authors are working on two natural follow-ups of the present research activity: integrating the proposed model with needs and requirements modeling as described in [7]; deriving from the integrated FBS model design prescriptions for preventing false intended affordances, misuses and failures.

Abbreviations

A	affordance
At	set including all the true affordances to the designer
Ai	true intended affordances
¬Ai	false intended affordances
Aun	true unintended affordances
EM	extended model
G	goal
Int	interacting interface
K_u	user's knowledge
M	manipulation
U	user
$(X)_u$	variable X observed from the user's perspective
$(X)_d$	variable X observed from the designer's perspective

References

1. Gero, J.S.: Design prototypes: A knowledge representation schema for design. AI Magazine 11(4), 26–36 (1990)
2. Gero, J.S., Kannengiesser, U.: The situated function-behavior-structure framework. Design Studies 25(4), 373–391 (2004)
3. Vermaas, P.E.: The Flexible Meaning of Function in Engineering. In: Proceedings of the 17th International Conference on Engineering Design (ICED 2009), Stanford University, California, United States, August 24-27, vol. 2, pp. 113–124 (2009)
4. Wang, L., Shen, W., Xie, H., Neelamkavil, J., Pardasani, A.: Collaborative conceptual design – state of the art and future trends. Computer-Aided Design 34, 981–996 (2002)
5. Chandrasekaran, B., Josephson, J.R.: Function in device representation. Engineering with Computers 16, 162–177 (2000)
6. Erden, M.S., Komoto, H., van Beek, T.J., D'Amelio, V., Echavarria, V., Tomiyama, T.: A Review of Function Modeling: Approaches and Applications. Artificial Intelligence for Engineering Design, Analysis and Manufacturing 22, 147–169 (2008)
7. Cascini, G., Fantoni, G., Montagna, F.: Reflections on the FBS model: proposal for an extension to needs and requirements modeling. Submitted to the International Design Conference - Design 2010, Dubrovnik - Croatia, May 17-20 (2010)
8. Gero, J.S., Kannengiesser, U.: Understanding Innovation as Change of Value Systems. In: Proceedings of the 3rd IFIP Working Conference on Computer Aided Innovation (CAI), Harbin, China 20-21, pp. 38–50 (2009)
9. Deng, Y.M., Britton, G.A., Tor, S.B.: Constraint-based functional design verification for conceptual design. Computer-Aided Design 32, 889–899 (2000)
10. Brown, D.C., Blessing, L.: The relationship between function and affordance. In: DETC2005-85017, Long Beach, California, USA (2005)
11. Norman, D.A.: The Psychology of Everyday Things. Basic Books, Inc. (1988)
12. Vermaas, P.E.: The physical connection: engineering function ascriptions to technical artefacts and their components. Studies In History and Philosophy of Science Part A 37(1), 62–75 (2006)
13. Kuipers, B.: Qualitative Reasoning: modeling and simulation with incomplete. The MIT Press, Cambridge (1994)
14. Keuneke, A., Allemang, D.: Exploring the no-function-in-structure principle. Journal of Experimental & Theoretical Artificial Intelligence 1, 79–89 (1989)
15. Chandrasekaran, B., Josephson, J.R.: Function in device representation. Engineering with Computers 16, 162–177 (2000)
16. Gaver, W.W.: Technology affordances. In: Proceedings of the SIGCHI Conference on Human Factors in Computing Systems: Reaching Through Technology, pp. 79–84. ACM Press, New York (1991)

17. Del Frate, L., Franssen, M., Vermaas, P.E.: Towards defining technical failure for integrated product development. In: Proceedings of TMCE 2010 Symposium, Ancona, Italy, April 12-16, pp. 1013–1026 (2010)
18. Becattini, N., Cascini, G., Rotini, F.: Correlations between the evolution of contradictions and the law of ideality increase. In: Proceedings of the 9th ETRIA/CIRP TRIZ Future Conference, Timisoara, Romania 4-6, pp. 26–34 (2009)
19. Umeda, Y., Tomiyama, T., Yoshikawa, H.: FBS Modeling: Modeling scheme of function for conceptual design. In: Proceedings of the 9th Int. Workshop on Qualitative Reasoning, Amsterdam, NL 11-19, pp. 271–278 (1995)
20. Vicente, K.J.: Cognitive Work Analysis, Toward Safe, Productive, and Healthy Computer-based Work. Lawrence Erlbaum Associates, Hove (1999)
21. IEC 61508-4 Functional safety of electrical/electronic/programmable electronic safety-related systems - Part 4: Definitions and abbreviations. International Electrotechnical Commission (1998)
22. IEC 60812 Analysis Techniques for System Reliability Procedure for Failure Mode and Effects Analysis (FMEA). International Electrotechnical Commission (2006)
23. Maier, J.R.A., Fadel, G.M.: Affordance-based methods for design. In: Proceedings of the ASME Design Engineering Technical Conference, vol. 3, pp. 785–794 (2003)
24. Dallard, P., Fitzpatrick, A.J., Flint, A., Le Bourva, S., Low, A., Ridsdill Smith, R.M., Willford, M.: The London Millennium Footbridge. Structural Engineer 79(22), 17–33 (2001)

A Formal Model of Computer-Aided Visual Design

Ewa Grabska and Grażyna Ślusarczyk
Jagiellonian University, Poland

This paper aims at contributing to a better understanding of essential concepts of inventive visual design. Towards this end, we first outline a framework of formal model of computer-aided visual design. Then, we define particular components of this model paying attention to the role of human visual perception treated as a dynamic process ("active vision"). Moreover, we present different types of logic models used in computer tools supporting the design process and consider an example of a graph-based data structure gathering information on which design knowledge can be based. Finally, the definition of the system of computer-aided visual design is presented. The approach is illustrated on examples of designing teapots.

Introduction

In the Internet age designers rely on cognitive tools to amplify their mental abilities. Almost half the brain is devoted to the visual sense and the visual brain is capable of interpreting visual objects in many different ways. Therefore the modern design process is characterized by the increased importance of the visualization of design concepts and tools. A sketch of a formal model, which gives us a base to develop computer tools supporting a visual design process, is proposed. This model can also provide insight in how humans solve problems in a way that uses active visual perception [1].

The designer has an internal world being a mental model of a design task that is build up of concepts and visual perceptions stored in his mind, and an external world composed of representations outside the designer [2]. Both drawings created by the designer and their internal representations can

J.S. Gero (ed.): *Design Computing and Cognition'10*, pp. 97–113.

be treated as situations in the external world build up outside the designer. The designer takes decisions about design actions in his internal world and then executes them in the external world. In this paper we assume that the designer's decision making process is supported by the computer-aided design system.

In the design process four types of actions: physical, conceptual, perceptual and functional, can be distinguished [3]. Physical actions consist of such operations like drawing, copying and erasing elements of design drawings. Nowadays these type of actions are usually aided by computer tools, for instance graphical editors, and the results of performing physical actions are displayed on the monitor screen.

In perceptual actions the designer discovers visual features of drawings, such as spatial relations between drawing elements, for instance closeness or neighbourhood, compares elements, for example searches for differences or similarities between them. His visual perception process is based on the analysis of drawings. Presently, this process is supported by the design system which is able to reason about design features on the basis of the internal representations of drawings, for instance in the form of graph data structures. The objective of functional actions is associating meaning with features discovered in the perceptual actions, relating abstract concepts to these features, and valuation of drawings. In conceptual actions new design goals and requirements are determined.

The paper considers a formal framework for computer-aided visual design, where human visual perception is treated as a dynamic process ("active vision") [1]. The proposed model of a design process is a modification of the model presented in [4], [5], [6]. It consists of three basic components:

- a domain $\boldsymbol{D}_T$ of *design tasks* related to formulation of design problems in terms of requirements,
- a domain $\boldsymbol{D}_A$ of *physical design actions*, and
- a domain $\boldsymbol{D}_V$ of a *computer visualization,* which consists of design drawings, data structures representing them, and a design reasoning mechanism.

The domain of design tasks is modified during the design process. At the beginning it contains only initial requirements, while later the devised requirements are added. Physical design actions of the second domain are related to the external world. The remaining design actions are constructed in designer's brain. They are based on the analysis of drawings and result in changes of requirements in the design task domain. We assume that in the domain of a computer visualization design drawings are automatically transformed into data structures being their internal representations which are essential in a reasoning process.

Moreover, this model is characterized by:

- an *operation method* that is a set of instructions specifying what types of physical actions can be taken under what circumstances, and
- an *active perception* which can be seen as a composition of a perceptual action and a functional one.

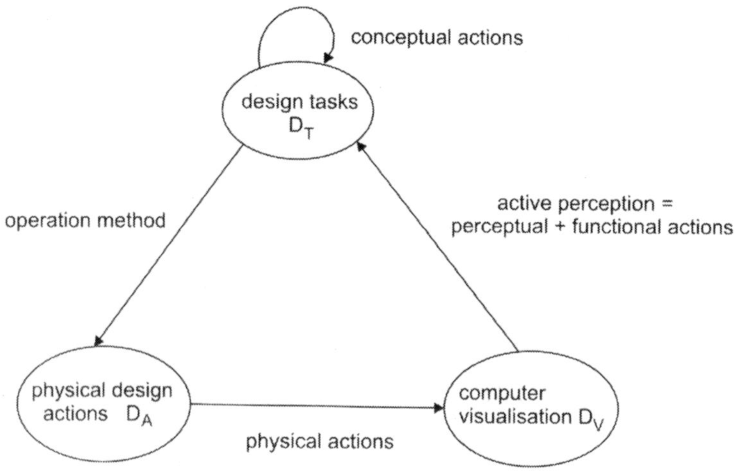

Fig. 1. Three domains of a design process

The essential aspect of a visual design process is devising new requirements which come into being as a result of composition of an active perception and a conceptual action. The relations among the three design domains are presented in Figure 1.

Our approach to design will be illustrated by examples of designing tea-pots.

Classifications and Logics for a Design Model

The domain of design tasks and the domain of design actions are characterized with the use of the notion of classification. The formal definition of a classification is as follows [7].

Definition 1. A **classification** is a triple $D = (O, \Sigma_o, \vdash_o)$, where:

- O – is a set of objects to be classified,
- Σ_o – is a set of types used to classify objects of O,

- $\vdash_o$ – is a binary relation between O and Σ_o that specifies which objects are classified as being of which types.

Entities of the domain of design tasks are classified by design requirements in the form of expressions of the *propositional logic.* In the domain of design actions only physical actions are classified. These actions can be classified using either structureless or structural objects depending on the type of generated drawings.

In the domain of a computer visualization the *first-order logic* is used as a reasoning mechanism. Information stored in the data structures corresponding to design drawings is translated to sentences of the first-order logic. In this process a problem-oriented relational structure, which assigns elements of data structures to entities of the specified first-order logic alphabet, is used.

In first-order logic we define a vocabulary $A = \{C, F, R\}$, where:

- C - is a set of constant symbols,
- F – is a set of multi-argument function symbols,
- R – is a set of multi- argument relations.

We assume that we have a set of variables written x and y, possibly along with subscripts. The set of *terms* is formed starting from constant symbols and variables and closing off under function application, i.e., if $t_1,\ldots, t_n$, n $\geq$ 1, are terms and $f\in F$ is an n-ary function symbol, then f($t_1,\ldots, t_n$) is also a term. An *atomic formula* is either of the form $r(t_1,\ldots, t_k)$, where $r\in R$ is an k-ary relation symbol and $t_1,\ldots, t_k$ are terms or of the form $t_1 = t_2$, where t_1 and t_2 are terms. The set of general logical formulas is built over atomic formulas using logical connectives and quantifiers, and closed under the consequence relation. The formulas contain variables universally quantified over appropriate component types. Formulas which do not have free variables are called *sentences*.

The semantics of first-order formulas uses relational structures a *relational structure* consisting of a domain of individuals and a way of associating with each of the elements of the vocabulary corresponding entities over the domain [8].

In our approach to the computer-aided visual design this structure is defined as follows.

Definition 2. A **relational *A*-structure *L*** consists of:

- a domain of a computer visualization D_V,
- an assignment of a k-ary relation $r^L \subseteq (D_V)^k$ to each k-ary relation symbol $r \in R$,
- an assignment of a n-ary function f^L: $(D_V)^n \rightarrow D_V$ to a n-ary function symbol $f\in F$, and

- an assignment of a $c^L \in D_V$ to each constant symbol c.

The next step to define the formal semantics of first-order formulas is specification of an interpretation of variables. A valuation υ on a structure L is a function from variables to elements of D_V. Given a structure L and a valuation υ on L, υ is inductively extended to a function that maps terms to elements of D_V. Let $\upsilon(c) = c^L$ for each constant symbol c and then the definition of υ is extended by induction on the structure of terms by taking $\upsilon(f(t_1,\ldots, t_n)) = f^L(\upsilon(t_1),\ldots, \upsilon(t_n))$.

Given a relational structure L with a valuation υ on L. $(L, \upsilon) \models \phi$ denotes that a formula ϕ is true in L under the valuation υ. The truth of the basic formulas is defined as follows:

- $(L, \upsilon) \models r(t_1, \ldots, t_k)$, where $r \in R$ is a k-ary relation symbol and $t_1, \ldots, t_k$ are terms, iff $(\upsilon(t_1), \ldots, \upsilon(t_k)) \in r^L$,
- $(L, \upsilon) \models t_1 = t_2$, where t_1 and t_2 are terms, iff $\upsilon(t_1) = \upsilon(t_2)$,
- $(L, \upsilon) \models \neg\phi$ iff $(L, \upsilon) \not\models \phi$,
- $(L, \upsilon) \models \phi_1 \wedge \phi_2$ iff $(L, \upsilon) \models \phi_1$ and $(L, \upsilon) \models \phi_2$,
- $(L, \upsilon) \models \exists x \phi$ iff $(L, \upsilon[x/a]) \models \phi$ for some $a \in D_V$, where $\upsilon[x/a]$ denotes the valuation with $\upsilon(x) = a$.

A Formal Model

As it has been considered in the introduction the proposed model of a design process consists of three basic domains: design tasks, physical design actions and a computer visualization.

A Design Task Domain

Design is a goal-directed activity that involves the decision making, exploration and learning. The target of designing is the object created by the designer during the design process. This object is formulated in terms of requirements. At the beginning only initial requirements are specified. During the design process the new requirements are added [3].

Definition 3. A domain of **design tasks** $D_T = (T, \Sigma_T)$ consists of a set T of objects to be classified, called *design situations* of D_T, and a set Σ_T of objects used to classify the situations, called the *types* of D_T.

If a design situation $t \in T$ is classified as being of type $\sigma \in \Sigma_T$, we write $t \vdash_T \sigma$ and say *t belongs to* σ.

Design situations of T are classified by design requirements of Σ_T in the form of expressions of the propositional logic. We assume that Σ_T is closed under the tautology $\equiv$, the negation $\neg$, the implication $\Rightarrow$, and the conjunction $\wedge$, with their usual truth-functional interpretations.

Example 1. Let the target of designing be a teapot with an interesting form. The set T of design situations is considered as the set of all possible teapot designs. It is classified by the following types:

- σ_1: a teapot of T stands firmly on a table,
- σ_2: all situations of T allows one to pour tea into a cup,
- σ_3: all situations of T can be lifted by hand, and
- σ_4: a teapot of T can be filed with water.

The design situation t corresponding to the teapot shown in Figure 2 belongs to types σ_2, σ_3, and σ_4 but it does not belong to σ_1 as this teapot cannot be firmly placed on a table. In other words, for the design situation t in the truth-functional interpretation the expression $\sigma_2 \wedge \sigma_3 \wedge \sigma_4 \equiv$ ***true*** and $\sigma_1 \equiv$ ***false***. The teapot shown in Figure 9 is an example of a design solution which belongs to all mentioned types, i.e., $\sigma_1 \wedge \sigma_2 \wedge \sigma_3 \wedge \sigma_4 \equiv$ ***true***, for this teapot.

Fig. 2. A design of a teapot

A Computer Visualization Domain

Designer's external world is associated with many drawings which allow the designer to interweave analysis with synthesis. Different surfaces are used for drawing, e. g., a sheet of paper or a monitor screen. Each of such surfaces along with a design drawing on it will be called a *visualization site*. In our model mainly a monitor screen with drawings is considered as a visualization site. A visualization site can be seen as a situation in the external world built up outside the designer.

Examples of visualization sites are shown in Figure 3. When using the monitor screen as a medium in visual design, usually the design process is started with an empty monitor screen (initial visualization site). Successive physical actions generate new visualization sites. It is worth noticing that each drawing being a step of a design solution leading to a final drawing is treated as a different visualization site and constitutes a different design situation.

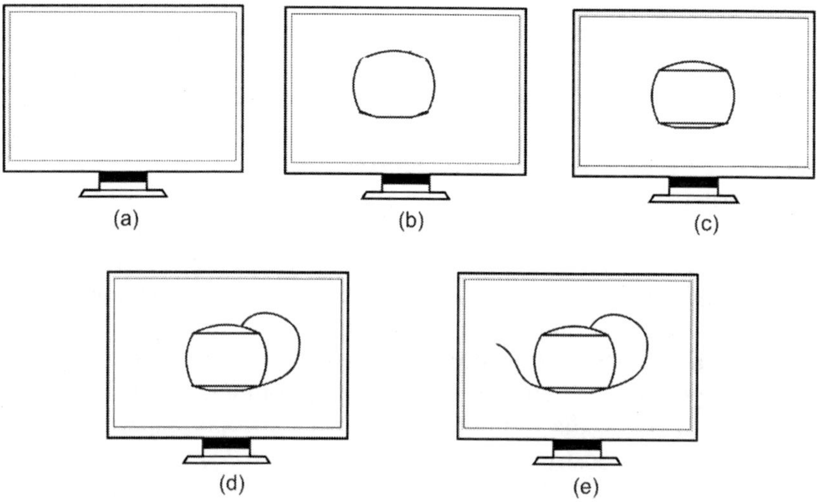

Fig. 3. Examples of visualization sites

Every visualization site belongs to a class of visualization sites that a collection of types classifies. In the considered computer-aided design system, the design drawings are automatically transformed into data structures being their internal representations.

Definition 4. Let I be a set of visualization sites and H be a set of data structures.

A domain of **computer visualization** $D_C = (I \times H, \Omega_I, \Psi_H)$ consists of:

- $I \times H$ - a set of pairs of the form (i, h), where i is a visualization site, h is a data structure representing a drawing of i,
- Ω_I - a set of types used to classify the visualization sites of I,
- Ψ_H - a set of first-order logic formulas supporting the classification of the visualization sites on the basis of data structures of H.

If a visualization site $i \in I$ is classified as being of type $\rho \equiv \omega \wedge \psi$, where $\omega \in \Omega_I$ and $\psi \in \Psi_H$, we write $i \vdash_I \rho$ and say *i belongs to ρ*.

Designer's classification of visual sites is supported by the computer-aided design system. The designer classifies visual sites using visual perception. This kind of classification is specified by means of a set Ω_I of types in the form of expressions of the propositional logic. We assume that Ω_I is closed under the tautology $\equiv$, the negation $\neg$, the implication $\Rightarrow$, and the conjunction $\wedge$, with their usual truth-functional interpretations. The types of Ω_I associated with visualizations sites are related to geometrical properties of drawings: appropriate geometrical objects and their transformations which allow for obtaining admissible components of design objects.

The second type of classification is done automatically by the design system using sentences of the first-order logic. These sentences are evaluated on the basis of internal representations of drawings and form design knowledge related to the drawings. Physical design actions, which result in modifications of drawings, automatically impose changes both in the data structures and design knowledge. Thus logic sentences form dynamic design knowledge, i.e., physical actions performed by the designer on drawings simultaneously modify this knowledge.

It is worth noticing that the proposed model can also be used to describe the design process without the support of a computer tool. In this case visual sites can be in the form of sketches on a sheet of paper and they do not have internal data structures, i.e., the sets H and Ψ_H are empty.

Our model mainly deals with computer-aided design systems, where the designer creates a design solution by drawing successive steps of the solution on a monitor screen. The drawings are made using a system editor and simultaneously the internal data structure of each drawing is automatically generated.

Example 2. Let us consider an example of designing a teapot. The successive drawing steps are based on the ICE language [9]. It serves to build shapes and forms along with their attributes. The basic visual elements in the ICE are lines, circles and curves. By means of transformations various shapes and forms can be obtained from these basic elements. The first shape on the visualization site is shown in Figure 4(a). Starting from this basic shape and using first a decomposition, Figure 4(b), and then a sum of shapes, a drawing of a teapot from Figure 4(c) is obtained.

The set I of visualization sites is classified by the types of Ω_I which describe the required shapes. The types ω_i, where $i = 1,\dots, 5$, specify which shapes can represent: a container (ω_1), a base (ω_2), a lid (ω_3), a spout (ω_4), and a handle (ω_5).

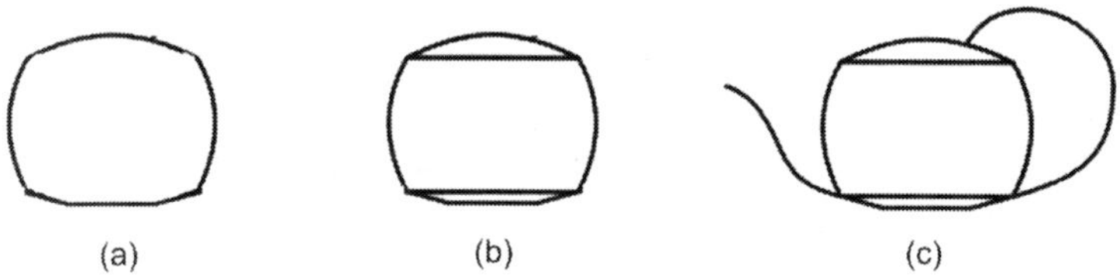

Fig. 4. The chosen steps of designing a teapot

The visualization site shown in Figure 4(a) is of the type ω_l, the site shown in Figure 4(b) is of the type $\omega_1 \wedge \omega_2 \wedge \omega_3$, while the type of the site shown in Figure 4(c) is the conjunction of all five types ω_i.

Each design drawing has its internal representation in the form of an *attributed hypergraph*. This type of structure allows us to represent multi-argument relations among design components. The proposed hypergraphs have two types of hyperedges, called *component hyperedges* and *relational hyperedges*. Hyperedges of the first type correspond to design components and are labeled by component names. In our example one component hyperedge labelled *C*, Figure 5(a), represents the basic shape presented in Figure 4(a). Hyperedges of the second type represent relations among fragments of components and can be either directed or non-directed in the case of symmetric relations. Relational hyperedges of the hypergraph are labelled by names of relations. After using a decomposition to the basic shape, three components representing a *lid*, *container* and *base* are generated, Figure 4(b). In a hypergraph representation this decomposition results in replacing one hyperedge shown in Figure 5(a) by three component hyperedges connected by two relational hyperedges (Figure 5(b)).

Component hyperedges are connected with relational hyperedges by means of nodes corresponding to common fragments of connected design components. Relational hyperedges representing the *supporting* relation (denoted by *sup*) are directed from the lower to the upper component. Adding a handle and a spout, Figure 4(c), the next two component parts of a designed teapot, is done by the application of the sum. As a result a hypergraph presented in Figure 5(c) is obtained. Two new component hyperedges connected by undirected relational hyperedges representing the *connectivity* relation (denoted by *con*) are added to the previous hypergraph, Figure 5(b).

To represent features of design components and relations between them attributing of nodes and hyperedges is used. Attributes represent properties (like shape, size, position, colour) of elements corresponding to hyperedges and nodes.

Let us come back to the semantics of the first-order logic formulas, which are used to reason about designs in an automatic way. The automatic reasoning requires an appropriate representation of design knowledge.

In the proposed visual design model the relational structure is in the form of a hypergraph corresponding to a design drawing on a visual site. It facilitates reasoning about important features of designs and enables the designer to trace changes in design knowledge resulting from physical actions applied during the design process.

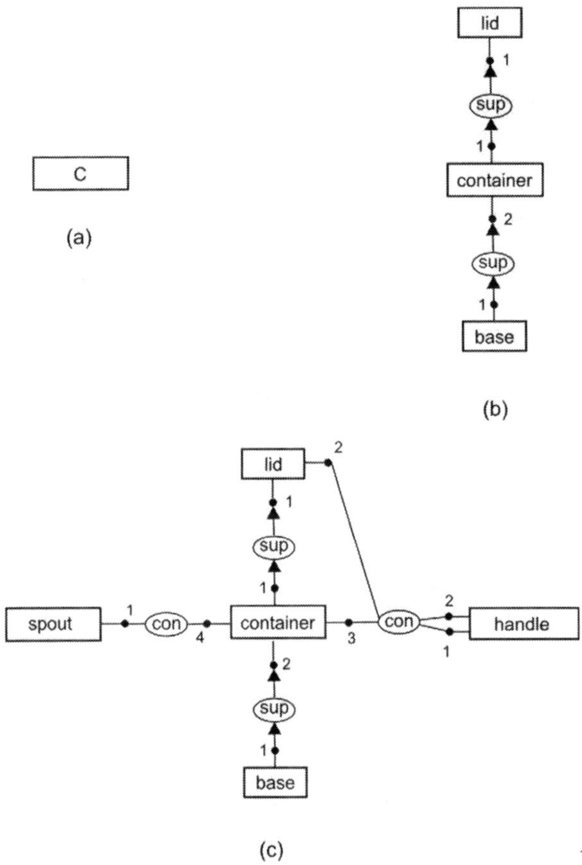

Fig. 5. The successive steps of generating a hypergraph corresponding to the designed teapot

As structures of design objects are represented by hypergraphs, the domain D_V includes:

- the set of component hyperedges, and
- the set of hypergraph nodes.

Relations between design components presented in the drawing are specified between fragments of these components, which correspond to hypergraph nodes. The interpretation of each relation is the hyperedge relation of the hypergraph such that there is a relational hyperedge coming from a sequence of nodes of at least one component hyperedge and coming into a sequence of nodes of other component hyperedges.

The two considered relations, *connectivity* and *supporting*, have at least two arguments. Each of these relations holds among design components if in the hypergraph there exist at least two nodes joined with component hyperedges corresponding to these design components and there exists a relational hyperedge labelled by the name of the relation, which connects these nodes. The connectivity relation is undirected, while the supporting relation is a directed one. Examples of the connectivity and supporting relations are presented in Figure 6(a) and 6(b), respectively.

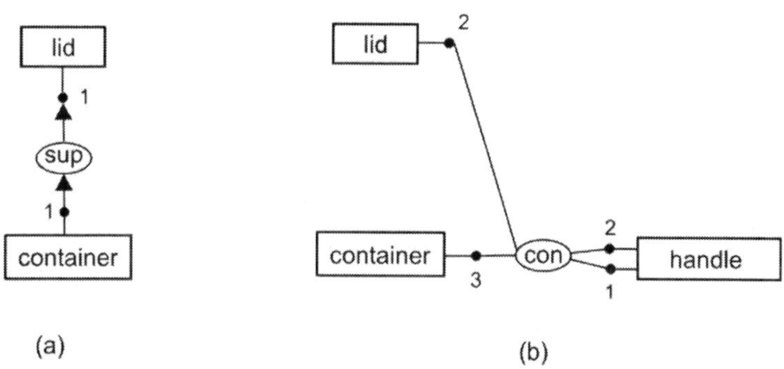

Fig. 6. Examples of the connectivity and supporting relations

Atomic sentences describing relations which hold among parts of the teapot presented in Figure 4(c) concern:

- connectivity between parts, and
- supporting relations between parts.

These atomic formulas constitute syntactic knowledge about the designed teapot obtained as the result of a design process. In this paper we omit the formal definitions of the formulas $sup(x_1,..,x_n)$ and $con(y_1,..,y_m)$ but we assume that both these formulas are specified in such a way that they are true in the given relational structure L under the given valuation υ, i.e., $(L, \upsilon) \models sup(x_1,..,x_n)$ and $(L, \upsilon) \models con(y_1,..,y_m)$.

Example 3. Let us consider the design of a teapot in Figure 4(c) and its hypergraph representation shown in Figure 5(c). The relations between teapot components are described by the following formulas:

- $\psi_1 \equiv$ *sup(base1, container2)* - the base supports the container,
- $\psi_2 \equiv$ *sup(container1, lid1)* - the container supports the lid,
- $\psi_3 \equiv$ *con(spout1, container4)* - the container is connected with the spout, and
- $\psi_4 \equiv$ *con(handle1, handle2, container3, lid2)* – the handle is connected with the container and the lid.

The visualization site in Figure 4(b) belongs to type $\rho_1 \equiv \omega_1 \wedge \omega_2 \wedge \omega_3 \wedge \psi_1 \wedge \psi_2$, while the visualization site in Figure 4(c) belongs to type $\rho_2 \equiv \omega_1 \wedge \omega_2 \wedge \omega_3 \wedge \omega_4 \wedge \omega_5 \wedge \psi_1 \wedge \psi_2 \wedge \psi_3 \wedge \psi_4$.

It should be noted that the handle is connected both with the container and with the lid. This information is described by means an atomic formula ψ_4. In the design knowledge related to design teapots, this formula does not belong to a set of basic atomic formulas which are to be satisfied for each teapot design. In such a case the design supporting system notifies the designer about the need to take it into consideration in the created solution (an appropriate shape of the lid or relocating one of the handle connections to touch the container).

A Physical Design Actions Domain

The last of the three design domains - the domain of physical actions is defined in the same design context. Physical actions are treated as a certain kind of events in the external world that start with an initial situation and result in another situation.

Definition 5. A domain of **physical design actions** $D_A = (A, \Delta_A)$ consists of a set A of physical actions to be classified, and a set Δ_A of objects used to classify the situations, called the *types* of D_A.

If a physical action situation $a \in A$ is classified as being of type $\delta \in \Delta_A$, we write $a \models_A \delta$ and say *a belongs to* δ.

Each action a has an input visualization site i_{in} and an output visualization site i_{out}, and as a consequence a tertiary relation $I \times A \times I$ is defined. We write $i_{in} \Rightarrow^a i_{out}$ and assume that each action has unique input site and output site.

Example 4. Let us consider the designing of a teapot in the context of the domain of physical design actions. The set Δ_A of types used to classify the situations contains constraints for actions leading to admissible component shapes of a teapot. The initial input visualization site i_0 is an empty monitor screen (see Figure 3(a)). The first action a_1 on i_0 results in the output visualization site i_1 presented in Figure 3(b). This action places the basic shape of a container and is classified as being of type δ_1, where δ_1 specifies actions which lead to appropriate shapes representing containers, for instance shapes which are closed curves or closed polylines. The successive physical actions are classified by types characterizing actions used to obtain the remaining teapot components.

The Active Perception

The designer obtains information about the design situation from a visual site using perceptual actions. He/she perceives a fact on a visual site *i*, that classifies this visual site. If a visualization site *i* is used to find a design solution *t* then we say that *i signals t*, ($i \rightarrow t$, where $\rightarrow$ is a binary relation from *I* to *T*). On the other hand, using functional actions the designer discovers the meaning of facts related to types that can classify design situations.

Design requirements can be treated as constraints on expected design solutions (design situations of D_T). Forms of visual constraints in the computer visualization domain D_V usually are different from forms in which the designer expresses requirements related to the design solutions of D_T. Therefore, types Ω_I of D_V that classify visual sites must be related to types Σ_T of D_T that classify design solutions. The connection between these types is expressed as a binary relation => from Ω_I to Σ_T, called a *semantic convention*. It relates constraints on graphical representations to designer requirements. In our design example the instance of the semantic convention is a relation $\omega_5 \Rightarrow \sigma_3$ between the design requirement σ_3: *the teapot can be lifted by hand* and the type ω_5: *the existence of an appropriate shape of a handle* which classifies the visualization site.

Thus the *active perception* can be seen as a combination of perceptual actions and functional actions. In our formal model the active perception is described by two relations, signaling and semantic convention, which together form a mapping from the computer visualization domain to the design tasks domain. The designer discovers information σ related to the design situation *t* from a visual site *i* only if $i \rightarrow t$ and there exists ω such that $i \models_I \omega$ and $\omega \Rightarrow \sigma$. In the running example, the information σ_3 can be discovered from the visual site i_3 (see Figure 3(d)), as $i_3 \rightarrow t$, where *t* is a teapot design, and there exists ω_5 such that $i_3 \models_I \omega_5$ and $\omega_5 \Rightarrow \sigma_3$.

The Operation Method

As it has been considered a method of operations is a set of instructions specifying what types of actions can be taken under what circumstances. The set of instructions can be defined by means of a set M of pairs (σ, δ) where $\sigma \in \Sigma_T$ and $\delta \in \Delta_A$. An individual instruction (σ, δ) allows one the following activity: if the design situation t belongs to σ ($t \vdash_T \sigma$), carry out any action a such that $a \vdash_A \delta$.

For example, in case of designing a teapot, the operation method contains an instruction of the form (σ_3, δ_3), where σ_3: *the teapot can be lifted by hand* and δ_3 specyfies actions which lead to appropriate shapes representing teapot handles. There are many sequences of actions belonging to the type δ_3, i.e., the shape of the handle can be designed in different ways as long as the applied physical actions describe situations belonging to σ_3.

The System of Computer-Aided Visual Design

After discussing three domains of visual design we can define the system of computer-aided visual design in the following way.

Definition 6. The **system** of **computer-aided visual design** is a 6-tuple

$S = (D_T, D_V, D_A, =>, \rightarrow, M)$, where:

- DT is a domain of design tasks,
- DV is a domain of computer visualization,
- DA is a domain of physical design actions,
- => is a semantic convention,
- → is a signaling relation,
- M is an operation method.

The system S allows us to define essential concepts of inventive design. One of these concepts is a notion of *emergence*.

The designer often obtains a drawing, which exhibits properties quite different from the mere summation of all components. During the active vision which is the dynamic process, the designer can discover and extract emergent shapes (the ones which had not been consciously constructed) in a generated drawing.

Example 5. In Figure 7(a) an initial drawing of a teapot is presented. An emergent shape, denoted by ω^*, which can be discovered by the designer in this drawing, is shown in Figure 7(b). The perceptual action allows the designer to notice this shape, while the functional action associates it with the stalk and the goblet of the flower. This association becomes a new inspiration in creating a form of a teapot (Figure 8) and enables the designer to formulate a devised requirement σ^* (a new type of Σ_T).

A three dimensional model of a teapot (Figure 9) corresponding to the last drawing from Figure 8 can be obtained by means of transformations of 2D basic shapes. This model is a design situation which belongs to σ^*.

Fig. 7. a) An initial teapot, b) an emergent shape

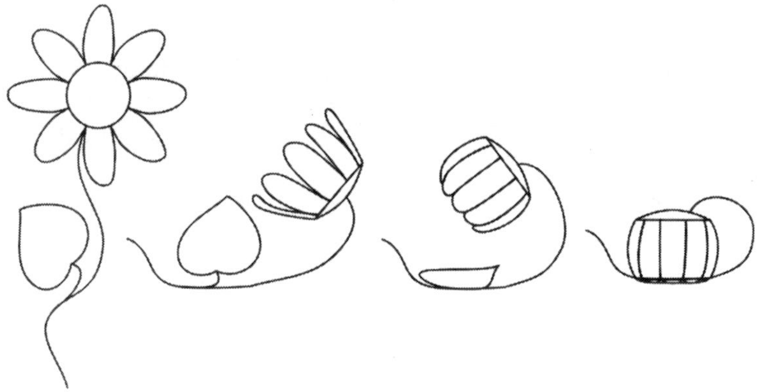

Fig. 8. A devised requirement – flower-shape form

Fig. 9. A new design of a teapot

It is known that emergent shapes elude a formal description. Our system enables us to handle the concept of the occurrence of emergence in a formal way.

Definition 7. Let *S* be a system of computer-aided visual design and $\sigma_1, \ldots, \sigma_n$ be types of Σ_T. We say that **emergence** occurs in *S* if:

- On the basis of the types $\sigma_1, \ldots, \sigma_n$ the method *M* admits a sequence $a_1, \ldots, a_m$ of the physical actions to be applied to visual site.
- Any subsequence of the sequence $a_1, \ldots, a_m$ of actions realizes a new fact ω^* on the visual site.
- According to the semantic convention (=>) an element ω^* of Ω_I can be transformed into a new type σ^* of Σ_T.

Conclusions

The main objective of this paper is to extend a formal model of computer-aided visual design system to include the method of automatic reasoning using data structures. The presented approach is based on the diagrammatic reasoning.

The logic model of reasoning, proposed here, uses data structures in the form of specific graphs called attributed hypergraphs. This structure is convenient to present designs as the elements of a visual language. Visual designing by means of shape grammars and structure-functional graphic editors require a different type of data structures. Such cases will be the subject of our studies in the future.

References

1. Ware, C.: Visual Thinking for Design. Elsevier, Amsterdam (2008)
2. Gero, J.S., Kannengiesser, U.: The situated Function-Behaviour-Structure framework. In: Gero, J.S. (ed.) Artificial Intelligence in Design 2002, pp. 89–104. Kluwer, Dordrecht (2002)
3. Suwa, M., Gero, J.S., Purcell, T.: Unexpected discoveries and S-invention of design requirements: Important vehicles for a design process. Design Studies 21(6), 539–567 (2000)
4. Shimojima, A.: Operational constraints in diagrammatic reasoning. In: Allwein, G., Barwise, J. (eds.) Logical Reasoning with Diagrams, pp. 27–48. Oxford University Press, Oxford (1996)
5. Grabska, E.: Computer-aided Visual Design (in Polish), EXIT, Warszawa (2007)

6. Arciszewski, T., Grabska, E., Harrison, C.: Visual thinking in inventive design: Three perspective (Invited). In: Proceedings of the First International Conference on Soft Computing in Civil, Structural, and Environmental Engineering, Madeira, Portugal (2009)
7. Barwise, J., Seligman, J.: The Logic of Distributed Systems. Cambridge University Press, Cambridge (1997)
8. Fagin, R., Halpern, J.Y., Moses, Y., Vardi, M.Y.: Reasoning About Knowledge. MIT Press, Cambridge (1995)
9. Akin, O., Moustapha, H.: Formalizing generation and transformation in design. In: Gero, J.S. (ed.) Design Computing and Cognition 2004, pp. 176–196. Kluwer, Dordrecht (2004)

Design Agents and the Need for High-Dimensional Perception

Sean Hanna
University College London, UK

Designed artefacts may be quantified by any number of measures. This paper aims to show that in doing so, the particular measures used may matter very little, but as many as possible should be taken. A set of building plans is used to demonstrate that arbitrary measures of their shape serve to classify them into neighbourhood types, and the accuracy of classification increases as more are used, even if the dimensionality of the space in which classification occurs is held constant. It is further shown that two autonomous agents may independently choose sets of attributes by which to represent the buildings, but arrive at similar judgements as more are used. This has several implications for studying or simulating design. It suggests that quantitative studies of collections of artefacts may be made without requiring extensive knowledge of the best possible measures—often impossible in real, ill-defined, design situations. It suggests a means by which the generation of novelty can be explained in a group of agents with different ways of seeing a given event. It also suggests that communication can occur without the need for predetermined codes or protocols, introducing the possibility of alternative human-computer interfaces that may be useful in design.

Introduction

Examination of the act of design by an individual agent, whether human or artificial, frequently involves an attempt to define the way in which that agent perceives the world. This paper suggests that the specific attributes an agent may perceive are relatively unimportant, but rather it is a high dimensionality of perception or input that is necessary.

J.S. Gero (ed.): *Design Computing and Cognition'10*, pp. 115–134.

In studying design or implementing an artificial agent, therefore, the attributes of a design artefact to be measured need not—indeed, should not—be determined a priori. While the suggestion that any set of attributes will do may seem counterintuitive, this paper will attempt to show that there is a more effective alternative strategy. This is to consider a large number of possible attribute dimensions, even if arbitrary, and allow the agent to select the relevant subset or subspace from these. This effectively allows for interpretation and reinterpretation on the part of the agent.

The strategy will be demonstrated with respect to a real set of design artefacts: building plans taken from various neighbourhoods. By taking a number of quantifiable measures of the shape of each, it is possible to classify the buildings such that each is identifiable as belonging to its particular neighbourhood. In brief, it will be shown that while some measures may be more or less useful in this, the correct identification of buildings improves as more measures are taken.

This has implications with respect to design creativity both at the level of the individual agent and of the group. For the individual, these concern the level at which symbolic representation occurs. Approaches to representation in Artificial Intelligence can be broadly positioned with respect to two extremes: a classical approach considering intelligence to be the manipulation of "physical symbol systems" directly representing the world [1], and a radically embodied one in which the world need not be represented at all [2], [3]. While the latter has strong merits, there are many aspects of design, from words to drawing conventions to standardised CAD representations, that appear strongly symbolic at least as far as communication is concerned. These symbolic elements are characterised by an interface that is clearly defined and comparatively low-bandwidth [4]—it is a reduction of the full dimensionality of possible measurements of the world. The classical assumption (famously made by Simon [5] in his description of an ant on the beach) is that this interface is identical to (or possibly external to) the boundary of the creative agent. What is suggested here, however, is that to the extent a symbolic interpretation or reduction of dimensionality exists, it must be internal to the creative agent. Perception is high-dimensional, then interpreted internally.

For a collective of many agents, this implies there may be at one time a variety of different interpretations of any observed event, a phenomenon that is arguably necessary for the generation of novel ideas. Many theories of creativity take the essential moment of insight as "seeing [something] as" something else [6] or changing "frames of reference" [7]; even within the extreme symbolic stance, Newell and Simon [1] mention the potential advantage of "moving from one representation to another". Clarke [8] explicitly notes from extensive archaeological data that novelty arises from small changes naturally inherent in the population; and reflective, hermeneutical [9] and systems [10] approaches to creativity or design likewise suggest that this novelty arises naturally, without being artificially imposed.

Hillier and Hanson [11] introduce the concept of *morphic languages*, in which the linguistic expressions may be the designed artefacts themselves, but the lack of a single, shared symbol system extrinsic to the agent raises a potential problem for communication. This paper aims to show that changes can happen as a result of different interpretations, as above, but communication is still possible. It will outline how agents can still make similar decisions due to patterns inherent in the observations, and demonstrate that this is possible for at least one set of data relevant to the design of architecture.

Clark and Thornton [12] make a distinction between two types of machine learning problems: type-1, in which the relevant patterns in data are immediately apparent; and the more difficult and complex type-2, in which any number of arbitrary patterns may be seen, and the data must be recoded before the relevant regularities are visible. The latter type are apparently far more prevalent in real world data (and interesting design situations), but by structuring our thought via language, social custom and other observation external to the data, humans demonstrate an ability to turn a type-2 problem into a tractable type-1. This paper will go a step further, to suggest that the data itself, in instances relevant to design, may gradually approach type-1 as more dimensions or attributes are observed. In this way, different agents may differ slightly in their independent judgements, yet overlap enough that communication via the morphic language of the artefacts themselves becomes possible.

Relevance

If it can be demonstrated that for many instances of design the particular choice of attributes/dimensions is of less relevance than the number used, this will impact at least three broad areas.

In the first case, it determines the possibility of quantitatively studying design via its artefacts without having to be sure about the validity of the particular measuring system used. If two significantly high-dimensional systems will converge on the same results, either one may be used effectively. This is particularly relevant as most real design situations deal with what Rittel and Weber [13] term “wicked” problems—a set of problems which can never be clearly defined and have unforeseeable implications and effects. Thus in studying design to make recommendations for real design practice, one cannot rely on knowing enough about the problem in advance to inform the particular choice about the most relevant attributes to measure.

In the second case, it appears necessary for a proper understanding of creativity, with respect to reinterpretation [6], [7], [14] and social interaction [9], [10], that the mechanisms for variance within a population be investigated. If creative leaps are ultimately rooted in small changes, a model that imposes these stochastically via straightforward random number generation (as occurs in “creative” models from genetic algorithms to populations of agents) may miss a crucial feature. Investigating how differences in interpretation occur may outline and quantify how much larger changes occur in a social system.

Finally, there is the very practical issue of how a designer can interact with the computer that is increasingly necessary in practice. Almost all current interfaces are constructed on the assumption that communication is based on predetermined protocols, often via agreed symbol systems, but this need not necessarily be the case. If two distinct agents can make similar judgements about an observed event via independently and arbitrarily chosen means of measuring it, then that event stands as effective communication. In the case of design, where an important element of communication is via sketching and similar methods that are both difficult to codify and easily reinterpreted, this may allow systems of interface with future design tools that are much more akin to the way designers interact with one another.

Example Systems: Observing Types in Architecture

The task of recognising and identifying distinct *types* of designed artefacts is taken as a primary subject of investigation relevant to design. Several approaches to type exist, sometimes distinguishing it from style in referring to objective matters of utility rather than subjective judgement [15]. As the main aim of this paper is to demonstrate that predefinition of relevant attributes is unnecessary, type will here refer broadly to all potential characteristics. In addition, the notion of a type is sometimes treated as clearly definable [16] and permanent [17], or sometimes recreated in every generation [18]. The latter view is taken again, for the same reason.

Clarke [8] sets out a clear working definition of type and demonstrates its effectiveness. An artefact type consists of a set of measureable attributes that is not monothetic, in that every member of the group displays all of the set of attributes, but polythetic, in that there is a looser overlap between attribute subsets. This correlation between individual members varies by context and scale, as Clark also uses the polythetic set to describe assemblages and cultural groups at higher levels. The effectiveness of this in an archaeological context is particularly relevant in that (as Clarke frequently notes) the attributes available to the archaeologist are necessarily limited and arbitrarily selected by the gap in time. This definition also lends itself to multivariate and computational methods, cluster analysis, and unsupervised learning.

The use of high-dimensional input has been shown effective in revealing types in artefacts at many levels of scale. For architectural and urban examples, spatial configuration is frequently represented topologically by a graph—the edit distance between these has been used, for example, to identify differences between Turkish and Greek house types in Cyprus [19]. At a larger scale, in a data set of 150 cities distributed around the world, the spectra of the entire street network graph was used to identify each as a vector in a 100 dimensional space, from which a subspace was extracted to represent the set [20]. The identification of a given city's geographical location was then found to be largely predictable purely by its form, Figure 1.

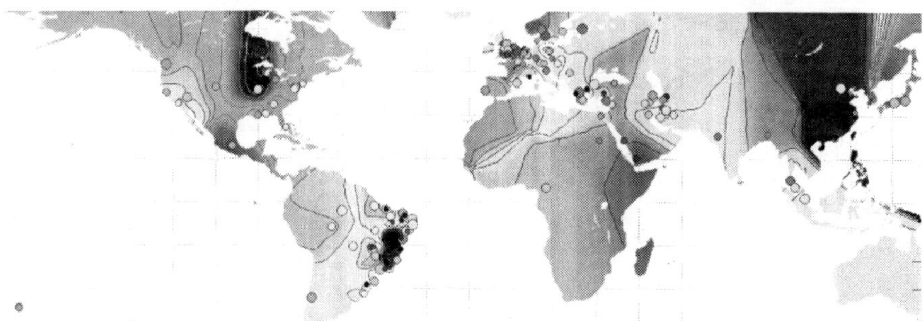

Fig. 1. Cities represented by their graph spectra can be placed geographically based on the form of their street network [20]

The use of such numerical type definitions has also been used to effectively guide a search in design generation or optimisation, by defining an objective function for a genetic algorithm to produce desk arrangements for the layout of workplace interiors [21]. Here, the objective is not set explicitly, but derived independently by a supervised learning algorithm based on a set of precedent examples. The algorithm derives the relevant features from the input set (e.g. convex groups of desks, clusters of a certain size) without any prior definition of these features, and generates plans to match these. In each of these cases, input to the algorithm is high dimensional, then reduced as required.

A similar set of types is used as the example for the work described here. The data is taken from a study of the properties of the building footprint and block configuration of four distinct neighbourhoods in

Athens, and one in London [22]. In this, it was demonstrated that a set of measurements of arbitrarily selected attributes of the plans of each block were sufficient to classify them by neighbourhood, even though the particular features relevant to this were not known. A set of thirteen individual measures were used, including topological features such as number of courtyard voids and geometrical features such as fractal dimension. Principal component analysis (PCA) of this thirteen dimensional space then revealed a distinct clustering of blocks by neighbourhood, Figure 2.

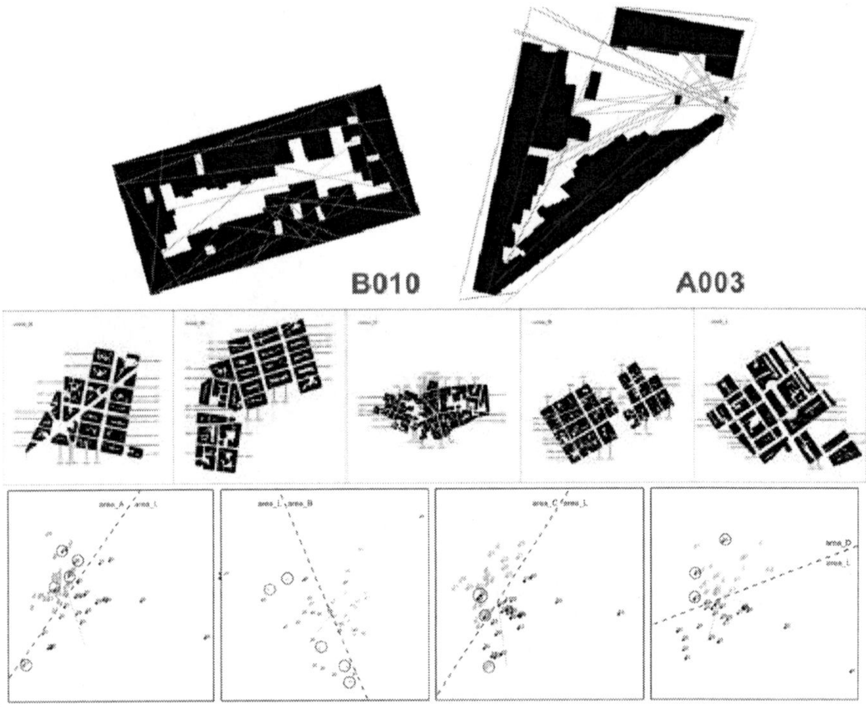

Fig. 2. A set of measurements taken of the shape of urban blocks (top) allow them to be clustered into distinct neighbourhoods. Image: Laskari *et al.* [22]

This example set of buildings has been chosen partly because its design scale is familiar. More importantly, while the dimensions of properties such as graph spectra are quite abstract, the particular measures used to describe the samples are each distinctly comprehensible, clear and distinct, even though their selection was arbitrary. The following section will investigate the reasons why such an arbitrary selection of attributes results in a correct classification into distinct neighbourhood clusters, and in particular the main hypothesis of this paper—that this is a result of there being a sufficiently high number of such attributes in the data set.

Method and Results

Under almost any circumstances, increasing the number of dimensions will permit an increased number of possible allowable classifications—this need not be tested. What will be tested here is classification on the particular space given by PCA. This is an unsupervised method which will yield a fixed subspace determined by the variance of the set as a whole, regardless of any class labels that might be assigned. The degree to which such an unsupervised (PCA) analysis of real samples allows classification into separate groups will therefore indicate the degree to which independent agents may observe the same phenomena without prior labelling.

Each sample is quantified by measurement of thirteen attributes, taken from [22] (the first four relate to changes in direct sight lines from different locations on the perimeter of the internal voids, see [23]):

1. mean connectivity value for the perimeter of the internal voids
2. mean distance between subsequent points of mean connectivity
3. vertical standard deviation of perimeter connectivity
4. horizontal standard deviation of perimeter connectivity
5. fractal dimension
6. perimeter of all voids internal to the block
7. number of voids internal to the block
8. total block area
9. building footprint area
10. number of buildings in the block
11. number of disjoint building clusters in the block
12. ratio of internal voids open to the street
13. number of vertices on the building contour

In each of the following sections, the method will select subsets of attributes, of sizes varying between $a=1$ and $a=13$. These will constitute the maximum possible input to a theoretical agent. Although far greater dimensionality would be possible, this range will suffice to show a clear trend of improved classification as more dimensions are used. PCA will then be performed on these to yield a feature space Φ of reduced dimensionality d (typically three-dimensional). It is within this space Φ that classification will be performed, and the effectiveness of this determined by the minimum linear classification error within this reduced PCA space.

Classification errors, Figure 3, will be calculated as the ratio of incorrectly classified samples within the total set of 125 samples (25 building groups in each of 5 classes). Errors will therefore be shown to 3 significant digits.

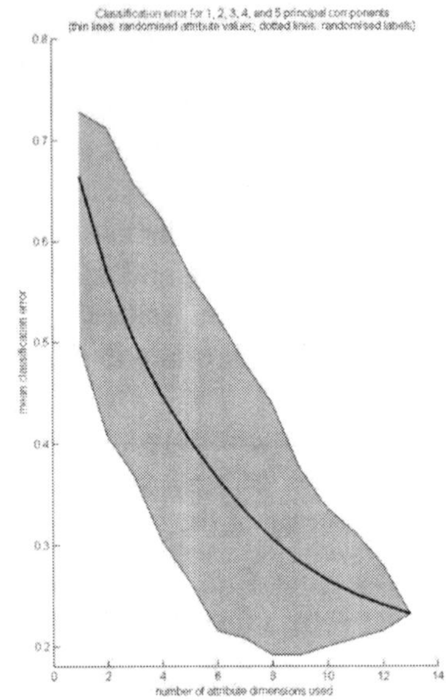

Fig. 3. The classification error decreases as the space in which samples are classified increases in dimensions (bold indicates mean error, grey the range from min to max error

Effect of Overall Attribute Dimensionality on Classification

Superficially, the number of dimensions used in any supervised classification task will have an obvious effect on the accuracy of the results—more dimensions yield a greater variety of hyperplanes for drawing distinctions between classes, and if all samples are appropriately labelled, the system has more opportunities to select the appropriate ones. This can be clearly seen in the data, Figure 3, with an expected decrease in error as dimensions increase from 1 to 13. (The mean error decreases monotonically while the minimum error reaches its optimum point at 8 dimensions, before the overall variance decreases due to a decreasing number of possible permutations.) What is not obvious, however, is whether an increase in the overall number of *available* attributes is of any further benefit beyond this. The above fact gives no reason to expect any improvement in classification whatsoever when:

- the examples are not labelled (unsupervised learning), as would be required for autonomous reinterpretation, etc. or;
- a subspace of constant, reduced dimensionality is available, as is always necessary in practice—an infinite set of attributes theoretically exists but is impossible to observe.

This section tests the hypothesis that the overall quantity of attributes is only relevant inasmuch as it provides a greater dimensionality in which arbitrary classification can take place (and therefore more possible classification hyperplanes), and finds it to be false. Rather, a pattern that may be considered intrinsic to the data set itself becomes progressively more evident as more attributes are used.

The effect of varying numbers of attribute dimensions within the data set was tested by taking the mean errors of classifications performed within a subspace of constant dimensionality, derived from principal components. A low dimensional subspace Φ of dimensionality $d=1$ to $d=5$ is taken from the principal components of the entire data set as specified by a given set of attributes $a \geq d$. Classification errors are then compared from linear discriminant analyses performed in Φ of equal dimensionality d. By varying the number of attributes a used in each subspace Φ, the effect of available attribute dimensionality can thus be compared independently of the classification dimensionality d. All possible permutations of a attributes are used from the total set of 13, classification performed on the resulting PCA subspaces Φ, then the overall mean error is recorded, Figure 4, for increasing sets of attributes $a=d$ to $a=13$.

Bold lines show the mean classification errors in Φ of dimensionality $d=1$ (top) to $d=5$ (bottom). Where $a=d$ (the minimum possible, with no reduction in dimensionality), classification is identical to that in the space of the original attributes and cannot be further improved. To the extent that the attribute dimensionality a is relevant only by virtue of increasing the dimensionality of the classification space φ, each of these mean errors should show no further improvement, and tests with randomly labelled data sets (dashed and thin lines, Figure 3) show this to be the case. However, all tests show improved classification. For a single component (top) this improvement is negligible, but for 3, 4 and 5 dimensions d there is significant improvement, approaching that of the optimal classification in the full dimensionality a of the original attributes.

This demonstrates that an increased number of attributes is clearly of value in describing the structure of the data, even when only a fixed number of components are used. As more attributes are used, the dimensions in which the data is naturally most varied overall more closely approximate the dimensions most useful for distinguishing the separate subclasses. This result is far from inevitable, as the attributes in question were chosen arbitrarily and so may have turned out to be redundant or conflicting.

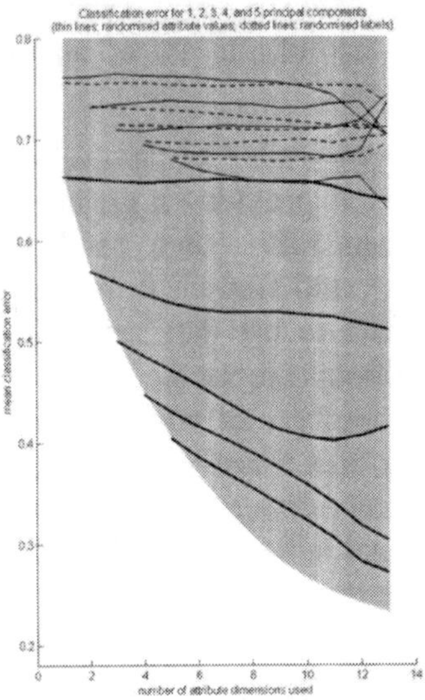

Fig. 4. Classification error decreases as more initial attributes are used, even if the space of classification remains of a constant dimensionality. Bold lines indicate spaces Φ of 1 to 5 dimensions. Thin and dashed lines show no improvement for random data sets

What this suggests is that the subgroups or clusters revealed by PCA are inherent to the data itself, rather than arbitrary designations imposed by the particular labelling scheme—they are gradually revealed as Clark and Thornton's [12] type-1 as more attribute dimensions are used. By contrast, the narrow lines in figure 4 indicate the effect of arbitrarily chosen classes, with dashed lines showing the mean errors for the same data in which only the labels were sorted at random, and solid lines showing the same for data in which each attribute value was resorted independently. In all cases the error rates are not only poor, but fail to increase as more attributes are used to define them.

This contrast indicates that the labelled clusters within the data set are intrinsically meaningful, in that they are discovered by unsupervised and unlabelled PCA, and are simply revealed by the measurement of larger sets of attributes. The following sections will unpack this observation by investigating whether a relationship is discernable for particular subsets of attributes.

Effect of Particular Sets of Attribute Dimensions

The possible subsets of attributes of any fixed dimensionality *a* are not equal in terms of accurate classification—there is some variance in the error rates from which the means above were taken. If constrained to a limited dimensionality of measurements to be taken, one would hope to be able to choose the optimal set of attributes to yield the best classification. This section examines whether the improvement in classification with larger sets above is a result of particular combinations of attributes, and whether these optimal combinations can be determined beforehand. It tests:

a) whether any *particular individual attributes* can be found to contribute to the overall reduction in misclassification error
b) whether there are similarities between *particular sets of attributes* that reduce the error
c) whether these *attribute sets can be determined prior to performing the classification* itself.

Particular individual attributes were found not to contribute significantly to the overall reduction in misclassification error. Contribution to errors for each attribute was calculated by taking the error rate (mean misclassified examples) in a constant subspace of dimensionality d=3 for all possible combinations of a=6 attributes that included the attribute in question. A significant variation in these would indicate specific attributes responsible for error or correct classification. However, while errors using subsets of a=6 attributes had a considerable range from 0.296 to 0.616 overall, the range in mean errors due to particular attributes was minor, as shown in Table 1.

Table 1 Contribution to errors: the mean errors for all sets of 6 attributes that include the attribute in question.

A1	A2	A3	A4	A5	A6	A7	A8	A9	A10	A11	A12	A13
.434	.464	.469	.459	.460	.465	.431	.457	.475	.447	.464	.451	.464

This very slight contribution of each attribute to error rates became more significant, however, when *particular sets of attributes* were considered. While there was a negligible correlation (R=0.16) found between the similarity between any two sets of attributes and their error rates, the very best sets resulting in the lowest errors (0.296–0.313) were found to contain four of the same attributes in common: [1 7 12 13], so the effect of limiting the sets to particular attributes was tested next. Attributes were ordered based on their contribution to error in table 1, and both the 'best' [7 1 10 12] and 'worst' [9 3 6 13] were progressively removed from the available attributes. The mean and range of error rates for the remaining combinations of attributes is shown in Figure 5. A noticeable change in the classification errors is evident here: rising monotonically when the 'better' attributes are unavailable and vice-versa.

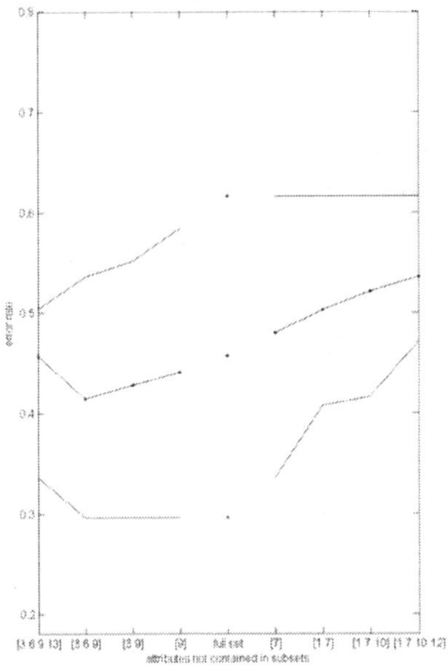

Fig. 5. Error rates of attribute subsets with specific attributes withheld. Error rates decrease when the 'worst' attributes are not used, and rise when the 'best' are withheld

In both tests above the optimal attributes for measurement were determined only by knowing the correct classification results—an impossibility if one is attempting a classification on unlabeled examples. The third test of attribute sets is whether these optimal subsets can be determined by any *measurable diversity within the data as a whole*, and therefore can be found before performing the actual classification itself. The increase in classification accuracy as dimensionality *a* increases (§4.1) suggests the hypothesis that the optimal attribute subsets are those that are most diverse, in terms of each attribute independently providing more information about each example. If this is true, it would both explain this improvement for large sets and suggest a method by which the appropriate attributes for any given data set can be found.

The simplest measurement of the independence of two attributes with respect to the data is the correlation between their respective measures of all the examples in the data set. A high product of these coefficients for a given set of attributes should indicate greater independence or diversity and therefore a lower classification error. This was found not to be the case, with almost no correlation between attribute dimension diversity and error rate ($r^2<0.0025$). A number of other measurements of attribute set diversity were also taken (product of minimum angles, sum or product of dot products, convex hull volume), using the projections of principle axes for each into the 3-dimensional classification space Φ, with similar results. Within each of these tests, neither the overall correlations nor the sets of attributes known to be error minimising revealed any discernable pattern—these latter, 'best' sets were sometimes of high diversity, but just as often medium or low.

These results suggest that while some subsets of attributes of a fixed dimensionality *a* are clearly superior in terms of providing an accurate classification, these do not have any relationship to the distribution of the data that can be found prior to classification itself. One therefore cannot identify a priori which ones to use for the classification.

Different Sets of Attributes for Different Classes

Classification errors for all five classes have been used together to examine the attribute subsets above, but the resulting suggestion that particular attributes (e.g. 1, 7, 10, 12) are optimal is somewhat misleading, as different classes may be better distinguished by different attributes. Table 2 shows the attributes that best classify each individual class in isolation (within a space φ of d=3, again found by PCA) for attribute sets of a=3, a=4, a=5 and a=6. The top row in each section shows the errors in each class for the optimal attribute set for Class 1, the second row for Class 2, etc.

Those that are optimal for one class are not for another, as evidenced by the greater classification errors for other classes (grey cells), and the fact that many of these errors even increase as more dimensions are used.

There is some overlap in the dimensions used—attributes 7 (no. of voids) and 8 (block area) occur repeatedly—but there is also a great deal of difference.

Table 2 The subset of attributes that best classify one class differ from those that best classify another. Bold indicates attributes that do not appear in the previous table

Best 3 attributes	Class 1	Class 2	Class 3	Class 4	Class 5
[7 8 10]	0.13	0.225	0.28	0.22	0.245
[3 9 12]	0.42	0.125	0.31	0.35	0.27
[7 8 13]	0.18	0.25	0.17	0.335	0.24
[3 8 10]	0.245	0.18	0.27	0.16	0.36
[1 7 12]	0.285	0.17	0.23	0.435	0.135

Best 4 attributes	Class 1	Class 2	Class 3	Class 4	Class 5
[**2** 7 8 10]	0.125	0.215	0.28	0.24	0.2

[**4 5 8** 12]	0.365	0.12	0.31	0.215	0.34
[**1** 7 8 13]	0.235	0.27	0.165	0.405	0.19
[3 **5 11 13**]	0.38	0.33	0.29	0.16	0.31
[1 **4** 7 12]	0.28	0.145	0.26	0.285	0.115

Best 5 attributes	Class 1	Class 2	Class 3	Class 4	Class 5
[2 **4** 7 8 10]	0.125	0.24	0.28	0.245	0.22
[**3** 4 8 **11** 12]	0.395	0.115	0.29	0.325	0.395
[1 8 **10 11** 13]	0.225	0.31	0.15	0.395	0.165
[**1 4** 5 11 13]	0.315	0.29	0.305	0.155	0.235
[1 **2 5** 7 **10**]	0.165	0.25	0.28	0.26	0.105

Best 6 attributes	Class 1	Class 2	Class 3	Class 4	Class 5
[2 4 **6** 7 8 10]	0.125	0.245	0.29	0.24	0.215
[**2** 3 4 8 11 12]	0.33	0.11	0.305	0.3	0.385
[1 **2** 8 10 11 13]	0.225	0.235	0.145	0.33	0.245
[1 **2** 4 5 11 13]	0.335	0.305	0.29	0.155	0.23
[1 2 5 **6** 7 10]	0.195	0.315	0.3	0.24	0.08

Worst attributes	Class 1	Class 2	Class 3	Class 4	Class 5
[**1 3** 6 **9 12 13**]	0.47	0.19	0.22	0.34	0.335
[**1 7 9 10** 11 **13**]	0.155	0.455	0.26	0.33	0.22
[1 2 **3 4 5 6**]	0.215	0.245	0.39	0.285	0.22
[1 **7 8 9** 11 13]	0.26	0.27	0.225	0.525	0.215
[**3** 6 **9 11 12 13**]	0.41	0.215	0.22	0.385	0.56

When limited to three attributes, [7 8 10] provide an excellent classification of Class 1 samples, but the best subset for classification of Class 2 samples, [3 9 12], contains none of the same attributes. Nor are these particular sets of much use in classifying the other classes within the data set: while [3 9 12] yields a successful result (error=0.125) for Class 2, it misclassifies nearly half of the Class 1 examples (error=0.42) and does little better for the others. The same effect can be seen for all classes and subset dimensionalities in the table.

There is also less consistency than might be expected as the dimensionality of the attributes a increases. Attribute numbers in bold show the new attributes added to the optimal subset for each class as the subset increases by one. In many cases, several new attributes replace the previous ones, indicating several of those providing the best classification when only three are allowed are no longer optimal when four are used. The final portion of Table 2 shows the *worst* performing attribute sets for each class, which reveals several attributes (not in bold) that appear also in the corresponding sets of optimal attributes immediately above.

These inconsistencies with respect to optimal attribute subsets for individual classes reinforce the result of section 4.2. While certain attributes or attribute subsets are particularly well suited for classification of specific classes, there appears no way to determine these without knowing the classification beforehand.

Mutual Classification and Communication

In considering communication between two artificial agents, or between a human subject and a computer system (or by extension even between two human subjects), the overall error rate resulting from a single attribute subset is less significant than the particular classifications that subset yields. For communication to be effective, it is necessary that the two agents in question make similar judgements on any given piece of data used for communication—that they see the world in a similar way. This section examines the way in which subsets of attribute dimensions matter with respect to the specific examples being misclassified, and the manner of their misclassification.

The mutual classification between pairs of attribute subsets was measured for subsets of a=3 to a=12. To determine the overall difference in how the pair perform, the total number of examples in the data set that the two classify differently is used; actual classification errors with respect to known labels are ignored. This difference in performance is then compared with the difference between the attributes themselves, where the similarity between any two attributes is calculated by the correlation between their independent classification of the data. The sum of these maximum correlations gives an overall measure of similarity between any two subsets.

Given the measure of the whole data set by one attribute α as $M(\alpha)=\{\mu_1,\ldots,\mu_n\}$, two attribute sets $[\alpha_1, \ldots, \alpha_a]$ and $[\beta_1, \ldots, \beta_a]$, and corresponding sets of classification results $[\kappa_{\alpha 1}, \ldots, \kappa_{\alpha n}]$ and $[\kappa_{\beta 1}, \ldots, \kappa_{\beta n}]$, the difference in performance is given by:

$$classDiff = \sum_{x=1}^{n} \kappa_{\alpha x} \neq \kappa_{\beta x}$$

and the similarity between attribute sets by:

$$\sum_{x=1}^{a} \max_{1 \le y \le a} corr^2\left(M(\alpha_x), M(\beta_y)\right)$$

Figure 6 illustrates that differences between attribute sets do result in approximately corresponding differences between sets of misclassified examples, but only for small *a*. As *a* increases, this correlation between attribute dimensions and classification results also decreases. For small subsets of dimensions (3, 4, etc.) there is a significant degree of correlation (approx r^2=50%), however this decreases to insignificant correlation as more dimensions are added, with r^2=12% for a=11 and r^2=0.3% for a=12.

With respect to the situation of two communicating agents, the prior agreement on the particular attribute dimensions used thus appears to matter greatly when the number of attributes is low, as evidenced by the strong correlations for low values of *a*. But this matters less the more dimensions we have. At the same time, increasing dimensions decreases the overall difference in performance between agents. If the general strategy for overall error reduction (as indicated earlier) is to increase *a*, this also appears to improve the likelihood of different agents making same distinctions (i.e. having similar φ) without prior agreement on the particular sets of attributes to use.

Conclusion

While one should generally expect to find a decrease in classification error as the number of dimensions of the classification space increases, there is no reason to expect this when a classification space of constant dimensionality is derived from arbitrarily varying sets of initial dimensions. As seen in §4.1, this is not the case for a randomly labelled set of data. Nevertheless, classification in spaces derived by PCA (with 1 to 5 components) was seen overall to steadily improve as more initial dimensions were used (§4.1), suggesting that for data sets with similar properties to the one under investigation it is generally beneficial to use as many dimensions of measurement as are available.

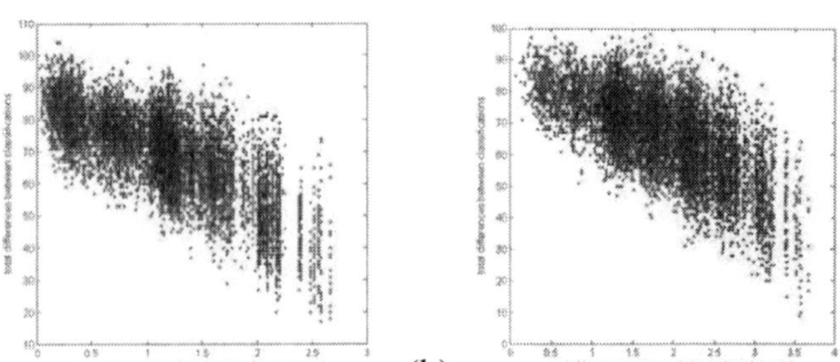

(a) (b)

(c) 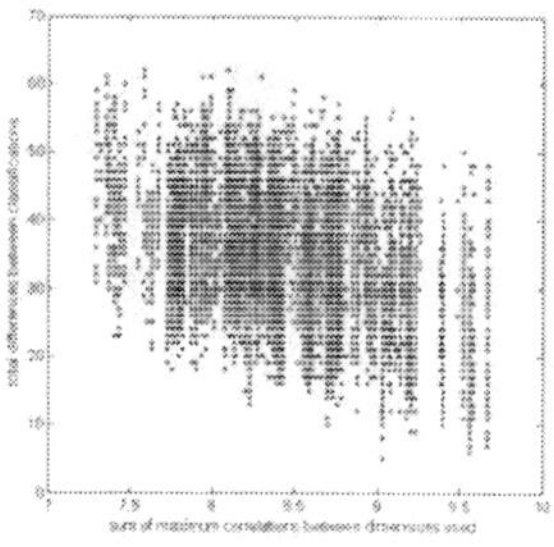 (d) 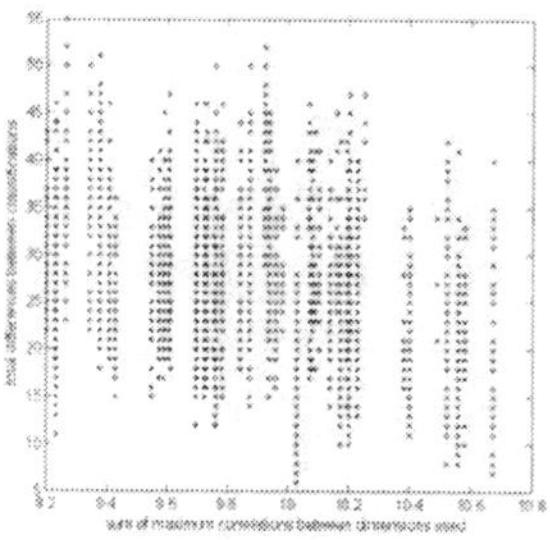

Fig. 6. Correlations between attribute subsets and classification results indicate the degree to which mutual understanding is dependent on the use of the same attributes by both agents. It is highly dependent for small subsets of 3 or 4, much less so for 11 or 12

There are, naturally, some particular small sets of attributes that yield a better classification than others, or better even than larger sets, but in practice there is no means for determining what these sets are. There appears to be no intrinsic relationship between these attributes (e.g. diversity) with respect to the data as a whole (§4.2), and the best subset of attributes for one class are unsuited to another (§4.3). In practice, in creating a system that is to evaluate any group of artefacts, the remaining viable strategy is not to carefully select the subset of features thought to best describe the relevant classes, but to use the full set of as many features as are available.

The reason larger sets result in better classification appears to be due to the nature of the examples themselves. Clark and Thornton's [12] distinction is useful: type-1 examples would contain regularities that are immediately apparent, while for type-2 the examples may afford a number of equally valid interpretations. The supposed way in which a culturally embedded person deals with the latter is by relying on a structure imposed externally by language or other cues to recode the observations to type-1. However, there may be intermediate possibilities between the two types. A hypothetical set of examples, measured in infinite dimensions, may be type-1, but appear as type-2 when only a limited subset of these are used. If the classes are determined by a polythetic set of attributes [8] this is almost certain to be the case, because the incomplete overlap between attributes would be less evident as fewer attributes are used. It would appear that the urban block data used here is of this nature, thus as it is viewed in more dimensions, the inherent type-1 pattern becomes gradually more evident.

This may also explain what is occurring when we are able to recognise as obvious distinct types in other sets of artefacts, even when we are unable to describe the specific criteria for the decision. The data set is type-1, but only when considered in many dimensions. If a continuum of intermediate possibilities is considered between type-1 and type-2, the degree to which classifications based on arbitrary subsets of attributes converge as those subsets increase in dimensionality offers a possible means of measuring this for a given set of data.

The observation that the overlap in attributes increases with larger sets implies that communication between distinct agents is possible without having to predetermine which particular dimensions are relevant. The strategy, again, of each agent using as many attributes as possible will ensure that each forms similar interpretations of a given event (§4.4).

However, the use of different subsets of attributes by different agents in a population may help to explain the generation of the novelty that is essential to the creative process in social models of creativity [10]. Many definitions of creativity hinge around coupled notions of novelty and utility [24], but while the latter is clearly justifiable for straightforward functional reasons, the need for novelty is less easily explained. If this novelty actually arises naturally, however, there need be no internal or external motivation required. It seems a fair conjecture that if slightly different sets of attributes can exist within a population that makes similar interpretations of events, then an ostensibly similar population may always contain some variance in its underlying choices of attributes, or ways of seeing. If an occasional event should arise that is interpreted differently among agents, an apparently novel difference of opinion will result. Such a variety of ways of seeing, or "frames of reference" is often stressed as the crucial component of creative insight [7][14], and appears in agent models of creativity [25][26]. In these, agents make design decisions by evaluating available options at any given point against a test criterion given by their own point of view. The result is that this difference in criteria easily drives collective innovation.

This also provides a notion of novelty that is not at odds with that of utility. Creativity is often described as a series of paradoxes [27], or explained as seeking a median point between "too similar" and "too different" [25], but the paradox does not exist if more dimensions are assumed. An interpretation of an event by an agent as perfectly normal may simply be seen by the next as highly unusual due to a measurement in a differing dimension. In human terms this conjecture needs further investigation, but it seems natural that we need only do what seems useful and appropriate to us, while others will always interpret things somewhat differently.

Acknowledgements

I would like to thank Anna Laskari for the initial analysis of building plans that made this research possible, and for the use of the data she has collected. This research has been supported by a Research Councils UK Academic Fellowship, EPSRC grant no. EP/E500706/1.

References

1. Newell, A., Simon, H.: Computer Science as Empirical Enquiry: Symbols and Search. Communications of the Association for Computing Machinery 19, 105–132 (1976)
2. Brooks, R.A.: Intelligence without representation. Artificial Intelligence 47, 139–159 (1991)
3. Dreyfus, H.: Why Heideggerian AI failed and how fixing it would require making it more Heideggerian. Artificial Intelligence 171, 1137–1160 (2007)
4. Haugeland, J.: The Nature and Plausibility of Cognitivism. Behavioral and Brain Sciences 2, 215–260 (1978)
5. Simon, H.: The Sciences of the Artificial, 3rd edn. MIT Press, Cambridge (1996)
6. Schön, D.A.: Displacement of Concepts. Tavistock, London (1963)
7. Akin, Ö., Akin, C.: Frames of reference in architectural design: analysing the hyperacclamation (A-h-a-!). Design Studies 17(4), 341–361 (1996)
8. Clarke, D.L.: Analytical Archaeology. Methuen & Co., London (1968)
9. Snodgrass, A.B., Coyne, R.D.: Is designig hermeneutical? Architectural Theory Review 2(1), 65–97 (1997)
10. Czikszentmihalyi, M.: Society, culture, and person: a systems view of creativity. In: Sternberg, R.J. (ed.) The nature of creativity: Contemporary psychological perspectives, pp. 325–339. Cambridge University Press, Cambridge (1988)
11. Hillier, B., Hanson, J.: The Social Logic of Space. Cambridge University Press, Cambridge (1984)
12. Clark, A., Thornton, C.: Trading spaces: Computation, representation and the limits of uninformed learning. Behavioral and Brain Sciences 20, 57–90 (1997)
13. Rittel, H.W.J., Webber, M.M.: Planning problems are wicked problems. In: Cross, N. (ed.) Developments in Design Methodology. John Wiley and Sons, Chichester (1984)
14. Koestler, A.: The act of creation. Hutchinson (1964)
15. Westfall, C.W.: Building Types. In: van Pelt, R.J., Westfall, C.W. (eds.) Architectural Principles in the Age of Historicism. Yale University Press, New Haven (1991)

16. Alexander, C., Ishikawa, S., Silverstein, M., Jacobsen, M., Fiksdahl-King, I., Angel, S.: A Pattern Language. Oxford University Press, New York (1977)
17. Rossi, A.: The Architecture of the City. The MIT Press, Cambridge (1982)
18. Colquhoun, A.: Typology and Design Method, Arena, 83: 11–14 Reprinted in Colquhoun A. Essays in Architectural Criticism: Modern Architecture and Historical Change. The MIT Press, Cambridge (1967)
19. Conroy-Dalton, R., Kirsan, C.: Small graph matching and building genotypes. Environment and Planning B: Planning and Design 35(5), 810–830 (2008)
20. Hanna, S.: Spectral comparison of large urban graphs. In: Koch, D., Marcus, L., Steen, J. (eds.) Proceedings of the 7th International Space Syntax Symposium. Royal Institute of Technology (KTH), Stockholm, Sweden (2009)
21. Hanna, S.: Defining Implicit Objective Functions for Design Problems. In: Proceedings of the Genetic and Evolutionary Computation Conference, GECCO 2007. ACM Press, New York (2007)
22. Laskari, S., Hanna, S., Derix, C.: Urban identity through quantifiable spatial attributes: Coherence and dispersion of local identity through the automated comparative analysis of building block plans. In: Gero, J.S., Goel, A. (eds.) Design Computing and Cognition 2008. Springer, Heidelberg (2008)
23. Psarra, S., Grajewski, T.: Describing shape and shape complexity using local properties. In: Third International Space Syntax Symposium, Atlanta (2001)
24. Boden, M.A.: The creative mind: myths & mechanisms. Weidenfeld & Nicolson, London (1990)
25. Saunders, R., Gero, J.S.: Artificial creativity: A synthetic approach to the study of creative behaviour. In: Gero, J.S., Maher, M.L. (eds.) Computational and Cognitive Models of Creative Design V, Key Centre of Design Computing and Cognition, pp. 113–139. University of Sydney, Sydney (2001)
26. Hanna, S.: Where creativity comes from: the social spaces of embodied minds. In: Gero, J.S., Maher, M.L. (eds.) Proceedings of HI 2005, Sixth International Conference of Computational and Cognitive Models of Creative Designs. University of Sydney, Sydney (2005)
27. Cropley, A.J.: Definitions of creativity. In: Runco, M.A., Pritzer, S.R. (eds.) Encyclopedia of Creativity, pp. 511–524. Academic Press, San Diego (1999)

A Framework for Constructive Design Rationale

Udo Kannengiesser[1] **and John S. Gero**[2]
[1]*NICTA, Australia, and University of New South Wales, Australia*
[2]*George Mason University, USA, and University of Technology–Sydney, Australia*

This paper proposes a framework for describing design rationale as a constructive notion rather than a fixed record of design reasoning. The framework is based on two views: an instance-based view of design rationale as an ordered set of decisions, and a state-space view of design rationale as a space of solution alternatives. The two views are connected with each other using the function-behaviour-structure (FBS) ontology. Constructive design rationale is defined and categorised based on reformulations of the function, behaviour or structure of the rationale. The drivers of the different reformulations are represented in the situated FBS framework.

Introduction

In design a typical ontology is concerned with either the object being designed or the processes of designing and hence views designing a forward moving activity; it is forward looking. ("Design" is used to indicate the output and "designing" the process that produces a design.) Design rationale is concerned with tracking the decisions that were taken to reach the design during the process of designing [1, 2]. In this it is backward looking in the same sense that history is backward looking. Design rationale can be equated to a truth maintenance system in that it aims to represent the beliefs behind the decisions and their dependencies.

One view of design rationale expands the elementary description given above to include not only the basis of the decision as listed but allows for any basis that produces the same decision. For example, in a closed world defined by a set of rules, this is conceptually similar to allowing for

J.S. Gero (ed.): *Design Computing and Cognition'10*, pp. 135–153.

potential multiple paths leading to a goal, where the paths model the rationale for the decision. As a consequence even if just part of or even the entire path taken that leads to a goal is retracted it does not necessarily follow that the goal is incorrect as there may be other paths that would support the goal. The same idea applies in an open world. It is consistent with the common observation that when designers are asked for the rationale underpinning their designs, they rarely produce pre-conceived or pre-recorded explanations but construct them on the fly [3]. The rationale they construct is adapted to the specific question being asked in the specific situation, which emerges from factors such as the presumed expertise and goals of the recipient of the rationale and expectations of accuracy and relevance. We refer to this notion as *constructive* design rationale.

In contrast, most computational models assume design rationale to be a fixed record of the design process. One of the problems associated with this "record-and-replay" [3] paradigm is that fixed rationale needs to be recorded as designing unfolds. Recording design rationale is a time-consuming activity that causes significant overhead for designers [4]. Recent approaches aim towards reducing this effort by providing integrated design environments that can automatically record design rationale [5]. However, over the course of several design projects, and even within the same design process, the relevance and applicability of a recorded design rationale decreases. This is because many design projects are unique, and rationale instances are based on assumptions [6] that frequently change during designing.

This paper proposes a framework of design rationale as a constructive rather than a fixed representation of designing. It provides an ontological basis for developing new design support systems that can generate different rationales for different situations.

An Ontological Representation of Design Rationale

The FBS Ontology

Modelling design rationale is facilitated by using an ontological framework that provides a common terminology with agreed meanings for a domain of discourse. The function-behaviour-structure (FBS) ontology [7, 8] provides such a framework for the design domain.

- *Function* (F) of an artefact is its teleology ("what it is for"). An example is the function "to wake someone up" that humans generally ascribe to the behaviour of an alarm clock.
- *Behaviour* (B) of an artefact is the attributes that can be derived from its structure ("what it does"). An example of a physical artefact is "weight", which can be derived directly from the product's structure properties of material and geometry.
- *Structure* (S) of an artefact is its components and their relationships ("what it consists of"). For physical artefacts, it comprises geometry, topology and material.

Humans construct relationships between function, behaviour and structure through experience and through the development of causal models based on interactions with the artefact. Function is ascribed to behaviour by establishing a teleological connection between the human's goals and measurable effects of the artefact. Behaviour is causally related to structure, i.e. it can be derived from structure using physical laws or heuristics. This may require knowledge about external (or exogenous) effects and their interaction with the artefact's structure. There is no direct relationship between function and structure.

Instance-Based and State-Space Views of Design Rationale

Design rationale can be conceptualised in three ways [9]: (1) as a historical record of designing, (2) as a set of claims about the properties embodied by an artefact, or (3) as a space of possible design alternatives.

The first two views can be combined into an instance-based view of design rationale as a history of specific decisions, about artefact properties or about the design process. Decisions on the decision alternatives are not independent of one another. A variety of dependencies can exist between design decisions [10]. During designing only a subset of these dependencies is known or anticipated in advance. A great deal of designing is based on the designer's assumptions that may be incomplete and incorrect, or that may change later in the process. Yet, the dependencies assumed by the designer are used to proceed from one decision to the next, thus generating a path through a network of possible design decisions, Figure 1. The instance-based view of design rationale can be understood as comprising three elements:

- *Starting point*: is an antecedent design decision
- *End point*: is a specific decision alternative selected for a consequent decision
- *Path*: is a sequence of intermediate decisions connecting the starting point and the end point. *Elementary* paths include only one intermediate

decision that is concerned with establishing a basis for assessing and then selecting specific decision alternatives. *Complex* paths include multiple intermediate decisions that create supplementary knowledge needed for reaching the end point. This includes, for example, decisions related to information-seeking activities and decisions on other design issues affecting the issue under consideration.

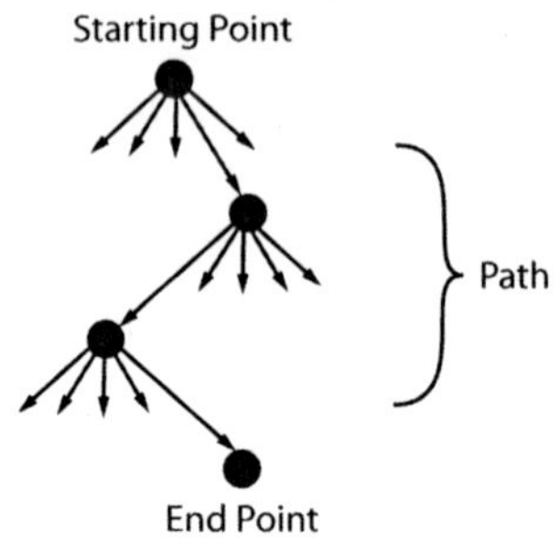

Fig. 1. A rationale instance described as a directed graph, with nodes representing decisions

In this paper, we refer to design rationale modelled in terms of these three elements as a *rationale instance*. Table 1 shows an example of a rationale instance in the context of a conceptual design process of a monitoring system for drilling tools.

Table 1 Example of a rationale instance for the conceptual design of a monitoring system for drilling tools. Selected decision alternatives are represented in italics.

Rationale element	Decision problem	Decision alternatives
Starting point	What level of automation?	*Automatic*, Semiautomatic, manual
Path	What data is monitored?	Tool wear, *Tool breakage*
	For what tool diameters?	> 10mm, > 5mm, > *0.8mm*
End point	What sensor type?	Force, Acoustic, *Laser*, Strain

Here, the end point decision on using laser sensors (rather than other types of sensors) is reached from the starting point decision on using automatic monitoring technology, and following a (complex) path that includes the intermediate decisions on monitoring tool breakage for tool diameters that may be as small as 0.8mm.

A state-space view of design rationale as a set of possible design alternatives can be understood as capturing classes rather than instances of design decisions. These generic classes of design decisions can be modelled as decision problems or issues associated with sets of alternative solutions, as shown in Table 1. This is consistent with the most common representations of design rationale. For example, in the QOC [11], IBIS [12] and DRL [9] approaches, issues are called "Questions", "Issues" and "Decision Problems", respectively. Alternative solutions are called "Options" in QOC, "Positions" in IBIS, and "Alternatives" in DRL.

Computationally, design decisions can be modelled as a state space in terms of variables and their ranges of values, where the variables correspond to issues and the ranges of values to a set of alternative solutions. A simple example of a design decision, taken from MacLean et al. [11], is the issue "how wide is the scroll bar" and the associated alternative solutions of "wide" and "narrow". Here, the variable is "width of the scroll bar", and its range of (qualitative) values comprises "wide" and "narrow". Design decisions can be more complex, consisting of multiple variables where every variable represents a decision problem on a finer level of granularity that can itself be represented in terms of variables and ranges of values.

An FBS View of Design Rationale

Individual design decisions may deal with the function, behaviour or structure of the artefact. However, what they all have in common is that they compose structured sets of decision variables. They can be viewed as abstract yet first-class artefacts, an idea that has found recent interest in the software design community [13]. We can apply the notion of structure (S) in the FBS ontology to describe design decisions as artefacts, and refer to the space of design decisions as the *structure state space* of design rationale.

Most approaches to representing design rationale include the notion of criteria that are used in design decisions as a basis for evaluating, comparing and selecting alternative solutions. In the scroll bar example, the criteria presented by MacLean et al. [11] include "screen compactness" and "ease of hitting with the mouse". Different alternative solutions fulfil these criteria to different extents. Criteria may be more or less formally defined, and may be qualitative or quantitative. In all instances, they represent the performance of a design decision, which in the FBS ontology is captured by the notion of behaviour (B). We call the set of criteria associated with a design decision the *behaviour state space* of design rationale.

Some accounts of design rationale include the strategies and goals underlying particular design decisions. This allows reasoning about the process or plan of designing, and establishes a basis for deriving criteria in accordance with particular goals [14]. The overall goal of a design decision is to create knowledge for advancing the state of the design [15]. The knowledge created is often needed to support other design decisions. Generally, the goals of a design decision include creating knowledge for refining or realising prior decisions and for enabling or guiding subsequent decisions. For example, the decision on using a scroll bar may have the goal of refining the prior decision on using a graphical user interface, and the goal of enabling the subsequent decision on scroll-bar width. The notion of function (F) in the FBS ontology can be used to capture the goals associated with design decisions. The set of functions for a design decision then establishes the *function state space* of design rationale.

The union of the function state space, the behaviour state space and the structure state space of design rationale is termed the *rationale state space*. We can establish connections between the instance-based view and the FBS state-space view of design rationale:

- The *starting point* of a rationale instance is covered by the function state space. This is because functions relate a decision to other decisions, including those that occur prior to that decision. The notion of issues in starting point decisions is covered by function variables. The notion of solution alternatives in starting point decisions is covered by ranges of function values.
- The *end point* of a rationale instance is covered by the structure state space. This is because the end point is a specific, targeted decision that is a point in the structure state space. The notion of issues in end point decisions is covered by structure variables. The notion of solution alternatives in end point decisions is covered by ranges of structure values.
- The *path* of a rationale instance, if it is elementary, is covered by the behaviour state space. This is because behaviour provides a link between function and structure [16] by forming a basis for assessing different structures oriented to achieving given functions. Behaviour variables (and their ranges of values) can then be viewed as path variables (and their ranges of values). Variables of a complex path that correspond to a set of additional intermediate decisions are not covered in the FBS state-space view of the rationale instance. They can be mapped onto the function, behaviour and structure variables of

those rationale instances that are associated with these intermediate decisions.

What Is Constructive Design Rationale?

Design rationale is termed constructive if there are reformulations of any of the three subspaces of the rationale state space. The reformulations affect a state space in terms of its variables or their ranges of values. According to this definition, changes of values within the boundaries of an original state space are not considered constructive.

State spaces are constructed for a current problem in a particular situation. As a result, reformulating a state space can be as simple as changing expectations about the current problem, by taking into account existing knowledge about potential issues and potential solutions. This is akin to a recombination of known concepts. In other instances, reformulating the rationale state space may involve new knowledge that has not existed before, leading to what may be called innovative or creative design rationale. We can use notions from research in creativity to categorise these differences in meaning of the word "constructive". Boden [17] draws a distinction between "historical" (or h-) creativity and "psychological" (or p-) creativity. H-creativity is the strongest form of creativity, where novelty is assessed in relation to the history of humankind. For example, the first steam engine was an h-creative design. P-creativity implies novelty with respect to the history of an individual. An architect designing a high-rise building using, for his or her first time, reflecting glass can be viewed as producing a p-creative design. H-creative designs must also involve p-creativity. This classification has been extended to include the notion of "situated" (or s-) creativity [18]. S-creativity is defined relative to the situation that pertains during the process of designing. A design or design feature is s-creative if it is the result of a change of the world within which designing operates. P-creativity must involve s-creativity.

The notion of constructive design rationale developed in this paper corresponds to the concept of s-creativity. By analogy, we may refer to it as "s-constructive" design rationale, and distinguish it from "p-constructive" and "h-constructive" design rationale. However, for reasons of simplicity in this paper, we will just use the term "constructive" and define it in the sense of "s-constructive".

Constructive design rationale allows producing rationale instances that have at least one element that is constructed: the starting point, the end point, or the path.

- *Constructed starting points* are based on novel issues or novel solution alternatives of antecedent decisions. They require reformulating the function state space in terms of its variables or ranges of values.
- *Constructed end points* are based on novel issues or novel solution alternatives of consequent decisions. They require reformulating the structure state space in terms of its variables or ranges of values.
- *Constructed paths* are based on novel issues or novel solution alternatives of intermediate decisions, and novel connections between intermediate decisions. For elementary paths, this requires reformulating the behaviour state space in terms of its variables or ranges of values. For complex paths, this may also require reformulating the individual rationale state spaces associated with intermediate decisions. In particular, reformulating the structure state space of an intermediate decision produces new intermediate decision variables or ranges of values. Reformulating the function state space of an intermediate decision produces new connections between intermediate decisions.

There are only seven combinations of constructed and non-constructed elements of constructive rationale instances, as shown in Table 2. Figure 2 shows graphically each of these combinations, Figures 2(b)-(h), contrasted with an instance of traditional, non-constructive rationale, Figure 2(a). Each of the seven combinations represents a type of constructive rationale. These types can be further elaborated, as non-constructed elements may be either fixed (i.e., their values remain unchanged) or variant (i.e., their values vary within the pre-defined ranges of the state space). Table 3 gives an overview of the 19 possible sub-types based on combinations of constructed, variant and fixed elements of constructive rationale instances.

Table 2 The seven possible types of constructive design rationale, based on different combinations of constructed and non-constructed elements

Type	End point	Starting point	Path
1	Non-constructed	Non-constructed	Constructed
2	Non-constructed	Constructed	Non-constructed
3	Non-constructed	Constructed	Constructed
4	Constructed	Non-constructed	Non-constructed
5	Constructed	Non-constructed	Constructed
6	Constructed	Constructed	Non-constructed
7	Constructed	Constructed	Constructed

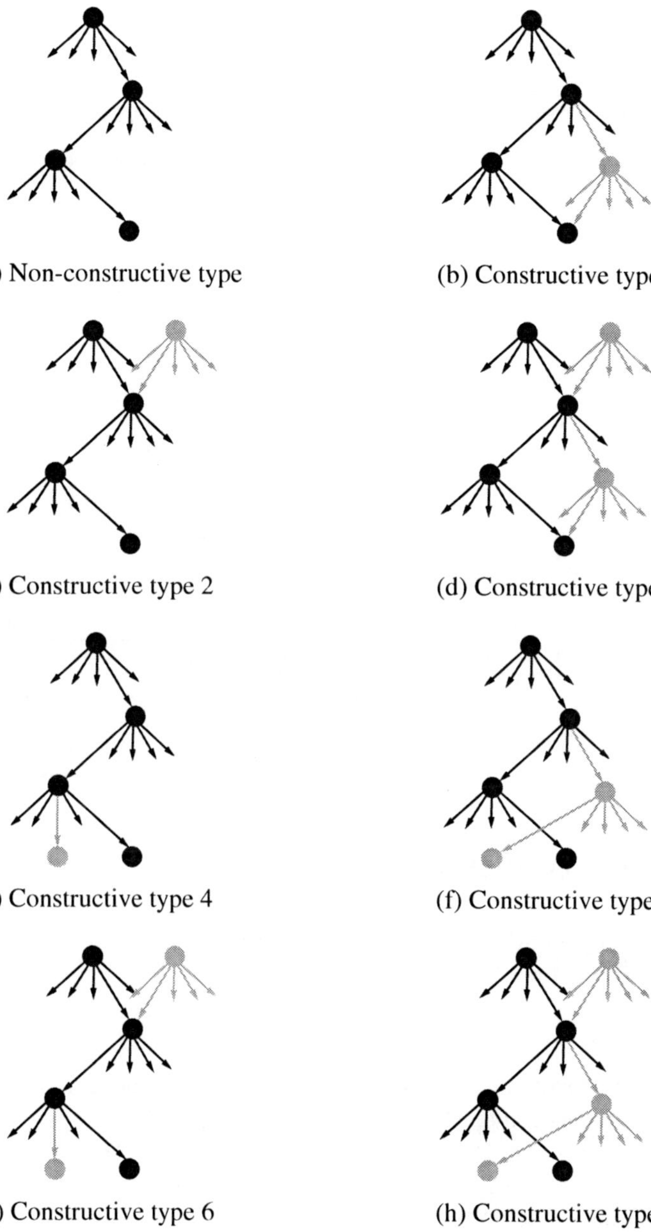

(a) Non-constructive type

(b) Constructive type 1

(c) Constructive type 2

(d) Constructive type 3

(e) Constructive type 4

(f) Constructive type 5

(g) Constructive type 6

(h) Constructive type 7

Fig. 2. Graph-based representations of rationale instances, including the non-constructive type (a) and the seven constructive types listed in Table 2 (b to h). Constructed elements are in grey; non-constructed elements are in black

Table 3 The nineteen possible sub-types of constructive design rationale, based on different combinations of constructed, variant and fixed elements

Type	Sub-Type	End point	Starting point	Path
1	1.1	Fixed	Fixed	Constructed
	1.2	Fixed	Variant	Constructed
	1.3	Variant	Fixed	Constructed
	1.4	Variant	Variant	Constructed
2	2.1	Fixed	Constructed	Fixed
	2.2	Fixed	Constructed	Variant
	2.3	Variant	Constructed	Fixed
	2.4	Variant	Constructed	Variant
3	3.1	Fixed	Constructed	Constructed
	3.2	Variant	Constructed	Constructed
4	4.1	Constructed	Fixed	Fixed
	4.2	Constructed	Fixed	Variant
	4.3	Constructed	Variant	Fixed
	4.4	Constructed	Variant	Variant
5	5.1	Constructed	Fixed	Constructed
	5.2	Constructed	Variant	Constructed
6	6.1	Constructed	Constructed	Fixed
	6.2	Constructed	Constructed	Variant
7	7.1	Constructed	Constructed	Constructed

Drivers of Constructive Design Rationale

This Section presents the drivers of constructive design rationale using the situated FBS framework [8].

The Situated FBS Framework of Designing

This Section provides a brief description of the situated FBS framework; for more information see Gero and Kannengiesser [8].

The basis for the situated FBS framework is a three-world model of designing interactions, Figure 3(a). The *external world* is composed of representations outside the designer or design agent. The *interpreted world* is built up inside the design agent in terms of sensory experiences, percepts and concepts. It is the internal representation of that part of the external world that the design agent interacts with. The *expected world* is the world imagined actions of the design agent will produce. It is the environment in which the effects of actions are predicted according to current goals and interpretations of the current state of the world.

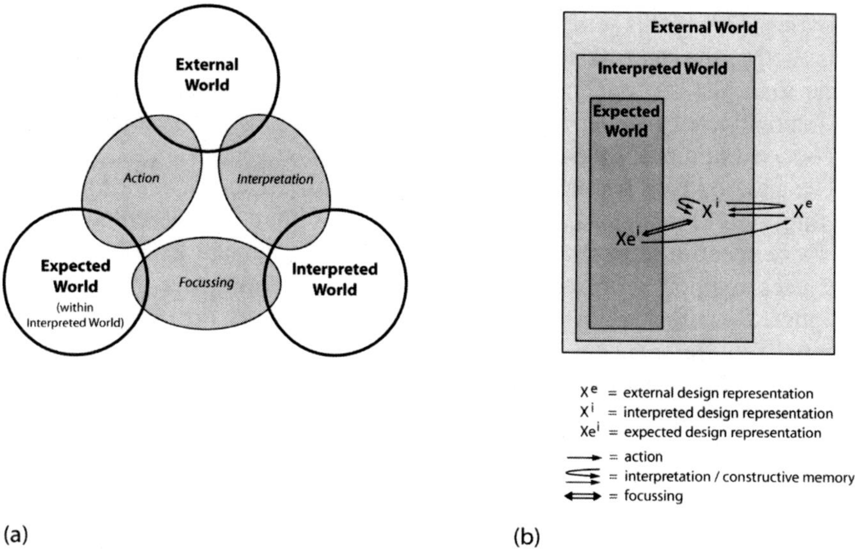

Fig. 3. Three interacting worlds: (a) general model, (b) specialised model for design representations

These three worlds are linked together by three classes of connections. *Interpretation* transforms variables which are sensed in the external world into the interpretations of sensory experiences, percepts and concepts that compose the interpreted world. *Focussing* takes some aspects of the interpreted world, and uses them as goals for the expected world that then become the basis for the suggestion of actions. *Action* is an effect which brings about a change in the external world according to the goals in the expected world.

Figure 3(b) specialises this model by nesting the three worlds and articulating general classes of design representations as well as the activity of reflection [19]. The set of expected design representations (Xe^i) corresponds to the notion of a design state space, i.e. the state space of all possible designs that satisfy the set of requirements. This state space can be modified during the process of designing by transferring new interpreted design representations (X^i) into the expected world and/or transferring some of the expected design representations (Xe^i) out of the expected world. This leads to changes in external design representations (X^e), which may then be used as a basis for re-interpretation changing the interpreted world. Novel interpreted design representations (X^i) may also be the result of *constructive memory*, which can be viewed as a process of interaction among design representations within the interpreted world rather than across the interpreted and the external world. Both interpretation and constructive memory are represented as "push-pull" activities [20]. This emphasises the role of individual experience in constructing the interpreted world, by "pulling" interpreted representations rather than just by "pushing" what is presented in the external world. It is the interaction of push and pull that may produce new representations that can be used to modify the design state space.

The situated FBS framework, Figure 4, combines the FBS ontology with the three-world model. Here, the variable X in Figure 3(b) is replaced with the more specific representations F, B and S. The situated FBS framework also uses explicit representations of external requirements given to the designer. Specifically, there are external requirements on function (FR^e), external requirements on behaviour (BR^e), and external requirements on structure (SR^e). However, we assume that there are no external requirements when applying the FBS ontology to modelling design decisions.

Drivers for Constructing Rationale Structure

Reformulation of structure (process 9 in Figure 4) covers constructing new end points of constructive design rationale. Two processes are potential drivers of this type of reformulation: the interpretation of an external structure (process 13), and the internal construction of an interpreted structure (process 6).

Interpretation of an end point decision (or external structure; process 13) is very common, as most rationale instances are generated based on a given decision that is available in the external world. Two of the most frequent scenarios that trigger the construction of a rationale instance based on a given end point include:

- Design justification: The design agent is given a particular decision (end point) and asked to communicate the reasoning that has led to this decision (starting point and/or path).
- Designing: The design agent interprets a potential decision on some aspect of a current design (end point) and reflects upon the reasoning that could lead to this or another decision (starting point and/or path).

In both scenarios, there is the potential for producing a different interpreted structure than originally intended by the design agent. This potential is enhanced by re-representations of the same external structure that may stimulate the emergence of new issues (or decision variables). Emergence is a process that makes implicit or unintentional design decisions explicit. Emergent design decisions often include visual forms and their potential consequences, although they are not limited to visual forms. They are based on the fact that producing designs, by means of sketching or modelling, necessarily imposes decisions on the organization and details of the design, not all of which are specifically intended by the designer. For example, sketching components of a design on a piece of paper produces a set of lines that compose shapes with intended spatial relations. Other spatial relations emerge when the designer inspects the sketch at a later point in time.

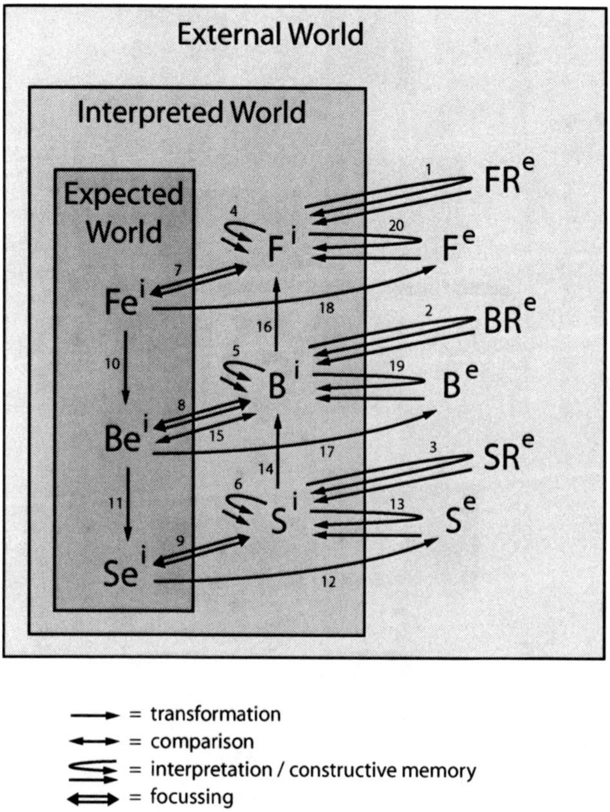

Fig. 4. The situated FBS framework

Take the layout of a set of buildings produced by an urban designer, shown in Figure 5(a). At the initial time of drawing the layout, the designer attends to the four buildings individually, leading to a set of independent decisions for each of them. Upon inspection of the layout, the designer becomes aware of a horizontal axis and an urban space between two buildings, as shown in Figure 5(b). These features are decisions on spatial relations that were implicit in the initial set of design decisions but are now made explicit. A rationale instance that comprises this constructed decision as an end point is of type 4 in Table 2.

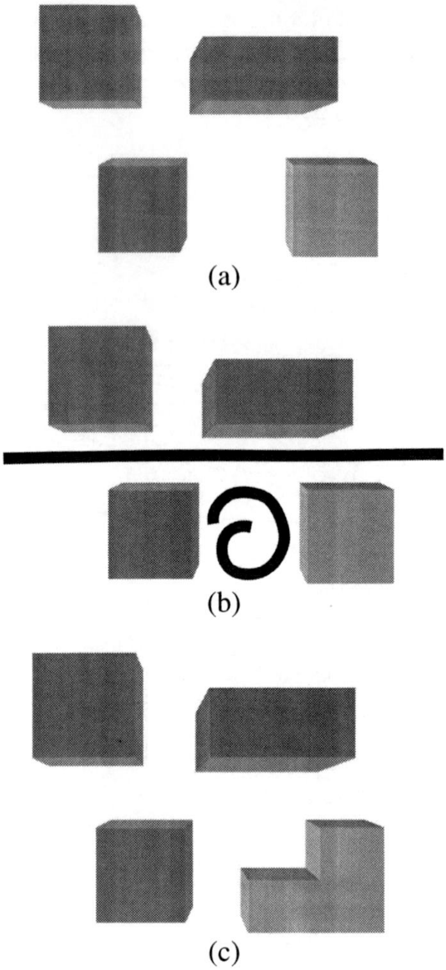

Fig. 5. A sequence of sketches of a town layout: (a) the initial layout; (b) the same layout highlighting an emergent urban space and horizontal axis; (c) subsequent change of the design as a consequence of the emergent urban space in (b)

Internal construction of new end point decisions (process 6) often occurs in the form of new solution alternatives that were not explicitly considered in the original decision-making process. This expands the ranges of values for decision variables. One benefit of this is that a design decision can be shown to remain valid or appropriate even when new decision alternatives, such as new technologies and competitors, come up later. And if one of these alternatives proves to be a better candidate decision, the design may be modified to incorporate it and to provide a closer fit with the design requirements.

Drivers for Constructing Rationale Behaviour

Reformulation of behaviour (process 8) covers constructing new elementary paths of constructive design rationale. New complex paths are covered by reformulation of the function and structure of intermediate decisions. Three processes are potential drivers of this type of reformulation: the interpretation of an external behaviour (process 19), the internal construction of an interpreted behaviour (process 5), and the derivation of interpreted behaviour from interpreted structure (process 14).

Interpretation can produce new interpreted behaviour (process 19) most commonly when a new external behaviour is also provided. The new behaviour may be in addition to or in conflict with previous behaviour. For example, the question "Does your decision on construction materials also consider recyclability besides strength?" represents a new, additional behaviour (i.e., recyclability). The question "Given we had to reduce our limit for material cost from $20 to $15 per unit, is our decision on suppliers still valid?" represents a new range of behaviour that is partially in conflict with the previous range of behaviour. The rationale instances resulting from additive and substitutive changes of behaviour are of type 1 in Table 2.

Internal construction of new behaviour (process 5) is often the consequence of inferring new interactions of end point decisions with exogenous effects. For example, let us assume an end point decision on using a particular part supplier. The path previously included cost per unit as a decision criterion. However, new government regulations (an exogenous effect) may require a minimum percentage of parts to be manufactured in a specific country, so the criterion of geographical location of the supplier must be constructed. This may add to or replace the previous cost criterion. The resulting rationale instance is of type 1.

Deriving new interpreted behaviour from interpreted structure (process 14) is usually the consequence of a reformulated structure. Returning to the example in Figure 5(b), the emerging decision on creating an urban space provides the basis for deriving decision criteria such as "support social interaction" and "provide a space for public events". As this corresponds to constructing a new path in addition to the new end point, this rationale instance is of type 5.

Reformulated behaviours may lead to subsequent refinements of end point decisions. For example, the urban designer may use the new criteria of "support social interaction" and "provide a space for public events" to more directly produce an urban space. Figure 5(c) shows how the emerged urban space is modified by changing the design of an individual building. This refined end point decision can be modelled in the FBS framework as the result of synthesis (processes 11 and 12), analysis (processes 13 and 14) and evaluation (process 15).

Drivers for Constructing Rationale Function

Reformulation of function (process 7) covers constructing new starting points of constructive design rationale. Three processes are potential drivers of this type of reformulation: the interpretation of an external function (process 20), the internal construction of an interpreted function (process 4), and the ascription of interpreted function to interpreted behaviour (process 16).

Interpretation can produce new interpreted function (process 20) most commonly when also a new external function is provided. This can occur when the robustness of an existing end point decision is assessed by relating that decision to a hypothetical starting point (e.g., during design review meetings): "What if we decide on using a different operating system; would we still use the same user commands?" In most cases, however, previous starting points become invalid because of changes in the external requirements on the product, the design process and the project. In non-constructive design rationale, this new starting point would invalidate all consequent decisions, including the original path and the original end point. In constructive design rationale, this need not be the case. An example is the decision to no longer outsource the manufacture of a physical part but to produce it in-house. The previous path included the decision on using a specific OEM parts catalogue; and the previous end point was a decision on a specific geometry of the part that is consistent with the catalogue. Based on considerations of maintainability and its associated principles of standardisation, the same path can be used leading to the same end point. This rationale instance is of type 2.

Internal construction of new function (process 4) can similarly produce new starting points, based on the designer's changed understanding of the design problem. Take the example from designing a distributed software system; an original starting point here is the decision to allow for extensibility of the system. The original path from this starting point includes a decision to use loose coupling, leading to an end point decision to use a Publish/Subscribe messaging model. Based on the designer developing a better understanding of the domain, the original starting point is modified to include a decision to allow for a high degree of system security. This leads to a modified path that includes a decision to use a mechanism that easily filters messages and then routes them according to their content. The Publish/Subscribe model still performs well under this additional criterion, and remains the chosen end point decision. This rationale instance is of type 3.

Ascribing new interpreted function to interpreted behaviour (process 16) is often triggered by reformulated behaviour. For example, the new behaviours associated with the decision on the urban space in Figures 3(b) and 3(c) may establish a basis for ascribing the new functions "to refine the decision to design for increased quality of urban life" (a new starting point) and "to guide the decision on what particular social activities should be supported in the urban space" (a new consequent set of decisions). The resulting rationale instance is of type 7, as all three elements of this instance are constructed.

Reformulated functions may lead to the formulation of new behaviours (via process 10), corresponding to constructing new elementary paths. The new behaviours can then be used to synthesise, analyse and evaluate refined end point decisions. New complex paths may need to be constructed by (re-) formulating the structure and/or function of intermediate decisions.

A new function and a new structure but with the same behaviour can occur, type 6 constructive rationale, when the designer's starting point has changed but leads to the same intermediate decision as previously existed and there is a change in the final decision of the rationale instance leading to a different structure than before. A new function, either as a result of an exogenous activity or as a result of emergence, changes the starting point of the rationale instance. The additional requirement that the artefact be collapsible may produce no change as that behaviour may already be embedded in its design. However, it will produce some different values for the variables that are propagated down the decision path. As a consequence it is possible that the final decision will therefore be different even though the same path as previously has been followed.

Conclusion

Design rationale can be understood either as a passive and fixed description of the history of designing, or as a dynamic act that constructs the assumptions underpinning the design decisions as they are needed in a current situation.

The first way of understanding rationale has benefits in supporting routine designing and activities such as auditing, learning and design maintenance. However, the remaining problems of capture and reuse of rationale in situations that are novel and dynamic, require a more subtle view of design rationale as a dynamic act that allows instances of design reasoning to be constructed on the fly. In the light of a new situation, a new line of reasoning can thus be generated that can provide new explanations for existing design decisions that may or may not lead to modifications of the design. When a set of antecedent decisions is invalidated or no longer available, a new set of decisions can be created without necessarily invalidating consequent decisions. In turn, new consequent decisions can be created without necessarily being in conflict with antecedent decisions. On the other hand, every new decision has the potential to affect other decisions, allowing for changes of both the design process and its outcomes to better adapt to different situations.

Our ontological framework can represent constructive design rationale as well as the drivers for constructing new decisions that form the starting points, the paths and the end points of rationale instances. The framework can be used for developing agent-based design rationale systems that not only capture and document rationale instances but also interpret them based on the agent's situation. This can produce different interpretations and thus different design rationale in different situations. Recent work on the relationship between design rationale and design creativity [21] can be supported by using constructive design rationale systems as testbeds for research hypotheses.

Future work includes validating our framework empirically. Studies need to capture initial rationale instances and their transformation as they are reconstructed by different designers or by the same designer at a later point in time. It would be interesting to establish which of the seven types and nineteen subtypes of constructive design rationale occur most frequently. The activities that drive the modification of rationale instances may be represented using an FBS-based coding scheme and then mapped onto the situated FBS framework.

Acknowledgments

This work is partly funded by the Australian Research Council Grant No: DP0559885 and by the US National Science Foundation Grant No. CNS-0745390. NICTA is funded by the Australian Government as represented by the Department of Broadband, Communications and the Digital Economy and the Australian Research Council through the ICT Centre of Excellence program.

References

1. Moran, T., Carroll, J. (eds.): Design Rationale: Concepts, Techniques, and Use. Lawrence Erlbaum, Mahwah (1996)
2. Dutoit, A.H., McCall, R., Mistrík, I., Paech, B. (eds.): Rationale Management in Software Engineering. Springer, Heidelberg (2006)
3. Gruber, T.R., Russell, D.M.: Generative design rationale: Beyond the record and replay paradigm. In: Moran, T., Carroll, J. (eds.) Design Rationale: Concepts, Techniques, and Use, pp. 323–349. Lawrence Erlbaum, Mahwah (1996)

4. Tang, A., Babar, M.A., Gorton, I., Han, J.: A survey of architecture design rationale. The Journal of Systems and Software 79, 1792–1804 (2008)
5. Burge, J.E., Brown, D.C.: Software engineering using RATionale. The Journal of Systems and Software 81, 395–413 (2008)
6. Brown, D.C.: Assumptions in design and design rationale. In: Burge, J.E., Bracewell, R. (eds.) Workshop on Design Rationale: Problems and Progress. Design Computing and Cognition 2006, The Netherlands, Eindhoven (2006)
7. Gero, J.S.: Design prototypes: A knowledge representation schema for design. AI Magazine 11, 26–36 (1990)
8. Gero, J.S., Kannengiesser, U.: The situated function-behaviour-structure framework. Design Studies 25, 373–391 (2004)
9. Lee, J., Lai, K.-Y.: What's in design rationale? Human-Computer Interaction 6, 251–280 (1991)
10. Kruchten, P.: An ontology of architectural design decisions. In: 2nd Groningen Workshop on Software Variability Management. Rijksuniversiteit Groningen, The Netherlands (2004)
11. MacLean, A., Young, R.M., Bellotti, V.M.E., Moran, T.P.: Questions, options, and criteria: Elements of design space analysis. Human-Computer Interaction 6, 201–250 (1991)
12. Kunz, W., Rittel, H.: Issues as Elements of Information Systems. Working Paper 131. Institute of Urban and Regional Development. University of California, Berkeley (1970)
13. Jansen, A., Bosch, J.: Software architecture as a set of architectural design decisions. In: 5th Working IEEE/IFIP Conference on Software Architecture, Pittsburgh, PA, pp. 109–120 (2005)
14. Lee, J.: Design rationale systems: Understanding the issues. IEEE Expert 12, 78–85 (1997)
15. Sim, S.K., Duffy, A.H.B.: Towards an ontology of generic engineering design activities. Research in Engineering Design 14, 200–223 (2003)
16. Qian, L., Gero, J.S.: Function-behaviour-structure paths and their role in analogy-based design. Artificial Intelligence for Engineering Design, Analysis and Manufacturing 10, 289–312 (1996)
17. Boden, M.A.: The Creative Mind: Myths and Mechanisms. Basic Books, New York (1991)
18. Suwa, M., Gero, J.S., Purcell, T.: Unexpected discoveries and s-inventions of design requirements: A key to creative designs. In: Gero, J.S., Maher, M.L. (eds.) Computational Models of Creative Design IV, pp. 297–320. University of Sydney, Australia (1999)
19. Schön, D.A.: The Reflective Practitioner: How Professionals Think in Action. Harper Collins, New York (1983)
20. Gero, J.S., Fujii, H.: A computational framework for concept formation for a situated design agent. Knowledge-Based Systems 13, 361–368 (2000)
21. Daughtry, J., Burge, J., Carroll, J.M., Potts, C.: Creativity and rationale in software design. ACM SIGSOFT Software Engineering Notes 34, 27–29 (2009)

DESIGN CREATIVITY

The curse of creativity
David Brown

Enabling creativity through innovation challenges: The case of interactive lighting
Stefania Bandini, Andrea Bonomi, Giuseppe Vizzari and Vito Acconci

Facetwise study of modeling activities in the algorithm for inventive problem solving ARIZ and evolutionary algorithms
Céline Conrardy, Roland De Guio and Bruno Zuber

Exploring multiple solutions and multiple analogies to support innovative design
Apeksha Gadwal and Julie Linsey

Creative and inventive design support system: Systematic approach and evaluation using quality engineering
Hiroshi Hasegawa, Yuki Sonoda, Mika Tsukamoto and Yusuke Sato

The Curse of Creativity

David C. Brown
Worcester Polytechnic Institute, USA

Computational design creativity is hard to study, and until fairly recently it has received very little attention. Mostly the focus has been on extreme non-routine cases. But there are hard sub-problems and others ways of moving towards creative systems that are worth considering. This paper presents three of the alternatives, discussing one in more depth: i.e., to look at what changes can be made to routine design systems in order to produce more creative outputs. This focuses on working "upwards" towards creativity, examining smaller, ingredient decisions that make a difference to the result. As the amount of creativity displayed by a design is a judgment made by some person or group, it should be possible to investigate the degree of impact of changes to routine design mechanisms. This will contribute to our understanding of less "extreme" reasoning that leads to judgments of increased creativity: i.e., the foundation on which other methods rest.

Introduction

It is common wisdom that people should be given tasks that computers can't do well, and computers should be given tasks that people can't do well. So in design computing why are we attempting to study computational design creativity?

The main answer is that the field (like many others) progresses by tackling simpler problems first and moving towards harder ones. Routine parametric design and design checking were starting points, moving gradually to Configuration and most recently to harder problems such as distributed/collaborative design and to creative design: from routine to non-routine [1]. One goal has always been to build working systems, while

J.S. Gero (ed.): *Design Computing and Cognition'10*, pp. 157–170.

another is to learn more about the knowledge and reasoning used for each type of design activity studied.

Computational design creativity is hard to study, and until fairly recently it has received very little attention, even though it is widely held to be very important both from intellectual and economic points of view. It has mostly been studied by looking at analogical reasoning and genetic algorithms: almost to the point of fixation. That is, the focus has been on extreme non-routine cases. There are hard sub-problems and other ways of moving towards creative systems that are worth considering. This paper presents three of the alternative approaches to computational design creativity research, discussing one in more detail.

Theoretical and Perceived Creativity

In Boden's theory of creativity [2], creative ideas must be new and valuable. In addition, the theory must be able to "distinguish first-time novelty from radical novelties". The former can be generated by a system, perhaps using rule-like knowledge that "underlies the domain and defines a certain range of possibilities". This resulting "conceptual space" defines what could be produced by a system, resulting in newness that is in some sense expected: i.e., each "new" design is just mapping out the possibilities defined by the system. However, the conceptual space needs to be changed by transformations in order to allow "radical originality": producing transformational creativity.

However, there is a difference between a formal theory of creativity, which attempts to define what might be called theoretical creativity, and how people detect and evaluate creativity: i.e., a performance-based view of computational design creativity, that we might call perceived creativity.

First, as creativity is judged, different individuals or groups may vary in their assessment of the product or concept. The scope of that judgment varies at least in the following (not independent) ways: how many people are judging (e.g., one person versus a group)); the depth of knowledge that this represents (e.g., professors or children); and the historical range represented (e.g., designs from this year only or since the beginning of time).

Second, people can judge degrees of creativity [3]. What's not clear is whether everyone judges in the same way. Boden [2] warns that "In general, one cannot assess creative ideas by a scalar metric". However, Ward et al. [4] hint at some idealized scale by noting "the possibility that the mundane and the exotic ... represent endpoints on a continuum of human creativity".

Besemer [5] has developed scales based on how people judge the creativity of products. The Creative Product Analysis Model (CPAM) [6] is the basis for a well-validated, practical, product creativity assessment instrument called CPSS [7], [8]. The model has three main dimensions (also known as factors): Novelty, Resolution and Style. Each of these factors has between 2-4 characteristics that further refine them: nine in total.

Rather than a simple scale, the scores for the nine characteristics represent a "fingerprint" of the product being evaluated, including, for example, an individual or group's judgment of the degrees of "surprise" or "elegance" that the product stimulates or displays.

However, these judgments are dominated by the Novelty dimension. This suggests that people would have little trouble viewing a product that was very novel as creative. In fact, the correlation between novelty and creativity is so widely recognized and strong that some writers actually confuse newness with creativity.

Besemer's statistical analysis of her data has led her to isolate Surprising and Original as characteristics of Novelty. An interesting research question is how much those characteristics correlate with transformational creativity. It is clear that this will vary greatly with the difference between the designer and the group judging, with regard to group size, depth of knowledge and historical range. However, by taking the mostly assumed default group as 'the world's professional designers of that type of product' and the range as 'from the beginning of civilization' the research question becomes more refined, and the standards for high creativity much tougher.

Current Approaches

It is not the goal of this paper to review current research into computational design creativity. However, this author believes that current creativity research tends to be based on the goal of transformational creativity. A lot of it appears to be based on or influenced by larger scale, general reasoning, such as Analogy [9], Genetic/Evolutionary Algorithms [10] and Conceptual Blending [11], [12].

The consequences of this goal are that:

1. researchers tackle a very hard problem "head on", making slow progress;
2. these powerful methods don't give a clear idea of what their limits are- knowing what a method can't do is as important as knowing what it can do;

3. the computational methods used don't always match what people can do, and therefore don't provide very good hypotheses about human creativity; and
4. detailed psychologically based hypotheses about what people might be doing tend to be ignored.

Some Research Alternatives

This section elaborates on three of the alternative approaches to computational design research. The first will be discussed in more detail in future sections.

New Wine in Old Bottles

The first alternative methodology is to take a well understood but not intentionally creative approach and see how it might be modified in order to produce results that people would be willing to say are creative, due to their novelty and other characteristics [5]. A secondary goal would be to determine whether the post-modification mechanisms could meet the criteria for transformational creativity.

This alternative addresses the four "consequences" given above. Although this too may produce slow progress, #1 is addressed by working on several smaller problems to examine the impact of their solutions. By using a routine design (RD) problem-solving method (PSM), #2 is addressed as we know the limits quite well. By picking a RD PSM which is already based on expert behavior we stand a better chance of addressing #3, and #4 can be addressed by focusing on using modifications based on hypotheses about the ingredients of creative reasoning that can be found in the psychological literature [13].

Of course, from this author's point of view, an obvious candidate for this first alternative is to look at what changes can be made to Design Specialists and Plans Language (DSPL) based routine design systems [14] [15] in order to produce more creative outputs. However, this isn't the only candidate.

This approach focuses on working "upwards" towards creativity, by examining smaller, ingredient decisions that make a difference to the result. It should be possible to investigate the degree of impact produced by changing the internal reasoning mechanisms in a DSPL system. This will contribute to our understanding of which less extreme reasoning mechanisms impact judgments of increased creativity. This is the foundation on which more extreme methods rest, as many authors agree that creativity is "...an outcome of subsets of ... processes acting in concert..." and not just a single reasoning mechanism [16].

Using Cognitive Science and Psychology

The second alternative is to look more carefully at what cognitive science and psychology tells us about creativity. Everyone agrees that "novelty" is a key ingredient of the production and evaluation of creativity in a designed product, while some others add "surprise". Novelty appears to be the principle component of all models of creativity, and all creativity metrics. Judging both originality and surprise appears to be quite difficult, and needs much more attention. Srinivasan & Chakrabarti [17], as well as others, have already made useful contributions to this problem.

Suggestions about the many ingredients of creative reasoning and its evaluation from the literature include:

A. Novelty: surprising and original; recognizing, evaluating and seeking it.
B. Domain Knowledge: having lots of it; being able to search it; finding relevant knowledge; rich interconnections; different representations; knowledge of its potential; similarities and differences; not just hierarchical representations.
C. Heuristic knowledge: having lots of it; for selecting ways to think (such as planning, simplification, analogy, etc.)
D. Constraints: being able to drop, weaken or invert them; having meta-knowledge about them to enable their modification.
E. Combinations: novel combinations of old ideas; combination of apparently unrelated ideas.
F. Associative reasoning: a quality of over inclusiveness; ability to associate the apparently unrelated.
G. Suppressing inhibitions: allows less relevant ideas/methods to "intrude" into the problem solving process.
H. Abstract and imprecise descriptions: such as for intermediate solutions and goals.
I. Alternative methods: for making decisions; for making goals more concrete.
J. Critical assessment: as an antidote to inclusiveness; identify misfits; heuristically eliminating very weak ideas and potential mistakes; resist pruning too strongly to just the routine ideas; resist too much novelty.
K. Problem recognition: error detection; recognition of product inadequacies; recognition leads to formulation of new goals.
L. Concept expansion: constructing, stretching, extending, modifying and refining concepts.

M. Analogical reasoning: far (cross domain) and near (same domain); depends on intentions and goals.
N. Visualization: mental simulation to examine existing things in new situations.
O. Meta-reasoning: breaking away from functional fixedness; abandoning old, unsuccessful problem-solving strategies; using meta-knowledge.
P. Least commitment: keeping options open as long as possible; suspending judgment; producing multiple partial solutions.
Q. Forgetting: productive forgetting; good mental management.

Products as Art

The third alternative is to focus on the role of artistic creativity evaluation [18], [19] in assessing the creativity of a product. Besemer [5] has detected "style" as one of the dimensions by which products are judged to be creative. The ingredient characteristics are "organic", "well-crafted" and "elegant". It is clear that many products with distinct style are close to works of art, and share many characteristics, such as attempting to manipulate the emotions of the viewer/user, for example. In addition, products that are highly related to established crafts (e.g., pottery) tend to be decorative, and some have "applied decorative design" [20] which move them closer to art.

As we have previously discussed, this is a very challenging area, as it isn't clear whether every ingredient of the evaluation of an artistic artifact for creativity can even be done reliably by a human. For example, product evaluation would include evaluating its intended function, and one would expect to be told it. From an artistic point of view, there might be a contribution to function, but more likely to the style dimension: there might also be intended (but undeclared) contributions to such purposes as "creating beauty", "entertainment", "healing", etc.

While studying this type of evaluation does avoid addressing the goal of transformational creativity, and does avoid tackling that very hard problem "head on", it may be substituting one very hard problem for another. However, considering product creativity with emphasis on the Style dimension is research that still needs to be done.

Ingredients of Routine Design Reasoning

From this point on we will concentrate on the first alternative: taking a well understood but not intentionally creative approach to see how it might be modified in order to produce results that people would be willing to say are creative.

Routine design means that everything about the design process is known in advance, including the knowledge needed. However, neither the resulting design nor the trace of use of the knowledge is known in advance. Typically, routine design knowledge is highly compiled: in the "knowledge compilation" sense of the term [21]. The DSPL language allows such routine design knowledge to be written down.

As previously presented [22], the ingredient types of reasoning supported by DSPL are:

1. Basic Synthesis
2. Criticism
3. Decomposition
4. Evaluation
5. Execution
6. Ordering
7. Patching
8. Planning
9. Recomposition
10. Retraction
11. Selection
12. Situation Recognition
13. Suggestion Making

Note that they are not independent, as some of these items involve other items, and are therefore at a different level of abstraction. The connection between this list and the mechanisms of DSPL are summarized in Table 1. We will use the terms presented in 1-13 above, but acknowledge that some have meanings that vary in the literature: e.g., "Synthesis" can also mean combining or generating, instead of calculating or selecting, which is why we use the modifier "Basic".

In DSPL, each Specialist contains Plans and plan selection knowledge. They each represent a subproblem, solving it by plan selection and execution. Plans are precompiled, ordered sequences of actions intended to provide the design for a subproblem. Each Plan provides a decomposition as well as sub-solution recomposition. Sponsors evaluate the suitability of a Specialist's plans for use in a particular situation, while a Selector picks the most suitable Plan.

Table 1 The Ingredients of Routine Design Reasoning

Type	*DSPL*	*Action*
Basic Synthesis	Step	Calculate, or select.
Criticism	Constraint	Values are tested/compared.
Decomposition	Plan Task Step	All three have sequences of actions.
Evaluation	Sponsor	Determine the quality of a plan.
Execution	Plan execution	Carry out the actions in a plan.
Ordering	Plan Task Step	All three have ordered actions.
Patching	Redesigner	Can change an attribute's existing value.
Planning	Plan Plan Sponsor Plan Selector	Hierarchically arranged collections of plans with plan selection produce a dynamically constructed design plan.
Recomposition	Plan	Each plan action adds its subproblem's solution to the overall design.
Retraction	Backtracking	One or more recent design decisions can be retracted and a re-design phase entered.
Selection	Plan Selector Step	The selector selects from amongst suitable plans, while a step selects from amongst suitable values for an attribute.
Situation Recognition	Plan Sponsor Step FHs	All three can make context sensitive decisions, based on recognizing patterns of previous actions or design decisions.
Suggestion Making	Suggestion	If any "agent" (e.g., a Constraint, or a Step) used by another fails, it passes suggestions (about how the failure might be fixed) back to the agent that called it from 'above'.

Steps are the building blocks of the design process, providing a value for an attribute of the design by calculation, or by selection using pattern matching. Tasks group Steps, and therefore define additional problem decomposition. Constraints test values and, on failure, make suggestions about patches. Redesigners attempt to patch the design, guided by suggestions, in order to correct a constraint failure. Failure Handlers (FHs) recognize failing situations that might be patchable, or can trigger suggestion-guided backtracking.

Modifications to Routine Design Reasoning

In this section we will examine some possibilities for modifying the ingredients of routine design systems in order to produce designs that are more likely to be judged to be creative.

Assumptions and Restrictions

We restrict possibilities by assuming that modifications are made without creating new agents (i.e., no additional reasoners are added), but that new mechanisms are allowed to be 'called' or added for exploiting meta-knowledge or meta-reasoning.

We assume that modifications are based on a RD knowledge-base (KB) constructed from DSPL, or something similar. We assume that the base system is doing configuration by selection between alternative pre-determined configurations.

As such an RD system is probably highly compiled, values will be constrained early to avoid failure later in the design process. The RD system could be considered to be very tight or loose, depending on how much earlier constraints restrict later decisions. One would expect tighter systems to be harder to modify to produce more creative results. This will require further study.

Note that the 13 ingredients of RD reasoning listed above allow the construction of not only an RD system, but also systems than can handle other types of tasks. For example, routine configuration, such as assignment or restricted layout, should also be easy to do [23]. However, as Situation Recognition plays a role at every level of an RD system, then it is possible to build a system that is dominated by that reasoning, where "design" decisions are actually flags that identify complex situations: thus allowing classification, the basis of much diagnosis. It may be possible to use this potential to enhance creativity.

Matching Creative and Routine Reasoning

In Table 2, the rows show the suggestions (A-Q) about creative reasoning from the literature, while the columns show the ingredients (1-13) of routine design reasoning. The table entries indicate places where relevant modifications might occur: others might be possible.

Table 2 Some possibilities for modifications

	Synth	**Crit**	Decomp	Eval	Exec	Order	Patch	Plan	Recomp	Retr	Sel	Recog	Sugg
Novelty	**y**	**y**		y			y				y	y	y
Domain	**y**	**y**	y	y			y	y			y	y	y
Heuristic		**y**	y	y			y	y			y	y	y
Constr.		**y**		y									
Combin.							y	y	y		y	y	
Assoc.		**y**						y	y		y	y	
Suppress		**y**	y	y				y	y	y	y	y	y
Abstract	**y**	**y**						y			y	y	y
Alt.	**y**		y			y	y	y			y	y	y
Assess		**y**		y				y			y	y	
Recog.		**y**					y				y	y	y
Expand									y				
Analogy	**y**		y			y	y	y					
Visualiz.		**y**											
Meta.	**y**	**y**	y			y	y	y			y	y	y
Least C.	**y**		y		y								
Forget		**y**											

Note that the first two columns (marked in bold) were considered in more detail and will be discussed below. Investigating the other 187 possibilities is more challenging and would require significantly more study. However, it's important to note how many opportunities there are for potentially interesting research into creative design systems given this 'humble' RD basis. The entries made in Table 2 were first done by considering each ingredient of routine reasoning in turn against all 17 of the creative reasoning suggestions, and then by considering it again in the opposite direction (i.e., for each of the suggestions, against all 13 of the ingredients).

Basic Synthesis & Criticism: Possible Modifications

This section will present some possible modifications to Basic Synthesis and to Criticism that should enhance the perceived creativity of a RD system's output.

Basic Synthesis

A basic synthesis step produces a value for an attribute using calculation or selection: for example, (set x to p + q) or (if a > b then set x to 5 else set x to 10).

Novelty might be enhanced by avoiding common values for attributes and also common combinations of values. It would be helpful to have knowledge of probabilities of values in successful designs, and be able determine the amount of deviation from a stereotype or from the mode. Pushing away from typical values towards the extreme values should produce novelty.

This sort of modification might be enhanced by having the system learn which attributes impact novelty the most, based on human feedback. A more detailed view might be obtained by an analysis of how the variation in novelty correlates with variation in attribute values.

Other possibilities include using other ways to calculate values-with less or more precision for example-and considering other ways to provide the values for selection process.

Domain Knowledge could be enhanced by adding models of existing designs, both in general and generated by this RD KB (similar to the rule models in Teiresias [24]). These models might include statistical records of the values of attributes, of configurations, and of complete designs, as well as correlations between each attribute value and others.

Combinations might be produced by selecting similar components to the "normal" one being considered at that point, based on the current partial design. Abstraction could be introduced by using less tight tolerances, or by using intervals or qualitative values. Alternatives methods in basic synthesis could be added by using alternative calculations or selections.

Analogical reasoning can be approximated by using CBR to determine an attribute's value. It might also be used to provide sets of values; i.e., including related attributes. Meta-reasoning can be supported by some of the domain knowledge described above. In addition, "creativity tolerance" might be used by system by keeping track of how many extreme choices have already been made during the design process and limiting the subsequent design actions if it has already gone too far. Lastly, Least Commitment in basic synthesis can be enhanced by producing multiple solutions, not just one.

Criticism

Criticism in an RD system is represented by Constraints, where intermediate values or attribute values are tested or compared. Constraint roles include: to detect design failure, to detect incompatible sub-solutions, to constrain in order to prevent design failure, and to check requirements.

Forms of constraints might include tests such as (x > 5) or comparisons such as (a < b).

Novelty might be increased by allowing values that only just fail (i.e., more extremes). Domain Knowledge to be added for constraints could include models of typical failure differences, value ranges, etc., as part of the test or comparison: i.e., don't make it a fixed test. It might be useful to keep track of how often and how much the constraint fails, and under what circumstances. Many of the constraints could be Heuristic, but it might be useful to know which are heuristic and which not, as this might allow those constraints to be flexed more safely.

Constraints themselves can be manipulated in many ways. It would be interesting to drop a constraint altogether, or move constraint until later in the reasoning (i.e., heuristic "de-compilation"). Constraints could be weakened in a variety of ways: change the test from "<" to "<="; change a constant, allowing (x > 1) to become (x > 0.9); allow tolerances, so as to change 1 to 1 ± 0.05; change the variable to one that is more inclusive; or "invert" the constraint, using failure handling to fix any negative consequences, if needed.

Associative reasoning might be enabled by using constraints from other similar components, while it is possible to be over inclusive by weakening constraints, as described above. Suppressing inhibitions might be equivalent to dropping constraints, or, more radically, by using constraints from other similar components.

Abstraction could be introduced by comparing types of values rather than actual values (e.g., "blue" instead of "navy"), or by converting values to qualitative values (e.g., "medium" instead of 10). It is clear that constraints play a key role in Critical Assessment, as they recognize actual and potential problems. They may also play a role in creativity tolerance.

Visualization could be introduced by replacing a compiled test by a simulation (e.g., object interference). This could be activated by Meta-Reasoning acting on meta-knowledge about the source/role of the constraint: i.e., a history of, and rationale for, its compilation. Other meta-knowledge might include records of the constraints' activities, such as success and failure counts/details, and when success leads to later failure.

Summary & Conclusions

The fine-grained analysis proposed by this paper is almost the antithesis of normal computational creativity research, in which the "blue skies" methodology is adopted: that is, to put it crudely, it isn't any good unless it appears impossible. While that approach has produced some great results in AI-autonomous vehicles for example-it tends to move researchers over and past more "mundane" problems, leaving them to be tackled later, with less prestige, or even left undone.

Creative systems may well be fuelled by important, large scale reasoning methods, such as analogy, that try to address the goal of transformational creativity, but they will be supported and enhanced by smaller scale reasoning such as has been presented here. It's important to note how many opportunities there are for potentially interesting research into creative design systems given this 'humble' Routine Design basis, with a focus on perceived creativity.

References

1. Brown, D.C.: Routineness revisited. In: Waldron, M., Waldron, K. (eds.) Mechanical Design: Theory and Methodology, pp. 195–208. Springer, Heidelberg (1996)
2. Boden, M.A.: What is creativity? In: Boden, M.A. (ed.) Dimensions of Creativity, pp. 75–117. The MIT Press, Cambridge (1994)
3. Amabile, T.M.: The social psychology of creativity. J. of Personality and Social Psychology 43, 997–1013 (1983)
4. Ward, T.B., Smith, S.M., Vaid, J.: Conceptual structures and processes in creative thought. In: Ward, T.B., Smith, S.M., Vaid, J. (eds.) Creative Thought: An Investigation of Conceptual Structures and Processes. American Psychological Association (1997)
5. Besemer, S.P.: Creating Products in the Age of Design. New Forums Press, Inc. (2006)
6. Besemer, S.P., Treffinger, D.J.: Analysis of creative products: Review and synthesis. J. of Creative Behavior 15, 158–178 (1981)
7. Horn, D., Salvendy, G.: Consumer-based assessment of product creativity: A review and reappraisal. Human Factors and Ergonomics in Manufacturing 16(2), 155 (2006)
8. O'Quin, K., Besemer, S.P.: The development, reliability, and validity of the revised creative product semantic scale. Creativity Research J. 2(4), 267–278 (1989)
9. Yaner, P., Goel, A.: From design drawings to structural models by compositional analogy. AI in Engineering Design, Analysis and Manufacturing 22(2), 117–128 (2008)
10. Koza, J.R.: Human-competitive machine invention by means of genetic programming, in special issue on Genetic Programming for Human-Competitive Designs. In: Spector, L. (ed.) AI in Engineering, Design, Analysis and Manufacturing, vol. 22(3), pp. 185–193. Cambridge University Press, Cambridge (2008)
11. Turner, M., Fauconnier, G.: Conceptual integration and formal expression. Metaphor and Symbolic Activity 10(3), 183–204 (1995)

12. Nagai, Y., Taura, T., Mukai, F.: Concept blending and dissimilarity: factors for creative concept generation process. Design Studies 30(6), 675–848 (2009)
13. Brown, D.C.: Guiding computational design creativity research. In: Gero, J.S. (ed.) Studying Design Creativity. Springer, Heidelberg (to appear, 2010), http://web.cs.wpi.edu/~dcb/Papers/sdc08-paper-Brown-25-Feb.pdf
14. Brown, D.C.: DSPL: Design Specialists and Plans Language (1983/1996), http://web.cs.wpi.edu/Research/aidg/DSPL.html
15. Brown, D.C., Chandrasekaran, B.: Design Problem Solving: Knowledge Structures and Control Strategies. Research Notes in Artificial Intelligence Series. Pitman Publishing, Ltd., London (1989)
16. Ward, T.B., Smith, S.M., Finke, R.A.: Creative cognition. In: Sternberg, R.J. (ed.) Handbook of Creativity. Cambridge University Press, Cambridge (1999)
17. Srinivasan, V., Chakrabarti, A.: Investigating novelty-outcome relationship in engineering design, in special issue on Creativity: Simulation, Stimulation, and Studies. In: Maher, M.L., Bonnardel, N., Kim, Y.-S. (eds.) AI in Engineering, Design, Analysis and Manufacturing, vol. 24(2). Cambridge University Press, Cambridge (2010)
18. Boden, M.A., D'Inverno, M., McCormack, J. (eds.): Computational Creativity: An Interdisciplinary Approach. In: Dagstuhl Seminar Proceedings 09291 (2009), http://drops.dagstuhl.de/portals/index.php?semnr=09291
19. Brown, D.C.: Artistic creativity and its evaluation, in computational creativity: An interdisciplinary approach. In: Boden, M.A., et al. (eds.) Dagstuhl Seminar Proceedings 09291 (2009), http://web.cs.wpi.edu/~dcb/Papers/Dagstuhl-paper.pdf
20. Jirousek, C.: Art, Design and Visual Thinking (1995), http://char.txa.cornell.edu
21. Goel, A.K., Bylander, T., Chandrasekaran, B., Dietterich, T.G., Keller, R.M., Tong, C.: Knowledge compilation: A symposium. IEEE Expert 6(2), 71–93 (1991)
22. Brown, D.C.: The reusability of DSPL systems, Workshop on Reusable Design Systems. In: Second International Conference on AI in Design, Carnegie Mellon University, Pittsburgh (1992)
23. Wielinga, B.J., Schreiber, G.: Configuration design problem solving. IEEE Expert, 49–56 (March-April 1997)
24. Davis, R., Lenat, D.B.: Knowledge-Based Systems in Artificial Intelligence. McGraw-Hill, New York (1982)

Enabling Creativity through Innovation Challenges: The Case of Interactive Lightning

Stefania Bandini[1], Andrea Bonomi[1], Giuseppe Vizzari[1], and Vito Acconci[2]
[1]*University of Milano-Bicocca, Italy*
[2]*Acconci Studio, USA*

This paper discusses a case in which an idea and a creative design for a reactive environment characterized by an adaptive lightning expressed by an artist was transformed into a prototype supporting the customization of the lightning effect. A specific configuration interface was realized to support the user in expressing and envisioning his creativity: by altering some specific parameters of the model implemented in the system, he/she can effectively change the behaviour of the adaptive lightning and immediately visualize the implications of his choice on parameters' values. This experience has been further developed towards the realization of a product based on the same model and approach: a configurable modular adaptive lightning system. The paper presents the starting scenario and its main characteristics, then the model supporting adaptive lightning is introduced. The model configuration and envisioning interface is described and the recent developments towards a product based on the model and configuration system conclude the paper.

Introduction

According to [1] the growing ease of satisfying material needs of common people has caused an increased focus and attention on immaterial, emotional needs. Stories and emotions have become a large part of what we consume, and we search them also in the products we buy, in the houses and environments we live in. Moreover, we want to have a deeper influence on the products we consume, we want to tailor them to our own needs, we want to find a self-actualization by marking them with our own

J.S. Gero (ed.): *Design Computing and Cognition'10*, pp. 171–187.

personal touch. In most cases, however, our personal choices are influenced by designs, objects, pieces of art that we have seen in the past and that we appreciated.

In this paper we will introduce an experience in the realization of a particular object, a modular interactive lightning system, able to support the user in the customization of its own adaptive behaviour. The basic idea for this particular object is derived by an original design defined by an artist for a specific installation, an adaptive lightning facility to be installed in a tunnel in the context of a renovation project. The illumination of the tunnel had to react to the presence of pedestrians, cars, bicycles, ect., to emphasize their presence, to 'light their way' in the tunnel, in addition to the basic functional illumination. Artists are often a powerful driver in technological innovation, since they often require combinations of features and performances that go beyond current technologies [2]. In this case, the artist's idea and desiderata in the given scenario stimulated the definition of a computational model and a simulator to turn the design into practice, through the realization of a prototypal implementation of the desired system. During the process, we realized that the model that we defined was able to generate a wide variety of adaptive ligthning effects by simply altering some of its parameters. A specific user interface was designed to support the exploration of the complexity of such variety by supporting the envisioning of the effects of the specification of a set of values for the model's parameters on the behaviour of the system. Finally, this experience led to the first steps towards the design and realization of a product embedding the same model for lightning adaptation into a modular tile able to react to the touch or to other stimuli.

The following section will describe the tunnel scenario that first triggered the project; the basic computational models that influenced the definition of the adaptive lightning model will be described in Section 3, while the proposed model for adaptive lightning will be described in Section 4. A description of the behaviour customization system will be described in Section 5; a brief introduction to the developments of this experience towards the definition of the modular interactive lightning tile and conclusions will end the paper.

The Indianapolis Tunnel Scenario

The Acconci Studio was founded in 1988 to help realize public-space projects through experimental architecture and public art efforts. The method of Acconci Studio is on the one hand to make a new space by turning an old one inside-out and upside-down; and on the other hand to insert within a site a capsule that grows out of itself and spreads into a landscape.

The Studio has recently been involved in a project for the renovation of a tunnel in the Virginia Avenue Garage in Indianapolis. The tunnel is currently mostly devoted to cars, with relatively limited space on the sidewalks and its illumination is strictly functional. The planned renovation for the tunnel comprises a set of interventions along the direction defined by the following narrative description of the project:

> *The passage through the building should be a volume of color, a solid of color. It's a world of its own, a world in itself, separate from the streets outside at either end. Walking, cycling, through the building should be like walking through a solid, it should be like being fixed in color.*
> *The color might change during the day, according to the time of day: pink in the morning, for example, becomes purple at noon becomes blue, or blue-green, at night. This world-in-itself keeps its own time, shows its own time in its own way.*
> *The color is there to make a heaviness, a thickness, only so that the thickness can be broken. The thickness is pierced through with something, there's a sparkle, it's you that sparkles, walking or cycling though the passage, this tunnel of color. Well no, not really, it's not you: but it's you that sets off the sparkle – a sparkle here, sparkle there, then another sparkle in-between – one sparkle affects the other, pulls the other, like a magnet – a point of sparkle is stretched out into a line of sparkles is stretched out into a network of sparkles.*
> *These sparkles are above you, below you, they spread out in front of you, they light your way through the tunnel. The sparkles multiply: it's you who sets them off, only you, but – when another person comes toward you in the opposite direction, when another person passes you, when a car passes by – some of these sparkles, some of these fire-flies, have found a new attractor, they go off in a different direction.*

Fig. 1. A visual elaboration of the desired adaptive illumination facility (the image appears courtesy of the Acconci Studio)

The above narrative description of the desired adaptive environment comprises two main effects of illumination, also depicted in a graphical elaboration of the desired visual effect shown in Figure 1:

- an overall effect of uniformly coloring the environment through a background, ambient light that can change through time, but slowly with respect to the movements and immediate perceptions of people passing in the tunnel;
- a local effect of illumination reacting to the presence of pedestrians, bicycles, cars and other physical entities.

The first type of effect can be achieved in a relatively simple and centralized way, requiring in fact a uniform type of illumination that has a slow dynamic. The second point requires instead a different view on the illumination facility. In particular, it must be able to perceive the presence of pedestrians and other physical entities passing in it, in other words it must be endowed with sensors. Moreover, it must be able to exhibit local changes as a reaction to the outputs of the aforementioned sensors, providing thus for a non uniform component to the overall illumination. The overall environment must be thus split into parts, proper subsystems.

However, these subsystems cannot operate in isolation, since one of the requirements is to achieve patterns of illumination that are local and small, when compared to the size of the tunnel, but that can have a larger extent than the space occupied by a single physical entity ("sparkles are above you, below you, they spread out in front of you, they light your way through the tunnel"). The subsystems must thus be able to interact, to influence one another to achieve more complex illumination effects than just providing a spotlight on the occupied positions.

Cellular Automata Models

Cellular Automata (CA), introduced by John von Neumann as an environment for studying self-replicating systems [3], have been primary investigated as theoretical concept and as a method for simulation and modeling [4]. They have also been used as computational framework for specific kind of applications (e.g. image processing [5], robot path planning [6]) and they have also inspired several parallel computer architectures, such as the Connection Machine [7] and the Cellular Automata Machine [8].

Asynchronous Cellular Automata

Cellular Automata have traditionally treated time as discrete and state updates as occurring synchronously and in parallel. The state of every cell of the automaton is updated together, before any of the new states influence other cells. The synchronous approach assumes the presence of a global clock to ensure all cells are updated together.

Several authors (e.g. [9, 10]) have argued that asynchronous models are viable alternatives to synchronous models and suggest that asynchronous models should be preferred where there is no evidence of a global clock. Nehaniv [11] has demonstrated an asynchronous CA model that can behave as a synchronous CA, due to the addition of extra constraints on the order of updating.

Cornforth, Green, and Newth argue that asynchronous updating is widespread and ubiquitous in both natural and artificial networks [12]. They identified two classes of asynchronous behavior: Random Asynchronous (RAS), and Ordered Asynchronous (OAS) updating. Random Asynchronous includes any process in which at any given time individuals to be updated are selected at random according to some probability distribution; Ordered Asynchronous includes any process in which the updating of individual states follows a systematic pattern.

Dissipative Cellular Automata

The Dissipative Cellular Automata (DCA) are a class of cellular automata that have been defined as dissipative, i.e., cellular automata that are open and makes it possible for the environment to influence their evolution [13].

The two main characteristics of the DCA are the asynchronous time-driven dynamics and openness. DCA are Asynchronous Cellular Automata: according to the asynchronous dynamics [14, 15], at each time, one cell has a probability of rate λ_a to autonomously wake up and update its state.

The above characteristics of modern software systems are reflected in DCA. can be considered as a minimalist open agent system (or, more generally, as a minimalist open software system). As that, the dynamic behavior of DCA is likely to provide useful insight into the behavior of real-world open agent systems and, more generally, of open distributed software systems.

Cellular Automata with Memory

Standard CA are ahistoric (memoryless): the cells have no memory of previous states, except the last one in the case the central cell is included in the neighborhood. Historic memory can be embedded in CA increasing the number of states and modifying the transaction function. Alonso-Sanz proposed to maintain the transaction rule unaltered, but make them act not only to the current state but weighted mean value of their previous states [16]. According to the author, CA with memory can be considered as a promising extension of the basic CA paradigm.

Adaptive Interactive Lightning Model

The proposed approach adopts an Asynchronous Cellular Automata with memory to realize a distributed control system able to face the challenges of the previously presented scenario. The control system is composed of a set of controllers distributed throughout the system; each of them has both the responsibility of controlling a part of the whole system as well as to collaborate with a subset of the other controllers (identified according to the architecture of the CA model) in order to achieve the desired overall system behavior. In the proposed architecture, every node is a cell of an automata that can communicate only with its neighbors, it processes signals from sensors and it controls a predefined set of lights associated to it. The approach is totally distributed: there is no centralized control and no hierarchical structuring of the controllers, not only from a logical point of view but also a physical one.

The designed system is an homogeneous peer system, as described in Figure 2: every controller has the responsibility of managing sensors and actuators belonging to a fixed area of space. All controllers are homogeneous, both in terms of hardware and software capabilities. Every controller is connected to a motion sensor, which roughly covers the controlled area, some lights (about 40 LED lights) and neighbouring controllers.

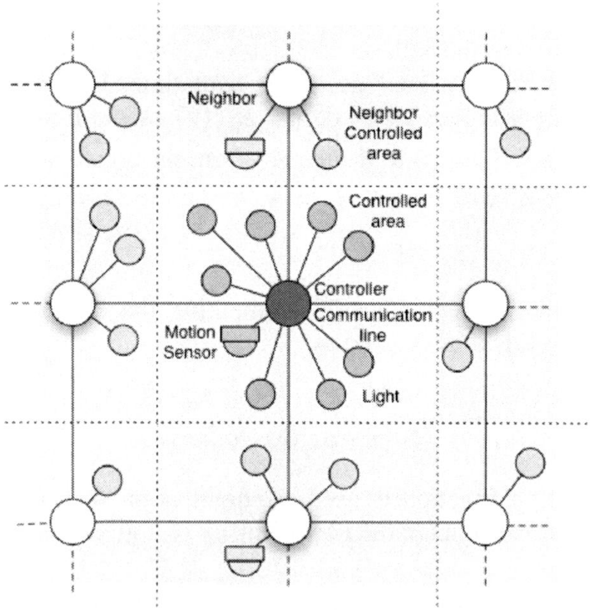

Fig. 2. The proposed architecture for the distributed control system to be managed through an Asynchronous Cellular Automata with memory

The state of the motion sensor influenced the internal state of the cell. The state of the sensor is represented by a single numerical value $v_s \in N_{8bit}$, where $N_{8bit} \subset N_0, \forall x : x \in N_{8bit} \Rightarrow x < 2^8$. The limit value was chosen for performance reasons because 8-bit microcontrollers are widely diffused and they can be sufficiently powerful to manage this kind of situation. The value of v_s is computed as $v_s(t+1) = v_s(t) \cdot m + s(t+1) \cdot (1-m)$ where $m \in R$, $0 \leq m \leq 1$ is the *memory coefficient* that indicates the degree of correlation between the previous value of v_s and the new value, while $s(t) \in N_{8bit}$ is the reading of the sensor at the time *s(t)*. If the sensor is capable of distance measuring, *s(t)* is inverse proportional to the measured distance (so, if the distance is 0, the value is 255, if the distance is ∞ the value is 0). If the sensor is a motion detector sensor (it able to signal 1 if an object is present or 0 otherwise) *s(t)* is equal to 0 if there is not detected motion, *c* in case of motion, where $c \in N_{8bit}$ is a constant (in our tests, 128 and 192 are good values for *c*).

The diffusion rule is used to propagate the sensors signals throughout the system. At a given time, every level 2 cell is characterized by an intensity of the signal, $v \in N_{8bit}$. Informally, the value of *v* at time *t*+1 depends of the value of *v* at time *t* and on the value of $v_s(t+1)$, to capture both the aspects of interaction with neighbouring cells and the memory of the previous external stimulus caused by the presence of a physical entity in the area associated to the cell.

The intensity of the signal decreases over time, in a process we call evaporation. In particular, let us define $\varepsilon_{evp}(v)$ as the function that computes the quantity of signal to decrement from the signal and is defined as $\varepsilon_{evp}(v) = v \cdot e_1 + e_0$ where $e_0 \in R^+$ is a constant evaporation quantity and $e_1 \in R,\ 0 \le e_1 \le 1$ is the evaporation rate (e.g. a value of 0.1 means a 10% evaporation rate).

The evaporation function *evp(v)*, computing the intensity of signal *v* from time *t* to *t*+1, is thus defined as

$$evp(v) = \begin{cases} 0 & \textit{if } \varepsilon_{\mathrm{evp}}(v) > v \\ v - \varepsilon_{\mathrm{evp}}(v) & \textit{otherwise} \end{cases}$$

The evaporation function is used in combination with the neighbours' signal intensities to compute the new intensity of a given cell.

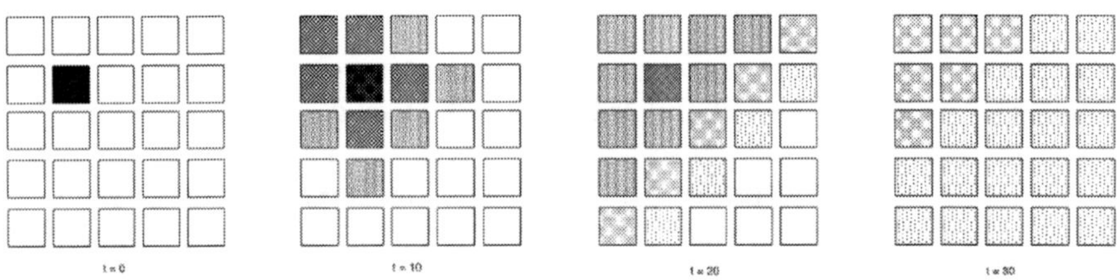

Fig. 3. An example of the dynamic behaviour of a diffusion operation. The signal intensity is spread throughout the lattice, leading to a uniform value; the total signal intensity remains stable through time, since evaporation was not considered

The automaton is contained in the finite two-dimensional square grid N^2. We suppose that the cell $C_{i,j}$ is located on the grid at the position *i, j*, where $i \in N$ and $j \in N$. According to the von Neumann neighbourood, a cell $C_{i,j}$ (unless it is placed on the border of the lattice) has 4 neighbours, denoted by $C_{i-1,j}$, $C_{i,j+1}$, $C_{i+1,j}$, $C_{i,j-1}$. For simplicity, we numbered the neighbours of a cell from 1 to 4, so for the cell $C_{i,j}$, N_1 is $C_{i-1,j}$, N_2 is $C_{i,j+1}$, N_3 is $C_{i+1,j}$, and N_4 is $C_{i,j-1}$.

At a given time, every cell is characterized by an intensity of the sensor signal. Each cell is divided into four parts (as shown in Figure 4), each part can have a different signal intensity, and the overall intensity of the signal of the cell is the sum of the parts intensity values. The state of each cell $C_{i,j}$ of the automaton is defined by $C_{i,j} = \langle v_1, v_2, v_3, v_4 \rangle$ where $v_1, v_2, v_3, v_4 \in N_{8bit}$ represent the intensity of the signal of the 4 subparts. $V_{i,j}(t)$ represents the total intensity of the signals (i.e. the sum of the subparts signal intensity) of the cell *i,j* at time *t*. The total intensity of the neighbours are denoted by V_{N1}, V_{N2}, V_{N3}, and V_{N4}. The signal intensity of the subparts and the total intensity of the cell are computed with the following formulas:

$$v_j(t+1) = \begin{cases} \dfrac{evp(V(t))\cdot q + evp(V_{Nj}(t))\cdot (1-q)}{4} & if\ \exists\ \mathrm{Nj} \\ \dfrac{\mathrm{evp(V(t))}}{4} & \text{otherwise} \end{cases}$$

$$V(t+1) = \alpha \cdot v_s(t+1) + \beta \sum_{i=1}^{4} v_i(t+1)$$

where $q \in R$, $0 \leq q \leq 1$ is the conservation coefficient (i.e. if q is 0 the new state of a cell is not influenced by the values of neighbours, if it is 0.5 the new value is a mean among the previous value of the cell and the neighbours value, if it is 1, the new value does not depend on the previous value of the cell but only from the neighbours). Of course the total intensity of the cell also consider the current state of the sensor cell, and parameters α and β are coefficients that can be used to fine tune the importance of neighbours compared to the local stimulus determined by the sensor cell. The effect of this modeling choice is that the parts of cells along the border of the lattice are only influenced through time by the contributions of other parts (that are adjacent to inner cells of the lattice) to the overall cell intensity.

In this project the actuators are LED lamps that are turned on and off according the state of the cell. Instead of controlling a single LED from a cell, every cell is related to a group of LEDs disposed in the same (small) area.

There are different approaches, called “coloring strategies”, to associate LED activity (i.e. being on or off, with which intensity) to the state of the related actuator cell. An example of coloring strategy consists in directly connecting the lights’ intensity to the signal level of the correspondent cell; more details on this will be given in the following Section.

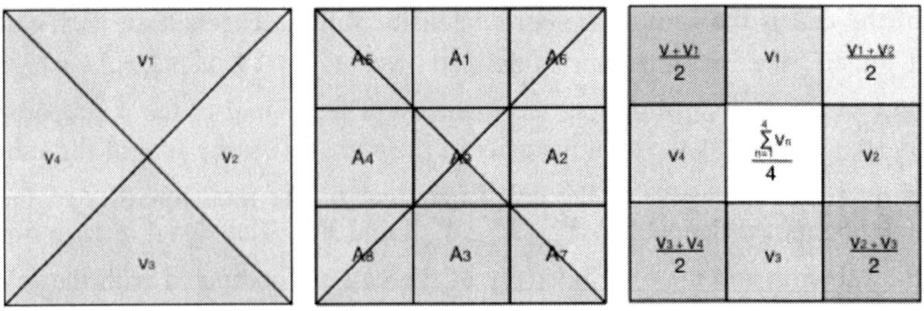

Fig. 4. Correlation between the upper layer cell subparts and the actuators layer cells

The Design Environment

The design of a physical environment (e.g. building, store, square, road) is a composite activity, comprising several tasks that gradually define the initial idea into a detailed project, through the production of intermediate and increasingly detailed models. CAD softwares, and also 3D modeling applications are generally used to define the digital models for the project and to generate photo realistic renderings and animations. These applications are extremely useful to design a lights installation like the one related to this scenario, but mainly from the physical point of view.

In order to generate a dynamics in this kind of structure, to grant the lights the ability to change illumination intensity and possibly color, it is also possible to “script” these applications in order to characterize lights with a proper behaviour. Such scripts, created as text files or with graphical logic editors, define the evolution of the overall system over time. These scripts are however heavily dependent on the adopted software and they are not suitable for controlling real installations, even though they can be used to achieve a graphical proof of concept. Another issue is that these tools are characterized by a “global” approach, whereas the system is actually made up of individual microcontrollers’ programs acting and interacting to achieve the global desired effect.

In this experience, our aim was to facilitate the user in designing the dynamic behavior of a lights installation by supporting the envisioning of the effects of a given configuration for the transition rule guiding lights; therefore we created an ad-hoc tool, also shown in Figure 5, comprising both a simulation environment and a graphical parameters configurator. This tool support the specification of the values for some of the parameters of the transition rule, affecting the global behavior of the overall system. The integrated simulation helps understanding how the changes of the single parameters influence the overall behavior of the illumination facility: every changed parameter is immediately used in the transition rule of every cell.

In the following paragraphs, the tool's main components are described. Ad the end of this section, some experimental configurations and the related dynamic evolution are presented.

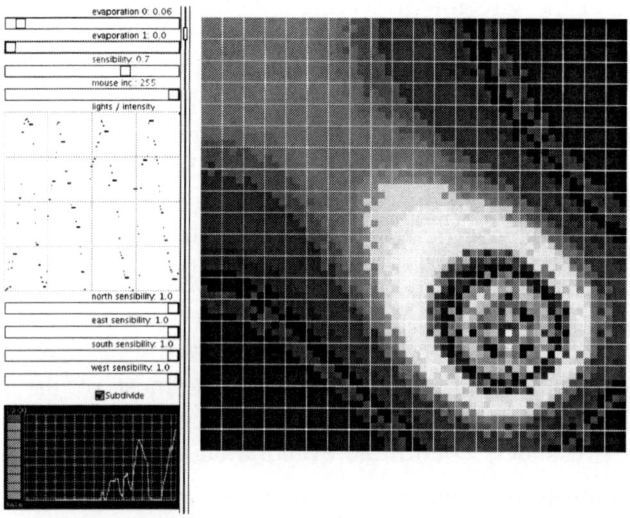

Fig. 5. A screenshot of the design environment. On the left, the system configurator panel and the global intensity graph, on the right the lights view

The Cells Simulator

The main component of the design environment is the simulator. This component simulates the dynamic evolution of the cell over the time, according to the transition rule. The simulated cells are disposed over a regular grid and each cell is connected to its neighbors according to the von Neumann neighbourood. By default, the tools is configured to simulate 400 cells, organized in a 20x20 grid. The grid is not toroidal, to better simulate a (portion of) the real installation space. Each cell has an internal state represented as an 8 bits unsigned number. In order to better simulate the real asynchronous system, an independent thread of control, that re-evaluates the internal state of the cell every 200 ms is associated to each cell. At the simulation startup, each thread starts after a small (< 1 s) random delay, in order to avoid a sequential activation of the threads, that is not realized in the real system. The operating system scheduler introduces additional random delays during both the activation and the execution cycle of the threads.

The Lights View

The aim of the Lights View is to realize an interactive visualization of the dynamic evolution of the system. In particular, the user can simulate the presence of people in the simulated environment by clicking on the cells and moving the mouse cursor. Each cell of the simulated system is associated an area of the screen representing a group of lights controlled by the cells. More precisely, it is possible to define at runtime if the area controlled by each cell is subdivided in 9 sub-areas (9 different lights groups) or if it is a single homogeneous light group. Each simulated group of lights is characterized by 256 different light intensity levels.

On the left of lights view, there is a graph showing the evolution over the time of the sum of all the cells intensity levels. This graph is particularly useful to set the coefficients of the evaporation function.

The System Configuration

Through this component, the user can define most of the parameters related to the transition rule of the simulated system.

The first two sliders control the evaporation coefficients e_0 and e_1, the next one controls the sensibility parameters q (see Section 4 for the parameters' semantics). The "mouse increment" slider defines the amount of the increment in the cell intensity when a user clicks on the cell: it represents the sensitiveness of the cell to sensor stimulus in the real system.

Under the four sliders there is a small panel that supports drawing the function that correlates the internal cell intensity value and the correspondent light group intensity value. The default function, represented by a diagonal segment between the (0,0) position and the (255,255) position, is the "equal" function (i.e. if the cell intensity has value x, the lights intensity has value x). It is possible to draw an arbitrary function, setting for each cell intensity value a correspondent light intensity value simply drawing the function over the graph.

The last four sliders control the sensitivity of each cell to the neighbors in the four directions (q_N,q_E,q_S,q_W); by keeping these values separated it is possible to configure the cell to be more sensitive to the cells in a specific direction (e.g. left or right).

Finally, there is a check-box to switch between 1 and 9 lights groups per cell.

Experimenting Different Configurations

This last section presents some consideration about the relations between the parameters value and the system behavior. This is not intended to be exhaustive analysis, it only presents some relevant usage examples of the design environment.

The first example, shown in Figure 6, describes 3 steps of evolution of the system configured with the default parameters ($e_0 = 0.75$, $e_1 = 0$, $q = 0.1$, mouse increment = 255, f=eq, $q_N=q_E=q_S=q_W=1$). The system, in this configuration, acts as a sort of spot-light around the stimulated areas. When the movement of a person on the space is simulated with a mouse input a light trace is generated following the mouse movement. The trace is not present anymore in Figure 7, with an increased evaporation coefficient ($e_0 = 1$, $e_1 = 0.1$, $q = 0.5$); on the contrary, in Figure 8 a long-persistent tail is produced with a very low evaporation level and no neighbours sensibility ($e_0 = 0.05$, $e_1 = 0$, $q = 0$). Figure 9 shows a different configuration with a high sensitivity to the southern neighbour ($e_0 = 0.01$, $e_1 = 0$, $q = 0.0$, f=eq, $q_N = 0.1$, $q_E = q_S = q_W = 0$). A sort of "smoke" arising from southern cells can be viewed. The last example, shown in Figure 10, is achieved through an ad-hoc intensity-light correlation function (shown in red line in the figure) and the following parameters: $e_0 = 0.6$, $e_1 = 0$, q=0.7, $q_N=q_E=q_S=q_W=1$.

It is interesting to notice how many different behaviors can be achieved by means of a different parameter specification of the same transition rule.

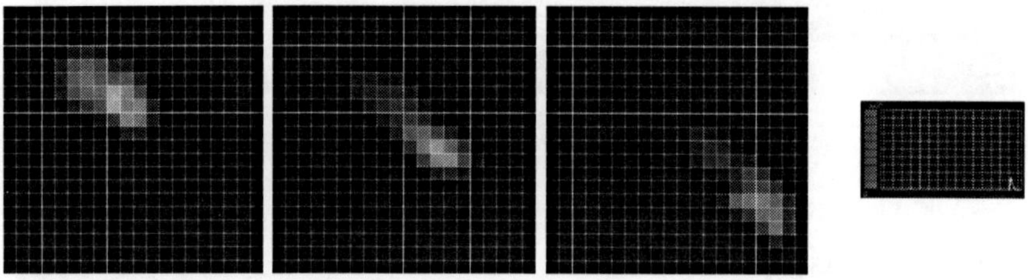

Fig. 6. Example 1: e0 =0.75, e1 =0, q=0.1, f=eq, qN =qE =qS =qW =1

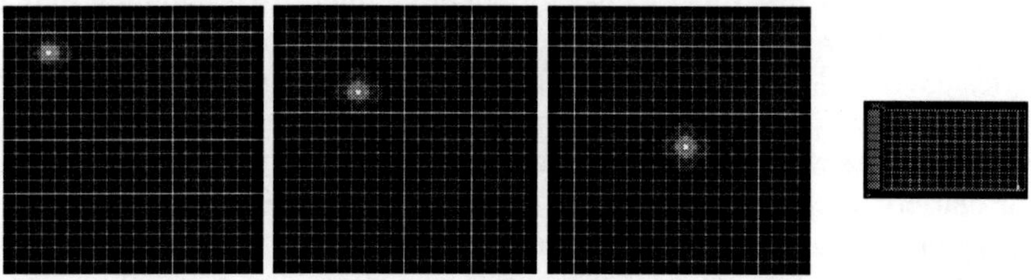

Fig. 7. Example 2 : $e_0 =1$, $e_1 =0.1$, q=0.5, f=eq, $q_N =q_E =q_S =q_W =1$

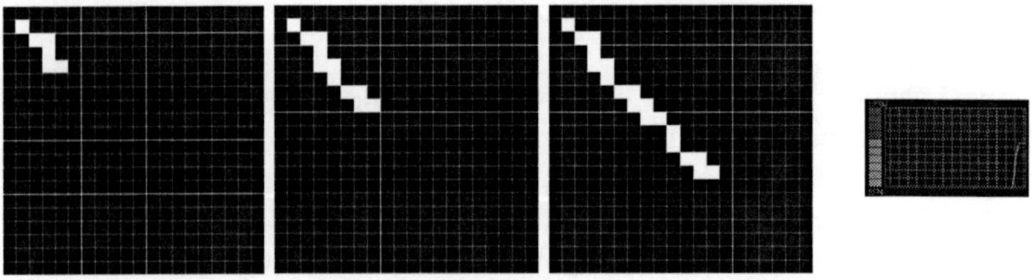

Fig. 8. Example 3: e0 =0.05, e1 =0, q=0.0, f=eq, qN =qE =qS =qW =1

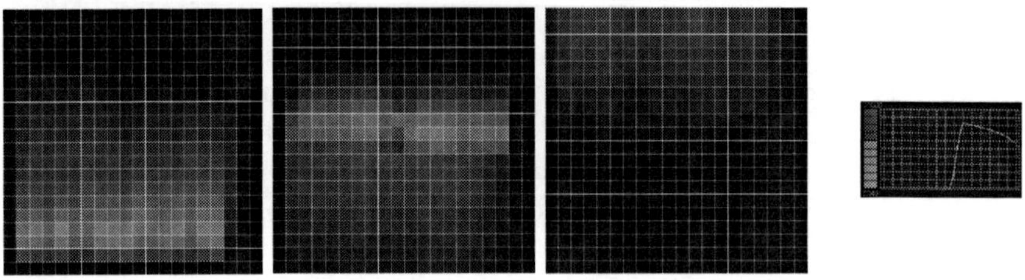

Fig. 9. Example 4: e_0 =0.01, e_1 =0, q=0.0, f=eq, q_N =0.1, q_E =q_S =q_W =0

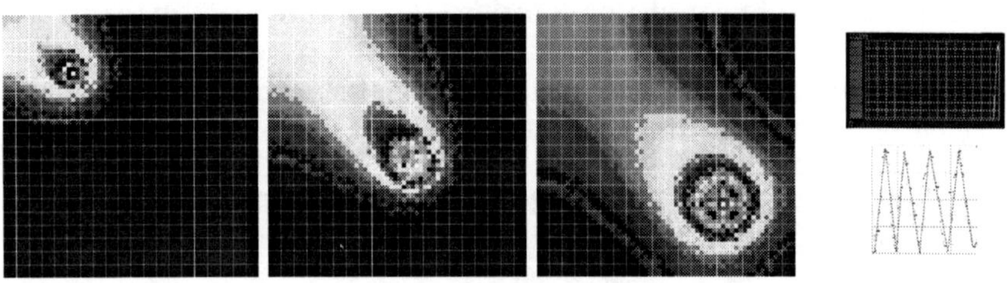

Fig. 10. Example 5: e_0 =0.6, e_1 =0, q=0.7, q_N =q_E =q_S =q_W =1

From a Prototype to a Product

After the above described experience we had the chance to effectively embed the above described model for adaptive interactive lightning into a specific hardware platform realized by an Italian company[1]. In particular, we designed a tile, also shown in Figure 11, comprising a set of proximity sensors (infrared devices able to perceive the presence, for instance, of a finger or a pen within few millimeters) and RGB leds (able to generate a wide variety of lightning effects). The tiles are provided with a 16 bit microcontroller and serial ports supporting both the programming of the tile and its interaction with other devices, including (and especially) other tiles.

[1] Egicon Srl - http://www.egicon.com/

Fig. 11. The front and rear views of a single tile of the modular adaptive interactive lightning system and a photo showing a sample two tile configuration

A single tile is associated to a composite cell of the above introduced CA (in particular it includes 16 sensors and 16 leds, but the structure of the cell is conceptually the same as the one presented previously) and it can be connected to other tiles to compose larger structures. These structures are able to exhibit an overall behaviour analogous to the one realized for the adaptive illumination of the Indianapolis tunnel. Howeverm, in this case, the tiles can be composed to realize arbitrary surfaces (e.g. wall decorations, floors) that are able to respond to the presence of people to generate self-organized adaptive illumination effects. Two different sample configurations characterized by a different assembly of the above tiles are shown in Figure 12.

Fig. 12. Two different configurations characterized by a different assembly of the adaptive interactive lightning tiles: a reactive carpet on the left and an adaptive wall decoration

Conclusions and Future Work

This paper has presented a successful case study in which the creative intuition of a designer and its narrative description have triggered the definition of both a specific computational model and prototype to effectively implement that intuition. In turn, this prototypal implementation lead to further research activities, for instance aimed at the comparison of alternative modeling approaches for self-organizing environments [17] and to the enhancement of traditional CAD systems by means of ad-hoc tools for the simulation of self-organizing systems dynamics, but also to the exploration of the wide variety of resulting behaviours that can be achieved altering the parameters of the specific adaptive illumination model that was defined. Both the designer and the end user can actually thus actually alter the behaviour of the system, tailor it to their own personal tastes and desires. While this possibility could be of interest for a large stable infrastructure like the Indianapolis tunnel, it surely represents an even more significant feature for smaller, personal objects or facilities. All these considerations lead to the realization of a prototypal implementation of elements of a modular lightning system based on the same conceptual and computational model. Current and future works are aimed, on one hand, at the definition of a language and supporting tools for the definition of self-organizing systems based on the introduced CA based model, and, on the other, at the realization of a specific application of the modular lightning system.

References

1. Mogensen, K.Æ. (ed.): Creative Man. The Copenhagen Institute for Futures Studies, http://www.cifs.dk/creativeman/CreativeMan.pdf (Last accessed May 2010)
2. Potts, J.: Art & innovation: An evolutionary economic view of the creative industries. UNESCO Observatory, Faculty of Architecture, Building and Planning, The University of Melbourne Refereed E-Journal, Multi-Disciplinary Research in the Arts 1(1) (2007)
3. von Neumann, J.: Theory of Self-Reproducing Automata. University of Illinois Press, Urbana (1966)
4. Weimar, J.R.: Simulation with Cellular Automata. Logos Verlag, Berlin (1997)
5. Rosin, P.L.: Training cellular automata for image processing. IEEE Transactions on Image Processing 15(7), 2076–2087 (2006)
6. Behring, C., Bracho, M., Castro, M., Moreno, J.A.: An algorithm for robot path planning with cellular automata. In: Bandini, S., Worsch, T. (eds.) ACRI, pp. 11–19. Springer, Heidelberg (2000)
7. Hillis, W.D.: The Connection Machine. MIT Press, Cambridge (1985)
8. Margolus, N., Toffoli, T.: Cellular Automata Machines. In: A New Environment for Modelling. MIT Press, Cambridge (1987)
9. Paolo, E.A.D.: Searching for rhythms in asynchronous random boolean networks. In: Bedau, M. (ed.) Alife VII: Proceedings of the Seventh International Conference, pp. 73–80. MIT Press, Cambridge (2000)
10. Thomas, R., Organization, E.M.B.: Kinetic logic: A Boolean approach to the analysis of complex regulatory systems. In: Thomas, R. (ed.) Proceedings of the EMBO Course Formal Analysis of Genetic Regulation, held in Brussels, September 6-16. Springer, Berlin (1979)
11. Nehaniv, C.L.: Evolution in asynchronous cellular automata. In: ICAL 2003: Proceedings of the eighth international conference on Artificial life, pp. 65–73. MIT Press, Cambridge (2003)
12. Cornforth, D., Green, D.G., Newth, D.: Ordered asynchronous processes in multi-agent systems. Physica D: Nonlinear Phenomena 204(1-2), 70–82 (2005)
13. Zambonelli, F., Mamei, M., Roli, A.: What can cellular automata tell us about the behavior of large multi-agent systems? In: Garcia, A.F., de Lucena, C.J.P., Zambonelli, F., Omicini, A., Castro, J. (eds.) SELMAS. LNCS, vol. 2603, pp. 216–231. Springer, Heidelberg (2002)
14. Buvel, R.L., Ingerson, T.E.: Structure in asynchronous cellular automata. Physica D 1, 59–68 (1984)
15. Lumer, E.D., Nicolis, G.: Synchronous versus asynchronous dynamics in spatially distributed systems. Phys. D 71(4), 440–452 (1994)
16. Alonso-Sanz, R.: The beehive cellular automaton with memory. Journal of Cellular Automata 1(3), 195–211 (2006)
17. Bandini, S., Bonomi, A., Vizzari, G., Acconci, V.: Simulation of alternative self-organization models for an adaptive environment. In: Proceedings of the Second Multi-Agent Logics, Languages, and Organisations Federated Workshops, Turin, Italy, CEUR Workshop Proceedings 494 CEUR-WS.org (2009)

Facetwise Study of Modelling Activities in the Algorithm for Inventive Problem Solving ARIZ and Evolutionary Algorithms

Céline Conrardy, Roland de Guio, and Bruno Zuber
INSA de Strasbourg, France
Lafarge Research Center, France

The aim of this paper is to contribute to a better understanding of modelling activities required to solve inventive problems. The scope encompasses both computer and cognitive computation. A better understanding of the nature of knowledge and models will provide information to help conducting inventive design process with high effectiveness (convergence) and efficiency. The contribution proposed in the following paper consists in developing a framework to compare some facets of modelling activities required by evolutionary algorithms and algorithm for inventive problem solving ARIZ. It aims to yield to practical guidance, insight and intuition of new approaches for computer aided innovation that reduce cost of modelling activities and increase inventiveness of solutions.

Conceptual Design Techniques Are Non Quantitative

Regular design procedure consists in starting with idea generation, which gives a first vague design embodiment. This first step is then followed by a detailed design stage using either experimental testing or numerical simulations. This second step is dedicated to validate choices or optimize parameters of the chosen design [1]. In such a process, first choices at conceptual design stages are mainly responsible of the product's performances.

Paper and pencil have been described as the best tools to support conceptual design stage [2]. Consequently, a lot of research has been

J.S. Gero (ed.): *Design Computing and Cognition'10*, pp. 189–207.

undergone to solve various problems linked with computing behaviour of systems drawn by CAD [3, 4]. However such processes cannot support any type of design embodiments, especially high level inventions using extra-domain knowledge. They are domain dependant: design of a new chemical reaction occurring in the matter of a mechanical device is not supported by CAD - computation chain although a sketch of system's behaviour can be drawn with a pencil on a paper.

Furthermore, such processes do not provide quantitative estimations to dictate how a CAD sketch should be redrawn in case of dissatisfaction with first design embodiment. It is always a posteriori computation of performances. On the contrary, pencil/paper sketches can be conduced by creativity or problem solving methods, but estimation of concept's accuracy relies on brain computation of experts. This process cannot provide valid quantitative estimations because inventive design often suppose going out of the scope of past expertise.

Aims

Conceptual and Detailed Design Stages Do Not Speak the Same Language

There is a contradiction in Computer Aided Innovation (CAI) between the ability to generate creative designs and easily transform modelling elements and the ability to get reliable quantitative estimation of the behaviour of a model under certain conditions. For example, improving skills of machines reduces those of designers [5, 6, 7, and 2]. In the construction sector this contradiction is often described as a cultural conflict between engineers and architects. However, the causes of this contradiction are deeply fixed by language's structure required to support both activities. The first language is that of functions and dynamic transformations and is better supported by sketches, drawings or schemes of problem models; the second is that of reliable computation within fixed structures supported by mathematical calculus, equations and theorems.

Co-operative frameworks to alternate from one language to the other [8, 9] and representations to rapidly switch from qualitative to quantitative [10] have been proposed. However, translation from numerical model to problem model and vice versa have already been proposed, but are restricted to problems of size, shape and topology and domain dependant types of problem-models [11].

Translation from One Language to the Other Is Required

This first attempt of translation is a step towards enabling conceptual and detailed design stages to be more interoperable in CAI. In order to go a step further, a better understanding of the nature of knowledge and models at each step is required. The present paper proposes to examine some characteristics of models used in two different design embodiment methods: evolutionary design [12] and algorithm for inventive problem solving (ARIZ) [13]. This article is the first a series dedicated to compare model creation and model manipulation through evolutionary computation and TRIZ based methods.

Significance

It is necessary to propose new approaches for computer aided innovation that reduce modelling and computing cost and increase inventiveness of solutions. It is required methods that enable starting design process with few models and provide guidance about choice of useful modelling extensions. Those extensions added step by step should be interoperable with the precedent models and be guided by a priori quantitative estimates of their potentialities. At last, they should provide solutions in a way that is controlled by designers. This will enable to use non connected and very scattered models in order to provide both quantitative estimations and new models choice guidance to get those estimated results.

Contrary to regular approach that begins with concepts of solution and try to refine them until they fit to the design requirements, the proposed synthesis activity aims at being able to start from something that is probably absolutely not a solution and to find step by step the design parameters and models that are required to fit design requirements in a parsimonious way.

Approach Followed in the Paper

A framework in which the modelling activities required by the two algorithms can be compared will be introduced. Its aim is to yield to practical guidance, insight and intuition of the matter at low learning cost. The development of the framework is based on several already existing frameworks among them C-K theory [14] and Hilbert space following the original purpose of this mathematical tool [15]. This framework is not dedicated to encompass the whole reality of the two algorithms, but it enables to study some characteristics of models used in the two approaches.

The study of ARIZ is restricted to some steps extracted from "Analysis of Initial Situation" (Part 0) and "Analyzing the problem" (Part I) that can be found in a translation of the original documents [13]. The interpretation of the text proposed in the paper is a personal approach developed through discussion with a TRIZ expert and application on several real problems. The study of modelling activities for EA has been realized through bibliography and discussion with an expert of the field on several problems.

Notions and Definitions

Hilbert Space: Search Strategy Driven by Objectives

Hilbert space as a design space that possesses the interesting property of having a number of dimensions equivalent to the total number of parameters necessary to describe all the objects it contains. Every possible object can be found by determining the points where its describing parameters intersect. It enables to think of the perfect object or system already being in existence and then to develop search strategies to find the parameters that describe it [15].

Steady State: Facets of the Notion in Various Scientific Fields

Steady states are parts of the design space where the type of physics remains the same, so that the system keeps acting in the same predictably way and is relatively unaffected by small changes. For the purpose of this article let us say that a steady state is when the same simple mathematical equations can be used to describe various solutions (the model is not changing, only the value of parameters). Of course, this notion depends on the degree of integration of several physics into more universal frameworks and the degree of precision required for the problem considered. Steady states notion leads to various appellations in different scientific fields: "type of design embodiment" [16] by mechanical designers, neighbourhood of local optima by mathematicians [17], mental highways [18] by cognitians, design concept [14], attractor [19], design or architectural paradigm [20, 21].

C-K: Distinction between Concepts and Knowledge

C-K proposes to describe design with two spaces: K – the knowledge space – is a space of propositions that have a logical status for designer (i.e. it exists mathematical solution or physical experiments that attest the logical status), and C – the concept space – is a space containing sets of parameters (concepts) that can have no logical status in K. This theory contributes to formalize a basic thinking process common to all design traditions and so facilitates the description of modelling activity during design.

Evolutionary Algorithms (EA)

Evolutionary algorithm (EA) uses iterative progress of a population of solutions based on some mechanisms inspired by biological evolution: reproduction, mutation, recombination and selection of the population. A simulation (most of the time performed on a computer thanks to a mathematical modelling of system behaviour) is able to evaluate the fitness function i.e. the level of performance (objective) and ability of survival (constraints) of each solution. This principle enables the algorithm to explore various steady states and to perform difficult optimization at a relatively low computation cost. It is particularly useful when facing optimization problems with a lot of parameters and local optima.

Algorithm for Inventive Problem Solving (ARIZ)

The algorithm for inventive problem solving (Russian acronym ARIZ) is a method based on TRIZ dialectical and systemic thinking to solve inventive problems. ARIZ and is composed of creativity tools (mental inertia breaker), problem modelling tools and very generic and more specific knowledge-base tools. The process of creating tools and organizing them included a kind of fitness evaluation by testing each modification of the algorithm with benchmark problems [22]. Alternation of analysis tools to model problem and synthesis tools (operators) to change the problem model into solution model enables to find solutions to inventive problems [23]. Based on TRIZ knowledge, several enhancement of ARIZ and knew methods have been proposed [24, 25, 26, 22, 27] in order to adapt method to rather simple or very complex problems. Some mechanisms of the thinking process proposed by ARIZ have already been reported in the literature [28], but no publication study the parsimony of knowledge and modelling activities with which an ARIZ expert is able to find a solution of an inventive problem.

Modelling Requirements in Evolutionary Algorithms

Inventive Problem Solving through Evolutionary Algorithm Is Restricted by the Lack of Dynamism in Models

In evolutionary algorithm, the analogy with natural process is restricted to the search process through design space. It does not encompass the way nature creates new steady states. In fact, the main limitation for invention is due to the architecture of numerical simulation: non progressive models, which leads to finite dimension search spaces. On the contrary biological systems have a flexible architecture, which enables their parameters to be continuously changed until they become optimally efficient [29, 30].

In order to overcome this limitation, some enhancements have been proposed consisting in non-finite dimension data structures: non-linear, non-fixed length, dynamic and indirect computing structures. Those structures enable the solution space to be potentially extended to an infinite number of dimensions (but countable) due to combination of a non restricted number of several unitary models. This kind of data structures can be trees like in genetic programming for example. Those techniques enable computer calculations to be competitive with human design performances [31, 32].

Even in construction sector, various data structures and enhancement of evolutionary algorithms have been applied to structural design problems and have produced solutions that offer significant improvements over traditional Genetic Algorithms based methods [33, 34, 35, 21, 8, 9 and 36]. Since EA paradigm has many ramifications, the illustrative example proposed below deals with one particular EA with flexible data structure. This should be understood as an illustration of a generic limitation in EA paradigm that remains valid for all its forms (to the knowledge of the authors and restricted to numerical parameters management).

Design Space Limits in Model Segmentation Approach

Let us consider a post with free the internal geometry i.e. every material can be placed everywhere in the post, enabling an infinite design space to be searched. It can be achieved for example if the post is constituted of elementary material elements. Each element has a non predefined size and a variable geometry (a set of different geometries is provided). A type of available material (or void) is attributed to each element. Rules about geometrical coherence between adjacent elements and linking mechanisms are additional models required to perform the computation. Since the size of elements is not predefined, the algorithm can theoretically use an infinite number of elements to refine the design of the post and explore original ways of assembling the elements. This non finite data structure is generated by a finite number of parameters type. Each element has a finite number of dimensions to describe it, but the whole post has a non finite number of design dimensions. Such a computation can lead to non expected shapes but does not provide solutions that are not a combination of the elementary models chosen to perform the computation. It does not solve the problem "what should be done when the programmer has no idea of elementary models that constitute the solution?" Elements (and its parameters) constituting the solution cannot be known a priori and are provided thanks to creativity of programmer.

In the previous case, the proposed dynamic data structure reposes on a possibility of spatial segmentation of computation of the behaviour of the whole object. Typical computable models changes are domain-specific and in restricted number [11], and even enabling all of them in the same computation provide no guarantee of requirements satisfaction.

Practical Limit of Model Segmentation

When freeing the structure of the computed object lots of modelling problems emerge in practice. Today, such generic simulation uses Finite Element Methods. When using this kind of computation techniques the number of element is not predetermined and some computation routines already exist to add, remove and connect various types of elements. The mechanical behaviour of each element constituted by a specific material can be computed and also the mechanical behaviour of each interface between different materials. However it exists no universal way to mathematically model and compute physical phenomena of various kinds at various scales. Each steady state needs a particular approach, so what occurs for FEM computation remains true with other computational approaches. In the post example, interfaces between materials could be very complex and the modelling approach has to change when dealing with very thin elements (foam structure for example). Moreover, material transitions in Nature are very progressive and use chemical overlapping of materials in order to prevent fragile behaviour [30], which cannot be computed in our approach. So, the result of such a computation is a priori expected to lead to poor practical relevance or infeasible design embodiments.

When segmenting models, every available physical model should be computable in order to maximize the chance to get more accurate and optimized results. So, computation strategy reaches fast limits such as exactness of codes or developing costs. How to find parameters, models and simulators useful for generation of new concepts with knowledge parsimony?

ARIZ applied on various technical fields has proven the human capacity to answer that question and could provide some directions of research for evolutionary algorithms. That is why ARIZ and the way it finds new steady-states will be examined now. The task consists in building a common representation framework of the two approaches in order to prepare their integration.

Modelling Requirements in ARIZ

Facetwise Modelling Reduces Complexity

Contradiction Statement

ARIZ proposes to understand problems through contradiction statement. When the actual state of a system is not satisfying (evaluation parameter EP1-), it should be searched a state of the system where actual bad value of one parameter can become satisfying (evaluation parameter EP1+); then it should be considered what gets worse (EP2-) and define what has changed to switch from one state to the other (action parameter AP and \AP) and vice versa. Both sides of the contradiction should be consistent, Figure 1. Contradiction solving leads to obtain both EP1+ and EP2+.

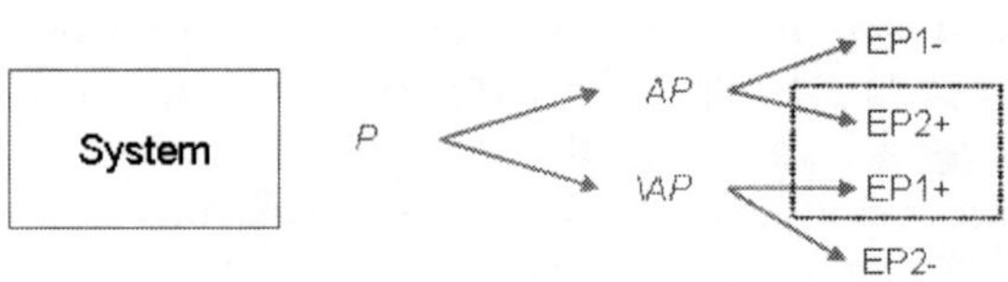

Fig. 1. Consistent contradiction

Let us consider storm water evacuation: if a regular concrete slab is poured on the ground, water needs to be evacuated through rivulets on its side, which leads to a system that is simple to implement but cannot evacuate a high quantity of storm water. On the contrary, if the whole surface is used for water evacuation (by use of a permeable concrete for instance), a huge amount of storm water can be evacuated, but the system cannot be simply poured on the ground, it requires experimented workers and several tools, Figure 2.

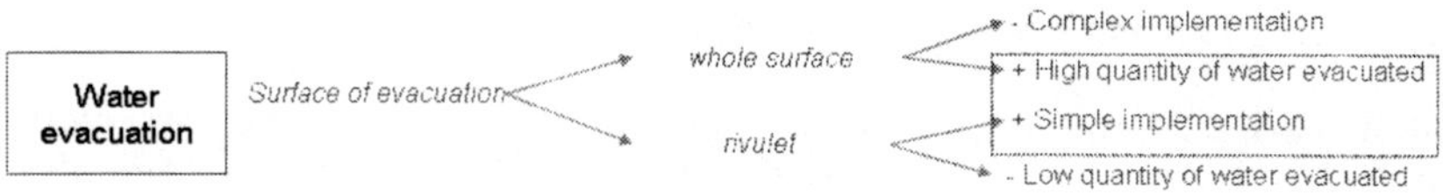

Fig. 2. Contradiction between simplicity of implementation and quantity of water evacuated

C-K framework proposes to study the cognitive mechanism in design by distinguishing concepts, that are sets of design parameters and knowledge that gives logical status to a set of parameters, Figure 3.

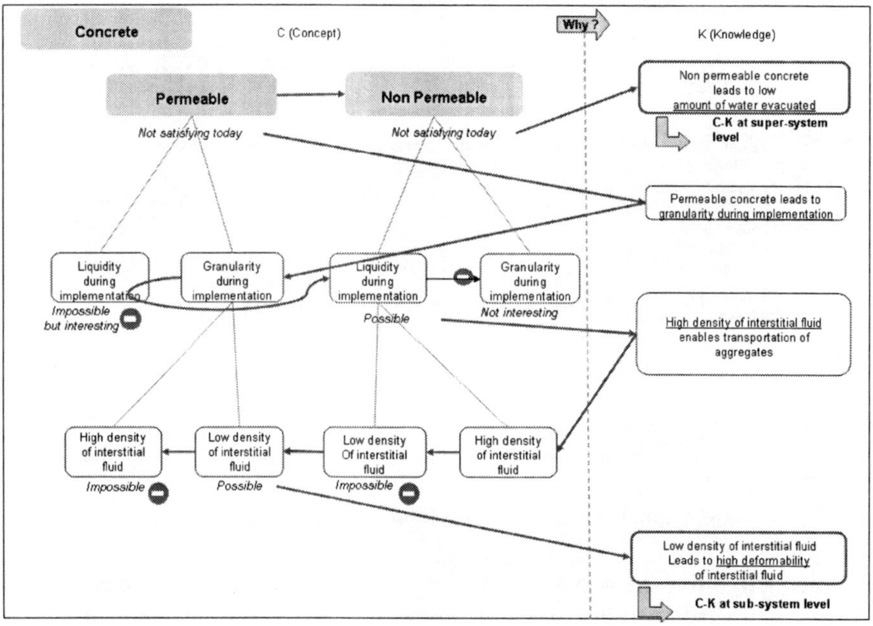

Fig. 3. Cognitive path of contradiction statement

For example, considering the permeable slab cited above, "granularity during implementation" induces expansive partition "liquidity during implementation" which seems impossible in mind of designer, but, if this parameter is added in the expansive partition "non permeable" previously generated by the parameter "permeable", then a logical explanation is known by the designer. It means that, when the knowledge of designer provides an answer to the questions "Why is it not satisfying?" or "Why is it possible?" a new parameter is added in C that induces an expansive partition.

Propagating Contradictions through System Levels

When solving a problem, we need to get a better understanding of what the problem is and how it is linked with other problems. Let's now detail what is going on at subsystem levels, Figure 4, suggested by the first C-K framework.

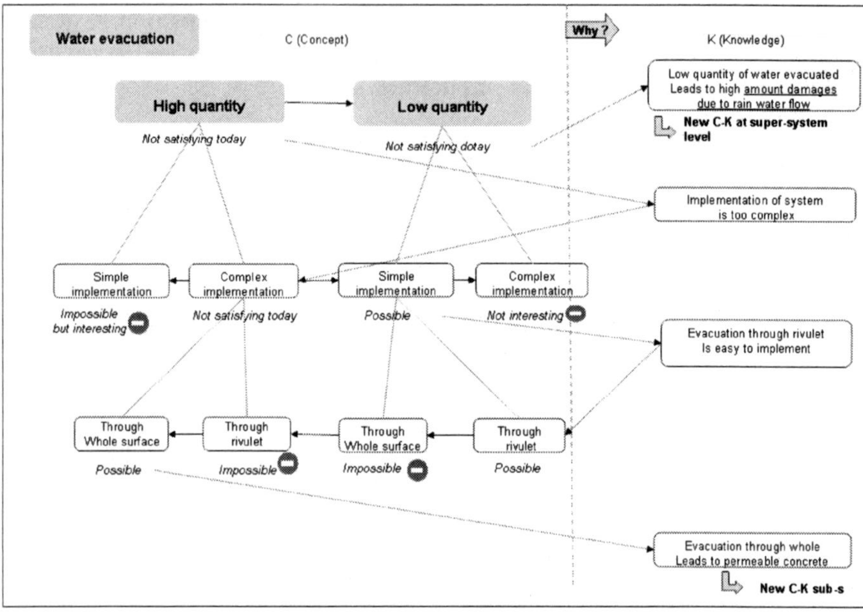

Fig. 4. Cognitive path of contradiction statement at sub-system level

Such cognitive mechanism can be pursued at higher or lower system levels. This elementary mental algorithm of systemic analysis is a cognitive pattern that repeats itself. Knowledge requirement to partition each set of concept leads to definition of evaluation and action parameters of the contradiction, elements and evaluation parameter of the super-system and sub-system.

Facetwise Modelling Reduces Complexity

Modelling Activities When Evaluating Logical Status of a Proposition

Let's examine more closely knowledge provided in K part of previous C-K representation. Why does an expert of the technical field consider that a set of parameters is impossible to achieve? How does he give a logical status to a set of parameter? How does he build a "logical" explanation? What are the models in his mind?

Logical status is given by consideration of the variables constituting a steady state, which is the expert's skill domain. The design variables are not explicit and constitute expert's implicit knowledge. Contradiction statement at this stage neither requires to state what all those variables are, nor to clarify any behavioural law. The subject of how the brain is managing such kind of implicit knowledge in reality should be further investigated. What is going on in other areas of design space? Why is there a lack of model in mind of expert that prevents him thinking beyond Pareto frontier? Partial answers could be:

- The expert ignores related physical laws and orders of magnitude, so he is not able to evaluate the result of possible variation of existing parameters.
- The expert has knowledge but should consider too much new parameters, so he does not have a clear vision of Pareto frontier, which leads to a too complex model that he cannot manage. This is a typical problem that can be solved by EA: the expert knows that changing the shape of the post will improve its performance, but it is too complex to be computed by his brain and traditional structural engineering techniques.
- The set of parameters he has in mind to compute (or predict) the behaviour of the system does not include a parameter able to describe the problem. If expert is trying to explain the non liquidity of permeable concrete with a background related to granular mechanics, he will explain that a chain of forces between aggregates holds them vertically together. He will never provide an explanation related to the nature of interstitial matter between aggregates, so he cannot think about reformulating contradiction at the system-level "interstitial fluid".

Conflicting Requirements of Two Steady States

The observation of impossible requirement when considering a single steady state "permeable concrete", leads to consider the expansive partition generated by the complementary of "permeable" i.e. "impermeable" (set-theory at the basis of C-K).

Since the modification of the first parameter enables a possible design, the expert answering the question "Why does it become possible?" searches an action parameter by comparison between the two contradictory design variable sets, that one of permeable concrete and that one of impermeable but liquid concrete. However, the expert of the first steady state is a priori not an expert of this second steady state. He can have heard about, or several domain experts can be participating to the problem solving. The action parameter then proposed by the expert, as "main explicative" parameter, was not necessary in the starting design sets of each steady state. It can rather be created by recombination of existing knowledge in a mental process that needs to be further investigated. For example, the density of interstitial fluid is not usually a parameter used to describe the behaviour of liquid concrete, but has been introduced in the purpose of explaining the difference between granular an liquid concrete behaviour.

Steady States Discontinuities and Integrated Framework

C-K does not provide a comprehensive framework to study the jump from one steady state to the other and does not focus the attention on the alternation and discontinuities in models at the basis of TRIZ thinking approach, Figure 5.

In order to merge the two paradigms in a comprehensive scheme, it should be represented not only the projection of design space on evaluation parameter axis in a 2D diagram but a flexible design set of valued parameters. In a Hilbert space, all considered parameters and the relations between their values as a result of applying knowledge on the specific resources of the problem (both physical laws of system and its environment) can be handled flexibly.

Principle of the description, Figure 6:

- Each line represents a parameter i.e. a dimension of the Hilbert Space. For commodity of representation, each axis is beginning at the value 0 and has only positive values (but we can imagine drawing the whole axis with its two opposites signs)
- Each point is the value of the parameter. It can be seen as unsatisfying (grimace) or satisfying (smile).
- The Hilbert space contains different volumes where parameters describing the same system can be found. In the permeable concrete example "permeable" is related to system "concrete" and "deformability" is a parameter linked to "interstitial fluid".

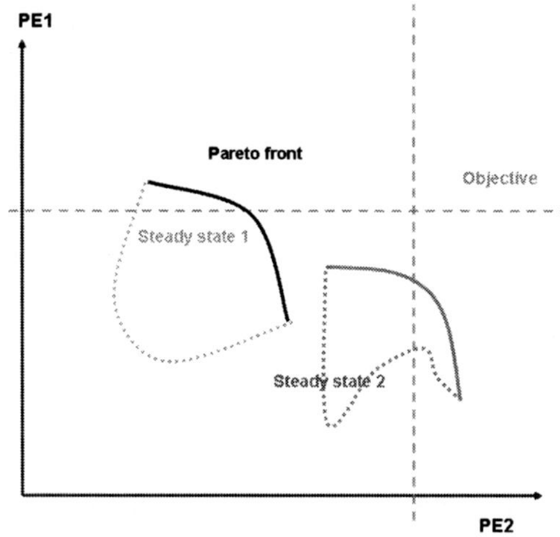

Fig. 5. Discontinuity in design space induced by two steady states

While contradiction is not solved, contradiction model parameters have two different values, i.e. two smiles have different colours on Figure 6 representing each side of the contradiction. For purposes of clarity, the parameters in the mind of the expert that are not related to contradiction statement are not represented.

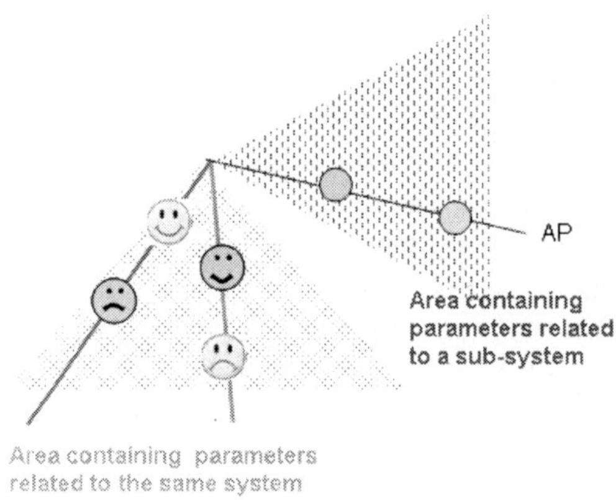

Fig. 6. Representation of a contradiction between two steady states in Hilbert space

Structuring Design Space through Multiple Contradictions Statements

Interconnecting Parsimonious Models Thanks to Systemic Approach

ARIZ proposes ways of complexity reduction thanks to a general structuring of problem that connects various very local (and so low cost) models into a coherent manner without creating unified models. The repetition of the same cognitive pattern in a quasi « automated way » at various system level leads to a precise structuring of various disconnected steady states. This structure minimizes the amount of knowledge considered or to be created during solving process.

A Model Structuring That Enables Travelling through Design Space

New design spaces are opened through changing the system level of problem statement. Contradictions enable travelling through design space. When changing system level for example, explanations are given by jumping into a new steady state (depicted as Pareto diagram), but parameters of the two diagrams are connected. Green (resp. pink) axis is related to each other through system levels on Figure 7.

The requirement of two opposite Evaluation Parameters at the beginning of ARIZ enables to start the search. ARIZ proposes a way to search design parameters of the ideal system by hurting contradictions, which can lead to other interesting contradictions, etc…

By use of several tools and knowledge base, ARIZ provides ways to either solve one or several of these contradictions in the specific design situation or to reformulate contradiction at another more fundamental level, letting new systems eventually appear in the problem description. This process of solution synthesis – reformulation converges to a set of systems, their respective sets of parameters and values, which contains no parameter with two distinct values. The convergence of this process will be further investigated in upcoming papers. In such a process, steady states bridging through establishment of a full new model occurs at the specific system-level where the solution is found. There is no need to bridge knowledge at each system level between the contradictory steady states studied to form a global computable model like in EA.

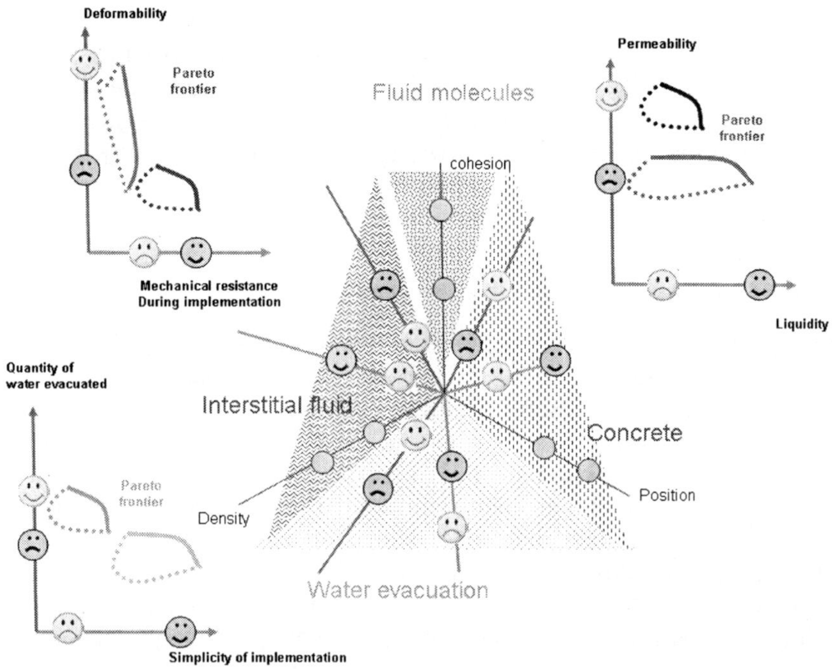

Fig. 7. Example of multi-level contradictions for permeable concrete

Discussion

Spatial Segmentation of Models vs. Propagation of Contradictions through System Levels

Evaluation of Modelling Cost

EA proposes to explore design space in the "sane part", where knowledge and models exists but are too abundant and complex for human brain to manage them. It manipulates knowledge elements in order to see if combinations of them can lead to a solution of design problem. The purpose of the decomposition is to loose the inventive character of the problem and to transform it into an optimization one. However, it requires modelling a lot and some tacit creativity in the segmentation in order to segment and dynamize models in a relevant manner. Evolution strategy remains a costly process where lots of steady states need to be explored (and mathematically described to be able of an evaluation).

TRIZ-based decomposition approach proposed in this paper explores design space in its “ill parts” and propagates the search of solution through discontinuities in models, like a material scientist using diffraction. In this second approach the propagation of contradiction through various systems limits the need of modelling activities by pointing directly towards interesting analytical decompositions. So contradiction statement participates to reduction of the search area when problem solving.

Towards Quantitative Estimates in Conceptual Design Stage

Although the present paper studies only one mechanism of analysis useful for inventive design, the new framework proposed enables to introduce quantitative evaluation useful to study convergence in inventive design algorithms. Following above logic, a first comparison criterion for knowledge parsimony can be stated in that form: “how many steady states (that do not solve the problem) should be visited before finding an appropriate steady state”?

This criterion leads to the following question: “Is there a method to straightforward split a problem into parts (analysis) in order to visit a minimum amount of steady states during problem solving?” or, how to build and tune a problem solving heuristics depending on the nature of knowledge available. It is expected that a combination of several mechanisms enables to reduce and control solution space.

Other Mechanisms of Problem Solving Used by the Two Approaches Needs to Be Studied

In order to go towards a full comprehensive framework of modelling activities useful for inventive design, ones still needs to understand the function (in terms of information processed) of other ARIZ tools: among them tools to perform analysis of system and its environment.

Conclusion

In order to investigate the contradiction between flexible parsimonious modelling and quantitative full models, some aspects of knowledge and models manipulated through EA and TRIZ-based method ARIZ have been studied in a Hilbert space framework. It has been explicated the difference between domain-dependant elementary models in EA and domain-independent contradiction pattern, that for each method contribute to structure the design space.

Expected Outcomes from Proposed Modelling Description

What is EA useful for TRIZ?

- Provide quantitative means of contradiction and mini-problem choice.
- Provide means of complex systems analysis, in order to study interaction between elements when expert's knowledge is lacking.
- rapid formulation of multi-level problem statement
- choice of a side of contradiction to be solved

What can TRIZ be EA useful for?

- reduce modelling complexity
- lower modelling work leading to low marginal payoff
- provide ways to split systems in order to enhance emergent properties appearance
- go out of global minima by model change

Details about those assumptions will be provided in further publications.

Acknowledgements

Pierre Collet for explaining us evolutionary algorithms and let us benefit from his skills. Denis Cavallucci, Sébastien Dubois and Dmitry Kucharavy for teaching us TRIZ and how to use ARIZ in practice. Lafarge for supporting this research.

References

1. Pahl, G., Beitz, W., Feldhusen, J., Grote, K.-H.: Engineering Design: A Systematic Approach. Springer, Heidelberg (2007)
2. Kara, L.B., Shimada, K., Marmalefsky, S.D.: An evaluation of user experience with a sketch-based 3D modeling system. Computers & Graphics 31(4), 580–597 (2007)
3. Foucaulta, G., Cuillièrea, J.-C., Françoisa, V., Léonb, J.-C., Maranzanac, R.: Adaptation of CAD model topology for finite element analysis. Computer-Aided Design 40(2), 176–196 (2008)
4. Thakura, A., Banerjeea, A.G., Gupta, S.K.: A survey of CAD model simplification techniques for physics-based simulation applications. Computer-Aided Design 41(2), 65–80 (2009)
5. de Bono, E.: Six Thinking Hats: An Essential Approach to Business Management (1985)

6. Yamamoto, Y., Nakakoji, K., Takadad, S.: Hands-on representations in a two-dimensional space for early stages of design. Knowledge-Based Systems 13(6), 375–384 (2000)
7. Rovira, N.L., Cueva, J.M., Silva, D., Gutierrez, J.: Automatic shape and topology variations in 3D CAD environments for genetic optimization. Interscience Publishers 30, 59–68 (2007)
8. Parmee, I.C.: Evolutionary and Adaptive Computing in Engineering Design. Springer, New York (2001)
9. Parmee, I.C.: Diverse evolutionary search for preliminary whole system design. In: Proceedings of the 4th International Conference on AI in Civil and Structural Engineering, Cambridge University, Cambridge (1995)
10. Huang, Z., Yip-Hoi, D.: High-level feature recognition using feature relationship graphs. Computer-Aided Design 34(8), 561–582 (2002)
11. Cugini, U., Cascini, G., Ugolotti, M.: Enhancing interoperability in the design process – The PROSIT approach. Trends in Computer-Aided Innovation, pp. 189–200 (2007)
12. Holland, J.H.: Adaptation in Natural and Artificial Systems. MIT Press, Cambridge (1992)
13. Algorithme de Résolution des Problèmes d'Invention: ARIZ-85C, © G.S. Altshuller (1956/1985), English Version to be found at `http://www.seecore.org/d/ariz85c_en.pdf`
14. Hatchuel, A., Masson, P.L., Weil, B.: Studying creative design: the contribution of C-K theory. In: Gero, J.S. (ed.) Studying Design Creativity. Springer, Heidelberg (to appear, 2010)
15. Small, P.: A biological way to think about information systems, people and collaboration, `http://www.stigmergicsystems.com/simpleexplain/biopaper2.html` (Last accessed May 2010)
16. Ashby, M.F.: Material selection in mechanical design. Technology and Engineering, p. 603 (2005)
17. Ahuja, R.K., Ergun, Z., Orlin, J.B., Punnen, A.P.: A survey of very large-scale neighborhood search techniques, vol. 123, pp. 75–102. Elsevier Science Publishers, BV (2002)
18. de Bono, E.: Lateral Thinking: Creativity Step by Step (1970)
19. Shayani, H., Bentley, P.J.: A more bio-plausible approach to the evolutionary inference of finite state machines. In: Genetic And Evolutionary Computation Conference. Proceedings of the 2007 GECCO Conference Companion on Genetic and Evolutionary Computation. ACM New York (2007)
20. Eeckhout, L., Bosschere, K.D.: How accurate should early design stage power/performance tools be? In: A case Study with Statistical Simulation, vol. 73, pp. 45–62. Elsevier Science Inc., Amsterdam (2004)
21. Arciszewski, T., Kicinger, R. (eds.): Structural design inspired by nature. Saxe-Coburg Publications, Stirling (2005)
22. Zlotin, B., Zusman, A.: Problems of ARIZ Enhancement (1991)

23. Zlotin, B., Zusman, A.: Managing Innovation Knowledge - The Ideation Approach to the Search, Development, and Utilization of Innovation Knowledge. J. of the Altschuller Institute for TRIZ Studies (1999)
24. Sickafus, E.N. (ed.): Unified Structured Inventive Thinking – How to Invent. Ntelleck, LLC (1997)
25. Horowitz, R.: ASIT, Méthode pour des solutions innovantes (2004)
26. Savransky, S.D.: Engineering of Creativity: Introduction to TRIZ Methodology of Inventive Problem Solving. CRC Press LLC, Boca Raton (2000)
27. Khomenko, N., Guio, R.D., Lelait, L., Kaikov, I.: A framework for OTSM TRIZ based computer support to be used in complex. Problem Management, vol. 30, pp. 88–104. Inderscience Publishers (2007)
28. Khomenko, N., Guio, R.D., Cavallucci, D.: Enhancing ECN's abilities to address inventive strategies using OTSM-TRIZ. International J. of Collaborative Engineering 1(1-2), 98–113 (2009)
29. Martelot, E.L., Bentley, P.J., Lotto, R.B.: A systemic computation platform for the modelling and analysis of processes with natural characteristics. In: Proc. of the 2007 GECCO Cconference Companion on Genetic and Evolutionary Computation. ACM, London (2007)
30. Benyus, J.M.: Biomimicry: Innovation Inspired by Nature. Perennial, HarperCollins (1998)
31. Koza, J.R.: Genetic Programming: On the Programming of Computers by Means of Natural Selection (1992)
32. Koza, J.R., Keane, M.A., Streeter, M.J., Mydlowec, W., Yu, J., Lanza, G.: Genetic Programming IV: Routine Human-Competitive Machine Intelligence. Springer, Heidelberg (2003)
33. Shaw, D., Miles, J., Gray, A.: Genetic Programming Within Civil Engineering. In: Adaptive Computing in Design and Manufacture Conference. Engineers House, Clifton, Bristol, UK (2004)
34. Perez, J.L., Monica, M., Rabuñal, J.R., Abella, F.M.: Applying Genetic Programming to Civil Engineering in the Improvement of Models, Codes and Norms. In: Proceedings of the 11th Ibero-American conference on AI: Advances in Artificial Intelligence, Springer, Lisbon (2008)
35. Rafal, K., Tomasz, A., Kenneth De, J.: Evolutionary computation and structural design. A survey of the state-of-the-art 83, 1943–1978 (2005)
36. Bentley, P.J., Corne, D.W.: Introduction to creative evolutionary systems, Creative evolutionary systems, pp. 1–75. Morgan Kaufmann Publishers Inc., San Francisco (2002)
37. Altshuller, G.S.: The Innovation Algorithm: TRIZ, systematic innovation and technical creativity (Technical Innovation Center, Inc. ed.), Worcester, MA (1999)
38. Hermetz, J., Clément, J.: L'optimisation multidisciplinaire, Maîtrise de l'optimisation (ONERA ed)
39. Goldberg, D.E.: The Design of Innovation: Lessons from and for Competent Genetic Algorithms. Kluwer, Dordrecht (2002)

Exploring Multiple Solutions and Multiple Analogies to Support Innovative Design

Apeksha Gadwal and Julie Linsey
Texas A & M University, USA

Idea generation and design-by-analogy is a core part of design. Designers need tools to assist them in developing creative and innovative ideas. Multiple solutions can be developed based on single analog and designers derive principles of design from the analogs (products) they experience. There is little research that discusses creating multiple solutions from a single analog or how multiple analogs can assist designers in mapping high level principles of design. This study explores two phases of design-by-analogy in which designers have difficulty, generating multiple inferences from a single source analog and the identification of high level principles given multiple example analogs in the presence of noise. Two hypotheses are proposed to explore the importance of analogies in design. 1. Multiple solutions can be generated from a single analog. 2. The mapping of high level principles increases with the increase in the number of example analogs and decreases with the amount of noise. The paper presents two laboratory experiments, "Multiple Solutions" and "Multiple Analogies" conducted to answer the proposed research questions and to understand how designers can become better analogical reasoners. The experiments are explained in detail with the methods for collecting data, metrics and the analysis. The results from the pilot experiments show that engineers, when directed to, can create multiple solutions from a single analog. This can allow designers to find better solutions and to evaluate their inferences. Results from the second experiment also indicate the mapping of high level principles increases with an increase in the number of analogs and decreases with distracters. A significant interaction is also observed between these two factors. The results indicate more future work with a greater sample size.

J.S. Gero (ed.): *Design Computing and Cognition'10*, pp. 209–227.

Introduction

Innovation is what drives new product development and engineering. Although some believe creativity cannot be invoked on demand, the presentation of appropriate stimuli greatly enhances the generation of concepts [1]. Analogy is one type of appropriate stimuli that aids in generating new ideas [2-6]. Analogy acts as a stimulus to generate new concepts and solve design problems. It can trigger breakthrough ideas [7]. Using an analogy to solve design problems helps generate innovative and creative solutions. There have been many instances where analogies have led to breakthrough innovations, for example Velcro was based on a burr and the Speedo swimsuit was inspired from shark skin.

Design problem solving is an integral component of engineering and understanding how engineering designers store and retrieve knowledge during the design process is very important. Casakin & Goldschmidt, Christensen & Schunn, demonstrate that designers frequently retrieve and use solutions from analogous designs to help them create innovative solutions to new problems. Although there have been many studies on analogies in design [7-10], there is more to explore in this filed. There are still many unanswered questions related to analogical design.

There are a number of cases where a single analog has inspired multiple innovations and often for the same problem (Figure 1). A Gecko's foot is the source analog and the different solutions developed based on this analog are a wall climbing robot, magnetic grips and tires. All these solutions solve the same design problem, i.e. a wall climbing device. So it is evident that a single analog can inspire multiple solutions but it is unclear how well a designer can derive multiple inferences. This observation leads to the first research question of this study: Can Multiple Solutions be drawn from a single source analog?

This paper explores research questions on Multiple Inferences and Multiple Analogies. Two separate experiments are conducted for this purpose. In the Multiple Solutions experiment subjects were given a design problem with a corresponding analog and asked to generate multiple solutions based on the analog. Prior work shows that creating more than one solution from a single analog is cognitively difficult [11] and participants do not typically attempt it [12]. The Multiple Analogies experiment tested the effect of multiple analogs and noise on high level principle mapping. The participants were given a set of products and a design problem. They were then asked to generate ideas for the design problem. The following sections first describe the foundations in the prior literature and then explain both the experiments and the results in detail.

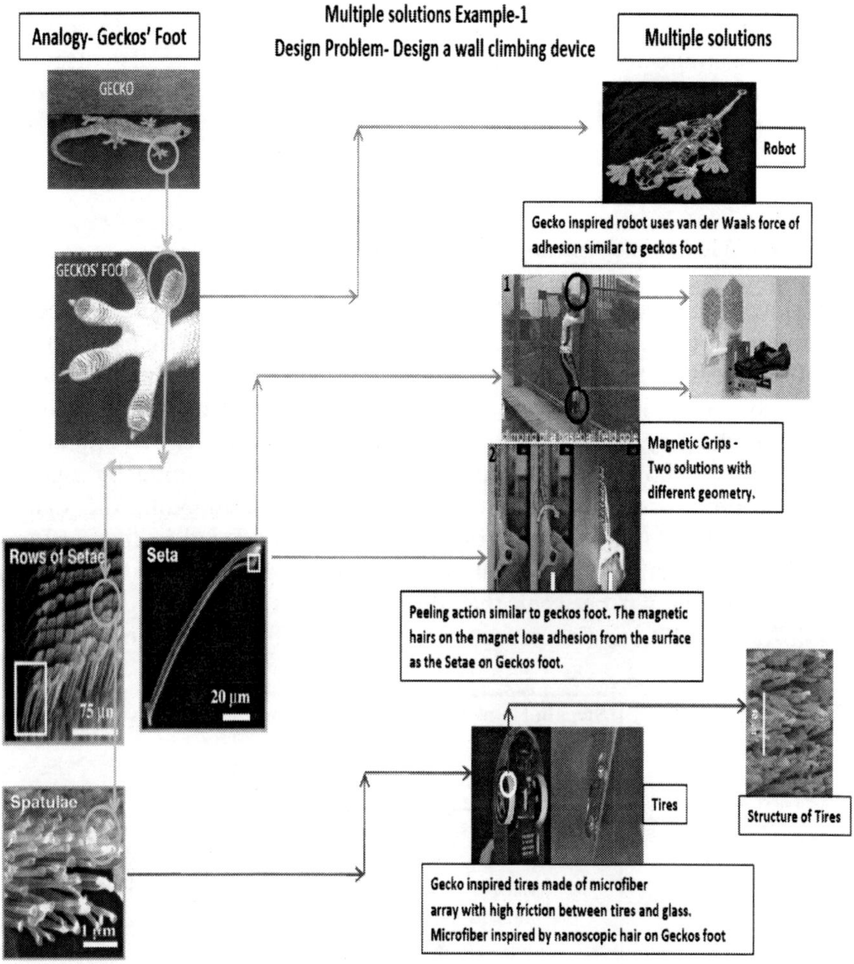

Fig. 1. Multiple Solutions example from a single source analog "Gecko's Foot"

Background

Analogical Reasoning as Basis for Innovative Design

Innovations resulting from analogies are ubiquitous in current trade journals, magazines and product offerings. Design-by-analogy is a powerful tool in creative design. The anecdotal and empirical evidence consistently demonstrates that professional designers often use analogies and it is a powerful tool for innovation [2, 4, 5]. Design teams also frequently use close-domain analogies in the form of references to past designs [13]. Eckert, *et al.* found designers use references to previous designs for more than just conceptual design. Designers also use past designs in a number of other phases of the design process including process planning, cost estimation, and evaluation of concepts for a new product [13].

Cognitive Models of Analogical Reasoning

Analogical reasoning is a widely used tool in design cognition. Analogical reasoning is a general human capacity [14] involved in many domains, mostly in creative fields like design, art and science [15]. "Analogy can be viewed as a mapping of knowledge from one situation to another enabled by a supporting system of relations or representations between situations" [16]. Figure 2 shows the various steps involved in reasoning by analogy [17]. The process of analogical reasoning starts with encoding the source. All the details or information from the source is encoded in memory and then, at a later time, suitable analogs are retrieved. Then relationships or mappings are drawn between the source and the target (design problem) and different solutions (inferences) are developed for the design problem.

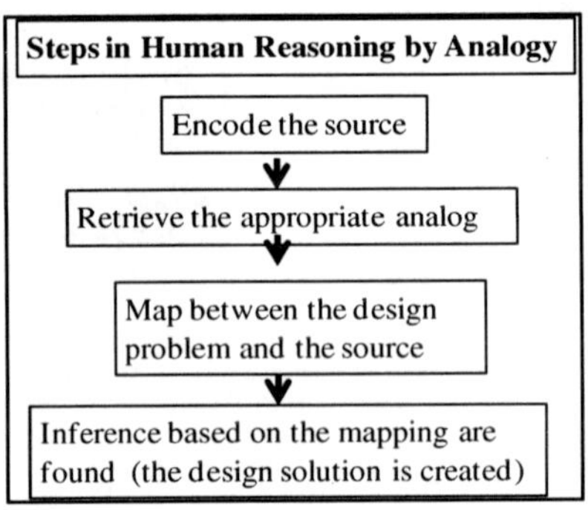

Fig. 2. Steps in reasoning by analogy

Structure Mapping Theory and Structural Alignment

Structural alignment is a more detailed description of the underlying process associated with models of analogical reasoning [16, 18, 19]. So in structural alignment the target domain is compared with the base domain based on their relational structure.

The structural alignment view of analogy shows an alignment of relational structure between the base and target domains. This alignment has three constraints to satisfy. 1) Structurally consistent: This implies that the alignment should have parallel connectivity i.e. it should be a one-to-one mapping between the base and target [19], 2) Relational Focus: This means that analogies should have relations in common between the base and target but need not have surface relations. 3) Systematicity: Analogies tend to match connected systems of relations [16, 18]. The structural consistency constraint and relational focus constraint are important for this present study as they talk about maintaining a parallel mapping between the source and target and focus on using relational features to map between the source and target. "A matching set of relations interconnected by higher order constraining relations makes a better analogical match than an equal number of matching relations that are unconnected to each other" [17]. Figure 3 shows structural alignment applied to an example. This figure is used for the purpose of illustration of the concept of structural alignment between existing solutions and the gecko. To create new solutions, designers structurally align the analog with the designer problem (not the solutions) and then inferences are drawn creating new solutions.

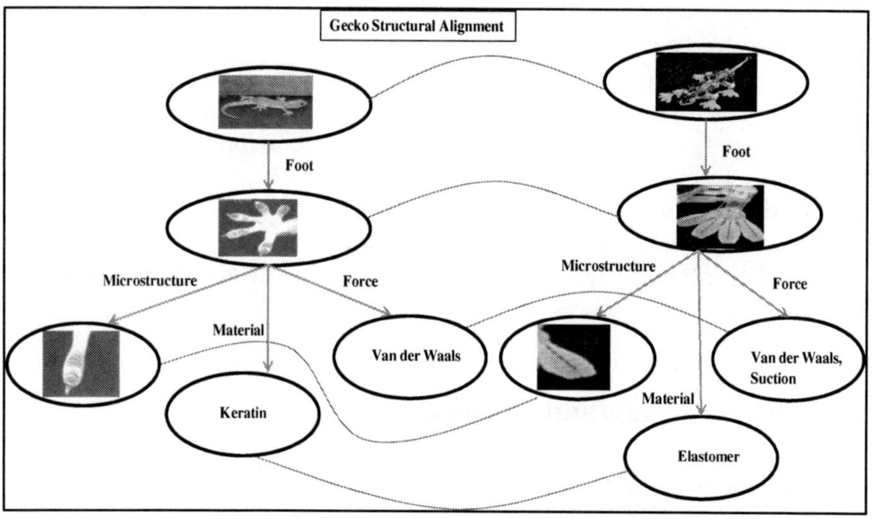

Fig. 3. Example for illustration only- Structural alignment process illustrated with a Gecko's foot and existing solutions. This does not illustrate new solution creation, only structural alignment between an analog and existing solutions

One-to-One Mapping Constraint

The fundamental purpose of analogy is to generate plausible and useful inferences [11]. In order to obtain useful inferences from analogical reasoning, analogical mappings have to be constrained, otherwise too many inferences are possible [11]. From structural alignment theory, one of the constraints of relational alignment is structural consistency. The alignment has to be parallel implying that only one-to-one mapping has to exist between the base and target domain. For example if we consider the above figure showing the relation structure for a Gecko's foot, the mappings between the source (Gecko's foot) and the target (Wall climbing robot) are parallel. There is only one feature or characteristic mapped between the source and the target. In Holyoak's theory the basic structural constraint is isomorphism. Isomorphism means finding structurally consistent mappings i.e. one-to-one mappings and then map elements as one-to-one correspondences [19]. This theory says that people generate multiple mappings (homomorphic mapping) but find it hard to generate multiple inferences. The one-to-one constraint is because inferences derived using a mix of incompatible element mappings are likely to be incoherent [18, 20, 21].

There have been theories in analogical reasoning which have assumed that one-to-one constraint discourages one-to-many mappings from base to target domain. But it has been found that even though people map one-to-many elements, they find it difficult to generate more than one inference. Multiple correspondences appear to arise from multiple isomorphic mappings, rather than from a single homomorphic mapping [11].

As the structural consistency requirement and one-to–one constraint restrict the use of analogies to draw multiple solutions, this leads to the first research question on generating more than one inference from a single analog. The research questions are described in detail in the following section.

Research Questions and Hypotheses

Based on the background, the following research questions are posed:

- Can engineers draw multiple solutions (inferences) from a single analog? If so, what types of inferences are typically made from the analog? What inferences are more and less likely to occur? Are relational (functional) or surface features more likely to be mapped?
- What causes the high level principles of an analog more likely to be mapped while in the presence of noise products?

Based on the prior literature and to investigate these questions three hypotheses are proposed.

Hypothesis 1: A lone designer is able to generate multiple inferences from a single source analog when instructed to do so.

This hypothesis is based on Holyoak's One-to-One Constraint in Analogical Mapping and Inference Theory [11]. The constraint theory restricts the mapping of one element from an analog in the base to multiple elements in the target. In this study people occasionally generate multiple mappings but find it hard to generate multiple inferences. Based on the one-to-one constraint, it is possible to generate multiple inferences but cognitively difficult. People generally generate multiple inferences from multiple isomorphic mappings rather than a single homomorphic mapping. In contrast for the field of design, more than one inference can be drawn from a single analog. The example of a Gecko's Foot shows that multiple products have been developed and the ability to generate multiple inferences is very useful. In order to see if engineers can generate multiple mappings when instructed to do so and how many they are able to generate, this study is conducted.

Hypothesis 2: The identification and mapping of the high level principle increases with multiple source analogs and decreases with the presence of noise products.

Hypothesis 3: The effects of multiple analogs will depend on the number of noise products present. Statistically, the number of source analogs and the number of noise products will interact to predict the identification and mapping of the high level principle.

Two experiments are conducted to evaluate the hypothesis and the results are analyzed.

Experiment 1- Generating Multiple Solutions

Overview

A single condition, requiring one hour, was conducted to evaluate the proposed hypothesis that Multiple Solutions can be generated from a single source analog. The participants were given a design problem and asked to generate Multiple Solutions based on the analog. Next, a feature listing task required participants to list the features they used from the analog to generate solutions. This task was conducted to identify patterns in the type and frequency of features used.

Method

Participants

The participants were graduate and undergraduate mechanical engineering students at Texas A & M University, seven graduates and one undergraduate. The undergraduate student was recruited from a Mechanical engineering senior design class and the graduate students were recruited through posted flyers. All the participants received paid compensation.

Procedure

Training

The participants were taught what Multiple Solutions were by being shown examples of Multiple Solutions, Figures 1 and 4. Participants received printed handouts with a definition of analogy, "Analogy in engineering design is used as a tool to solve design problems. Use of an analogy to solve design problems gives innovative solutions."

The next part explained the definition of Multiple Solutions, their use in engineering design and presented a detailed explanation of the examples. The definition was "Multiple solutions mean finding more than one solution from a given analogy by using various features from the analogy. Multiple solutions are very useful in the process of design as there would be more than one solution available for a design problem and hence designers can select the best solution from the various options available". The Gecko's foot example was developed from a literature review of innovations based on the gecko [22-25]. The Air Mattress example was developed based on the participant solutions from a previous analogy study [12, 26]. The entire training lasted ten minutes.

Idea Generation

The training session was followed by the idea generation task. In this task, the participants were given an analog and design problem Figures 5 and Figure 6. The design problem was used in a prior analogy experiment and the participants in that experiment did not spontaneously generate multiple solutions [12, 26]. The participants were then asked to generate Multiple Solutions based on the analog and to describe their ideas using sketches and/or words on 11" by 17" paper.

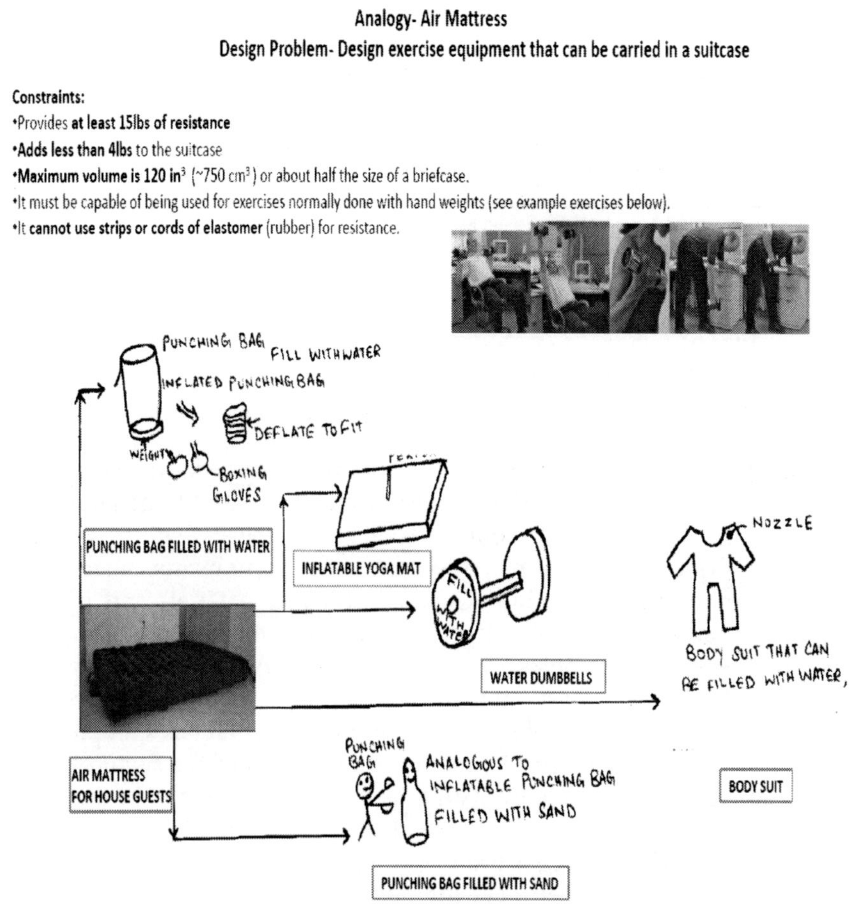

Fig. 4. Multiple Solutions example-2 "Air Mattress"

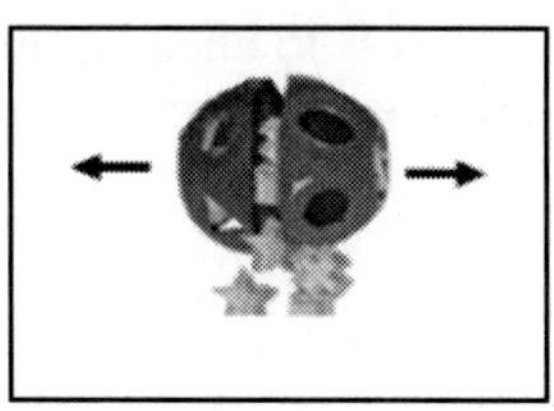

Fig. 5. Analog

The idea generation task lasted thirty minutes. Multiple colors of pens were changed at five, ten and twenty minutes to record when ideas were generated. A five minute warning was given before the end of the activity.

Design Problem:

Design a kitchen utensil to sprinkle flour over a counter.

- **The only material that is available to build the kitchen utensil from is various thickness of stainless steel wire.**
- **The entire kitchen utensil must be made from only one thickness of wire.**
- **The kitchen utensil must be manufactured by bending and cutting the wire only.**
- **The kitchen utensil must be capable of containing the powdered substance and carrying the powdered substance 1 meter without losing the powdered substance.**

Create as many solutions as possible based on the analogy.

Fig. 6. Design problem - flour sifter

Feature Listing

The next task was feature listing. In the feature listing task, the participants had to list features from the given analog which they used to generate their solutions. They were also required to identify these features on their solutions. Prior to beginning this task, the participants were given examples of features including Geometry/Shape, material, and physical principles such as Van der Waals force, friction, and adhesion. They were also given specific examples of features for the Air Mattress, Figure 4, such as inflate/deflate, easy storage and use of available substance. The purpose of this task was to identify the patterns in the features, i.e. the type of features (functional or surface) and the frequency of the features being used. The results of the feature listing task are not presented in this paper and will be used in future work to evaluate specific predictions made by the One-to-One Mapping Constraint.

At the end of the experiment the participants were asked to fill out a survey and participate in a five minute interview. The survey measured demographic information and previous design experience. In the interview the participants were asked few questions regarding the experiment. The main purpose of the interview was to check if all the instructions, material and tasks were clear. The entire experiment lasted for fifty minutes.

Metrics

The results from a previous analogy study showed that participants did not create Multiple Solutions when asked to generate ideas for the same design problem and analog as in this experiment [12]. So the main aim of this experiment was to see whether Multiple Solutions could be created from a single analog and how easy it was for the participants to do this. In order to evaluate this, the number of ideas each participant generated based on the source analog was measured.

Results and Discussion

Hypothesis: A lone designer is able to generate multiple inferences from a single source analog when instructed to do so.

The data from the eight participants was analyzed and showed the average number of Multiple Solutions generated was three, Figure 7. This is consistent with Holyoak's theory and shows that when instructed to do so participants can create Multiple Solutions based on a single analog. Given the average was only three ideas in a half hour; this is clearly a very difficult task for which engineers need support.

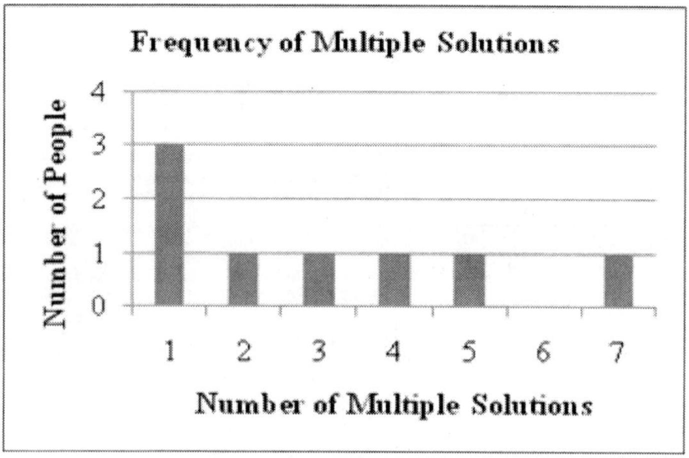

Fig. 7. Frequency of multiple solutions generated by participants.

Table 1 Summary of results for the Multiple Inferences Experiment

Total number of ideas based on functional features	17
Average number of Multiple Solutions per participants based on functional features	2
Total number of ideas based on surface features	7
Average number of Multiple Solutions based on surface features	1

Experiment 2- Learning Design Principles from Multiple Analogs

Overview

To further explore the analogical reasoning process, a between-subjects factorial experiment evaluated the effects of Multiple Analogs (one or five) and amount of noise (none or 3 noise products per analog) on idea generation and the mapping of a high level design principle (energy storage through elastic material deformation). Participants were given a design problem, a set of products, Figure 8, and then asked to generate solutions.

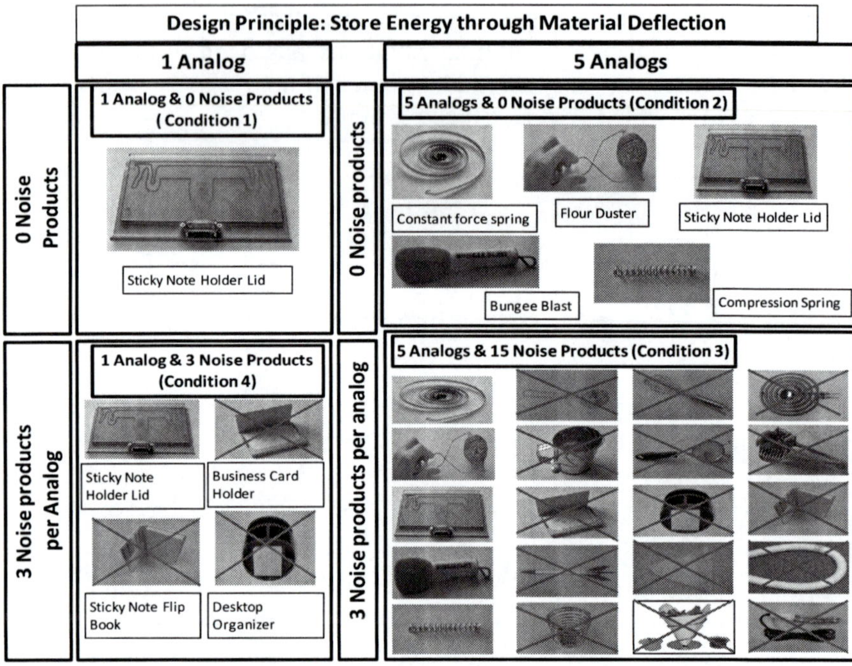

Fig. 8. Analogs and associated noise products

Method

Participants

The participants in this experiment were undergraduate and graduate mechanical engineering students from Texas A & M University. There were a total of thirty-four senior undergraduate students and thirteen graduate students for the two hour study. There were forty-one male and six female participants with an average of eight months engineering work experience. Twelve graduate and thirty-four undergraduate students were recruited from their mechanical engineering design classes. One graduate student was recruited through posted flyers. There were twelve students per condition except in condition four which had eleven students. There were three graduate students in each condition, except in condition three which had four. Most of the participants received class credit as compensation. Two undergraduate and one graduate student received paid compensation.

Design problem

The participants were given a printout of the design problem along with a hand sketch (Figure 9 and Figure 10).

NASA astronauts are on a mission to Mars and a critical component has broken down; "Door Pin Lock" as shown in the handout. NASA engineers are anticipating this situation. They want to design features into the parts ahead of time allowing astronauts multiple avenues to provide temporary solutions to this problem.

NASA is looking for innovative solutions to fix this problem. So, your task is to provide a temporary fix to this problem satisfying the following condition.

- The door pin must automatically return to the locked position even when there is no electricity.

Since the parts are still being designed, you can add or remove features to the parts.

NASA will send supplies to the space station with the astronauts. The supplies will consist of a wide variety of materials and tools but NASA has not decided what materials and tools will be needed for this problem. It costs them millions of dollars per pound, so they want to send as little material as possible. So your solutions will help to determine what supplies to send.

Constraints:

- You cannot use a metal coil spring. NASA is aware of this solution and needs others.

Your task is to design a temporary mechanism to move the pin back to the locked position.

Generate as many solutions as possible for the given design problem.

Fig. 9. Design problem for the Multiple Analogies Experiment

Procedure

The overall procedure was same for all the conditions. The products and distracters (noise products) were described and demonstrated briefly before the idea generation. The participants were given a set of products according to their condition and then asked to generate ideas for the design problem for forty minutes. They were instructed that the given products may or may not help them to generate ideas. The participants were asked to sketch and/or use words to describe their ideas and to describe one idea on each sheet of paper.

Again, different colors of pen were used. Pen colors were changed every 10 minutes and also after the first 5 and 10 minutes. The aim of pen change was to keep track of when the ideas were generated.

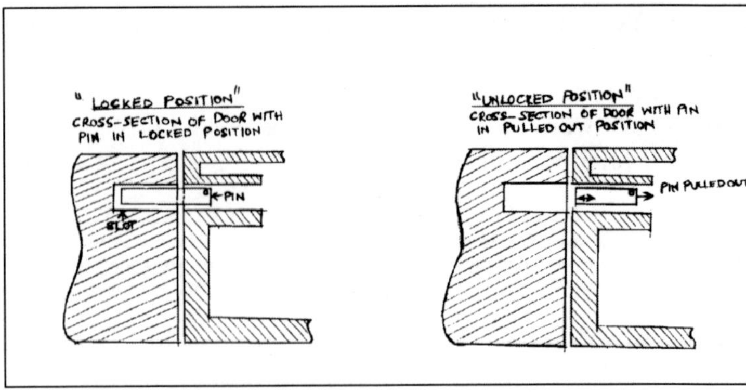

Fig. 10. Design problem sketch for Multiple Analogies Experiment

In the next task the participants were given an example of an analogy, Velcro being based on a burr, and they were asked to cross out the ideas that did not use the given products as analogies. This was followed by a feature listing and similarity rating task. In this task, the participants were asked to list all the features that they had used from the given products to generate each of the ideas. The details of this task are not presented here nor are the results. This task lasted a total of thirty minutes.

The next two tasks were to identify the high level principle from the given products. In the first stage, requiring five minutes, the participants were asked to identify the high level principle that many of the given products shared and mark their ideas with a star that used this principle. In the second stage, requiring ten minutes, participants were told which products shared the same high level principle which solves the design problem and then asked to list the principle. They also circled any ideas already generated based on the principle and then generated more ideas for the design problem.

At the end of the experiment, the participants were asked to fill out a survey and participate in a five minute interview asking questions about their demographic information and previous design experience. The purpose of the interview was to check if all the instructions and activities were clear. The entire experiment lasted for one hour and fifty minutes.

Metrics

To evaluate the hypothesis that the mapping of high level principle increases with an increase in the number of products and decreases with an increase of noise, identification of the high level principle was used as the main metric. All the four conditions were compared with the corresponding input and the output was evaluated.

Results and Discussion

Hypothesis: The mapping of high level principle increases with an increase in the number of products and decreases with the amount of noise.

There was a clear increase in the mapping of the high level principle when the number of products was increased from one to five, Figure 11 and Figure 12. Results are shown for the first stage of principle identification for all conditions. There was also a significant decrease in the mapping of the high level principle when noise products were presented. From Figure 11 there is a clear interaction between the number of products and number of distracter products. It is interesting to note that five examples are better than one, even when those five examples are in the presence of noise (distracter products) and the one example is presented alone.

These results are statistically significant. The outcome variable is binary (yes or no) so a logistic regression model is appropriate. A two predictor model (analogs and noise products) shows one of the factors to be statistically significant (analogs: Wald= 6.085, $p = 0.014$, distracters: Wald= 1.941, $p= 0.164$). A quasi complete separation is observed in the statistical test, so the interaction effect is not considered. The line graph above shows a clear interaction between the analogs and distracters which implies that the distracters also have an effect in predicting the high level principle mapped.

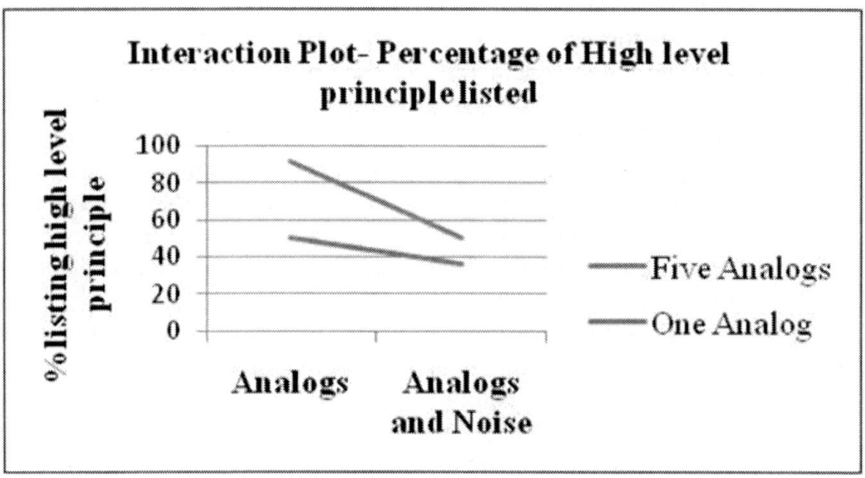

Fig. 11. Interaction plot showing percentage of high level principle listed in each of the four experimental conditions

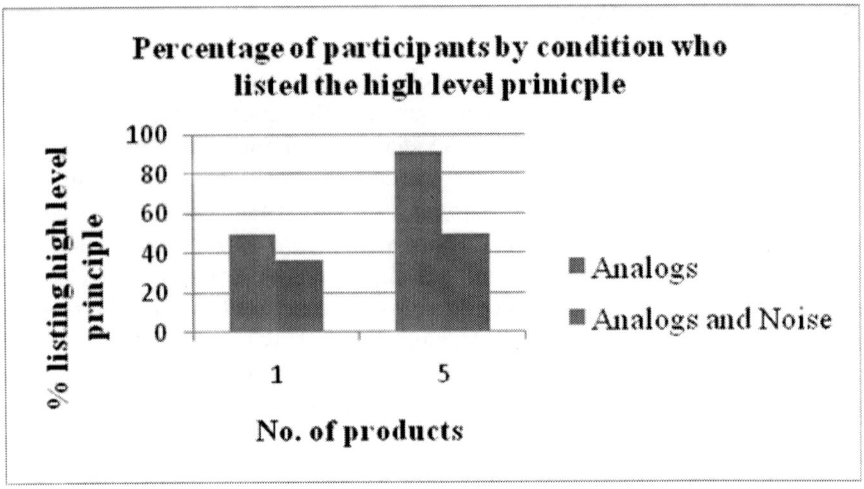

Fig. 12. Percentage of participants in each condition who were able to list the high level principle from the products

Conclusion

Designers innovate through analogy and numerous opportunities for enhancing this process are present. Often for a given design problem, a single analog provides multiple paths for solutions. It is often difficult to generate multiple solutions but this provides the opportunity to identify the optimal solution and evaluate the analogical inferences. In addition, designers frequently derive principles of design from sets of analogs. This study explores both of these aspects of design-by-analogy to provide a basis for improved design methods and tools to support the design process.

The use of analogies aids engineering idea generation process. The results from Experiment 1 indicate that it is possible to generate more than one solution from a single source analogy. So, this would help designers have multiple solutions for the design problem and they could select the best possible solution available from the various options. The other preliminary result from Experiment 1 showed that more functional features are mapped from the analogy than surface features. This supports the definition of an analogy- "Analogy is a comparison between two items in which their relational, or causal structure, but not the superficial attributes match [17] So, we can state that analogy aids in generating Multiple Solutions from a single source.

The results from the second experiment also show the importance of analogies in design. Analogies help to solve design problems. The results from the second experiment indicate that the use of multiple analogies helps designers identify the high level principle that can be used for solving given design problems and it also shows the effect of noise on the idea generation process. Noise in presence of analogies distracts the designers and negatively affects the mapping of the high level principle.

A bigger sample size and more work needs to be done on the Multiple Solutions experiment to make a strong conclusion.

Therefore, from the preliminary results of both the experiments it is clear that analogy acts as a tool to solve design problems during idea generation process and hence design by analogy is important in the process of concept generation.

Future Work

Future work will focus on developing approaches and methods to assist designers in identifying multiple solutions and high level principles of design. Further exploration of the data and additional experiments will also be completed. From the results of experiment 1, it is clear that further study is required on the Multiple Solutions experiment. A sample size of at least 20 is needed to make a proper conclusion of the results from the pilot studies and to find out whether or not Multiple Solutions can be created from a single source analog. Also we need to find out if this method can be used universally given any design problem and a corresponding analog.

The second experiment is planned as a 3X4 factorial experiment (products=1, 2, 3, 5 and distracter ratio=0, 1, 3). As an initial experiment, only four of the twelve conditions were run. Also the data from the follow up activities of feature listing and similarity rating needs to be analyzed.

Acknowledgements

Partial support for this work was provided by the Aggie 100 Scholars Program. We would like to thank members of the IDREEM lab at Texas A&M University, especially Ms. Kelsey Brawner for their support throughout this work. We would also like to thank all the students who participated in the experiments.

References

1. Mak, T., Shu, L.: Abstraction of biological analogies for design. CIRP Annals-Manufacturing Technology 53(1), 117–120 (2004)
2. Leclercq, P., Heylighen, A.: 5,8 analogies per hour. In: Gero, J.S. (ed.) Artificial Intelligence in Design 2002, pp. 285–303. Springer, Heidelberg (2002)

3. Ahmed, S., Christensen, B.T.: An in situ study of analogical reasoning in novice and experience design engineers. J. of Mechanical Design 131(11) (2009)
4. Christensen, B.T., Schunn, C.: The relationship of analogical distance to analogical function and pre-inventive structures: The case of engineering design. Memory & Cognition 35(1), 29–38 (2007)
5. Casakin, H., Goldschmidt, G.: Expertise and the use of visual analogy: implications for design education. Design Studies 20(2), 153–175 (1999)
6. Dunbar, K.: How scientists think: On-line creativity and conceptual change in science. In: Ward, T.B., Smith, S.M., Vaid, J. (eds.) Creative Thought: An investigation of Conceptual Structures and Processes. American Psychological Association, Washington (1997)
7. Herstatt, C., Kalogerakis, K.: How to use analogies for breakthrough innovations. International Journal of Innovation and Technology Management 2(3) (2005)
8. Kalogerakis, K., Herstatt, C., Lüthje, C.: Generating innovations through analogies- an empirical investigation of knowledge brokers. Technische Universität Harburg (2005)
9. Linsey, J., Wood, K., Markman, A.: Modality and representation in analogy. AI EDAM 22(02), 85–100 (2008)
10. Daugherty, J., Mentzer, N.: Analogical reasoning in the engineering design process and technology education applications. Journal of Technology Education 19(2) (2008)
11. Krawczyk, D., Holyoak, K., Hummel, J.: The one-to-one constraint in analogical mapping and inference. Cognitive Science 29(5), 797–806 (2005)
12. Gadwal, A.: Exploring two phases of design-by-analogy Multiple Solutions and Multiple Analogies. In: Mechanical Engineering, Texas A& M University, College Station, p. 70 (2010)
13. Eckert, C.M., Stacey, M., Earl, C.: References to past designs. In: Gero, J.S., Bonnardel, N. (eds.) Studying Designers 2005. Key Centre of Design Computing and Cognition, pp. 3–21. University of Sydney (2005)
14. Holyoak, K., Thagard, P.: Mental Leaps: Analogy in Creative Thought. MIT Press, Cambridge (1996)
15. Christensen, B., Schunn, C.: The relationship of analogical distance to analogical function and preinventive structure: The case of engineering design. Memory and Cognition 35(1), 29 (2007)
16. Gentner, D.: Structure-mapping: A theoretical framework for analogy. Cognitive Science 7(2), 155–170 (1983)
17. Gentner, D., Markman, A.: Structure mapping in analogy and similarity. American Psychologist 52, 45–56 (1997)
18. Falkenhainer, B., Forbus, K., Gentner, D.: The structure-mapping engine: Algorithm and examples. Artificial intelligence 41(1), 1–63 (1989)
19. Holyoak, K., Thagard, P.: Analogical mapping by constraint satisfaction. Cognitive Science 13(3), 295–355 (1989)

20. Gentner, D.: Are scientific analogies metaphors? Bolt Beranek and Newman, Cambridge, MA (1981)
21. Markman, A., Gentner, D.: The effects of alignability on memory. Psychological Science 8(5), 363–367 (1997)
22. Autumn, K., et al.: Evidence for van der Waals adhesion in gecko setae. Proceedings of the National Academy of Sciences 99(19), 12252–12256 (2002)
23. Campolo, D., Jones, S., Fearing, R.: Fabrication of gecko foot-hair like nano structures and adhesion to random rough surfaces (2003)
24. Sitti, M., Fearing, R.: Synthetic gecko foot-hair micro/nano-structures for future wall-climbing robots (2003)
25. Menon, C., Murphy, M., Sitti, M.: Gecko inspired surface climbing robots (2004)
26. Linsey, J.: Design-by-analogy and representation in innovative engineering concept generation. Mechanical Engineering, The University of Texas at Austin, Austin (2007)

Creative and Inventive Design Support System: Systematic Approach and Evaluation Using Quality Engineering

Hiroshi Hasegawa[1], Yuki Sonoda[2], Mika Tsukamoto[1], and Yusuke Sato[3]
[1]Shibaura Institute of Technology, Japan
[2]NTT DATA, Japan
[3]Ricoh IT Solutions, Japan

In recent years, designers have been required to design creative products in a short period of time. In addition, the problem definition and conceptual design phases significantly change the direction of product attraction. These phases depend heavily on the creativity of the designer or development team and dictate the quality of the product concept. In this situation, it is difficult for designers to quickly create a host of solutions without some kind of support for their creative thinking processes. To overcome this difficulty, we used a systematic approach to develop the thinking process of the Creative and inventive Design Support System (CDSS). Moreover, its validity was examined by using the design of experiments (DOE) and the Taguchi method (TM). These evaluation results confirmed that creative design solutions can be explored using the support of the CDSS thinking process.

Introduction

The problem definition and conceptual design phases significantly change the direction of product attraction. These phases depend heavily on the creativity of the designer or development team and dictate the quality of the product concept. Additionally, it is necessary to shorten the product development cycle. Moreover, to find one big-hit product, it is necessary to explore over 3,000 ideas [1]. In this situation, it is difficult for designers to quickly create a host of solutions without some kind of support for their creative thinking processes.

J.S. Gero (ed.): *Design Computing and Cognition'10*, pp. 229–248.

Support methodologies for the creative thinking process have been studied by many researchers in various application areas. In particular, computer-aided engineering design, which was the target of our research, is classified broadly into two types. The first is a support system based on utilizing systematized and integrated design knowledge. The other type is a system that supports the thinking process for exploring design solutions. The former includes the study of the framework for the systematization of functional knowledge based on ontological engineering [2], the fundamental strategy of the Universal Abduction Studio (UAS) [3], etc. Our study belongs to the latter, which includes the theory of inventive problem solving (TIPS) [4-12] and the engineering design methodology (EDM) [13-18].

In TIPS, G. S. Altshuller's TRIZ (referred to as Classical TRIZ) is the most prominent theory [4]. He defines TIPS, not as "*Problem Solving by Trade-Off*," but as "*Contradiction Solving*" (it is used to solve the contradictions that worsen other attributes when an attempt is made to improve a particular attribute). Classical TRIZ develops a design solution using problem solving techniques such as the principle of invention and the contradiction solving matrix, using ARIZ (Algorithm for Inventive Problem Solving) [5], [6]. However, ARIZ is often said to be a complicated and a difficult process that only a TRIZ specialist can use [7]. To improve this problem, Systematic Innovation (referred to as Mann's TRIZ) [8] and USIT (Unified Structured Inventive Thinking) [9] have been proposed. In contrast to classical TRIZ, Mann's TRIZ introduced many problem definition and solving methods from the point of view that "*90% of a problem is defining what the problem is*" [8]. However, since this theory expanded the range of choices for the problem solving method, a specific model is needed for every method. As a result, it may easily fall into the situation of a "*Paradox of More Models with Less Effectiveness*" [11]. On the other hand, unlike Mann's TRIZ, USIT accessibly simplifies its thinking process by using an object, an attribute, and a function as basic analysis elements for the problem analysis, and introducing the closed world and qualitative change in the problem characteristic conditions of SIT (Structured Inventive Thinking) as a sufficient condition. Its theory makes it easy to understand the path of the thinking process. Nevertheless, beyond the designer's knowledge and know-how, support methods—for example, a principle of invention, or the effects of physics and chemistry —for combining with other knowledge do not exist in USIT.

In EDM, Pahl *et al.* proposed a systematic approach (referred to as the P&B method) [13], and Suh proposed a thinking process that uses an axiomatic theory as Axiomatic Design [14], [15]. Additionally, Shai et al. proposed Infused Creativity, which applies the graph theory and solves problems on the generalized problem domain [16-18].

The process of the P&B method determines the function structure of the target problem, explores a physical function to realize the lowest level of the function (i.e., the design solution principle), combines this adequately, and then draws on the overall function. This process requires the principle knowledge in combination with other knowledge, along with contradiction solving. However, methods for supporting these necessary items do not exist, and all of the items depend on the designer's capability. This makes it difficult, which has also been pointed out in the case study of Kawamo et al. using the P&B method [19]. On the other hand, Axiomatic Design and Infused Creativity solve a problem through abstracting it by using the axiomatic or graph theory, which simplifies these processes compared to the P&B method. However, these also do not support a thinking process for exploring functions and solving contradictions. Therefore, as with the P&B method, it is dependent on the designer's capability.

In this study, the creative thinking process is defined as the creation of a new combination of existing elements (functions or technologies), which is believed to have an edge on others, by a user (a designer or development team) [7]. It needs to mobilize the user's knowledge and know-how, as well as other things beyond them. The solution that results from this process is defined as a creative design solution. Moreover, this process must be supported without depending on the user's attributes or knowledge. Therefore, the common generalized items, which are capabilities required for this process are as follows: A capability to (1) obtain a creative design solution, (2) widely mobilize other elements beyond a user's knowledge and know-how, and (3) provide a systematic approach that beginners can use to obtain a creative design solution. However, item (3) has not been satisfied since the problem solving procedure of classical TRIZ—ARIZ-85—which is hard to understand, and the paradoxical situation of Mann's TRIZ has been brought up in Ref. [11]. Moreover, USIT, the P&B method, Axiomatic Design, and Infused Creativity do not support item (2). As a result, the methodologies for TIPS and EDM do not satisfy all of the requirements.

To satisfy these requirements, in this paper, we develop the thinking process of the Creative and inventive Design Support System (CDSS), and evaluate the validation of CDSS's process by using quality engineering.

Systematic Approach for the Creative and Inventive Thinking Process

Proposal of the Thinking Process

To realize requirements (1) and (2), as outlined in the former section, "the definition of contradiction understanding and solving processes" and "the derivation by the principle of invention of TRIZ" are applied for the mobilization of other elements beyond a user's knowledge and know-how. As for item (3), "the simplification of the thinking process" for CDSS is applied. Because, the determination of "What method should be chosen?" and "What kind of sequence should be followed?" cannot be judged, to be overly prepared in the thinking procedure and problem solving method, like TRIZs. In addition, for this simplification, a two-dimensional matrix and standardized diagrammatic representation are introduced for the problem understanding and problem solving techniques. This are used because, in cognitive psychology, it is believed that although we cannot make to do thinking which operates a logical expression, we can examine various situations using a visual image as a model [20].

Next, the proposed thinking process of CDSS is shown in Figure 1. This process consists of a combination of two processes—the problem understanding and problem solving processes—that allow beginners to obtain a creative design solution by following a simple thinking process. These two processes can be explained as follows.

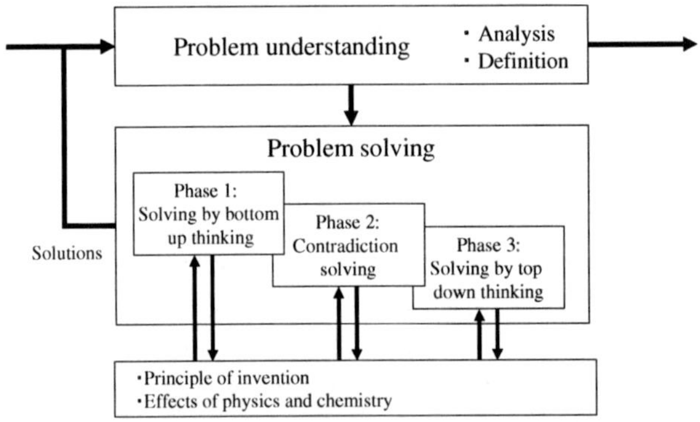

Fig. 1. Proposed thinking process for CDSS

Problem understanding process: Before the problem solving process, "requirement (demand and constraint are called requirement) without design solution" and "contradiction of design solution" are precisely defined and broken down via this process. For this reason, this process is not only made a problem definition process, but is defined as a problem understanding process since it combines problem definition and analysis.

Problem solving process: The problem solving process is constituted by three phases bottom up thinking (Phase 1), contradiction solving (Phase 2), and top down thinking (Phase 3). Each phase supports the deriving of a design solution beyond the user's knowledge and know-how by combining a principle of invention or the effects of physics and chemistry. These are freely selected and combined by the user. Next, each phase is simply summarized as follows:

- **Solving by bottom up thinking (Phase 1):** This approach analyzes in detail the situation where a requirement occurs, and synthesizes its result to obtain a design solution.
- **Contradiction solving approach (Phase 2):** This approach obtains a design solution based on Altshuller's definition [4].
- **Solving by top down thinking (Phase 3):** A design solution (lower conception) is derived from the dominant conception called an Ideal Solution (IS). Although the definition of the Ideal Final Result (IFR) of TRIZ is that the solution contains all of the benefits and none of the costs or harms [4-7], the IS is not this strict, but is a solution where the user considers the situation or assumption as ideal.
- **Degree of freedom in selecting thinking:** To draw on a creative solution without relation to thinking process likes and dislikes, these phases can be freely selected and combined in the problem solving process, as shown in Figure 2. Additionally, if a designer is stuck when creating a design solution, there is a degree of freedom to select another thinking phase.

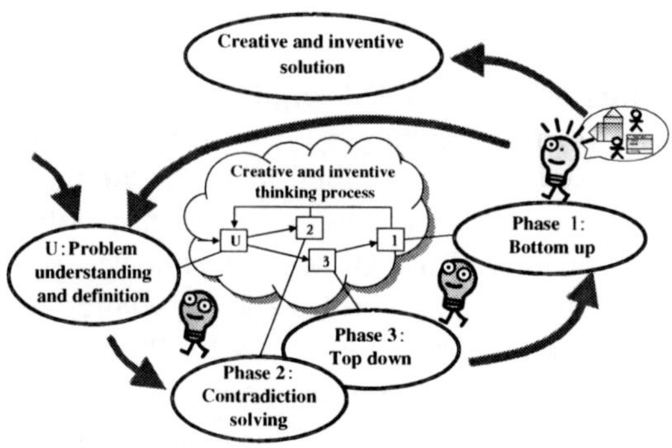

Fig. 2. Combination image of phases for CDSS

In the proposed thinking process, an explored design solution is fed back to the problem understanding process, and it is grasped again whether the solution satisfies the other requirements. If there is a contradiction, the process of finding a new design solution in the problem solving process is repeated, as shown in Figure 1.

Problem Understanding Process

In this process, the problem definition and analysis (a vague problem is clarified as a requirement, and contradictions of requirements are distinguished from clarified requirements) are carried out, not by documentation like TRIZ or USIT, but by using a two-dimensional matrix of quality function deployment (referred to as the QFD matrix). This process follows the next four steps.

1. **Extraction and definition of problem:** This extraction is performed using the class diagram shown in Figure 3. This class structure makes a requirement an abstraction class, and also consists of four subclasses: physical requirements, functional requirements, human requirements, and economic requirements. The requirements used as values for the attributes of these subclasses are clearly listed as required items in the QFD matrix of Figure 4.
2. **Finding a solution:** Both considered solutions and solutions used with existing products, similar products, etc. are extracted as initial design solutions. The QFD matrix of Figure 4 is completed by listing these solution items. There is a high probability that a solution in the QFD matrix at the initial step is an average design solution (referred to as a Common-Sense Solution, CSS) for the corresponding requirement. However, if a contradiction of the design solution of CSS is clarified and this is solved, this design solution will turn into a creative design solution [21].
3. **Problem analysis:** If a solution is a design solution for a requirement, a "O" is used, whereas, if there is a conflict, an "X" is placed in the QFD matrix. From this result, a solved requirement, a contradiction of a design solution, and a requirement with no solution (three items) can be clarified. In addition, this contradiction understanding becomes a vital step since inventive problem solving is a contradiction solution. This matrix becomes an index that shows the balance, logical consistency, and contradiction of design solutions for the whole requirement. In addition, this matrix can manage the quality for the creativity step by step.
4. **Transition to problem solving process:** The result of the problem analysis—the analysis for contradictions and determining a requirement with no solution—using the QFD matrix is delivered to the problem solving process. If a design solution is obtained in the problem solving process, it returns to step 2 and the QFD matrix is updated by this solution, as shown in Figure 5.

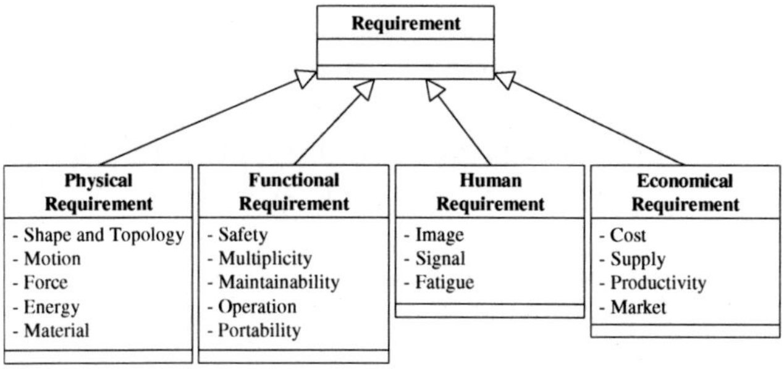

Fig. 3. Requirement class. This includes demand and constraint

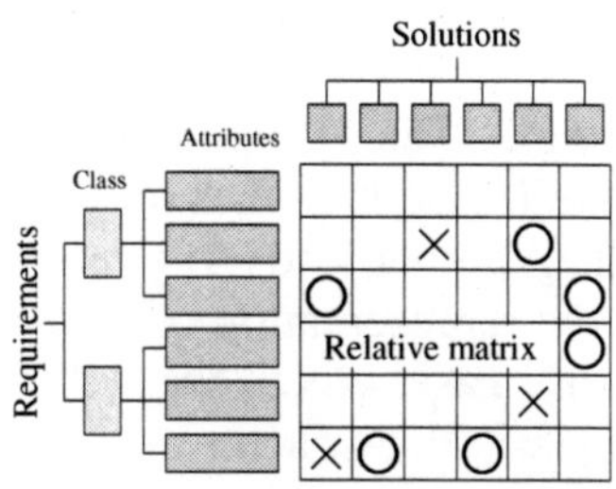

Fig. 4. QFD matrix

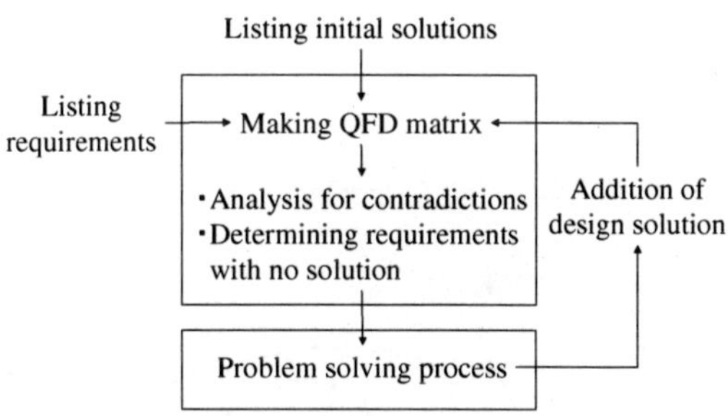

Fig. 5. Problem definition and analysis process for problem understanding process

Problem Solving Process

Solving by Bottom Up Thinking (Phase 1)

For a requirement with no solution, an analysis of the problem conditions for this requirement is carried out, and a design solution is explored via the analysis results in this thinking phase. An object-oriented model is applied in this phase. An object has an attribute, function, and response. Its object's response becomes two kinds of responses, i.e., the response to another object's behavior (external factor) and to the object's own behavior (internal factor). In the case of USIT, although the object-oriented concept is also applied to the problem analysis process, it is not the concept that was standardized by OMG (Object Management Group). Rather, it is based on an original concept and is not suitable for the generalization and simplification of a thinking process. In this study, the Unified Modeling Language (UML), which was standardized by OMG and has been used for the analysis of the business process, was applied in this phase. For this reason, a guarantee of the generality and schematization could be accelerated. Phase 1 follows the steps shown below.

(I) Analysis step

1. As shown in Figure 6, the situation that a requirement of the QFD matrix yields—a problem situation leads to a requirement—is enumerated as a use case of the use case diagram, like use cases A, B, and C, through the position of an actor. The use case is described using the format of "verb + noun." The UML symbol for the actor is the stick figure shown in Figure 6, and becomes an object when involved in use cases. A problem situation that leads to a requirement is made easier to understand by using the representation of the use case diagram.
2. The verb part of the use case, which we will focus on, is defined as a function (F) of an object (O). An actor that carries out its function, i.e., that becomes the subject of its requirement, is extracted as an object, as shown in Figure 6. In this figure, the object that focuses attention on "use case B" is "actor A," and from the description of its use case, its function becomes "function A ( )."

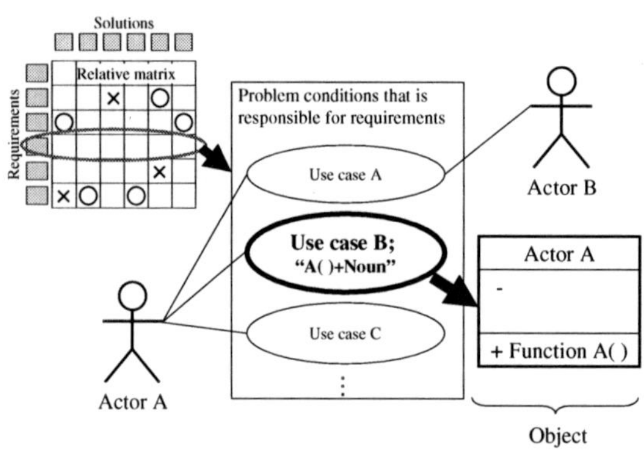

Fig. 6. The use case diagram and object

3. Next, the attribute (A) where an attribute value changes by the function (F) of the object (O) in the problem situation is extracted. Moreover, the response (R) shows a situation where the changed attribute values are enumerated. Additionally, since the response shows a situation it is described by the verb format. Table 1 shows the attribute (A) from which

the attribute value changed by function A( ) of actor A and its response (R). This extracted relationship is called the O-A-F-R relation.

4. The core of the problem is selected. This is the independent O-A-F-R relation among the requirements, which is extracted when the cause of the focused requirement was generated. For example, as shown in Figure 7, the response to function A( ), which generates the cause of the focused requirement, analyzes whether it has been similarly generated in other functions. In fact, although response A or C is generated by functions B( ), C( ), and D( ) besides A( ), response B responds only to A( ). For this reason, if response B can come under control, it is thought that the cause of its requirement will not occur. The O-A-F-R relation, i.e., actor A - attribute B - function A( ) - response B, is obtained via this analysis. This relation that was documented as a requirement becomes the core of the problem.

Table 1 Pick out the O-A-F-R

Object	Attribute	Function	Response
Actor A	A	A( )	A
	B		B
	C		C

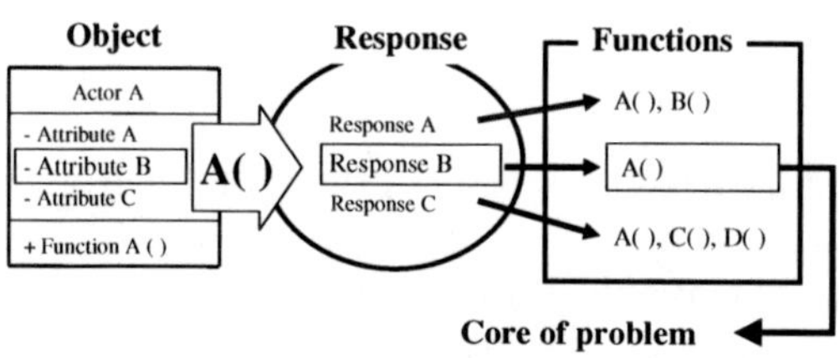

Fig. 7. Select the core of the problem

(II) Solving step

In this step, the best design solution to reduce changes in an attribute value is explored for extracting the core of the problem. This draws on a solution by focusing on the response of the change in the attribute value. Three points of the view are introduced for this solving method: "deal with an action in advance," "deal with an action after the fact," and "make the situation to act with sufficient convenience."

1. **Preparing in advance:** From the viewpoint of "dealing with an action in advance," the thinking method of "preparing in advance" is applied. This draws through invention principles of classic TRIZ as follows:

   - A solution that negates a non-desirable situation by predicting the non-desirable situation in advance and making the reverse situation beforehand (invention principle 10, preliminary action, of classic TRIZ),
   - A solution that cancels the bad influence was produced beforehand by its interaction, when a certain interaction is required (invention principle 9, preliminary anti-action, of classic TRIZ).
2. **Preparing a protection resource:** From the viewpoint of "dealing with an action after the fact," the thinking method of "preparing a protection resource" is applied. A solution that applies a protection resource by predicting the non-desirable situation (invention principle 11, beforehand cushioning, of classic TRIZ) is used as the design solution.
3. **Changing into the usual situation from the situation after an alteration:** This method develops a design solution from the viewpoint of "the situation is made to act conveniently." Although it is non-desirable to shift from the "usual situation" to the "situation after an alteration" if the "situation after an alteration" acts conveniently, this does not worsen a situation. Therefore, a solution is found by using the "situation after an alteration" in reverse. This corresponds to the "blessing in disguise" or "turn lemons into lemonade" (invention principle 22 of classic TRIZ).

Contradiction Solving Approach (Phase 2)

For the contradictory design solution of the QFD matrix, the "separation in space" or the "separation in time" are carried out, and a design solution is found by using an inventive principle obtained via the contradiction method in Figure 8. Here, the "separation" of the contradiction solving method is used. This is the separation in space and time that is used in the uniqueness of USIT or the contradiction matrix of TRIZ. In this phase, it judges which shall be used between the separation in space and time by clarifying the characteristics of a contradictory design solution. The following questions are used for this judgment.

Q1: Which parts desire the characteristics of design solution 1?
Q2: Which parts desire the characteristics of design solution 2?
Q3: What kind of case desires the characteristics of design solution 1?
Q4: What kind of case desires the characteristics of design solution 2?

When the answers to Q1 and Q2 differ, the "separation in space" is adopted, and when Q3 differs from Q4, the "separation in time" is used.

	Solution 1	Solution 2	Solution 3
Requirement A	○	×	×
Requirement B		○	
Requirement C	×	○	

Contradiction

Contradiction solution routes	Solving methods
Separation in space	Segmentation
	Taking out
	Local quality
	Nested doll
	Other way around
	Curvature
	Another dimension
Separation in time	Prior action
	Beforehand cushioning
	Dynamics

Fig. 8. Contradiction solving method

Solving by Top Down Thinking (Phase 3)

Phase 3 performs the problem solving by the top down thinking under the precondition of "having an ideal solution." In CDSS, the personification method is used to draw on a specific design solution using this ideal solution. This is an improved version of a similar method used in USIT or TRIZ, and is called particles. Particles have dynamic behaviors. However, it is difficult to find a design solution from its behavior because its behavior is not defined in UIST or TRIZ.

In this phase, its behavior is defined by using six behaviors, i.e., "increasing," "decreasing," "separating," "moving," "transmitting," and "staying." The effect acquired from activating these behaviors is related to the effects and principles of physics and chemistry shown in Table 2.

As this result, the effect and the principle can be searched as the reverse resolution from particles. Additionally, we prepare only six behaviors rather than numerous behaviors to avoid the situation of having too many choices [22].

For example, when particles activate the behavior of "transmitting," the effects of "transmitting a force/torque," "transmitting heat," or "transmitting light" can be drawn on by "force/torque," "heat," and "light" (these are objects) from the reverse resolution of Table 2. If "transmitting a force/torque" was focused on, a realization action such as "transmitting by friction or a wave" is obtained. From this action, a specific design solution is found.

Table 2 Reverse resolution for effects

Behavior of particles	Object	Effect	Realize an behavior
Increasing	Force/Torque	Increase a force/torque	High-pressure
			The principle of leverage
			Thermal expansion
			Centrifugal force
			Impactive force
	Volume	Increase a volume	Thermal expansion
	Temperature	Increase a temperature	Friction
			Heat pipe principle
			Vibration
			Infrared ray
			Electromagnetic induction
Decreasing	Force/Torque	Decrease a force/torque	The principle of leverage
	Volume	Decrease a volume	Thermal expansion
	Temperature	Decrease a temperature	Joule-Thomson effect
			Capillary tube and porous material
			Peltier effect
			Seebeck effect
Separating	Mixture	Separate a mixture	Electromagnetic field
			Centrifugal force
			Osmotic pressure
			Adsorption
	Object	Break down an object	Impactive force
			Discharge
			Sympathetic vibration
			Thermal expansion
			Supersonic
Moving	Object	Move an object	Magnetic field
			Vibration
			Thermal expansion
	Liquid/Gas	Move a liquid/gas	Pressure
			Capillary phenomenon
			Osmotic pressure
Transmitting	Force/Torque	Transmit a force/torque	Friction
			Wave
	Heat	Transmit a heat	Heat radiation
			Heat conduction
	Light	Transmit a light	Optical reflection
Staying	Temperature	Stay a temperature	Phase transition
			Bubble
			Evaporation
			Electromagnetic induction
	Object	Stay a position or a motion of an object	Electromagnetic field
			Gyro effect
			Reaction
			Friction
			Magnetic fluid

Quantitative Evaluation for the Creative and Inventive Thinking Process of CDSS

In this study, to quantitatively evaluate the effectiveness of the proposed thinking process, a questionnaire survey was carried out for the problem understanding process. For the problem solving process, a Design of Experiment (DOE) [23], [24] was carried out, and a necessary condition for a creative design solution in the evaluation of the problem solving process became a best design solution for a designer. Therefore, as an evaluation criterion, we used the question "Could a best design solution for oneself be found?" by using the proposed thinking process. Since the aim of the support by CDSS is to draw on a creative design solution without depending on the characteristics and knowledge of the user, the test subject was a beginner who was not familiar with inventive problem solving methods like TRIZ and USIT.

Next, the design problem for this evaluation was "a problem involving the improvement of a vacuum cleaner that presents an improvement solution in relation to functional improvement, cost reduction, environmental problems, etc." Before carrying out this design problem, in order to understand the structure and function of the vacuum cleaner, the test subject assembled and took apart two kinds of vacuum cleaners: the stick and the general type. The function structure diagram of the P&B method [13] was created by using the prior exercise. Then, the design problem experiment was performed. The test subjects were 72 students, who evaluated "whether a design solution was the best" on an absolute scale, where the questionnaire entries of the problem understanding process using the evaluation criteria of this five-grade evaluation: "1: think so very much," "2: think so," "3: have no preference," "4: don't think so," and "5: don't think so at all." In addition, the evaluation is higher when this evaluation value is lower.

Evaluation for the Problem Understanding

The problem understanding process involves "listing a requirement item," "derivation of an initial design solution for a requirement item," and "a contradiction understanding of design solutions" for the QFD matrix. Because of this, this section shows how three evaluation criteria related to these items were prepared as questions. The usefulness of this process is examined via the results obtained from the absolute evaluation. In addition, the evaluation of the answers was used to calculate an average value for each question by using the value of the evaluation criterion. For an overall evaluation value for the problem understanding process, the average of the evaluation values for three questions was calculated and used.

- Q1: It is easy to summarize a requirement item.
- Q2: It is easy to think of a design solution to a requirement item.
- Q3: It is easy to describe "O: a solution of a requirement" and "X: a conflict solution" to the QFD matrix.

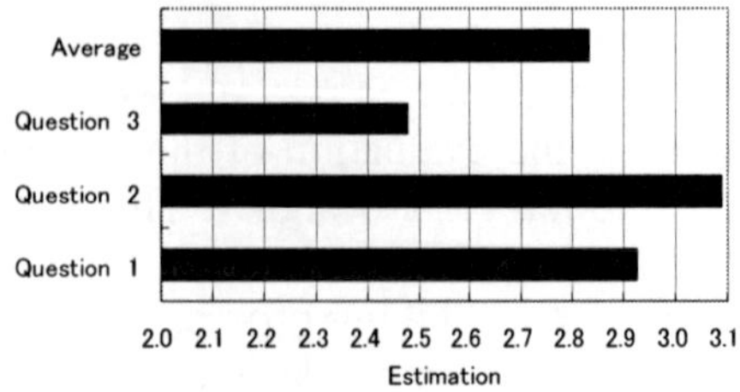

Fig. 9. Results of three questions for problem understanding process

The evaluation values for the three questions and the average value for the overall evaluation are shown in Figure 9. First, an overall evaluation of the usefulness of the problem understanding process is performed by using the average evaluation value. Figure 9 confirms that the average value is lower than "3: have no preference." Therefore, the overall evaluation confirms that the problem understanding process is a useful process. However, some of the individual evaluation values are larger than "3: have no preference." Thus, we considered every question and confirmed the following items. In question 1, since the evaluation value becomes a little lower than "3: have no preference," it is found that summarizing a requirement in the QFD matrix is not difficult. In question 3, since the evaluation value is lower than the middle value of "2: think so" and "3: have no preference," it is confirmed that the contradiction understanding can be performed easily using the QFD matrix. However, in question 2, the evaluation value becomes larger than "3: have no preference." From this result, we understand that it is difficult for a test subject to list initial design solutions (the possibility of a Common-Sense Solution is high), even if assembly and disassembly tasks and the creation of a function structure diagram were performed beforehand. Of course, if the derivation of a design solution is easy, it is not necessary to support the problem solving process. Therefore, we believe that this result is appropriate.

Validation of the Problem Solving Using the Design of Experiment

In order "to check the usefulness of the three solving phases (Phases 1, 2, and 3)" and "to analyze a desirable combination for the solving phase," the evaluation is performed using DOE. This evaluation assigns the three solving phases (Phases 1, 2, and 3) and Phase 1-1 (a combination used to examine the effect in the case of combining the same solving phases) to the control factors. The factor uses two levels (1: enforcement and 0: non-enforcement) to define whether the solving phase is carried out. Therefore, the orthogonal array L8(2^7) is applied as shown in Figure 10. The combination was used in the cases from trial 1 to trial 7 and assigned to the orthogonal array of Figure 10. Trail 0 is set to 5.0 as the maximum evaluation value, because it becomes a case where no solving phases are carried out. Moreover, as an example, in the case of the thinking process that combined Phase 3 and 1, it is written as Phase 3-1 in this paper. A test subject drew on a design solution and evaluated it using the assigned trial case. A test subject's evaluation results were used to calculate the mean for all the data, and this value was used for the DOE response value.

Trial	Phase 1	Phase 2	Phase 3	Phase 1-1
0	0	0	0	0
1	0	0	1	1
2	0	1	0	1
3	0	1	1	0
4	1	0	0	1
5	1	0	1	0
6	1	1	0	0
7	1	1	1	1

Trial 1	Phase 3, Phase 3-1	Trial 5	Phase 3, Phase 1-3
Trial 2	Phase 2, Phase 2-1	Trial 6	Phase 1, Phase 1-2
Trial 3	Phase 2, Phase 2-3	Trial 7	Phase 1-2-3-1
Trial 4	Phase 1, Phase 1-1		

Fig. 10. The orthogonal array; L8(2^7)

Validation of the Usefulness of the Solving Phase

This section verifies the usefulness of each phase in the problem solving process. The main effects obtained as the evaluation results of these phases are shown in Figure 11. Figure 11 shows that the test subjects understood that it was better to perform, rather than not perform, all of the solving phases in the problem solving process. This result shows that the best design solution can be found with the support of these phases in the problem solving process.

Validation for the Combination of the Solving Phases

The problem solving process of CDSS assumes that the combination of the solving phases will be used within the process. In this section, in order to examine whether this view of the combination is appropriate, an evaluation is made to determine the most desirable type of combination pattern. For this evaluation, we use the interaction diagram of DOE to answer the questions, “Do solving phases have a dependency or not?” and “Does an evaluation value improve by combining or not?” The interaction diagrams for combining the phases are shown in Figure 12.

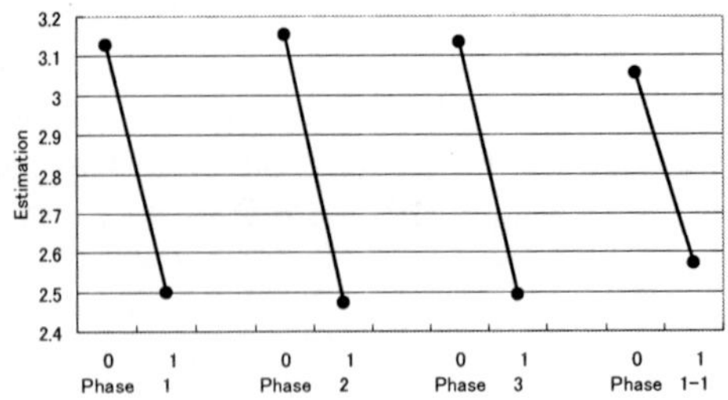

Fig. 11. The main effect of phases

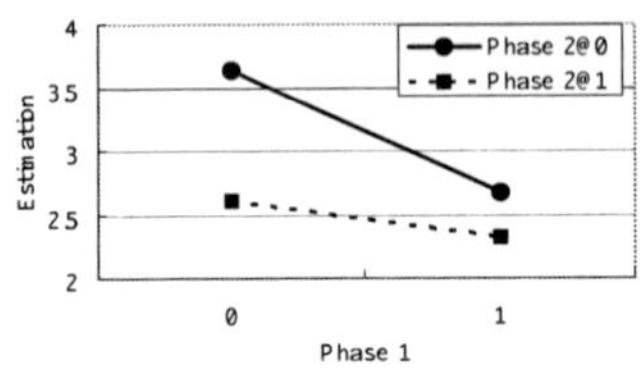

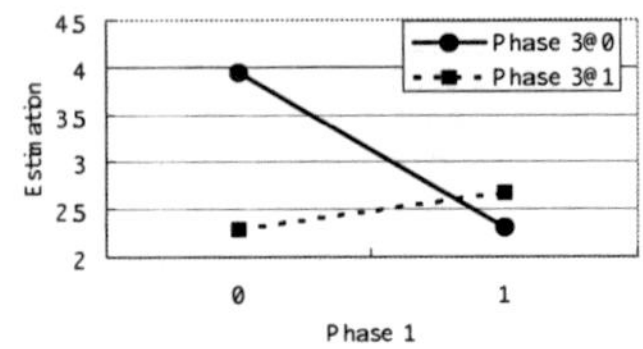

(a) Interaction of Phases 1 and 2; (b) Interaction of Phases 1 and 3

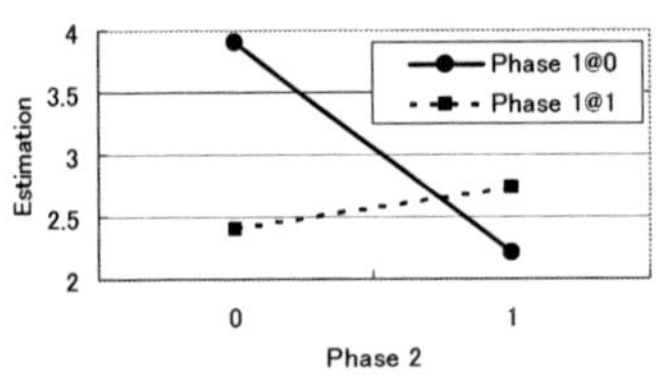

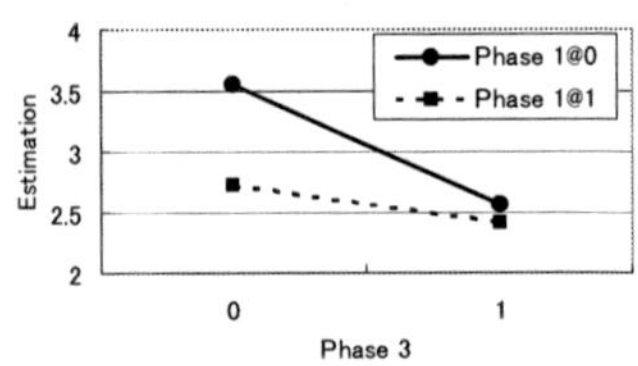

(b) Interaction of Phases 2 and 1; (d) Interaction of Phases 3 and 1

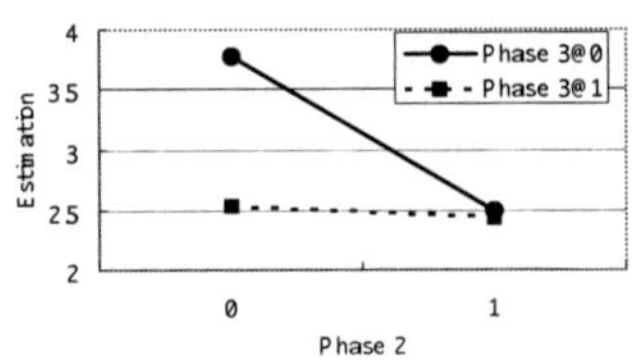

(e) Interaction of Phases 2 and 3

Fig. 12. Interaction of phases

From these figures, it is understood that the solving phases depend on combining the phases. Moreover, it is confirmed that Phase 1-2 and Phase 3-1 are interactions that are not changed by the superiority or inferiority of a level. Therefore, these evaluation values show that an evaluation value has been equaled or improved, compared to the case where each phase is carried out independently.

Next, the evaluation values for combining the solving phases are shown in Figure 13. This figure shows that the evaluation value becomes high with the order Phase 1-2, Phase 3-1, Phase 2-3, Phase 1-1, Phase 1-2-3-1, Phase 1-3, and Phase 2-1. Moreover, it is confirmed that it is better to use Phase 1-2 and Phase 3-1 rather than using a solving phase independently. This result is also supported from the result based on the interaction shown previously. Therefore, it is understood that the degree of freedom for using a solving phase combination is effective for supporting the derivation of a creative design solution. However, Figures 12 and 13 confirm that the best design solution is not necessarily obtained when using Phase 2-1, Phase 1-3 (reverse combination of Phase 1-2 and Phase 3-1), or a combination of numerous phases, like Phase 1-2-3-1.

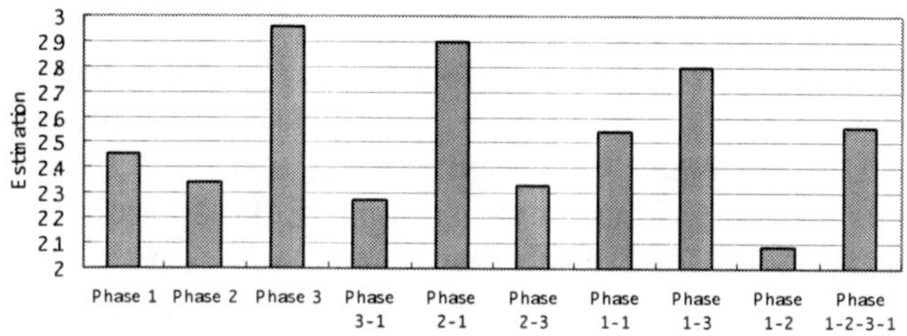

Fig. 13. Estimations for phases and combinations of phases

Evaluation of the Robustness Using the Taguchi Method

The robustness evaluation of the problem solving process was carried out in consideration of the variation in the test subjects' attributes. The Taguchi method (TM) [25] was used as the method for this evaluation. From this evaluation result, we can obtain the most robust solving phase for the robustness of a test subject's attributes and knowledge.

The Taguchi Method

In this section, a test subject's attributes are assigned as an error factor to the orthogonal array $L8(2^7)$ of Section 3. For the test subject's attributes, we presuppose "Creativity; imagination is rich," "System thinking; system engineering is a favorite," and "Mechanical knowledge; knowledge about mechanical engineering is abundant." Its ability is estimated using the "0: No" or "1: Yes" replies from the test subject. Our TM uses the Smaller-is-Better characteristic as a static characteristic because the evaluation is higher when the evaluation value of the phase is lower. The SN ratio for an evaluation of robustness is shown in the following equation (1).

$$SN\text{ ratio} = -10\log_{10}\left[\frac{1}{n}\sum_{i=1}^{n} y_i^2\right]^2 \quad (0 \le y \le \infty) \tag{1}$$

where *y* denotes the response value. When the *SN* ratio becomes increasingly large, it is not influenced by the error factor. In the case of the Smaller-is-Better characteristic, when the *SN* ratio approaches 0.0, it is not influenced by the error factor since the *SN* ratio denotes a minus value.

Solving Phase with Robustness

The three attributes of a test subject were individually assigned as the error factors, and the *SN* ratio results were obtained by applying the TM to each phase, as shown in Table 2. The results show that Phase 2 has the largest *SN* ratio for each attribute in the table. Therefore, it is confirmed that Phase 2 is a solving phase with robustness. Moreover, it makes assignments to its outer orthogonal array by using all of the attributes as error factors, and the result obtained by TM is shown in Figure 14. From this result, it is understood, similar to Table 2, that Phase 2 is a phase where the *SN* ratio is the largest of all the attributes. On the other hand, when the *SN* ratios for all of the attributes related to Phase 3 are compared, the *SN* ratio of creativity is larger than the ratios for the other attributes, and slightly exceeds the *SN* ratio of creativity of Phase 1. However, when all of the attributes are used as error factors, there are few rates of change for the *SN* ratio of Phase 3, as shown in Figure 14. For this reason, it is understood that Phase 3 is easily influenced by a test subject's attributes. From the above, Phase 2 is the phase that is the easiest to use in CDSS. Additionally, since inventive problem solving is a contradiction solution, it is a very desirable result that it is easy to use this phase. By recommending this to beginners, it seems that it becomes an easy-to-use thinking process.

Table 2 *SN* ratios

	Phase 1	Phase 2	Phase 3
Creative	-7.96	-7.29	-7.92
System Thinking	-7.85	-7.4	-8.21
Mechanical Knowledge	-7.82	-7.24	-8.74

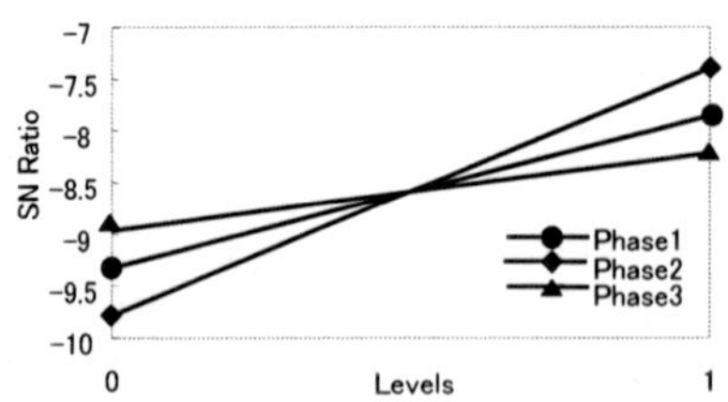

Fig. 14. *SN* ratios of all the attributes

Conclusion

In this study, as a system to support an existing conceptual design process, the theory of inventive problem solving and the engineering design methodology were analyzed. We newly systematized its system as a creative and inventive thinking process based on the analysis results and proposed it. Moreover, the proposed thinking process was evaluated quantitatively to determine its effectiveness using quality engineering. As a result, it was confirmed that it is possible to find a creative design solution (the best design solution for the user) by using the thinking process of CDSS. Moreover, it was possible to obtain desirable combinations during the phases of the problem solving process (Phase 1-2 and Phase 3-1) and the easy-to-use solving phase was robust (Phase 2). By using these obtained phases, it seemed that CDSS was an easy-to-understand thinking process for a designer by presenting a user beforehand the easy-to-use solving phase and combination of the phase in the problem solving process.

References

1. Stevens, G.A., Burley, J.: 3000 raw ideas = 1 commercial success. Research Technology Management 40(3) (1997)
2. Kitamura, Y., Mizoguchi, R.: A Framework for Systematization of Functional Knowledge based on Ontological Engineering. Transaction of the Japanese Society for Artificial Intelligence 17(1), 61–72 (2002) (in Japanese)
3. Shimomura, Y., Yoshioka, M., Takeda, H., Tomiyama, T.: Fundamental Plot for a Designer Support Environment based on Abduction. Transactions of the Japan Society of Mechanical Engineers, Series C 72(713), 274–281 (2006) (in Japanese)
4. Altshuller, G.: Creativity as an Exact Science. Translated by Williams A. Gordon and Breach Science Publishers (1984)
5. Altshuller, G., Zlotin, B., Zusman, A., Filatov, V.: Tools of Classical TRIZ. Ideation International (1997)
6. Savransky, S.D.: Engineering of Creativity, Introduction to TRIZ Methodology of Inventive Problem Solving. CRC Press, Boca Raton (2000)
7. Kawamo, K., Suga, M.: Methodology of Creative Engineering Design – Principle of new Design and Manufacturing, Yokendo (2003) (in Japanese)
8. Mann, D.L.: Hands-on Systematic Innovation. CREAX Press (2002)
9. Sickfus (ed.): Unified Structured Inventive Thinking. Ntelleck, LLC (1997)
10. Mann, D.L., Dewulf, S., Zlotin, B., Zusman, A.: Matrix 2003 Updating the TRIZ Contradiction Matrix. CREAX Press (2003)
11. Nakagawa, T.: Overall Dataflow Structure for Creative Problem Solving in TRIZ/USIT. In: The 7th Altshuller Institute TRIZ Conference (2005)

12. Hororwitz, R., Maimon, O.: Creative Design Methodology and the SIT Method. In: Proceedings of DETC 1997: ASME Design Engineering Technical Conference (1997)
13. Pahl, G., Beitz, W., Feldhusen, J., Grote, K.H.: Engineering Design a systematic approach, 3rd edn. Translated by Wallace K and Blessing L. Springer, Heidelberg (2007)
14. Suh, N.P.: The Principles of Design. Oxford University Press, Oxford (1990)
15. Suh, N.P.: Axiomatic Design: Advances and Applications. Oxford University Press, Oxford (2001)
16. Shai, O., Reich, Y.: Infused design: I theory. Research in Engineering Design 15(2), 93–107 (2004)
17. Shai, O., Reich, Y.: Infused design: II practice. Research in Engineering Design 15(2), 108–121 (2004)
18. Shai, O., Reich, Y., Rubin, D.: Infused Creativity: An Approach to Creative System Design. In: Proceedings of ASME DETC2005-85506 (2005)
19. Kawamo, K., Ishikawa, N., Fon, T.: Conceptual Design by Using a Systematic Approach Engineering Design and TRIZ. Journal of Japan Society for Design Engineering 35(4), 119–126 (2000) (in Japanese)
20. Ichikawa, S.: Science for thinking. Chuokoron-shinsha (1997) (in Japanese)
21. Altshuller, G., Shapiro, R.V.: About a Technology of Creativity. Questions of Psychology (6), 37–49 (1956) (in Russian)
22. Sheena, S.I., Mark, R.L.: When Choice is Demotivating: Can One Desire Too Much of a Good Thing? Journal of Personality and Social Psychology 79(6), 955–1006 (2000)
23. Taguchi, G.: Design of Experiment, 3rd edn., vol. 1. Maruzen (1967) (in Japanese)
24. Taguchi, G.: Design of Experiment, 3rd edn., vol. 2. Maruzen (1977) (in Japanese)
25. Taguchi, G.: Mathematics for Quality Engineering. Japanese Standards Association (1999) (in Japanese)

LINE, PLANE, SHAPE, SPACE IN DESIGN

Line and plane to solid: Analyzing their use in design practice through shape rules
Gareth Paterson and Chris Earl

Interactions between brand identity and shape rules
Rosidah Jaafar, Alison McKay, Alan de Pennington and Hau Hing Chau

Approximate enclosed space using virtual agent
Aswin Indraprastha and Michihiko Shinozaki

Associative spatial networks in architectural design – Artificial cognition of space using neural networks with spectral graph theory
John Harding and Christian Derix

Line and Plane to Solid: Analyzing Their Use in Design Practice through Shape Rules

Gareth Paterson and Chris Earl
The Open University, UK

Design practice is complex and multifaceted. Designing the form of three-dimensional objects, for example, involves the creation of a number of distinct, but related, design descriptions. This paper focuses on relations between the descriptions used to create three-dimensional designs, and analyses their use through shape rules derived from empirical studies of practice.

Introduction

Stiny [1] observes that attempting to define a design by the process used to create it is unproductive. While this approach can reveal the methods and behaviour which create designs, it can do so without necessarily saying what designs *are*. Attempting to define a design by the products of the design process instead is similarly unproductive. In addition to the designed object itself, the products of the design process typically also consist of a number of distinct design descriptions, such as multiple drawings, models and statements of functional requirements, rather than a single entity. Rather than product or process, however, Stiny suggests that designs can be characterized instead by the relations that exist *between* these design descriptions. It is the relations, inherent and emergent, between multiple drawn views of a building, or a product design, for example, which allow its three-dimensional form to be described sufficiently clearly for physical objects to be made from them. It is in the individual usages of a multiplicity of descriptions therefore, and the relations that exist between them, that designs are to be found.

This paper will show that close observation of a series of design protocols, undertaken by the first named author, provides strong evidence

J.S. Gero (ed.): *Design Computing and Cognition'10*, pp. 251–267.

for Stiny's claim that a design is '...an element in an *n*-ary relation among drawings, other kinds of descriptions, and correlative devices as needed.' [1]. Further, the paper demonstrates that the framework of an *n*-ary relation among geometrical shape descriptions leads to explicit and objective accounts of the design process. These accounts are especially useful in gaining a better understanding of how three-dimensional designs are created. Stiny notes that, generally, formal, functional, rational, and historical accounts of how design occurs are incomplete as practice is inevitably more complex than they suggest. More specifically, viewing designs as *n*-ary relations calls into question programme, context, technology, and material as the exclusive generators of form [1].

The benefit of multiple descriptions to design practice in general has been examined by several authors including Soufi and Edmonds [2], Kilian [3], and Li [4]. Of these Li has so far engaged with the nature of the relations between multiple *geometrical* descriptions in his shape grammar for teaching the Yingzao fashi architectural style. Li's linking of related elements across plans and elevations in his grammar, however, touches on a point raised by Knight [5] that the lines and other geometrical elements contained in these descriptions are representationally ambiguous. They can be employed to represent abstract divisions of space in an architectural *parti*, or as more concrete spatial elements such as walls and floors. This ambiguity should be borne in mind when considering relations between descriptions; Knight's analysis indicates that this multiple referencing by spatial elements is one of the many ways in which representational ambiguity is pervasive throughout design [5].

While representational ambiguity is implicit in the graphical elements operated on in shape grammar implementations, Knight also notes that the focus of research into shape grammars has so far largely been on their part ambiguity; that is the ambiguity in how parts of a design are perceived and used [5]. This is understandable as the ability of part ambiguity to create emergent shapes from existing elements is a unique and powerful feature of the shape grammar formalism. However, many shape grammar interpreters do not yet take into account the use of multiple descriptions, or support many of the features of design practice that enable 'designerliness' such as parameterization or weights [4]. The emphasis of research on the relations of elements within a single graphical description in particular, rather than on the relations of these elements across the multi-dimensional *n*-ary of design descriptions, can therefore limit the applicability of shape grammars to three-dimensional designs. This is especially the case when three-dimensional designs are conceived, during the design process, in terms of several related descriptions.

This paper analyses detailed cases of such design processes when creating three-dimensional designs. Generally, understanding design processes can be assisted by seeing how relations among multiple design descriptions are used. More specifically, creating shape grammars to assist in these kinds of design processes is held back by a lack of knowledge of the relations between design descriptions [1].

Aims and Significance

The broad aim of this paper, therefore, is to deliver a greater understanding of how three-dimensional designs are created through empirical observations of practice.

Creating three-dimensional designs usually entails more than manipulating a collection of abstract linear elements in a single drawing. Multiple drawings are often required to establish three-dimensional relationships, and models may also be employed as drawings can be insufficient in themselves to describe the shape of a design fully. To this end the paper focuses on the relations between multiple drawn views, and between drawn views and three-dimensional descriptions (such as CAD or physical models), as they are employed in creating three-dimensional designs.

Design practice is, of course, complex and multifaceted, and the processes involved in creating even the simplest of design descriptions are both voluminous and ephemeral. These processes can be captured on video tape, but creating a description of them (to allow the relations between elements across the *n*-ary of design descriptions to be analyzed) requires an appropriate structure to be derived from the video record. A more specific aim of the paper, therefore, is to employ shape rules for this purpose. Although shape rules have in the past been used either for the synthesis of new designs, or for analysis of the *results* of the design process in the form of existing designs, here they are used instead as a means of analyzing the *processes* involved in creating designs.

The general significance of the findings is that they highlight the necessity of considering relations between descriptions, if the end result of any design process is a three-dimensional object. The use of shape rules in the analysis of the protocols also relates the findings directly to computational design. If a shape grammar, for example, is intended to result in a three-dimensional object and is implemented in a single two-dimensional design description, shape rules applied in it may result in designs which are inconsistent in three dimensions. The findings show that it is beneficial in design to consider the relations between design elements *across* descriptions, such as related views, or other descriptions, such as models, either when drawing design proposals or when applying shape rules. There are practical reasons, therefore, '...to connect many devices of many kinds to make a design' [1].

Method

The raw data for the enquiry is comprised of video recordings of a series of design protocols. In these a number of participants from a design, or design related background undertake a three-dimensional design task.

The task itself is designed to encompass a meaningful chunk of practice; one which is large enough to include the possibility of employing multiple design descriptions, while still small enough not to swamp the enquiry in data. Furthermore, by allowing the participants as much freedom as possible in their choice of descriptions, the protocols are intended to provide a record of their actions which, given the constraints imposed by any enquiry, is a representative sample of their normal form generation practices.

An artificial task was chosen in preference to observing designers on live projects for two reasons. Firstly, while observing designers *in situ* would undoubtedly produce data which genuinely reflects design practice, disseminating the results of these observations would be problematic. Observing live projects inevitably raises the issue of protecting the intellectual property of the participants. Furthermore, the complex phenomena that arise from the interaction of participants with a number of physical and virtual design descriptions, over an extended period of time, also has the potential to generate enormous volumes of data.

Participants, Protocol Locations, and the Design Task

Seven protocols were recorded in a period spanning from November 2006 to March 2008. Wherever possible these were recorded in the participants' normal working environment. The participants themselves were drawn from an initial pool of eleven individuals which contained both practicing designers and design students. No selection of the participants took place other than on the basis of their availability within the timeframe of the enquiry. The detailed results on protocols, and their analysis through shape rules and schemas is presented in the first named author's PhD thesis [6].

In each one hour protocol the participants were given an identical design task which required them to generate a proposal for the handle area of a domestic clothes iron. An iron was chosen as the subject of the design brief as it would potentially result in a more organic form, that all participants would be familiar with the subject, and that its design would be equally amenable to physical modeling, virtual modeling, or drawing. At the start of each protocol participants were presented with the design brief, three copies each of three full-size orthogonal views, a perspective view and a physical model of the iron base.

A significant aspect of the protocols, that participants were allowed complete freedom in their choice of design descriptions, raised the possibility that many participants, when faced with one hour in which to generate a design proposal, would only produce drawings. However, rather than presenting the task as one specifically geared to creating physical or three-dimensional descriptions, the participants were instead invited to produce any form of description which would allow *ergonomic* assessment of their proposals. While this might be seen as weighting the design protocols toward using physical three-dimensional descriptions, violating the intention to allow participants complete freedom in their choice of description, it ultimately proved to be a successful approach in generating a range of descriptions, and strategies for their use. While one participant did go on to produce drawings alone as might be expected, and another produced a virtual three-dimensional description, both did come up with alternative strategies to allow them to assess the ergonomic aspects of their designs in the course of their protocols.

The design descriptions employed in the protocols ranged from purely two-dimensional (physical) drawings to fully three-dimensional virtual geometry. In addition to a number of drawings, four participants produced blue foam models, one sculpted a half-scale clay model, and one produced a virtual three-dimensional design description.

Of the seven protocols recorded, only the four undertaken by experienced designers have so far been analyzed. The reasons for focusing exclusively on the experienced designers are twofold. The first is that their protocols are more likely to be representative of professional design practice. The second is that the analysis process, described in the next section, is a lengthy and painstaking one. The least complicated protocol (which encompassed 25 minutes of drawing) contained 59 distinct states, while the most complicated (which encompassed the whole hour and included making a blue-foam model) contained 321.

Deriving Shape Rules from Segmentation Schemes

Extracting these distinct states from the video record requires a segmentation scheme which can be rigorously applied, regardless of the description making actions employed by participants. Where actions are easily differentiable, such as making a mark on paper with an individual stroke, a state is defined as the modification made to the description by that single action. Where compound actions occur (such as rehearsing the stroke of a pencil before committing a mark to paper, repeatedly tracing over the same line to reinforce it, or applying a number of sanding strokes to a block of foam) regarding individual strokes of a pen, or a sheet of glass-paper as distinct states would result in fragmenting the designers' actions to the point of unintelligibility, without being any more representative of their intentions. A state in these cases is defined instead with reference to Schön and Wiggins 'see-move-see cycle' [7]. Here multiple strokes are regarded as a single design '*move*', which is bounded by the preceding and subsequent pauses in the participants' activity (equivalent to '*sees*') where they stop momentarily to assess the effect of their actions on the design description.

States are recorded by taking stills from the video record showing the beginning and end of the physical action(s) which create them. The elements contained in the design description at these points are then recreated in CAD, by working directly over scans of the participants' drawings. Transitory shapes are extracted from particular actions either by observing them directly in the video record as they are created, or by examining the video record for individual drawing strokes (or other making actions) which match the shapes still present in the final design description. The resulting images are then collected into an initial statement of a '*shape rule*', Figure 1.

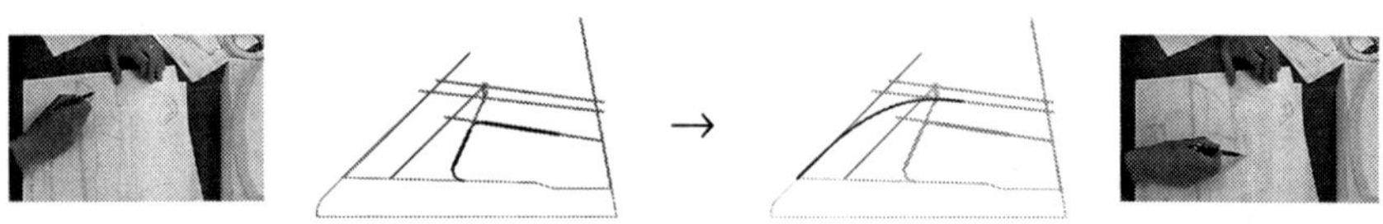

Fig. 1. The initial form of shape rule A013 (the 13th rule application in Andrew's protocol): an example of how modifications to the design description are initially captured and recorded

Shape rules are employed here as they lend themselves to formalizing predominantly visual generative processes [8], and are defined by Stiny and Gips [9] in the following procedure:

1. Find the subshape of the given shape that is geometrically similar to the left side of a shape rule.
2. Find the Euclidean transformations (translation, rotation, scale, mirror image) that make the left side of the shape rule identical to the corresponding subshape of the given shape.
3. Apply these transformations to the right side of the shape rule.
4. Substitute the resulting shape for the occurrence of the subshape in the given shape.

The initial statement of a shape rule in Figure 1 does not conform strictly to Stiny and Gips' definition, as no attempt is made to identify which element(s) of the design description (subshape/shapes) is responsible for triggering it. Instead, the state of the design description immediately prior to the transformation contained in the rule is given on the left side, while the transformation, i.e. the rule itself, is laid over a ghosted version of that state on the right side. The right side of the rule, as it has a visible result on the design description, can of course be stated explicitly at this stage.

Having identified the right side of a rule, the left side subshape(s) can now be inferred by noting explicit spatial relations between the new element created by the shape rule and pre-existing elements. In the case of shape rule A013, as the new element connects tangentially with two lines describing the outer envelope of the forward end of the side elevation, it is reasonable to assume that these, or some portion of them, are responsible for triggering this particular rule application. In the full statement of rule A013 the modification contained in the initial statement is now matched up with left side trigger conditions and organized into a rule schema, in this case under the heading of 'tangential connect', Figure 2.

Fig. 2. Shape rule A013, now shown with the shape required to trigger it. The rule schema is a tangential connection between individual lines in the left side shape

This rule sets a frame for subsequent rules and its right hand side result is retained through subsequent moves. Relating this to the segmentation scheme, and referring once again to Schön and Wiggins 'see-move-see cycle', the preceding *'see'* is equivalent to the left side of the rule while the *'move'* is now equivalent to its right.

Individual shape rules are then collected into *episodes*. When shape rules are applied to a two-dimensional design description an episode is defined as commencing with the first shape rule applied to that design description, and terminating with the first shape rule applied in a different type of design description, or another instance of the same type of design description.

While this definition suffices for the initial stages of each protocol where participants employed two-dimensional design descriptions, it is less obviously applicable in the latter stages where participants would often work on a *single* instance of a three-dimensional design description for considerable periods of time. However, their *use* of this single description would vary during this time and could be divided into more or less distinct episodes. The key to defining how the use of a three-dimensional design description varies lies in disentangling the spatial degree of the description itself from the spatial degree of the shape rules applied to it.

Each design description is equivalent to an instance of the algebra, U^*, with some extension of the terminology to allow weighted curves rather than points, lines and planes alone. To differentiate episodes by use when three-dimensional descriptions are employed it is necessary to identify the space that a design element, acted on in a shape rule, is transformed in. Stiny's array of algebras of shapes, U_{ij} [10], indexes U^* by the spatial degree of the element itself, i, and the space it is transformed in, j (Fig3). Lines on paper, for example, as linear elements which can be transformed in a two-dimensional space, are in U_{12}.

$$\begin{array}{cccc} U_{00} & U_{01} & U_{02} & U_{03} \\ & U_{11} & U_{12} & U_{13} \\ & & U_{22} & U_{23} \\ & & & U_{33} \end{array}$$

Fig. 3. The array of algebras of shapes [10], U_{ij}, where i is the spatial degree of the element transformed in a shape rule, and j is the degree of the space it is transformed in.

Both shape rules, and algebras of shapes, are useful concepts for understanding the processes involved in creating three-dimensional designs. Mapping the space that elements of a design description are transformed in, rather than the space that a design description itself occupies, is a significant distinction; if a designer consistently transforms elements of a design description in a spatial degree lower than that allowed by the design description, it is reasonable to infer that this lower spatial degree corresponds to the spatial degree of their internal representation. In turn, identifying the left side of shape rules can also reveal where elements from more than one design description, i.e. parallel computations, are required to define a rule. In the following section examples of parallel computations, extracted from Andrew's protocol, are used to make the otherwise implicit relations between design descriptions explicit. The nature of the correlative devices which enable these relations to propagate across multiple design descriptions will then be examined with reference to algebras of shapes.

Results

At the point in Andrew's protocol when rule A013, Figure 2, is applied there is only a single instance of a design description to work on. The trigger conditions for rule application, a particular subshape or collection of subshapes in the left side of the rule, are therefore contained in the same design description they are applied in. When, a little further into Andrew's protocol (after the 18th rule application) he produces further instances of two-dimensional, and then three-dimensional design descriptions, the possibility now exists that the shape(s) or subshape(s) that trigger a rule application may be contained in a separate design description.

Parallel Computations across Design Descriptions

Shape rules A030 and A031 from Andrew's protocol can be used to illustrate this. Both rules are invoked to create lines which connect the upper edges of the supplied base drawing. The initial form of shape rule description, Figure 1, which records the state of the design description immediately before, and then immediately after, rule application shows these lines as straightforward additions to the design description, Figure 4.

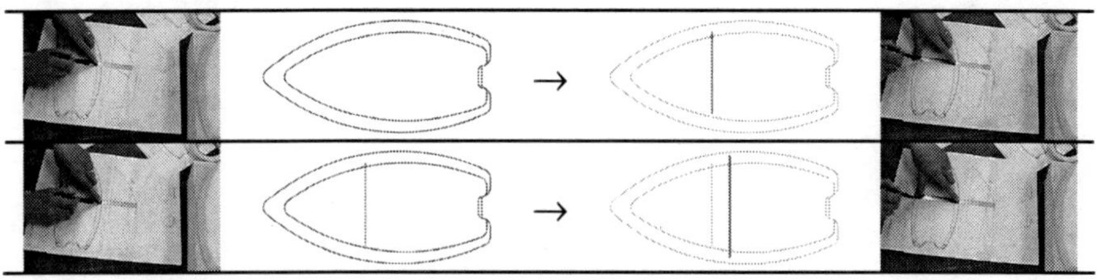

Fig. 4. The first shape rules applied in the plan view, A030 and A031

Inferring the left side of these shape rules, by noting relations between elements contained within the plan view, suggests that A030 employs a 'connect' schema while A031 employs a combination of 'connect' and 'offset'. The 'offset' schema in A031 applies to the line just created in rule A030, Figure 5.

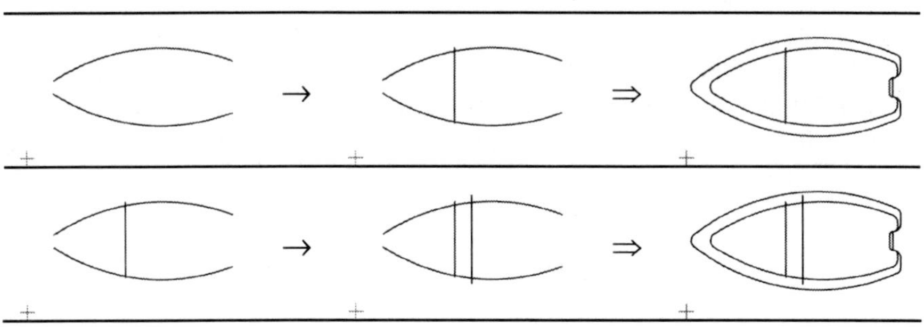

Fig. 5. A030 and A031, with left side shapes from the same design description

Where more than one design description is employed, however, participants may deliberately align one with the other, using one description as a reference to guide modifications made to the other. An example of this can be seen in the video stills contained in Figure 4. Here Andrew creates explicit spatial relations between elements in the plan view and the side elevation, by first aligning a ruler over them, and then lightly tracing marks connecting them, prior to making the marks contained in rules A030 and A031.

The additional trigger shape, the forward edge of the handle opening in the side elevation, is itself decomposed from a collection of shapes previously created in rules A008, A009, and A011, Figure 6.

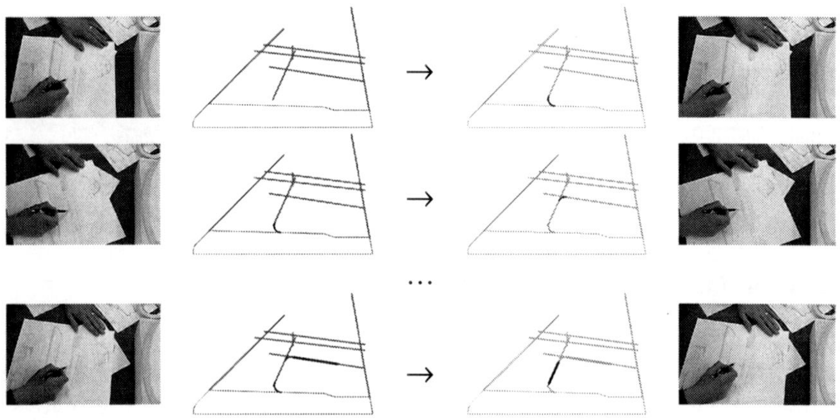

Fig. 6. The three shape rules which create the forward edge of the handle opening

Taking this additional shape into account, the left side of rule A030 now comprises the shape from rules A008, A009 and A011 in the side elevation, and the upper edges of the base in the plan view. These are related by a tangential connection between the foremost edge of the element in the side elevation, projected down across both descriptions to intersect with the upper edges of the base in the plan view, Figure 7, upper.

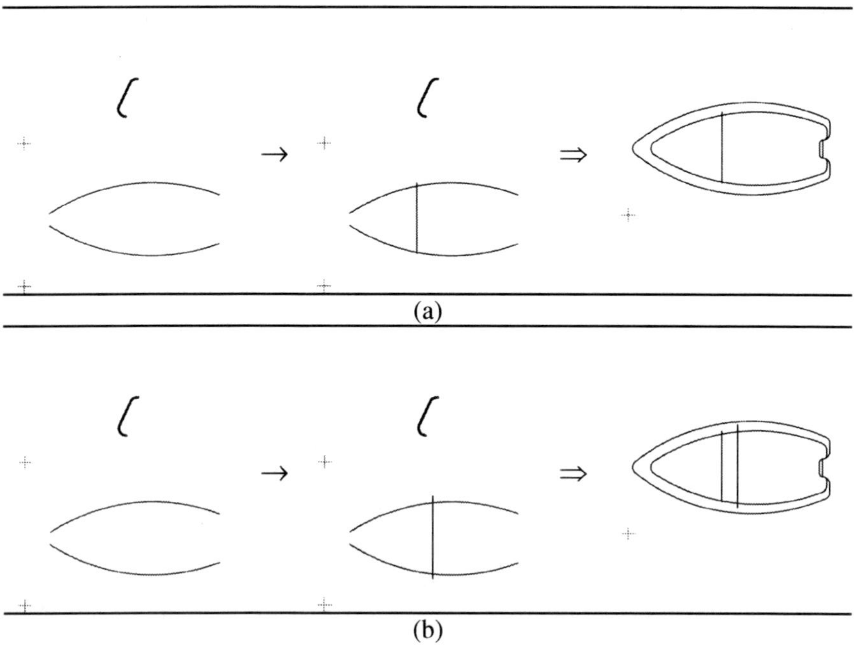

Fig. 7. Rules (a) A030 and (b) A031, as parallel computations with two design descriptions on the left side

Similarly, in rule A031, Figure 7(b), the rearmost end of the trigger shape in the side elevation is now projected down to intersect with the upper edges of the base in plan view. Including the shape from the side elevation in the rule definition also changes the nature of the rule schemas employed in rules A030 and A031. A030 can now be seen to consist of tangential connect and connect, while rule A031 consists of three connect schemas.

Spatial Relations between Design Descriptions

The relationship between the plan view and side elevation, imposed by their adjacency on Andrew's desk top in rules A030 and A031, has apparently produced a workable result by treating two separate instances of design descriptions as one. It is immediately apparent, however, if the axes of each view are labeled, that the transformational space represented by each drawing plane is different, Figure 8. The transformational space of the plan view extends along the X-Y axes, while that of the side elevation extends along the X-Z axes instead.

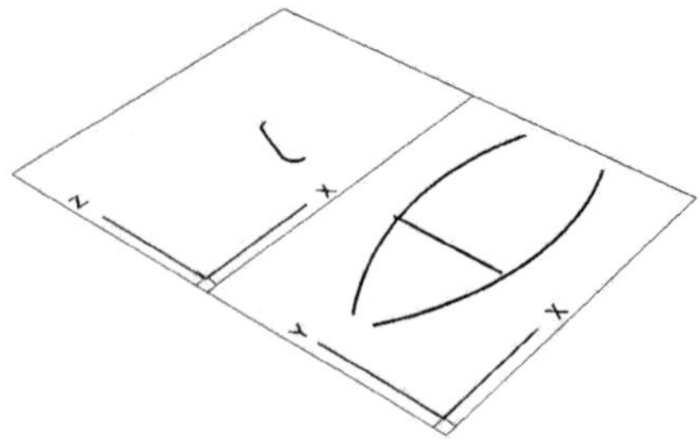

Fig. 8. The side elevation occupies an instance of U^* in x-z drawing plane, the plan another instance of U^* in the x-y drawing plane

Bringing these two instances of separate design descriptions into their correct spatial relationship, Figure 9, highlights the differing representational purpose, not only of the two design descriptions themselves, but also of the true relations between the elements contained in them. The correlative device of the 'vertical' line which Andrew draws in the side elevation, Figure 4, is only vertical while it is contained in the side elevation. When this single element, created by a single rule (or design move) continues onto the plan view it becomes, not so much representationally ambiguous as representationally 'schizophrenic'. The same element which previously represented a vertical line in the plan view now represents a 'horizontal' line instead.

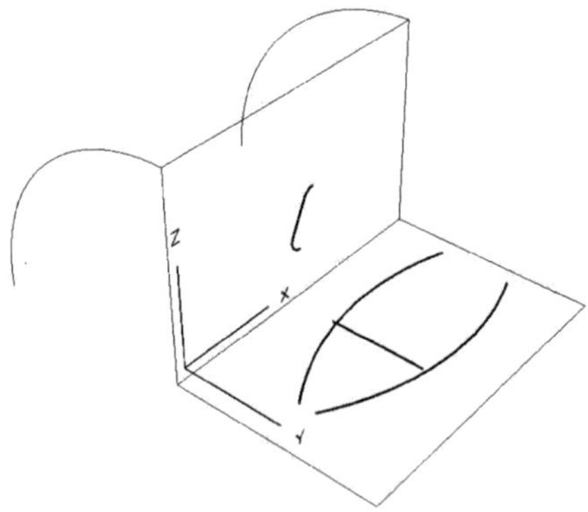

Fig. 9. The result of the parallel computation of shape rule A030, in U^*U^*, when they are placed in their correct spatial relation

As a result, while the spatial relations between the elements involved in triggering rule A030 are apparent if both design descriptions are treated as a single one, or as two of the same type in a parallel computation, when they are treated instead as a parallel computation involving descriptions in differing spatial orientations it is no longer clear how one acts on the other.

Medial Lines as Correlative Devices

An answer may lie in a further aspect of representational ambiguity, above and beyond that already noted in the introduction. As well as the ambiguity that points, lines, and planes can be used to represent physical or spatial features of a design, there is a further form in which connected linear elements have a 'medial' quality, where they can be seen as representing the boundary of a plane [5], Figure 10.

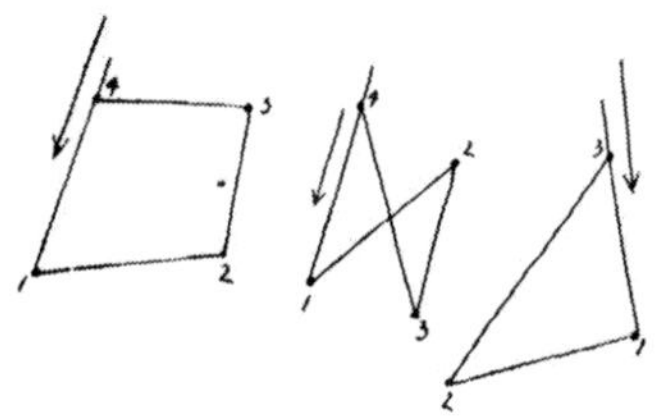

Fig. 10. Examples of medial lines from Klee's Pedagogical Sketchbook [11]

While the design descriptions employed in rule A030, as related orthogonal views of the same object, use the implicit relations between elements in these views to negate their potential representational ambiguity, an extension of the representational ambiguity that gives rise to medial lines can be used to explain how the descriptions in rule A030 may act on each other. The orthogonal projection itself is a special case of the perspective projection, where its vanishing point is set at infinity. Lines drawn in orthogonal views (representing the profiles of parts of the design) can therefore also be seen as the intersection of the drawing plane with an implicit surface projecting at right angles from it, Figure 11. The medial quality of these lines arises from their linear/planar ambiguity, where a linear element can also be seen as the intersection of two planes.

Proposing implicit surfaces as the correlative devices which relate two-dimensional descriptions of three-dimensional objects to each other, and two-dimensional descriptions to three-dimensional ones, arises from evidence, in this and other protocols, that they form part of design cognition in practice.

In all protocols which employed a three-dimensional description, physical or virtual, the transition from two-dimensional descriptions to three-dimensional ones included a distinct orthogonal phase. In this phase the initial rules invoked to create the three-dimensional representation would be used to project profiles, previously drawn in the two-dimensional descriptions, as implicit surfaces running through the three-dimensional description.

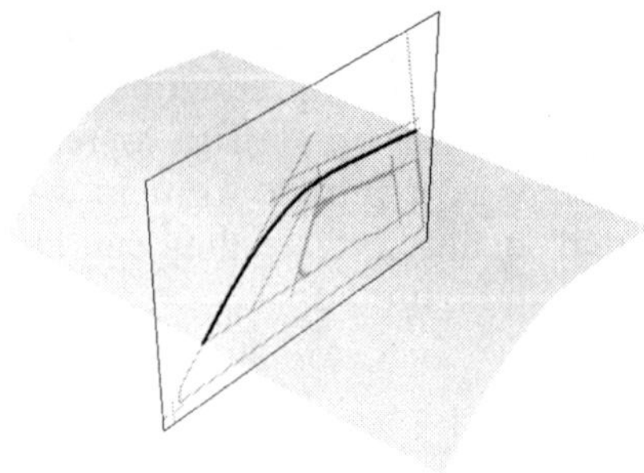

Fig. 11. A linear element (U_{12}) in shape rule A018, as it is contained in an orthogonal view, is also a medial line which lies at the intersection of the drawing plane with an implicit surface (U_{23}) extending at right angles to the drawing plane

Rule A084 from Andrew's protocol can be used to show this in action. Here, in the first shape rules applied to the three-dimensional solid model, A080 – A084, Figure 12, the shape shown in Figure 11 reappears as an implicit surface which is used to create the model's upper surface.

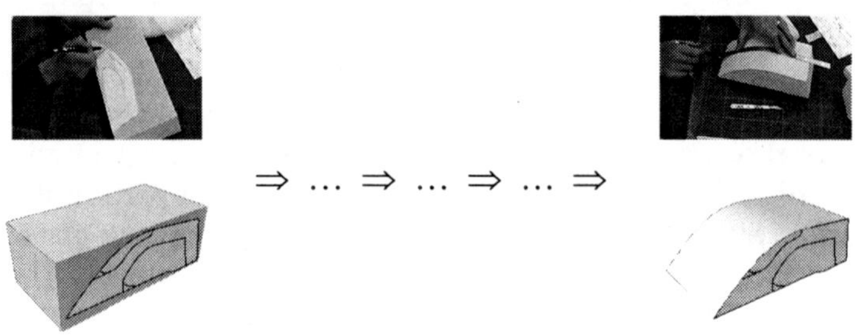

Fig. 12. Examples of implicit surfaces in the transition from two-dimensional descriptions to three-dimensional ones

Returning to rule A030, if implicit surfaces (U_{23}) are projected from linear elements (U_{12}) in design descriptions placed in their correct spatial orientation, their intersection creates an additional U_{23} surface which defines both the location of the linear element created by the rule in a three-dimensional transformational space (U_{13}), and its projection onto the plan view, Figure 13.

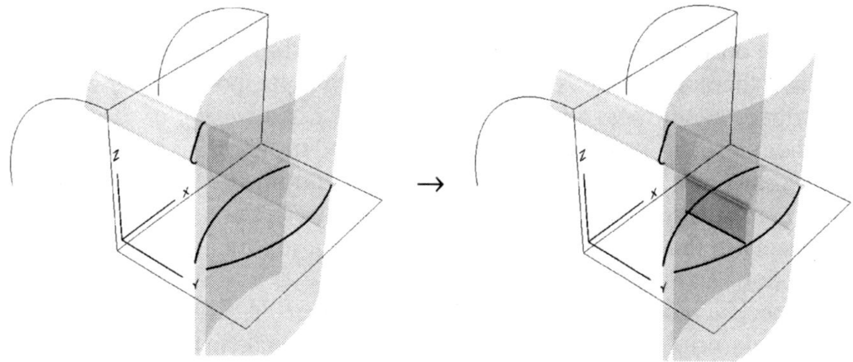

Fig. 13. Shape rule A030 as a parallel computation in $U^*U^*U^*$. Implicit surfaces ($U_{23,}$ U_{23}) act as correlative devices between the first instance of U^* in the side elevation, and the second instance U^* in the plan view. These are used to guide the creation of additional geometry at the intersection of these surfaces, in a further instance of U^*, and to generate the (highlighted) linear element (U_{13})

Implicit surfaces, functioning in this way as projections of two-dimensional linear elements into three-dimensional space, are therefore proposed as an explanation of how the relations between two-dimensional design descriptions are mediated, and which allow a number of descriptions (plus associated relations) to generate three-dimensional designs. By generating further linear elements in a three-dimensional space (U_{13}) at their intersections, Figure 13 illustrates how, in the parallel computation contained in shape rule A030, and in the visual and spatial reasoning of experienced designers, (medial) lines and (implicit) planes can lead to three-dimensionally consistent solids.

Discussion and Conclusions

This research provides examples of how relations work across geometrical design descriptions in practice; either between multiple instances of descriptions of the same type (such as numbers of drawings of the design from different viewpoints), or between different types (such as when two-dimensional descriptions are used to create three-dimensional ones).

A specific aim of the enquiry was to devise a means of analyzing the phenomenological flow of a series of design protocols. The outcome presented here is a segmentation scheme based on Schön and Wiggins' 'see-move-see' cycle. In turn, as this scheme notes the state of the participants' design descriptions at each pause for visual and/or tactile assessment, it is possible to derive a series of shape rules directly from empirical observations of designers' form generation activity.

When these shape rules are married up to a corresponding series of left side trigger conditions, a further outcome is the uncovering of examples of left side shapes in one design description triggering right side rule applications in a related one. The theoretical and practical outcome of this is the confirmation of the value of multiple descriptions in design, and of Stiny's claim that a design is an element in an *n*-ary relation among design descriptions and correlative devices.

A specific contribution made by this paper is, when the relations between descriptions of elements of a design are taken into account, that lines in related orthogonal views of a design have an additional medial quality. They can be seen as a linear elements on a planar surface (as it might be presented in a formal geometrical model) or as the intersection between the drawing plane and a planar surface extending from it (as might be used to define a curve through relations between higher dimensional elements). By expanding the concept of medial lines as correlative devices that relate linear elements in separate drawings (disjoint sets of elements, U^*U^*) it is possible to explain the relations between design descriptions. It is these relations expressed explicitly which allow designers to turn two-dimensional design descriptions into three-dimensional artefacts.

The aggregates of elements such as points, lines (including curves), and planes (including surfaces) are also considered as elements in an *n*-ary relation of descriptions. These *n*-ary relations offer the scope for individual shape grammars, each representing moves on descriptions containing elements such as points, lines, and planes, to be combined to represent moves in a richer world. These moves take place in three-dimensional solids, implemented in physical solids or in a CAD counterpart, which lead to the transition from descriptions with lines and planes, to new descriptions detailing solids. The shift from line and plane to solid is facilitated through *n*-ary relations of constituent descriptions. The generative moves on constituent descriptions during a design process, connected through *n*-ary relations, deliver the integrative moves from line and plane to solid to which the title refers.

The significance of this enquiry to shape grammar implementations, as well as to design practice in general, is that undertaking design in two-dimensional descriptions alone can result in designs which are inconsistent in three dimensions. In those areas of design which are concerned with creating three-dimensional objects it is essential to consider the relations between design elements, and *across* a number of descriptions, such as related views, or other descriptions, such as physical models, when generating the form of that object. Grammar implementations in particular should ideally work on two scales: on the scale of graphical elements in individual instances in one design, and on the scale of those descriptions as elements in an *n*-ary relation of descriptions.

References

1. Stiny, G.: What is a design? Environment and Planning B: Planning and Design 17, 97–103 (1990)
2. Soufi, B., Edmonds, E.: The cognitive basis of emergence: Implications for design support. Design Studies 17(4), 451–463 (1996)
3. Kilian, A.: Design Exploration through Bidirectional Modeling of Constraints. PhD thesis submitted to the Department of Architecture, Massachusetts Institute of Technology, Massachusetts (2006)
4. Li, A.J.-K.: A Shape Grammar for Teaching the Architectural Style of the Yingzao Fashi. PhD thesis submitted to the Department of Architecture, Massachusetts Institute of Technology, Massachusetts (2001)
5. Knight, T.: Computing with ambiguity. Environment and Planning B: Planning and Design 30, 165–180 (2002)
6. Paterson, G.: Form Generation in Design. PhD thesis submitted to the Department of Design, Development, Environment and Materials, The Open University, Milton Keynes (2009)
7. Schön, D.A., Wiggins, G.: Kinds of seeing and their functions in designing. Design Studies 13(2), 135–156 (1992)
8. Antonsson, E.K., Cagan, J.: Preface. In: Antonsson, E.K., Cagan, J. (eds.) Formal Engineering Design Synthesis. Cambridge University Press, Cambridge (2001)
9. Stiny, G., Gips, J.: Algorithmic Aesthetics: Computer Models for Criticism and Design in the Arts. University of California Press, Berkeley and Los Angeles (1978)
10. Stiny, G.: Shape: Talking about Seeing and Doing. MIT Press, Cambridge (2001)
11. Klee, P.: Pedagogical Sketchbook. Faber and Faber, London (1925)

Interactions between Brand Identity and Shape Rules

Rosidah Jaafar, Alison McKay, Alan de Pennington, and Hau Hing Chau
University of the Leeds, UK

Brand and the maintenance of brand identity are key drivers in consumer product development activities. This paper aims to give early results of research that sits between marketing and shape computation technologies. An initial review of brand identity from a marketing perspective is presented along with an analysis from student designers in identifying brand characteristics of consumer products. Results of early experiments with a prototype computational design synthesis system are reported.

Introduction

Initial design concepts are typically produced at the early stages of a product development process. When designing new products within a range, often the company or brand owner needs to preserve aspects of shape style so that the brand identity is maintained. It is the task of product designers to preserve aspects of style across existing and new product designs. How to define and capture style within the same brand remains an open question. The long term goal of the research reported in this paper is to support product designers in synthesising design shapes more effectively. Building on earlier studies of shape grammars as a potential design tool to support design synthesis, this paper reports an experiment where shape aspects of a common brand were identified, defined in a shape grammar and then used in a prototype software tool to generate new designs in the style of the brand. The resulting designs were compared with designs generated during a shape grammar workshop by a Masters level Product Design student.

J.S. Gero (ed.): *Design Computing and Cognition'10*, pp. 269–284.

Many products from well-known brands have made many changes to their shapes within the group style as the brand evolves and develops. For example the shapes of Coca-Cola bottles have been through a number of changes since they were first manufactured in the form of glass bottles. Although the bottle designs have been altered and refined over the years, their bottle shapes retained certain characteristics [1]. Some of the changes can be explained by the need to respond to new customer needs, new production processes and materials, or the need to compete with the same product from others brand products in the market. A key challenge for product designers lies in capturing a brand style (and so the brand identity) while creating new designs in a given range. The long term goal of the research reported in this paper lies in establishing systematic methods for capturing shape aspects of brand identity. In addition to supporting the design of new products that communicate the brand identity through their shapes, the use of such systematic methods could also contribute to improving the product development process as a whole, for example by making it faster, more efficient and more responsive to the needs of a number of stakeholders including brand owners and retailers.

Background

This paper draws together research from four distinct areas: shape design, branding, shape grammars and the role of computational design synthesis tools in product development processes. Reviews of current work in each of these areas are provided in the following sections.

Product Shape Design

The visual appearance of a product is a critical determinant of consumer response and product success [2] and, as such, plays a key role in both establishing the product in a market and in maintaining its competitiveness. A framework to communicate consumer response to product appearance was developed by Crilly [3] who asserts that visual appearance of products plays a significant role and influences commercial success and quality-of-life. Bloch [2] also noted that product appearance is a key component in defining product-person relationships and significantly affects commercial success. As such, product appearance should be integral to the product concepts that are developed in product development processes.

Design function plays an important role in defining the physical forms of the product to meet customer needs, which includes mechanical design and industrial design [4]. Sometimes design is equated with integrating and reconciling the nature and characteristics of materials and components, their quality, price and ergonomic [5]. Berkowitz give some examples of familiar packaged food items that involve minor shape modifications such as a small cube or ball of chocolate instead of an elongated bar, crinkle cut potato chips instead of flat ones and cut green beans instead of whole ones. In this context, the innovation strategy is to be consistent with social and cultural trends for the potential increase share in an existing market. Berkowitz also noted that design is related to creation of a product or brand identity. For this reason, it is important to develop a product shape consistently with current consumer life-styles and for commercial success of the product.

Product designer produces sketches as a significant proportion of one's output. A product designer typically produces a range of varied initial ideas before converging to a chosen solution principle and develops it further [6]. For each idea, it usually starts from a preliminary layout with more details being added to it gradually. In the process maturing of a concept towards an initial idea, replacing an existing feature with a new one, one at a time, is a norm that can be observed by inspecting a sequence of developing sketches. The pedagogical value of find-and-replace approach is concurred by product design educators [7], [8] and [9].

Brand Identity

Brand is the promises and expectations that reside in each customer's mind about a product, service and company [10]. Successful brands give good promises to customers and can lead to brand loyalty for as long as the promises are maintain. People buy branded products mostly because of the way they can fit those brands into their own lives and because they have an affinity with the personalities projected by them [10]. From a marketing perspective, brand is important for its ability to attract and maintain consumer attention. For this reason it is essential for the brand to have its own identity that differentiates it from competitors. Aaker [11] defined brand identity as a unique set of brand associations that the brand strategist aspires to create or maintain. The association represents what the brand stands for and implies a promise to customers. Meanwhile, Wheeler [12] defined brand identity as the visual and verbal expression of a brand, which supports, expresses, communicates, synthesizes and visualizes the brand. For example, brand can be conveyed by their shape such as the Apple iPod; graphics devices such as Kodak's yellow; words such as Coke and Coca–Cola, Kleenex or Pampers; and people/characters such as Ben and Jerry, Mickey Mouse or Hello Kitty.

A brand should have the ability to stand apart from its competitors. When a company produces products that are consistent with the brand strategy, then all aspects of the brand work in unison to compete effectively in the marketplace [13]. It is important also for a brand to stay relevant to a large consumer market segment in order to maintain or increase market share. To maintain brand identity, Park [14] proposes measuring the similarity of appearance between the new and existing designs based on product level features and the consistency between the product and brand concepts. In determining feature similarity, Person [15] reports a study on the influence of market environment. Person found that it is more important for the new product to be similar to the existing product portfolio in the maturity phase than in the growth phase of the product/brand life cycle.

Shape Grammars and Shape Rules

Design is the process of transforming an initial set of requirements into an explicit tangible form that meets those requirements [16]. The process of transformations involves activities such as adding details, modifying or adding new components to the current design. A design grammar is a language that relates to the representation structure and transformation mechanisms. In this context, design using a grammar involves the application of transformation rules to an initial design until the final design satisfies the design requirements. The types of designs that result from a given transformation process depend on the particular mechanism (for example, the group of design rules) used and the point from which the transformation process was initiated.

The first published paper on shape grammars by Stiny and Gips [17] illustrated shape grammar as an original language of paintings. An approach for creating new grammars from scratch was first proposed in 1980 by Stiny in his kindergarten grammar [18]. Shape grammars were defined by Stiny [19] as a production system of shape which accommodates the property of emergence. Stiny and Gips also noted that shape grammars were defined over alphabets of shapes in which one can generate n-dimensional shapes. This means that a shape can be generated using a shape grammar by applying a series of shape rules.

The initial exploration of shape grammar by Stiny was focused in the architecture field, for example, the Chinese lattice design [20], the Mughul garden [21] and the Queen Anne house [22]. More recently shape grammars have been used in engineering design, industrial design, and product design. For example, Agarwal and Cagan [23] explored the use of shape grammars in product design, and defined a coffeemaker grammar, while Pugliese and Cagan proposed the Harley Davidson motorcycle shape grammar [24]. Shape grammars have also been demonstrated as tools to generate products consistent with a brand, for example, McCormack and Cagan [25] captured the brand characteristics of Buick styling for Buick vehicles in a shape grammar.

Stiny and Gips defined a five stage programme for creating new design languages: a vocabulary of shapes, spatial relations, shape rules, initial shape and shape grammars. A shape grammar consists of a finite set of shapes, a finite set of labels, a finite set of shape rules and an initial shape. The shapes in the shape rules and the initial shape can be linked with labels to constrain the shape generation process. The shape generation process starts with an initial shape and proceeds through a series of shape rule applications. Any Euclidean transformation of a rule, such as mirror or rotate, is regarded as equivalent to the transformed rule itself. An example is illustrated in Fig. 1 for a simple shape grammar in generating new shape.

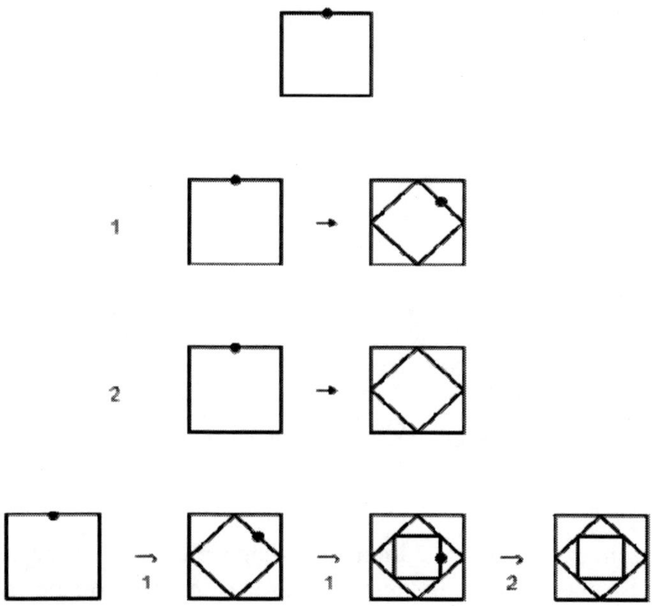

Fig. 1. Generation of a shape from the shape rules, reproduced from [19]

The Role of a Computer Aided Design Synthesis System

Product development capabilities are important as the basis for successful competition of products in the market. Improving product lines is crucial for the success for any company or organization in order to maintain competitive in the market and increase the sales. Successful product development processes typically include mechanisms to meet customer needs, reduce waste and minimize time-to-market, so bringing products to market earlier in order to be competitive in the market. One way of achieving improvements lies in implementing design technology in process design effectively. Ulrich and Eppinger [4] noted that success of product development consists of producing quality products that meets with customer needs, time development should response quickly to technology development and competitive market, product cost development should be less and development capability for new product should be more effectively. Therefore, in product design activities, it is essential to perform and respond to the criteria of the success product development so that it gives good feedback to the design activities. Any approaches to organizing and enhancing product development processes will help to reduce of product and process design cycle time. Figure 2 illustrates the interrelation between design practice and design technology.

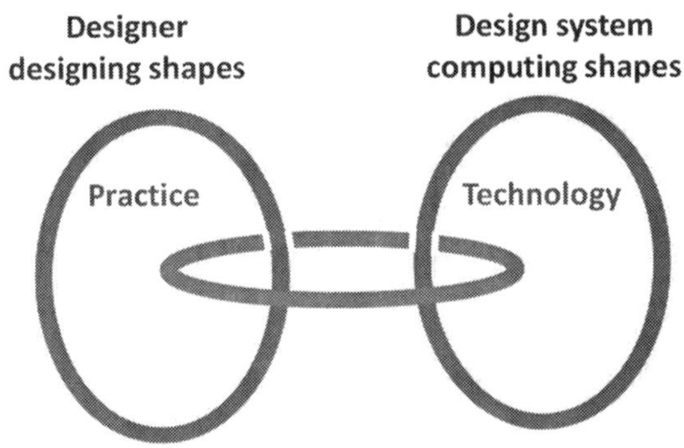

Fig. 2. Interrelation between design practice and technology (from [26])

Method

The research explored the capture of shape elements of product brand identity with the goal to create a design system that can be used in the synthesis of new designs in a given brand. Based on an earlier study of shape grammar and its use in the design of consumer products, this research has the potential to support the creation of new product designs with the style of a given brand. The designs that are of interest for this research are consumer products in a three dimensional shape.

The research process is summarized in Fig. 3. It can be seen that the process started with the selection of a corpus of designs; in this case, a selection of products in the same brand. Analysis of the products was used to inform the derivation of brand characteristics that were used to inform the definition of a shape grammar. This grammar was implemented in a prototype computational design synthesis system and used to generate new design shapes. Finally, the design shapes were evaluated. The research process, overlaid on the content of Fig. 2, is shown in Fig. 4.

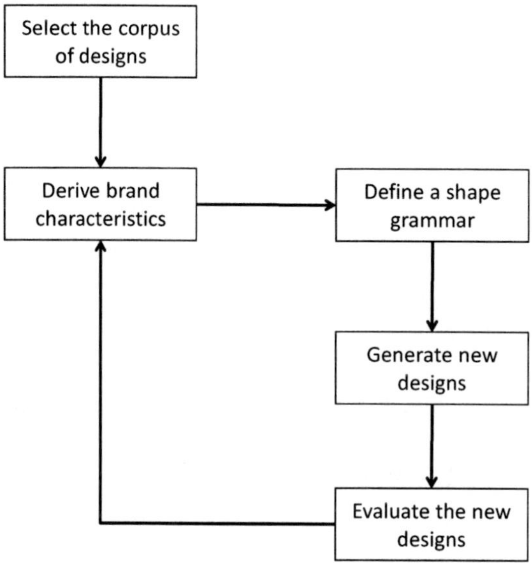

Fig. 3. Experimental method

Results

The student designers completed the project reported in this paper as part of a Masters level design research module; the project was one of three 50 hour design research projects that form the module. It began with four teams of three or four students (15 students in total) being assigned the task of identifying a brand and purchasing one product within the brand per student. Students then analysed these products to identify brand characteristics and each created a page of new design shapes in the style of the brand. These designs were used as input to a shape grammar workshop where students were introduced to shape grammars as a way of articulating design languages for their selected brands. Finally, during and after the workshop, they applied ideas from shape grammars to produce new design shapes in the style of their selected brand.

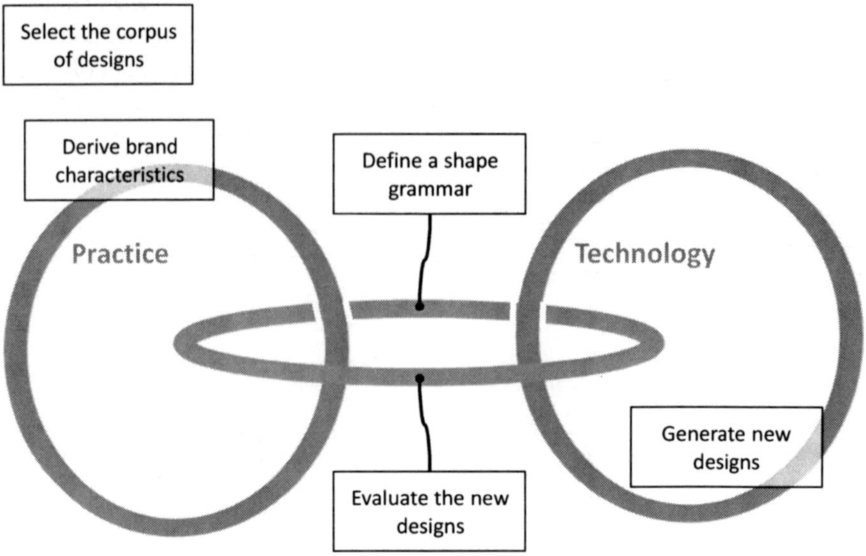

Fig. 4. The experimental method in the context of design practice and technology

Selection of the Corpus of Designs

The students were assigned the task to select a brand and then go shopping and buy four products in the brand (one per team member). They were to buy products whose shapes were as different to each other as possible but all belonging to the same brand. Cost may have influenced their choice because the students were told to choose a brand that contained products they could afford because they would not be reimbursed for the costs of these products. The brands chosen were Heinz (ketchup products), Domestos (domestic cleaning products), Garnier (personal care products) and Apple Macintosh (ipod products).

Having selected the brands, the students produced an initial page of sketches of selected product; The Heinz student group analysed Heinz tomato ketchup bottles (plastic and glass) and a salad cream bottle. The Heinz tomato ketchup bottle was chosen for the study reported in this paper because it is a recognized brand, the brand has a number of clear shape characteristics common across bottle content and material, and, from the students' work, the application of shape grammars resulted in a wide range of novel design ideas. Fig. 5 shows the sketches of one student for the Heinz brand. At this stage, the students had some experience in generating design concepts but not in using shape grammars. They were not told that the project related to shape grammars until after they had submitted these sketches.

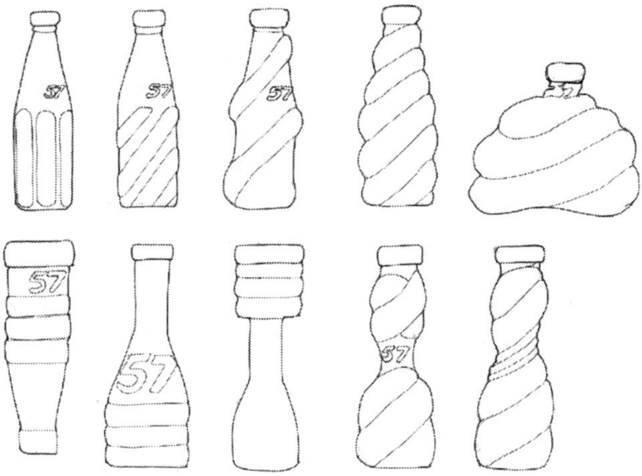

Fig. 5. Initial sketches

The original Heinz Tomato Ketchup bottle is shown in Fig. 6(a) and the outer line of the bottle shape together with outer line of label is shown in Fig. 6(b).

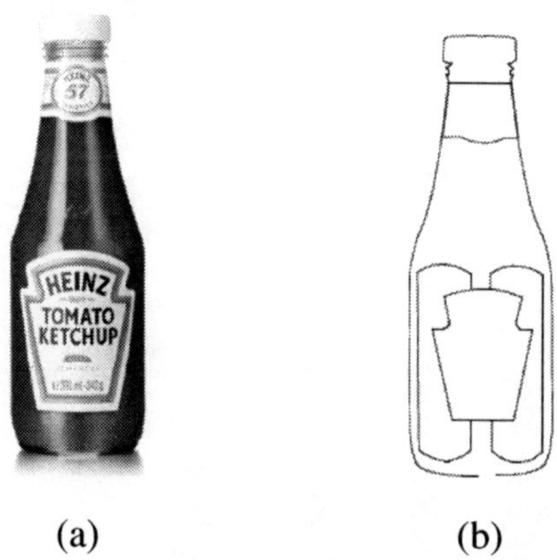

Fig. 6. Heinz Tomato Ketchup bottle

Derivation of Brand Characteristics

Once purchased, the students were asked to analyse their selected brand and the products they had bought to identify key characteristics of the brand shape.

Schön [27] outlines design as a process of seeing, moving and seeing again which can be described as a series of 'design moves'. Schön also states that there are four kind of seeing in design: visually appreciating what is there to see, detecting the consequences of a move, judging the quality of a configuration, which may be produced as consequences of a move, and recognising a feature, or quality, regardless of whether or not we can explain why. Based on this outline, the analysis was carried out by a group of student designers to identify brand characteristics from their sample products. Based on these analyses, brand characteristics were grouped into four categories: shape, colour, label and graphics, and surface finish. Shape is one of the brand characteristics that is essential in maintaining brand identity through the bottle design. For marketing purposes, which needs to attract women or men products, the shape of the bottle can be personified into masculine or feminine look.

Color can trigger an emotion and evoke a brand association [12]. Distinctive colors can build brand awareness and can apply to make obvious color difference between lid and main body of the bottle. The colors also can be repeated in graphics as key product color. The label or graphics image is recommended to have relevant connection to the type of product of what it is represented and go along with the colors chosen for that product. The label and graphics should have bright graphics with recognisable font and high contrast between label and product. The concept and idea of the student designer is shown in Fig. 7.

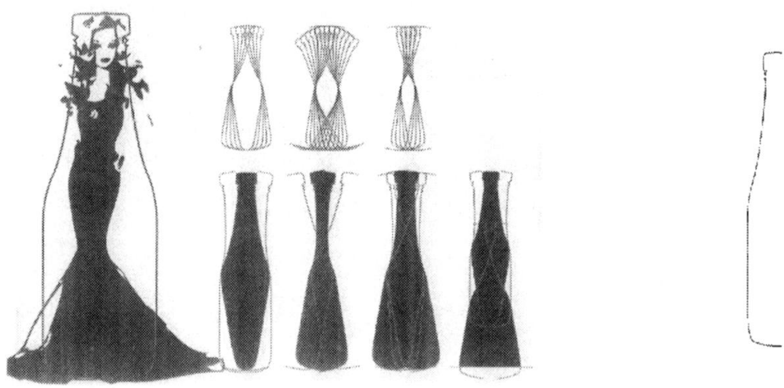

(a) (b)

Fig. 7. (a) The concept and idea of student designer for Heinz Tomato Ketchup design. (b) Half of bottle profile

The student designer developed the shape shown in Fig. 7 using hand-drawn sketches and Adobe Illustrator for the final presentation. It can be seen that the student designer experimented by taking half of the profile of the original bottle, as shown in Fig. 7(b), and transformed it by rotating around a midpoint several times, reflected and positioned at three different intersection distances to see what possible new shapes were created. The outer bottle shape was furthered analysed and combined several lines to form a number of new shapes of the bottle. The combination of several lines still involved the original part or lines from the original bottle so that the brand identity of the bottle still maintain. This is an example in which transformation in shape grammars can be applied to create a new shape using an Euclidean transformation, t, which includes translation, rotation, reflection and scale.

Definition of a Shape Grammar

Based on an interview and observation of how the student designer manipulated shapes, a collection of shape rules was defined. These are illustrated in Fig. 8. It can be seen that half of outer bottle shape from Fig. 6 was selected for use in the definitions of shape rules for the Heinz Tomato Ketchup bottle. Firstly, shape rule 1 was defined by rotating the half profile around a midpoint twice by 5° in a counter clockwise direction. Meanwhile in shape rule 2, the half profile was transformed by rotating it around a midpoint twice by 5° in a clockwise direction.

The mechanism of the application (c.f. creating a shape rule) of a shape rule does not recognise any Euclidean transformation. The application of a shape rule, one step of shape computation $C' = [C - t(A)] + t(B)$ using a shape rule $A \rightarrow B$, denotes (i) to move, to rotate and to scale shape A such that there is a matched occurrence in the working shape C; (i) to remove shape A, which is on the left hand side of the shape rule; and (ii) to replace shape B, which is on the right hand side of the shape rule, to from a resultant shape C'.

Nevertheless, when creating a shape rule, one might use Euclidean transformation to capture a design intent. For example, shape Rule 1 in Figure 1 can be seen as (i) retaining the original square while removing the labeled point; and (ii) taking another copy of the original square and labeled point, and rotating them by 45° and scaling down by $1/\sqrt{2}$ by the centroid.

Centroid and symmetry planes are obvious reference geometries for creating a new shape rule. Krishnamurti [28] introduced the idea of registration points, which included all intersection points of each and every pair of basic elements in a shape. Tapia [29] and Chau [30] applied registration points in their respective shape grammar implementations as centres of rotation and other shape transformation tasks.

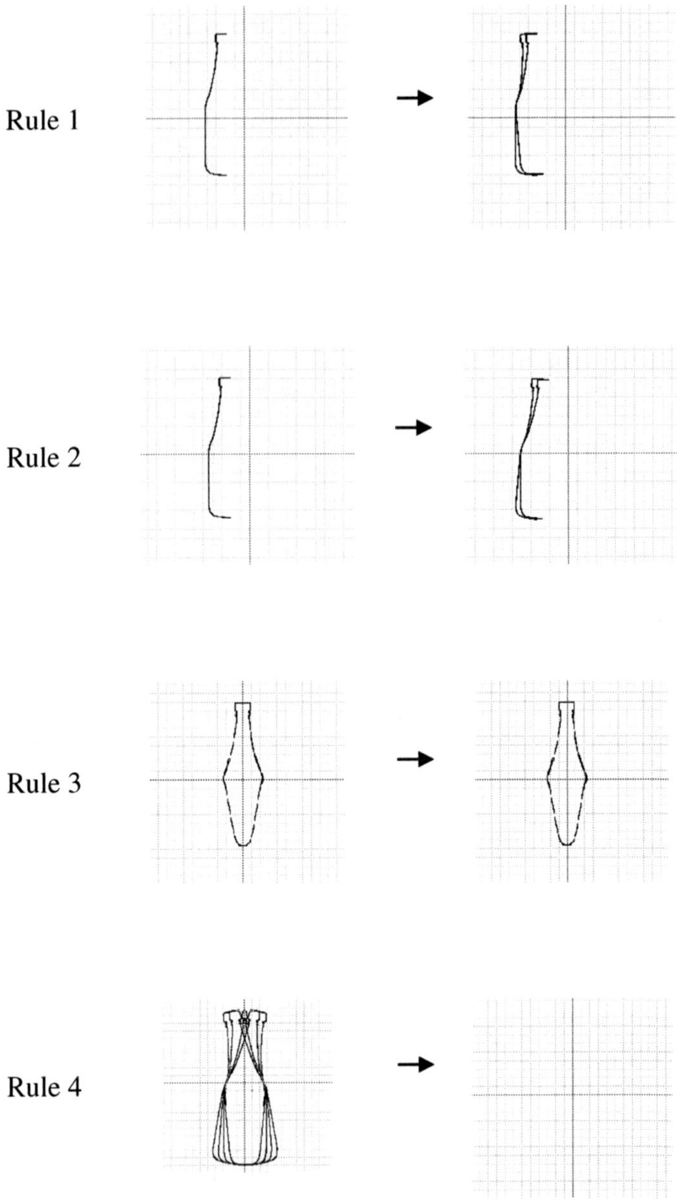

Fig. 8. Shape rules in the Heinz Ketchup bottle grammar

Shape rule 3 denotes an identity rule [31] which detects a resultant shape of the interest of this shape computation. The current working shape is then decomposed as a lattice [32] using the shape identified by Rule 3 as one part. The complement shape [33] can be identified which forms the left hand side of Rule 4.

Generation of New Designs

The Heinz Ketchup bottle grammar was used to generate a series of designs using a prototype computational design synthesis system [26] and starting from the initial shape as shown in Fig 9.

Fig. 9. Initial shape

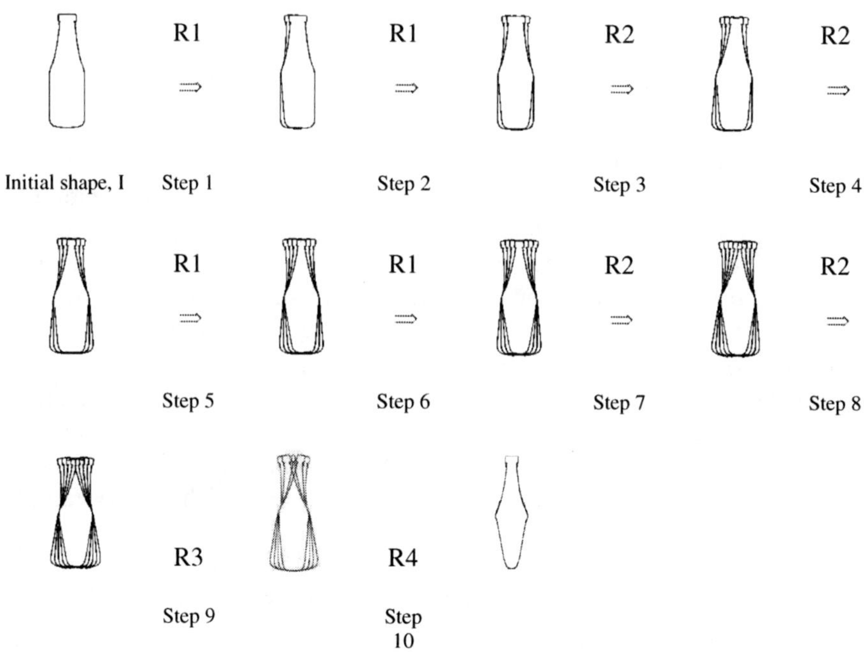

Fig. 10. An example of shape derivation

Evaluation of the New Designs

Examples of the resulting designs are shown in Fig. 11.

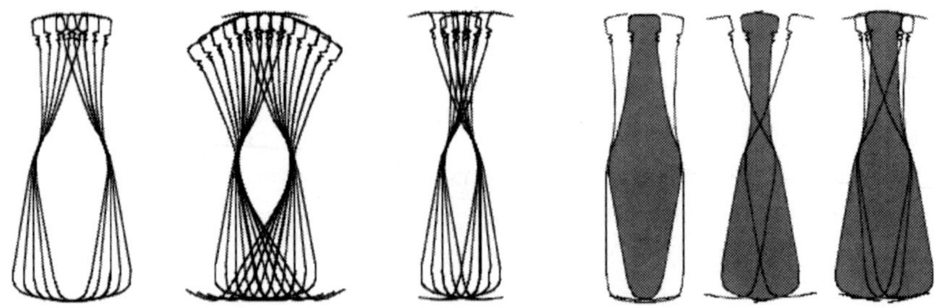

(a) (b)

Fig. 11. Transformation of profiles and resulting new bottle shapes

It can be seen that bottle designs similar to those produced by the student, Fig. 7(a), can be extracted from the computer generated shapes shown in Fig. 11(a). Examples of such shapes are those highlighted in red on Fig. 11(b).

Conclusion

In this paper we have reported an early experiment where brand characteristics were identified by analysing a corpus of designs within a common brand. These characteristics were captured in a shape grammar that was implemented in a shape grammar-based design synthesis system that was used to generate new designs. The resulting designs were evaluated through comparison with the original student generated designs. These early results provide promise that shape grammars might be used as a method to capture brand identity and inform new design shapes.

Further exploration in applying shape rules in the design of branded consumer products is now in progress. The focus of the research is on the support of real design scenarios in the context of a whole product development process. Key issues include establishing systematic approaches that can be used to extract brand style, integration within product development processes without impinging on designers' creative processes, and more building quantifiable means of evaluating generated designs so that the most viable designs can be identified quickly.

Acknowledgements

Rosidah Jaafar is on secondment from Universiti Teknikal Malaysia Melaka studying for a PhD; her research is sponsored by the Ministry of Higher Education, Malaysia. The authors also wish to thank Dr Iestyn Jowers who provided advice on the use of the software prototype and Ms Sarah Stevens, the design student whose work is used in this paper.

References

1. Chen, X.: Relationships between product form and brand: A shape grammatical approach in School of Mechanical Engineering. PhD: University of Leeds (2005)
2. Bloch, P.H.: Seeking the Ideal Form: Product Design and Consumer Response. The J. of Marketing 59(3), 16–29 (1995)
3. Crilly, N., Moultrie, J., Clarkson, P.J.: Seeing things: Consumer response to the visual domain in product design. Design Studies 25, 547–577 (2004)
4. Ulrich, K.T., Eppinger, K.D.: Product Design and Development. Tata McGraw Hill, New York (2003)
5. Berkowitz, M.: Product Shape as a Design Innovation Strategy. J. of Production Innovation Management 4, 274–283 (1987)
6. Liu, C.: Innovative Product Design Practice: Carl Liu Design Book. CYPI Press, London (2007)
7. Krisztian, G., Schlempp-Ülker, N.: Visualizing Ideas: From Scribbles to Storyboards. Thames & Hudson, London (2006)
8. Pipes, A.: Drawing for Designers: Drawing skills, Concept sketches, Computer systems, Illustration, Tools and materials, Presentations, Production techniques. Laurence King Publishing, London (2007)
9. Eissen, K., Steur, R.: Sketches: Drawing Techniques for Product Designers. BIS Publishers, Amsterdam (2007)
10. Upshaw, L.B.: Building Brand Identity: A Strategy for Success in a Hostile Marketplace. John Wiley & Son, Inc., Chichester (1995)
11. Aaker, D.A.: Building Strong Brands. Free Press, New York (1996)
12. Wheeler, A.: Designing Brand Identity: A Complete Guide to Creating, Building and Maintaining Strong Brands. John Wiley, Hoboken (2006)
13. Cagan, J., Vogel, C.M.: Creating Breakthrough Products: Innovation from Product Planning to Program Approval. Financial Times Prentice Hall, Englewood Cliffs (2002)
14. Park, S.W., Milberg, S., Lawson, R.: Evaluation of Brand Extensions: The role of product feature similarity and brand concept consistency. The J. of Consumer Research 18(2), 185–193 (1991)
15. Person, O., Schoormans, J., Snelders, D.: Should new products look similar or different? The influence of the market environment on strategic product styling. Design Studies 29, 30–48 (2008)
16. Brown, K.: Grammatical design. IEEE Intelligent Systems 12(2), 27–33 (1997)

17. Stiny, G., Gips, J.: Shape grammars and the generative specification of painting and sculpture. In: IFIP Congress. North Holland Publishing, Amsterdam (1971)
18. Stiny, G.: Kindergarten grammars: designing with Froebel's building gifts. Environment and Planning B 7, 409–462 (1980)
19. Stiny, G.: Introduction to shape and shape grammars. Environment and Planning B 7, 343–351 (1980)
20. Stiny, G.: Ice-ray: a note on the generation of Chinese lattice designs. Environment and Planning B 4, 89–98 (1977)
21. Stiny, G., Mitchell, W.J.: The grammar of paradise: on the generation of Mughul gardens. Environment and Planning B: Planning and Design 7, 209–226 (1980)
22. Flemming, U.: More than the sum of parts: the grammar of Queen Anne houses. Environment and Planning B: Planning and Design 14, 323–350 (1987)
23. Agarwal, M., Cagan, J.: A blend of different tastes: the language of coffeemakers. Environment and Planning B: Planning and Design 25, 205–226 (1998)
24. Pugliese, M., Cagan, J.: Capturing a rebel: modeling the Harley-Davidson brand through a motorcycle shape grammar. Research in Engineering Design 13, 139–156 (2001)
25. McCormack, J.P., Cagan, J.: Speaking the Buick language: capturing, understanding, and exploring brand identity with shape grammars. Design Studies 25, 1–29 (2004)
26. McKay, A., Chase, S., Garner, S.W., et al.: Design Synthesis and Shape Generation. In: Designing for the 21st Century-Volume 2: Interdisciplinary Methods and Findings, pp. 304–321. T. Inns, Ashgate Publishing Limited (2009)
27. Schön, D., Wiggins, G.: Kinds of seeing and their functions in designing. Design Studies 13(2), 135–156 (1992)
28. Krishnamurti, R.: The construction of shapes. Environment and Planning B: Planning and Design 8(1), 5–40 (1981)
29. Tapia, M.: A visual implementation of a shape grammar system. Environment and Planning B: Planning and Design 26(1), 59–73 (1999)
30. Chau, H.H.: Preserving Brand Identity in Engineering Design Using a Grammatical Approach. PhD thesis, School of Mechanical Engineering, University of Leeds (2002)
31. Stiny, G.: Useless rules. Environment and Planning B: Planning and Design 23, 235–237 (1996)
32. Stiny, G.: The algebras of design. Research in Engineering Design 2, 171–181 (1991)
33. Stiny, G.: Shape: Talking about Seeing and Doing. MIT Press, Cambridge (2006)
34. Stiny, G.: Shape rules: closure, continuity, and emergence. Environment and Planning B: Planning and Design 21, 49–78 (1994)

Approximate Enclosed Space Using Virtual Agent

Aswin Indraprastha and Michihiko Shinozaki
Shibaura Institute of Technology, Japan

In agent-based pedestrian model, steering movement is driven by position of attractor or goal, and a graph of their relations. Our work studied on constructing relationship between spatial cognition and enclosed space using virtual agent. Instead of focusing on location-based goals, we are investigating enclosed space as primary factor for locomotion. Our contribution on the identification of enclosure enhances the artificial model of spatial cognition. This is significant for the development of agent-based simulation with spatial cognition to determine and to measure space in architectural design model. We present our approach using three stages of methods. First, we constructed object detection algorithm on agent line of sight. Second, by decomposing detected objects as set of points, we analyzed their attributes and properties to define center of enclosed space. Third, as point of enclosed spaces determined, we classified them into L-shaped space and U-shaped space using simple arithmetic algorithm. Finally, computational results of points represent goals for navigational purpose.

Enclosed Space

Space or enclosure is a medium or interfaces of how we interact and occupied the design of architecture. As one of high level emotional needs, a space is an object of necessity that manifest into three mental needs: stimulation, security and identity [1], [2], [3]. In addition to this, Barker (1968) elaborated notion of behavioral setting to describe how our behavior is influenced and even constrained by space settings.

Furthermore, space settings are one of important component to develop locomotive ability that conceptualized by spatial knowledge [4]. This ability is necessary for the study of way finding behavior and path learning. Based on this, spatial information related to geometry features of

J.S. Gero (ed.): *Design Computing and Cognition'10*, pp. 285–303.

environment plays important role for determining enclosed space that lead to the further investigation on behavior-based simulation and emergent shape recognition.

Despite of enormous numbers of possibilities to define space by its elements, there are six defined architectural spaces [5]: by linear elements (column), by single vertical plane (i.e. wall), L-shaped plane, parallel planes, U-shaped plane and four planes to define full closure. We are investigating two of these characteristics: L-shaped space and U-shaped space defined by walls.

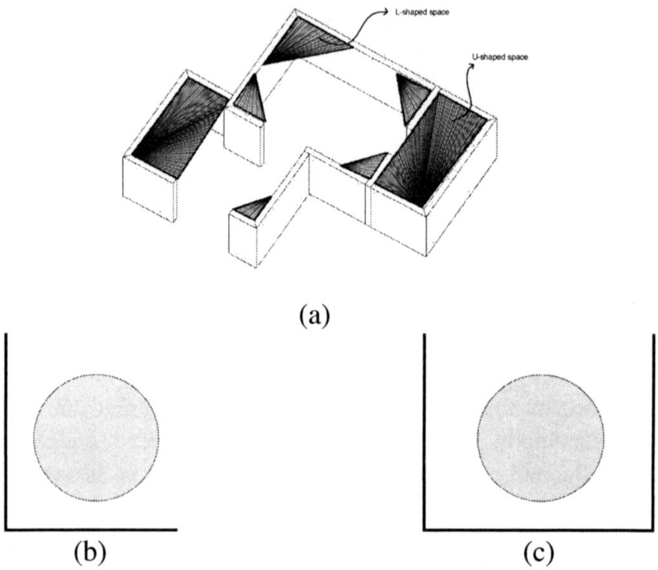

Fig. 1. The diagrammatic model of enclosed space. (a) L-Shaped space; (b) U-shaped space; (c) an example in architectural design

Previous Simulation on the Interaction between Architectural Design and Virtual Agent

Virtual simulation on the interaction between virtual agent and their environment has reach significant result in term of their methods and findings. Depend on focus and goal of simulation, various attempts have been conducted particularly to:

a. Visualize and analyze their spatial relationship [6], [7], [8]
b. Modeling interactive objects and their interplay with agent behavior [9],[10].

In case of architectural design, Narahara [11] explored possibility to develop virtual agent that *aware* of their physical setting by using programmable modeling environment tool with variables of constraints. These constraints represent set of rules for the interaction between agent and their architectural features. There are four rules to control the behaviors:

1. Attractor-based reaction: assigned to particular element to attract agent's visual attention in a time-based action.
2. Variable dependent reaction: internal states of each agent that can influence their decision for particular stimulation.
3. Visibility-based reaction: assigned to agent's vision to access certain reaction if they encounter with visible objects.
4. Agent to agent reaction: different degree of social ability to develop agent to agent interaction.

Those four basic rules contribute important step for modeling relationship between architectural features and basic behavioral patterns.

Wei Yan and Yehuda E. Kalay [12] stepped forward by integrating statistical model as decision tool into a cognitive model. Their cognitive model is based on four components which are able to handle dynamic problems that are not predictable in advance:

1. Knowing: Information related to the geometrical properties of all elements in the environment is provided as the basic knowledge
2. Finding: Set of computational process to define path toward the destination. The modification of A* algorithm is developed to search the optimized path to the goal
3. Seeing: Modeling social spaces as an influential space for avoidance and recognition behaviors.
4. Counting: Compute duration of certain action based on statistical analysis and set this as threshold for upcoming action.

On both system and methodology, we can see that the process of acquiring information from environment features is a primary factor for developing behavior-based simulation. However, it still remains question how does the agent can detect '*space*' or enclosed space which is formed by spatial configuration of architectural elements.

Geometric Modeling of Enclosed Space

Enclosed space is a product of geometrical relationship between architectural features. To understand how an enclosed space is formed and for the purpose of this research, we first examine geometric properties of wall and the relationship of it lines and points. All architectural elements can be decomposed into set of lines and further, set of vector points. By examining these points through their co-linearity, we can determine convex area and non-convex area for further classification of enclosed space. Each method for the experiment is explained in the following section.

Co- linearity of Three Points

Detecting co-linearity of three points can be achieved by applying simple algebra using their determinant property. If there are exist three points P (x_p, y_p), $Q(x_q, y_q)$, and $Z(x_z, y_z)$, the co-linearity of PQR is computed by using determinant equation as in Figure 2.

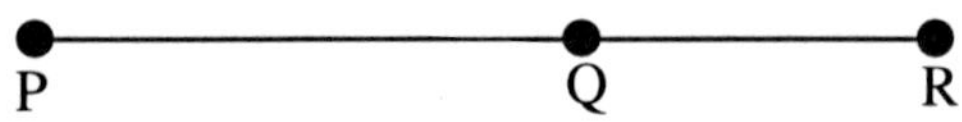

$$\Rightarrow \frac{1}{2}\begin{vmatrix} X_p & Y_p & 1 \\ X_q & Y_q & 1 \\ X_z & Y_z & 1 \end{vmatrix} = 0, \Rightarrow \begin{vmatrix} X_p & Y_p & 1 \\ X_q & Y_q & 1 \\ X_z & Y_z & 1 \end{vmatrix} = 0$$

$$\Rightarrow (X_p(Y_q - Y_z) - X_q(Y_p - Y_z) + X_z(Y_p - Y_q)) = 0$$

Fig. 2. Co-linearity test of three points

Boolean algorithm then is applied to define co-linearity of PQR. *True* value indicates co-linearity. If it resulted in *false,* those points must constitute an area that can be either convex area or non-convex area. The term *convex* and *non convex* in the experiment is relative to the position of the observer. An area is defined as *convex* if location of observer is *inside* or at the border line area formed by those points.

Determination of convex area is important step as logical consequence for defining an enclosed space based on agent position.

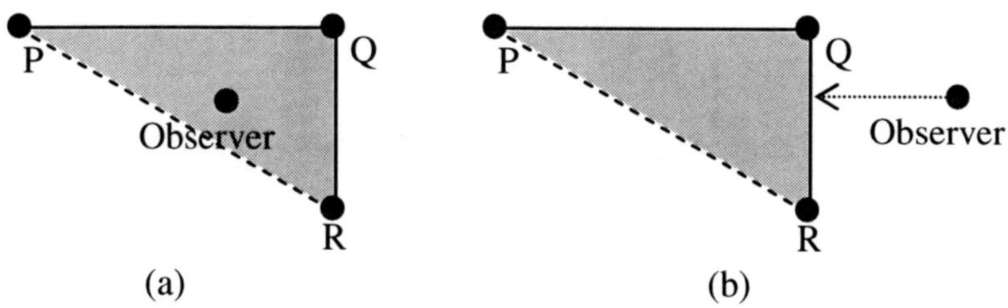

Fig. 3. (a) Convex area; (b) Non-convex area in the context of observer location

The Determination of Convex Area

Following the procedure of co-linearity, we employed theory of convex polygon and *winding number* [13] to determine convex area based of the position of the agent. The concept of winding number is to determine whether a certain point O is inside closed polygon area L by counting *number of times a loop* wind of L traverses around O. We assume our convex polygon is categorized as simple loop since it does not intersect itself. In the most cases of our experiment, the enclosed space is formed as either in form of triangle or rectangular. Figure 4 shows concept of winding number applied in a convex area.

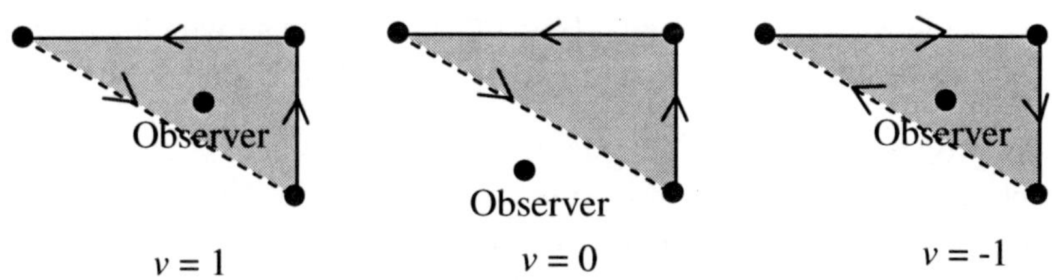

Fig. 4. Procedure of Winding Number

By employing winding number method, we can determine convex area that depends on agent position. On later section we present application of this theory and method into algorithm.

Define Centroid

The last procedure of defining enclosed space is to determine centroid of convex area which is based on those two previous results. The centroid is center of the convex area and will be set as a goal for agent.

The existence of centroid is established if and only if the convex area is determined. Therefore, in order to define coordinate of centroid (C_x, Cy) we will use previous findings and compute them using this formula:

$$A = \frac{1}{2}\sum_{i=0}^{n-1}(X_iY_{i+1} - X_{i+1}Y_i)$$

$$Cx = \frac{1}{6A}\sum_{i=0}^{n-1}(X_i + X_{i+1})(X_iY_{i+1} - X_{i+1}Y_i)$$

$$Cy = \frac{1}{6A}\sum_{i=0}^{n-1}(Y_i + Y_{i+1})(X_iY_{i+1} - X_{i+1}Y_i)$$

A= area of convex polygon
C_x=coordinate on-X axis of centroid
C_y=coordinate on-Y axis of centroid

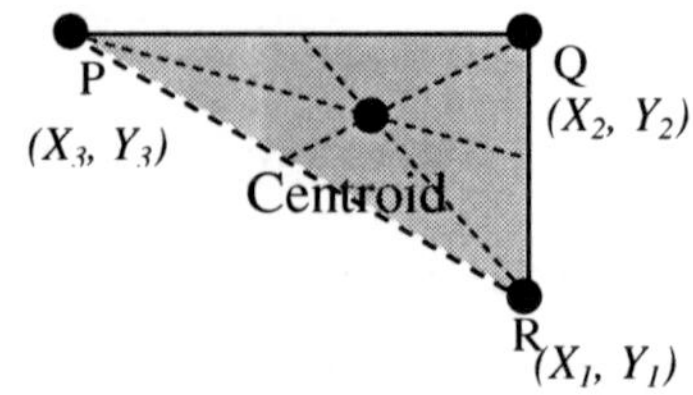

Fig. 5. Define centroid of a convex area

Method

Figure 6 illustrates overall framework consists of:

- visible objects detection
- convex polygon analysis by co-linearity test and winding number
- centroid determination of L-shape space and U-shaped space

As proposed, result on computation of the centroid is set as goals for agent navigation. The visible objects detection is obtain by mean of perception model. We used single line sight to represent visual direction. As agent rotates, any objects that intersect with the line considered to be visible. As consequences, there is possibility of unseen area as depicted in Figure 7.

Therefore, we develop procedure to step by step computing centroid based on visible objects as follows:

1. At initial position, detect objects by self-rotation and record objects into an array A. After all possible centroid determined, record all centroid into array B and rank centroids based on the distance to the agent
2. Go to the highest distance and repeat detecting and recording process. Update array A and array B and go to the next goal in array B.

In the final result we get maximum possibility of visible objects and the centroids.

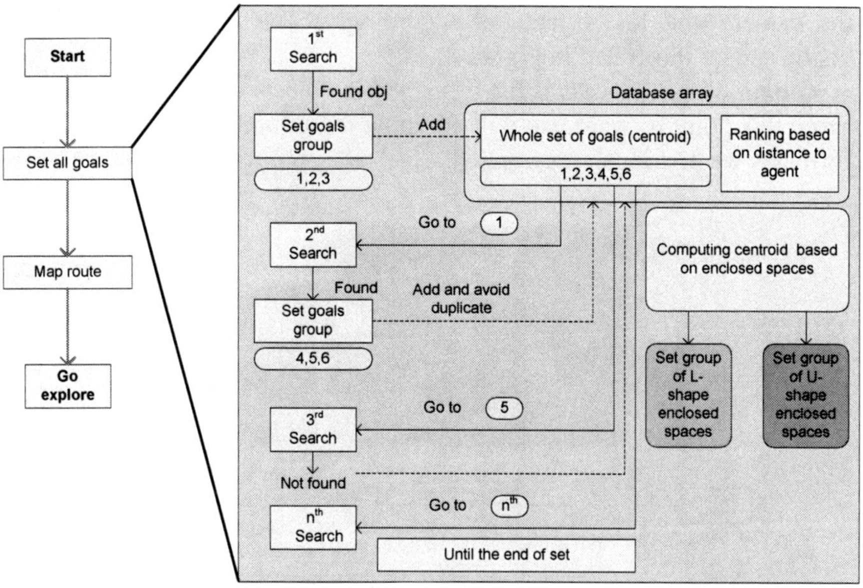

Fig. 6. Overall experiment framework

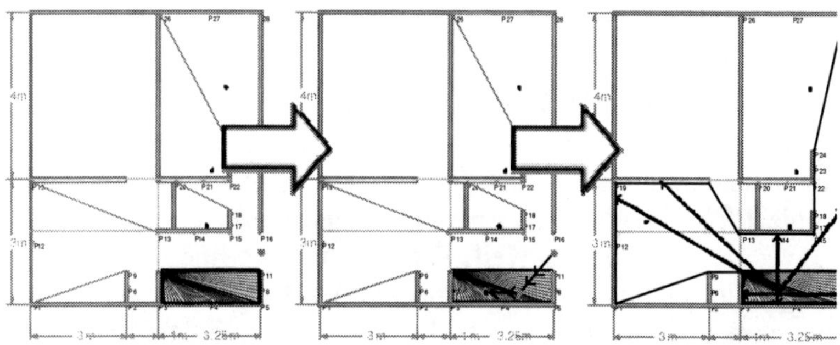

Fig. 7. Conceptual step by step process

Acquiring Information of the Environment

Initial process of the experiment is to acquire information related to the geometry properties. Instead of providing agent with knowledge of location-based attractor or set of network path, we first let agent examine their environment by using their vision model. A simple approach to detect and recognize object is by accompanying agent with camera (as their *eyes*) and line (as their line *of sight*). As head and body change their direction, the camera and line orientated accordingly. The virtual agent and his vision can be illustrated in Figure 8.

Output of this process is list of viewable objects from agent standpoint. From this point forward, we will compute these inputs using mathematical formulas we previously mentioned.

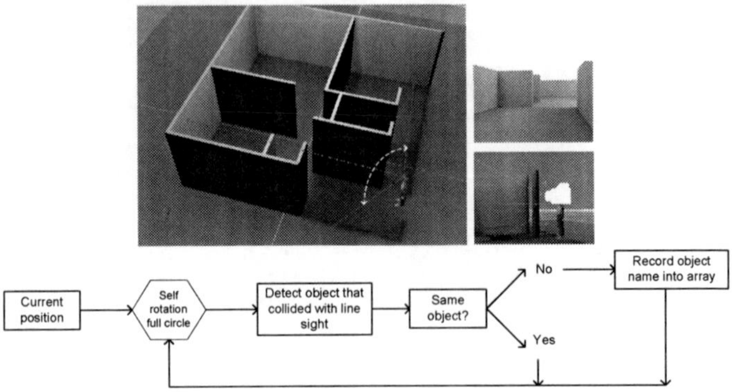

Fig. 8. 3D Environment and detection procedure

Decomposing Object into Set of Points

By the input of 3D CAD model, the output of detection process gives us geometrical properties such as: centroid location of object (in 3D vector), maximum and minimum vector points (the edge of its boundary box). Each object may have arbitrary maximum and minimum vector points and normal vector orientation as well. Thus, we have to find solution to reduce the complexity of this 3D information into 2D in which we can perform further computation. Figure 9 illustrates concept to transform 3D information into 2D information.

As in figure 9, in order to perform convex area analysis, we are only counting surface of each detected object. The procedure of decomposing objects is as following: First, we decompose each object by its boundary box to obtain *Maximum* and *Minimum* vector (pair vectors). Second, we compute distance of each pair vectors to the agent so the lowest distance must be at the intersected surface. Third, we transform maximum point (in figure 9, Original Max A) so that it co-linear with previous point.

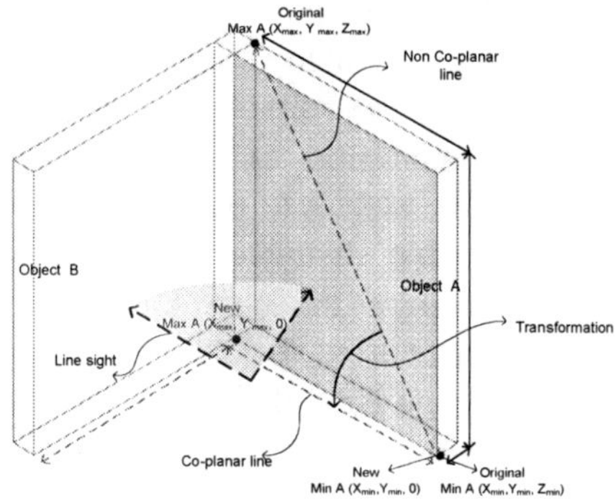

Fig. 9. Decomposing detected object (vector 3D) into surface (vector 2D)

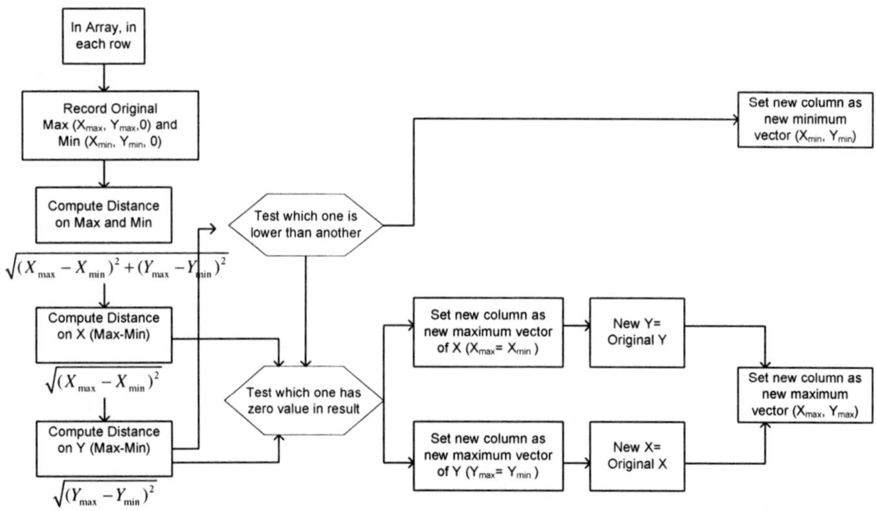

Fig. 10. Procedure to determine ne surface (vector 2D) coordinate

Determining Convex Area and Centroid

After successfully set 2D vector of each detected surface, we perform computational procedure to determine which pair of vectors is forming convex area.

Let us define two consecutive objects: A and B. Since our agent detect object *sequentially*, we assume that object A and B is connected to each other. In our case, this relation is in form of co-linear or non-co linear. If object A is defined by *Max A* $({}^{A}X_{max}, {}^{A}Y_{max})$and *Min A*$({}^{A}X_{min}, {}^{A}Y_{min})$, object B is define by *Max B* $({}^{B}X_{max}, {}^{B}Y_{max})$and *Min B* $({}^{B}X_{min}, {}^{B}Y_{min})$then there are six combination at which a point has the same location (A=B).

Regarding the *Max* or *Min* points is arbitrary, the method to seek this point (joint point) is to compute the distance of six combinations of two points with the agent and find out the same value to determine the joint point. Figure 11 shows implementation to determine joint point out of six possible combinations. Next step is to search similar value on each column of possible states. The same distance value of two consecutive points indicates that those points are at the same location. Up to this point we are able to determine two consecutive surfaces (of objects) that have common joint. The remained challenge is to analyze these objects of their convex or non-convex area.

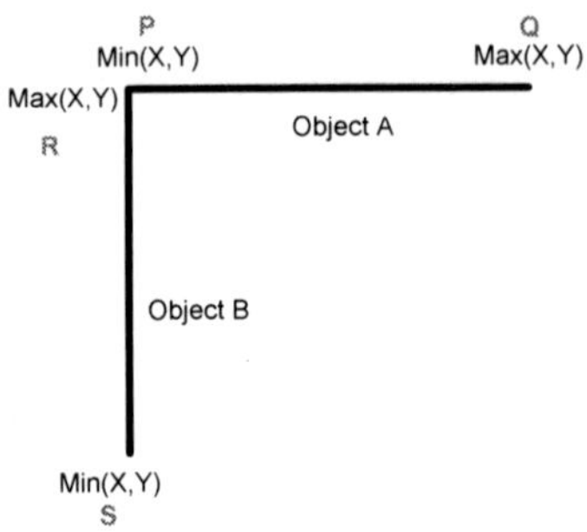

Determine possible combination (un-ordered) of two distinct pair from P, Q, R, and S using:

$$_{n}C_{k} = \frac{n!}{k!(n-k)!}$$

$$\Rightarrow n = 4, k = 2;$$

$$\Rightarrow \frac{4!}{2!(4-2)!}$$

$$\Rightarrow \frac{12}{2} = 6$$

21 : PQ	22 : RS	23 : PR	24 : QS	25 : PS	26 : QR
0	0	0	0	0	0
0	0	0	1	0	0
0	0	0	1	2	0
0	0	0	0	2	0
0	0	4	0	0	0
0	0	4	0	0	0
0	0	6	0	0	0
0	0	7	0	0	0
0	0	7	0	0	0
0	0	0	0	0	9
0	0	0	0	0	9

Fig. 11. (Above) Six possible combinations of pairs; (Below) Results after computation show pair of objects

One approach for analysis is to perform Graham's Scan algorithm [14] on each pair of objects, Figure 11. Graham's algorithm utilizes principle of winding number to determine convex area from list of vector points.

On each of pair, there are four points (each two from each object). Following Graham's algorithm we determine arbitrary first point. The second point obtained through computation of the determinant. This computation is explained as follows:

1. Calculate determinant of first point to each remained three points using *cross product*:

$$axb = \det\begin{vmatrix} a_x & a_y \\ b_x & b_y \end{vmatrix}$$

$$\Rightarrow (a_x b_y) - (a_y b_x)$$

1. Choose lowest determinant value to determine second point.
2. Choose second lowest determinant value to determine third point.

	Un-ordered list				Ordered List	
Points	X	Y	Det		X	Y
First	3.8	-2.9			3.8	-2.9
2	0.5	-2.9	-9.3		0.4	-2.9
3	0.5	-2	-6.1	⇨	0.5	-2.9
4	0.4	-2.9	-9.5		0.5	-2
Loop	3.8	-2.9	0		3.8	-2.9

Fig. 12. Transforming list of points based on the cross product of the first point

As result we obtain ordered list of points for further computation on area and its centroid.

We compute area and then centroid based on the ordered list of vectors as mentioned earlier. Entire computation process is undertaken on each of six columns explained earlier. On each array of ordered list of points, first we compute the sum of its determinant based on the X and Y values. The result will be divided by 2 to obtain area (A). Second, to compute X component of centroid ($_cX$) we add X value on each row and X values on next row (X_i and X_{i+1}). The result will be multiplied by their *cross product*. The sum of this value will be divided by (6A) to obtain $_cX$. The same method is applied to obtain Y component of centroid ($_yC$) except we use Y value (Y_i and Y_{i+1}).

Figure 13 shows overall procedure of determining area and centroid coordinate.

Experiment

We use Virtools Educational Version 5.0 for the simulated environment of this modeling and programming tool. The entire 3D geometry objects is modeled in 3DStudio Max.

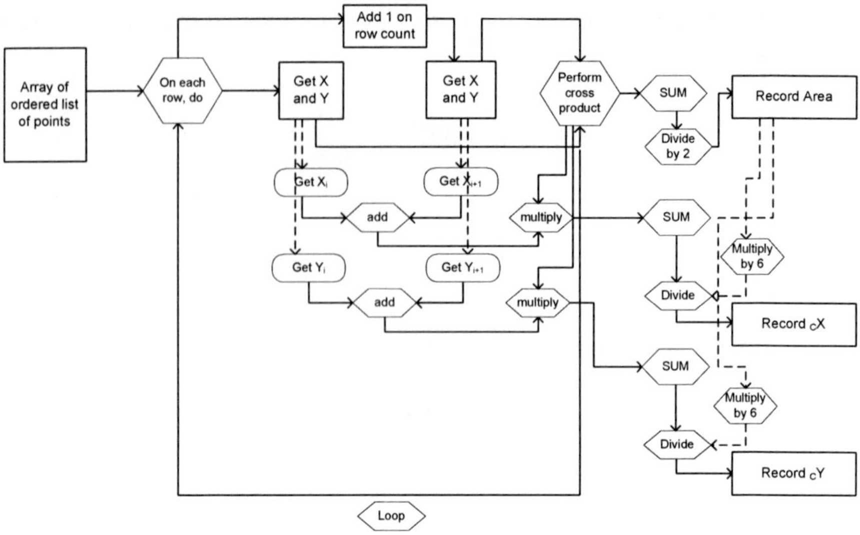

Fig. 13. Procedure to determine area and centroid coordinate

Constructing Agent and Environment

By default, Virtools comes with humanoid character equipped with animation clips which can be embedded to the character and programmed to be triggered by events, objects or procedures. This feature is significant since one approach to get character move is by triggering animation sequence (e.g. *walking*) with a particular procedure (e.g. *go to object*).

Figure 14 illustrates animation controller and trigger method of character.

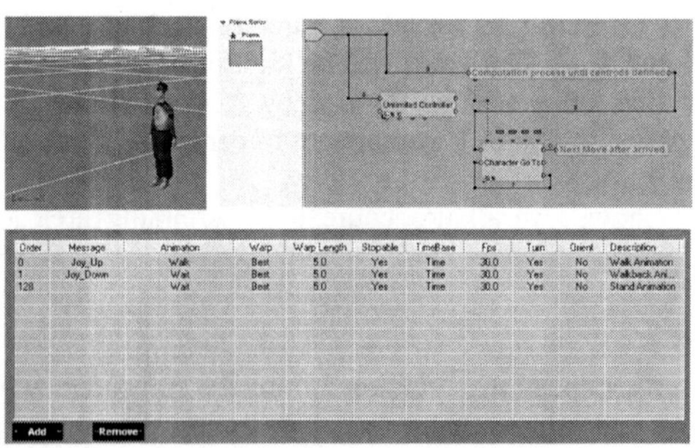

Fig. 14. Agent and his basic animation property. (Above Right) General procedure for moving agent towards goals

For the experimental architectural plan, we modeled partial plan from *Greenhouse House*, Connecticut by architect John M. Johansen (1973-1975) [5], Figure 15.

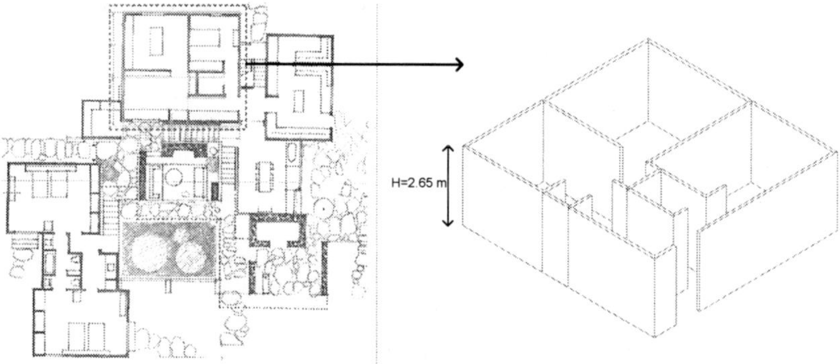

Fig. 15. Experimental architectural plan and 3D axonometric view

The standard metric unit (meter) is used to sync with default unit of the agent in Virtools. Particular emphasize given on the modeling of 3D wall into consideration that the detection process will only distinguished different objects by their name. To address this consideration, we modeled each wall following nature of building construction as illustrated in Figure 16.

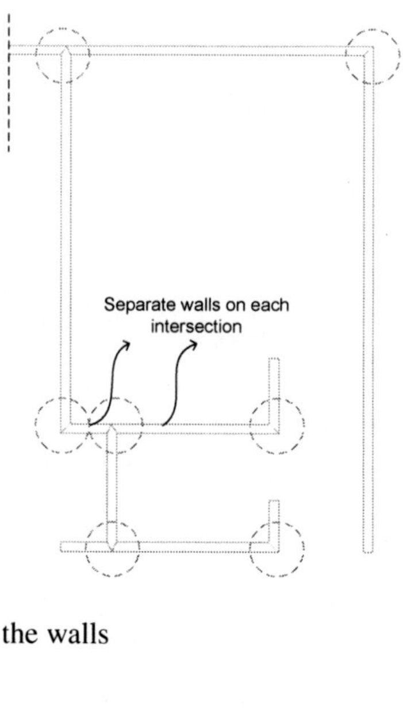

Fig. 16. Intersection of the walls

Additional properties such as texture and lighting are finished in 3D Studio Max using baked texture technique in order to be properly applied in Virtools. Figure 17 shows complete initial setup in Virtools.

The basic procedure of *collision avoidance* is available in Virtools as *attribute*. This attribute is attached to the related object according to their function, as example:

1. *Floor attributes*: applied on floor object to indicate object to be treated as base for the agent
2. *Fixed Obstacle:* applied on walls to indicate object as obstacle and considered not to be collided or passing through

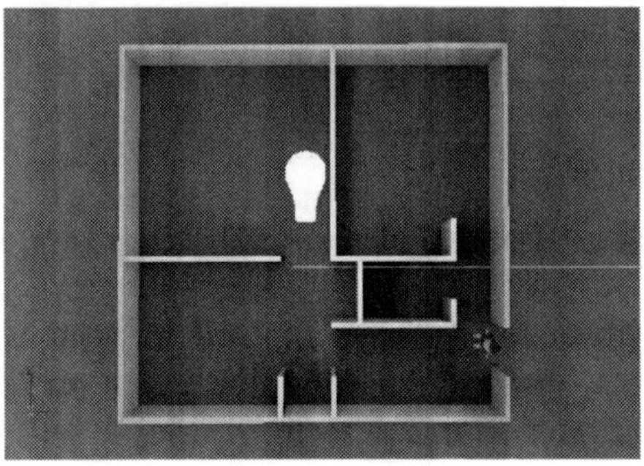

Fig. 17. Overall initial construction of environment

Computational Process and Result

As explained in methodology section, we equipped agent with camera as his eye and one line sight parallel to his direction. The whole computation processes are coded as scripts in these following components:

1. Initiation script (attached to level): to attach camera to the character's head and translates according to the character translation, Figure 18.

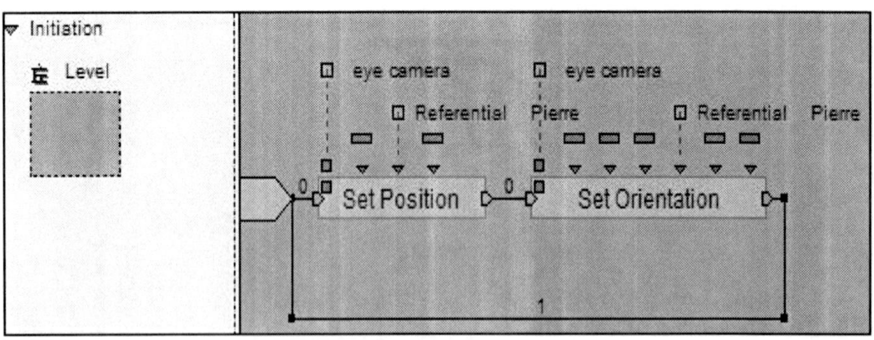

Fig. 18. Initial script at level

2. Character script (attached to character): this script consists of:
   a. *line of sight*: to attach line of sight to the camera so its position and orientation refers to the camera
   b. *rotate and detect objects*: using time-based translation technique to set amount of time as duration of complete rotation. At run time, it records delta progression (from 0 to 1= complete) and it is used as multiply factor for angle of rotation. As result, we can set up time-based 360 degree rotation for agent. The longest time, the slower agent rotates and more accurate it detects objects. For detection procedure, we applied ray intersection method. As character rotates, the ray intersection that attached to the line of sight detects objects and records them as group, Figure 19.

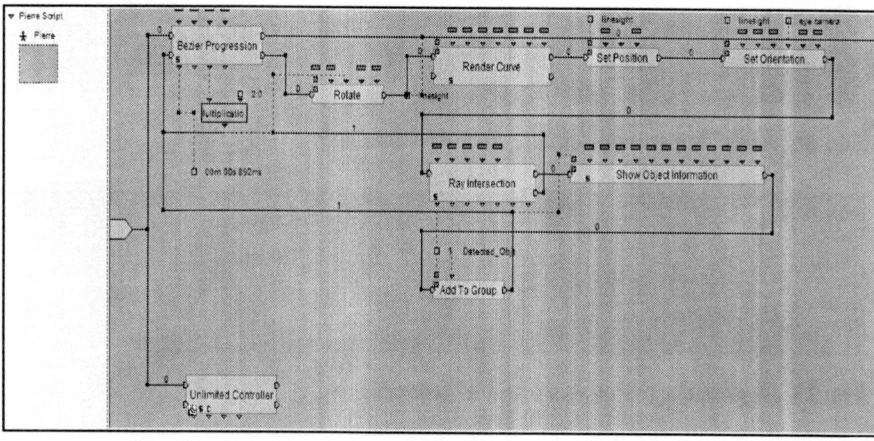

Fig. 19. Procedure of rotation and detecting object

   c. *record objects*: while character rotates and detects objects, it records each object into group. After rotation finished, the script goes to get the member of the group, which are the detected objects and record them into database array.
   d. *character to go*: this script is activated after computation process in array finished. Basically it will conduct agent to choose centroid determine by previous process.
3. Computation in array script (attached to level) : following the methodology, this script consists of:
   a. determine *maximum* and *minimum* 2D vector points on each surface of detected object.
   b. determine and detecting pair of non co-linear objects by the joint points.

c. computing convex area and centroid coordinate of each pair of object

Figure 20-23 show snapshots of computation process to determine convex area and centroids.

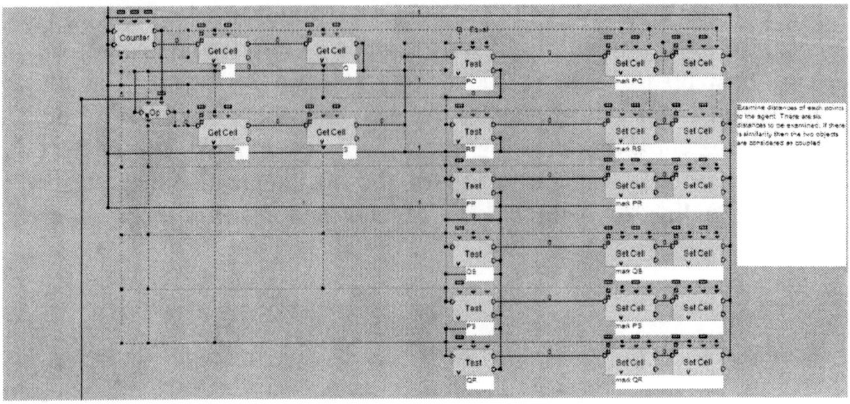

Fig. 20. Assigning computation results into database

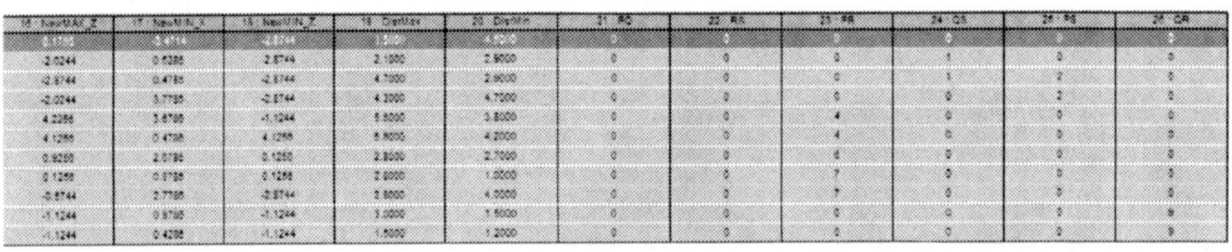

Fig. 21. Database array shows result of pair objects

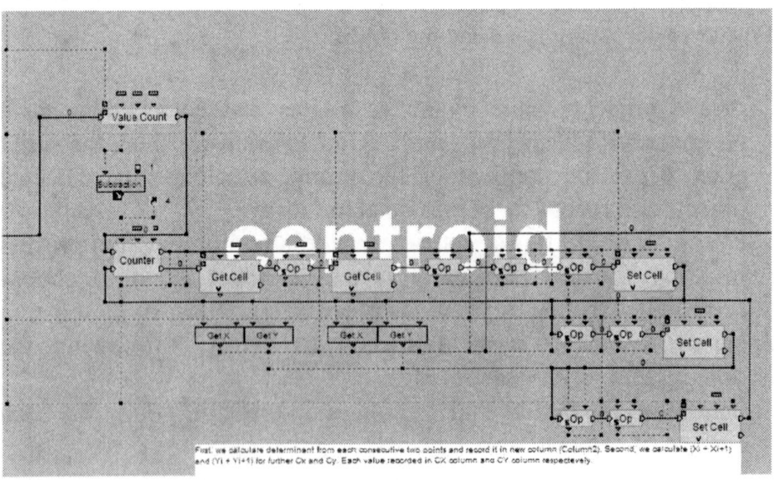

Fig. 22. Procedure to determine centroid

	0 : Coordinate	1 : Det_temp	2 : Det	3 : CX	4 : CY
0	3.7,-1.1	-15.7200	19.3500	145.1250	58.0500
1	3.8,4.1	-13.5300	0.7900	5.9250	6.5570
2	3.7,4.2	-13.0700	13.0700	54.8940	108.4810
3	0.5,4.1	0.0000	-15.7200	-66.0240	-47.1600
4	3.7,-1.1	0.0000	0.0000	0.0000	0.0000

Fig. 23. An example of pair object to be calculated for the centroid

Figure 24-25 shows snapshots of determined centroids in first, second and third steps respectively. As illustrated, the approximate number and position of enclosed spaces are determined according to the detection process of visible objects.

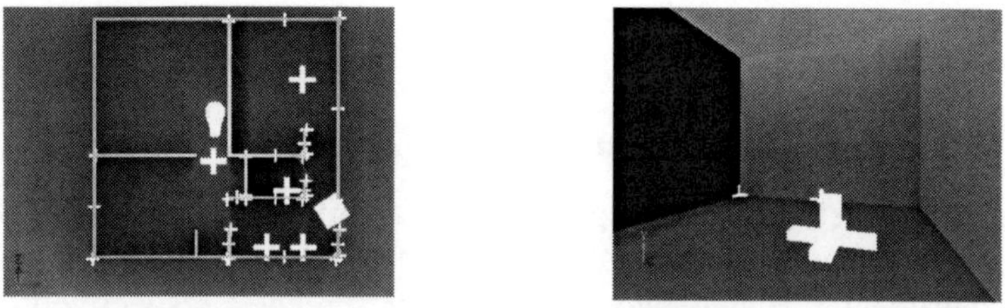

Fig. 24. (Left) First thread result indicating the whole set of detected object (small crosses) and the possible centroid of enclosed spaces (big crosses). (Right) Agent's view at one of the goal

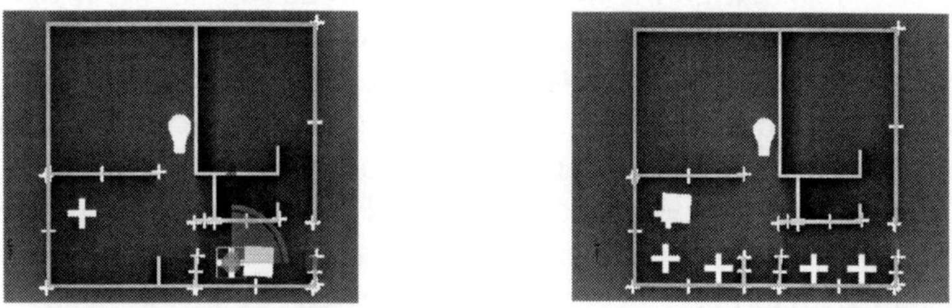

Fig. 25. Second and third thread simulation result

In experiment, the most significant factor to determine accuracy of approximate enclosed space is the detection objects. It is indicating that the more accurate agent detects objects, the more accurate it will determines the existence of enclosed spaces.

The relationship between objects properties used in simulation and execution times required for simulation is illustrated in figure 26.

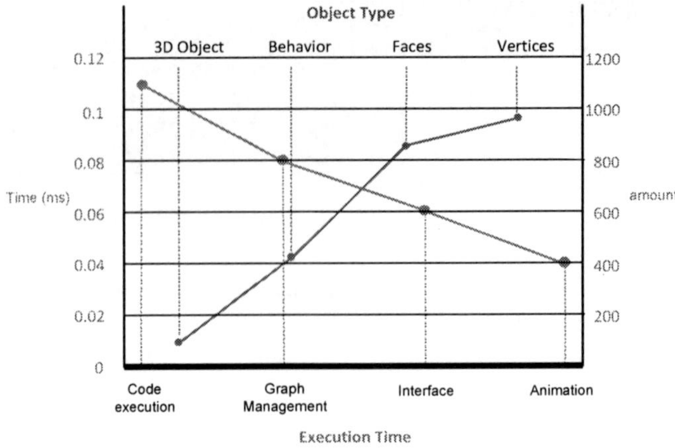

Fig. 26. Object properties versus execution time used in experiment

Concluding Remarks and Future Development

We present methods and algorithm to approximate the existence of enclosed spaces using computation of geometrical properties. The result of this experiment revealed two essential points related respectively to the navigation and interaction process of the agent and the environment.

We argue that this finding would have contribution to the development of behavioral-based simulation for architectural design where agent can aware of the spatial configuration and utilize this information to interact with other agents and/or other architectural elements in their environment. As one approach to model relationship between architectural design and behavioral pattern, enclosed space detection has significant input towards one aspect of quantitative understanding of design quality.

In the future development, we would investigate the property of personal space and its relationship with the enclosed space in a simulation to visualize interplay between goal-directed behavior and reactive behavior.

References

1. Lawson, B.: The Language of Space, 5th edn. Architectural Press, Elsevier (2007)
2. Rasmussen, S.E.: Experiencing Architecture. MIT Press, Cambridge (1959)
3. March, L., Steadman, P.: The Geometry of Environment. MIT Press, Cambridge (1971)

4. Golledge, G., Reginald, R., Stimson, J.: Spatial Behavior: A Geographic Perspective. Guilford Publication, New York (1997)
5. Ching, F.D.K.: Architecture: Form, Space and Order, 3rd edn. John Wiley and Sons, New Jersey (2007)
6. Wiener, J.M., Franz, G.: Isovists as a Means to Predict Spatial Experience and Behavior. In: Freksa, C., Knauff, M., Krieg-Brückner, B., Nebel, B., Barkowsky, T. (eds.) Spatial Cognition IV. LNCS (LNAI), vol. 3343, pp. 42–57. Springer, Heidelberg (2005)
7. Turner, A., Penn, A.: Encoding Natural Movement As an Agent-based System: An Investigation into Human Pedestrian Behavior in The Built Environment. Journal of Environment and Planning B: Planning and Design 29, 473–490 (2002)
8. Mankyu, S., Gleicher, M., Chenney, S.: Scalable Behaviors for Crowd Simulation, Eurographic. Blackwell Publishing, Oxford (2004)
9. Thalmann, D., Musse, S.R., Kallman, M.: From Individual Human Agent to Crowds, Informatik (2000)
10. Kallman, M., Thalmann, D.: Modeling Behaviors of Interactive Objects for Real Time Virtual Environment, Computer Graphics Lab, Swiss Federal Institute of Technology (2004)
11. Narahara, T.: Enactment Software: Spatial Designs Using Agent-based Models. Complex Interaction and Social Emergence (2007)
12. Wei, Y., Kalay, Y.E.: Geometric, Cognitive and Behavioral Modeling of Environmental Users. In: Design Computing and Cognition 2006. Springer, Heidelberg (2006)
13. Needham, T.: Visual Complex Analysis. Clarendon Press, Oxford (1997)
14. Graham, R.L.: An Efficient Algorithm for Determining the Convex Hull of a Finite Planar Set. Information Processing Letters 1, 132–133 (1972)

Associative Spatial Networks in Architectural Design: Artificial Cognition of Space Using Neural Networks with Spectral Graph Theory

John Harding[1] **and Christian Derix**[2]
[1]*University of Bath / Ramboll, UK*
[2]*University of East London /Aedas R&D, UK*

This paper looks at a new way of incorporating unsupervised neural networks in the design of an architectural system. The approach involves looking the whole lifecycle of a building and its coupling with its environment. It is argued that techniques such as dimensionality reduction are well suited to architectural design problems whereby complex problems are commonplace. An example project is explored, that of a reconfigurable exhibition space where multiple ephemeral exhibitions are housed at any given time. A modified growing neural gas algorithm is employed in order cognize similarities of dynamic spatial arrangements whose nature are not known a priori. By utilising the machine in combination with user feedback, a coupling between the building system and the users of the space is achieved throughout the whole system life cycle.

Introduction

In 1953, the cyberneticists Gordon Pask and Robin McKinnon-Wood designed and constructed the so called 'Musicolour Machine' [21]. This device assisted in the production emergent musical 'designs' as an output of a complex interaction between a human composer and computer hardware. Here a structural coupling was formed which resulted in a composition that whilst being neither chaotic nor boring, took on a direction that could not be predicted in advance. The composer was adaptive to the machine, and vice-versa after each had cognized each other's output.

J.S. Gero (ed.): *Design Computing and Cognition'10*, pp. 305–323.

Similar thinking to producing built form in space can be applied to architectural designs. If the environment and its people are considered dynamic entities, how can architecture acknowledge and utilise this to the advantage of all parties? As Pask remarked himself, architecture is a discipline that fits the bill insofar as the abstract concepts of cybernetics can be interpreted in architectural terms [8].

Part of an architect's role is to organise of space in three dimensions that follows from a cognition of a complex set of requirements. The architect thus digests this information, and attempts to reduce a complex problem into a traditionally fixed form of bricks and mortar or whatever else he has at his disposal. Here the required actions are performed at a lower dimensional level themselves but are dependant on making sense of high dimensional inputs.

If the requirements of an architectural problem change after construction however, a building must be capable of adaptation. Cedric Price's collaboration with John Frazer in the Generator project [7] is one such example of an 'intelligent structure' whereby the goals of the system are not defined by an initial project brief. Instead the building continually adapts to the changing client requirements that are in turn influenced by the current building configuration. Such a system is completely different to goal orientated optimisation whereby the problem requirements are always somewhat fuzzy. Kalay [17] notes that such non-deterministic processes need an unsupervised mechanism to pick out results that are truly emergent and suggests the use of unsupervised neural networks to achieve this. Here the machine must continually understand changing inputs if it is to advise on the morphology of form. The design approach described in this paper thus employs artificial neural network that takes spatial data in high dimensional feature space and though a process of machine cognition, helps generate new informed designs. These realised designs in turn influence the users of the space who reinvest information to the system and hence redefining the requirements for the next cycle of spatial organisation.

Artificial Neural Networks in Architectural Systems

The use of neural networks form one method of using artificial cognition in the design process. Early examples include Petrovic and Svetel's computer algorithm that assisted the architect in making design decisions - the 'PDP-AAM' program [23]. Picture and words were associatively compared via a neural net design system in order to generate informed house designs. Part of the self-organising system also included understanding the attitudes and preferences of the inhabitant via neural network learning.

More recently, Langley [19] has used a form growing neural network to generate representations of dynamic activity in existing urban environments, although the generation of new designs based on this learning was not explored. Derix and Thum [2] have deployed an unsupervised self-organising map [18] to absorb large amounts of urban data in order for the machine to form its own interpretation of space and hence make interventions using its own internal comprehension. One can compare these last two examples to the first in that the learning algorithm becomes part of the system itself over the whole design lifecycle, as with the system presented here. This differs from using learning algorithms to assist in the production of new designs and then constructing them in a frozen form in real space.

Overview of the Design Problem and Approach

The experiment presented in this paper uses a two stage neural network as a spatial pattern recognition tool towards the development of an emergent two dimensional plan form – a large exhibition hall that is used to house various ephemeral exhibitions, which in turn contain exhibits to be arranged on the plane. Figure 1 shows the system components.

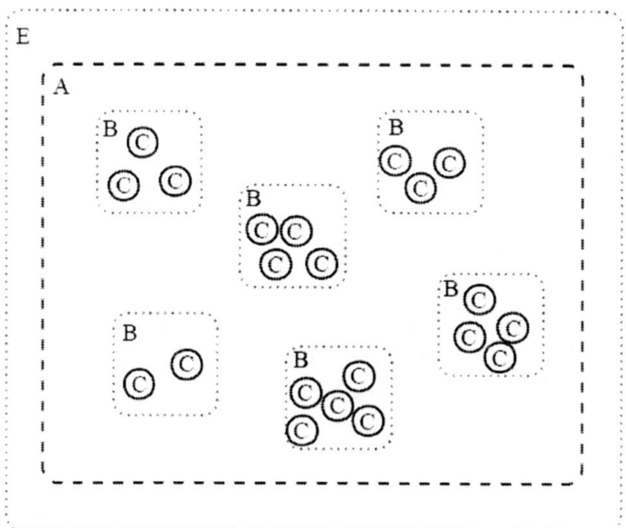

Fig. 1. Exhibition System: Environment 'E' contains exhibition hall 'A' which contains exhibitions 'B' which contain exhibits 'C'

As there is limited space in the hall, similar exhibitions 'B' must be found so that they will be able to share the same space for a given time once the hall is reconfigured for a given time period. As the characteristics of the exhibits change over time, so do the exhibitions and hence their relationship to each other.

In order to understand the system, firstly we employ a self-organising map (SOM) [18] in order to arrange the exhibits on the plane by comparing their differences. This associative map is then converted into a spatial plan that represents the exhibition. Secondly, we compare these plan graphs so as to find clusters of similar spatial topologies. A common graph is then found for the cluster and realised in built form. Finally as the exhibitions house participant users, feedback is collected in order to assess the various qualities of the exhibits themselves in a subjective manner. At the end of the time period, the exhibition hall reconfigures to suit a change in the exhibitions and hence the overall design requirements. The proposed design is thus a dynamic system that uses machine learning to find patterns in the exhibits and then the exhibitions according to qualities influenced by the human users.

Generation of Spatial Graphs for the Exhibitions

Each exhibition that is to be housed contains a number of exhibits that each have particular qualities. Table 1 shows such qualities for an exhibition of famous chairs resulting in a six-dimensional 'synaptic' vector for each. Here the features to measure and their values are defined by a single human (the authors of this paper), however methods to remove these preconceptions are discussed in section 4.4 whereby user feedback influences the choice and value of the qualities. This is very important as the designer must attempt to remove his/her own biased values.

For our experiment, we have selected thirty such separate exhibitions on various topics. For each exhibition the exhibits must be arranged in real space and this requires the cognition of high dimensional data.

Mapping of Exhibits Using Dimensionality Reduction

The goal of the mapping is to find differences between exhibits so that there is some associative logic in the way the exhibition is spatially arranged and hence experienced. So for example, two similar exhibits could be placed adjacent to each other in real space. In order to do this, a self organising map (SOM) [18] is used to reduce the dimensions to $\mathbf{R}^2$ while retaining topological associations. The self-organising map is a type of artificial neural network first described by Teuvo Kohonen that uses an unsupervised learning process. It has the ability to reduce the dimensionality of inputs whilst retaining the topological similarities between them.

Table 1 Characteristics of various chair designs that form their synaptic vectors – in this example the choice of features and their values have been chosen by the authors.

	foldable	4 legs	adjustable	armrests	cushioned
Red Blue	0	1	0	1	0
Barcelona	0	1	0	0	1
Butterfly	1	1	1	0	0
Aalto Stool	0	0	0	0	0
Eames	0	0	1	1	1
Bubble	0	0	0	0	1
Deckchair	1	1	1	1	0
Wassily	0	0	0	1	0

As the size of the feature space is known, the parameters for an accurate SOM mapping can be determined. Such associations between exhibits are then maintained spatially by applying a plan graph. The plan graph is generated by finding the nearest neighbour from each node and joining with an edge. Such graphs have been shown by Eppstein et al. [6] not to violate planarity for a point set in two dimensions. By joining any other nodes that are a similar distance away to the nearest neighbour, planarity has been found to hold so long as this tolerance is kept relatively small.

Planar forms are preferable due to the physical reality of spatial morphologies (movable walls) being easily incorporated in $\mathbf{R}^2$. Normalised weights are also associated to the graph edges which represent the difference between two adjacent exhibits. These weights are used later in the process when the exhibition hall is realised in space. By maintaining the weights in our graphs we exploit the generalised mapping that a self-organising map creates. This generalised property also ensures that not only similarities between certain exhibits are expressed, but also differences between certain ones as well, i.e. exhibits that are furthest away from each other in space are the most different.

The plan graph formed is a topological representation of spatial adjacency as described by Jupp and Gero [16] and Steadman [24]. Other graph applications in the architectural analysis of spaces have been studied by Hiller & Hanson [15]. These space syntax mappings are able to find a representation of space from the perspective of the participant observers. In this project described here, although simple plan graphs are used, because the values of the exhibits are influenced by the users they also include influence from their perspective, Figure 2. It is important to note that such a mapping technique will not always result in exactly the same distribution due to the nature of the SOM algorithm and the random initial distribution of exhibits - however similar types of graph structures do appear for each exhibition each time a mapping takes place simply by the nature of the differences for the exhibit synaptic vectors.

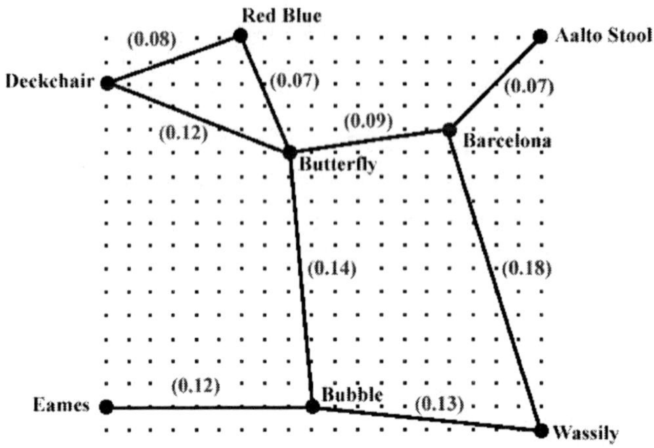

$$s = \begin{bmatrix} 5.562 \\ 3.414 \\ 3.414 \\ 3.000 \\ 1.438 \\ 0.586 \\ 0.586 \\ 0.000 \end{bmatrix}$$

Fig. 2. A graph is found following a self-organised mapping on the plane for an exhibition of chairs. The weightings are the associated edge lengths that have been normalised. The shorter the length, the stronger the association. The graph spectrum is shown on the right and is discussed in the next section

Spectrum Representation of Graph Features for Synaptic Vectors

We now have the plan graphs of the exhibits and their associated adjacency matrices. However, the next stage in the process involves using a meta-level neural network to find differences between exhibitions. In order to do this, we find a unique graph spectrum for each exhibition. This spectrum thus forms a synaptic vector representation in feature space and makes comparisons between graphs much easier to conduct.

Successful applications involving comparisons of graph spectra have been previously described by Hanna [13]. These include image recognition techniques, and the subsequent generation of face designs by a computer forming its own concepts for objects. Reducing a graph to its spectra is not always full proof mapping, but in most cases good enough for the task required. For example, some graphs may share the same spectrum (co-spectral) but are not isomorphic. For a heuristic algorithm the technique still gives sound results however.

The accuracy of the process is greatly improved by finding the spectrum for the Laplacian Matrix which is a simple deviation from the adjacency matrix.[1] Zhu & Wilson [25] have analysed and confirmed the general effectiveness of using Laplacian spectra in comparing different graphs.

One important part of the Laplacian spectrum is the number of zeros it contains as this represents the connectedness of the graph. Such zeros at the end of the feature vector change the dimensionality of the spectrum and hence a cognition tool able to adapt to varying vector dimensions is required at the next stage. In our experiment, thirty spectra for each exhibition are derived each with a varying amount of connectedness and hence dimensionality. Once found, we can now use another artificial neural network to cognise and hence find clusters of similar plan graphs.

Mapping a Modified Growing Neural Network

Our intention is to find clusters of similar plan graph structures so that they will be able to share a configured space at the exhibition hall. Here we use the spectrum from each graph and use it as a new feature vector for a neural network to learn.

Hanna has shown that principle component analysis can be successfully used in the comparison of axial map graphs of known buildings [13],[14] whereby clusters are found by visual inspection. Gero and Jupp have successfully compared the complexity of plan graphs in order to correlate distinct architectural styles [16]. In Ireland's poly-dimensional living model [3] a meta-level self-organising map compared various other 'association networks' from self-organising maps themselves that representing living patterns in order to design a living space.

In these examples, knowledge of the input data is known before analysis is conducted. In our design problem however we have no way of knowing the nature of the spectra to be presented for analysis. The ephemeral nature of the exhibitions themselves mean that a system capable of adapting change in the graph spectra must be used – this includes the possibility that the spectra to be compared are of varying dimensions in feature space due to the connectedness of the graphs. As Derix states [4], one of the key problems for a good match between the dimension of the network and the dimension of the signal space in many applications is that the signal space is not known yet. We therefore require a learning algorithm that includes the following capabilities:

- The network can adapt to the size of the input signals dynamically.
- Clusters can be identified by the machine automatically.

[1] A detailed description of finding a graph's Laplacian spectrum is given by Hanna [13].

- The network can continue until a certain number of clusters is found.

To meet these requirements, a *growing* neural network was developed by the authors very similar to the 'growing neural gas' (GNG) algorithm by Bernd Fritzke [11]. The GNG is the ability to automatically find a problem specific network structure through a growth process. The insertion and deletion of neurons causes the network to adapt to the data presented to it. In contrast to the self-organising map and the original neural gas algorithm by Martinez and Schulten [20], all parameters are constant including the width of the neighbourhood around the best matching unit (winner) where adaption takes place [10]. It has been shown that a GNG can adapt to a signal distribution which has different dimensionalities in different areas of the input space [11] which makes it suitable to handle varying spectra dimensions caused by differing graph connectedness.

In order to visualise the clustering that the GNG finds, we have extended the algorithm to include a disc embedding approach similar to that used in Fritzke's growing cell structures algorithm [9]. This is achieved by assuming each topological connection in feature space is a simple linear elastic string of a given spring constant in $\mathbf{R}^2$. A test mapping was conducted by the authors and is shown in Figure 3 where three clusters are correctly found from signals produced in 3-dimensional feature space. The network is then mapped into two dimensions for visualisation.

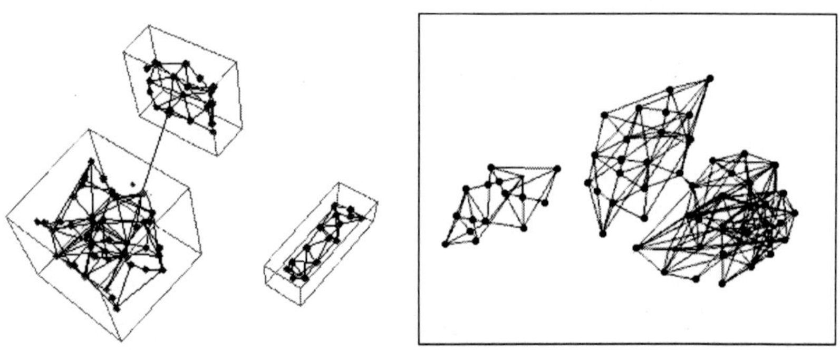

Fig. 3. Left: a growing neural gas mapping in three dimensional space - the topology of three cubes. Right: the real-time embedding using a particle spring system in two dimensions

Clustering of Graph Types Using a Growing Neural Gas

The use of a growing neural gas has an advantage over the self-organising map in that clusters are identified automatically. Size of the clusters can be altered by allowing the system to run a certain number of iterations to suit the size of the exhibition space. As mentioned by Fritzke, the GNG is able to continue learning, adding units and connections, until a performance criterion has been met [10]. Here the machine is instructed to finish when 16 separate clusters have been identified. Such an instruction is related to the finite size of the exhibition hall (the boundary conditions of the site). It is important to note that although the GNG loses some of the generalised mapping qualities that are possessed by Kohonen's self-organising map, for finding a discrete number of clusters for similar graphs such a capability is not necessary.

Results

Figure 4 shows the growing neural gas running when presented with thirty exhibition topologies. After 1500 signals, 16 separate clusters have been identified by the algorithm and hence the machine selects these onto the next stage.

The resulting clusters are shown in Figure 5. We can see from our own observations that the clusters formed by the algorithm are quite sound. For example, the large set {2,6,20,23,28} contains all connected graphs of a similar structure. Similarly, set {5,14} contains sparse graphs, and set {3,18} has the same amount of connectivity, with each connected sub-graph being similar by visual inspection. We can also identify unique graphs with only one member in the cluster. For such a heuristic method, the results are fit for purpose and as the graphs used here are relatively simple, visual inspection is adequate. With more complex graph structures, better verification of the algorithm is recommended, similar to that previously discussed [25].

Investigation of Dynamic Inputs

The mapping of dynamic inputs can also be achieved if the spatial graphs are subject to change throughout the cognition process. This would result in a dynamic mapping similar to that used by Langley [19]. As Fritzke comments, self-organising neural networks are rarely considered for tracking non-stationary distributions since many of the models use decaying adaption parameters (e.g. self-organising map [18], neural gas [20] or the hypercubical map [1]).

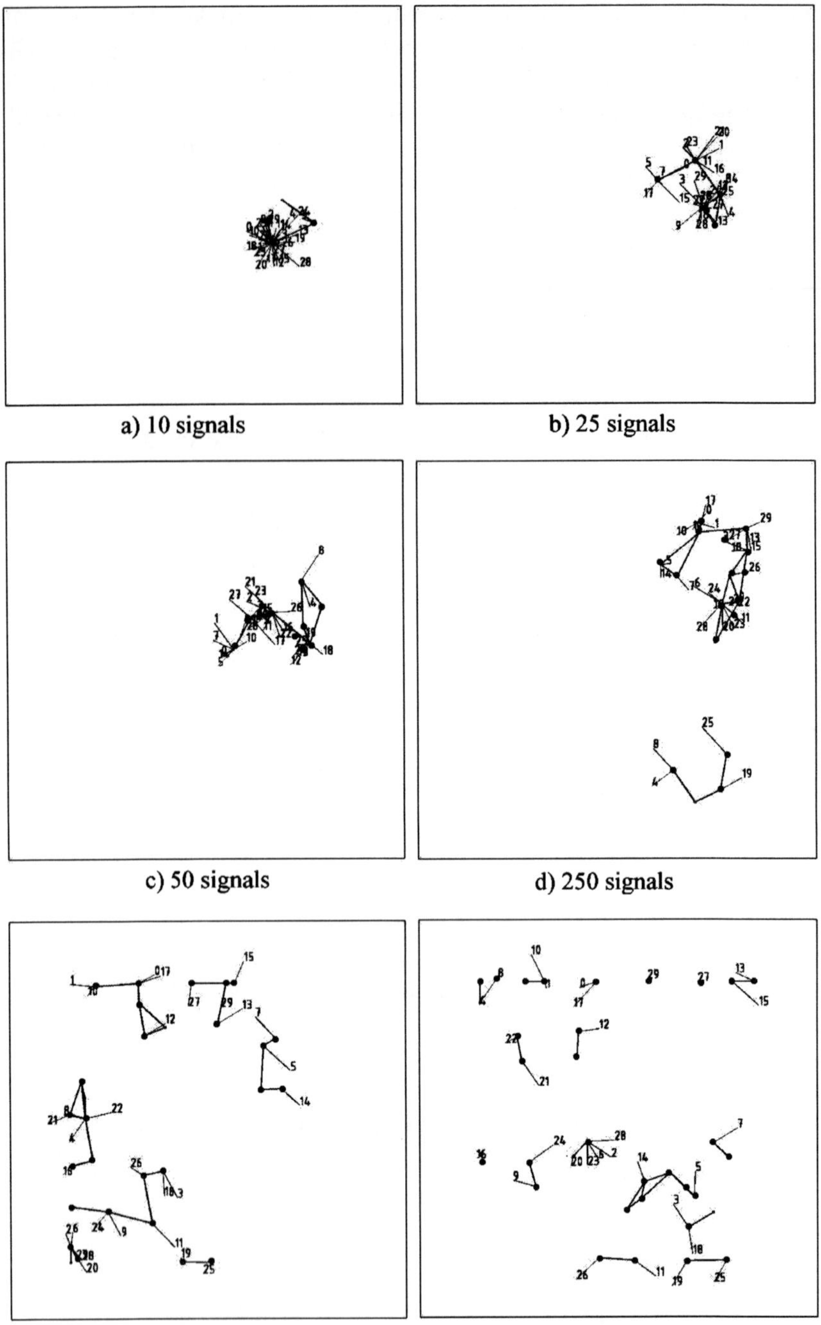

a) 10 signals b) 25 signals c) 50 signals d) 250 signals e) 750 signals f) 1500 signals

Fig. 4. The growing neural gas finds clusters of similar graph structures

Similar studies were conducted for this particular project but such a modified growing structure was deemed unnecessary due to the large time-spans between the morphology of the exhibition hall thus meaning sufficient solutions can be found by simply re-executing the GNG algorithm when any changes in the graphs take place.

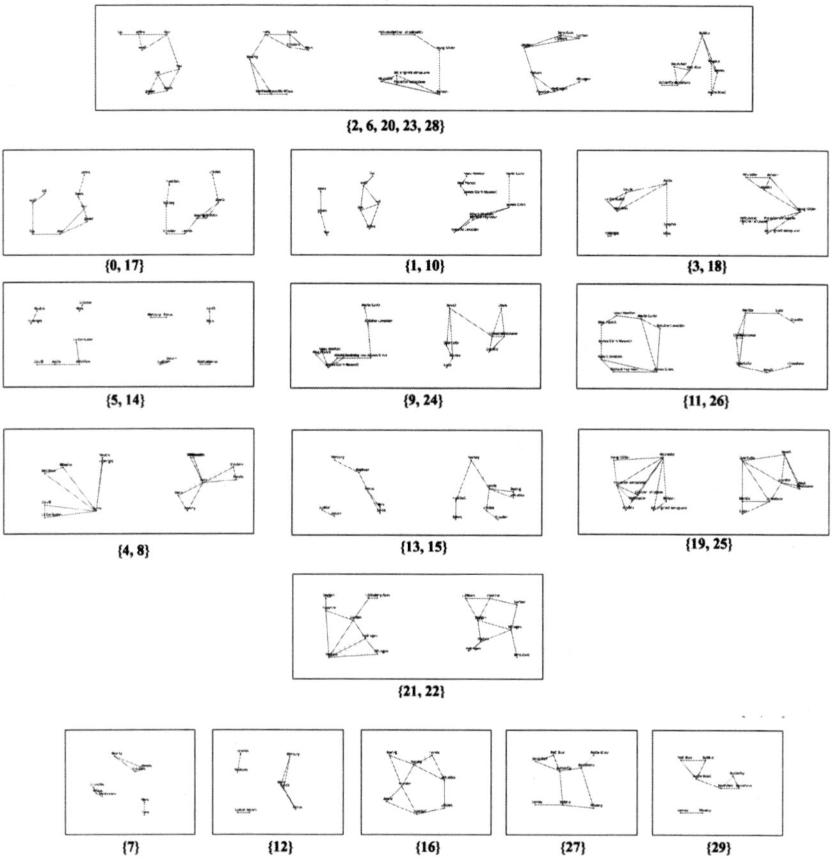

Fig. 5. Results from the modified growing neural gas. Each cluster is shown bounded. Thirty different exhibitions were analysed by the growing neural network

Generation of Spatial Layouts – 'An Artificial Curator'

Once clusters of similar graphs have been identified by the growing neural gas, the exhibition hall can now be reconfigured. This clustering therefore creates the timeline for when particular exhibitions are housed due to several exhibitions being able to share one configured space.

If a cluster involves several graphs, which most do, there needs to be a way to find the 'average' graph for the configuration of the space, Figure 6. This is done again by looking at the spectra of the graphs. The spectra vectors for all the graphs in the cluster are summed up and an average taken. Each graph is then measured against this average using the Euclidean distance in feature space. The closest graph selected is then translated into form. If there are only two graphs in the cluster then one is selected at random.

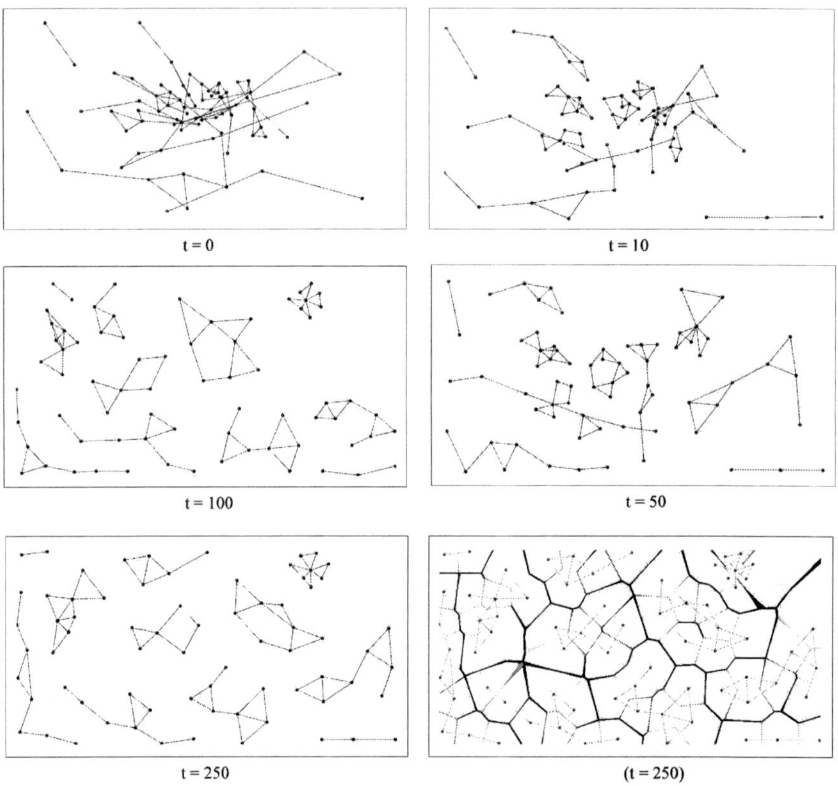

Fig. 6. A repulsion algorithm distributes the graphs evenly over the exhibition hall. Topological connections are simulated as springs to help keep the graph nodes adjacent.

Spatial Realisation of Graphs Using a Particle Repulsion Algorithm

Once we have a selection of average graphs, we must then morph our form based on these spatial topologies. For the two dimensional case, we have various plan graphs that are arranged on the plane, and movable walls that can be easily constructed due to the planarity of the graph structures.

We do this using a simple repulsion algorithm combined with a Voronoi diagram. Figure 7 shows an example realisation.

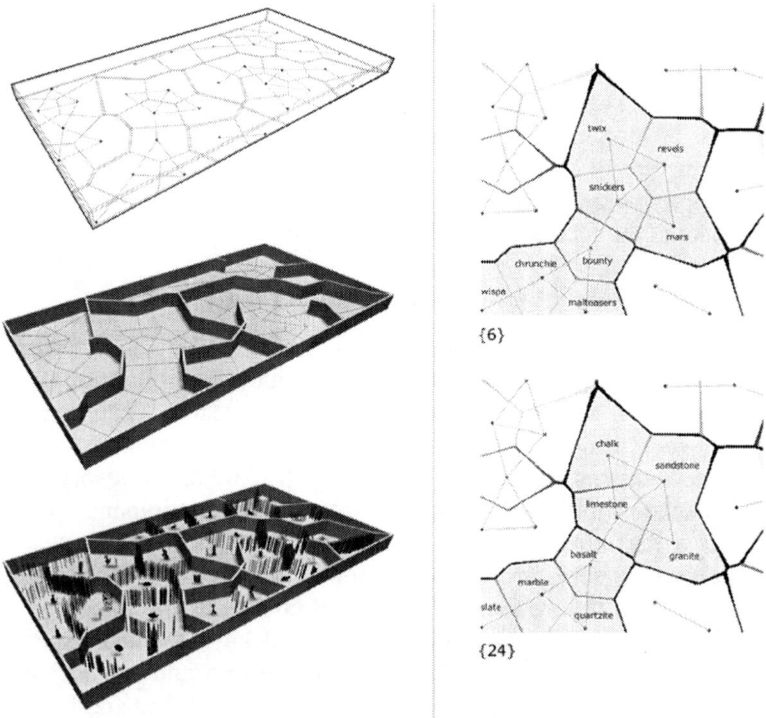

Fig. 7. Left: a small scale realisation at a particular time. Although the example shown here is a simple 2d plan extrusion, three-dimensional form could be incorporated. Right: two exhibitions that formed a cluster share the same spatial topology. The permeability strength of the boundary walls can change

Each graph shakes itself until equilibrium is reached, with the resulting layout used for the Voronoi tessellation. The boundary walls of the Voronoi cells are an equal distance from each adjoining exhibit and are assigned a permeability based on the weighting from the original SOM plan graphs as described previously. This permeability changes for each specific exhibit housed in the same spatial topology. Once the form has been configured, the exhibition hall is open to the public for a given time period. As the exhibitions may each run for a different length of time, it is possible at this stage to run the GNG algorithm again for a sub-set of the graphs, although this is not shown here. Instead, a set period of for example one month is allowed for before the exhibition space reconfigures itself for either new exhibits, or modifications of the existing ones.

Meta-cognition of Exhibition Spaces by Users

The iteration between certain exhibitions, their sequence of spaces and the types of exhibits they house would topologically be similar. Therefore the users can start to create types of 'meta-associations' between the exhibitions themselves and attribute qualities to spaces that derive from the exhibits. The user creates expectations through the coupling and generation of cognitive domains of what he/she will find. Here, the subjective nature of the human mind cannot be predicted by the authors, only an appreciation that the system must facilitate any associations that he/she make at the feedback stage of the process.

Integrating Participant Users Using Unsupervised Goals

So far the qualities for the exhibits have been defined by an external hand (i.e. the authors), this was the case for the chairs exhibit example given earlier. However, as the designer becomes a system designer, he/she should try to remove personal assumptions from solutions by initiating only the mechanisms and setting a context [5]. If we are to respect that the exhibition space should form a relationship with its participants then some attempt at a part-closure of the system via user feedback is necessary. Given a closed system, inside and outside exist only for the observer who beholds it, not for the system itself [21]. Of course, as the inhabitants lead lives outside of the system proposed here and have learned experiences themselves, complete internal closure is not desired and would lead to a non-adaptable architecture that would deny outside cultures and influence.

One approach to incorporate user feedback is to have the hardware the neural network itself whereby interactions with the environment (e.g.: humans) are directly cognised. This has been instigated by Derix in an experimental project whereby a physical unsupervised neural network exhibit derives its own associations from how people interact with it, Figure 8. Exactly what semantic interpretations a human participant will form as a response to the exhibit is not known beforehand. The same thinking applies to the exhibition hall and its users – we cannot tell in advance who will be there and how their respective brains are *and will be* wired!

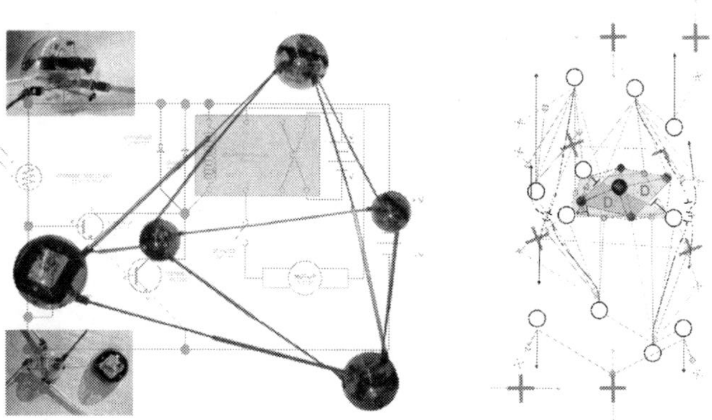

Fig. 8. "Analogue Embedded SOM" by Derix as displayed at the digital intuition - the design space of artificial learning exhibition, London, 2009

For this project then there are two key bits of information that we require in order to try to remove author assumption:

- The quality criteria used to assess differences between exhibits.
- The relative scoring of these qualities for each exhibit.

One proposal would be a scoring system that is given to the users of the exhibition hall after a visit in order to influence change the qualities that belong to the exhibits. This in turn would influence the exhibitions themselves. It is important that any subjective information is fed-back into the qualitative mappings at the SOM stage where associative mappings take place, Figure 9. This differs from the GNG whereby a discrete mapping of clusters leads to a loss of generalisation. We must respect that subjective data needs to be collected and hence more descriptive methods may be required. It is also clear that a combination of subjective and objective qualities must be maintained, for example with an exhibition on animals a horse has 4 legs and a monkey has 2 legs. At first these seem not to be qualitative attributes but objective facts – but what is a leg anyway? It is the human's value of what makes animals *different* from each other that still applies and must be captured – but this opinion can be influenced by the machine's learning in that hidden connections are revealed that are not necessarily obvious by human cognition alone. Likewise, changes in human opinion are able to influence how the machine understands the exhibits and exhibitions.

Strategies to capture subjective feedback in this context is a subject of ongoing research. Whichever strategy we employ, we can still appreciate that the qualities of the exhibits that are compared must be to some extent represented as variables so that new spatial graphs are generated. The unsupervised nature of the neural network thus becomes crucial to the process.

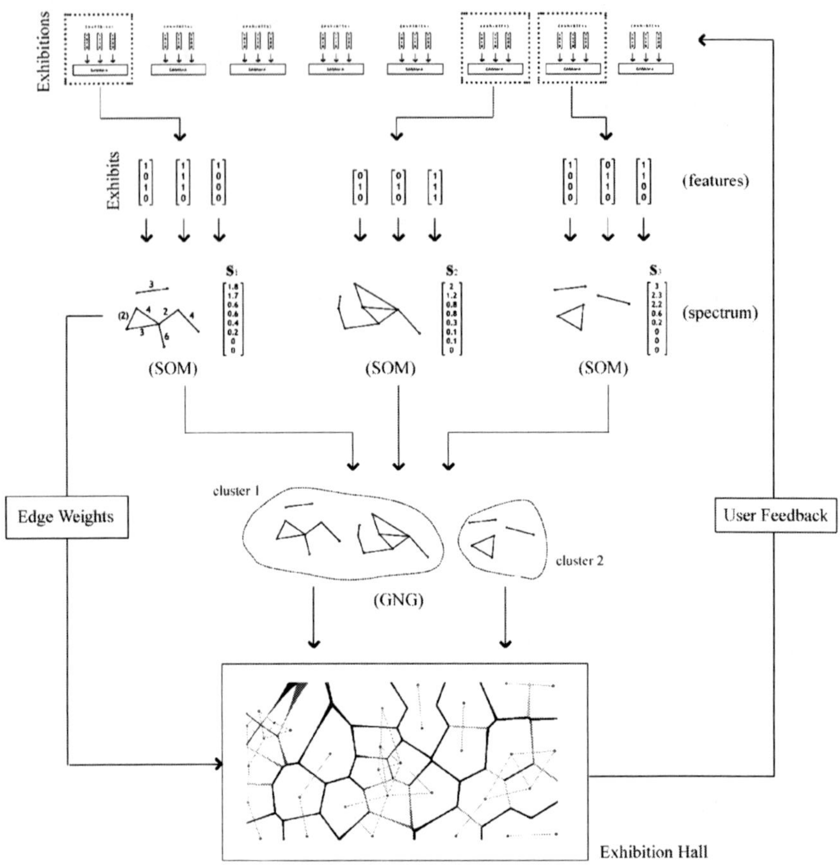

Fig. 9. An overview of the design system. Spatial clusters are found and a cluster average realised in built form. Feedback from the users influences the exhibitions themselves for the next cycle.

Conclusion

In this paper we have attempted to show that dimensionality reduction by a machine can be used in the design process to generate designs informed by its environment – in this case in the design of an exhibition hall whereby both the exhibitions and the users are intrinsic to an adaptable system. In our experiment, the spatial graphs used have been relatively simple in order to verify the heuristic method by visual inspection, but there is no reason why more complex graph structures of even hundreds of elements could not be compared. Three dimensional realisations of form are also possible, Figure 10, although the reconfiguration of the space is likely to be more problematic in reality.

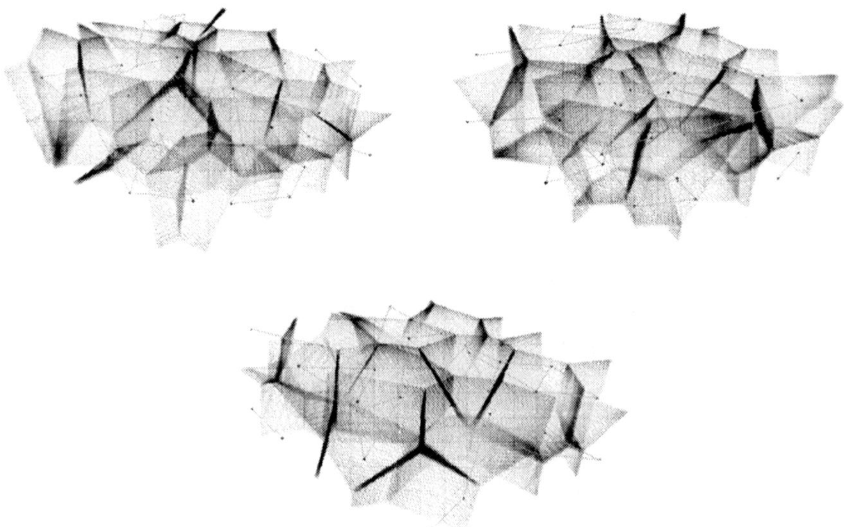

Fig. 10. 3-dimensional Voronoi would not require a planarity condition on the graphs, but would be harder in reality to reconfigure

Further investigation is required in terms of different sized exhibitions. With a variation in exhibit numbers, the length of the spectrum changes and hence the GNG should be well suited to cope with such variation. In other situations, if a plan graph contains two disconnected sub graphs, these could potentially form independent exhibitions themselves. Likewise, the merging of initially separate exhibitions is yet to be explored.

In summary, we have shown a potential application of machine learning in the context of a whole building lifecycle, not just the initial generation of a frozen form. As the environment in which our buildings are constructed becomes ever more dynamic, so we must look at integrating the potential for similar approaches to design by considering architectural interventions as complex adaptive systems coupled to their environment.

Acknowledgements

The authors would like to thank Paul Coates and Manos Zaroukas for helpful discussions throughout this work. This research has been supported by the Industrial Doctorate Centre in Systems, Ramboll and the Engineering and Physical Sciences Research Council, UK.

References

1. Bauer, H.U., Villmann, T.H.: Growing a Hypercubical Output Space in a Self-Organizing Feature Map. International Computer Science Institute, Berkeley (1995)
2. Derix, C., Thum, R.: Artificial Neural Network Spaces. In: International Conference on Generative Art, Milan
3. Ireland, T., Derix, C.: An analysis of the Poly-dimensionality of living - An experiment in the application of 3-dimensional self-organising maps to evolve form. In: 21st eCAADe Conference Proceedings, Graz, September 2003, pp. 449–456 (2003)
4. Derix, C.: Approximating Phenomenological Space. In: Proceedings of Intelligent Computing in Engineering and Architecture, Ascona, Switzerland, pp. 136–146. Springer, Heidelberg (2006)
5. Derix, C.: Genetically Modified Spaces. In: Littlefield, D. (ed.) Space Craft - Developments In Architectural Computing, pp. 22–26. RIBA Publishing (2008)
6. Eppstein, D., Paterson, M.S., Yao, F.: On nearest-neighbor graphs. Discrete and Computational Geometry 17(3), 263–282 (1997)
7. Frazer, J.H.: An Evolutionary Architecture. Architectural Association, London (1995)
8. Frazer, J.H.: The cybernetics of architecture: A tribute to the contribution of Gordon Pask. Kybernetes 30(5), 641–651 (2001)
9. Fritzke, B.: Growing cell structures, a self-orgainising network for unsupervised and supervised learning. International Computer Science Institute, Berkeley (1993)
10. Fritzke, B.: Kohonen feature maps and growing cell structures, a performance comparison. In: Advances in Neural Information Processing Systems (1993)
11. Fritzke, B.: A growing neural gas network learns topologies. NIPS, Denver (1994)
12. Fritzke, B.: A Self-Organizing Network that can follow Non-Stationary Distributions. In: International Conference on Artificial Neural Networks, pp. 613–618. Springer, Heidelberg (1997)
13. Hanna, S.: Representing Style by Feature Space Archetypes: Description and Emulation of Spatial Styles in an Architectural Context. In: Gero, J.S. (ed.) Design Computing and Cognition 2006, pp. 3–22. Springer, Heidelberg (2006)

14. Hanna, S.: Automated Representation of Style by Feature Space Archetypes: Distinguishing Spatial Styles from Generative Rules. International Journal of Architectural Computing 1(5), 1–23 (2007)
15. Hillier, B., Hanson, J.: The Social Logic of Space. New edition. Cambridge University Press, Cambridge (1989)
16. Jupp, J., Gero, J.S.: Towards computational analysis of style in architectural design. In: Argamon, S. (ed.) IJCAI 2003 Workshop on Computational Approaches to Style Analysis and Synthesis, IJCAI, Acapulco, pp. 1–10 (2003)
17. Kalay, Y.: Architecture's New Media: Principles, Theories, and Methods of Computer-Aided Design. MIT Press, Cambridge (2004)
18. Kohonen, T.: Self-organizing maps, 3rd edn. Springer, Berlin (2000)
19. Langley, P., Derix, C., Coates, P.: Meta-Cognitive Mappings: Growing Neural Networks for Generative Urbanism. In: Generative Arts conference, Milan (2007)
20. Martinetz, T.M., Schulten, K.J.: A neural-gas network learns topologies. In: Kohonen, T., et al. (eds.) Artificial Neural Networks, pp. 397–402 (1991)
21. Maturana, H.R.: Biology of Language: The Epistemology of Reality. In: Miller, G.A., Lenneberg, E. (eds.) Psychology and Biology of Language and Thought: Essays in Honor of Eric Lenneberg, pp. 27–63. Academic Press, New York (1978)
22. Pask, G.: A Comment, a Case History and a Plan. in Cybernetic Serendipity. In: Reichardt, J., Rapp, C. (eds.) Reprinted in Cybernetics, Art and Ideas, pp. 76–99. Studio Vista, London (1971)
23. Petrovic, I., Svetel, I.: From Number Cruncher to Digital Being: The Changing Role of Computer in CAAD. In: Architectural Computing from Turing to 2000, eCAADe Conference Proceedings Liverpool, vol. 15(17), pp. 33–39 (1999)
24. Steadman, P.: Architectural Morphology: An Introduction to the Geometry of Building Plans, Pion, London (1983)
25. Zhu, P., Wilson, R.C.: A study of graph spectra for comparing graphs and trees. Pattern Recognition, Pergamon 41(9), 2833–2841 (2008)

DECISION-MAKING PROCESSES IN DESIGN

Comparing stochastic design decision belief models: Pointwise versus interval probabilities
Peter C Matthews

The redefinition of the paradox of choice
Michal Piasecki and Sean Hanna

Rethinking automated layout design: Developing a creative evolutionary design method for the layout problems in architecture and urban design
Sven Schneider, Jan-Ruben Fischer and Reinhard König

Applying clustering techniques to retrieve housing units from a repository
Álvaro Sicilia, Leandro Madrazo and Mar González

Comparing Stochastic Design Decision Belief Models: Pointwise versus Interval Probabilities

Peter C. Matthews
Durham University, UK

Decision support systems can either directly support a product designer or support an agent operating within a multi-agent system (MAS). Stochastic based decision support systems require an underlying belief model that encodes domain knowledge. The underlying supporting belief model has traditionally been a probability distribution function (PDF) which uses pointwise probabilities for all possible outcomes. This can present a challenge during the knowledge elicitation process. To overcome this, it is proposed to test the performance of a credal set belief model. Credal sets (sometimes also referred to as p-boxes) use interval probabilities rather than point-wise probabilities and therefore are easier to elicit from domain experts. The PDF and credal set belief models are compared using a design domain MAS which is able to learn, and thereby refine, the belief model based on its experience. The outcome of the experiment illustrates that there is no significant difference between the PDF based and credal set based belief models in the performance of the MAS.

Introduction

Modern trends in product development and production have seen a shift from products being fully developed in-house to products being developed in collaboration with ever increasing numbers of external partners. Along with this shift, there has been the need to complement the change in design with a change in the enterprise structure. Where product development previously could be undertaken by a small co-located team generating a design that would be manufactured in-house, it is now more common to see large disperse teams being composed from multiple organisations along with the manufacture of the product being undertaken in a similar multi-site manner [1, 2].

J.S. Gero (ed.): *Design Computing and Cognition'10*, pp. 327–345.

The complexity of modern design and manufacturing introduces a new source of uncertainty [3]. A component of this uncertainty is how well potential partners are able to work together. For example, two organisastions can have a very good (tacit) understanding of their mutual capabilities and constraints and therefore are able to work well together. On the other hand, another pairing of organisations could completely lack this mutual understanding and place unrealistic demands on each other resulting in poor collaboration. It is this ability to identify suitable partners for collaborative work that this paper addresses. There have been efforts on selecting partners through capability profiling [4], however there remain issues on how the individual capability scores are determined. These capability scores are subjective to the capability supplier rather than to the purchaser. There is therefore a need to devise capability metrics that are subjective to the purchaser.

The challenge is how can an agent determine if a collaboration will be successful. To address this, there are a number of assumptions: (1) it is not appropriate to represent capability by a point score, rather a distribution-based representation should be used; (2) for the purposes of simulation based experiments, an abstract measure of success should be used; and (3) the actors within this experiment can be modeled using agents, and the interaction between these can be modeled using a Multi-Agent System (MAS). The experiments will test how well the agents are able to 'learn' suitable probability distribution functions (PDFs) for 'potential successful interaction'.

An important problem in learning PDFs is the uncertainty about the accuracy of the PDF. Specifically, how accurate is the probability that a certain outcome will occur? To address this, two approaches for representing this stochastic information will be compared. The first will be the well known PDF representation, where probabilities for all the outcomes are represented by pointwise values. The second will be the use of credal sets [5] where the probabilities of outcomes are represented by a probability interval. The rationale behind this is that the PDF provides a well known benchmark, with classic learning algorithms and simple summary statistics to be used for decision making. The credal set approach provides greater flexibility and supports a richer representation of an agent's understanding of outcome probabilities. For the credal set approach, similar learning algorithms and decision algorithms will be required. The credal set approach would also enable the knowledge elicitation process to work with interval probabilities rather than pointwise, which are easier to elicit from a domain expert. For example, credal sets have been used to design a pressure vessel with a new material (representing uncertain design conditions). In this pressure vessel design case, it was shown that where the imprecision is large, the credal sets outperformed the classic pointwise probability distribution representation method [6].

The remainder of this paper is structured as follows: The next section introduces the mathematical concepts and representations required to implement both a PDF and credal set belief model. The following section describes the learning algorithm implemented for both belief models. This is followed by the presentation of how these models are implemented in a multi agent system. Next section describes the empirical trial environment. Finally, the results are presented and discussed.

Stochastic Representations

Stochastic decision support requires a stochastic domain model. This paper sets out to compare two different approaches for implementing the stochastic domain model. The first will be the well understood probability distribution function (PDF). The PDF will be used as a benchmark to compare the credal set approach. Both of these approaches are expanded on in this section.

Two different approaches to representing stochastic information will be considered. The first is the classical probability distribution function (PDF) representation. Let Ω be the set of all possible outcomes for a random variable, X. A PDF is a function, f(x) = P(X = x) from the set of outcomes, x in that maps onto the probability interval [0;1], subject to the condition that:

$$\sum_{x \in \Omega} f(x) = 1 \qquad (1)$$

For the purposes of this paper, only discrete valued outcome spaces will be considered. The arguments remain valid for continuous valued spaces as well, the summations simply need to be replaced by integrals.

Under the PDF approach, the probability of an outcome is precisely defined. For example, consider the outcome of a certain collaboration between two agents. The possible outcomes for the collaboration are: fail, poor, fair, excellent. The probability of each outcome can then be presented as: P(C = fail), P(C = poor), P(C = fair), and P(C = excellent). Each of these represents the probability that the collaboration has the stated outcome. While the outcome of the collaboration is unknown, the PDF representation for this case is precise, e.g.:

$$P(C = \text{fail}) = 0{:}1$$

$$P(C = \text{poor}) = 0{:}2$$

$$P(C = \text{fair}) = 0{:}4$$

$$P(C = \text{excellent}) = 0{:}3$$

Given the above values for this system, an observer would expect to see 10% of collaborations fail, 20% of collaborations be poor, 40% of collaborations be fair, and 30% of collaborations be excellent. However when parameterising this case, there might not be such certainty on these values. For example, the expert might not have total confidence in the pointwise values being assigned to both outcomes. A very simple method for overcoming this is to represent the probabilities as intervals rather than as points [5, 7].

The use of intervals for probability outcomes provides a natural extension to the pointwise approach. The length of the interval is related to the confidence, or certainty, of the probability value of an outcome. Specifically, the probability of an outcome is specified as a range of values. Using the above example, it becomes possible to say P(C = fair) = [0:3;0;5]: i.e. the probability that the collaboration will be successful lies between 0.3 and 0.5.

Functional Distribution Representation

In classical probability, the well understood and used method for representing the probability of various outcomes is the probability distribution function (PDF). This can be easily plotted in two dimensions for visualisation using the outcome space as the horizontal axis and the probability of each outcome occurring as the vertical axis. Figure 1 illustrates two PDFs on the same set of axis, illustrating how characteristics (e.g. the distribution mode) can be identified from this representation. For the scope of this work, it is sufficient to consider the discrete case. Let Ω be the set of all possible outcomes for some random variable X (for simplicity of the mathematical notation, it is assumed that Ω is an ordered set). The PDF for this random variable is then defined as:

$$
\begin{aligned}
&f_X(x) = P(X = x) \\
&\text{s.t.} \sum_{x \in \Omega} f_X(x) = 1
\end{aligned}
\tag{2}
$$

An equivalent representation is the cumulative distribution function (CDF). This is simply the sum (or integral in the continuous case) of the PDF in Equation 2 along the horizontal axis, and can therefore also be visualised in two dimensions. As the PDF is bounded between 0 and 1 and sums to unity, the CDF is a monotonically increasing function from 0 to 1. The CDF is defined as:

$$
F_X(x) = \sum_{i \le x} f_X(i) \tag{3}
$$

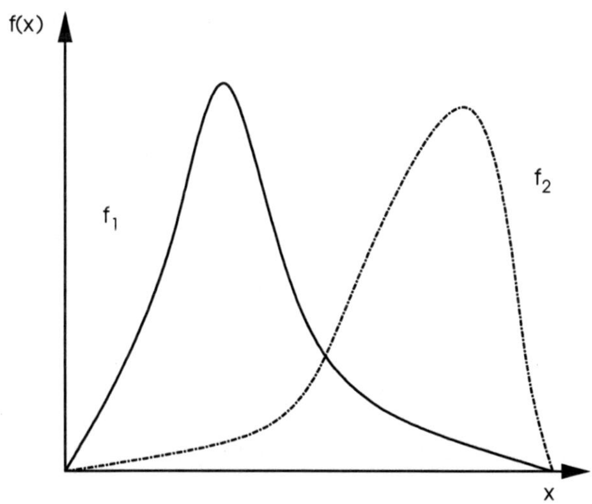

Fig. 1. Two probability distribution functions: f1 represents a distribution with a low valued mode and f2 represents a high mode

Functional Representation of Interval Probabilities

A slightly different approach is required to represent the stochastic information in the interval case. In the interval probability case, the CDF representation can be used to define the full range of possible (pointwise) distribution functions for a given variable. By extending the concept of interval probabilities for each outcome, two CDFs can be defined: firstly the CDF defined by all the lower probability ranges, F(X = x) and secondly the CDF defined by all the upper probability ranges, F(X = x).

These two CDFs provide the envelope for all possible CDFs that could represent the distribution function for the given variable.

The p-box [8], or credal set [9], representation for uncertainty is based on an envelope that defines the range of possible cumulative distributions. This is bounded by the maximal ($\overline{F}$) and minimal ($\underline{F}$) values taken by all possible distribution functions for the given uncertainty. This credal set is then formally defined as:

$$M = \{F : \underline{F}(i) \leq F(i) \leq \overline{F}(i), \forall i \in \Omega\} \tag{4}$$

As the area within the envelope defines the range of possible distributions, it follows that the larger this area is the greater variety of distribution functions that exist to represent the variable. Therefore it is possible to numerically define and measure the uncertainty about the distribution for the given variable. Numerically, the uncertainty for a given credal set M will be defined as:

$$\text{Unc}(M) = \tfrac{1}{\Omega} \sum_{i \in \Omega} (\overline{F}(i) - \underline{F}(i)) \tag{5}$$

From this equation, it can be seen that where the credal set 'narrows' to the limit, Unc(M)= 0 and conversely where the credal set contains all possible distribution functions, Unc(M)= 1.

To illustrate the credal set, consider again the random variable C representing the success of a collaboration between two agents. Now, the probabilities of each outcome is represented by an interval.

P(C = fail)=[0:0;0:2]
P(C = poor)=[0:1;0:3]
P(C = fair)=[0:3;0:5]
P(C = excellent)=[0:1;0:4]

These intervals define the range of distribution functions that represent the probabilities for each possible outcome, subject to the probabilities summing to unity. Within this range there are infinitely many possible distribution functions. The credal set has the property that for any two distributions taken from the credal set say F_1 and F_2, all linear combinations of the form $\alpha F_1 + (1-\alpha)F_2$, for a. [0;1] will also be a member of the credal set.

Agent Learning

The agents within this system learn (gain experience) through interaction and observation. The learning is based on a pair of agents initiating a collaboration. This collaboration will have some measurable degree of success, and this evidence is used by the agents to learn about their mutual ability to collaborate. Specifically, the learning process modifies the agent's stochastic information held about the observed variable (in this case the ability to successfully collaborate). If the current stochastic belief is given by F and the observed evidence is given by E, then the updated stochastic belief is determined as follows:

$$F' = (1-\gamma)F + \gamma E \tag{6}$$

where $\gamma \in [0,1]$ is the learning rate. Where $\gamma = 0$, no learning takes place and at the other extreme, $\gamma = 1$, the updated stochastic belief is completely determined by the last piece of evidence seen.

This abstract learning function requires further detail in both the PDF and credal set cases. This is developed below.

PDF Updating

In the PDF case, the evidence must be transformed into a PDF as well. The evidence will have been the observation of a single event, say xe. In the discrete event case, this evidence PDF can be encoded as:

$$e(x) = \begin{cases} 0 : x \neq x_e \\ 1 : x = x_e \end{cases} \tag{7}$$

Then the updated PDF that is used by the agent for future decision making is given by:

$$f'(x) = (1-\gamma)f(x) + \gamma e(x) \tag{8}$$

It is worth noting that the constraint. f (x)= 1 remains satisfied after applying this learning function.

Credal Set Updating

In a similar manner to the PDF approach, in this case the observed evidence must be transformed into a CDF. Again, if the observed evidence is given by xe, the evidence CDF can be encoded by:

$$E(x) = \begin{cases} 0 : x < x_e \\ 1 : x \geq x_e \end{cases} \tag{9}$$

This produces a step-function, with the step rising at the evidence point. Note that this is simply the integral of the PDF version (Equation 7).

In the credal set case, there are now two CDFs that need to be considered, the lower and upper bounds of the credal set. In a similar approach, the learning algorithm is simply applied to both boundaries. The rationale for this is that if the same piece of evidence (or observation) was presented at each learning cycle, the credal set must converge to this evidence function.

$$\overline{F}'(x) = (1-\gamma)\overline{F}(x) + \gamma E(x) \tag{10}$$

$$\underline{F}'(x) = (1-\gamma)\underline{F}(x) + \gamma E(x) \tag{11}$$

Note that the updated functions are a linear weighted combination of two other CDFs and therefore they are also CDFs. Hence, these properly define a credal set.

Agent Implementation

Agents are used to model a set of individuals observing and acting within an environment. Ultimately, the aim is to test how well they are able to identify, through evidence based learning, suitable other agents for collaborating on a given task. For the purposes of this paper, the aim will be the slightly simpler task of learning to identify which other agents within the system an agent should prefer to collaborate with through forming a direct network link with.

Each agent forms a view (or belief) of the characteristics of all the other agents within the system. This belief is updated with evidence as and when it is observed by the agent. The agent uses the belief of the other agents' characteristic to determine which is most appropriate to interact with. If there are N agents within the MAS, then the data structure for each agent i is given by $(B_{i1}, B_{i2}, \ldots, B_{iN})$, where B_{ij} is agent i's belief of agent j's characteristic distribution. The belief component can be either a PDF or a credal set, and these will be compared in the empirical section of this paper.

To compare the PDF and credal set belief models, the UCI Car design database is used as a design domain model [10]. Here, each design variable is determined by an agent, and the agents need to identify suitable agents to interact with in order to complete a car design.

UCI Car Design Database

The UCI Car design database [10] provides a simplified representation of an automotive design domain. This domain consists of ten variables with a known hierarchical rule structure (see Figure 2). The design variables can be categorised as design parameters (controlled by the designer, shown in boxes) or design characteristics (function of the design variables, shown in ovals). As a corollary, the design parameters are the nodes of the rule structure. The database contains all the 1728 possible (legal) designs within this domain.

In this work, each variable was 'assigned' to a unique agent. The agents were not provided with any prior information about the rule structure, and therefore prior to any learning were completely unbiased to which agents they would prefer to collaborate with. Each agent was given a utility function that mapped variable state with 'cost' of moving to that state. These utility functions were not part of the original database, but needed as part of the MAS bidding system.

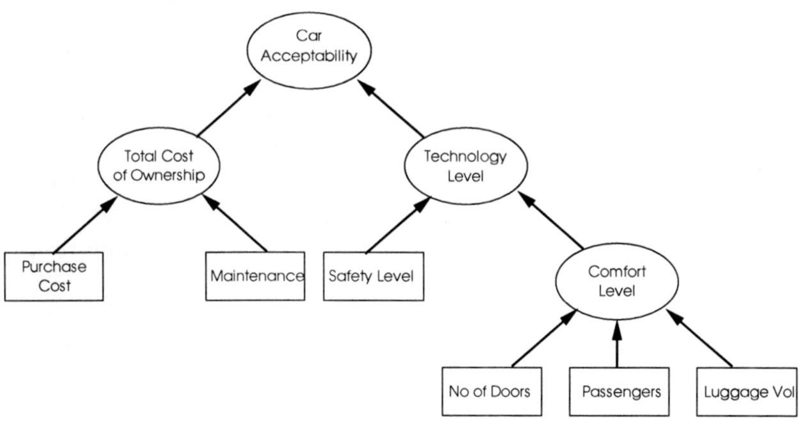

Fig. 2. Rule structure of the conceptual car domain [10]

The database was used to generate design tasks. This was accomplished by randomly selecting a design from the database, thereby ensuring a legitimate design, and then transforming it into a design task. The design task was created by randomly blanking out a preset number of the design variables. The remaining set variables (four were left set for these simulations) represented the 'design task'. The aim for the MAS was to then complete this design, i.e. define the blanked out design variables. It is possible that for any given design task, there were multiple possible complete satisfying designs, and so the aim was not to recreate the original randomly sampled design.

The design process was undertaken by a set of agents. Each agent had 'control' over a single design variable. The agents were able to observe each other through a blackboard approach [11]. The design task provided a global goal for the MAS. Within this, local goals were set for the individual agents in the form of the target variable setting for each agent. Individual agents must collaborate to be able to set or modify their own variable. Initially, the agents are given no information on the design variable structure and therefore must learn this through action and observation. When a pair of agents attempt to collaborate (i.e. change their variable settings), the overall success of this collaboration is used to update the belief of success for future collaborations.

The success of an agent-agent collaboration was measured using an abstract quality metric. For any collaboration, there were four possible outcomes, ranging from most successful to least: (1) the collaboration is a total success – both agents are able to move to the new variable setting; (2) the initiating agent is successful, the supporting agent is not; (3) the supporting agent is successful, the initiating agent is not; and (4) neither agent is successful in changing state. It is this collaboration outcome that is used to update an agent's belief in its own suitability to collaborate with the other agent.

To reflect the logistical challenges that are inherent to physical design, each agent was augmented with a cost function. These cost functions were individually tailored for each agent and effectively represented the difficulty for an agent to achieve a specified variable state.

Each individual design task can be measured against a set of metrics. These metrics were used specifically to measure how well the MAS was learning as a result of the interactions involved in each design task. The metrics were:

Cost each design task incurred a cost as a result of the sum of the 'interactions' that occurred between the agents during the design;

Score each agent interaction was scored based on the degree of change in the design, ranging from 0 (no change) to 1 (both agents successfully changed state) and provides a measure of 'quality of collaboration';

Task completion was the number of design variables set. As the MAS was not able to backtrack the design process (for example by clearing previously set design variables) it could therefore find itself in a 'dead end', hence this provided a measure of how successful the MAS was in completing designs; and

Number of agent interactions any individual interaction was not necessarily going to be successful in modifying the design, and therefore the total number of interactions for a given design task measured how efficient any given design process was.

In addition to these basic metrics, two further metrics were derived:

Mean cost per variable set this is the average cost incurred in determining each design variable for a given design and

Mean interactions per variable set this is the average number of interactions required to set each design variable.

The mean cost metric is important due to the MAS not being able to backtrack. It is therefore possible that design cost reduces as a result of task completion reducing. By measuring the mean cost per variable set, this is able to illustrate if this is indeed the case. Note that it is not necessary to measure mean cost per interaction, as this is determined by utility function, and therefore will be independent of any learning and constant throughout the simulation.

The mean number of interactions per variable provides another measure to how efficient (in terms of 'effort' due to collaboration) the MAS is at any point. This basically measure how often an agent on average attempts to set a design variable before succeeding.

Empirical Trials

The empirical trials seek to compare the MAS performance characteristics using a PDF belief model against the credal set belief model. This comparison will be undertaken using the UCI Car design domain. Due to the stochastic nature of these experiments, each individual trial must be repeated several times to get meaningful results.

The UCI Car domain has added complexity to reflect the complexities that are involved in real product design. The goal is to identify other agents that are suitable collaborators based on observed past performance. These observations arise as a result of attempted collaborations.

For each experiment, there is a small set of parameters. Principally, these parameters set the learning rate (γ) and the duration or the complexity of the task (N). For the Car design domain this is determined by the number of variables that were specified prior to the design task, and therefore the larger N, the fewer free design variable are left to be determined by the MAS.

The hypotheses that will be tested are:

H0 There is either no significant difference between using the credal set belief model over the PDF belief model;

H1 The credal set belief model and the PDF belief model are significantly different.

The car design experiments used a series of tasks. Each task specified the same number of design variables, but the exact variables and target values were randomly sampled. Each design task requested that the MAS identify a design solution with certain design variables set to given values. Each design task was generated by randomly selecting a design from the car database and then randomly selecting a subset of design variables.

Each agent has a belief of how well it is able to collaborate with each and every other agent. As there are 10 agents in the car design domain, each agent has 9 belief models. An agent will use their belief models to determine which other agent is the most likely to result in a successful collaboration. A collaboration is initiated by one agent who wishes to achieve a given result and a collaborating agent who will offer to help the initiating agent in obtaining that result. After an initiating agent has selected a collaborator and attempted to collaborate with that agent, it is able to see the result of this collaboration. There are four possible outcomes of the collaboration (in descending order of overall quality of outcome): (1) both agents successfully change; (2) the initiating agent is able to change, but not the collaborator; (3) the collaborator is able to change, but not the initiator; and (4) neither agent successfully changes. This outcome is mapped onto a numerical 'quality of collaboration' score which is in turn used as evidence for the initiating agent to update their belief model.

The key experimental variable was the belief model learning rate, γ. The key experimental outcomes that were measured were: (1) the average cost of setting each variable, (2) the average number of interactions per variable set, and (3) the average task completion level. In terms of 'optimal' outcomes, lower is better for cost and number of interactions while higher is better for task completion.

For this experiment, each independent design trial was run for 120 iterations. This run allowed the agents to refine their belief models. Each trial was repeated 30 times, and the average value is reported.

Results

Figures 3 to 9 represent how the car design MAS performs against the three metrics as γ ranges from 0.0 to 1.0 in increments of 0.1. These graphs clearly suggest that for most values of γ (the learning rate), there is little difference in performance of the car design MAS. At $\gamma = 0$, both systems undertake no learning and therefore this effectively represents the performance of the MAS at its initial state. From the graphs it can be seen that its performs poorly in terms of cost and number of interactions. Oddly, this case has the best (highest) task completion of all. For the remaining values of γ, there is little trend to be seen, and the systems appear to be stable for all . values.

Figure 3 compares the convergence of the average cost per variable set of the PDF and credal set belief models. Initially, the credal set-based belief model performs worse than the PDF-based belief model, but by iteration 70 both approaches have converged in terms of this metric. Further, they both continue to perform at this level for the remainder of the learning process, suggesting that both systems have stabilised. This learning run was performed at the MAS datum parameter settings.

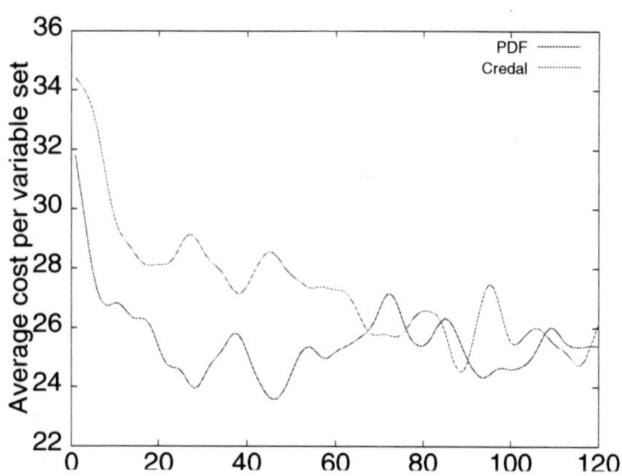

Fig. 3. Comparing convergence between the PDF and credal set representations of the average cost per variable set as learning progresses ($\gamma = 0{:}2$)

In a similar vein, Figure 4 compares the average number of interaction each agent performs per variable set. Under this metric, both the PDF and credal set-based systems initially perform at the same level. From iteration 50, a divergence is seen with the credal set-based system performing worse, however this appears to be temporary. It is also worth noting that the difference between these two systems is small: at the greatest divergence (ca iteration 80), the difference between them is 3 interactions.

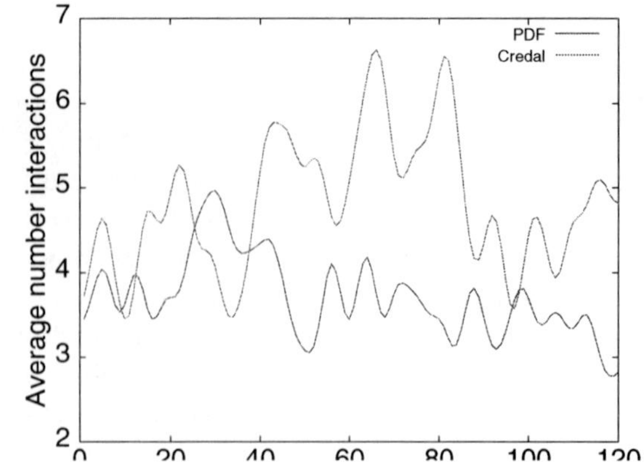

Fig. 4. Comparing convergence between the PDF and credal set representations of the average number of interactions per variable set as learning progresses ($\gamma = 0{:}2$)

A similar result is seen in Figure 5, the level that the design tasks have been completed as learning progresses. Both systems perform similarly, with the credal set system completing about one more design variable than the PDF system throughout the learning process.

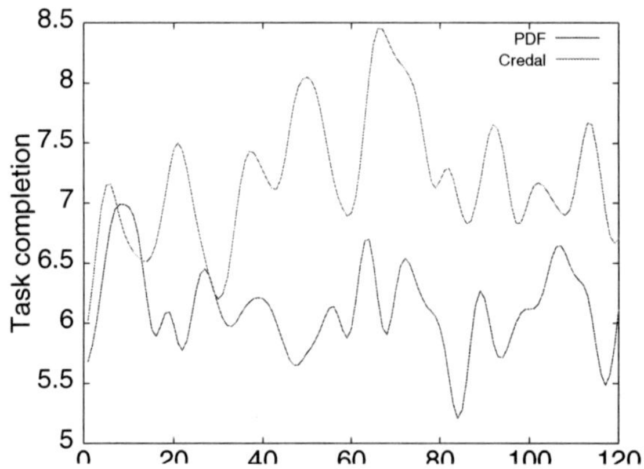

Fig. 5. Task completion as learning progresses ($\gamma = 0{:}2$).

The following set of results compare the performance of the two systems as a function of the learning rate, γ. Within these results, it is worth noting that at the one extreme $\gamma = 0$, no learning occurs, whereas at the other extreme, $\gamma = 1$, the learning is completely based on the previous iteration's observation. In both these extremes, the two systems will perform identically. The first of these results, Figure 6 compares the average cost of setting each design variable against the learning rate.

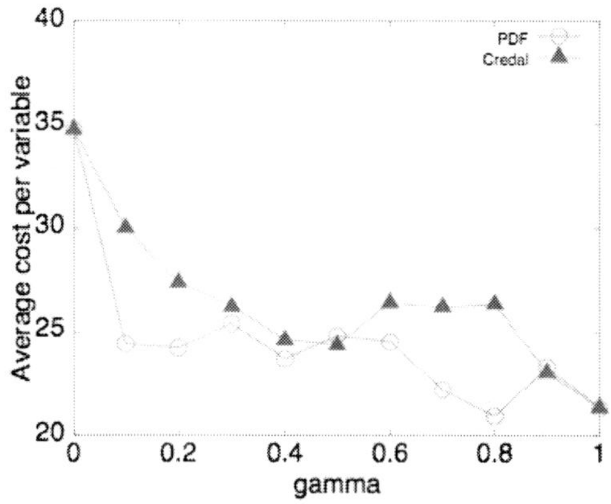

Fig. 6. Comparing average cost per variable set after learning for different values: PDF v Credal

On average, the credal based system performs slightly worse. However it is more consistent throughout the learning rate range, whereas the PDF based system rises again in the middle of the γ range.

Figure 7 illustrates the task completion level for both systems against the learning rate. Similar characteristics are again seen, with both systems converging for the middle of the γ range. Towards either extreme, the PDF based system does perform slightly better than the credal set system.

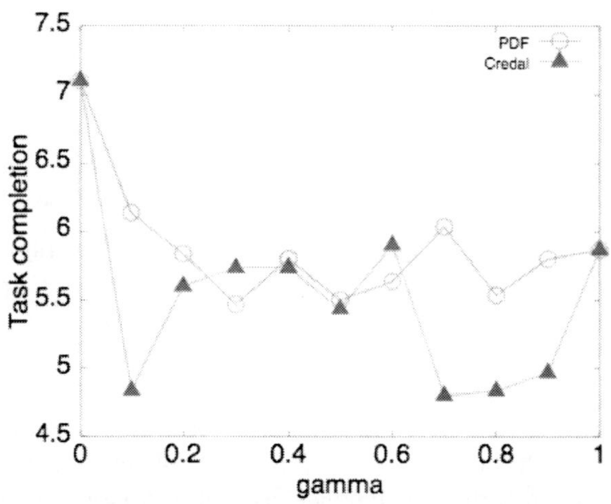

Fig. 7. Task completion against learning rate (γ)

The final set of empirical trials were to compare how the two systems responded to different design task settings. These are independently illustrated for the PDF case, Figure 8, and the credal set case, Figure 9.

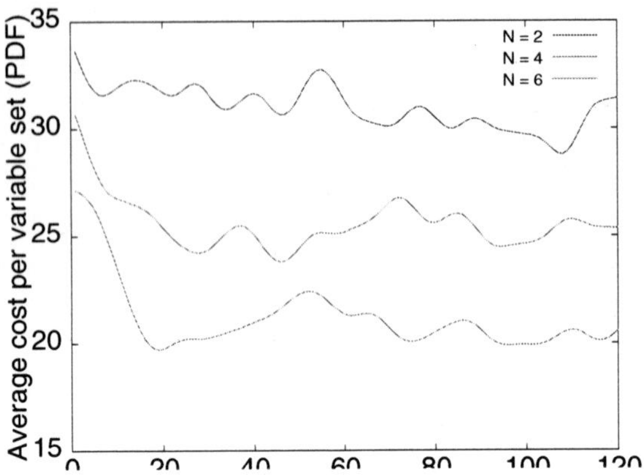

Fig. 8. Average cost per interaction as learning progresses for different design task complexities given by N, the number of variables that have been predetermined (PDF case, γ = 0:2)

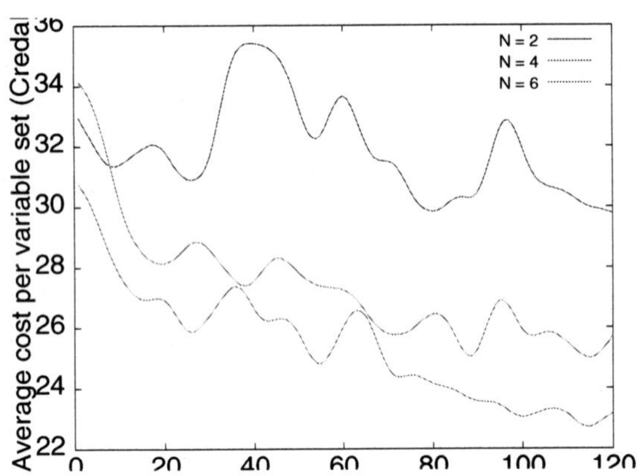

Fig. 9. Average cost per interaction as learning progresses for different design task complexities given by N, the number of variables that have been predetermined (Credal set case, γ = 0:2)

The three values of N and the number of design variables that are predetermined (or how many variables are set as part of the design 'specification') were tested. In both cases (PDF and credal set) it can be noted that the performance, in terms of average cost per variable set, decreases as there are fewer design variables to set. Again, in both cases, both systems converge to similar performance levels for the same task complexity.

Discussion

The empirical work compared how a PDF based belief model performed against a credal set based belief model. The key metrics that were used to base the comparison on were: (1) how well the design MAS could complete the design task, (2) the mean cost for setting each design variable, and (3) the mean number of interactions required to set each design variable.

The experiments were based on the assumption that there was no prior knowledge to initialise the belief model. Therefore, in both cases the belief models were initialised uniformly. In the PDF case, this meant a uniform distribution model across all outcomes. On the other hand, in the credal set case, the belief models were initialised to be the full interval for all outcomes. Both cases used a similar learning algorithm, where evidence was used to modify the belief models to increase the probability of the recently seen evidence.

The key conclusion that can be drawn from this comparison is that a stochastically based decision support algorithm can use either the PDF or credal set representation of belief. In terms of the hypothesis presented earlier, H0 would be accepted. Specifically, it means that where a PDF representation has been used in the past, this is able to be replaced by a credal set representation. As credal sets are more readily elicited from domain experts than PDF [7], this represents an important advance in decision support systems instantiation.

Conclusion

The key question this paper set out to investigate was to compare the effectiveness of pointwise versus interval probabilities in a collaborative engineering design process. To tackle this question, an abstract design environment was used themed on a multi-agent car design process. The car design was divided into a small number of loosely coupled sub-design tasks, and each of these tasks was undertaken by an independent agent. These agents needed to collaborate to be able to successfully complete the design.

The design agents were initially given no information as to which other agents within the system they could collaborate with. Through interaction with each other, the agents could gather evidence and learn which other agents they most successfully were able to collaborate with. Specifically, the evidence was used to update the agents' belief models of each others' capabilities.

The belief models were implemented using the two stochastic representations: pointwise and interval (credal set) probabilities. The computational experiments demonstrated that the credal set approach performed no worse that the (classic) pointwise representation. Given the benefits to be gained from using a credal set approach, such as ease of elicitation from domain experts, it can therefore be argued that there is a call to replace the use of pointwise stochastic models with credal sets in engineering contexts.

Further work is required to determine how sensitive the credal belief model is to the interval size given for each outcome. Clearly, as the interval of the credal set narrows, the credal set in the limit converges into the PDF. Therefore, it is important to know what the lower interval limit for an effective credal sets is. This will support the belief elicitation process in providing guidance as to how much information is required.

Acknowledgements

This research was undertaken while the author was visiting the University of Amsterdam Informatics Institute. Thanks must be given to the Royal Academy of Engineering and Thales Research Netherlands for supporting this visit. The conceptual basis for this work was developed while the corresponding author was a Fellow at the Institute of Advanced Study, Durham University.

References

1. Maier, A.M., Kreimeyer, M., Hepperle, C., Eckert, C.M., Lindemann, U., Clarkson, P.J.: Exploration of correlations between factors influencing communication in complex product development. Concurrent Engineering-Research and Applications 16(1), 37–59 (2008)
2. Maier, A.M., Kreimeyer, M., Lindemann, U., Clarkson, P.J.: Reflecting communication: A key factor for successful collaboration between embodiment design and simulation. Journal of Engineering Design 20(3), 265–287 (2009)
3. Chalupnik, M.J., Wynn, D.C., Clarkson, P.J.: Approaches to mitigate the impact of uncertainty in development processes. In: Norell Bergendahl, M., Grimheden, M., Leifer, L., Skogstad, P., Lindemann, U. (eds.) Proceedings of the 17th International Conference on Engineering Design, Stanford, vol. 1, pp. 459–470 (2009)
4. Armoutis, N.D., Maropoulos, P.G., Matthews, P.C., Lomas, C.D.W.: Establishing agile supply networks through competence profiling. International Journal of Computer Integrated Manufacturing 21(2), 166–173 (2008)

5. Tonon, F.: Some properties of a random set approximation to upper and lower distribution functions. International J. of Approximate Reasoning 48(1), 174–184 (2008)
6. Aughenbaugh, J.M.: The value of using imprecise probabilities in engineering design. J. of Mechanical Design 128(4), 969–979 (2006)
7. Guo, P., Tanaka, H.: Decision making with interval probabilities. European Journal of Operational Research 203(2), 444–454 (2010)
8. Ferson, S., Hajagos, J.G.: Arithmetic with uncertain numbers: rigorous and (often) best possible answers. Reliability Engineering & System Safety 85 (1-3), 135–152 (2004)
9. Levi, I.: The Enterprise of Knowledge: An Essay on Knowledge, Credal Probability, and Chance. MIT Press, Cambridge (1980)
10. Asuncion, A., Newman, D.J.: UCI machine learning repository. University of California, Irvine, School of Information and Computer Sciences (2007)
11. Tan, G.W., Hayes, C.C., Shaw, M.: An intelligent-agent framework for concurrent product design and planning. IEEE Transactions on Engineering Management 43(3), 297–306 (1996)

A Redefinition of the Paradox of Choice

Michal Piasecki and Sean Hanna
University College London, UK

Barry Schwartz defined the paradox of choice as the fact that in western developed societies a large amount of choice is commonly associated with welfare and freedom but too much choice causes the feeling of less happiness, less satisfaction and can even lead to paralysis. The paradox of choice has been recognized as one of the major sources of mass confusion in context of the B2C online mass customization. We propose to redefine the paradox of choice with an emphasis on the meaning of choice in conjunction with the amount of available options, rather than just the quantity of choice. We propose that it is the lack of meaningful choice, rather than an overwhelming amount of choice, that can cause customers' feelings of decreased happiness, decreased satisfaction and paralysis. We further propose that since users themselves are often not able to explicitly define what constitutes a meaningful choice, the task they face belongs to the category of ill-defined problems. The challenge for mass customization practitioners is thus not to limit the scope of choice, as has been suggested in previous literature, but to provide users with choice that is relevant to them. We further discuss two computational approaches to solving problems related to the redefined paradox of choice in the context of the B2C mass customization. The first is based on recommender systems and the second is an implementation of artificial selection in genetic algorithms. We present findings of an empirical comparison of genetic algorithm and parametric product configurators. We find that the genetic algorithm tools, which allow users to move through a solution space by recognition of meaningful options rather than their definition, appear to be more popular among the users when it comes to browsing through solution spaces with larger number of dimensions.

Introduction

It is a common assumption in western developed societies that the more choice people have the more freedom they have and the more freedom they

J.S. Gero (ed.): *Design Computing and Cognition'10*, pp. 347–366.

have the more welfare they have. A syllogism thus suggests that the more choice people have the more welfare they have. Yet it was shown in multiple experiments that increasing amount of choice is related to increased feeling of satisfaction only up to a certain point. Beyond this threshold growing amount of choice starts to have a negative impact on the decision-maker. Schwartz called this contradiction the "paradox of choice" [1], [2].

We are going to propose that it is not necessarily the overwhelming amount of choice what makes people unhappy, but rather a lack of meaningful choice. We will propose a redefinition of the term "paradox of choice" with a decreased emphasis on quantity of choice and increased emphasis on meaning of the attributes to choose from. We will argue that current definition of the paradox of choice does not fully acknowledge the complexity of decision-making process. It assumes that one deals with well-defined problems, that are the problems defined explicitly. It thus implicitly assumes that the rational choice theory is a valid model of decision-making process even though it has been argued it is overly simplified [3], [4]. Schwartz's definition of paradox of choice fails to model the task well in the case of ill-defined problems, where 'ill-defined' refers to problems in which the formulation changes while the person trying to solve them moves through the space of potential solutions [5], [6]. These types of problems constitute the vast majority of problems the decision-makers described by Schwartz encounter.

Schwartz's definition of the paradox of choice refers to every aspect of modern life, from shopping through relationship and career related choices up to medical care decisions. However, this discussion will concern a successful implementation of mass customization, understood as the producers' ability to provide users with exactly what they need at mass production cost [7]. Configurators used for mass customization of products over the Internet on the business to customer axis (B2C) will be the main context of this enquiry.

We will discuss computational means of solving problems related to the redefined paradox of choice. We will also provide findings from an empirical comparison of two types of B2C configurators. One of them featured an implementation of artificial selection in genetic algorithms and thus allowed the users to move through the solution space by recognition of meaningful options rather than their definition. The other featured parametric configurators, which required an explicit setting of attribute levels and thus demanded a definition of meaningful solutions.

We outline a definition of the terms "paradox of choice" and "mass confusion", already proposed in the literature. The next section is a discussion of means of solving problems related to the paradox of choice and mass confusion in the context of these definitions. We propose to redefine the term "paradox of choice" and provide an outline of computational means of solving problems related to the redefined paradox of choice. We discuss implementation of recommender systems algorithms and artificial selection. A discussion of an empirical experiment on an implementation of artificial selection in B2C configurators is provided, summarized with a conclusion.

The Paradox of Choice and Mass Confusion

The Paradox of Choice

Schwartz notes that an increased amount of choice can be observed in nearly every aspect of lives of the citizens of western developed countries [1]. Increasingly larger choice is available during shopping for any kinds of goods, be they groceries, gadgets, knowledge or entertainment. Furthermore, for the past couple of decades, people have been exposed to new kinds of choices, such as the choice of utility providers, health insurance, retirement plans or even types of medical treatments. Schwartz argues that the variety of choice in modern developed countries is commonly associated with freedom. Freedom in turn is associated with welfare. Thus increased amount of choice is understood as a sign of welfare and it is expected to have a positive influence on how people feel. However, Schwartz's argument is that the opposite might often be true. An overwhelming amount of choice can result in dissatisfaction and even more serious drawbacks. In order to describe this phenomena Schwartz proposed that there is a certain threshold in a relation between the amount of choices available to the chooser and their satisfaction. He argues that the level of satisfaction rises together with the number of available options only up to a certain point. Beyond this point an increasing amount of choice results in a decreased feeling of satisfaction. Schwartz describes empirical evidence suggesting that a too large a scope of choice can lead to dissatisfaction both with the decision-making process and with the results of the decision itself, even if this decision is objectively good. Even more serious consequences, such as depression and customer paralysis, were also confirmed empirically. Paralysis might be of significant importance to producers, as it may mean that customers will not decide to purchase any of the available products if they are confronted with an unbearable scope of choice. Schwartz concludes that a wide scope of choice is understood as a sign of the freedom and welfare, while in fact a scope that is too broad can lead to numerous drawbacks and general feeling of insecurity. He proposed to call this phenomenon the "paradox of choice" [1], [2], [8].

Wundt Curve Representation of the Paradox of Choice

Schwartz has conceptualized a diagram depicting the relation between satisfaction and amount of choice. The diagram very closely resembles a curve often called the "Wundt curve". Berlyne [9] adopted the curve from Wundt's description of measures of human and animal arousal response to stimuli. It illustrates the hypothesis that beyond a certain point the stimulus no longer causes pleasure. Instead it begins to contribute to unpleasant feelings. In Schwartz's case the shape of the curve before the threshold depicts increased amount of choice resulting in increased satisfaction. Beyond the threshold increased amount of choice begins to result in decreased satisfaction, Figure 1.

Fig. 1. A Wundt curve-like relation between amount of choice and subjective feeling of satisfaction

In his definition, Schwartz discusses the amount of choice explicitly, but Keller and Stealin suggested an existence of U-shaped relationship between both quantity and quality of choice and the choice effectiveness [10]. They proposed that an increasing quantity of information available during the decision making process, while the quality of information is fixed, will lead to decreasing decision effectiveness. On the other hand increased quality of information without a change to its quantity will result in increase of the choice effectiveness. Keller and Staelin did not reference the Wundt curve directly, but it could well be used as an illustration of their hypothesis.

Saunders and Gero used the Wundt curve as a reference to the amount of novelty recognized as a meaningful contribution by a creative community [11]. Their thesis is that both too little novelty and too much novelty causes lack of understanding and disapproval from the side of the community. Thus a threshold should exist, which marks the most appropriate amount of novelty recognizable by creative peers. Saunders and Gero modeled the behavior of a creative community with agents capable of generating art works with an aid of a genetic algorithm. Their curve is based on Berlyne's original diagram of the sum of two non-linear functions. One marks the amount of reward and the other the amount of punishmenta particular agent receives, depending on how his art works are judged by the rest of the agents.

Holbrook and Gardner referenced Berlyne's proposal of the relationship between arousal and stimulus directly [12]. They have empirically demonstrated phenomena conforming the Wundt curve relation between the duration of listening to particular piece of music and the arousal caused by it. They have suggested that a peak listening time will occur in case of medium arousal, which constitutes a similar assumption to the one proposed by Saunders and Gero.

Mass Customization and Mass Confusion

Mass customization (MC) was first anticipated by Toffler [13], however the term itself was coined by Davis and popularized couple of years later by Pine [7], [14]. Mass customization can take place both on business-to-business axis (B2B) and business-to-customer axis (B2C). There are multiple definitions of MC functioning simultaneously in the literature, but the one provided by Pine enjoys the widest recognition. Pine proposed that mass customization aims at fulfilling all the needs of every customer at the mass production cost [7]. In recent years the rapid development of computer aided manufacturing and e-commerce technologies have caused new MC commercial initiatives to flourish.

Online tools provided by the producers to customize goods have been referred to as "toolkits for user innovation and design", "choice-boards", "design systems" and "co-design platforms", but the term "configurators" has been used most recently [15], [16], [17], [18]. The customizable features of products are referred to as product attributes and various options within a particular attribute are called attributes levels [18], [19]. An example of an attribute is color or size and attribute levels are understood as all colors and sizes available to choose from. The multidimensional space where all the levels of all the attributes can be mapped has been called a "search space" or "attribute space", but "solution space" is the currently accepted term [16], [18], [20].

To satisfy a potentially large amount of configurators' users, companies harnessing mass customization have to provide them with a large variety of options. Thus companies have to allow users to customize large amount of product attributes. Further on, each of the attributes should expose a variety of levels to choose from. However it is an understandable assumption that if the paradox of choice takes place in case of non-customizable products, it is even more likely to take place in case of MC. Users of MC configurators must not only choose which configurator to use, but also have to specify a considerable amount of levels of product features. It has been recognized in MC literature that interaction with some of the product configurators might lead to the feeling of insecurity. This feeling was called "mass confusion" and Piller distinguished three main sources of it [16], [21], [22], [23]:

- Burden of choice.
- Customers' inability to match their needs with product attributes.
- Lack of information about manufacturer's behaviour.

Burden of choice refers to an overwhelming amount of choice, which can result in users' paralysis and lack of any decision at all. Thus it is another term for what Schwartz referred to as the paradox of choice. Customers' inability to match their needs with product features to customize is directly related to the amount of choice. Companies are likely to fulfill customers' needs by allowing them to customize a large amount of attributes, but they face the threat of exposing their users to the burden of choice. On the other hand when the choice of product attributes is limited, customers are likely to find out that the product attribute they wish to customize is not customizable. Lack of information about manufacturer's behavior is a problem of communicating a company's identity over the Internet and is outside of the scope of this paper.

Means of Solving Problems Related to the Paradox of Choice and Mass Confusion

Approaches to solving problems related to the paradox of choice, proposed by Schwartz and researchers from the field of MC, can be grouped into three categories:

- Approach 1: Limitation of the solution space according to averaged useds' needs [1], [18], [21].
- Approach 2: Nourishing online communities [21].

- Approach 3: Learning from the users with an aid of recommender systems.

Later we will propose a 4th approach, based on an implementation of artificial selection in genetic algorithms. This approach takes into consideration the redefined paradox of choice by enabling the users to recognize valid solutions without the need of identifying the attributes levels contributing to them.

Limitation of Solution Space According to Averaged Users' Needs

Schwartz proposed 11 ways in which people can reduce the negative effects of the paradox of choice [1]. Two points extracted from his proposition are outlined below:

- Omit the choice when it is not really important.
- Choose based on reflection on what is important in this decision.

These propositions are underpinned by an assumption that the users encounter well-defined problems and thus they are able to define which attributes constitute a meaning to them. Piller et al. provided similar propositions arguing for limiting the size of the solution space [21]. They have proposed that of the necessary capabilities of MC practitioners is an ability to identify the product attributes along which users needs diverge the most. However this proposal bears a danger of bringing mass customization practice close to the realm of mass production, where an average is derived from overall space of users' needs to satisfy the majority. This approach is not valid in the light of Pine's definition of MC, which explicitly assumed the fulfillment of *all* the needs of *every* user [7].

Nourishing Online Communities

A different approach to solve problems related to mass confusion was proposed by Piller et al., who argued for creation and nourishing of online communities, where users share their experience and expertise in interaction with configurators [21]. They have suggested that the advice provided by experienced users might be highly useful for other users who are new to a particular configurator.

Learning from the Users with the Aid of Recommender Systems

An approach that has attracted a considerable amount of attention in relation to mass confusion is the use of recommender systems. Salvador et al. proposed "assortment matching", which they explained as a software adjusting configurators' solution space according to a model of user preferences [18]. "Elicitation" was a term proposed by Zipkin to describe "leading customers through the process of identifying exactly what they want" [23]. Both propositions assume an existence of a user's profile, which is built over time and updated with that user's online activities such as purchase and browsing patterns. Both propositions also assume an existence of an algorithm able to narrow down the size of the available solution space, based on the profile information. The algorithm is responsive to individual user's preferences directly, instead of responding to global patterns of preferences of all users as the first approach suggested.

One of the approaches mentioned by Schwartz as a solution to the paradox of choice was "customization of customers' experience". He further argued that this approach is especially relevant in case of e-commerce and mentioned an algorithm, which "knows the customer well enough" to be able to propose appropriate and manageable choice [2]. Thus Schwartz, similarly to Salvador et al. and Zipkin argued for the usage of recommender systems.

Recommender system algorithms are already widely adopted in e-commerce of non-customizable products, but their implementation in B2C mass customization of products over the Internet remains rare [24], [25].

Redefinition of the Paradox of Choice

Schwartz's definition of the paradox of choice discussed above does not fully recognize the complexity of the choice problems. Both in the case of customizable and non-customizable products, the problems which users face during a typical decision-making process are not well-defined problems argued for by the rational choice theory, but rather ill-defined problems similar to those referred to as 'wicked problems' by Rittel and Webber and 'design problems' by Maher and Poon [3], [5], [6]. We thus propose a following redefinition of the paradox of choice:

- Lack of meaningful choice rather than an overwhelming amount of choice can cause customers' feelings of decreased happiness and satisfaction as well as customers' paralysis.
- Lack of meaningful choice is generally accommodated by providing larger amount of options.
- Providing more choice is not guaranteed to solve the problem because meaningful choice is ill-defined, that is the users themselves are often not able to explicitly define what constitutes a meaningful choice.

Thus we propose that the solution to the paradox of choice is to consider both amount of choice and the *meaning* of the choice to the individual user. We additionally argue that the level of customers' satisfaction is likely to be directly influenced by the availability of meaningful choice.

Lack of Meaningful Choice

The 2-dimensional Wundt curve model proposed by Schwartz as a relationship between the amount of choice and customers' satisfaction is an unnecessarily simplified one, because it assigns the same weight to all of the product attributes. Importance of the attributes is likely to vary - consider attributes such as "color" and "safety" when buying a family car. On the other hand, for a user buying a coffee table shape and color might be of significant importance, while they are less likely to think about safety of the piece of furniture. The definition of the paradox of choice proposed by Schwartz takes into consideration the quantity of choice only, but fails to consider the quantity of possibilities in conjunction with their quality. Quality in this case refers to the attributes' meanings perceived by every individual user.

Inability to Express What a Meaningful Choice Is

Meaningfulness of the scope of choice is already inscribed in Pine's definition of mass customization. The assumption that the producer is able to provide the user with the precise thing that they need implies that the user himself knows what he needs. This however is an assumption of rational choice theory, which, as has been argued, does not constitute a proper model of most decision-making situations. Rational choice theory attempted to explain human choice mechanisms assuming that people posses knowledge about costs and benefits coming from every feature of every option [3]. Thus people were thought to be able to compare the options on a single dimensional scale to choose the best one. Problems where such comparison is possible belong to the category of well-defined ones. Design and engineering literature commonly proposes that looking for a best solution for these kind of problems should be referred to as optimization.

Simon argued that optimization is unlikely to be possible in real life situations in the 1950s [4]. He suggested that this is due to the complexity of human surroundings and human cognitive limitations. Later, Rittel and Webber proposed the term "wicked problems" as a reference to problems in which potential solutions evolve together with formulation of the problem itself [5]. Such problems were also referred to as "ill-defined". Maher and Poon provided a similar definition of a design process in which the term “design” refers to the process of simultaneous definition of the solution as well as the problem [6]. An argument along similar lines in mass customization literature was provided by Franke and Piller, who noticed that customers approach configurators without a clear idea of what they need [15]. It thus seems that both in the case of choice among customizable products and non-customizable options users are faced with “wicked problems” or “design problems”. Thus the complexity of problems faced by designers, configurator users and people choosing non-customizable products varies, but all of these problems belong to the same category of ill-defined tasks. If the users are not able to explicitly state what constitutes a meaningful choice to them, the MC practitioners task to provide them with such choice becomes an ill-defined problem as well.

Schwartz’s argument on that matter is also worth mentioning. During the discussion of features of “maximizers” - the type of consumers who are not satisfied with a “good enough” option but instead are on the quest of looking for the best option of all. Schwartz proposed that they aim at “optimizing their choice”. However, he further suggested that goals of this “optimization process” are often unclear [26]. Thus Schwartz used the term “optimization” not according to the use common in engineering and design literature but as a reference to solving ill-defined problems. Nevertheless a discussion of these problems did not became an inherent part of his definition of the paradox of choice and thus the redefinition proposed above.

Multidimensional Wundt Curve Model of the Paradox of Choice

Schwartz argued that his definition of the paradox of choice and his proposal of a threshold in the function of amount of options and user satisfaction can be represented by a diagram close to a 2-dimensional Wundt curve model [1], [2], [8]. However such a model is an unnecessarily simplified one in the context of the redefinition of the paradox of choice, because it fails to recognize that the importance of each attribute can vary across users. A model to better correspond with our proposition might be a multidimensional Wundt curve, where each product attribute is mapped onto a separate dimension. In such a model, the threshold marking the point before which a user’s satisfaction increases with an increasing amount of options and beyond which it begins to decrease is specific for each attribute. It is positioned further in case of more important attributes and closer in case of less important ones. Thus the shape of the curve in every dimension will vary from user to user. A multidimensional Wundt curve model would acknowledge Schwartz's contribution highlighting that a large amount of choice influences users' feelings of decreased satisfaction, but at the same time it extends this proposition, in that that it is not only the quantity, but also the meaning of choice, that affects users' feeling of satisfaction.

Solution to the Paradox of Choice

The level of customers' satisfaction will not increase if they are confronted with a bearable amount of attributes that are irrelevant to them. Thus the task of MC practitioners becomes not to provide an ideal *number* of customizable attributes, but to provide each user with exactly the attributes they wish to customize.

The proposed solution lies in the prospect that while a customer may be unable to define this set of attributes explicitly, they are adept at recognizing a complete bundle of attributes that meets their needs when they see them embodied in a product. They are adept judges who know what they like. This task can be aided with computational means, for example the implementation of recommender systems and artificial selection in genetic algorithms. The latter proposition allows users to recognize a bundle of meaningful attributes without the need of defining them explicitly. Both propositions are further discussed in section 5 and an empirical experimentation on the latter is discussed in section 6.

Computational Approach to the Paradox of Choice

Both Schwartz and mass customization researchers acknowledge the complexity of the problems decision-makers face by recognition that recommender systems might be an aid for the paradox of choice [2], [18], [23]. We will discuss a possible implementation of recommender systems as well as our proposition of an implementation of artificial selection in genetic algorithms.

Learning from the Users with an Aid of Recommender Systems

Artificial intelligence algorithms that aid decision-making process are commonly called recommender systems. Collaborative filtering and content based filtering are two most popular and best-documented examples of these algorithms [24], [27], [28]. Content based filtering works by looking for attributes similar to the attributes of products a user has already bought or viewed. Collaborative filtering compares browsing patterns of different users and suggests choices made by one user to another if they appear to have similar interests.

Recommender systems are already very common in e-commerce. Examples from the domain of non-customizable products are Amazon and Zafu [29]. From the information domain Daily Perfect and Last.fm can serve as examples [30], [31]. A recent survey by the authors showed that the usage of recommender systems in product configurators is currently a very rare approach. Among 55 commercial configurators reviewed, only 1 featured a recommender system. Moreover it was the simplest kind of recommender system, called "dynamic queries", which does not require building the user's profile [25]. Felfernig et al. recognized that recommender systems developed especially to aid configuration of products could help to tackle mass confusion [24], but MC practitioners have not exploited the potential of these algorithms yet.

Given the redefined paradox of choice which we have proposed, future applications of recommender systems in MC configurators of products will aim at finding patterns in attributes' importance to particular users.

Recognition of Meaningful Options with the Aid of Evolutionary Algorithms

Another possible aid for the paradox of choice is an evolutionary approach, based on the application of artificial selection in genetic algorithms. Users interacting with genetic algorithm configurators experience a choice-over approach that is they are not forced to explicitly define values of all the parameters available to customize. Instead they express preferences among a couple of different options. This proposition can be considered as a fourth approach to solve problems related to the paradox of choice, in addition to the three approaches outlined earlier.

Genetic algorithms (GAs) were first conceptualized by Holland as search algorithms, where every potential solution is represented by a genotype and a phenotype [32]. A classical genotype is a string of binary numbers, which constitute encoded values of the parameters. These can be used to generate the phenotype using a given parametric definition. A typical GA operates on a population of potential solutions. It creates a series of genotypes, maps them into phenotypes and then uses a series of formulas such as selection, crossover and mutation to create new genotypes, thus next generations. Natural selection in GAs has been widely recognized as an optimization algorithm that is as a good engine for search for solutions of well-defined problems. In this type of selection, the parents of next generation are selected automatically, based on how well do they fulfill explicitly stated, quantifiable fitness criteria. Given that problems faced by configurators' users belong to the category of ill-defined problems, natural selection in a GA does not bare remarks of a promising application in the context of MC. However Todd and Latham have suggested that natural selection is not the only type of selection available for implementation within GAs [33]. They proposed the use of artificial selection, where parents are selected manually in an application used for iterating through different versions of sculptures. They also suggested different applications of such breeding tools producing a range of output from jewelry to business plans. Cho has also referred this type of selection as an "interactive genetic algorithm" while discussing applications in fashion design and emotion based image retrieval [34]. According to the authors' recent review, artificial selection has not yet been implemented in B2C configurators operating commercially [25].

In genetic algorithm MC configurators, users can navigate through the solution space based on non-explicit features of the products, which are not expressed by a single attribute but rather by a bundle of attributes. Our assumption guiding the experiment in the following section is that the shape and size of such bundle of attributes constituting a meaning to the user is likely to vary among users and that a GA configurator is a tool robust enough to accommodate these differences. An additional advantage of this approach is that the designer of the GA configurator does not need to posses the knowledge about the shape and size of these bundles thus they do not need to know what the users of the tool find relevant.

Experiment Setup

We have conducted an empirical comparison of six B2C product configurators, belonging to two distinctive categories. The first category featured three configurators with artificial selection in GAs. These tools will be further referred to as "genetic algorithm" or "GA" configurators. The second contained three configurators featuring a menu-based navigation through a solution space, which will be further referred to as "parametric configurators".

The purpose of the experiment is to test the new definition of the paradox of choice, that is whether users will be likely to accept a larger scope of choice in case if it is more meaningful to them. A second purpose is to test the genetic algorithm-based solution to the paradox of choice. The hypothesis is that configurators with artificial selection by GA are expected to handle larger and more diverse bundles of attributes. Thus these configurators are expected to perform better in case of solution spaces with larger amount of dimensions.

Types of Configurators Subject to Experiment

In the GA configurators, attribute levels are not set explicitly, but rather through manual selection of meaningful solutions and breeding of subsequent generations of products. Each population in GA configurators consisted of 6 individuals and the users where able to select each as a parent of the next generation multiple times.

Parametric configurators, on the contrary, required the users to explicitly define each attribute levels and thus they called for a definition rather than a recognition of meaningful solutions. The users where able to set the attribute levels through a slider-bar menu and were presented with a single visualization of the current instance of the product. Configurators belonging to this category represent a type of a menu-based configurator, similar to the vast majority of configurators currently available online [25].

A subject to customization was a parametric definition of a NURBS surface. It was meant to represent a piece of furniture, but it's function was not specified.

The configurators where divided into three pairs. Each contained a tool from both categories with the same number of solution space dimensions. The number of dimensions for each pair were 4, 18 and 38. Names of the configurators used in the further discussion are based on whether the configurator belongs to the GA or the parametric category and on their dimensionality. They are outlined below:

- 4P – 4 attributes, parametric configurator (see Figure 2A).
- 4GA – 4 attributes, generic algorithm configurator (see Figure 2B).
- 18P – 18 attributes, parametric configurator (see Figure 2C).
- 18GA – 18 attributes, generic algorithm configurator (see Figure 2D).
- 38P – 38 attributes, parametric configurator (see Figure 2E).
- 38GA – 38 attributes, generic algorithm configurator (see Figure 2F).

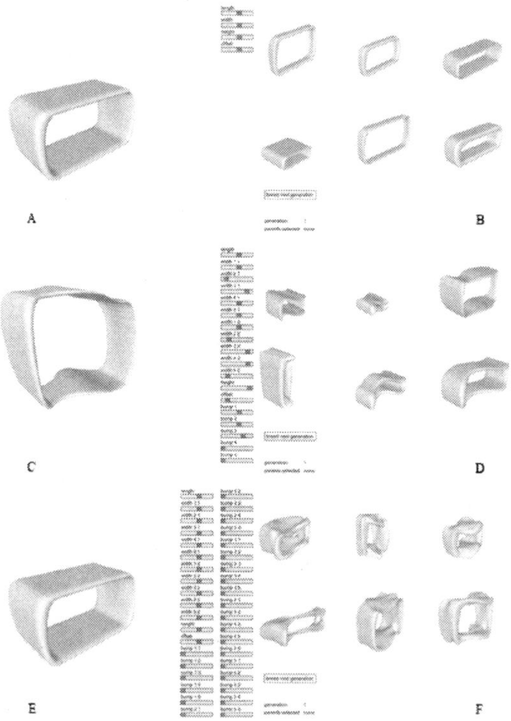

Fig. 1. Six configurators subject to the experiment

Experimental Method

The experiment was conducted over the Internet. Each configurator was an applet written in Processing, an open-source programming language initiated by Reas and Fry with designers in mind [35]. A user choosing to begin the experiment was randomly directed to one of the six configurators and was not presented with any description of the purpose or method of the experiment. The user was then able to browse through different configurators with an applet, using links available on each site. The links directed to a configurator of the same dimensionality, but from a different category and to the configurators from the same category with smaller and larger number of dimensions, if those where available. For example: 18P was linked to 18GA, 4P and 38P, but 38P was linked only to 38GA and 18P, since there was no configurator with larger dimensionality. These links were meant to ease the definition of the user's preferences of the category of the configurators as the function of the solution space dimensionality. The users where not asked to provide any information such as via a questionnaire, but patterns of their interactions with each tool where saved as .txt files onto the hosting server. The experiment was conducted over 29 days between 20th of January and 17th of February 2010.

Findings

The findings are based on the data from 313 user interactions with the configurators, which occurred in the course of 29 days of the experiment. The data was collected on the number of sessions with every configurator and on an average number of options reviewed.

It was found that as the number of solution space dimensions increases, the level of voluntary user interaction with the GA configurators increases approximately linearly, while a threshold exists for the parametric configurators beyond which the number of interactions steadily drops. Thus it appears that the Wundt curve relation of users satisfaction and the scope of choice applies to parametric configurators, but not to GA configurators, Figure 3.

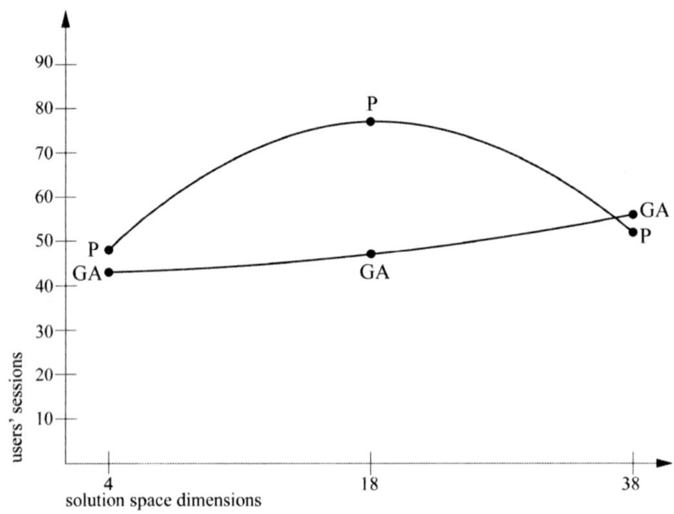

Fig. 3. Amount of Interactions with each Configurator.

The actual valyes for number of solutions reviewed is not directly comparable between tool types, since in the case of the GA configurators the number of *generations* bred is counted while in the case of the parametric configurators it is the number of *changes* applied to attribute levels. However what can be compared is the shape of the curves depicting the relation of these numbers to the solution space dimensionality. It appears that in case of this data as well, a threshold exists for parametric configurators while the number of generations bred with GA tools continues to steadily increase, Figure 4.

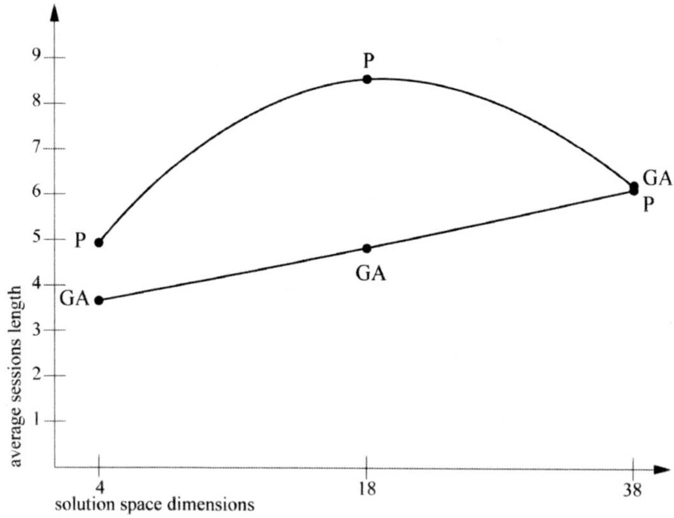

Fig. 4. Average amount of possible solutions and generations reviewed using each out of six configurators

Conclusion

We have rephrased Schwartz's proposition of paradox of choice, putting an emphasis on the meaning of the available options rather than just on their amount. We have also suggested that the problems that people encounter during the decision-making process belong to the category of ill-defined ones. Users choosing non-customizable products and interacting with product configurators face problems of a different complexity than do designers, but all of these problems belong to the same category. The formulation of this kind of problem changes while a person trying to solve them moves through the solution space. We have further argued that the Wundt curve model of relation between satisfaction and the scope of choice may be an appropriate one only if it is considered in multiple dimensions, where each dimension represents the significance of a single attribute. We have suggested that the threshold beyond which more choice begins to contribute to dissatisfaction is positioned further on the curve in the case of meaningful attributes and sooner in case of less relevant ones. Additionally, different attributes are likely to be meaningful for different users.

We have further discussed several computational means of providing meaningful choice. Implementation of recommender systems is one such option and a use of artificial selection in genetic algorithms is another. We have suggested that because artificial selection in GAs enables the users to navigate through a solution space by recognizing rather than defining meaningful options, it is likely to be a possible solution to the redefined paradox of choice.

The findings of an empirical comparison of six B2C product configurators appear to confirm this hypothesis. Beyond a certain number of solution space dimensions the popularity of parametric configurators appears to decrease, while the popularity of genetic algorithms configurators appears to grow continuously. The same is true for the average number of sessions reviewed with each tool. Thus we conclude that a non-explicit recognition of attribute bundles with an aid of artificial selection in GAs appears to be an attractive means of browsing through solution space with a large number of dimensions. We further conclude that this is because what contributes to users satisfaction is not only an ideal scope of choice but also the availability of a meaningful choice.

Acknowledgements

We would like to thank Alasdair Turner for his advice and help in setting up the online experiment described above.

References

1. Schwartz, B.: Paradox of Choice. HarperCollins Publishers, New York (2004a)
2. Schwartz, B.: Paradox of Choice – Discovery Expert Series interview by Furier, J. (2004 b), http://74.125.155.132/scholar?q=cache:WvWuU3APF30J:scholar.google.com/&hl=en&as_sdt=2000 (Accessed November 2009)
3. Von Neumann, J., Morgenstern, O.: Theory of Games and Economic Behaviour. Priceton University Press (1944)
4. Simon: A Behavioural Model of Rational Choice. Quaterly Journal of Economics 59, 99–118 (1955)
5. Rittel, H.W.J., Webber, M.M.: Dilemmas in General Theory of Planning. Policy Sciences 4, 155–169 (1973)

6. Maher, M.L., Poon, J.: Modelling Design Exploration as Co-Evolution. Computer Aided Civil and Infrastructure Engineering 11(3), 195–209 (1996)
7. Pine, J.B.: Mass Customization - The New Frontier in Business Competition. Harvard Business School Press, Cambridge (1993)
8. Schwartz, B.: Self-Determination. The Tyranny of Freedom (2000)
9. Berlyne, D.E.: Aesthetics and Psychobiology. Appleton-Century-Crofts, New York (1971)
10. Keller, K.L., Staelin, R.: Effects of Quality and Quantity of Information on Decision Effectivness. The Journal of Consumer Research 14(2), 200–213 (1987)
11. Sauders, R., Gero, J.S.: Artificial Creativity: A Synthetic Approach to the Study of Creative Behaviour. In: Gero, J.S., Maher, M.L. (eds.) Computational and Cognitive Models of Creative Design V. Key Centre of Design Computing and Cognition, pp. 113–139. University of Sydney, Sydney (2001)
12. Holbrook, M.B., Gardner, M.P.: An Approach to Investigating The Emotional Determinants of Consumption Durations. Journal of Consumer Psychology 2(2), 123–142 (1993)
13. Toffler, A.: Future Shock. Bantam Books, New York (1970)
14. Davis, S.: Future Perfect. Addison-Wesley, Reading (1987)
15. Franke, N., Piller, F.: Value Creation by Toolkits for User Innovation and Design: The Case of the Watch Market. The Journal of Production Innovation Management 21(6), 401–415 (2004)
16. Piller, F., Schubert, P., Koch, M., Moslein, K.: Managing High Variety: How to Overcome the Mass Confusion Phenomenon of Customer. In: Deppery, D. (ed.) Proceedings of the EURAM 2003 Conference, Milan (2003)
17. Schreier, M.: The value increment of mass-customized products: An empirical assessment and conceptual analysis of its explanation. Journal of Consumer Behavior 5(4), 317–327 (2006)
18. Salvador, F., de Holan, P.M., Piller, F.: Cracking the Code of Mass Customization. MIT Sloan Management Review 50(3), 71–78 (2009)
19. Franke, N., von Hippel, E.: Satisfying Heterogeneous User Needs via Innovation Toolkits: The Case of Apache Security Software. Research Policy 32(7), 1149–1291 (2003)
20. Swait, J., Adamowicz, W.: The Influence of Task Complexity on Consumer Choice: A Latent Class Model of Decision Strategy Switching. Journal of Consumer Research 28(1), 135–148 (2001)
21. Piller, F., Schubert, P., Koch, M., Moslein, K.: Overcoming Mass Confusion: Collaborative Customer Co-Design in Online Communities. Journal of Computer-Mediated Communication 10(4) article 8 (2005)
22. Teresko, J.: Mass Customization or Mass Confusion. Industry Week 243(12), 45–48 (1994)
23. Zipkin, P.: The Limits of Mass Customization. MIT Sloan Management Review 42(3), 81–88 (2001)

24. Felfernig, A., Friedrich, G., SchmidtThieme, L.: Guest Editors' Introduction: Recommender Systems. IEEE Intelligent Systems 22(3), 18–21 (2007)
25. Piasecki, M., Hanna, S.: Review of B2C Online Product Configurators. In: MCPC. Fifth World Conference on Mass Customization and Personalization Proceedings, Helsinki, Finland (2009)
26. Schwartz, B.: Maximizing Versus Satisfacing: Happiness Is a Matter of Choice. J. of Personality and Social Psychology 83(5), 1178–1197 (2002)
27. Maes, P.: Agents that Reduce Work and Information Overload. Communications of the ACM 37(7), 30–40 (1994)
28. Schmidt-Thieme, L.: Compound Classification Models for Recommender Systems. In: Proceedings of the Fifth IEEE International Conference on Data Mining, pp. 378–385 (2005)
29. Amazon (2010), `http://www.amazon.com` (Last accessed: January 2010)
30. Daily Perfect (2010), `http://www.dailyperfect.com` (Last accessed January 2010)
31. Last.fm (2010), `http://www.last.fm/home` (Last accessed: January 2010)
32. Holland, J.H.: Concerning Efficient Adaptative Systems. In: Yovits, M.C., Jacobi, G.T., Goldstein, G.D. (eds.) Self-Organising Systems, pp. 215–230. Spartan Books (1962)
33. Todd, S., Latham, W.: The Mutation and Growth of Art by Computers. In: Bentley, P.J. (ed.) Evolutionary Design by Computers, pp. 221–251. Morgan Kaufmann Publishers, San Francisco (1999)
34. Cho, S.B.: Towards Creative Evolutionary Systems with Interactive Genetic Algorithm. Applied Intelligence 16, 129–138 (2002)
35. Reas, C., Fry, B.: Processing – A Programming Handbook for Visual Designers and Artists. MIT Press, Cambridge (2007)

Rethinking Automated Layout Design: Developing a Creative Evolutionary Design Method for the Layout Problems in Architecture and Urban Design

Sven Schneider[1], Jan-Ruben Fischer[2], and Reinhard König[2]
[1]*Technical University Munich, Germany*
[2]*Bauhaus-University Weimar, Germany*

The research project presented in this paper deals with the development of a creative evolutionary design methodology for layout problems in architecture and urban planning. To date many optimisation techniques for layout problems have already been developed. The first attempts to automate layout were undertaken back in the early 1960s. Since then, these ideas have been taken forward in various different manifestations, for example shape grammars, CBS, cellular automata and evolutionary approaches. These projects, however, are mostly restricted to very specific fields or neglect the creative, designerly component. Since pure optimisation methods are of little practical use for design purposes, there have been no useful attempts to derive a universally applicable method for computer aided layout design. For this we need to be aware that designing is a process that occurs at different levels and degrees of abstraction. The solution space is explored in the realm between intuition and rationality in a variety of ways. Good solutions can only arise through an intensive and fluid dialogue between the designer and the generating system. The goal of our project is to develop an adaptive design system for layout problems. To this end we examine different approaches to achieving the best possible general applicability of such a system and discuss criteria that are crucial for the development of such systems.

Introduction

Layout tasks in architecture and urban design are of central importance. The aim is, at different scales, to arrange lots, buildings, rooms or building elements in a creative and functionally sensible way, Figure 1. Their arrangement significantly determines the quality and sustainability of

J.S. Gero (ed.): *Design Computing and Cognition'10*, pp. 367–386.

buildings, neighbourhoods or even entire cities. Depending on the project and context, a plethora of different requirements must be met. To synthesise these specifications at the beginning of the traditional design process, design hypotheses are established and then constantly reviewed and refined during the design process. A prerequisite for this approach is a significant reduction in the complexity of a design task with the help of heuristics or models. Such an approach can be considered as a top-down design strategy. The goal of the efforts in computer-aided layout design is to support the designer in undertaking complex tasks with bottom-up strategies to generate adequate solution criteria.

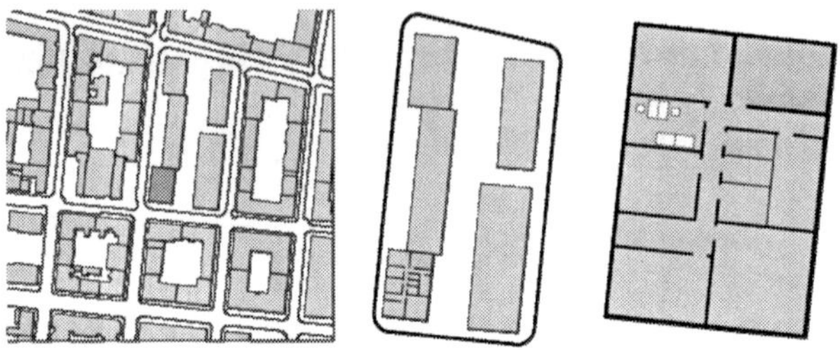

Fig. 1. Different levels of scale of layout objects

Computer-based solutions for layout problems have already been tested that employ various optimisation techniques [1], [2]. However, in these studies the crucial creative component is very often missing. Most projects aim to make design processes more rational, faster and more effective. It is questionable, however, whether increased efficiency in the design process actually leads to better designs. Because pure optimisation methods are of little practical use for design tasks, there have been no useful attempts to derive a universally applicable method for computer-aided layout design. This paper aims to explore this issue and to find methods and techniques for such a universal system. In the following we will first look in detail at the problems that arise in conjunction with an automated system for the resolution of design problems. We then discuss various projects and methods in the domain of layout automation. In the last chapter we define the objectives of an adaptive system for layout problems.

Between Intuition and Rationality

As a prerequisite for successfully dealing with computer-based design support, we must distinguish between operational and non-operational problems. While most operational problems can be solved algorithmically with the help of a computer, non-operational problems require human interpretation. A problem is operational if it can be described so accurately that one can specify the steps necessary to solve it. This is achieved by analysing and breaking down a complex problem into sub-problems, which can be solved independently and their solutions then merged together to find the final solution. The goal of the analysis of a design is a description that is so accurate that it contains the solution [3], definable and tangible criteria for describing problems are referred to as operational criteria.

By contrast, non-operational problems are "vaguely defined [...] and significant elements of the task are unknown or cannot accurately (quantitatively) be defined. Their solution criteria are not clearly formulated, and the decision-making process is less concerned with finding solutions, but rather with the specification and demarcation of the problem as well as closing open constraints" [4]. For these problems, there is no clear right or wrong solution. Architectural design problems are usually non-operational problems and accordingly differ from most problems in other engineering disciplines [5]. The answer to such questions invariably depends on intuitive, subjective and contextual aspects. Creative decisions are always a response to poorly defined situations and solutions are always a product of both operational and non-operational issues [4]. To fully formalise such non-operational design problems, thereby in essence turning them into operational problems, would be tantamount to "measuring the link between the environment and the neuronal activity of the organism" [6]. The comprehensive collection of all the necessary empirical data is, despite all the advances in cognitive science, all but impossible and the option of performance simulation similarly offers only limited potential.

The problems a design has to react or respond to are for the most part not predefined, but must be discovered and identified during the design process. Instead of being presented with problem situations one is required to discover problem situations [7]. Before we can start searching for solutions, the problems need to be discovered and structured, a process that requires imagination and divergent thinking. In this context, Michael Polanyi [8] distinguishes between explicit and hidden or tacit knowledge. Tacit knowledge is non-verbalized or non-formalized pre-existing knowledge of as yet undiscovered and non-explicit connections. It is the essential basis for heuristic procedures. Computer-based search techniques for problem solving can only be used once non-explicit and pre-existent patterns have been recognised or generated imaginatively and intuitively, i.e. once the problem has been structured. The real problem when designing therefore lies in formulating the wicked problem [3]. With the first step of explaining the problem we have already determined the nature of the solution [9]. Lawson [10] also concludes that problem and solution are interdependent because analysis, synthesis and evaluation often occur simultaneously as complex mental processes. Therefore, software that should act like a designer can only be applied in a "highly restricted situation, a narrowly defined chunk of a design process, where the design world employed by designers can feasibly be assumed as given and fixed" [11]. Designing is a process that takes place at different levels and degrees of abstraction. Between intuition and calculation, a range of approaches is explored in a solution space. Schön, therefore, concludes that "practitioners of Artificial Intelligence in the design world would do better to aim at producing design assistants rather than knowledge systems phenomenologically equivalent to those of designers" [11].

In which form computer-based methods can be integrated as a tool for solving non-operational problems in the circular process of creative design needs to be clarified. But first, we will examine some previous approaches to dealing with computer-based layout problems.

Revisiting Automated Layout Design

Initial attempts to automate layout design have existed since the early sixties. These have been regularly picked up and taken forward by new methods (e.g. L-systems, shape grammars, constraint-based systems, Figure 2, cellular automata and agent-based systems). Currently it appears that the combination of generative processes with evolutionary methods is the most promising approach. Below the most important methods for the automation of layout and design are presented and evaluated.

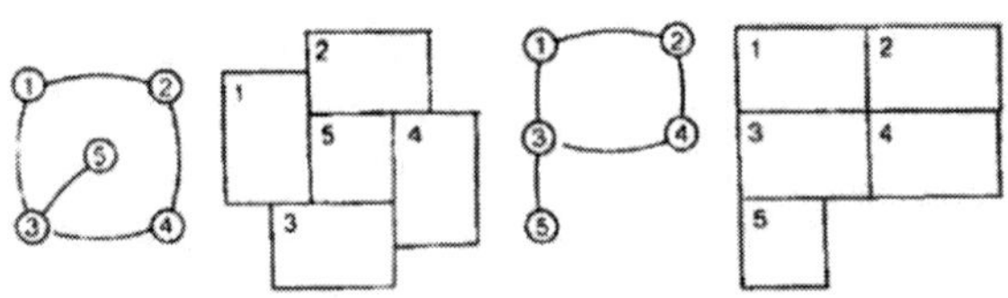

Fig. 2. Graph and topological plan [12]

Constraint-Based Systems

In a Constraint-Based System (CBS) the conditions (constraints) for a design are explicitly defined at the outset. Using this basis we can solve the design or layout problem by means of mathematical modeling and nonlinear programming. The main conditions include information about the dimensions of the elements and their functional relationships [13]. These constraints are typically reported as numeric or Boolean parameters or in the form of a relationship matrix. To input the element relationships a matrix is usually defined. A far more elegant method is to draw a graph, Figure 3 [14].

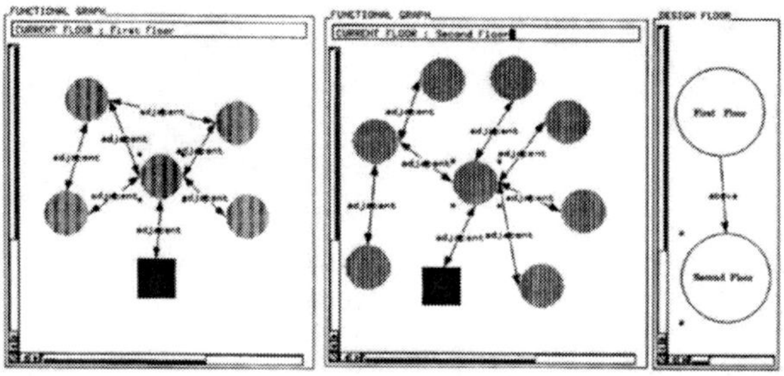

Fig. 3. Archiplan, functional diagram [14]

In the examples of generated floor plans by Li, Frazer, and Tang [13] or Medjdoub and Yannou [14] they have shown the possibilities and limitations of the CBS optimisation strategy. The rooms can all be arranged within a given perimeter and fulfil the requirements with regard to relative sizes, orientations and spatial relationships. A disadvantage of the CBS method is that it is a linear process starting with a definition of the problem by the user and proceeding to the final computed solution – the influence of the user on the results is limited to the initial definition of constraints. Additionally the computation time for more complex systems increases exponentially.

Cellular Automata and Agent-Based Systems

Applications of cellular automata (CA), and agent-based systems are usually found on the urban and regional planning level. Cellular automata consist in their simplest form of a grid cell whose cells change their states depending on the states of their neighbouring cells, Figure 4 [15].

Based on the rules for the change of state of a cell, spatial arrangements of certain elements can be generated [16], [17], [18]. Agent-based systems consist of autonomous units (agents) that can exchange and share information with each other and with their environment. In this way it is possible to generate road structures in a landscape [19], exchange rates between settlements [20] or place buildings depending on the urban context [21]. Often the agent-based system is combined with a CA for the representation of a landscape or city. A restriction of CA is that they can only represent certain geometric structures - mostly regular cell grids, Figure 5.

Fig. 4. Cellular automata [15]

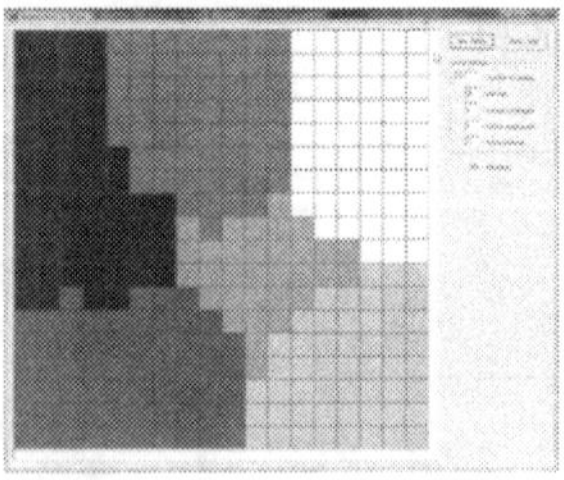

Fig. 5. Agent-based system in a cellular space

When transferred to an irregular cell system [22], geometric constraints still always remain since the neighborhood relationships have to be clearly defined. Alternatively, a stochastic cellular automaton has to be used, but their results are difficult to control. Although agent-based models allow for greater geometric freedom [23], [24] they can only be controlled indirectly through the interaction rules of the agents. Furthermore, they must be combined with other reproductive methods to produce complex geometries.

Shape Grammars

Shape Grammars (grammars of form) consist of a set of basic shapes and symbols as well as syntactic rules. The rules are used to transform one form or collection of forms into a new form. Recursively applied to an initial form, the rules result in structures, which belong to a (Shape) Language [25], [1]. The theory and applications of Shape Grammar has been collected comprehensively in the books by Mitchell [26] and Stiny [27]. Despite their promising possibilities, so far no systems for supporting design using Shape Grammar have been developed for practical use. The problem of this method lies in the fact that architectural design cannot be reduced to producing graphical representations or imitating styles. Shape Grammars have been used mainly for analytical purposes in the context of the generative description of styles [28]. The intention was primarily to split existing entities into their constituent parts, to reproduce these based on their syntactic rules or to try new combinations. Design drawings, however, also have a semantic meaning, for example, derived from the function that an element of a building or settlement structure should have. But functions in architecture are strongly dependent on the context, which makes it almost impossible to describe such design problems with this method. The work of Duarte [29] extends the use of shape grammars to a model that is an extension of the grammar formalism by including a shape grammar, a description grammar, and a set of heuristics that allows the generation of a design solution that matches requirements given a priori. This model is called a discursive grammar and leads to good results but is still restricted by a set of given pre-implemented rules, Figure 6.

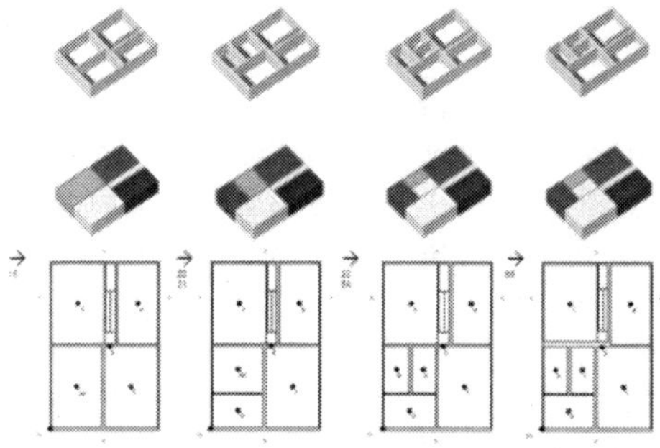

Fig. 6. Jose Duarte, Customizing Mass Housing [29]

Physically-Based Systems

Physically-based systems try to automate layout organisation by implementing physical properties such as repulsion and attraction, Figure 7 [30]. The designer assigns force fields to plan elements which affect their topological relationship on the one hand and change their geometric shape on the other hand. This responsive design process is aimed at facilitating a very natural, interactive, intuitive and flexible way of designing. The design intention can be translated into a mass-spring system, which is imbalanced in its initial state and aims to achieve a state of equilibrium during the simulation process, Figure 8.

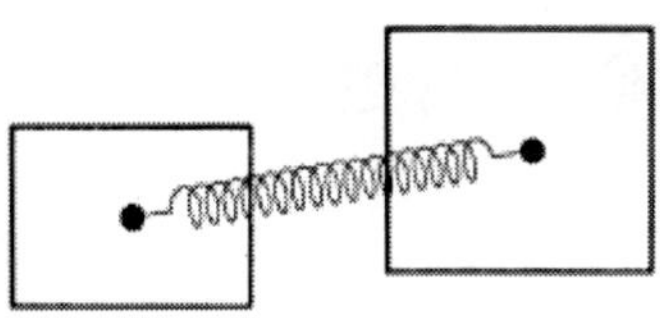

Fig. 7. Adjacency objective [30]

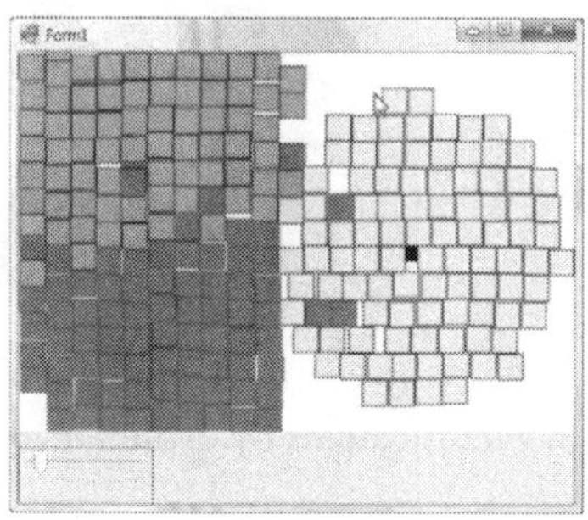

Fig. 8. Physically based space entities

A disadvantage of physically-based systems is that the quantity of topological design goals is limited, and that design goals which cannot be translated into physical processes cannot be realised. Another problem is that the elements block one another as a result of their physical features as they attempt to reach a balance of forces. Hence it is usually impossible to identify how well a solution has fulfilled the design goals.

Evolutionary Algorithms

Evolutionary algorithms (EA) are so-called heuristic methods. They do not guarantee the solution of a problem but reduce the time required to solve the problem considerably. EA, which can be understood as an emulation of biological evolution, are the most creative of all the known processes in the current state of computer-based research. They are the only method available with which to implement a means of tackling a poorly-defined problem. This allows us to find new solutions which are not even contained in the specifications. In the field of architecture, the first experiments with EA were published by Frazer [31], [32]. However the results of these studies are still very abstract in nature and are of purely academic interest. The works that were created in the context of Paul Coates in the 90s at the CECA [33], [34] result in abstract spatial structures which can serve as inspiration for the further development of design solutions, Figure 9.

The first convincing examples that use an evolutionary approach in the field of computer-based layout development include those by Jun and Gero [2] as well as [35]. However, these early examples are still based on a rectangular grid, so the possible geometries are very limited. Among the many computer-based layout development systems that have arisen in recent years, the work of Elezkurtaj [6] which uses an additive process, is in many respects one of the most well elaborated, Figure 10.

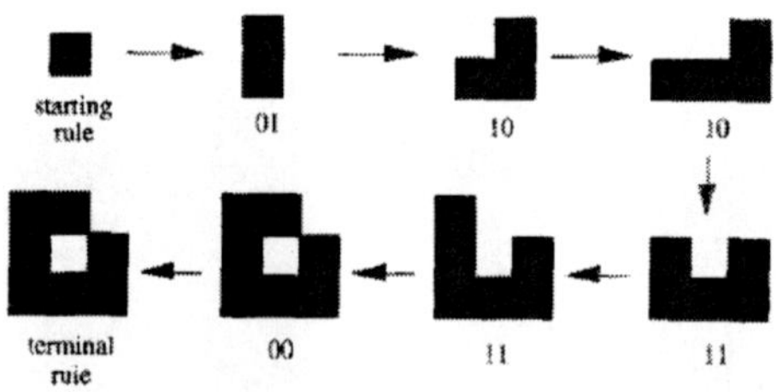

Fig. 9. Modifications by evolutionary algorithms [2]

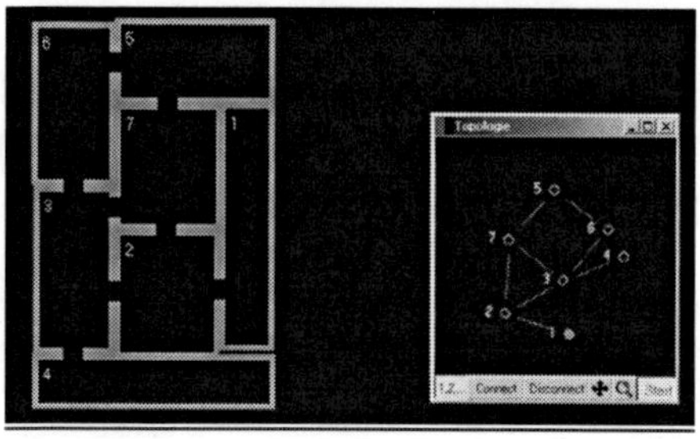

Fig. 10. Interactive Layout System based to evolutionary algorithms [6]

The system calculates design proposals in real time, providing the ability to meaningfully interact with the generative software and allowing a continuous interplay between the designer and computer.

Conclusion

The above overview of existing projects shows different methods and their potential as well as limitations for use as a design system. However, the biggest problem with all software that has been developed to serve this task is that they obstruct creativity in the design process more than they support it. There are various reasons for this. Firstly, all the aforementioned methods deal with operational problems. However creative solutions most often arise in the context of poorly-defined situations. A solution can only be called creative when it does not derive directly from a detailed description of the problem. This illustrates why the methods discussed above are not suitable on their own for supporting the creative design process. In addition, there is a lack of universality in the above projects – each deals with certain sub-problems. However these sub-problems often arise in the course of the design process and even then they are hard to identify as such. "Of course the human design process in architecture is not a process of suboptimisation. So the computer as 'oracle' has not so far proved to be helpful and is not likely to do so." [10]. Finally in this context it should also be mentioned that most of the common computer-based design systems follow a linear principle: after the user executes a specific function, an automatic search for solutions is undertaken without the possibility of further interaction until satisfactory results have been found or the process is aborted.

A good approach in supporting the heuristic way a designer works can be seen, for example, in the projects by Elezkurtaj and Franck [6], Arvin and House [30] and in the work of the Aedas R&D Group. Here one can see a first example of circular interaction between the user and the generative system. These approaches need to be further enhanced and be made more flexible. The challenge is not only to develop new and better optimised algorithms, but especially to incorporate computational methods in the designerly way of doing. On the one hand it is important to generate creative solutions using the computer and on the other to support the creativity of the user in this process. A balance needs to be found between the creative power of the computer and the individual creativity of the designer.

In order to work out how to optimise the system in this direction we will leave the technical level for the moment and take a closer look at the interaction between designer and tool. For the development of tools one is directly involved in the creation of action and possibility spaces in which a designer works. That means that when you develop a tool to support design processes, the tool provides a possibility space in which a designer is able to explore and develop – ideally without restrictions – his own individual design world. The aim of design supporting systems is to create possibility spaces which allow one to explore the problems one is working on as flexibly as possible. The quality of such software is not so much dependent on what the programmer makes possible, but rather on its potential to engender a dialogue with the user (conversation) and its openness (experience of surprise) [11] has also detailed a few criteria for how such action spaces should be designed in relation to the development of artificial intelligence systems in architecture. These include the simplest unit of design experimentation (designer's seeing-moving-seeing), appreciating design qualities and setting design intentions and problems, storing and developing prototypes, and communicating across divergent worlds. A computer-based design system should also provide opportunities for surprise and puzzlement to enable new insight and accept vague and incomplete, sketchy information to provide opportunities for personal ways of seeing. How these requirements can be implemented conceptually is discussed below.

Developing an "Adaptive" Layout-Design System

"Again, it is worth stressing that such systems are not intended to replace people, but increase productivity and creativity by allowing people to explore more and a wider variety of solutions than they could without such computer systems" [36]. Here we should additionally note that designing is not just about searching through the solution space, but also about continuously revising the solution space. One can think of this as the ongoing adaptation of the solution space and the direction of one's investigation while designing. This continuous process of adaptation does not follow a predefined procedure. The goal of our project is to develop a layout-design system that is applicable to various design problems. Top-down and bottom-up strategies should be combined seamlessly. This means that one should be able to start with both a minimal description of the problem as well as a highly-restricted situation. For this we need to treat the layout problem at the most general level and use an explorative approach. Instead of searching for optimal solutions to a parameterised problem, the system must allow the user to define a flexible search space, change it and explore different paths.

For the development of this system we have considered different sub-aspects. These are shown in Figure 11 and include the graphical representation, how one interacts with it, the generative system and the internal representation of the layouts. These sub-aspects are closely linked with each other. At the centre of this is the generative mechanism, which is continually generating solutions and optimising them according to evaluative criteria. The user always has the possibility to intervene based on the graphical representation of the data generated using this mechanism. In the following three sections we describe the criteria and requirements for the proposed adaptive system.

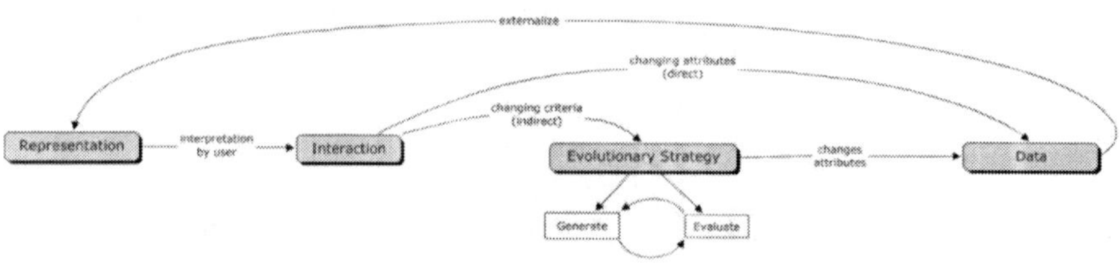

Fig. 11. Project structure

Definition of Layout

Scale-Independent Layout

Designing is a process that takes place simultaneously at several scale levels. A system supporting this process should therefore not be confined to one level only but must operate adaptively at different levels. In the context of the topic discussed here we first attempt to define layout in a way that is as universally applicable as possible for various problems. Generally speaking, in architecture and urban design layout describes the evident arrangement of various elements such as lots, buildings, rooms, areas, components, furniture and so on at different scales. As we change scale, the definition of the elements changes and with it their respective demands as well as those they place on the system. Because we effortlessly switch back and forth between the different meanings of elements while designing [10] a too narrow definition of the space of design possibilities would end up being too restricting in a certain direction. This limits the designer in his scope to tackling only predefined problems as given by the definition of what should be arranged. We therefore need to find a definition for the central issue of layout that is as universal as possible. To build a data structure, that can be used in different contexts, we define layout here (solely) as the arrangement of elements in elements in elements, etc. The elements themselves need to remain largely undefined. They are geometric elements whose semantic meaning is left open to begin with. The user himself should ascribe meaning to the elements used. The description of the layout is not determined, and always remains variable during the design process. This lack of definition allows the user to construct his own design world without being bound by predefined descriptions of the problem in terms of scale or meaning, Figure 12.

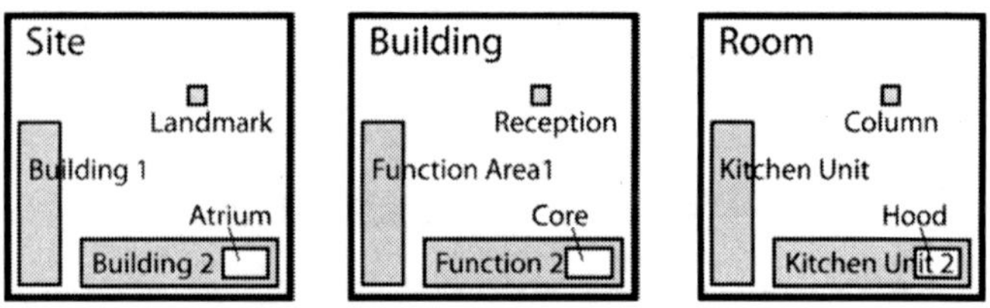

Fig. 12. Same layout with different interpretations at different scales

Interconnected Hierarchies

If one understands layout as a complex arrangement of nested elements, then a hierarchical order is inevitable. Both homogeneous as well as heterogeneous elements can exist on a single hierarchy level. As a result it is possible that when designing, heterogeneous elements may mix almost arbitrarily at various levels, Figure 13.

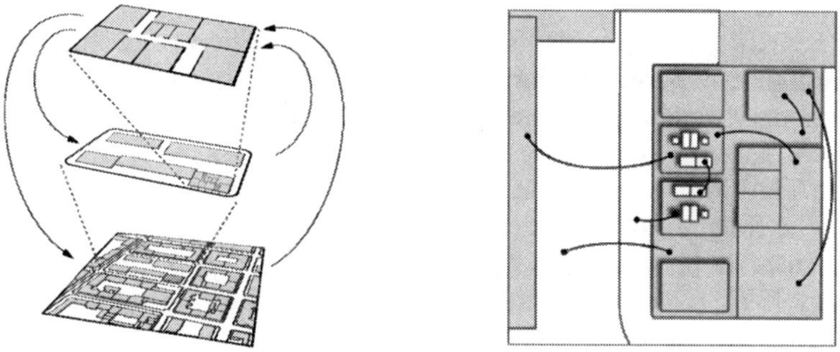

Fig. 13. Hierarchical relationships between interconnected hierarchies

Therefore it must be possible to create different relationships between elements, including connections that extend across layers and hierarchical boundaries. For example, a bathtub could have an abstract connection to the kitchen because of the desire to locate them in close proximity to one another, without it being necessary to combine the kitchen and bathroom facilities. Several elements could also be combined into functional zones and arranged alongside other functional areas in an enclosing space or in individually defined spaces. The goal of such interconnected hierarchies, is to be able to connect different levels of scale seamlessly with one another. This makes it possible, for example, to evaluate the suitability of the form of a building to see if it first into the context of the built environment or whether it can accommodate a specific spatial programme. The corresponding datastructure has to be object-orientated and describes the single layout elements as objects with properties (name, form, etc.), its individually definable constraints (minimal size, proportions, topology, etc.) as well as their relations to sub- and superordinated elements (parent, child).

Generative Mechanism

Since the user's perception of design problems and his response to them occur as parallel processes [37], design problems cannot be programmed ad hoc with concrete and prefabricated formulas. Many problems are first discovered during the design process. The system must always facilitate the possibility of being able to redefine and refine the problem. During the search for solutions the designer moves between fields of problem definition, formal elaboration and evaluation and optimisation of the result. Instead of following a strictly solution-oriented, linear approach, we therefore pursue a strategy that upholds the exploratory approach. The basis for this is a recursive circular trial and error system using generative and evaluative mechanisms that are based on evolutionary algorithms.

Evolutionary algorithms are used for the evaluation mechanism because their heuristic nature lends themselves to the solution of design problems: "...through experimentation and analysis we have learned that evolutionary techniques have excellent abilities as general-purpose problem solvers. Indeed, as Goldberg states, the genetic algorithm is 'a search algorithm with some of the innovative flair of human search' " [38]. The function of evolutionary algorithms will not be discussed further here, since it is well known. It is, for example, described in detail by Bentley [39]. A crucial factor here is that the direction during the search for solutions of evolutionary algorithms changes depending on the evaluative function. These evaluate the generated results with regard to certain criteria. However, as these need to be explicitly defined, the evaluation is necessarily based on operational criteria. The difficulty is that on the one hand, these criteria vary with each designer or design task and on the other that it is not necessarily true that a design that has been optimised for a certain criteria is better than one that has not been optimised. At best, these criteria can be used to search the solution space in certain directions in order to learn more about the object being designed. It is crucial that the evaluation function can be flexibly defined. For this purpose, a multi-criteria fit-function will be implemented which can be weighted flexibly during the evaluation process.

As the criteria for evaluation depend on the intentions of the designer or the interpretation of the presented layouts (depending on what meaning is ascribed to them), we propose the use of criteria that are as neutral and multifunctional as possible and can therefore be applied at different scales. Possible criteria include: overlapping, topological relationship, orientation, proportion, size and screening. These criteria must be assigned to the various elements flexibly and can be controlled by parameters. In addition one can also use operationally-specific assessment criteria for evaluating layout. Specific criteria are capable of being applied specifically to only one level of scale, for example, land use characteristics, density, building line, overshadowing and connectivity, etc.

We have investigated two approaches to achieving these goals.

The first approach focuses on the evaluation mechanism that assesses the quality of the elements (weak AI), the latter focuses on the implementation of reasonable rules for the dividing and generative mechanism (strong AI).

The prototype developed for the first approach uses the principle of packing to generate solutions. It uses a combination of evolutionary strategies for packing and genetic algorithms for the topological dependencies. Both mechanisms work in opposition in a co-evolutionary process, but after several iterations they usually find a satisfactory optimum for the packing and topological dependencies [6]. These prototypes also implement hierarchical structures that show how elements can be arranged at different levels of scale while offering the advantage of faster, more efficient computation, since the calculations can be performed in hierarchically staggered smaller groups. This approach makes it possible to influence solutions across different scales. For example forms can arise at an urban planning level which in turn influence the underlying configuration of the floor plans, Figure 14.

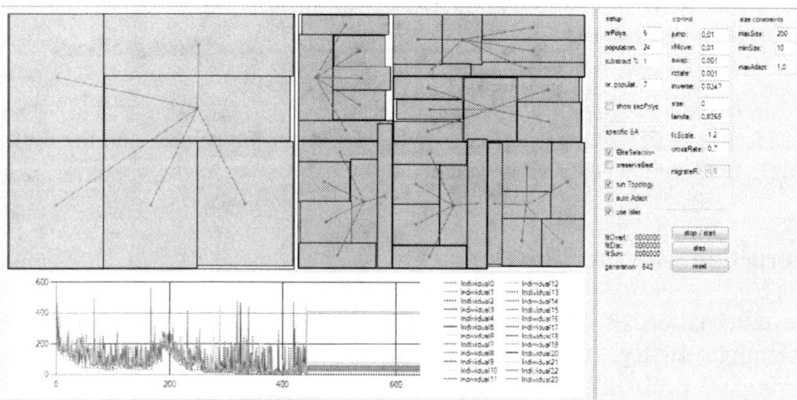

Fig. 14. Prototype using hierarchies with a packing algorithm

A second approach is to subdivide elements with a similar goal. This follows the logic of division, according to which an output element is divided recursively until a stop rule is satisfied and the result can be evaluated.

Different algorithms come into play that generate different solution sets. Initially these prototypes only support rectangular shapes to ensure that they work properly. As a result, the search space is limited to a linear range (bold line Figure 15. In further investigations, the aim will be to expand this subspace within the search space by implementing free polygons or even entirely free forms. We need to ascertain how much of the entire search space can be covered.

The challenge here is to be able to create a large variety of shapes on the one hand, while on the other to be efficient enough to make user interaction possible in real time (see next section).

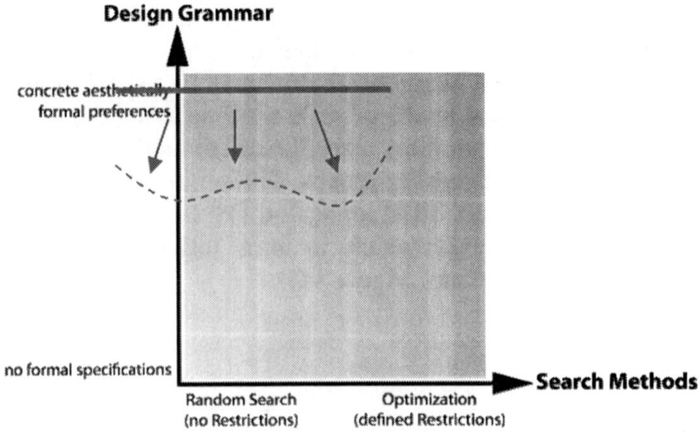

Fig. 15. The search space is defined by the active prototype area and the desirable flexible field for the generative system

Interaction with the Situation

The automation of design tasks presents two fundamental problems for designing: firstly, the difficulty described above of reducing non-operational problems to operational issues so that they can be formulated formally, and secondly, the fact that problems often change or even first arise in the process of developing a design. The integration of human capabilities is, therefore, a crucial factor for such a system.

It should be noted that the process of designing "combines slow reflection with intense periods of very rapid mental activity as the designer tries to keep many things in mind at once." [10]. The designer must be able to see the consequences and impact of his actions immediately. This means that the system must function in real time as much as possible so that users can see the effect of changes in criteria and parameters immediately and are able to react to them. Here we can distinguish between direct interaction, which refers specifically to the geometry and the attributes of the elements, and indirect interaction with the criteria and parameters of the evaluation mechanism. Through these indirect interventions the attributes of the elements (phenotype) change, which can be modified again by means of direct manipulation. The indirect (non-operational) and direct (operational) interaction therefore complement one another.

Direct Manipulation

To give the user a permanent and immediate opportunity to intervene directly in the optimisation process, the design system must provide a function that allows the user to select, move, resize, change the outline and anchor elements in place. It should be possible to add new elements, delete existing ones or merge them together. This direct manipulation addresses the mainly non-operational problems, since this intervention is done in the graphic representation directly, intuitively and playfully.

To address non-operational problems adequately which arise during the process and which were discovered by the user, it is important that this intervention works on several levels. A hybrid approach is desirable to connect between the generic, which is drawn by the user himself, the parametric, which deals with simple relations, and the generative computer-generated solutions. The user must be in a position to be able to draw shapes and elements, to define parametrical divisions for these elements, or let certain elements be generated within certain evaluation criteria. Associated with this is the set of partial solutions. Since humans are only able to consider a limited number of elements simultaneously, it is important, particularly for complex projects, that certain parts of solutions can be fixed. Such partial solutions can, for example, be elements that are to remain for some reason at a certain position, or a combination of several elements that correspond in size and topological relationship to each other to meet certain requirements.

Additionally the question arises as to how to indicate which assignments on one hierarchical level can serve as a restriction for another level. A typical scenario is the definition of the outline of a building, within which the organisation of rooms will occur. In the process, it may also be necessary to modify the outline, to accommodate the needs of the rooms within. Conceivable is also the definition of certain scopes, within which a solution can adapt to meet the constraints set at another level.

Indirect Manipulation

In the case of indirect manipulation, the user does not interact directly with the graphical output, but changes the evaluation criteria of the evolutionary algorithm (EA). The user controls the direction in which the system will search the solution space. Such modified EAs are also known as interactive or collaborative EAs [36]. Because of the need to formally describe these parameters, this indirect manipulation addresses the operational criteria. The impact of changes to the performance criteria must be visible immediately so the user can learn from immediate feedback and have the opportunity to examine what impact the optimisation of certain parameters (light, compact form, etc.) will have on the layout. The control of these parameters, e.g. overlay, should be similar to a performance-synthesizer. Here the multi-criteria fitness function can be flexibly defined and weighted, and, if necessary, also loosened. After all, creative solutions are often characterized by the lifting of specific restrictions to enable another path to be taken through the space of possibilities that may lead to a better solution than the most direct route. Accordingly, the evaluation may be made on purely aesthetic criteria, or in terms of the straightforward optimization of certain criteria.

Conclusion / Outlook

Design is a highly complex process, which often takes place simultaneously at different scales and levels of abstraction. If digital tools are to support complex design tasks, such as layout design, they must be able to become an integral part of this complex process.

Until now systems for computer-aided layout design are still grappling with the integration of their methods into a suitable workflow that can actually assist the user in designing. Several reasons for this have been discussed, concerning designing as a reflective practice.

In this paper, we present a highly process-oriented, exploratory system, which involves the user more actively in the generative mechanism. At present prototypes are being developed for several sub-aspects of the system in order to study the feasibility of individual requirements. The ultimate goal is to integrate these prototypes into a collective application.

We have shown how the basic conditions for such a design system can be implemented and how the supporting layout generation can work with rectangular shapes at hierarchically staggered scales. We have used the principle of packing, which produces meaningful solutions using a combination of evolutionary algorithms. The next step is to investigate the influence of the degree of freedom of the shapes used (rectangles, polygons, bubbles). Further linked connections across different levels of hierarchy have to be explored along with the impact this may have on the generated solutions.

Acknowledgements

The project is funded by the German Research Council (DFG).

References

1. Stiny, G., Gips, J.: Shape Grammars and the Generative Specification of Painting and Sculpture. In: IFIP Congress 1971, Amsterdam (1972)
2. Jo, J.H., Gero, J.S.: Space layout planning using an evolutionary approach. Artificial Intelligence in Engineering 12, 149–162 (1998)

3. Rittel, H.W.J., Webber, M.M.: Dilemmas in a General Theory of Planning. Policy Sciences 4, 155–169 (1973)
4. Röpke, J.: Die Strategie der Innovation: Eine systemtheoretische Untersuchung der Interaktion von Individuum Organisation und Markt im Neuerungsprozess. Mohr Siebeck, Tübingen (1977)
5. Simon, H.: The Science of the Artificial. MIT Press, Cambridge (1996)
6. Elezkurtaj, T., Franck, G.: Algorithmic Support of Creative Architectural Design. Umbau 19, 129–137 (2002)
7. Getzels, J.W., Csikszentmihalyi, M.: Scientific creativity. Science Journal 3, 80–84 (1967)
8. Polanyi, M.: The Tacit Dimension. Doubleday, Garden City, New York (1966)
9. Rittel, H.W.J.: On the Planning Crisis: Systems Analysis of the First and Second Generations Studiengruppe für Systemforschung e.V., Heidelberg (1976)
10. Lawson, B.: How Designers Think: The Design Process Demystified, 4th edn. Architectural Press, Oxford (2006)
11. Schön, D.A.: Designing as Reflective Conversation with the Materials of a Design Situation. Research in Engineering Design (1992)
12. Whitehead, B., Eldars, M.Z.: An approach to the optimum layout of single storey buildings. The Architects Journal, 1373–1380 (June 1964)
13. Li, S.-P., Frazer, J.H., Tang, M.-X.: A Constraint Based Generative System for Floor Layouts. In: Fifth Conference on Computer Aided Architectural Design Research in Asia, Singapore (2000)
14. Medjdoub, B., Yannou, B.: Dynamic space ordering at a topological level in space plannung. Artificial Intelligence in Engineering 15 (2001)
15. Toffoli, T., Margolus, N.: Cellular Automata Machines: A New Environment for Modeling. MIT Press, Cambridge (1987)
16. Batty, M., Xie, Y.: From Cells to Cities. Environment and Planning B: Planning and Design 21(7), 31–48 (2004)
17. Koenig, R., Bauriedel, C.: Computer-generated City Structures. In: Generative Art Conference, Milan (2004)
18. Koenig, R.: Generating urban structures: A method for urban planning and analy-sis supported by cellular automata in Research on New Towns: Second International Seminar in Druck II. International New Town Institute, Almere (2007)
19. Schweitzer, F.: Wege und Agenten: Reduktion und Konstruktion in der Selbstorganisationstheorie. In: Krug, H.-J., Pohlmann, L. (eds.) Selbstorganisation, pp. 113–135. Duncker & Humblot, Berlin (1997)
20. Batty, M.: Cities and Complexity: Understanding Cities with Cellular Automata, Agent-Based Models, and Fractals. MIT Press, London (2005)
21. Coates, P., Schmid, C.: Agent Based modelling. In: CECA - The centre for computing & environment in architecture. University of East London School of Architecture, London (2000)

22. O'Sullivan, D.: Exploring spatial process dynamics using irregular cellular automaton models. Geographical Analysis 33, 1–18 (2001)
23. Braach, M.: Kaisersrot - computergestützter individualisierter Städtebau Werk Bauen Wohnen, vol. 4 (2002)
24. Coates, P., et al.: Current work at CECA, Three projects: Dust, Plates & Blobs. In: Generative Art International Conference, Milan (2001)
25. Stiny, G.: Pictorial and Formal Aspects of Shape and Shape Grammar. Birkhäuser, Stuttgart (1975)
26. Mitchell, W.J.: The Logic of Architecture.: Design, Computation, and Cognition, 6th edn. MIT Press, Cambridge (1998)
27. Stiny, G.: Shape: Talking about Seeing and Doing. MIT Press, Cambridge (2006)
28. Fleming, U.: The Role of Shape Grammars in the Analysis and Creation of Designs. In: Kalay, Y.E. (ed.) Computability of Design, pp. 245–272. Wiley, New York (1987)
29. Duarte, J.P.: Customizing mass housing: a discursive grammar for Siza's Malagueira houses. Faculty of Architecture. MIT, Cambridge (2000)
30. Arvin, S.A., House, D.H.: Modeling architectural design objectives in physically based space planning. Automation in Construction 11, 213–225 (2002)
31. Frazer, J.: Reptiles. Architectural Design, 231–239 (April 1974)
32. Frazer, J.: An Evolutionalry Architecture. Architectural Assocoation Publications, London (1995)
33. Broughton, T., Tan, A., Coates, P.: The Use of Genetic Programming In Exploring 3D Design Worlds: A Report of Two Projects by Msc Students at CECA UEL. CAAD Futures. München Kluwer Academic, Dordrecht (1997)
34. Coates, P., Hazarika, L.: The use of Genetic Programming for applications in the field of spatial composition. In: Generative Art Conference (1999)
35. Rosenman, M.A.: The generation of form using an evolutionary approach. In: Dasgupta, D., Michalewicz, Z. (eds.) Evolutionary Algorithms in Engineering Applications. Springer, Heidelberg (1997)
36. Bentley, P.J., Corne, D.W.: An Introduction to Creative Evolutionary Systems. In: Bentley, P.J., Corne, D.W. (eds.) Creative Evolutionary Systems, pp. 1–76. Morgan Kaufmann, San Francisco (2002)
37. Schön, D.A.: The Reflective Practitioner: How Professionals Think In Action. Basic Books, New York (1983)
38. Goldberg, D.E.: Genetic Algorithms in Search, Optimization and Machine Learning, 1st edn. Addison-Wesley, Boston (1989)
39. Bentley, P.J. (ed.): Evolutionary Design by Computers. Morgan Kaufmann, San Francisco (1999)

Applying Clustering Techniques to Retrieve Housing Units from a Repository

Álvaro Sicilia, Leandro Madrazo, and Mar González
ARC Enginyeria i Arquitectura La Salle, Spain

The purpose of BARCODE HOUSING SYTEM, a research project developed over the last four years, has been to create an Internet-based system which facilitates the interaction of the different actors involved in the design, construction and use of affordable housing built with industrialized methods. One of the components of the system is an environment which enables different users – architects, clients, developers – to retrieve the housing units generated by a rule-based engine and stored in a repository. Currently, the repository contains over 10,000 housing units. In order to access this information, we have developed clustering techniques based on self-organizing maps and k-means methods.

Introduction

Nowadays, the possibility of carrying out design processes collaboratively on the Internet offers different users the opportunity to participate in the design of mass customized housing [1], [2], and [3]. This requires the design of environments which support user interaction by means of appropriate interfaces while at the same time taking advantage of computer programs' capacity to generate and evaluate design solutions.

BARCODE HOUSING SYSTEM creates housing blocks as aggregations of housing units which have been automatically generated using a rule-based system. The overall process of designing a housing block, from the housing unit to the overall building, is open to the participation of the different stakeholders (architects, builders, manufacturers, occupants, facilities managers), which lead the design process by providing inputs at different stages using the interfaces accessible on-line. A housing system has been specifically created which is based on a combination of horizontal and vertical space bars. The housing system is represented in a graph which contains all possible floor plans [4].

J.S. Gero (ed.): *Design Computing and Cognition'10*, pp. 387–401.

At the outset, one possible way to generate a housing unit is by specifying its individual characteristics. A second way it is by specifying the generic characteristics of a type of housing unit so that the system returns a series of floor plans. In the first case, there is the risk that system is unable to return a valid solution, or that it takes a lot of time to find it. With the second procedure, both problems can be avoided. The trade-off in this case is that it is necessary to have a background generation of the housing units which are stored in a repository. To search the housing units in the repository, an information retrieval system is necessary. To improve the performance of the retrieval, we have applied clustering techniques which facilitate the interaction between users and the design system.

Clustering techniques have been used to solve classification problems in information systems in different domains, such as in financial [5] and traffic data analysis [6], to classify massive document collections [7], and to analyze protein sequences in the field bioinformatics [8]. Some applications can also be found in the domain of architecture, such as data mining techniques to analyze design information stored in a case library [9], or techniques which are integrated into an information system to retrieve floor plans [10]. Also, clustering techniques have been used along with shape grammar systems to categorize shape rules and designs in order to facilitate design exploration [11].

In this paper we describe the search tools which users employ to retrieve the repository of housing units previously generated. Two clustering techniques – Self-Organizing Maps and k-means – have been implemented and their performance has been compared.

BARCODE HOUSING SYSTEM: A Generative System for Housing Design

BARCODE HOUSING SYSTEM consists of interwoven working spaces in which the different actors (architects, developers, manufacturers, occupants) participate synchronously and asynchronously throughout the entire process of design and construction of housing units and of the buildings resulting from their aggregations.

The work spaces and their functionalities are the following:

- PROJECT DEVELOPMENT. In this working space, developers, architects and building managers specify site properties (area, size), the number and type of housing units, building and planning regulations (building volumes and height) and environmental conditions (climate, orientation). Alternative building solutions (massing, location) can then be explored for the given site conditions and brief.

- HOUSING LAYOUTS. In this space, architects select a set of units that will later be used to generate a building. The units have been generated by the system in batch processing and are stored in the system database. The selection becomes a "discovery" process as the architect finds the housing layouts while navigating through the space of solutions – using clustering techniques – which the system has generated in previous project developments. Should an adequate layout not be found from the pool of the existing solutions, the architect can request the generative system to create alternative layouts that conform to the desired criteria (surface, number of rooms, number of bathrooms, open or closed kitchen). The new solutions are stored in the database, thus enhancing the previously existing pool of solutions.
- HOUSING CONFIGURATION. Occupants describe their housing program (number of family members, usage of spaces, lifestyles) working with user-friendly interfaces that represent housing units and layout in a graphic language that can be understood by lay people (schematic plans, photographs depicting activities in spaces, bubble diagrams). The system returns the housing units which most closely correspond to the criteria defined by the users and they select those that most closely meet their needs. Then, the selected units are used in the generative process that creates the housing block. Once the housing units have been assembled, there is a process by which occupants and architects collaborate using a 3-D environment to define the arrangement of a living unit (finishing, partitions and furniture).
- HOUSING ASSEMBLY. In this environment, the architect defines the design criteria for the assembly of housing units, including: degree of compactness of the housing block, the degree of optimization of building services, the minimum distances to access cores (staircases, elevators), the material of the structural skeleton and so on. Once the design values are set, a generative process creates the solutions that satisfy these criteria.
- BUILDING COMPONENTS CATALOGUE. A XML-based product modeling catalogue enables manufacturers to enter descriptions of their products, which will be then selected by the team in charge of the project development. Based on this selection, the future occupant chooses the components (doors, windows, partitions) and inserts them in the 3-D depiction of a dwelling.

In the following sections, we will introduce the retrieval processes in the HOUSING LAYOUTS work space where the clustering has been applied.

Housing Layout Workspace

As Steadman observed, there are two basic approaches to automatic floor plan generation: to generate one or few plans that satisfy a set of specified constraints, or to produce all the possible plans which cover all the requirements [12]. In the second case, we avoid the high cost of generating a possible solution by shifting the computational power to the search process. In this way, it is possible to facilitate the process of finding a solution by guiding the search for designs that suit the specific set of criteria.

We have opted for separating the rule-based generation of design solutions from the search in the design space.

In the Housing Layouts work space, architects select and/or generate a set of housing units according to the project specifications: floor plan dimensions and area, and access type. These specifications have been previously set in the Project Development workspace by the different agents – developers, architects and building managers – involved in the project.

Housing Layout encompasses two environments (Figure 1): Housing Generation and Housing Selection. In the former, the floor plan layouts are created through a generative process. The spatial structure of a housing unit is represented by a graph. The nodes of the graph are cell spaces and the edges represent the connections between them. To minimize the computational cost of generating a layout by searching through all connections in the graph, a constraint list has been implemented. For example, the constraint list contains the information about the sizes and proportions of the spaces or the type of entrance to the dwelling (by staircase or walkway). The constraint list contains information about the graph and it is separate from it. In this way, the information contained in the list can easily be edited so that the user can guide the generative process of the layout. At the end of the generative process, the designs are stored in the Housing Repository, Figure 1.

The search on the designs previously generated takes place in the Housing Selection environment. At the start, a clustering process is launched to classify the designs. Then the floor plans subsequently generated are classified using the current clusters configuration. These clusters help the architect user to identify the housing design he or she needs for a particular housing project. Also, the groups previously created by other users can provide insights that facilitate the search. Furthermore, by adding tags to the discovered designs users participate in creating a metadata layer, thus bringing their individual knowledge into the system. These tags also can be used in future searches by other users.

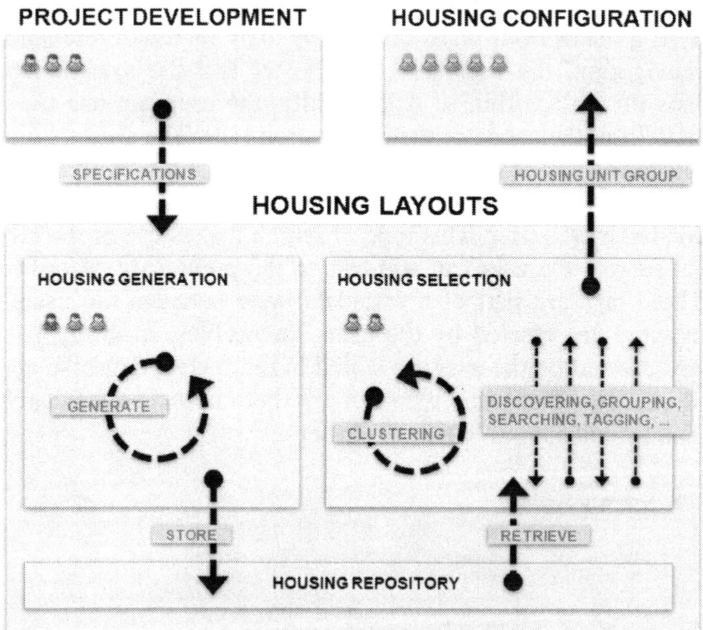

Fig. 1. Structure of the system and workspaces relations

The Housing Configuration workspace is enabled when the architect has created a collection of floor plans. Later on, the future resident contributes to the design process by describing the characteristics of the dwelling through three interactive interfaces. Tenants choose their living units from among the collection previously selected by the architect. Afterwards, the tenants can customize their dwelling with the help of the architect interacting with a three-dimensional model. Later on, once the building has been constructed, residents can still have access to the 3-D model and add tags to describe their experience after living in the unit. These tags can inform the search process of architect users in later projects.

Housing Selection Workspace

In the Housing Selection workspace, the architect user can search and collect designs through a discovery process. This environment provides a variety of tools – attribute search, cluster navigation, similarity search, social search and group navigation – to assist the user in this process, Figure 2.

With searches, the user creates a query using the attributes of the housing designs, giving a specific value for each attribute (e.g., area, room type). Also, the user can specify the weight of the attributes. The output of the query is a list of floor plans ordered by their degree of relevance. In the cluster navigation, the user selects a cluster and the system returns the floor plans that fall within it. Additionally, the user can use the attribute search to delimit the scope of the exploration [13]. The clusters are created by the clustering system, which will be described later. With the similarity search, a collection of designs are retrieved which are similar to a floor plan provided by the user. This type of search makes use of the clusters. In the social search, the user can add tags to the previously described search types. These tags are part of a metadata layer between the users and the repository and are created by the users themselves. In group navigation, the groups created by the user are stored in the system database and can be accessed later by other users who can use them in their own search. These groups can be considered 'custom' clusters.

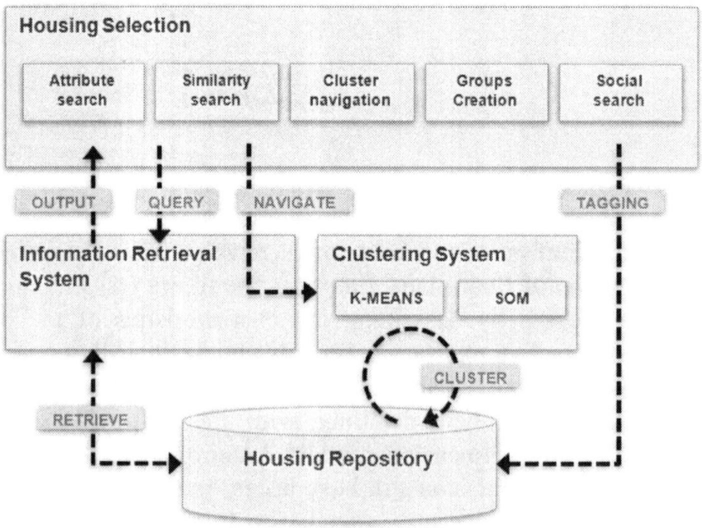

Fig. 2. Selection Housing Units structure

The Housing Selection workspace is composed of the Information Retrieval System and the Clustering System, Figure 2. The Information Retrieval System (IRS) is a search engine based on an enhanced vector model adopted from information retrieval science (IRS) [14]. IRS parses the user queries, retrieves the elements from the Housing Repository and sorts the outputs according their relevance. The query is defined as a list of the floor plan attributes. These attributes can be architectural, Table 1, tags, or associations of clusters and user groups. In this way, the IRS can meet the requirements of the different search tools described above. To perform a query, the IRS needs a list with the weights of each attribute.

Then, the elements having at least one attribute are retrieved from the repository and the IRS calculates their degree of relevance, Figure 3. Finally, the outputs are ordered, situating the elements with a high relevance value at the top of the list. Moreover, IRS makes use of a cache memory to speed up the response time.

$$\sum_i W_i Att_i$$

$$Att_i : \begin{cases} 1, element\ has\ the\ attribute\ i \\ 0, element\ hasn't\ the\ attribute\ i \end{cases}$$

Fig. 3. Calculation equation of relevance degree

The Clustering System (CS) is responsible for clustering the Housing Repository. As the architect user generates new floor plan layouts, the CS assigns a cluster to them. Should the new layouts not fit the existing cluster configuration, the CS will cluster the entire repository content from scratch. The CS implements two clustering methods – Self-Organizing Maps and k-means – and it can run them with different configurations. In the case of Self-Organizing Maps, it sets up the grid dimension and the learning coefficient. The configuration parameters of the k-means method are the number of clusters, the distance function and the cluster initialization method.

The Housing Repository is organized as a multilayer structure, Figure 4. At the bottom layer are the housing floor plans. On top of them are the architectural attributes that describe the floor plans which are extracted by the generative process. The cluster data is created by the Clustering System accessing the architectural attribute layer. The cluster layer is formed by links to the housing floor plan layer. The tag layer is composed of metadata generated by users and includes links to the other layers.

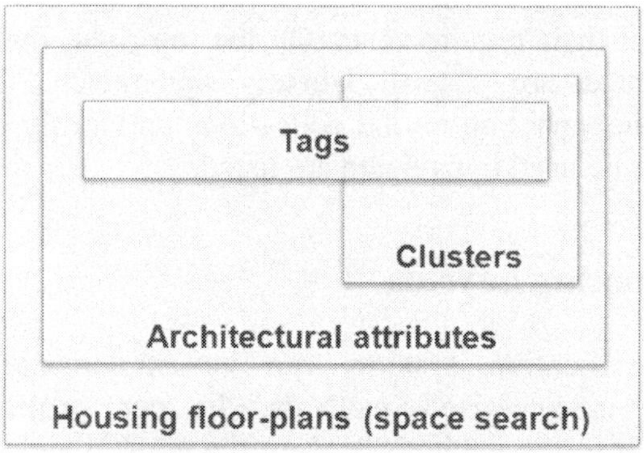

Fig. 4. Housing Repository data structure

The interface for architect users is depicted in Figure 5. Using the previously described search tools, the user can search for housing layouts with, for example, a 70 sqm area, two bedrooms and one bathroom. These values of attributes and tags are introduced in the lower gray window. The output of the query is shown in the large window. On the top left corner, the most relevant layouts are shown, with the less relevant ones in the opposite corner.

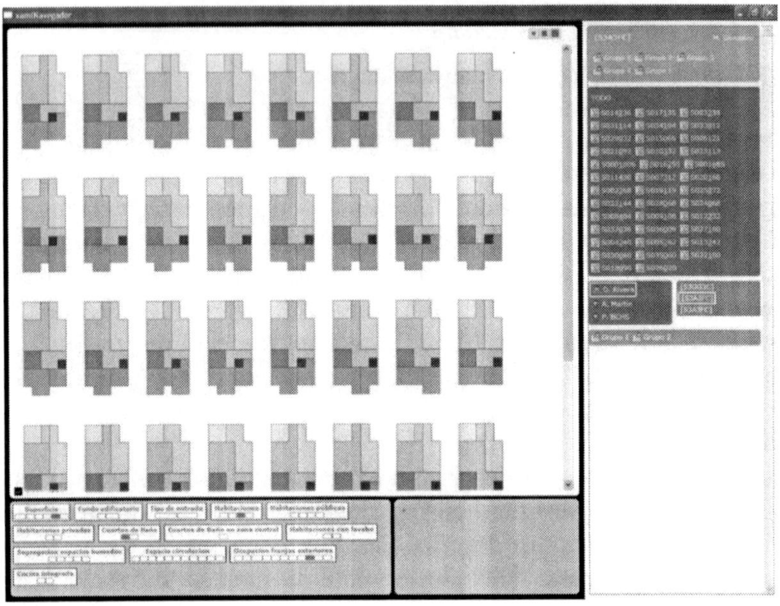

Fig. 5. Architect user interface to select housing units

When the relevance is far below the maximum value, the figure is dimmed. In this way, the users can see at a glance the housing layouts which best meet their requirements. On the top right, the collections that are being created are shown. Directly underneath, the clusters are displayed. At the very bottom the collections previously created by other users which can be used to navigate are listed.

Clustering Housing Layouts

Automatic classification systems can be implemented using either supervised or unsupervised methods. In our project, we applied unsupervised clustering methods because the number and types of samples to be clustered is increasing over time, as more layouts are being generated and stored in the repository. However, there is no guarantee that the optimum results will be achieved with such methods [15]. Therefore, the project aims to apply known cluster methods to our data and compare their outputs.

At the time of writing, the housing generation processes have created over 10,000 layouts. There are sixteen characteristics that describe a layout in terms of space, circulation and services, Table 1.

Table 1 Attribute list that characterize a housing layout

Attribute name	**Description**
Surface	Surface of the housing unit in square meters
Rooms	Number of rooms
Private rooms	Number of private rooms
Public rooms	Number of public rooms
Water-closet	Number of water-closets
Toilets	Number of toilets
Water-closet at the center	Yes/No
Balcony	Yes/No
Room extensions	Yes/No
Building depth	Depth of the building
Entrance type	Staircase, walkway
Wet spaces segregation	Distance value from toilettes to other rooms
Rooms with water-closet	Yes/No
Circulation space	Surface of the circulation space of the apartment
Exterior spaces	Surface of exterior spaces (galleries, balconies)
Kitchen integrated in living rm	Yes/No

The types of attributes are numeric and Boolean, and most of them are numeric with a known range. The attributes listed in Table 1 are used in different ways. For instance, cluster algorithms use them as dimensions of the input elements, and the architect user uses them to perform queries.

We have opted for Self-Organizing Maps and k-means because they are two well-established techniques that are unsupervised, easy to understand and implement, have a linear time complexity, and are successfully used to cluster large amounts of data [16]. To compare them, we have used the following quality measures:

- Q: This takes into account the intra-cluster and inter-cluster distances [15]. Also, the weight of a cluster (e.g., its size) is used in the measurement. In Figure 6, d_i is the inter-cluster distance where d_{ij} is the minimum distance between the samples in the cluster and all the other samples in the remaining clusters. The variable $M_{i,k}$ is the value of the attribute i of the sample k; and w_i is the value of the attribute i of the centroid. s_i is the distance intra-cluster, where k_i is the number of elements of the cluster. δ_i is a weight indicator proportional to the size cluster.

$$d_i = \min_j(d_{ij}), where\ d_{ij} = \min_{h,k}(\|M_{i,k} - M_{j,h}\|)\ j,h,k = 1,2 \ldots K$$

$$s_i = \frac{1}{k_i}\sum_{j=1}^{k_i}\|M_{ij} - w_i\|^2 \qquad \delta_i = \frac{k_i}{N}, where\ N = \sum_i k_i \qquad Q = \sum_{i=1}^{K}\delta_i\frac{s_i}{d_i}$$

Fig. 6. Equations of Q quality measurement

- Qn: This measurement indicates how close a sample is to its cluster with regard to the others clusters. In other words, it indicates the cohesion of the clusters. The variable d_i is the distance of the element i to its cluster and r_i is the distance to the closest remaining cluster.

$$Qn = \sum_i^N \frac{r_i - d_i}{\max(r_i,\ d_i)}$$

Fig. 7. Equation of Qn quality measurement

- QNN: This measurement indicates the cohesion of the samples. The equation is similar to the previous one, but it uses the distance between samples instead of distances between clusters. The variable $\check{d}_i$ is the distance of the sample i to the closest sample from the same cluster; and $\check{r}_i$ is the distance to the closest sample from a different cluster. To make it less sensible to noise, a variant of this measure uses the mean of several samples instead of only one.

$$Qnn = \sum_i^N \frac{\check{r}_i - \check{d}_i}{\max(\check{r}_i,\ \check{d}_i)}$$

Fig. 8. Equation of Qnn quality measure

Clustering Algorithms

Self-Organizing Maps (SOM) is a classification technique based on a type of unsupervised, competitive neural network with a regular distribution grid that can discover the underlying relationships between data. This technique aims to reduce the number of data dimensions using neural networks [17]. One of its most remarkable advantages is that it can show the quality of the results graphically. Another significant characteristic is that it can show data similarities. We have implemented the original algorithm which has less computational load [7]. We have used a Gaussian neighborhood function in a rectangular grid with variable dimensions. The number of executions is set at twice the input samples. The learning rate has been set to be sensitive enough to get a maximum number of clusters with the largest cohesion possible. The visualizing power of this technique is not relevant to our case.

We have implemented an automatic process to label the nodes and merge the similar ones based on the U-Matrix [18] without information on the input classes. This iterative process follows these steps:

1. First, the process selects a neuron with the minimum similitude value.
2. If there are input samples for which this neuron is the winner, then it is labeled.
3. Afterwards, it searches for other neurons with the same similitude within a small threshold; these neurons are labeled like the first one.
4. Go to step 1 if there are unlabeled neurons.

The k-means technique is a type of cluster analysis whose goal is to minimize the quadratic function error [19]. The inputs of the algorithm are the samples to be clustered and the number of partitions. It is important to accurately choose the proper number of partitions, the centroid initialization method and the appropriate distance function. We have taken the number of clusters value from the SOM results as input, and we have tested four centroid initialization methods – random values, random domain values, random sample and D2 weighting – and three distance functions: Euclidian, Manhattan and Hamming. The D2 weighting method [20] uses probabilities to improve the original k-means. It is an iterative process, where first a centroid is chosen as a sample from an input samples list. Then, the next centroid is calculated by selecting the farthest sample with the proportional probability to the minimum distance between the element and the previous centroid.

Results

We have executed the algorithms 200 times with different configurations and compared the results. Both techniques use the architectural attributes of the floor plan layouts. We have tested three different k-means configurations and one SOM configuration. The statistics in Table 2 show that the different methods behave similarly. Taking into account the standard deviation, we can see that the SOM method is more stable than others according to the executions.

Table 2 Statistics generated after 200 executions

		Q	Qn	Qnn
K-MEANS	Existing elements	1,112 ± 0,387	0,530 ± 0,020	0,997 ± 0,0006
K-MEANS	Domain values	1,454 ± 1,076	0,534 ± 0,038	0,998 ± 0,0004
K-MEANS	D2 Weighting	1,230 ± 0,914	0,532 ± 0,031	0,997 ± 0,0006
SOM	20x20	1,114 ± 0,114	0,520 ± 0,013	0,996 ± 0,0004

In order to test the significance of the measurements, we have plotted the quartiles of the results of the different methods. The quality measurement Q relates the cohesion of the samples and the spread between clusters: the lower the value the better the results. Figure 9a shows that there are no significant differences between the results obtained by each method. Therefore, we cannot compare these results. The Qn measurement determines the unity of the input samples with their clusters, in this case, the closer the value is to 1 the better. We can see that the k-means method with domain values has the best result, but Figure 9b shows there are no significant differences between methods. Finally, for the measurement Qnn which evaluates the cohesion of the samples, we can see that the behavior of all methods is significantly different, so we can conclude that for this measurement the k-means with domain values is the best option.

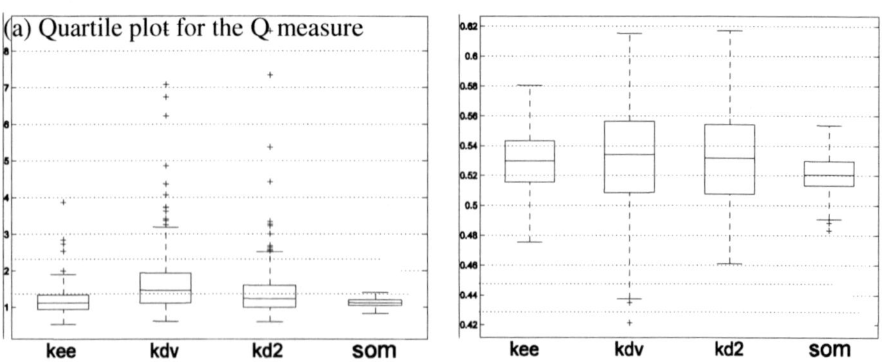

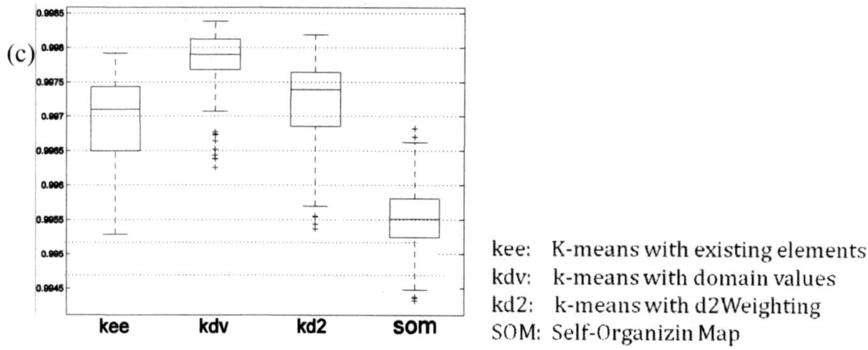

Fig. 9. Quartile plot for the different quality measurements

Once we have assessed the results, it is difficult to choose the best method from this data because they all perform similarly, and some quality measurements cannot be used to compare them. Because of this, we have turned to the best execution of all the methods to cluster the floor plans. The number of clusters is high enough to express all of the variations. With the help of the interface, Figure 5, of the Housing Selection, we have manually checked the quality of the clusters. As seen in Figure 10, the final clusters have a high degree of cohesion and are homogeneous.

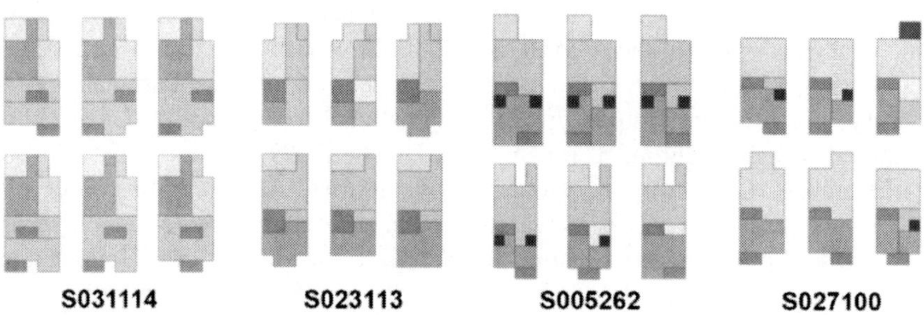

Fig. 10. Clusters generated with a SOM algorithm

Conclusions

We have successfully integrated a rule-based generative process which generates housing floor plans with an information retrieval system which returns a series of plans with the help of clustering techniques. The application of the search tools has demonstrated that they provide proper housing units, thus facilitating the navigation through the repository. With the data we have used, there have been no substantial differences in the performance of the two clustering techniques. Further developments of the visualization capacities inherent to SOM techniques and the interfaces that support them would enhance the cognitive potential of clustering techniques.

Acknowledgements

This research project was carried out with the support of grant BIA2005-08707-C02-01 from the Spanish National RDI Programme, 2005-2008. We would like to thank Francesc Teixidó, professor in the Computer Science department at Enginyeria La Salle, for his advice and support in interpreting the results.

References

1. Chien, S.F., Shih, S.G.: A Web Environment to Support User Participation in the Development of Apartment Buildings. In: Special Focus Symposium on WWW as the Framework for Collaboration, InterSymp., Baden-Baden, Germany, pp. 225–231 (2000)
2. Gerzso, J.M.: Automatic generation of layouts of an Utzon housing system via the Internet. Reinventing the Discourse - How Digital Tools Help Bridge and Transform Research, Education and Practice in Architecture. In: 21st Annual Conference of the ACADIA, Buffalo, New York, pp. 202–211 (2001)
3. Huang, J.C., Krawczyk, R.: A Choice Model of Consumer Participatory Design for Modular Houses. In: 25th International Conference Aided Architectural Design in Europe, Germany, pp. 679–686 (2007)
4. Madrazo, L., Sicilia, A., González, M., Martin, A.: Integrating floor plan layout generation processes within an open and collaborative system to design and build customized housing. In: Tidafi, T., Dorta, T. (eds.) Joining Languages, Cultures and Visions: CAADFutures, pp. 656–670 (2009)
5. Deng, Q.: Combining Self-Organizing Map and K-Means Clustering for Detecting Fraudulent Financial Statements. In: IEEE International Conference on Granular Computing, GRC 2009, pp. 126–131 (2009)
6. Chen, Y., Zhang, Y., Hu, J., Yao, D.: Pattern Discovering of Regional Traffic Status with Self-Organizing Maps. In: Intelligent Transportation Systems Conference, ITSC 2006, pp. 647–652. IEEE, Los Alamitos (2006)
7. Kohonen, T.: Self organization of a massive document collection. IEEE Transactions on Neural Networks 11(3), 574–585 (2000)
8. Zhong, W.: Improved K-Means Clustering Algorithm for Exploring Local Protein Sequence Motifs Representing Common Structural Property. IEEE Transactions on NanoBioscience 4(3), 255–265 (2005)

9. Lin, C., Chiu, M.: Smart Semantic Query of Design Information in a Case Library. Digital Design: Research and Practice. In: 10th International Conference on CAADFutures, pp. 125–135 (2003)
10. Inanc, B.S.: Casebook. An Information Retrieval System for Housing Floor Plans. In: CAADRIA 2000, 5th Conference on Computer Aided Architectural Design Research in Asia, Singapore, pp. 389–398 (2000)
11. Lim, S., Prats, M., Chase, S., Garner, S.: Categorisation of Designs According to Preference Values for Shape Rules. In: Gero, J.S., Goel, A.K. (eds.) Design Computing and Cognition, pp. 41–60. Springer, Heidelberg (2008)
12. Steadman, J.P.: Architectural Morphology. Pion Limited, London (1983)
13. Quintarelli, E.: Facetag: Integrating Bottom-up and Top-down Classification in a Social Tagging System. Las Vegas IA Summit (2007)
14. Baeza-Yates, R., Ribeiro-Neto, B.: Modern Information Retrieval. ACM Press, Addison-Wesley, New York (1999)
15. Savaresi, S.: Cluster selection in divisive clustering algorithms. In: 2nd SIAM ICDM, Arlington, VA, USA, pp. 299–314 (2002)
16. Jain, A.K.: Data clustering: A Review. ACM Computing Surveys 31(3) (1999)
17. Kohonen, T.: Self-Organizing Maps. Springer, New York (1995)
18. Ong, J.: Data Mining Using Self-Organizing Kohonen maps: A Technique for Effective Data Clustering & Visualization. In: International Conference on Artificial Intelligence (IC-AI), Las Vegas (1999)
19. MacQueen, J.B.: Some Methods for classification and Analysis of Multivariate Observations. In: 5th Berkeley Symposium on Mathematical Statistics and Probability, pp. 281–297. University of California Press, Berkeley (1967)
20. Arthur, D., Vassilvitski, S.: K-Means++: The advantages of careful seeding. In: Bansal, N., Pruhs, K., Stein, C. (eds.) 18th Annual ACM-SIAM Symposium on Discrete Algorithms, New Orleans, Louisiana, pp. 1027–1035 (2007)
21. Singhal, A.: Modern Information Retrieval: A Brief Overview. Bulletin of the IEEE Computer Society Technical Committee on Data Engineering 24(4), 35–43 (2001)
22. Aghagolzadeh, M.: Finding the number of clusters in a dataset using information theoretic hierarchical algorithm. Electronics, Circuits and Systems. In: ICECS. 13th IEEE International Conference, Nice, France, pp. 1336–1339 (2006)
23. Michalski, R., Stepp, R.: Learning from observation: Conceptual clustering. Machine Learning: An Artificial Intelligence Approach, pp. 471–498. Morgan Kaufmann, Los Altos (1986)
24. Baçao, F., Lobo, V., Painho, M.: Self-organizing Maps as Substitutes for K-Means Clustering. In: 5th International Conference Computational Science - ICCS, Atlanta, GA, USA (2005)
25. Nguyen, Q.H., Rayward-Smith, V.J.: Internal quality measures for clustering in metric spaces. Int. J. Business Intelligence and Data Mining 3(1), 4–29 (2008)

KNOWLEDGE AND LEARNING IN DESIGN

Different function breakdowns for one existing product: Experimental results
Thomas Alink, Claudia Eckert, Anne Ruckpaul and Albert Albers

A general knowledge-based framework for conceptual design of multi-disciplinary systems
Yong Chen, Ze-Lin Liu and You-Bai Xie

Learning concepts and language for a baby designer
Madan Mohan Dabbeeru and Amitabha Mukerjee

Organizing a design space of disparate component topologies
Mukund Kumar and Matthew I Campbell

Different Function Breakdowns for One Existing Product: Experimental Results

Thomas Alink[1], Claudia Eckert[2], Anne Ruckpaul[1], and Albert Albers[1]
[1] *Karlsruhe Institute of Technology, Germany*
[2] *The Open University, UK*

This paper describes the findings of an experiment on how different engineers understand the notions of function and functional breakdown in the context of design by modification. The experiment was conducted with a homogenous group of 20 design engineers, who had all received the same education. The subjects were asked to analyze how a hydraulic pump works and summarize their understanding in a function tree. The subjects were given either the hydraulic pump itself (with part of its casing removed), or a maintenance drawing that showed a section cut of the pump. This paper shows typical outputs of the designers and discusses the differences between the subjects' approaches and resulting function trees; and points to typical mistakes the subjects made.

Introduction

Analysis of existing products is at the heart of new product development in two ways, as inspiration from which the designers of the new products can learn and as a starting point for the development of a new design. With very few exceptions complex technical products are designed by modification from existing products, which are changed to meet new needs and requirements. While these changes can be substantial and allow scope for radical innovation, many products carry components or solution principles across from their predecessor designs. It is therefore important for designers to understand how the old product works, what its strengths and weaknesses are, and what problems have occurred with it. This paper reports on an experiment, where 20 designers were asked to analyze a given product in terms of the functions that occur in the product, and provide a summary in the form of a classical function tree as a common format that allows a comparison of their understanding.

J.S. Gero (ed.): *Design Computing and Cognition'10*, pp. 405–424.

Developing a degree of shared understanding of the problem and the product is important for an effective design process (see for instance Hacker [1], and Hinds and Weisband [2]). This can be problematic when participants with different backgrounds and interests need to communicate across object worlds – the differing sets of concepts and interpretations of terms that people gain from education and experience in particular fields (see Bucciarelli [3]). Shared understanding has mainly been studied from the perspective of computer support for distributed teams (e.g. Arias et al. [4]) or eliciting the degree of shared understanding in teams (e.g. Hill et al [5]).

Functions are a fundamental, albeit in practice elusive, part of many prescriptive design methodologies, such as the Pahl and Beitz methodology. Pahl and Beitz [6] define a function as the relation between an input and an output of a system. The Pahl and Beitz definition represents the class of transformative notions of functions. However, transformative approaches are claimed to fail for the human-centered aspects of how the product is used (Warell, [7]). Crilly [8] states that the designers' notions about the term 'function' depend on the situation they are faced with.

The Function Behavior Structure (F-B-S) model of Gero and Kannengieser [9] is closer to the everyday use of English words, than flow or transformation notations of function. F-B-S sees design as the transformation of a set of functions into a set of descriptions of structure (S). The function of an object is defined as its teleology, i.e. what the product is made for. They introduce the term 'behaviour' to describe what the system is expected to *do* (behaviour expected, Be) or what the system actually does when it is designed, i.e. the behaviour (Bs), derived from structure (S). Designing is defined as the transformation of those variables. With the F-B-S ontology Gero clearly distinguishes different terms for what e.g. the Pahl and Beitz describe with a single term. Based on the work of Pahl and Peitz the German Engineers Association (VDI) Guidelines were created. The VDI Guidelines [10] do not distinguish between function and behaviour. However, the VDI Guidelines are the basis for the education of most engineering designers in Germany. Design researchers are still debating the concept of function (e.g. Vermaas and Houkes [11]; Kirschman and Fadel [12]) because the concept of function is fundamental to understanding designing and to improving methods that build on this concept.

This paper reports on an experiment, where 20 designers were asked to analyze a given product and express their understanding in a function tree. The experiment has removed many of the factors that negatively affect teams by giving a very homogeneous group of people the same product and time to analyze its functions. The way designers approached the task and the function trees they produced were strikingly different. Designers chose different representations and approaches to analyzing the product and drew on different notions of functions. The paper includes an analysis of the different function trees, comparing for all subjects the written and verbally expressed functions, as well as comparing them against a detailed function tree generated by the researchers to assess the completeness of the different functional analyses. It addresses the mistakes that different subjects have made, before the findings and further work are discussed in the conclusions.

Research Design

The experiment has been carried out as part of a systematic evaluation of the Contact and Channel Model (C&CM) approach, developed at the IPEK in Karlsruhe (Albers et al [13]). C&CM is an approach to describing, analyzing and designing products in terms of their functions. The functions are mapped onto the form of the product through working surface pairs (WSP) and channel and support structures (CSS). For the fulfilment of any function at least 2 WSP and a CSS connecting the WSPs must be designed. So far the method does not distinguish between function and behaviour or between wanted and unwanted functions. This broad definition allows the application of a simple modelling approach, but requires experience to apply it efficiently. C&CM has the potential to be employed in design synthesis. However, its main application has been so far in the analysis of existing systems as a preparation for a synthesis step.

Before the experiment the second author interviewed 14 of the subjects about their experience with C&CM and their notions of function. The experiment was designed to compare the analysis of an existing product by engineers familiar with C&CM with those by engineers less familiar with C&CM. 20 subjects took part in the experiment over a period of one month in the summer of 2009. The subjects (18 males, 2 female) were all design engineers and graduates from the University of Karlsruhe, thus all have received roughly the same design education. All subjects are working as researchers at the IPEK. Three subjects have more than 5 years post graduate experience. Two of them are experienced engineers who have been teaching C&CM to the other subjects. Twelve subjects have between 2 and 5 years post graduate experience and five subjects have less than 2 years.

The task of the experiment was to analyse a hydraulic pump that is employed in off road vehicles for lifting or moving heavy parts, for example the arm and shuffle of a digger. The product was selected as an example of a product, which is highly optimized for robustness and durability, and sufficient complex to reflect components that are designed by teams in industry. The subjects were divided into two groups. 11 subjects were given a physical model of the pump, which has been prepared for teaching as shown in Figure 1 by replacing a motor with a manual handle and cutting the top cover to expose the pumping mechanism. The subjects were allowed to manipulate the pump, but could not unscrew its components. The second group of 9 subjects was provided with a 2 D maintenance drawing (see Figure 2 on the right), which was obtained from the publically accessible product documentation. The subjects were given identical briefing notes, which asked them to imagine that they were a new engineer at the pump developing company and needed to find ways to improve the pump. They were asked to analyze this pump to assess the potential for improvement and summarize their understanding in a function tree.

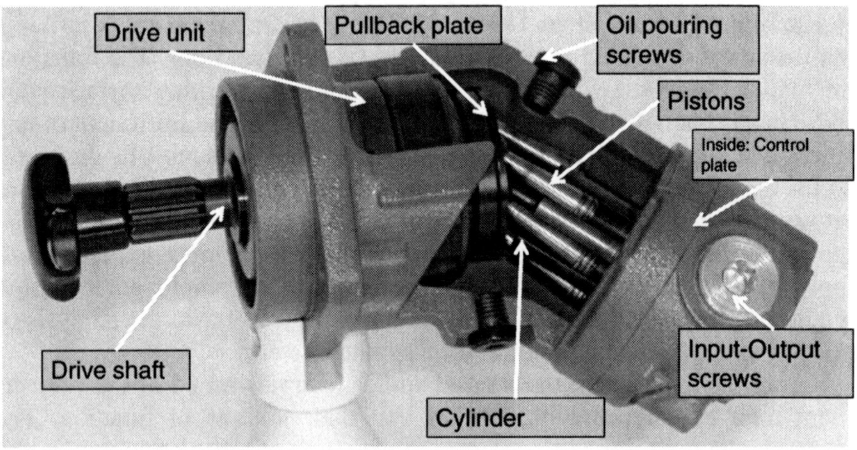

Fig. 1. Pump used in the experiment

All experiments where observed by the first two authors, recorded with two video cameras. The subjects were asked to explain the pump to one of the experimenters. To gain a protocol, that was as complete as possible the experimenters asked the subjects what they were thinking or looking at in order to keep them talking and asked them targeted questions to encourage them to fully explore the pump. While the experimenters might have biased the activity compared to a classical concurrent verbalization approach, the experiments yielded very rich protocols and the subjects commented that they experienced their explanations as a natural dialogue similar to a student coaching session.

The experimenters discussed the outcomes after each experiment and provided feedback to the subjects at the end of the experimental period.

The audio recordings were transcribed. The third author is analyzing the results as part of a diploma thesis. She has listened to the recordings, reviewed all sketches and completed the function trees with functions that were mentioned, but not written down. The assignment to a location in the function tree sometimes had to be extrapolated from the context in which they were expressed. The experiments were conducted in German. Component and function names are translated by the authors.

Results of the Experiment: Representations, Notions and Approaches

The brief only asked for the subjects to summarize their understanding in a function tree, but did not prescribe the form of representations. The subjects were provided with A1 paper and different coloured pens of different thickness. One subject requested post-it notes, so that he could rearrange the functions as his understanding grew. None of the subsequent subjects made use of the sticky notes. This section aims to provide a sense of the range of data produced, starting with two examples of typical representations, before discussing the ways functions were expressed and components are named. While an indication of the range of formal and informal notions of the functions is provided, an in-depth analysis of how different notions of the functions affect the analysis of the product is outside the scope of this paper. The section will conclude with an overview of the approaches people took to analyzing the product.

Typical Representation

The most obvious difference between the two groups was that the subjects who were given the 2D schematic drawing did not sketch, whereas ten out of eleven subjects in the other group drew sketches to understand or illustrate parts of the pump. Nineteen of the subjects drew traditional function trees, as illustrated in Figure 6; one subject did not draw any tree. Function trees arrange the functions in a hierarchical manner, where lower level functions contribute to the fulfilment of a higher level one. All but two subjects mentioned functions that they did not include in the function tree. The subjects using the maintenance drawing pointed out and marked the subsystems that carried out the functions they were speaking about.

Figure 2 shows a typical output from an experiment where the subject was given the maintenance drawing. On the left side in the figure the function tree can be seen as a box representation. Sub-functions are arranged inside the box of the main function. The Working Surface Pairs

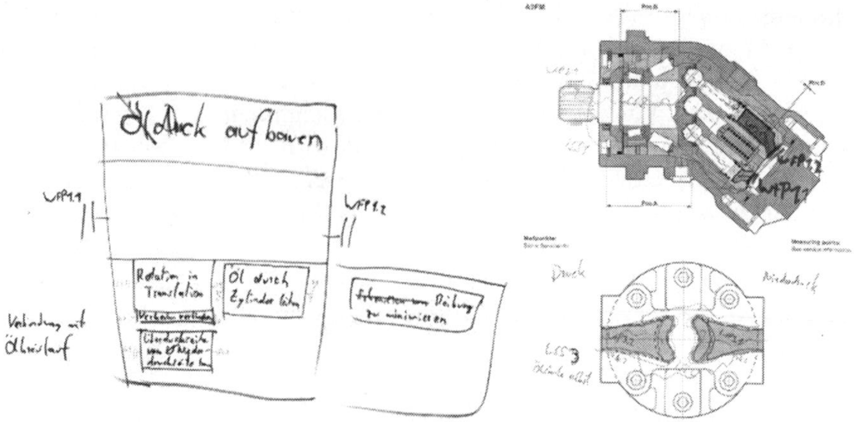

Fig. 2. Typical representation of a maintenance drawing subject

(WSP) and Chanel and Support Structures (CSS) are included in this function tree as part of a C&CM analysis. The right side shows the annotations the same designer added to the maintenance drawing.

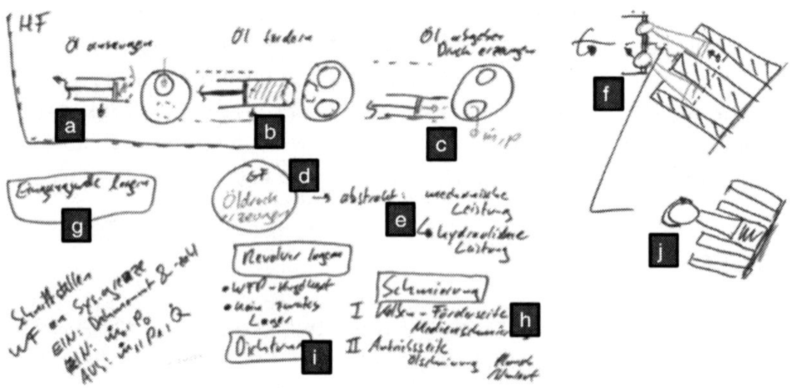

Fig. 3. Example of an analysis by subject using the physical pump

Figure 3 shows typical notes produced by one pump subject, who created additional drawings to clarify aspects of the pump. On the upper left side of the figure the main function of the system (HF = Hauptfunktion, i.e. main function) is broken down into three sub functions. These functions correspond to the sequence of the pumping operation: *"draw in oil"*(**a**), *"convey oil"* (**b**) and *"dispense oil/create pressure"*(**c**). To understand the piston movement, he created a sketch of one piston in different positions (**a**, **b**, **c**). As soon as the sketch was created, he identified further functions at a lower level of detail, for which he did not produce more sketches, but wrote down the functions (**g** = "*position drive unit*", **h** = "*lubrication*"). **f** shows the arrangement of the pistons that "*draw in oil*" and "*dispense oil*", two functions that occur sequentially, which makes him aware of the need to position the piston cylinder. **j** refers to the function of "*support the revolver*". **i** refers to lubrication, which is a function on a lower level in the function hierarchy, as it supports all other functions of the pump. The arrow from the upper drawing (see **f** to **j** in Figure 3) indicated a detailing of the analysis.

Different Notions and Expressions of Function

The homogeneous group of subjects had different notions of functions. The interviews indicated that these were partly derived from everyday use of language and partly from the different methodological approaches they had been exposed to. In consequence they chose different formats to express functions and were also working on different levels of abstraction. None of the subjects drew an explicit distinction between function and behaviour, as German language design methodologies do not include a distinction, in particular the VDI Guidelines [10]. A more detailed analysis of notions of function and how they affect the analysis is ongoing; however this section provides a flavour of the variability of the data.

Notions of Functions

Some subject had been asked what they understand by the term 'function' and how they would define it during the initial interviews. The remaining subjects were asks after the trail to avoid biasing before the analysis. The following Table 1 gives an impression of the range of answers. The subjects were asked to elaborate on their definition, but were not challenged on it. The left columns list the key concepts and the right columns elaborate the flavour of the answer. The partial nature of the answer reflects the subjects' statements.

Table 1 Variation in the notion of function

Concept	Given definitions
Goals	Utility of a system; Must fulfil an utility; Functions often also are requirements; Allows to fulfil the goal of the design process
Flow	Flow of energy, material and information
Syntactic	Need a verb; First noun, verb then Has subject
Transformation	Transformations of an input into an output; change in state;
Property of the function	Must be quantifiable; Must be related to form
Others	exist on the level of principal solutions; solution neutral description of what the system does

Expression of Functions

During their degrees the subjects had exposure to a variety of engineering methods, such as QFD or FMEA methodology, which implicitly compose different conceptualizations of the term function. In the teaching of C&CM functions were introduced through examples rather than a formal definition. The students learn to apply C&CM in the context of practical engineering tasks.

Some subjects expressed functions in very concrete terms, while others used very abstract descriptions. However, often the same subject used different levels of abstraction for the same function or rephrased a function in more abstract or more concrete terms later. The subject in Figure 3 even wrote down two different expressions. He started by expressing the main functions very concretely as *"draw in oil"*, *"convey oil"* and *"dispense oil/create pressure"*, which he summarized as the main function *"generate oil pressure"* (marked as **d** in Figure 3).

He then rephrased this on a more abstract level as "transform *mechanical capacity into hydraulic capacity"* (**e**), when he thought explicitly about what would be an appropriate way to phrase the function. Other subjects expressed the functions without any reference to the specific pump or the components that carry out the function. For example *"Transform rotation into translation"* does not provide any impression of which subsystem carries out the function. Some subjects state explicitly that they are trying to write functions as a *"solution neutral description of what the system does"*. All subjects agreed that the main function of the pump was pumping the oil. Two subjects additionally mentioned that the pump could also be a motor that *"transforms the flow of oil into a rotation"*. Table 2 illustrates the range of phrasing and the level of abstraction for the main function of the product they used.

Table 2 Abstract and concrete formulation of the main function

Level of abstraction	**Outcomes**
Concrete verbalization: (12)	Convey fluid, oil (6); Provide oil pressure, (4); Provide oil pressure for the system (1); Acceleration of fluid, providing pressure (1)
Abstract verbalization: (6)	Convert mechanical to hydraulic energy (2); Convert mechanical rotation in hydraulic pressure and flow (2); Convert power, Provide of a volume flow from A to B
Mixed: (1)	Providing oil pressure/oil flow; convert mechanical into hydraulic energy

Diverging Terminology for Parts

Subjects often pointed to components rather than naming them, or vaguely referred to generic terms like “thing” or “this part”, or names of very broad categories like “disk”. The oil was described on different levels of abstraction according to the designers’ views on functional descriptions, as oil, hydraulic fluid or medium. This appears to be related to the designer’s notation of function. Several commented that they would call the oil fluid or medium to be as abstract as possible, because they felt functions needed to be described as abstract as possible.

A complete analysis of the terminology used by the subjects has not yet been carried out. However, Table 3 gives a flavour of the variation of component names in translation. The left column uses the terminology of the English maintenance instructions. Several subjects used the same word for different components with very different functions, e.g. using disk for both the pullback plate and the control plate. Without a detailed analysis of the context when the words are mentioned three distinct naming strategies have emerged:

- Geometrical names, where the subject use the shape is a guide. Examples are: “plate” or “disk” for the pullback plate and “cone” for the pistons
- Functional names, where an assumed functions provides the name, e.g. “sealing ring” for the control disk.
- Analogical names, where they had encountered a similar looking objects in a different context, such as “pistons” in engines.

Many words could be interpreted as a geometric or as a functional term, as the term “Kolben”, the German word for piston. However without a detailed analysis of protocol, it is difficult to distinguish which meaning the subjects had in mind. Some commented on analogies when they were struggling to name the part.

Table 3 Variation in terminology for components

Components	**Word used**				
Oil	Oil	Fluid	Medium	Liquid	
Piston	Piston	Conelet	Cone	Plunger	Stamp
Cylinder	Cylinder	Cylinder-block	Socket	Cylinder-case drum	Revolver
Centre pin	Piston	Centre piston	Centre bolt	Plunger	Bearing pin
Pull back plate	Disk	Plate	Swash plate	Ball head fastener	
Control plate	Control disc	Disc	Valve plate	Plate	Split-washer
Shaft	Shaft	Input shaft	Drive shaft	Partial shaft	

Different Approaches

Subjects adopted different strategies to analyze the product. A more in depth analysis is on-going, which will look at their effectiveness. The following strategies have been identified so far:

- **Top-down (4 of 20 subjects):** Subjects analyze the overall system and proceed to further levels of detail. The focus becomes more detailed step by step.
- **Important things first (2 of 20 subjects):** Subjects focus immediately on what they perceive as the main function of the most relevant subsystem. The criteria for the decision are not explicit. The focus is chosen arbitrarily or based on experience.
- **Issue driven (7 of 20 subjects)**: First the subject creates an overview and then details specific issues, which seem obviously important or they find problematic. They write things down as they go along.
- **Power flow throughout the system (7 of 20 subjects):** They follow the power flow from the source of the power to interfaces with other parts of the system. The flow starts, where torque and rotational motion come into the system through the drive shaft. The rotation is then used to move the pistons in order to draw in and dispense the hydraulic oil.

Results: Different Functional Breakdowns

This section concentrates on the function trees provided by the subjects. For each experiment the third author went through the transcripts and drew out the functions that were mentioned but not written down. The authors generated a complete function tree based on all the functions mentioned. This was uses as a benchmark of completeness for the subjects' function trees. The section will also analyze the overlap between different function trees. The focus of this section is quantitative.

The Function Trees

The basic statistics of the functions produced by subjects is summarized in Figure 4. On average the subjects identified 11.3 functions (written plus verbalized). The largest number of functions written down was 17.

This subject, who had nearly 4 years experience, did not verbalize further functions. The other extreme was one subject, who wrote down only 1 function but verbalized a further 9 functions. He had just finished his degree, and had problem understanding the mechanism. This picture is characteristic for the groups. With experience the number of function written down increased. Subjects who were confident about their analysis, were more willing to commit it to paper. People using the pump also generated more functions than those who used the maintenance drawing.

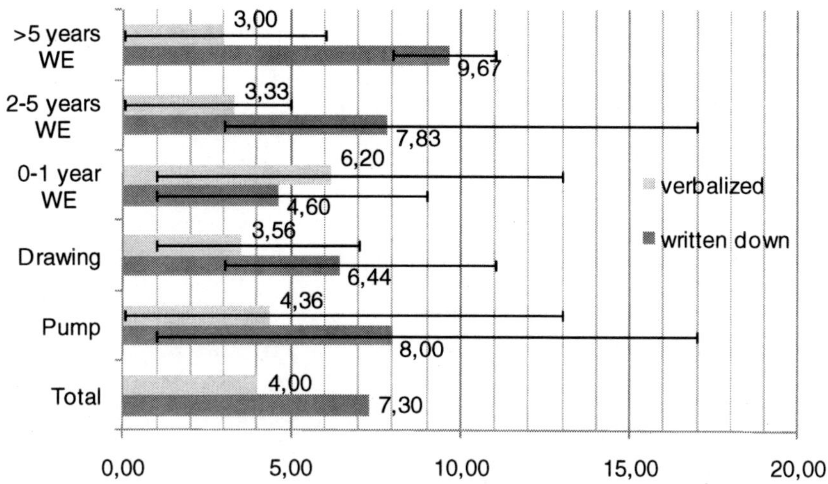

Fig. 4. Average number of written down and verbalized functions with their statistical spread

People using the pump also generated more functions than those who used the maintenance drawing. When the authors later created a more complete model, taking all the analyses into consideration and not limiting the time, they identified 46 functions.

A Comprehensive Model of the Pump

A comprehensive model, Figure 5, was generated from all the functions mentioned by the subjects, discarding those that were clearly false. The authors generated a detailed C&CM analysis to think through the logic of the product and identify the functions associate with each component.

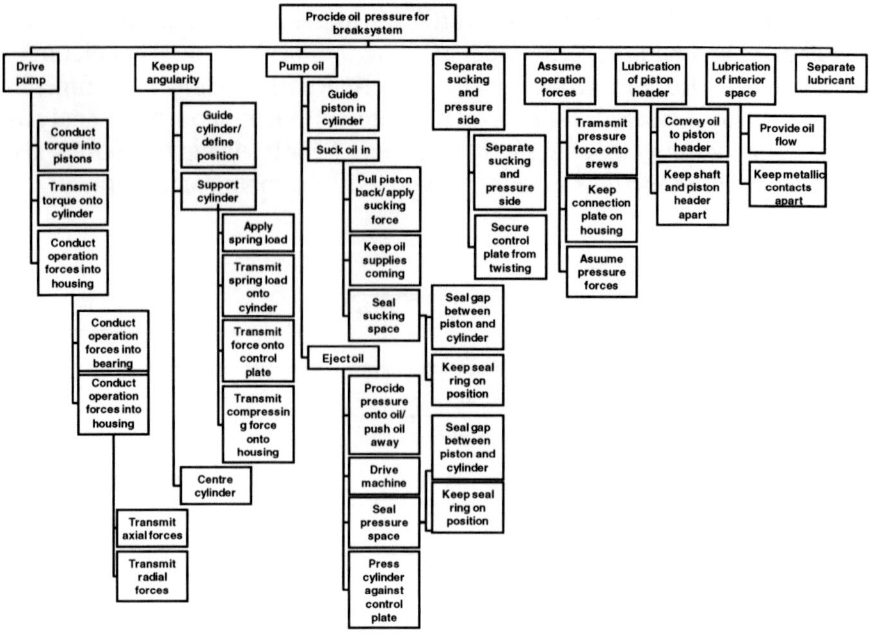

Fig. 5. Functions of the reference model

They used both the pump and the maintenance description. Thus they had a richer data set than the subjects. While many subjects commented they had understood the product sufficiently, none reached this level of detail.

Written versus Verbalized Functions

Figure 6 and Figure 7 show typical function trees. All function trees were first converted from the format they were written down, e.g. the box notation in Figure 2, into a tree structure, and displayed in a standardized format to allow visual comparison. All subjects mentioned more functions than they wrote down. The written functions are shown in black, connected with continuous lines. Verbalized functions are added to the function tree at the most suitable point, are shown in grey and are connected with dashed lines. There was considerable variation in the number of functions that were only verbalized, both for very rich descriptions and for fairly coarse descriptions.

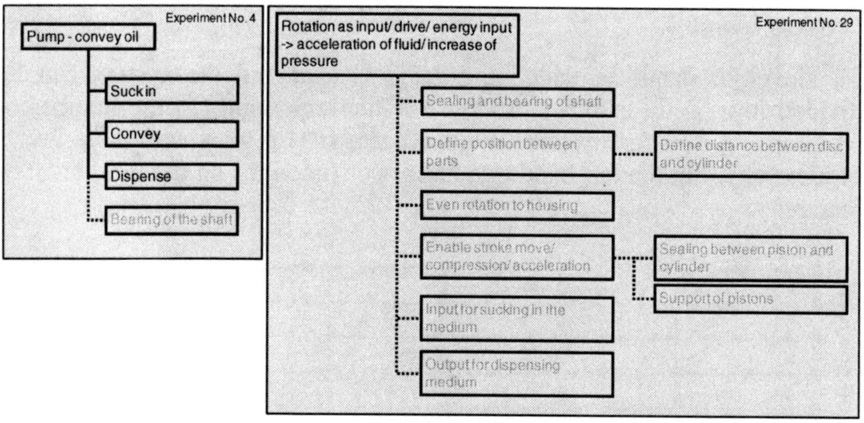

Fig. 6. Function tree on two levels and function tree with verbalized functions

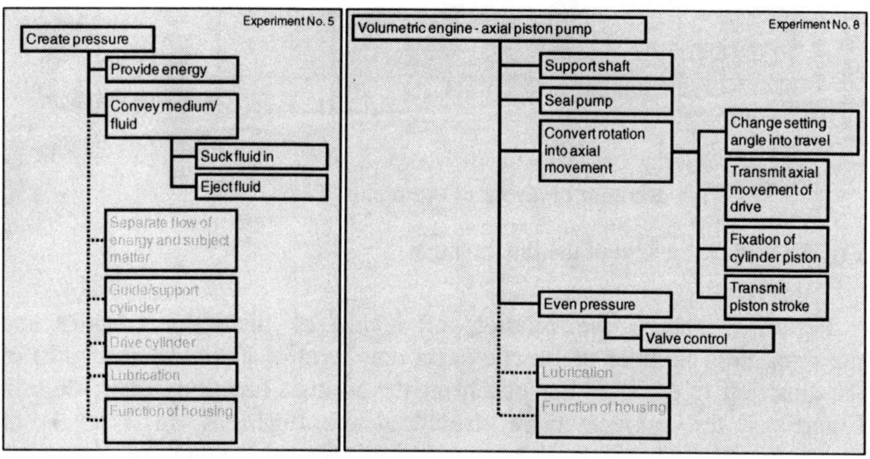

Fig. 7. Two three level function trees

Figure 6 shows two coarse descriptions. In the right hand example, only the main pumping operation was written. The subject really struggled. He was a fairly new staff member, and might have been lacking in confidence. He identified a key function using a mixture of abstract and concrete descriptions as "Rotation as input/drive/energy input → acceleration of fluid/increase of oil pressure". He then mentioned what he thought the key components did, sometimes in detail, but did not aim to summarize them as a tree.

Levels of Hierarchy

The level of detail in which the subjects analyzed the system can be divided into (i) the number of levels of hierarchy and (ii) the number of functions on a hierarchy level. Figure 7 shows two trees with three levels of hierarchy, but the left hand tree has fewer functions on the third level of hierarchy.

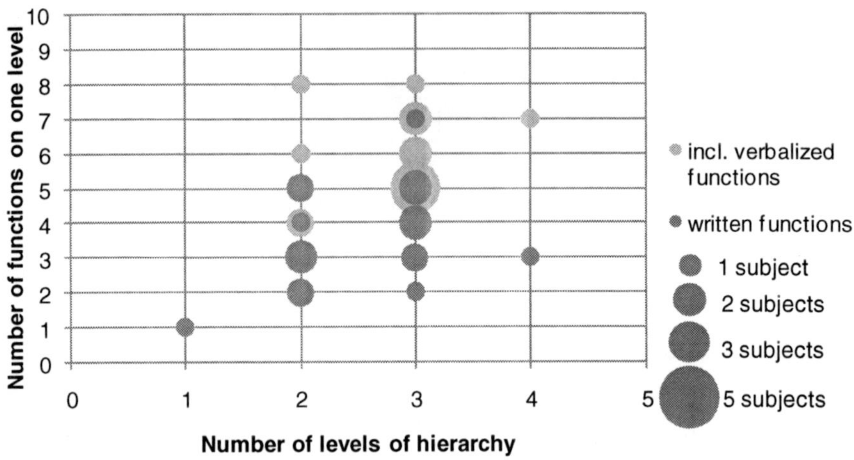

Fig. 8. Depth to breadth of the function tree

Figure 8 shows the relation of levels of hierarchy (depth) and the maximum number of functions on one level of hierarchy (breadth) of the function trees. Looking purely at the written functions (dark dots in Figure 8), ten subjects have structured the functions on 3 levels of hierarchy. Eight subjects elaborated the details on 2 levels of hierarchy. Only one subject chose a breakdown on 4 levels and only one subject wrote down only the main function. If additionally the verbalized functions are considered fifteen subjects break down the functions on 3 levels of hierarchy (bright dots in Figure 8) of whom eight have a maximum breadth of 5. Four subjects remain with 2 levels of hierarchy (bright dots in Figure 8). The functions on the third and fourth levels of hierarchy are predominantly "auxiliary functions" like *"Lubrication of part X"*, *"Pivoting of cylinder"*. The variability of the maximum number of functions on each level of hierarchy is higher than the number of levels of hierarchy. Most functions are assigned to the second level of hierarchy. Further detail on the third level of hierarchy was only provided for 1 or 2 second level functions (as in Figure 7). The lack of completeness on the third level of detail can be ascribed to (i) limited time in the experiment,

(ii) subjects finding that they understood the system's function to their satisfaction, or (iii) subjects being unable to advance because they could not develop sufficient understanding of how the pump worked on a more abstract level to generate a function tree as the case in Figure 6 on the right. The generation of the authors' own model required several iterations and restructuring of assigned functions, which was not possible within the time given to the subjects. The author's model has 5 levels of hierarchy, with 8 second level, 18 third level, 13 fourth level and 6 fifth level functions (see Figure 5).

Similarities in the Layout of the Function Trees

While the function trees look quite different at first glance there are some patterns of similarity amongst the trees.

Physical Arrangement of the Function Tree

While only seven subjects followed the power flow as a deliberate strategy, eleven subjects built a function tree that refers to the power flow through the system. The functions of the drive shaft and the bearing unit (combined as "drive unit" in Figure 1) are arranged on the left side. The flow of power then led the subjects to the right side, to the functions of the pump mechanism. In the example shown in Figure 9 the main stages of the power flow are drawn left to right. Lubrication is added, because it seemed important.

The physical pump, as well as the maintenance drawing also provides a standard alignment. The way the pump was prepared for teaching only affords a close look from one direction. The maintenance drawing has words and sentence on it, which also determines how the drawing must lie in front of the subjects.

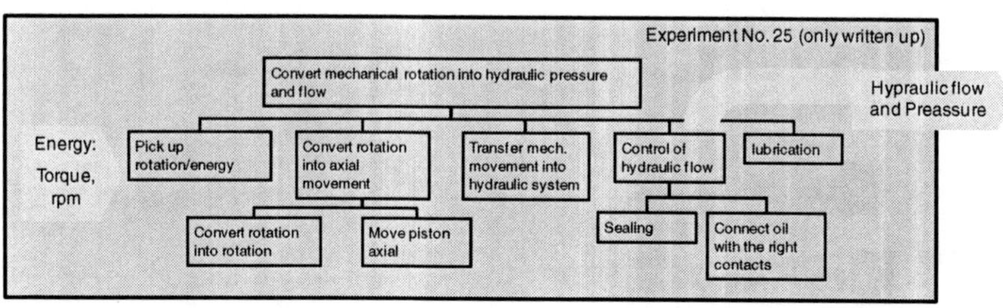

Fig. 9. Power flow reflected in physical layout of a function tree

As described in the section on “different approaches” seven subjects built up the function trees in an issue-driven way. They focused on subsystems that for them were obviously important. Four subjects positioned these important issues at the left or at the top of the function tree

Lubrication and sealing

Lubrication and sealing are often considered as auxiliary functions, which are required in order to maintain other functions, or rather to maintain the components which carry them out in running order. Half of the subjects included lubrication or sealing in their function trees. Only three linked them to the function tree below another function. The other seven subjects stated that they considered them as a different kind of function. They arranged them at the right side or the bottom of the function trees (six subjects, e.g. in Figure **7**) or they were marked as auxiliary functions (one subject) and thus were excluded from the function tree.

Structuring the Function Tree into a Sequence

Five subjects structured the function tree by describing the functions that happen at the same time in different pistons as a sequence (as in Figure 10: (**a**) *“drawing in oil”*, (**b**) *“conveying oil”* and (**c**) *“ejecting oil”*) for a single piston. That is the perspective of a small volume of oil flowing through the pump. Those descriptions are arranged top down in the function tree (and not from left to right) in four out of five cases.

This way of arranging functions in a sequence in order to model functions that are fulfilled at the same point in time is part of the C&CM approach, which the subjects had learned in a recent tutorial.

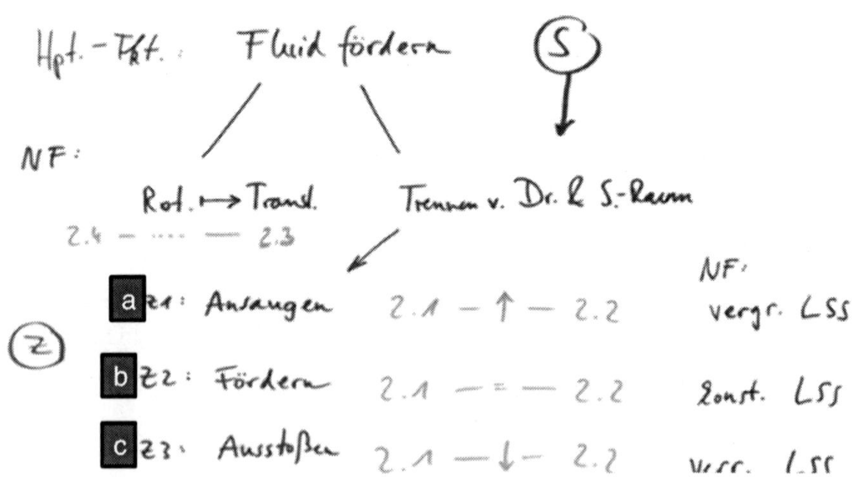

Fig. 10. Structure of functions in a sequence

Mistakes in the Trees

The maintenance drawing provided two cross-sectional views of the product. However, putting the two sections together was a challenge for all subjects. Only if they mentally map these two drawings together can they understand how the oil flows in and out of the pump. Several subjects assumed that the oil would flow in through the lubrication screws. Similar issues occurred around the central rod (see Figure 3 **j**), which holds the cylinder block in place. This was clearly visible on the drawing, but difficult to see on the pump. Logically it was clear that the location of the cylinder block had to be fixed, but many designers mistook it as the central rod for another piston.

While none of the functions trees provided a comprehensive analysis, nine trees did not include factual mistakes in the description. As Figure 11 shows, seven out of eleven subjects who used the pump made mistakes, and four out of nine who had used the drawing. Analysis of the causes of mistakes is ongoing; however the physical pump did not allow the designers to see important details, such as the control plate and the piston heads, so that the designers had to speculate. The less experienced designers were, on the whole they found fewer functions. They were mainly listing the three phases in the pumping sequence, leaving themselves less room for error.

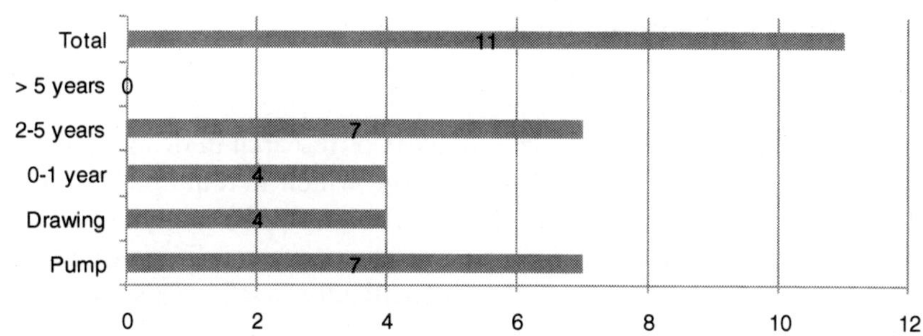

Fig. 11. Number of persons who made mistakes

Seven subjects made wrong assumptions based on the view of the product they were presented with. In particular the physical pump was cut open to provide a free view of the pumping system, i.e. the cylinder block, the casing, and the screws closing the lubrication inlet and outlet. In its normal state the pistons are gliding inside the cylinder. Being cut open it could not build up pressure and thus function as a pump. Several subjects only remarked on problems with understanding the system when they were asked about their assumptions how the pressure was built up by the experimenter. Only subjects working with the physical pump made that wrong assumption about the casing containing the liquid. However several subjects, using both the pump and the drawing, got confused about the oil inlet, which they interpreted as the place where the hydraulic fluid entered. For example Figure 12 shows on the left side the sketched shape of the control plate which *"separates the high pressure side from the low pressure side"*, which has kidney shaped inlets. Discrete holes would both limit the amount of oil drawn in and the amount of oil pushed out the cylinder. Two subjects working with the physical pump made this particular wrong assumption. One subject wrote the wrong understanding down possibly due to lack of attention, the other corrected himself.

Fig. 12. Wrongly assumed shape of the control plate and actual shape

Conclusions and Implications

Design is classically described as an iterative cycle of problem analysis, synthesis and evaluation [14].The emphasis of research in design cognition has been on synthesis, rather than analysis, which is required throughout the cycle, but predominantly during the problem analysis and the evaluation phases. Analysis of existing products is a very important part of many design processes and is not systematically taught. This experiment points to the variability of approaches and results in the analysis of products. The range of function trees and the differences in quality and level of detail have interesting implications for the shared understanding that a team might have. Most of the subjects left the experiment with the sense that they had understood the product to their satisfaction, but their understanding was very different. This could potentially cause problems in a joint design project, where each of them would bring their own divergent understanding without realizing that others might interpret the product in a different way. This is particularly critical as this group was far more homogeneous than many design teams, being graduates of the same university with very similar experiences.

The paper also provides empirical evidence for the relevance of the ongoing debate on the nature of functions. All the subjects in the experiment struggled with defining functions, and each of them had a slightly divergent definition of functions. None of the academic definitions has been adopted universally. This also sheds light on the challenges of introducing methodologies into industry which require a coherent understanding of functions.

The goal of this paper is to alert the wider design community to some of the challenges designers face in the analysis of existing products and in using the concept of function. The paper provides an overview of the findings of the hydraulic pump experiment. Detailed analyses of the data are ongoing, and future work will look in detail at the processes by which products are analyzed, the challenges the designers encountered, and the help they require in the future.

Acknowledgements

The authors would like to thank all people who participated as subjects in the experiment and also the student research assistants who were transcribing the records with great stamina.

References

1. Hacker, W.: Improving engineering design Ð contributions of cognitive Ergonomics. Ergonomics 40(10), 1088–1096 (1997)
2. Hinds, P., Weisband, S.: Knowledge Sharing and Shared Understanding. In: Gibson, C., Cohen, S. (eds.) Virtual Teams That Work Creating Conditions for Virtual Team Effectiveness. John Wiley & Sons, Jossey-Bass (2003)
3. Bucciarelli, L.L.: Designing Engineers. MIT Press, Boston (1996)
4. Arias, E., Eden, H., Fischer, G., Gorman, A., Scharff, E.: Transcending the individual human mind-creating shared understanding through collaborative design. ACM Transactions on Computer-Human Interaction (TOCHI) 7(1) (2000)
5. Hill, A., Song, S., Dong, A., Agogino, A.M.: Identifying Shared Understanding in Design Using Document Analysis. In: Proceedings of the 2001 ASME Design Engineering Technical Conferences, Pittsburgh, PA, DETC2001/DTM-21713 (2001)
6. Pahl, G., Beitz, W.: Engineering design: a systematic approach, 2nd edn., translated by Wallace, K., Blessing, L. and Bauert, F. Springer, London (1996)

7. Warell, A.: Introducing a use perspective in product design theory and methodology. In: Proceedings of the 1999 ASME Design Engineering Technical Conferences, DETC99/DTM8782, Las Vegas, NV (1999)
8. Crilly, N., Good, D., Matravers, D., Clarkson, P.J.: Design as communication: exploring the validity and utility of relating intention to interpretation. Design Studies 29(5), 425–457 (2008)
9. Gero, J.S., Kannengiesser, U.: The Situated Function-Behaviour-Structure Framework. In: Gero, J.S. (ed.) Artificial Intelligence in Design 2002, pp. 89–104. Kluwer, Dordrecht (2002)
10. Ingenieure, V.D.: VDI Richtlinie 2223, Methodisches Entwerfen technischer Produkte. Beuth, Berlin (2004)
11. Vermaas, P.E., Houkes, W.: Technical functions: a drawbridge between the intentional and structural natures of technical artefacts. Studies in History and Philosophy of Science 37(1): 5(18) (2006)
12. Kirschman, C.F., Fadel, G.M.: Classifying functions for mechanical design. Journal of Mechanical Design 120(3), 475–482 (1998)
13. Albers, A., Alink, T., Deigendesch, T.: Support of design engineering activity – The Contact and Channel Model (C&CM) in the context of problem solving and the role of modelling. In: Proceedings of the International Design Conference 2008, Dubrovnik, Croatia (2008)
14. Asimov, M.: Introduction to Design. Prentice-Hall, Englewood Cliffs (1962)

A General Knowledge-Based Framework for Conceptual Design of Multi-disciplinary Systems

Yong Chen, Ze-Lin Liu, and You-Bai Xie
Shanghai Jiao-Tong University, P.R. China

Designers are encouraged to explore in wide multi-disciplinary solution spaces for finding novel and optimal principle solutions during conceptual design. However, as cultivated in limited disciplines, they often don't have sufficient multi-disciplinary knowledge for fulfilling such tasks. A viable solution to this issue is to develop an automated conceptual design system so that knowledge from various disciplines can be automatically synthesized together. Since conceptual design is often achieved through reusing and synthesizing known principle solutions, this paper proposes a knowledge-based framework for achieving automated conceptual design of multi-disciplinary systems, which comprises three primary parts, i.e. a flexible constraint-based approach for representing desired functions, a situation-free approach for modeling functional knowledge of known principle solutions, and an agent-based approach for synthesizing known principle solutions for desired functions. A design case demonstrates that the proposed framework can effectively achieve automated conceptual design of multi-disciplinary systems. Therefore, designers can then be automatically guided to explore in multi-disciplinary solution spaces during conceptual design.

Introduction

Conceptual design is responsible for generating suitable PSs (abbreviated as PSs) for desired functions [1]. Here, PS means the basic physical mechanism of a system or sub-system for achieving a desired function. During the conceptual design stage, designers are encouraged to explore in wide multi-disciplinary solution spaces to find novel and optimal PSs for desired functions [1]. Therefore, they should have sufficient multi-disciplinary knowledge to fulfill conceptual design tasks, which is often a

J.S. Gero (ed.): *Design Computing and Cognition'10*, pp. 425–443.

big challenge for them since they have been merely cultivated in a single or few limited disciplines.

A possible solution to this issue is to develop an Automated Conceptual Design (abbreviated as ACD) system so that knowledge from various disciplines can be automatically synthesized together for the conceptual design of multi-disciplinary systems, which is just the primary motivation for this research.

There are still two other significant issues that motivate this research. One is that engineering design research discloses that human designers prefer to use familiar solutions to solve design problems [1]. Design preferences as such can often prevent designers from searching for optimal solutions in wide multi-disciplinary solution spaces. An ACD system can assist designers in overcoming this issue since it can generate PSs from wide multi-disciplinary spaces for them. The other is that new technical components are ceaselessly invented in a fast speed in the current knowledge-explosive era. Since these new components can either exhibit better performance or can achieve new functions, it is desirable that they can be immediately utilized for developing new systems or improving existing systems. However, since human designers often can't learn such components in time, system innovations as such are often delayed. This issue can also be overcome as long as the knowledge about such new components can be timely input into the knowledge base of an ACD system.

A general knowledge-based framework for achieving ACD of multi-disciplinary systems is proposed here. The premise of this research is that known PSs in various disciplines can be abstracted from existing systems as building blocks for future design synthesis. Since function often plays a crucial role in design synthesis, how to represent desired functions and the functional knowledge of known PSs is studied here. An agent-based design synthesis approach then briefly introduces how the proposed functional representation approaches can be employed to achieve ACD synthesis.

Representing Desired Functions

Since conceptual design is a kind of function-centered activities, functional representation is critical for developing an ACD system. Here, function refers to the general relationship between the input and the output of a system aiming at performing a task [1]. Functional representation in this paper is classified as two types, i.e. the representation of desired functions and the representation of functions achieved by known PSs. In this section, we will focus on how to represent a desired function.

There can be two primary approaches for representing a desired function, i.e. the verb-noun pair approach and the input-output flow approach. For conceptual design of multi-disciplinary systems, designers are particularly interested in how to transform flows from one discipline to another. Therefore, the latter approach is more eligible for representing desired functions here. However, this approach also has some limitations. For example, it can't represent the detailed features of related flows.

Here, we propose to represent a desired function as a set of constraints on the input and output flows. Obviously, the representation of a flow is its foundation, which therefore is elaborated as below at first.

The Representation of Flows

The first kind of information for representing a flow is its name. It is often the truth that a flow used in different disciplines or by different experts may be called different names. Therefore, a taxonomy approach is employed here to standardize the names of multi-disciplinary flows. Here, flow names are first collected from various disciplines. These names are then classified into different classes based on their physical meanings. The synonyms of the same kind of flows are then detected and removed.

Note that here we prefer to use the names of standard physical variables to indicate the names of the corresponding flows since it can easily smooth away the ambiguities during the standardization process. For example, instead of using the word "rotation" to indicate a flow's name, we use "Angular_Velocity" or "Angular_Displacement" to indicate the names of two different kinds of flows. As a result, a set of standard flow names are then developed, together with a flow taxonomy. A snapshot of the taxonomy is shown in Figure 1. Note that only the items in rectangles in the figure can be employed to indicate the name of a flow.

The flow name alone is not sufficient for representing a flow explicitly. Besides that, a flow also has some detailed features. For example, for the to-and-fro translation output by a slider-crank mechanism, the name of the flow is "Linear_Velocity", while its features include the to-and-fro feature, the orientation, etc. When representing the features of flows, a key point must be remembered, i.e. the flows in different disciplines may have different features. For example, the features of an electrical current flow are obviously different from those of a rotational flow. Therefore, the feature-representing mechanism should be flexible enough to allow the customization of different feature sets for flows in different disciplines.

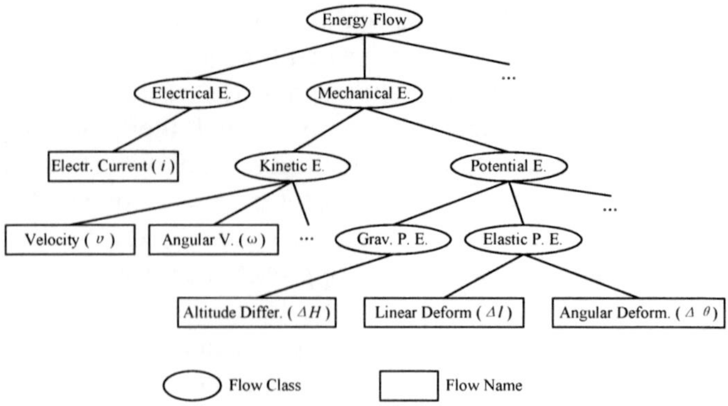

Fig. 1. A snapshot of the flow taxonomy

Here, the attribute-value approach is employed to represent the detailed features of a flow. Based on this approach, the features of a flow are represented as a set of attributes and values. For example, the to-and-fro feature of a translational flow can be represented as a combination of the attribute "*direction*" and the value "To-and-Fro". Standard attributes and values are provided here so that designers can represent the features of flows with unified terms. Note that the attributes of flows only have qualitative values here since conceptual design primarily deals with qualitative synthesis.

Based on the standard name and the related attribute-values, a flow can then be explicitly represented. Figure 2 shows the schemes for representing two flow classes from different disciplines, *Linear_Velocity* and *Electrical_Current*. In these schemes, the attributes are expressed in lowercase and italic font in the braces (e.g. "*orientation*"), while their possible qualitative values are expressed in the brackets with the first letters in uppercase font (e.g. "Constant").

Flow_Name: Linear_Velocity	*Flow_Name*: Electrical_Current
{ *stability:* [Constant, Variable];	{ *Type:* [AC, DC];
orientation: [X, Y, Z, X-Y, Y-Z, X-Y-Z];	*stability:* [Constant, Variable];
direction: [Positive, Negative, To-and-Fro];	*direction:* [Positive, Negative, To-and-Fro];
intermittence: [Continuous, Intermittent] }	*intermittence:* [Continuous, Intermittent] }

Fig. 2. Two cases of the flow-representing schemes

Note that the two representation schemes in Figure 2 are general ones and don't refer to a concrete flow. When representing a concrete flow, a designer should select a suitable value for each of its attributes according to its features. For example, a translational motion in a specific situation can be represented as "Linear_Velocity {*stability*: Variable; *orientation*: X; *direction*: To-and-Fro; *intermittence*: Continuous}".

The Representation of Constraints on Flows

A constraint on a flow denotes what feature it should have. Based on the attribute-value representation, a constraint can be represented as an equation. For example, the constraint that a translation should have a to-and-fro feature can be represented as "*direction* (Linear_Velocity) = To-and-Fro", where translation is regarded as a "Linear_Velocity" flow, "*direction*" is the related attribute, and "To-and-Fro" is the value that it should have.

It is also possible that the attribute of a flow in a constraint can have multiple values. For example, when the orientation of a Linear_Velocity flow is constrained to "X" or "Y", the corresponding constraint can then be represented as "*orientation* (Linear_Velocity). = X || Y", where "||" denotes the logical OR relation between the values.

Note that it is not allowed to represent a constraint as an inequality here since it will make an ACD system unable to process the constraint information effectively. For example, a designer can't represent a constraint as "*direction* (Linear_Velocity) ≠ To-and-Fro". Instead, such inequality-based constraints should be converted to equation-based constraints in advance. For example, the above inequality constraint can be converted as "*direction* (Linear_Velocity) = Positive || Negative".

The Representation of Desired Functions

A desired function here is represented as a set of constraints on the input and output flows. Here, the constraints on the input flows indicate what flows can be input into the system to be developed, while those on the output flows refer to the goal flows desired.

A schema is proposed here for representing desired functions, as shown in Figure 3. Here, the "Semantic_Description" is provided for designers to define desired functions, which will not be processed by an ACD system. Note that the constraints on the flows here are conceptualized as a set of binary groups, (Attr_Name, Constrained_Values), since this kind of representation can be easily implemented in a relational database system. Together with the schema, how to represent the function, to convert a rotation in X- or Y- orientation into a to-and-fro translation in Z orientation, is also shown in the figure as an illustrative example.

As seen in the above example, a desired function can be explicitly represented with the aid of the proposed functional representation schema. Our approach has two major advantages. One is that the attribute-value representation enables designers to define the related flows in various disciplines with more details. The other is that the constraint-based representation makes them to define the flows' features more flexibly.

FUNCTION
{ *Semantic_Description*
Input_Flow_Name
/**A list of constraints on the input flow**/
[(Attr_Name, Constrained_Value*)]*
Output_Flow_Name
/**A list of constraints on the output flow**/
[(Attr_Name, Constrained_Value*)]* }

(a) Schema for representing desired functions

FUNCTION
{ *Semantic_Description:* "Convert rotation to translation"
Input_Flow_Name: "Angular_Velocity" ;
[(*magnitude,* "Constant")
(*axial_orientation,* "X" || "Y")
(*direction,* "Clockwise" || "Anticlockwise")
(*intermittence,* "Continuous")]
Output_Flow_Name: "Linear_Velocity" ;
[(*magnitude,* "Variable")
(*orientation,* "Z")
(*direction,* "To-and-Fro")
(*intermittence,* "Continuous")] }

(b) A case of the representation of a desired function

Fig. 3. The schema for representing a desired function

Representing the Functional Knowledge of Known PSs

It is generally agreed that designers often reuse known solutions for design synthesis [14, 15]. Note that since designers seldom reuse the total PS of a total system during conceptual design, PS here primarily refers to the PS of a sub-system, which can achieve an independent function. Therefore, the total PS of a product often should be decomposed into multiple PSs to facilitate their reuse. For example, multiple PSs (such as *AC-Motor*, *Slider-Crank-Mechanism,* etc.) can be abstracted from a punching machine.

As is well-known, conceptual design is a kind of function-centered activities. In order for an ACD system to reuse known PSs, these PSs should be indexed with their functions in the knowledge base. Therefore, how to represent the function(s) of a known PS is the critical issue here. A possible approach is to represent the function(s) of a known PS as the input-output flow pair together with their features. For example, based on the attribute-value representations of the Angular_Velocity and Linear_Velocity flows, the function achieved by a slider-crank mechanism in a specific situation can be represented as: (Angular_Velocity {*stability*: Constant; *orientation*: X; *direction*: Clockwise; *intermittence*: Continuous}, {*stability*: Variable; *orientation*: Y; *direction*: To-and-Fro; *intermittence*: Continuous}).

Representing the function of a known PS as above has a major pitfall, i.e. it merely denotes the function achieved by a known PS in a particular situation. It is often the case that a known PS can be deployed in different situations for achieving different specific functions. For example, besides the above function, a slider-crank mechanism can also be employed to achieve another function, (Angular_Velocity {*stability*: Variable; *orientation*: Y; *direction*: Clockwise; *intermittence*: Continuous}, Linear_Velocity {*stability*: Variable; *orientation*: Z; *direction*: To-and-Fro; *intermittence*: Continuous}).

Therefore, a situation-free functional representation approach should be developed so that a known PS can be reused for design synthesis in multiple situations. For clarity, the functional information represented with this approach is called the functional knowledge of a known PS. The functional knowledge of a known PS involves four parts, i.e. the input-output flow name pair, the constraints on the input flow, the constraints on the output flow, and the attribute-mapping rules between the input and the output flows, which will be elaborated separately as below.

The Input-Output Flow Name Pair

The input-output flow name pair is employed here to represent the transformation of the flow's name from input to output achieved by a known PS. It can be represented as a binary group (I_F, O_F), where I_F and O_F denote the input flow's name and the output flow's name, respectively.

With the aid of the input-output flow's name pair, the corresponding functional knowledge of a slider-crank mechanism can then be represented as (Angular_Velocity, Linear_Velocity), where "Angular_Velocity" denotes the name of the input rotation while "Linear_Velocity" that of the output translation.

Obviously, the input-output flow name pair is a general and rough description of the function achieved by a known PS. It doesn't involve the change of the specific features of the related flows.

The Constraints on Input Flows

As described before, a flow's attribute can have multiple values. However, not all values are acceptable for the input flow of a known PS. For example, the attribute *type* of an Electrical_Current flow can have two possible values, i.e. "AC" or "DC". Here, "AC" and "DC" refer to alternative current and direct current, respectively. However, only the value "DC" is acceptable for the known PS *DC-Motor*.

Therefore, it is necessary to use the constraints on the input flow to denote what values are acceptable for a known PS. Constraints on an input flow can also be represented as equations here, just as the constraint representation mentioned in section 3.2. For example, the above constraint on the attribute *type* of the input Electrical_Current flow can be represented as "*type* (Electrical_Current) = DC". When multiple values are acceptable for a flow's attribute, they can be linked with the symbol "||", which indicates the logical OR relation between them. For example, since the Angular_Velocity flow of a slider-crank mechanism can be either a clockwise or an anticlockwise one, the corresponding constraint can then be represented as "*direction* (Angular_Velocity) = Clockwise || Anticlockwise".

The Constraints on Output Flows

Similar to the input flow of a known PS, not all values of a flow's attribute are meaningful for the output flow of a known PS. For example, although the attribute *direction* of a Linear_Velocity flow can have three different values, "Positive", "Negative" and "To-and-Fro", only the value "To-and-Fro" is meaningful for the output translation of a slider-crank mechanism since this PS can merely output to-and-fro translation.

Therefore, the constraints on an output flow are employed here to denote the meaningful values of the output flow's attributes for a known PS, i.e. what values an attribute can have for its output flow. The constraints on an output flow can also be represented as an equation. For example, the above constraint on the output Linear_Velocity flow can be represented as "*direction* (Linear_Velocity) = To-and-Fro". Similarly, the constraint on the attribute *orientation* of the output Linear_Velocity flow of this mechanism can be represented as "*orientation* (Linear_Velocity) = X || Y || Z" since its value can be any one of "X", "Y" and "Z".

The Attribute-Mapping Rules

The function of a PS is to transform its input flow into an output flow desired by a designer. Such a transformation often involves not only the names of the related flows, but also the values of their attributes. For example, the function achieved by a slider-crank mechanism in a particular situation involves not only the flow name transformation from "Angular_Velocity" to "Linear_Velocity", but also the transformation of the related orientations. Obviously, such kind of transformation knowledge should also be represented as a part of the functional knowledge of a PS.

Here, we employ production rules to represent such kind of transformation knowledge. As such rules deal with the attribute's mapping relations between the input and output flows of a known PS, they are called attribute-mapping rules here. According to the artificial intelligence research [15], a production rule can be represented as: *IF* (*precondition*) *THEN* (*action*), where the *precondition* part denotes under what situation the rule can be activated, while the *action* part indicates what actions will be taken.

For an attribute-mapping rule of a known PS, the *precondition* part denotes what value the related attribute of the input flow should be in order for the rule to be activated, while the *action* part indicates what value(s) the corresponding attribute of the output flow can have after the execution of the rule. For example, when a solar battery transforms continuous sunlight (a Visual_Light flow) into continuous current, an attribute-mapping rule about the continuity can then be written as: "IF (*intermittence* (Visual_Light) = Continuous), THEN (*intermittence* (Electrical_Current) = Continuous)", where *intermittence* is an attribute, while "Continuous" is the corresponding value.

Note that it is possible that one value of an input flow's attribute can correspond to multiple values of the related attribute of the output flow in an attribute-mapping rule. For example, for a slider-crank mechanism, when the input rotation's axial-orientation is "X", the orientation of the output translation can then be either "Y" or "Z", depending on what orientation the mechanism is deployed in. Such mapping knowledge can be represented as: "IF (*axial_orientation* (Angular_Velocity) = X), THEN (*orientation* (Linear_Velocity) = Y || Z)".

It should be pointed out that either the *precondition* part or the *action* part of an attribute-mapping rule is restricted to involve only one attribute. This restriction is based on the following logical foundation: since the attributes of a flow are usually irrelevant to each other, the attribute-mapping rules about one attribute can be independent of those about another one. For example, since the attributes *orientation*, *intermittence*, and *direction* of the flow *Linear_Velocity* are irrelevant to each other, the related rules can then be represented independently.

The General Functional Knowledge Representation Schema

Based on the above research, a general schema can then be developed for representing the functional knowledge of a known PS, as shown in the left part of Figure 4. Note that the *precondition* part and the *action* part of a known PS here are also both represented as binary groups, *(Attr_Name, Constrained_Values)*, to facilitate its implementation.

In Figure 4, how to represent the functional knowledge of a slider-crank mechanism is also given as an example. For this PS, the input flow and the output flow are regarded as "Angular_Velocity" and "Linear_Velocity", respectively. The constraint on the output flow, *(magnitude, "Variable")*, means the magnitude of the output flow can't be constant. Rule *1* means that if the value of the attribute *axial-orientation* of the input Angular_Velocity flow is "X", the value of the attribute *orientation* of the output Linear_Velocity flow can be either "Y" or "Z". Rule *m* means that if the input flow is continuous, the output flow will also be continuous.

Function
  Principle_Solution
  Input_Output_Flow_Name_Pair
    (*InFlowName, OutFlowName*)
  Constraints_On_Input_Flow_Nane
    {(*Attr_Name, Attr_Values*)}
  Constraints_On_Output_Flow_Name
    {(*Attr_Name, Attr_Values*)}
  Production_Rules
    {[Rule ID:
      IF (*Precondition*) THEN (*Action*)]
    }

(a) The functional representation schema of a PS

Function
  Principle_Solution: *Slider_Crank_Mechanism*;
  Input_Output_Flow_Name_Pair
    (*Angular_Velocity, Linear_Velocity*)
  Constraints_On_Input_Flow_Name
  { (*magnitude*, "Constant" || "Variable"),
    (*axial_orientation*, "X" || "Y" || "Z"), ···}
  Constraints_On_Output_Flow_Name
  { (*magnitude*, "Variable"), ···}
  Production_Rules
  { [Rule 1: IF (*axial_orientation*, "X")
      THEN (*orientation*, "Y" || "Z")]; ...
   [Rule m: IF (*intermittence*, "Continuous")
      THEN (*intermittence*, "Continuous"] ; ···}

(b) A functional representation case of a PS

Fig. 4. The functional representation schema of a known PS

Based on the above functional representation schema, the functional knowledge of a known PS can then be explicitly represented in a situation-free manner, i.e. its functional knowledge is independent of any concrete situation it is deployed in. As a result, it is then possible for an ACD system to reuse these known PSs for design synthesis.

An Agent-Based Design Synthesis Approach

As seen above, the desired functions and the functional knowledge of known PSs have been represented with different models. Therefore, it is then difficult for the ACD system to judge whether a PS can match with a desired function, which means that it is difficult to use traditional search technologies to achieve automated conceptual design synthesis here. Based on the agent-based search technology, we develop an intelligent approach for achieving automated conceptual design of multi-disciplinary systems. Due to limited space, only its fundamental process is introduced here.

The Agent-Based Design Synthesis Process

According to the artificial intelligence research, the basic working mechanism of an agent-based system can be regarded as an iterative sense-action process [15]. At first, an agent senses its surrounding environment, and then it selects a suitable action to change its state. The sense-action process will continue until its new state reaches the desired goal.

When the agent-based approach is employed here to fulfil the conceptual design synthesis process, the known PSs in the knowledge base can be regarded as the agent's action tools for transforming flows, the input flow of a desired function can be regarded as the agent's initial environment, the output flow can be regarded as the goal environment that it wants to reach, and the agent's aim is to find a set of PSs to transform it from the current environment to the goal environment.

Based on the agent-based problem-solving approach, the fundamental conceptual design synthesis process can be briefly described as below:

- Step 1: A designer inputs a desired function represented with the constraints on the input and output flows and prescribes the maximal search depth;
- Step 2: According to the constraints on the input flow, the ACD system constructs a set of flows and puts them into its environment.
- Step 3: Through sensing its environment, the system selects an environmental flow that has never been explored before and that doesn't exceed the maximal search depth; if successful, set it as the current flow; else, go to step 6
- Step 4: The system then selects all suitable PSs that can act on the current flow, and use their functional knowledge to act on the current environmental flow one by one, resulting in some new output flow (s).
- Step 5: The system puts the new output flow(s) into its environment, and then return to step 3;
- Step 6: According to the constraints of the desired function on the output flow, the system selects an flow that satisfies these constraints flow from its environment; if successful, set it as the current flow for backtrack, and continue; else, go to step 9 ;
- Step 7: The system traces the flow-transforming path back from the current flow, with a result of a sequence of flows and a set of PSs that enable such transformations;
- Step 8: The system then remove the current flow from its environment and return to step 6;
- Step 9: If there have been combinatorial PSs generated, exit with success; else, exit with failure.

Note that the exhaustive search strategy is employed here to develop the agent-based synthesis process in order that the ACD system can generate optimal PSs in wide multi-disciplinary solution spaces. To prevent the ACD system from falling into an endless cycle, the above process uses the predetermined maximal search depth to control the search cycle. In addition, since the agent can only sense a concrete flow, flows represented with constraints should be converted into concrete flows before being put into the environment. This rule also fits for the result of a PS's action on an environmental flow.

An Illustrative Case

Assume that a designer wants to design a toy boy that can bow iteratively when it is turned on. As environmental issues are becoming more and more important, he tends to use solar light energy to drive the toy. Obviously, this conceptual design task deals with knowledge from multiple disciplines, which is suitable for demonstrating the above agent-based design synthesis approach.

Defining the Desired Function

At first, the designer should define the desired function for the ACD system. Here, the input flow of the desired function is solar light, while the output flow, i.e. the bowing behaviour of the toy, is a sway motion. Based on the proposed functional representation approach, the above desired function can then be represented in a form shown Figure 5.

```
FUNCTION
{  Semantic_Description: "Convert solar light into a to-and-fro
                          translation in the X-or Y-orientation" ;
   Input_ Flow_Name: "Visual_Light" ;
      [(stability, "Constant" || "Variable" )
      (intermittence, "Continuous" )
      (type, "Hot_Light" ) ]
   Output_ Flow_Name: "Angular_Velocity" ;
      [(magnitude, "Variable" )
      (axial_orientation, "X" || "Y" )
      (direction, "Bi-Direction" )
      (intermittence, "Continuous" ) ]                }
```

Fig. 5. The input and output flows of the desired function

Note that based on the standard flow name vocabulary, solar light is regarded as a Visual_Light flow here, while the sway motion is regarded as an Angular_Velocity flow with a to-and-fro feature.

The PS Knowledge Base

The PSs in the knowledge base comes from existing engineering systems. To illustrate this case, ten PSs from several disciplines are given here, which will serve as the PS knowledge base of the ACD system.

The first PS is *Solar-Battery*. Its function is to convert solar light into electrical current. Note that its output is DC, other than AC. The second PS is *DC-Motor*. Its function is to convert electrical current to rotational motion. The third PS is *AC-Motor*. Similar to *DC-Motor*, it can also convert electrical current to rotational motion. From the viewpoint of function, the primary difference between them is that *AC-Motor* takes AC as its input while *DC-Motor* takes DC as its input. The fourth PS is *DC-to-AC-Inverter* from the electronics discipline, which can convert AC to DC. The fifth PS is *Rack-Pinion*, a mechanism that can either transform rotational motion into translational motion, or transform translational motion into rotational motion. The sixth PS is *Crank-Rocker*, which can transform rotational motion into to-and-fro sway. The seventh PS *Parallel-Pulley-Belt* and the eighth PS Spur-Gear-Pair are usually for increasing or decreasing the rotational speed. The tenth PS is *Fluorescent-Lamp,* which can transform AC into light. Due to limited space, we are sorry that we can't present the pictures of these PSs here.

The ACD Process

Based on the proposed agent-based design synthesis approach, the primary conceptual design process is as follows.

First, the ACD system transforms the constraint-based representation of the input flow into a set of environmental flows with detailed features. Since only the attribute *stability* of the Visual_Light flow has two values (i.e. "Constant" and "Variable"), two environmental flows can then be constructed, i.e. "{*stability*: Constant; *intermittence*: Continuous; *Type:* Hot_Light}" and "Visual_Light {*stability*: Variable; *intermittence*: Continuous; *Type:* Hot_Light}".

Second, the ACD system begins to sense its environment, resulting in that the first environmental flow, "Visual_Light {*stability*: Constant; *intermittence*: Continuous; *Type:* Hot_Light}", is detected as the current flow. The system then analyzes the constraints of each known PS on its input flow to find suitable PSs for acting on the current flow. As a result, the PS *Solar-Battery* is then selected as the eligible PS.

Third, the ACD system employs the functional knowledge of the selected PS to act on the environmental flow. Based on its input-output flow name pair, the system can know that this PS will output an Electrical_Current flow. The primary constraints on the output flow of this PS are that its direction must be positive and that the current type must be

DC. For the current flow, this PS also have two executable attribute-mapping rules, i.e. “IF (*stability* (Visual_Light) = “Constant”), THEN (*stability* (Electrical_Current) = “Constant”)”, and “IF (*intermittence* (Visual_Light) = “Continuous”), THEN (*intermittence* (Electrical_Current) = “Continuous”)”. Based on these two rules, the ACD system can know that the values of the attributes *stability* and *intermittence* for the output Electrical_Current flow are “Constant” and “Continuous”, respectively. As a result, the system can generate an output flow for this PS, “Electrical_Current {*stability*: Constant; *intermittence*: Continuous; *direction:* Positive; *Type*: DC}”, which will be further put into its environment.

The ACD system will continue to sense its environment until all environmental flows have been explored, i.e. it will get the environmental flow next to the current one, set it as the current flow and find suitable PSs to act on it again. For example, when the system senses the above flow just generated by the PS *Solar_Battery*, it can then select *DC_Motor* to act on it. As a result, some new output flows will be generated again, such as, “Angular_Velocity {*stability*: Constant; *intermittence*: Continuous; *axial-orientation*: X; *direction:* Clockwise}”, “Angular_Velocity {*stability*: Constant; *intermittence*: Continuous; *axial-orientation*: Y; *direction:* Clockwise}”, etc. And when the flow, “Angular_Velocity {*stability*: Constant; *intermittence*: Continuous; *axial-orientation*: X; *direction:* Clockwise}”, is selected as the current flow for further exploration, the PS *Crank-Rocker* can then be selected to act on it. As a result, a new output flow, “Angular_Velocity {*stability*: Variable; *intermittence*: Continuous; *axial-orientation*: X; *direction:* To-and-Fro}” will be generated.

Finally, when the above search process ends, the ACD system will verify whether there are some flows in its environment that can satisfy the constraints on the output flow of the desired function, and then output the combinatorial PSs. For example, when the flow, “Angular_Velocity {*stability*: Variable; *intermittence*: Continuous; *axial-orientation*: X; *direction:* To-and-Fro}”, is verified as a satisfying output flow, the system will then get the corresponding previous flows through a backtracking process and chain them as a flow sequence, e.g. “Angular_Velocity {*stability*: Variable; *intermittence*: Continuous; *axial-orientation*: X; *direction:* To-and-Fro} ← Angular_Velocity {*stability*: Constant; *intermittence*: Continuous; *axial-orientation*: X; *direction:* Clockwise} ← Electrical_Current {*stability*: Constant; *intermittence*: Continuous; *direction:* Positive; *Type*: DC} ←Visual_Light {*stability*: Constant; *intermittence*: Continuous; *Type*: Hot_Light}”. Corresponding to this flow-transforming chain, the combinatorial PS is “*Crank-Rocke*r ← *DC-Motor* ← *Solar-Battery*”.

The conceptual design synthesis process with more details is shown in Figure 6. Here it is assumed that the maximal search depth is 3. For this maximal search depth, only one possible combinatorial PS can be generated with the available PSs in the knowledge base, i.e the combinatorial PS mentioned above.

However, if the designer is unsatisfied with the above combinatorial PS, s/he can increase the maximal search depth and launch the conceptual design synthesis again. For example, the maximal search depth is set as 4, more combinatorial PSs can be generated, such as "*Solar-Battery* → *DC-Motor* → *Spur-Gear-Pair* → *Crank-Rocke*r", "*Solar-Battery* → *DC-Motor* → *Crank-Slider* → *Rack-Pinion*", "*Solar-Battery* → AC-To-DC-Inverter → *AC-Motor* → *Crank-Rocker*", etc.

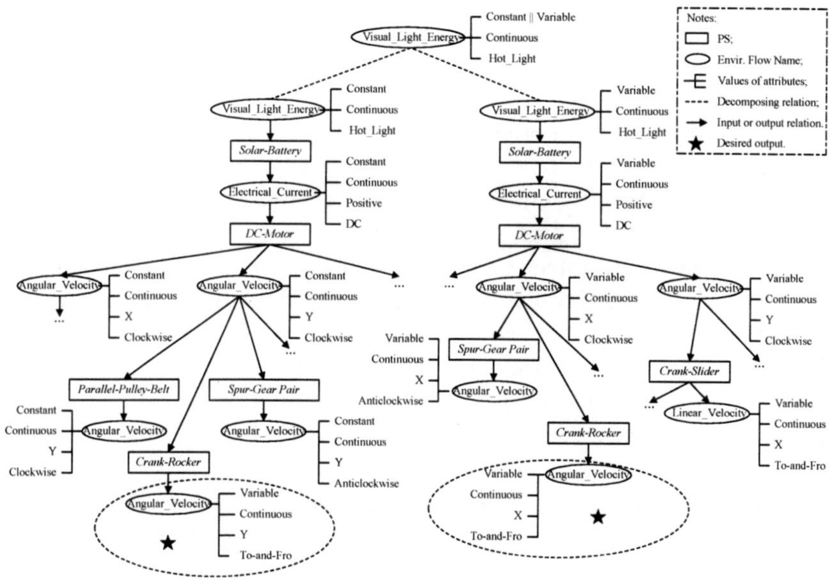

Fig. 6. Illustration of a conceptual design synthesis process

The above design case demonstrates that the general knowledge-based conceptual design framework developed here can employ known PSs from various disciplines to generate suitable PSs for a desired function effectively. For example, the combinatorial PS generated in the above case, "*Solar-Battery* → AC-To-DC-Inverter → *AC-Motor* → *Crank-Rocker*", is composed four PSs from completely different disciplines. During the conceptual design synthesis process, the proposed framework can not only consider whether different PSs can match with respect to the flow names, but also can identify whether they are compatible with respect to their specific features. For example, the ACD system can know the functional difference between *AC-Motor* and *DC-Motor*, and can employ the correct PS for design synthesis. As a result, designers unfamiliar with multi-disciplinary knowledge can obtain more reliable support.

Discussion

Existing computer-aided conceptual design systems primarily falls into two categories, i.e. interactive conceptual design systems and automated conceptual design (ACD) systems. An interactive conceptual design system first retrieves one or multiple existing design solutions for the design problem at hand from its knowledge base according to a designer's inquiry, and then the designer selects a suitable solution and adjusts it according to the current design problem. Typical systems of such a kind are reported in Ref. [3, 4]. A major drawback of such systems is that they can are prone to design bias towards familiar PSs since the function-decomposing process and the PS-searching process largely depend on the knowledge and design experience of designers.

Different from interactive conceptual design systems, ACD systems can independently fulfill the functional reasoning and solution-searching tasks. Typical systems of this kind can be found in Ref. [5-12]. Compared with the former kind of systems, ACD systems have two major advantages. One is that they don't require that designers should have full knowledge about the future PS. The other is that they can achieve automated design synthesis without any biases. Obviously, both advantages are of great value for the conceptual design of multi-disciplinary systems. Therefore, this paper also focuses on how to achieve ACD with a knowledge-based approach.

Compared with the existing ACD approaches, our knowledge-based conceptual design approach has multiple advantages. We briefly compare it with two primary kinds of existing ACD research as below.

One is the research on how to achieve ACD of mechanical devices [5-7, 16]. A significant feature of this kind of work is that it usually employs domain-specific approaches to represent and reason about functions. For example, to achieve automated synthesis of mechanisms, Kota and Chiou [5] have proposed a motional matrix approach for representing functions and a matrix-decomposing approach for reasoning about them. Obviously, such approaches are domain-specific and merely suitable for the conceptual design of mechanisms. Different from such kind of work, the approach proposed in this paper is independent of any specific discipline. As seen in the case, the ACD can use PSs from various disciplines for design synthesis, which is therefore suitable for achieving ACD of multi-disciplinary systems.

The other is the research on how to achieve ACD of multi-disciplinary systems [8-11]. Since these approaches usually employ the bond graph theory to represent the function (behaviour) of a component, they can merely be employed to fulfil conceptual design tasks dealing with scalar variables [12]. For example, since the bond graph theory can't consider the orientation information, these approaches can't consider the orientation change achieved by a perpendicular bevel gear pair during the design synthesis process. Obviously, our approach is more general than theirs, and can achieve conceptual design synthesis with more detailed features. In addition, there are still four significant differences between our approach and theirs.

First, our approach employs the attribute-value approach to represent the detailed features of the input and output flows, which makes the flow represented more explicitly and more flexibly. Since this approach is independent of any specific discipline, knowledge engineers can then customize different sets of attributes and values for different kinds of flows in different disciplines to show their differences. For example, two different sets of attributes have been used to represent the features of the Angular_Velocity and Electrical_Current flows. However, in the previous research, the flows have to be represented with a uniform set of features, e.g. the 7-nary group model for representing flows in Ref. [9]. Such kind of fixed representation models obviously can't be adjusted for specific flows in different disciplines.

Second, our approach employs not only the input-output flow name pair, but also the constraints on the related flows and the attribute-mapping rules to represent the functional knowledge of a known PS in a situation-free manner. As a result, known PSs can be distinguished at a more detailed level from the viewpoint of function, resulting in that the ACD system can use them more exactly for design synthesis. For example, although the PSs *AC-Motor* and *DC-Motor* both take electrical current as input, the system can distinguish them according to the different values of the *type* attribute they require for the flow *Electrical_Current*; as a result, the system can combine *Solar-Battery* and *DC-Motor* directly in the previous design case, other than *Solar-Battery* and *AC-Motor*. However, in the previous conceptual design synthesis approaches, such a difference between *AC-Motor* and *DC-Motor* often can't be considered.

Thirdly, our approach primarily employs known PSs to achieve design synthesis, which is different from the previous approaches using the basic elements to achieve the task. For example, our approach uses *Slider-Crank-Mechanism* as a known PS for design synthesis, while the previous approaches usually have to use the basic elements of this PS (e.g. slider, crank and rod) for design synthesis. Obviously, our approach can lead to more effective reuse of design solution knowledge in various disciplines.

Finally, our agent-based reasoning approach for design synthesis is domain-independent, unlike the domain-specific approaches developed in previous research. Based on the agent-based approach, the ACD system can simulate the action of known PSs on environmental flows and generate suitable output flows. In contrast, the traditional approaches are usually based on the domain-specific search and match mechanism. Obviously, our agent-based approach is more effective and more intelligent.

Conclusions

Designers are encouraged to explore in wide multi-disciplinary solution spaces for finding novel and optimal principle solutions during conceptual design. However, as cultivated in limited disciplines, they often don't have sufficient multi-disciplinary knowledge for fulfilling such tasks. To address this issue, we develop a general knowledge-based framework for achieving ACD of multi-disciplinary systems. It is primarily composed of a constraint-based approach for representing desired functions, a situation-free approach for representing the functional knowledge of a known PS and an agent-based approach for automated design synthesis. An illustrative design case demonstrates that the proposed framework can effectively achieve ACD of multi-disciplinary systems using the known PSs from various disciplines. As a result, designers can then explore in wide multi-disciplinary solution spaces during conceptual design, even if s/he doesn't have a full knowledge about them.

Acknowledgements

The research work introduced in this paper is supported by Ministry of Science and Technology of China (Granted No. 2008AA04Z108), Natural Science Foundation of China (Granted No. 50975173, 50935004, 50821003) and Science and Technology Commission of Shanghai Municipality (Granted No. 09QA1402800). The authors are also grateful to the anonymous reviewers for their valuable comments.

References

1. Pahl, G., Beitz, W.: Engineering design: A systematic approach. Springer, New York (1996)
2. Wang, L., Shen, W., Xie, H.: Collaborative conceptual design- State of the art and future trends. Computer-Aided Design 34, 981–996 (2002)

3. Umeda, Y., Ishii, M., Yoshioka, M., et al.: Supporting conceptual design based on the function-behaviour-state modeler. AI EDAM 10(4), 275–288 (1996)
4. Prabhakar, S., Goel, A.K.: Functional modeling for enabling adaptive design of devices for new environments. Art. Int. Eng. 12, 417–444 (1998)
5. Kota, S., Chiou, S.J.: Conceptual design of mechanisms based on computational synthesis and simulation of kinematic building blocks. Research in Engineering Design 4, 75–87 (1992)
6. Chakrabarti, A., Bligh, T.P.: An approach to functional synthesis of mechanical design concepts: Theory, applications and merging research issues. AI EDAM 10, 313–331 (1996)
7. Li, C.L., Tan, S.T., Chan, K.W.: A qualitative and heuristic approach to the conceptual design of mechanism. Engineering Applications of Artificial Intelligence 9(1), 17–32 (1996)
8. Ulrich, K., Seering, W.: Synthesis of schematic descriptions in mechanical design. Research in Engineering Design 1(1), 3–18 (1989)
9. Welch, R.V., Dixon, J.R.: Guiding conceptual design through behavioral reasoning. Research in Engineering Design 6, 169–188 (1994)
10. Bracewell, R.H., Sharpe, J.E.E.: Functional description used in computer support for qualitative scheme generation- Schemebuilder. AI EDAM 10, 333–345 (1996)
11. Campbell, M.I., Cagan, J., Kotovsky, K.: Agent-based synthesis of electromechanical design configurations. Journal of Mechanical Design 122, 61–69 (2000)
12. Rosenberg, R., Karnopp, D.: Introduction to physical system dynamics. McGraw-Hill, New York (1983)
13. Gero, J.S.: Design prototypes: A knowledge representation schema for design. AI Magazine 11(4), 26–36 (1991)
14. Sivaloganathan, S., Shahi, T.M.M.: Design reuse: an overview. Journal of Engineering Manufacture 213, 641–654 (1998)
15. Nilsson, N.J.: Artificial Intelligence: A New Synthesis. Morgan Kaufmann, CA (1998)
16. Chen, Y., Feng, P.E., He, B., et al.: Automated conceptual design of mechanisms using improved morphological matrix. Journal of Mechanical Design 128, 516–526 (2006)

Learning Concepts and Language for a Baby Designer

Madan Mohan Dabbeeru and Amitabha Mukerjee
Indian Institute of Technology Kanpur, India.

We introduce the "baby designer enterprise" with the objective of learning grounded symbols and rules based on experience, in order to construct the knowledge underlying design systems. In this approach, conceptual categories emerge as abstractions on patterns arising from functional constraints. Eventually, through interaction with language users, these concepts get names, and become true symbols. We demonstrate this approach for symbols related to insertion tasks and tightness of fit. We show how a functional distinction - whether the fit is tight or loose - can be learned in terms of the diameters of the peg and the hole. Further, we observe that the same category distinction can be profiled differently - e.g. as a state (clearance), or as a process (the act of insertion). By having subjects describe their experience in unconstrained speech, and associating words with the known categories for tight and loose, the frequencies of words associated with these can be discriminated. The resulting linguistic labels learned show that for the state profile, the words "tight" and "loose" emerge, and for the action, we get "tight" and "easy". Once an initial grounded symbol is available, it is argued that knowledge-based systems based on such symbols can be sanctioned by its semantics, as well as its syntax, leading to more flexible usage.

Symbols and Design Reasoning

Machine design systems have been used for encoding the final design, and for downstream functions such as analysis or manufacturing. In the attempt to generalize it to conceptual design, it is tempting to define a set of symbols and rules for modeling a domain (knowledge-based systems) [5, 6, 7, 12, 17 and 30]. However, the word "symbol" as used in the context of computers has a far narrower interpretation from that in human usage, which can lead to considerable inflexibility. In computers, symbols are defined formally, i.e. only in terms of other symbols, and lack the connection to domain experience underlying flexible usage among human designers. If we may present an analogy, computer usages of symbols are

J.S. Gero (ed.): *Design Computing and Cognition'10*, pp. 445–463.

like the understanding of a colour symbol like "red" by a blind man; he knows that it is an instance of something called "colour", and that "green" and "blue" are other colours, and maybe even that "crimson" and "vermilion" are shades of "red", but his understanding is dramatically different from that of a sighted `person, because the semantics is not connected to direct experience.

When human designers use symbols, their usage is flexible and even for very abstract terms, the semantics is well-grounded and they can easily come up with detailed instances of the idea. Computers using symbols may be able to provide instances for basic symbols (based on programmer definitions), but not for symbol compositions. Further, the semantics defined by programmers cannot take into account many contextual factors that may change the interpretation of a symbol; indeed, rules that apply over a general domain often need to be modified to fit the problem - "general rules never decide concrete cases" [28]. While symbolic reasoning systems have been used successfully in the context of design for very limited situations, we argue that they may prove difficult to scale up for the following reasons:

- Designers often differ widely in what they mean by any term; the meaning of any term is rarely independent of its context. Thus attempts at defining a "standardized" vocabulary may not fructify.
- Formal semantic models, typically based on "intension", i.e. a set of rules defined on other formal symbols (e.g. [24]) provide a very narrow, inflexible interpretation.
- Cognitively, related terms are often organized in a loose, hierarchy that is defeasible, i.e. memberships can be overruled in exceptional situations (a "bird" may be an animal that can fly, but it may also be an toy resembling the animal). On the other hand hierarchies in computational models (ontologies) are rigid, leading to unforeseen failures in novel situations.
- Crucially, symbols defined in terms of other symbols alone are un-grounded, like the blind man's "red". This implies that every possible relation with other symbols, and all possible consequences for actions must be explicitly encoded. Thus, it must know that after a wing injury, a "bird" may no longer fly, but still remain a bird. The number of such axioms is potentially unbounded. Grounded symbol semantics avoids this problem because the model of birds that don't fly, or the roles of wings in flight would also have emerged at some stage of experience, leading to a graded inference ("birds fly" is a rule, but it can be overruled).

Learning Symbols

Here we propose to treat the term "symbol" in machine design as it is understood by human designers, and not as in formal algebras. Thus a symbol would constitute a close coupling of a term or label and a semantic representation or "image schema" [18]. In the design context, the image schema may be viewed as a set of constraints abstracted from different design experiences. Good designs repeatedly reveal certain inter-relations among the design variables; we propose to learn these as image schemas (e.g. "clearance", "oscillating motion", etc.) Such image schemas need to be discovered from functional associations during design exploration, since the functions defining them are often too complex to be modeled by simple rules, and in any event, the definitions may change considerably based on context. This may be why designers find it difficult to define terms they use regularly, and why it is felt that much of their knowledge may be implicit [28]. This discovery process must start with the simplest concepts and build up, which is why we call our approach the "baby designer enterprise".

In earlier work, we have explored the emergence of image schemas in a baby designer through computational simulations [23]. Here we focus on learning the labels for a schema, so the resulting label-schema pair becomes a true symbol. We demonstrate the process by learning the simple distinction between tight and loose fits. The baby designer is given a set of explicit performance measures and show how graded, grounded schemas can emerge based on this functional distinction.

Subsequently, we consider the baby designer as interacting with a human expert, who describes the different situations using language. In this interaction, we assume no knowledge of language or grammar; all we assume is that the system is that the human narratives is available as word-separated text rather than raw speech data (human infants show good word-boundary separation skills starting around 9 months of age). Then we show how our computational baby can learn some labels merely by considering the frequency of words that are associated with these conceptual distinctions. Thus, the system now has a symbol in terms of both its label and its grounded semantics. The semantics for it may now broaden with further exposure. This approach has also been attempted in other domains [14, 27 and 29].

We observe now that the semantics of a symbol is much more than just its referent. The semantics also encodes many subjective aspects, such as how the referent is being viewed - how it is being *profiled* [19]. For example, entering a space may be profiled either as an action (process, "enter") or as a whole unit (atemporal relation, "into"). Though the conceptual structure is largely the same, there is a subtle difference caused by the focus on different aspects of the same structure. Thus the process view would accept temporal modifiers, the state view would not. This led us to conduct two different experiments for collecting language narratives. In the first, we ask the users to describe, in unconstrained English, their experience with an already assembled peg and hole. In the second we ask them to describe the experience of putting the peg into the hole. The word associations are significantly different, but one word does appear in both contexts - thus, its semantics is already enriched by these two meanings.

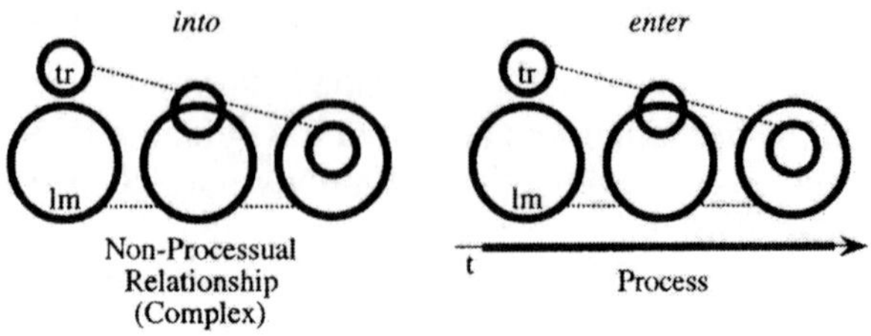

Fig. 1. *Profiling.* The same conceptual structure may be viewed as a relational complex or a process (action). Langacker [9]

At the end of this process, the semantics of these words is grounded in one limited context, so to our baby, the words mean have this limited meaning, but as its experience broadens, she will generalize the meaning to new situations. Note that the semantics learned is not a "primary" or "core" meaning – indeed, whether such privileged senses exist is itself uncertain. The sense learned is just one in a continuum of possible interpretations, many of which will be learned with further exposure, including possible metaphorical extensions to other domains.

Are Designs Emergent?

The approach here focuses on a situation where the image schema is available before the label is known. For human designers, this assumption may be questioned, since designers (e.g. students) learn many concepts by being told – i.e. the concept arises after its name is given. For a human learning language, beyond a small initial inventory of symbols, the vast majority of words are learned through its correlation with other words [4]. Nonetheless, the early inventory of symbols is crucial, for it provides grounding for the compositions that define later symbols. Only in this manner would grounding be available for the new concept. In the design scenario, the need for experiencing a domain directly is all the more crucial, which is the basis for hands-on approaches in design pedagogy, as opposed to other didactic disciplines.

A second reason for symbols to emerge in a design context is because the designer faces a challenge far greater than mere problem-solving or search, since the very constituents of his problem are ill-defined. One of the first tasks the designer must do is to discover representations at the right level for encoding and formulating the problem. Some of these representations, if they keep recurring, may become symbols. This process is related to the discovery of *chunks*, an abstraction formed from many input variables, and a topic often associated with design expertise. For example, we note that a designer who is an expert in a particular domain is "confident of immediately choosing a good [design] based on experience" [13]. Clearly, some sort of simplification of the problem space has occurred for the expert. While models of expertise have focused on chunks as they arise among trained designers, the process applies to all learning, and forms the backbone of symbol learning from our "baby designer" to the very best designers. This is why we feel this mode of learning is scalable -- from the very early stages demonstrated in this paper, to far more complex situations encoding large swathes of domain knowledge.

Related Work: Discovering Patterns in Design Spaces

An early attempt at discovering patterns in the design space of shapes may be seen in relation to 2D shapes [25]. Another approach to discovering chunks in design operates within the tradition of formal ontologies [22]. Here a learning layer, operating as a manager (M-agents), is added to a system being used to create designs. The M-agents consider the good designs that have come up, and try to identify some patterns which eventually become chunks that are added to memory. Further, the effectiveness of a new chunk can be tracked in subsequent designs to ascertain its utility. In design problem re-formulation mechanism suggested by Sarkar *et al.* [26], designers can identify latent relationships among different design variable groups by using Singular Value Decomposition on the co-occurrent variable matrix. This helps designers to redefine the design problem by re-representing it with a possible reduction in dimensionality.

None of these proposals however learn the semantics underlying the symbol in a grounded manner, using the feedback received from exploring the space of "good designs". An optimization-based approach towards this problem can be seen in [8]. Here the results of multi-objective optimization are analyzed manually to propose the possible inter-relationships between design variables necessary for "good design". The approach proposed here automates this process for both linear and non-linear relationships. Also, the discovered "dimensions" are proposed as "chunks' that are the putative basis for symbol discovery.

Baby Designer Enterprise

Our approach differs from the above in that it discovers structures in the input space, and posits these as proto-symbols, for which labels are then determined through language association. Our "baby designer" is at the apprentice level that has no prior knowledge about domains, though it knows many machine learning algorithms and has a bias towards shorter, information-condensing representations. It is given a set of raw descriptors (design variables) and a set of performance metrics defined on these variables so it can be evaluate different design instances. By exploring the design space it first identifies the pattern of good solutions, the Functionally Feasible Regions or *FFRs*. Next, it seeks to determine if the FFR is embedded in a lower-dimension sub-space, reflecting some constraints that holds among design variables among the "good" solutions. These inter-relations are proposed as one of the mechanisms for discovering chunks in design.

Discovering chunks results in a set of priors which encode domain-specific knowledge, but these are not the same as symbolic rules, because they are not explicit. For example, the system cannot provide a justification for such decisions. We may compare these with decisions that a human designer knows are good but finds difficult to justify, e.g. by saying "looks right" [1]. However, if similar chunks are observed repeatedly, especially in different domains, one may become conscious of the pattern, a process called *reification*. These reified chunks are more stable, and are sometimes called perceptual symbols in cognitive science [2]; in this work we refer to these as image schemas.

Subsequently, if the system interacts with a language user and finds that a string is strongly correlated with the image schema, then it may identify this string as the label, and the resulting structure may become a true symbol, with both a semantics (the image schema) and a linguistic label. Any priors that were earlier implicitly known will now get mapped explicitly into symbolic rules. These symbolic structures are situated, in the sense that they map the semantics in the "right" context at the "right" level of detail. Eventually, this will enable the system to reason with the symbol more flexibly, generate exemplar instances, and to justify its decisions in symbolic terms.

Learning Containment

Now we present the computational baby designer with the task of inserting a peg into a hole. A very early discovery for human infants is that pegs must be narrower than holes, i.e. the hole-width w must be greater than the peg thickness t, Figure 3(a). For the human designer, such primitives lie at

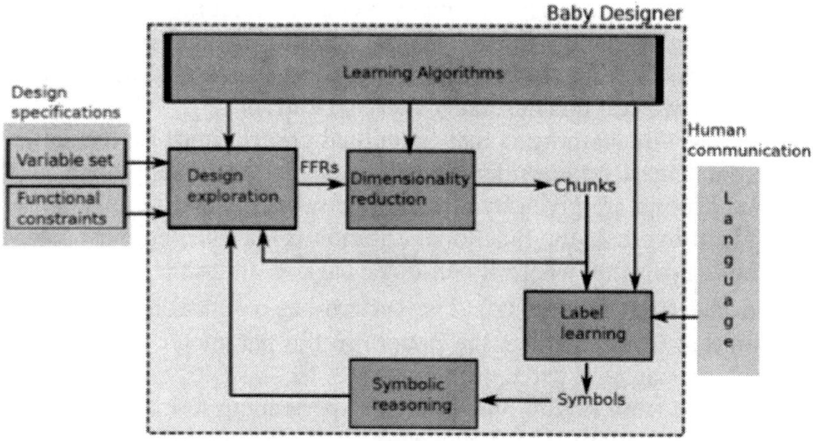

Fig. 2. *Architecture of the Baby Designer:* The baby designer learns patterns as in an apprenticeship situation. It is presented a design problem with design variables and a set of functional constraints that map design instances to performance metrics. While exploring the design space, it uses an inventory of domain-general learning algorithms to discover patterns that hold among the better designs, which may become chunks if they recur often enough. Labels for these may be learned after exposure to language, when they become true symbols. Implicit associations (priors) become encoded as domain rules in this symbolic space, thus enabling symbolic reasoning

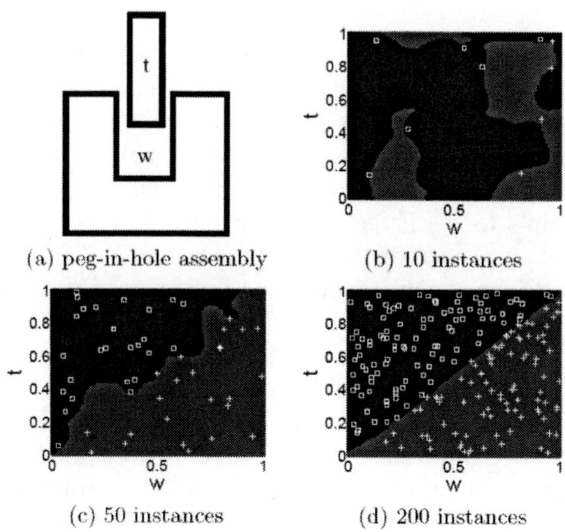

Fig. 3. Learning through experience that inserted-object-must-be-smaller-than-container ($w > t$). After a few instances, the experience is unstable, but the pattern converges after sufficient instances

the heart of the semantics for many symbols relating to containment, assembly, dimensioning, fit, etc. Thus, learning design symbols cannot start as an adult - it must start with our earliest conceptual achievements, which is why we call this the "baby designer enterprise".

The key to this learning is that functional criteria must be available. In these initial stages, we consider the baby designer as an apprentice, so that functional criteria are given by a mentor or some other external source. In this first learning task, the functional criterion is that the peg must enter the hole. Those instances where it can enter (+) constitute the FFR, which is distinguished from failures (□). The system has a function generalization algorithm that can generalize the pattern in the parameter space- here we use a back-propagation perceptron.

We observe from Figure 3 how, after experiencing just a few instances, Figure 3(b), the pattern is inchoate, so the baby keeps trying to insert some fat pegs into smaller holes, but this exploration itself keeps filling up the negative (black) area of the figure, Figure 3(c) and (d). Eventually the defining boundary becomes sharper, and at some point it does not change as much with new experiences, so that it feels it may have discovered a stable pattern, at least implicitly. This knowledge can be thought of as a simple prior defined on the w, t space, an implicit version of the rule that w must be greater than t.

Learning about Clearance

In the next step, let us consider that our baby designer has is exploring different types of successful insertions - particularly, those involving instances of tight fit and loose fit. Given a functional description of these, it learns the corresponding FFRs, Figures 4(a), (b), FFRs in gray. These functions also become sharper with greater experience.

Next, it attempts to see if the good instances may be lying along some low-dimensional sub-space of the design space $<w,t>$. To begin with, we may use linear dimensionality reduction. Trying out Principal component analysis [3], we find that the results on the early learning after 20 samples results are not very clear, but after 100 samples, the first eigenvalue is clearly dominant (33.60,0.11), and the first eigenvector (-0.72, 0.69) is along a 45 degree line in the w, t space (Fig. 4a, bottom). Thus, though the design space had two dimensions, we discover that the distribution of good instances of tight fit is largely distributed along one dimension. A parallel eigenvector is found to be dominant in the loose fit case. These two 1-D lines constitute the basis for the categories tight (C_T) and loose (C_L). The invariant along either line is the quantity $w-t$, which becomes the learned "chunk"; its value eventually may form part of the semantics for the symbol "clearance".

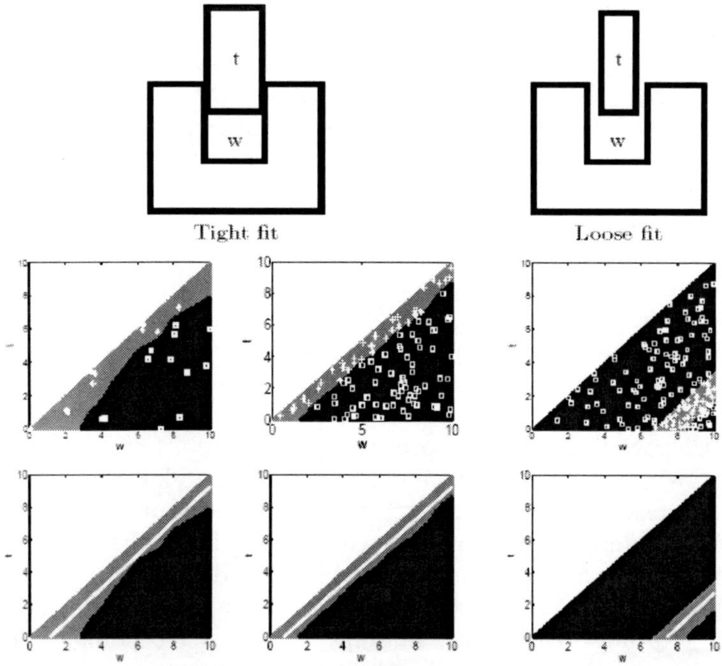

(a)e1 = (-0.72, 0.69) (b) e1= (0.70, -0.70) (c) e1= (0.72, 0.70)

Fig. 4. Emergence of chunks for fit: “tight" vs. “loose": Given a task which requires a “tight’ fit, the Functionally Feasible regions (FFRs) learned after 20 instances is diffuse (a), but improves within 100 trials (b). For loose fits, a well-learned stage is shown (c). These FFRs are modeled using PCA, resulting in a 1-D characterization where the principal eigenvector represent invariance in *w-t*. Thus, when considering the concept of clearance, the number of parameters involved reduces from (*w; t,* 2-D) to the emergent chunk *w-t*, a single parameter (1-D). This learning process results in two chunks, C_T and C_L but the system does not have any names for these yet.

Such correlations, which are embedded as lower-dimensional manifolds in the high-dimensional design space, may be rather common in design. For example, if strength is to be maximized while minimizing weights, then many dimensions need to be balanced – they would rise (or fall) in tandem. Thus, for good designs, these inter-relations may result in a single chunk. Discovering these interdependences is a first step towards the process of creating semantically rich models of design. While the example here deals with only linear subspaces, we have elsewhere dealt with non-linear manifold discovery [23].

Language Mapping

At this stage, we have an implicit notion of the categories C_T and C_L, but we cannot relate this idea to other concepts because there is no handle or label with which to refer to it. The label is a crucial part of the symbol - without it, the chunk or image schema that has been learned cannot be related to a broad host of other concepts. Implicit rules where these chunks play a role cannot be used explicitly to justify decisions, or to reason about the effects of decisions. More importantly, a label such as "clearance" stabilizes the semantics through social convention; without it the semantics may drift with new, similar experiences.

In order to learn a label, we obtain human commentary on the same C_T - C_L distinction which has been already learned. We provide human subjects with a simple apparatus - several flat pieces of wood with a hole, and some cylindrical pegs that fit in these holes with different types of clearance, Figure 5. We give different combinations of these to human subjects and have them describe their experience with them in unconstrained English. Then we would associate individual words appearing in these descriptions with the concepts, and see if any good labels would emerge.

As discussed above, the same insertion task may be viewed under different profiling distinctions. In some situations the complete task is viewed as a whole (atemporal), while elsewhere one may consider its evolution over time (temporal). We also explore how these distinctions may lead to differences in the symbol used to refer to the same concept.

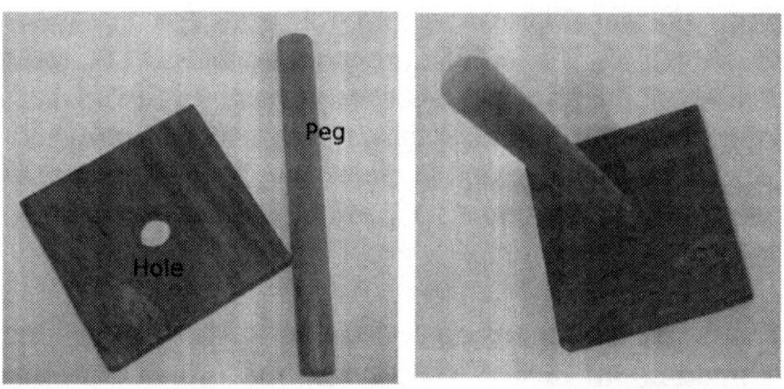

(a) peg and hole (b) peg-in-hole assembly

Fig. 5. Peg-in-hole assembly : A, B, C with three hole sizes (22.5, 17.1 and 12.74) average diameter respectively) and pegs 1 through 6 (22.4, 21.2, 16.9, 14.5, 12.5, and 10.3) are used. A:1, B:3 and C:5 are tight fits, A:2, B:4 and C:6 are loose.

For the human designer also, profiling distinctions are important. For example, "easy to insert" takes a process view of function, while "loose fit" relates more to an atemporal view, although both may be talking about the same design. In different design contexts, a designer may use one symbol or another and no amount of standardization can do away with such differences. Indeed, this reveals an important aspect of symbol use contrasting with the formal view - a symbol is not merely the object being referred to (referent), or even a class definable by a set of attributes and associations, for it also encodes a subjective view of the referent. As a cognitive linguist has colourfully said, the same person may be referred to as "eminent linguist" and "blonde bombshell" [10], expressions that highlight different profilings for the same referent.

To test the effect of profiling in greater detail, we designed two different scenarios for collecting user data. Although these are not experiments in the traditional sense (there is no hypothesis to be validated), we refer to these as experiments for want of a better term. In the first experiment, we placed already inserted peg-hole assemblies in front of our subjects, and they were instructed not to pick up the block with the hole, so they could not practice insertion actions. In the second experiment, we gave them the block with the hole and the peg separately and had them play around while inserting them. In either case, we asked them to describe the interaction between the peg and hole in plain, unconstrained English, and did not correct any grammar errors etc, though we did transcribe them into written texts. No constraints were imposed on the language, and at least in one instance a subject held forth at length about the colour and smoothness of the apparatus and other aspects – this was the only narrative excluded from our analysis, though it does not substantially affect the result.

Also, before collecting data, users were permitted two trial rounds, one with a tight fit and one with a loose fit, without telling them anything about these.

Associating Linguistic Labels

Next, we outline the process used to discover a linguistic label for the image schema that has been learned. The association of a word with a concept *C* can be measured in many ways; the machine translation literature poses many association measures. One of the simplest is based on conditional probability, but here there is some debate re: the direction of causality - should we consider the conditional of word *w* given *C* or *C* given *w*? Let n_T and n_L be the number of words in the C_T and C_L narratives, and let word *w* have a count of k_L and k_T in each narrative. Then one may estimate the conditional probability $p(w/C_T)$ as k_T/n_T. For the conditional probability of *C* given w we may adopt Bayes' rule, which gives us

$$p(C/w) = \frac{p(w/\tilde{C})p(C)}{p(w)}$$

Now, since the prior word frequency *p(w)* is independent, we have

$$\frac{p(C_L/w)}{p(C_T/w)} = \frac{p(w/C_L)}{p(w/C_T)}\frac{p(C_L)}{p(C_T)}$$

Furthermore, the number of C_L and C_T instances in the training data is roughly the same so $p(C_L)$ and $p(C_T)$ are equal. Hence the ratio of the two conditionals are equal in both directions. Thus, if we find the w,C pair that maximizes this ratio, then either conditional would be maximized.

So our objective is to compute $\max_i \frac{p(w_i/C_L)}{p(w_i/C_T)}$ so that w_i would have the strongest association with C_L. Similarly, the inverse ratio is to be maximized for the strongest association with C_T.

For this task, the user narratives are first transcribed and all the narratives relating to each fit situation (tight or loose) are combined. The word counts n_T, n_L, k_L, k_T are used to compute the conditional ratio and the top five correlations in all four Concept-Profile are presented in Table 3. We also observe that subjects use many morphological variations - e.g. for "tight" (count=26), we also have "tighter" (3) "tightly" (4) etc. We may use stemming [15] (discard common afffixes) to count only the roots of such words. We contrast below the results with and without stemming – and find that even without stemming, correlations are quite strong.

Another step frequently adopted in NLP is to remove the frequent words which occur in many diverse contexts, so their relevance in a particular task may be less. These include particles and grammatical markers like *the, a, an, of, in, to, is, am*, etc. However, we found good results even without removing these words. Correlations discovered without these steps implies minimal assumption of linguistic knowledge for the word association process.

Experiment: STATE: The purpose of this first experiment is to focus on state distinctions - i.e. collect the spoken English data for the situation where the subjects is given a peg already inserted into the hole. We then collect their unconstrained English descriptions.

Method: *Apparatus*: six wooden pegs (1...6) and three blocks A,B,C as shown in Figure 5.

Participants: Eighteen IIT Kanpur graduate students, both male and female, of age 18-24, participated in generating narratives. Students had back-grounds in physics, mechanical engineering, biology, electrical engineering, chemical engineering and design. Level of competence in spoken English varied somewhat across the group sentence structures were retained as spoken, even if they were ungrammatical.

Procedure: Each participant is presented with the following instruction:
"This is a peg and this is a hole. The peg is already inserted into the hole. Play with the assembly, but please do not lift the block with the hole from the table. Describe the interaction between the peg and hole in English."

Note that no reference was made to the tight-loose distinction; many participants reported on many other aspects such as the relative sizes of the peg and hole pairs, their shapes, the kind of construction, etc. In state-profiled trials a peg pre-assembled into blocks A, B, and C was placed on the table and the subject could not to lift these blocks from the table – thus they could experience the assembly more as a whole, rather than the insertion task as a process. Each subject was given alternating tight and loose assemblies in the order (A:1) - (A:2) - (B:3)- (B:4) - (C:5) - (C:6). Sample narratives for two (A: 1) and (A: 2) of a speaker is given in Table 1.

Analysis and Discussion: The narratives for tight {(A: 1), (B: 3), (C: 5)} and loose {(A: 2): (B: 4) and (C: 6)} cases result in a small sample corpus (1099 words for C_T, and 904 words for C_L). Many words appearing in one set do not appear in the other, but fortunately, the top twenty-five words appearing in either set were also present in the other, so the ratios could be computed for these frequent words.

Table 1 Transcribed narrative: State profiled. [...] indicates pause

A1	This one is very sticky kind of thing [...] it is not moving in the either of directions and its firmly hold it to the basement and [...] I am not able to rotate or pull it to the any side of block.
A2	this piece of art the later piece can be moved easily in all directions [...] and can be removed out of the block and can be rotated either directions.

To ensure the relevance of these words, we tried to ensure that the probability of the word appearing in this context is higher than its prior in general usage; i.e. $p(w/C)$ should be greater than $p(w)$. Priors were estimated from a spoken English corpus (based on TV scripts) with 29 million words [16]. For all the top 25 words the conditional was higher than the prior. Table 2 shows the top five words for the tight corpus (C_T), in descending order of the ratio $p(w/C_T)$ by $p(w/C_L)$. Thus, the strongest correlations for C_T are "tight" , "to", "into", "not", "rotate". Now, the prior probability of words such as "to" and "not" are several orders of magnitude higher - i.e. these are most likely used in a wide variety of situations. Thus the more appropriate words for this situation are "tight", "into", and "rotate". In the unstemmed case, we find "cannot" and "am" in addition to "tight", "into", "to". The term "tight" is the most likely key word in both stemmed and unstemmed cases. The word "rotate" does not appear without stemming since morphological variants such as *rotating* (5 times) and *rotation, rotations* (1 each) are lost. The word *rotate* itself appears only in the state profile, probably because subjects were instructed not to remove the peg so the only action they could try was to rotate it in the hole.

A different set of eighteen graduate students, both male and female, of age 18-24, studying physics and mechanical engineering, participated in generating narratives. Each participant is presented with the following instructions:

"This is a peg and this is a hole. The peg can be inserted into the hole. Describe the interaction between the peg and hole in English."

Table 2 State profiled: [tight] and [loose] corpora.

State profiled: [tight] corpus								
Term	fr	$p\hat{}()$	f_L	$p\hat{}(\frac{w}{C_L})$	$f_{T.L}$	$p\hat{}(w)$	$p(w)$(T.V)	$\frac{p'(\frac{w}{C_T})}{p'(\frac{w}{C_L})}$
Without stemming								
tight	26	0.02366	3	0.01447	29	0.01447	0.0004	7.14
cannot	7	0.00637	1	0.00399	8	0.00399	0.000121	5.76
to	33	0.03003	8	0.02046	41	0.02046	0.02810	3.40
into	12	0.01092	3	0.00749	15	0.00749	0.00093	3.29
am	9	0.00819	3	0.00599	12	0.00599	0.001294	2.47
With stemming								
tight	34	0.03094	3	0.01846	37	0.01846	0.00004	9.33
to	33	0.03003	8	0.02046	41	0.02046	0.02810	3.4
into	12	0.01092	3	0.00749	15	0.00749	0.00093	3.29
not	19	0.01729	7	0.01297	26	0.01297	0.00660	2.24
rotate	23	0.02093	12	0.01747	35	0.01747	0.00000	1.58
State profiled: [loose] corpus								
Term	fr	$p\hat{}()$	f_L	$p\hat{}(\frac{w}{C_T})$	$f_{T.L}$	$p\hat{}(w)$	$p(w)$(T.V)	$\frac{p'(\frac{w}{C_T})}{p'(\frac{w}{C_L})}$
Without stemming								
loose	27	0.02983	4	0.01547	31	0.01547	0.00003	8.2
much	10	0.01105	2	0.00599	12	0.00599	0.00117	6.07
can	28	0.03094	12	0.01996	40	0.01996	0.00912	2.83
quite	8	0.00884	4	0.00599	12	0.00599	0.00018	2.43
easily	10	0.01105	7	0.00848	17	0.00848	0.00002	1.73
With stemming								
loose	30	0.03315	4	0.01697	34	0.01697	0.00003	9.11
much	10	0.01105	2	0.00599	12	0.00599	0.00117	6.07
can	28	0.03094	13	0.02046	41	0.02046	0.00912	2.62
easy	17	0.01878	9	0.01297	26	0.01297	0.00023	2.29
in	15	0.01183	13	0.01397	28	0.01397	0.00966	1.4

Experiment: ACTION: In this experiment the subjects are permitted to actively insert the peg into the hole, using the same apparatus as above.

As in the previous experiment, each subject is given alternating tight and loose fit situations, and asked to describe their experience in spoken English (sample narratives in Table 3).

Table 3 Transcribed narratives: Action profile

Peg-in-hole Assy.	Description in spoken English
B3	This is another hole and this is another peg [...] this is also going very tightly [...] it is not going completely inside the hole
B4	This is another one this is very loose [...] in this hole and we can pass it through the hole very easily

The top five word associations for C_T are presented in Table 4 sorted by the ratio of conditional probabilities.

Table 4 Action Profile: [tight] and [loose] corpora: Top five words by conditional ratio

Action profiled: [tight] corpus							
Term	f_T	$\hat{p}(\frac{w}{C_T})$	f_L	$\hat{p}(\frac{w}{C_L})$	$f_{T,L}$	$\hat{p}(w)$	$\frac{\hat{p}(fracwC_T)}{\hat{p}(\frac{w}{C_L})}$
Without stemming							
tight	26	0.02050	4	0.00462	30	0.01406	4.43
first	13	0.01025	3	0.00347	16	0.00750	2.96
not	30	0.02366	8	0.00925	38	0.01782	2.56
a	11	0.00868	3	0.00347	14	0.00656	2.5
i	30	0.02366	9	0.01040	39	0.01828	2.27
With stemming							
tight	33	0.02603	4	0.00462	37	0.01735	5.63
first	13	0.01025	3	0.00347	16	0.00750	2.96
not	30	0.02366	8	0.00925	38	0.01782	2.56
i	30	0.02366	9	0.01040	39	0.01828	2.27
of	27	0.02129	11	0.01272	38	0.01782	1.67
Action profiled: [loose] corpus							
Term	f_L	$\hat{p}(\frac{w}{C_L})$	f_T	$\hat{p}(\frac{w}{C_T})$	$f_{T,L}$	$\hat{p}(w)$	$\frac{\hat{p}(\frac{w}{C_L})}{\hat{p}(\frac{w}{C_T})}$
Without stemming							
easy	13	0.01503	2	0.00158	15	0.00703	9.53
second	10	0.01156	2	0.00158	12	0.00563	7.33
easily	18	0.02081	4	0.00315	22	0.01031	6.6
can	13	0.01503	4	0.00315	17	0.00797	4.76
very	21	0.02428	11	0.00868	32	0.01500	2.8
With stemming							
ease	35	0.04046	6	0.00694	41	0.01922	5.83
second	10	0.01156	2	0.00231	12	0.00563	5.00
can	13	0.01503	6	0.00694	19	0.00891	2.17
very	21	0.02428	11	0.01272	32	0.01500	1.91
diameter	13	0.01503	9	0.01040	22	0.01031	1.44

It is observed that both with and without stemming, “tight”, and “first” are most relevant associations for action profiled C_{T_1} , though after stemming, the term “tight” emerges stronger. For loose fits when profiled as a process, “easy” is found to have the strongest association both with and without stemming.

Discussion

These experiments demonstrate that finding the linguistic labels for image schemas like tight and loose is possible without much prior knowledge, of either the domain or of language. Initially we had thought we may need to use standard techniques like stemming and stop word removal, and the results - that “tight” and “loose” came up so readily in the state profiled experiment - were a pleasant surprise. Further, in the action profiled case, the word for describing the action where the clearance is high, in English, is not “loose”, but “easy” - an association that some may find reasonable. The other interesting fact is that the same word, “tight” is used as the label for two related but somewhat different semantics - the tight fit case profiled as a state, and also an action. This is not very surprising, since languages must encode, in a finite inventory of units, the unbounded phenomena present in the universe.

This type of polysemy, which is sometimes called lexical polysemy (as distinct from accidental polysemy or homonymy), is extremely widespread, e.g. the word “motor”, which may indicate the engine of a car or an electric motor. Without taking contextual cues into consideration, symbol meanings are impossible to define.

Conclusion

This work aims to introduce a new paradigm for reasoning based on grounded symbols, as opposed to narrow human-defined symbols. The present demonstration is necessarily very limited, as indicated by the “baby designer” appellation. A simple demonstration of learning symbols as label-meaning pairs with minimal prior domain knowledge, along with our understanding of the human learning process, argues for scalability of such an approach.

At this stage, our goal is to present grounded acquisition of symbols as an alternative to traditional knowledge-based systems for design. At first glance, symbols defined formally appear capable of capturing a large part of high-level design knowledge, but without being structured on top of our very basic, earliest experiences, such systems face severe limitations in generalizing to new domains. On the other hand, the system proposed here, is grounded and likely to be scalable and more flexible in its deployment.

As of now, the system learns patterns only in the apprenticeship mode, which works only for well-understood domains. The symbols learned thus can be used to create flexible knowledge structures for these known domains. Thus, the first models emerging out of a grounded knowledge discovery paradigm in design may apply to systems that at least humans understand well. It is only after such knowledge with a rich semantic basis has become familiar that it would be possible to attempt other challenges such as novel domains or creative designs.

More than the results itself, we feel that the key contribution of this work is to open up several new avenues for further integration of cognitively motivated approaches into computational systems for design. A baby's symbols are initially over-specialized, limited to the contexts where they learn it. Similarly, the baby designer's symbols will initially have narrow, specific meanings. However, once the initial symbols are available, one may merge other experiences sharing the same label, to broaden the semantics. How such experiences would be amalgamated into a more general conceptual structure remains a key question for further work. Another question that arises is the compositions of symbols. New symbols often arise as compositions - “thick flange”, “airy hallway”, “pinwheel arrangement” etc. instantiate or blend some aspects from each of its constituents, and which to choose and which to ignore remains a serious challenge for any model of semantic composition.

Another avenue opened up by this work is a possible rapproachment between formal or systematic approaches to design [5, 12, 24] and the situative view which argues for a more creative view of design[28, 25]. The symbols that are learned here are initially emergent, but may eventually be used to formalize design knowledge and lead to coherent theories for different domains. This may provide a basis for unifying the so-called systematic and creative camps in design theory.

How does the baby designer become an useful designer? Clearly, the system will need to be exposed to many more situations (after all, a human designer goes through a fifteen-year childhood, and then a many-year apprenticeship). Also, we will also have to develop much of the learning structures needed to consolidate these experiences. The paradigm proposed here makes only a small start, and like the interaction of a baby with its world, it opens up more questions than answers. We hope that others will also work on similar ideas, and that we may discover the contours of some of these answers, and that these may then illuminate the potentialities of this approach over the coming years.

References

1. Ahmed, S., Wallace, K.M., Blessing, L.T.: Understanding the differences between how novice and experienced designers approach design tasks. Research in Engineering Design 14(1), 1–11 (2003)
2. Barsalou, L.: Perceptual symbol systems. Behavioral and Brain Sciences 22, 577–660 (1999)
3. Bishop, C.: Pattern recognition and machine learning. Springer, Heidelberg (2006)
4. Bloom, P.: How children learn the meanings of words. MIT Press, Cambridge (2000)
5. Bohm, M.R., Stone, R.B., Szykman, S.: Enhancing virtual product representations for advanced design repository systems. Journal of Computing and Information Science in Engineering 5(4), 360–372 (2005)
6. Campbell, M.I., Cagan, J., Kotovsky, K.: Agent-based synthesis of electromechanical design configurations. Journal of Mechanical Design 122(1), 61–69 (2000)
7. Chakrabarti, A., Sarkar, P., Leelavathamma, B., Nataraju, B.S.A.: Functional representation for aiding biomimetic and artificial inspiration of new ideas. AIEDAM 19(2), 113–132 (2005)
8. Deb, K., Srinivasan, A.: Innovization: innovative design principles through optimization. Tech. Rep. Kangal, 2005007. IIT Kanpur (2007)
9. Ericsson, K.: Expertise. MIT Encyclopedia of Cognitive Science (1999)
10. Evans, V., Green, M.: Cognitive Linguistics: An Introduction. Edinburgh University Press (2006)
11. Gero, J.S., Fujii, H.A.: Computational framework for concept formation for a situated design agent. Knowledge-Based Systems 13(6), 361–368 (2000)
12. Gorti, S.R., Sriram, R.D.: From symbol to form: a framework for conceptual design. Computer-Aided Design 28(11), 853–870 (1996)
13. Gross, M.D.: Design as Exploring Constraints. PhD thesis, Department of Architecture. MIT, Cambridge (1986)
14. Guha, P., Mukerjee, A.: Language Label Learning for Visual Concepts. Discovered from Video Sequences. Springer, Heidelberg (2008)
15. Jurafsky, D., Martin, J., Kehler, A.: Speech and language processing: an introduction to natural language processing, computational linguistics, and speech recognition. MIT Press, Cambridge (2000)
16. Keffy. Wiktionary: Frequency lists for TV and movie scripts (2006), `http://en.wiktionary.org/wiki/Wiktionary:Frequency_lists` (accessed February 10, 2010)
17. Kurtoglu, T., Campbell, M., Gonzales, J., Bryant, C., Stone, R.: Capturing empirically derived design knowledge for creating conceptual design configurations. In: Proceedings of the ASME Design Engineering Technical Conferences And Computers In Engineering Conference. DETC2005-84405, Long Beach, CA (2005)

18. Langacker, R.: An introduction to cognitive grammar. Cognitive science 10(1), 1–40
19. Langacker, R.: Cognitive Grammar: A Basic Introduction. Oxford University Press, USA (2008)
20. Lawson, B.: Schemata, gambits and precedent: some factors in design expertise. Design Studies: Expertise in Design 25(5), 443–457 (2004)
21. Martinetz, T.M., Berkshire, S.G., Schulten, K.J.: Neural gas network for vector quantization and its application to time-series prediction. IEEE Transactions on Neural Networks 4, 558–569 (1993)
22. Moss, J., Cagan, J., Kotovsky, K.: Learning from design experience in an agent-based design system. Research in Engineering Design 15(2), 77–92 (2004)
23. Mukerjee, A., Dabbeeru, M.M.: The birth of symbols in design. In: Proceedings of DETC 2009, ASME Design Engineering Technical Conferences (2009)
24. Nanda, J., Thevenot, H., Simpson, T., Stone, R., Bohm, M., Shooter, S.: Product family design knowledge representation, aggregation, reuse, and analysis. AIEDAM 21(02), 173–192 (2007)
25. Park, S., Gero, J.: Qualitative representation and reasoning about shapes. In: Visual and Spatial Reasoning in Design, Sydney, Australia, vol. 99, pp. 55–68 (1999)
26. Sarkar, S., Dong, A., Gero, J.S.: Design optimization problem reformulation using singular value decomposition. Journal of Mechanical Design 131(8), 081006–1–10 (2009)
27. Satish, G., Mukerjee, A.: Acquiring linguistic argument structure from multimodal input using attentive focus. In: 7th IEEE International Conference on Development and Learning, ICDL 2008, pp. 43–48 (2008)
28. Schoen, D.A.: Designing: Rules, types and words. Design studies 9(3), 181–190 (1988)
29. Steels, L.: Evolving grounded communication for robots. Trends in Cognitive Science 7(7), 308–312 (2003)
30. Yaner, P., Goel, A.: Analogical recognition of shape and structure in design drawings. AIEDAM 22(2), 117–128 (2008)

Organizing a Design Space of Disparate Component Topologies

Mukund Kumar and Matthew I. Campbell
University of Texas at Austin, USA

In a previous DCC paper, the authors presented an approach to generate a large space of conceptual designs by a set of grammar rules. These results indicated a large number of topologically unique solutions that could be created from a single black box that consists of a simple description of the function of the product and the input and output flows. The problem remains as to how to efficiently organize and search this space to find the best design for a given set of user preferences. In this paper we present new results that organize the candidate space using clustering methods such as K-means algorithm that group a large number of points in space based on a certain spatial property. Candidate component topologies, referred to as Component Flow Graphs (CFGs), are categorized using this method based on properties that physically distinguish them. From a theoretical and computational standpoint, this is an open research question as the CFGs may be vastly different graph topologies and the nodes and arcs of the graph may represent many different types of components and component connections. This paper details an experiment wherein ten products are designed from function structure to CFG. A space of over 8000 candidate solutions is developed. From this large set, clustering algorithms are employed to organize the space, and eventually aid an automated or interactive search algorithm that can find a best candidate solution for a particular user. A vast space of clustered concepts would allow an interactive process to query the user about particular CFGs and gauge whether the user would like to see more similar CFGs (i.e. from the same cluster) or is more interested in different ones (i.e. ones from other clusters). Such an interactive tool would be useful in mass customization.

Introduction

Automating the conceptual design process is a challenging task on several levels. In 2008, Kurtoglu et al. [1] present an approach where a graph

J.S. Gero (ed.): *Design Computing and Cognition'10*, pp. 465–485.

grammar is developed for generating numerous conceptual designs from a black box. In that research, the ultimate challenge was searching the resulting tree of solutions. Ideally each solution in this immense tree would be evaluated to find the best solution. As will be discussed later, evaluation of a generic design for several performance requirements is a complex process. As a result, a universal ranking scheme that can rate these candidate designs is not available for this search. By leveraging human expertise in evaluating design problems, we can bypass the complication of using an automated evaluation approach and at the same time address the needs and requirements of the user. However, the time commitment required of the user in rating thousands of design solutions is not practical.

Several approaches may be employed to realize this concept and there are many questions that need to be answered. On what criteria do we pick the sample candidates from the thousands of designs? What is the minimum sample space that is enough to represent the design space? How do we identify a group of designs of similar properties? On what basis do we categorize the designs so that we are able to present a set of solutions similar to any particular solution? How do we go about the grouping process?

In this research we try to answer these questions and provide a solution for organizing the design space. We use products such as hair dryers and water pumps and attempt to redesign them in different ways using components from different products to create a large space of unique designs. We have investigated how clustering approaches organize the designs by grouping them based on characteristic parameters. It has also been shown that clustering can prove to be an efficient approach to organizing large sets of design solutions and can aid downstream processes of evaluation using an automated approach or using human expertise.

Background

The use of graphs in engineering design is unavoidable. From electric circuit and chemical process diagrams to function structures and flowcharts, engineers have found graphs as a quick and clear way to describe what needs to be designed. In this work, we create a graph grammar for transforming between two different types of graphs: the function structure [2] and the component flow graph, which is simply the components in a product with arcs indicating the flows between the elements. This two-stage approach can be seen to be a simplification of the FBS approach [3] which formalizes the relations among function, behavior, structure to retrieve design information to conduct analogy-based design. Other than the graph-rewriting approach adopted here, typical examples of computational synthesis applications start with a set of fundamental building blocks and some composition rules that govern the combination of these building blocks into complete design solutions. Hundal [4] designed a program for automated conceptual design that associates a database of solutions for each function in a function database. Ward and Seering [5] developed a mechanical design "compiler" to support catalog-based design. Bracewell and Sharpe [6] developed "Schemebuilder," a software tool using bond graph methodology to support the functional design of dynamic systems with different energy domains. Chakrabarti and Bligh [7] model the design problem as a set of input–output transformations. Structural solutions to each of the instantaneous transformation are found, and infeasible solutions are filtered according to a set of temporal reasoning rules. The A-Design research [8] is an agent-based system that synthesizes components based on the physical interactions between them.

Graph grammars have been adopted by the design community for their ability to create sweeping topological changes to a graph in a rigorous fashion [9]. The approach used here sees each grammar rule modification as a three-step process: recognition, choosing, application. From a given host, all rules are checked to find which contain valid transformations (recognition). Presented with the list of valid transformation or option, some decision-making agent, either human or computer, must choose which one to invoke. Through the application of the chosen grammar rule on a particular location of the graph, elements of the host are deleted, added or modified to create a new state. Using this formalism, researchers have created grammar rule sets for a number of engineering applications such as: function structures [10], coffee makers [11], truss structures [12], sheet metal [13], mechanical clocks [14] and gear trains [15, 16].

The approach taken here is unique as the grammar rules are not specific to a particular problem domain, but to an array of electro-mechanical products. As a result of being so broad in defining the rules, it is difficult to search and evaluate entities in the space. Thus, this paper also incorporates algorithmic innovations from the field of clustering. A good review of clustering techniques can be found in [17].

What we need is a method that could summarize this design space into a manageable smaller set of solutions that represent the range of all the designs generated. This will enable us to obtain a general picture of the types and quality of the designs generated. From this sample set, we can use human expertise to judge the best solution. If there existed a method where given a particular design, a set of designs that are topologically similar are easily found, then this process could be carried out in an iterative manner with a human designer evaluating candidate solutions.

Generation of Designs

In the previous research [1], grammar rules were created based on taking apart existing products and building grammar rules from the extracted design knowledge. In Fig. 1, two graphs are shown for a disassembled hair dryer – both are screenshots from the GraphSynth software [18], which is developed to create and execute generative grammars. The graph on the left represents a function structure following the component basis and the graph on the right is a component flow graph (CFG) which is first presented in Kurtoglu *et al.* [1].

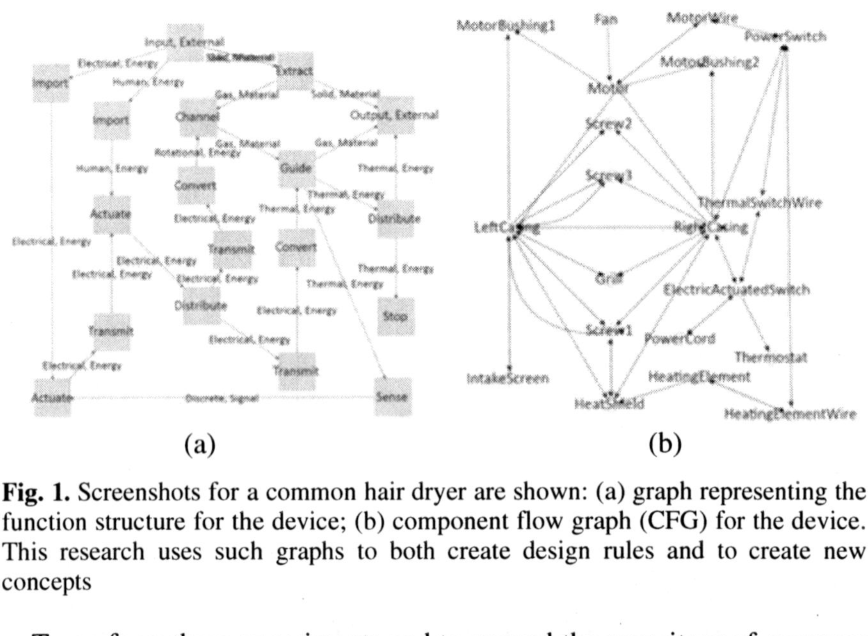

(a) (b)

Fig. 1. Screenshots for a common hair dryer are shown: (a) graph representing the function structure for the device; (b) component flow graph (CFG) for the device. This research uses such graphs to both create design rules and to create new concepts

To perform these experiments and to expand the repository of grammar rules, 10 products such as a leaf blower, submarine water pump, a hair dryer etc. were selected. These products were disassembled and its internal components and their connectivity were studied and laid out in the form of CFGs. The functional structure of these products such as the flow of energy and materials and how they are processed and utilized as the product is being used was also constructed. The data for the function structures and the CFGs can be found online at [19]. In Table 1, a summary of the 10 products is shown.

Table 1 Number of candidate designs generated from each function structure

ID	Product Name	# functions in FS	# components in CFG	# generated Candidates
A	Common Alternator	13	36	210
B	Hair Dryer	18	20	768
C	Hydraulic Jack	18	64	1000
D	Lycoming T53 Turbine	14	20	1000
E	Portable Air Compressor	25	64	1000
F	Proctor Toaster	17	54	1000
G	Squirt Gun	21	63	1000
H	Stanley Staple Gun	26	60	928
I	Troy Bilt Leaf Blower	10	19	480
J	VW Bug Carburettor	20	58	1000
Average		18.2	46	838.6
Min		10	19	210
Max		26	64	1000

There was an average of 18 functions and 46 components with the smallest being the Troy Bilt Leaf Blower (10 functions and 19 components) and the largest being the Portable air compressor (25 functions and 64 components). The rule set used for this study was created from these 10 products themselves. An automated grammar rule generation process compares the function structure graph and the CFG of each of these products and extracts modules of components from the CFG that satisfy discrete functions in the Function structure. The module of functions is then assigned as the left-hand-side of the rule (L) and the components as the right-hand-side (R) of the grammar rule. Using this rule generation process which will be presented in the future, 91 rules are created from these ten products. These ten products also served as our test bed for the experiments presented in this paper and acted as seeds for the generation process. By providing the 91 rules, CFGs are recreated for the products. As can be seen from the last column in Table 1, hundreds of unique CFGs are found for every product. While the search should recreate the original solution, it also creates countless others by borrowing elements from other products. As an example, a solution to the hair dryer function structure involves elements from a portable air compressor, squirt gun, hydraulic jack and a toaster as shown in Fig. 2. As previously mentioned, without a ranking scheme these candidates cannot be evaluated without human expertise. To aid future interactive methods, we seek a way to organize these candidate solutions in order to make the process more streamlined.

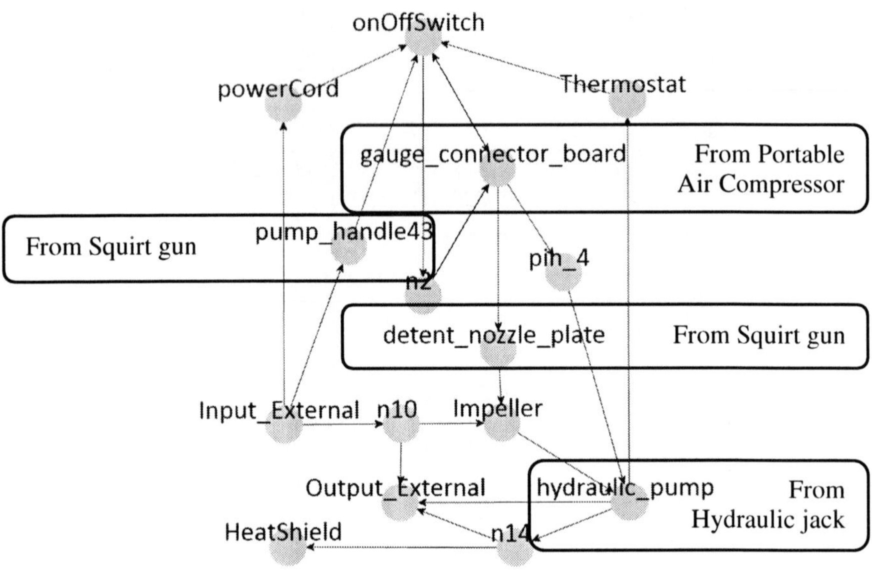

Fig. 2. One of the candidates generated from the function structure of a Hair dryer. This includes components from other products such as a Squirt gun, Proctor Toaster, Hydraulic jack etc.

Reducing the Design Space through Confluence and Matrix Comparison

Confluence between rules leads to the generation of duplicate designs in a search process. Confluence is defined as independence in the recognition and application of rule options. Confluent options can be applied in parallel or in any order and the result is the same. In this particular problem domain, the instances of confluence between the 91 rules and 10 seeds were found to be very high. Out of millions of designs created, only a few thousands of them were unique solutions. This causes the search process to be extremely time consuming. Given current computational limits, the elimination of confluent designs is a necessary step to proceed.

The identification and removal of confluent options has been implemented following the Recognize step of the tree search process even before the designs are generated. The recognition process checks to see if each rule is a sub-graph of the host graph. The check is not a Boolean but zero, one, or several distinct locations where the sub-graph is found in the host. When two options are recognized at completely different locations in the host graph, they essentially cannot affect each other's output. If there are two confluent options, *a* and *b*, *b*'s applicability is unaffected by *a* since the elements that *a* will potentially delete or modify are not considered in *b*'s recognized location. The same holds true for the effect *a* has on *b*. What this means is that options *a* and *b* can be applied in any order and they will produce the same designs. By identifying the set of options that are recognized at independent locations, a list of "confluent options" is created.

From this list of confluent options, all but one is deleted thus reducing the branching factor and size of the tree. This does not eliminate any unique designs since the deleted options will be re-recognized following the application of the first confluent option. It should be noted that the current function to detect confluence is not perfect. Due to possible changes in neighboring arcs, the algorithm is conservative in predicting confluence. There are many cases in which rule options are indeed confluent but the algorithm is unsure, and errors on the safe side by keeping these false-negative options.

As a result of erring on the side of caution, there are still many repeat candidate solutions that are created. Therefore, a second step in eliminating duplicate designs after generation is developed. This is a post-processing step that compares the current candidate with all other unique candidates that are generated thus far to determine whether it is a duplicate or not. The time consumed in this step is quite high as for every design created, it has to be compared with the full set of unique designs that were generated before that. To speed up this process, instead of matching entire graphs, graphs are translated into Design Structure Matrices (DSM) [20, 21] which is a common design approach to capture the interconnectivity between elements in a product. A DSM is an adjacency matrix that simply stores the connectivity between the nodes of a graph. As a square co-occurrence matrix an identical order of the product's components is stored to correspond with both the rows and columns of the matrix and a non-zero value in a cell represents an arc connecting between the row component and the column component. Since different graphs may have a different amount and type of nodes, it is important to normalize the DSM so that all graphs or CFGs are the same size and have identical ordering of rows and columns. In order to solve this, we use an alphabetized list of component types as described in the component basis research [22]. Therefore if two separate CFGs have a motor attached to a gear than both of the resulting DSMs will have a '1' in cell $[i, j]$ where the ith row corresponds to motor and the jth column corresponds to gear.

With these DSMs for each candidate solution, one DSM can be subtracted from another to create a matrix that indicates what elements and connectivity is different between the two. If the difference matrix is only filled with zeros, then we can be confident that the two CFGs are identical. And, as a result, only one of the CFGs needs to be stored. This is a form of reduction of the design space purely based on the topology of the graph. The design need not be a part of a tree for this property to exist. This matrix subtraction approach can also be used to prove confluence as the same designs will have the same DSM irrespective of the path of creation. This process is faster than comparing the entire design graphs. One subtle drawback of this method is that a DSM of a design graph is not unique to it and there is a possibility of the same DSM getting generated for two different CFGs.

Along with these two methods of confluence reduction and DSM comparison to identify the unique solutions in the tree, practical limits exist that require us to further reduce the tree. For the results shown later in the paper, we limit the space to a 1000 unique candidates for each problem. Even with this truncation of results, the following efforts represent more than one hundred hours of computation time.

Organizing the Design Space Using Clustering Methods

In an ideal search process, every candidate solution would be generated and evaluated to identify the one most suitable or optimal solution for the posed performance parameters or customer needs. In addition to challenges in handling such a large set of solutions, there is a limitation in evaluating conceptual design such as the CFGs presented here. Conceptual solutions do not provide details on standard component dimensions, shapes or materials. If non-standard components are used, these are even more ill-defined since CAD models and manufacturing process are not provided. In order to perform common mechanical analysis like dynamics, stress analysis or heat transfer, such details are required. Additionally, conceptual design problems often have more than one performance parameter and handling multi-objective problems adds an additional layer of complexity to the problem. Given the differences in concepts, an automated analysis would be challenged with interpreting the variety of configurations, identifying proper boundary conditions, and determining objective values of similar accuracy across the candidate solutions.

Because of these reasons, evaluation of individual designs is not pursued. Another option is to leverage user expertise by creating an interactive search process. This proposed method would be similar to the interactive stochastic search discussed in [23]. Since user ratings are time-consuming and the limited due to user fatigue, it is not possible to present the user with all the generated designs to obtain preference ratings. So, the approach taken here is to present the user with the minimal number of designs that spans the variety present in the entire design space. In order to determine this, we group the generated designs and pick a single design from each group to be presented to the user. This process of grouping the designs is done by using clustering algorithms. Designs are classified into clusters based on their topology or a parameter that represents the CFG of the design which is compared against other CFGs to measure the difference. This characteristic parameter should satisfy the following properties.

1. The parameter should be a property of the graph and representative of its topology. Two different graphs should not have the same value.
2. The distance between the parameter of the two graphs should increase as the graphs become more and more different.
3. The parameter should be equal or the distance between them should equal to zero when the corresponding graphs are identical.

There were three parameters found that could be used for clustering. The first is based on the Hamming distance [24] between the unordered lists of rules used to arrive at this graph from the seed. Since each rule represents a unique contribution to the design, it seems reasonable to simply count how many rules are different between two CFGs. Due to the sheer amount of confluent rules this method seems practical and has in fact been the approach used before [1]. However, with the improvement in detecting confluence, it is apparent that this method is not accurate, and thus it fails to solve the first and third rules shown above. A second choice for the parameter is to set it to the Hamming distance of comparing ordered list of options used to arrive at the CFGs. The option numbers represents the exact path taken in the creation of a graph from a particular seed. The distance between two parameters will represent the hamming distance in the tree search where it was created. If all confluent options were correctly detected and eliminated, this method would be accurate, but given the false-negative produced by the method, it seems that the third rule will be violated often.

The above two options have the disadvantage that they are strictly tied to the generation process which depends on the rules used and the seed function structure. They do not directly represent the topology of the graph and they cannot evaluate graphs that were created outside the search tree. Confluence is also an issue that restricts the usage of both. The solution is to return to the Design Structure Matrix approach described above. Extensive research was completed to translate the CFGs into a common

DSM. This is important since different CFGs may have different number and type of nodes.

To find the distance between two DSMs, a matrix subtraction is performed followed by the taking the 2- norm of the resulting matrix (square root of the sum of the squares). This reduces it to a single dimensional number for comparison purposes. For example, if we have two designs that differ by just a single connectivity, the difference matrix of the two DSMs will have all zeros except for a single element which will be 1 that represents the difference between the two candidates. The 2-norm of this quantity will just be 1 which is the distance between the two candidates.

For a large set of solutions, many differences have to be calculated. For example, for a hundred candidate solutions to be organized using clustering, we will have the corresponding 100 DSMs. In the clustering procedures that will be discussed, it will be a repeated activity to find the distance between every pair of candidates. Calculation of the distance between two DSMs is a time consuming process, and needless repetition of this calculation can be eliminated by creating the D-DSM matrix beforehand and then initiate the clustering process. The D-DSM is a symmetric square matrix of the size of the number of candidates and the value at a particular row *i* and column *j* gives the distance between the DSMs of the candidates *i* and *j*. Thus the diagonal of this matrix which represents the distance between the same candidates is always 0. Calculation of the D-DSM is one of the major time-consuming activities in the entire clustering process. For 1000 candidates it takes between 20 and 30 minutes to calculate this matrix; for a ten-thousand it would take nearly two days. Fortunately, the clustering methods employed are such that recalculation of this matrix is minimized or never needed.

Using the DSM as the design parameter, we can now cluster the design space by choosing the correct clustering algorithm. Given the multidisciplinary application of clustering, there are numerous algorithms that are tuned for its various applications. For our problem of clustering designs using their DSM, it is important that the clustering method function in multidimensional space. If the DSM is considered as a coordinate dimensions of a candidate, then the candidates are present in an n^2 dimensional space, where n is the number of rows in the DSM. Also, the DSM are matrices comprised of integer values, if the clustering method requires a characterization of the center of a cluster, it will be difficult and meaningless to create an artificial DSM to characterize the centroid's coordinates. Therefore, clustering methods that do not require a mean or centroid solution are preferred.

The k-means algorithm [17] is the most popular approach used for clustering. The aim of this algorithm is to minimize the sum of the distances of each candidate from its corresponding cluster center. This is done in the following fashion:

1. Randomly select the cluster center positions.
2. Calculate which candidates will fit best with which cluster depending on their distance from the cluster centers.
3. With the list of candidates present in a particular cluster, recalculate the cluster center by finding the point that is equidistant from all the candidates.
4. Go back to step 2.

The iteration continues until it reaches a stable situation where none of the candidates are reassigned to a new cluster. There are some shortcomings of directly using this algorithm in this process. The D-DSM matrix takes a considerable amount of time to be recalculated and in this method it has to be calculated for every iteration because the distance of all the candidates from their respective cluster means needs to be calculated. Additionally, as stated above, this average candidate will be an imaginary candidate of no practical significance. A small modification to this approach is to take the median as the center of a cluster instead of the mean. This approach is called the K-medoid algorithm. Because of this modification the DDSM need not be calculated during every iteration.

A major disadvantage of the K-means and the K-medoid algorithm is the sensitivity of the results on the random choosing of the initial cluster centers and the arbitrary prior specification of the number of clusters. The global-k-means algorithm [25] solves this problem by repeating the method n times (where *n* is the required number of clusters) each time with the addition of a new cluster point. The randomness element in this method is very minimal compared to the original approach. The trade-off here is that there needs to be *n* trial runs for a single problem.

What we are using in this paper is a modified version of Global K-means algorithm, where, to overcome the problem of calculating the D-DSM for each iteration, instead of the mean, the median is used as done in the K-Medoid algorithm.

In order to measure the efficiency of our methods, the following criterion is used as the clustering error.

$$E(m_0, m_1, m_2, \dots, m_M) = \sum_{i=1}^{N}\sum_{k=1}^{M} I(x_i \in C_k)\|x_i - m_k\|^2$$

$$E(m_0, m_1, m_2, \dots, m_M) = \sum_{i=1}^{N}\sum_{k=1}^{M} I(x_i \in C_k)\|x_i - m_k\|^2$$

Eq 1

The above equation measures the distance of each of the candidate (x_i) from its cluster center (m_k). The distance between two candidates is calculated by the distance between their DSMs represented by the norm difference term ($\|x_i - m_k\|$). To ensure that the candidate distances are measured from their corresponding cluster medians, the function I(x) is used which is 1 if the condition is true and 0 otherwise. The cluster sets are represented by C_1, C_2, C_3 ... C_M where M is the number of clusters. The candidates, m_1, m_2, m_3 ... m_k are the corresponding medians of the M clusters.

If the candidates in the design space form dense groups of clusters that are widely separated from other cluster then this criteria will be produce of low error, E. In other words, if the error is found to be very low, then the design space can be and has been clustered effectively into groups of designs.

Results and Discussion

From the store of 8,386 CFGs, we invoke the Global K-Medoid clustering method on each of the 10 problems to cluster the design space. The number of clusters is tested from one to half the number of candidates in the design space. While the number of clusters could be increased, half the number is a practical upper limit as clustering into more clusters will force clusters to have just a single design. On average, around three hours is required to complete clustering on one of the ten problems.

Fig. 4 shows the CFGs of 3 toaster candidates from cluster 1 and 3 candidates from cluster 2 when the algorithm was run for 15 clusters. It can be seen that there are certain characteristics pertaining to the elements within a cluster and some distinguishing features of elements from the other cluster. For example, all three candidates in the second cluster have a pin connection between the switch and the nozzle plate. This is not present in the elements of the first cluster. Also the candidates in the second cluster do not get input to the nozzle plate directly, but through a component like the pressure tube or the adjustable knob lock spring for the cases shown.

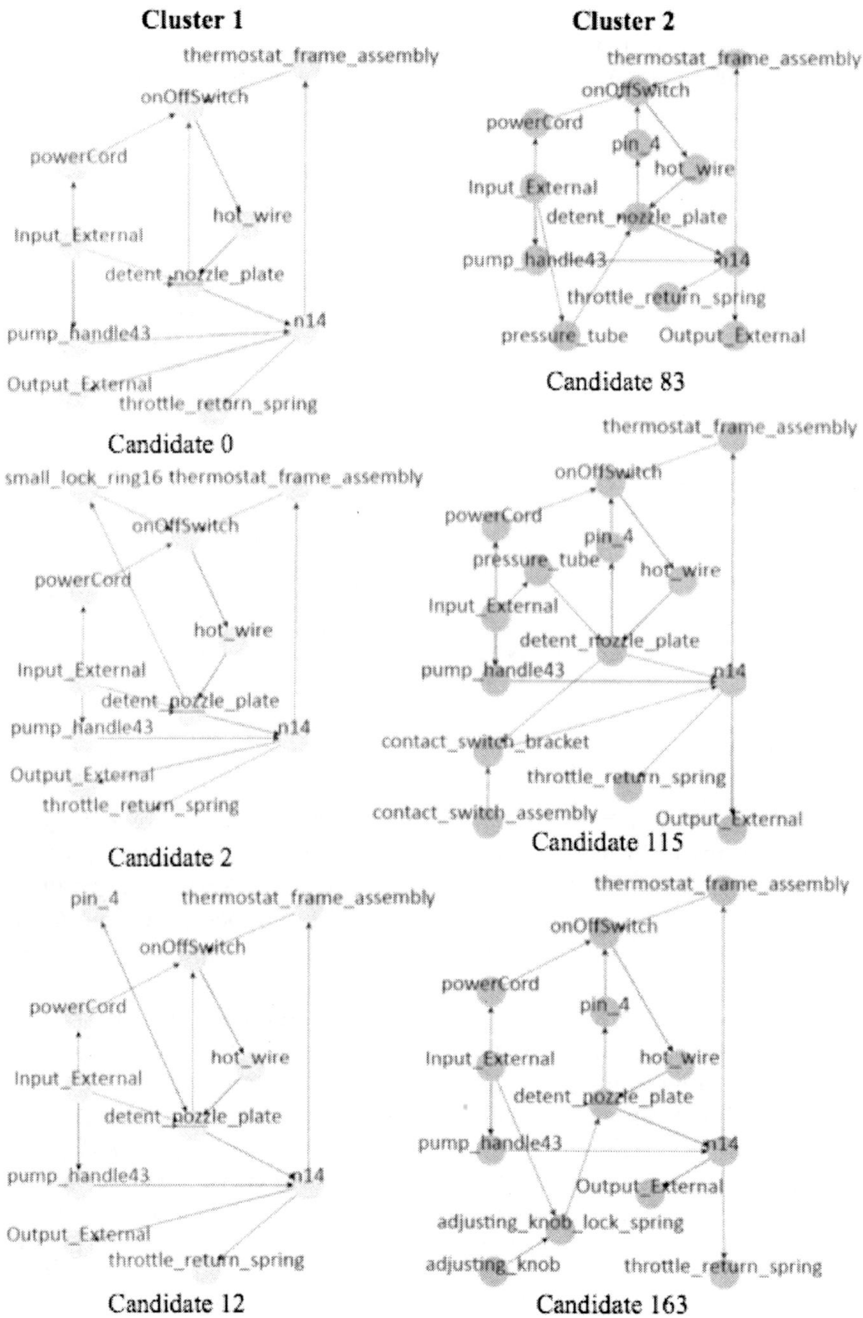
Cluster 1
Cluster 2
thermostat_frame_assembly
onOffSwitch
powerCord
Input_External
hot_wire
detent_nozzle_plate
pump_handle43
n14
Output_External
throttle_return_spring
Candidate 0
thermostat_frame_assembly
onOffSwitch
powerCord
pin_4
hot_wire
Input_External
detent_nozzle_plate
pump_handle43
n14
throttle_return_spring
pressure_tube
Output_External
Candidate 83
small_lock_ring16
thermostat_frame_assembly
onOffSwitch
powerCord
hot_wire
Input_External
detent_nozzle_plate
pump_handle43
n14
Output_External
throttle_return_spring
Candidate 2
thermostat_frame_assembly
onOffSwitch
powerCord
pin_4
pressure_tube
hot_wire
Input_External
detent_nozzle_plate
pump_handle43
n14
contact_switch_bracket
throttle_return_spring
contact_switch_assembly
Output_External
Candidate 115
pin_4
thermostat_frame_assembly
onOffSwitch
powerCord
hot_wire
Input_External
detent_nozzle_plate
pump_handle43
n14
Output_External
throttle_return_spring
Candidate 12
thermostat_frame_assembly
onOffSwitch
powerCord
pin_4
Input_External
hot_wire
detent_nozzle_plate
pump_handle43
n14
Output_External
adjusting_knob_lock_spring
adjusting_knob
throttle_return_spring
Candidate 163

Fig. 3 shows trend of the clustering error (Equation 1) as a function of the number of clusters. Recall that the clustering method functions by grouping all candidates into a single cluster and then introducing new clusters. It can be seen that the clustering error decreases with increasing number of clusters. It reaches the least value only when there are an equal number of clusters as the number of candidates in which case it will become zero. However, this does not present a meaningful solution to have hundreds of clusters for a design space of a thousand candidates.

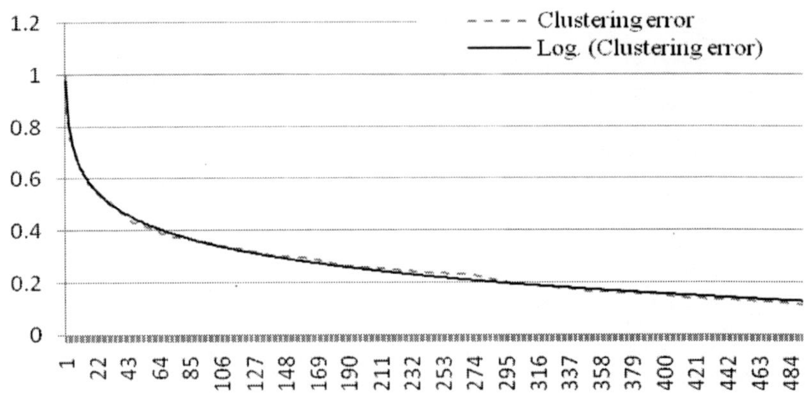

Fig. 3. Clustering error vs. number of clusters

What can be observed from the graph is that there is only a marginal decrease in the error as the number of clusters increase. As the number of clusters increase, we are adding more complexity to the organization of the design space. A meaningless increase in the number of categories of design will slow down the analysis of downstream processes. Thus we need to identify a particular number of clusters that provides an optimum against two opposing criteria: minimize the number of clusters, and minimize the clustering error. We choose to identify the "elbow" point or the sharpest point of the curve. A point where the curve starts to flatten out at a much faster rate is seen to be a good stopping point in the clustering process.

The clustering error graphs that are created have very high noise in them and this does not allow us to clearly see the general trend. A logarithmic trend-line is found to represent these graphs accurately as shown in Fig. 3.

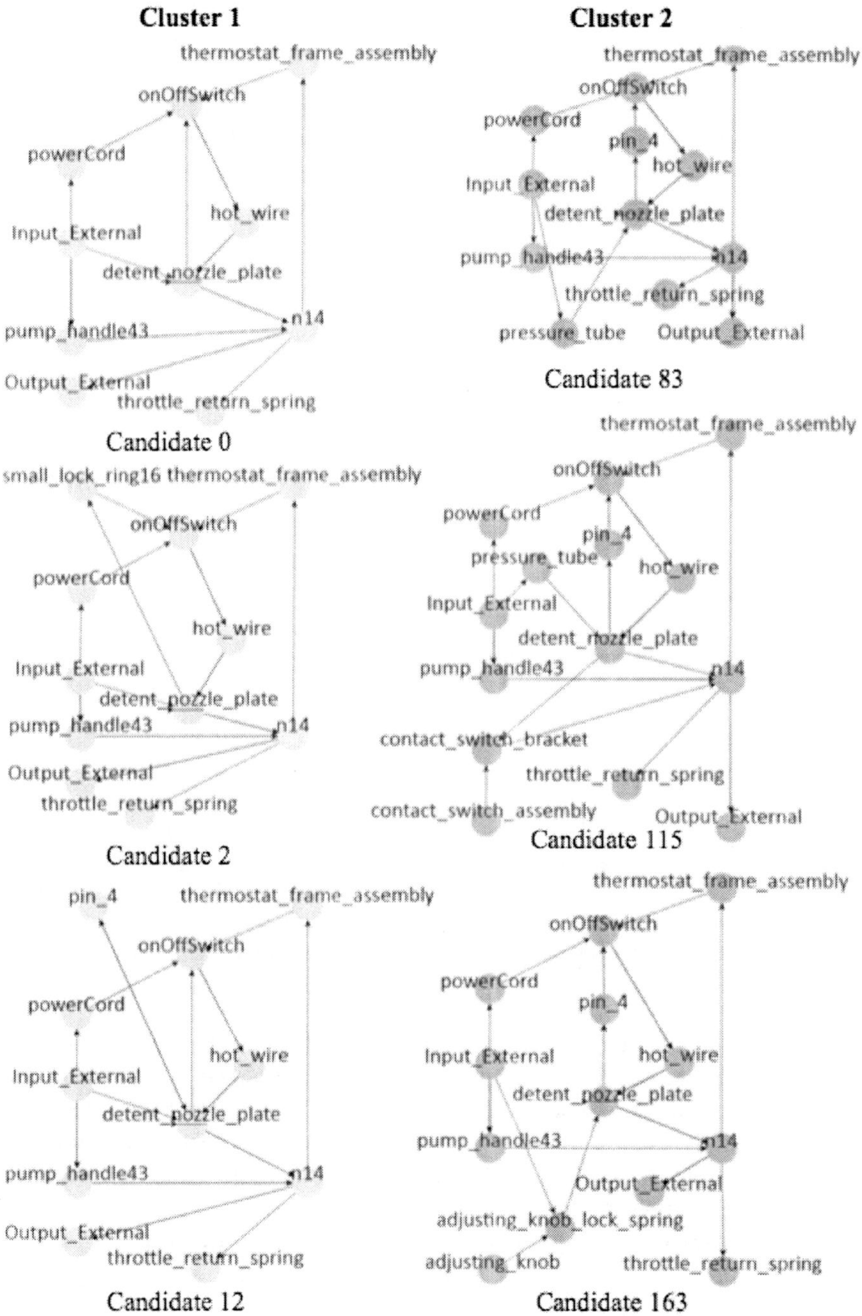

Fig. 4. CFGs of some candidates from cluster 1 and cluster 2 with the proctor toaster as the seed function structure

Also a logarithmic equation helps us to easily compare and analyze the curves. The logarithmic trend-line is created for all the 10 products and the following

Table 2 derives certain observations from the plot. Listed in the table are: the parameter m of the equation $y = m \log x + c$ (which is the generic form of the logarithmic trend-line equation), the number of candidates, whether an exhaustive search was completed, and an approximate range of the elbow point that is obtained by observation of the graphs.

From Table 2 we can make the following observations. It is shown that seven designs were stopped before the search process completed due to hardware limitation and three were allowed to complete. The sharpness of the logarithmic trend-line or the presence of a clear elbow point reduces as the number of candidates in the design space decreases, Figure 5.

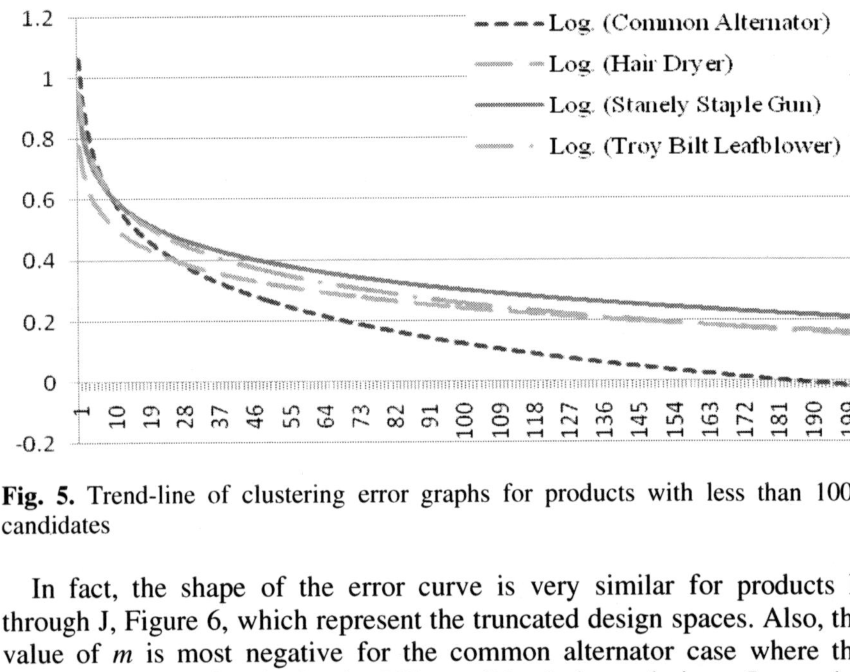

Fig. 5. Trend-line of clustering error graphs for products with less than 1000 candidates

In fact, the shape of the error curve is very similar for products E through J, Figure 6, which represent the truncated design spaces. Also, the value of m is most negative for the common alternator case where the entire gamut consisted of only 210 possible design solutions. It remains within the range of 0.12 to 0.14 for candidates that had 1000 designs even though the entire search tree was not scanned.

The parameter m is highly interesting as it shows the variation of the clustering error parameter and summarizes the shape of the graph in a single number. A low value of m indicates that the curve is blunt or there is no clear elbow point.

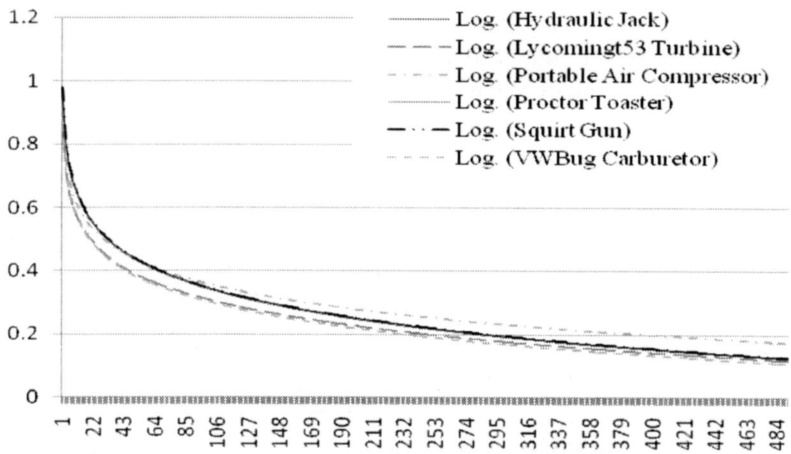

Fig. 6. Trend-line of clustering error graphs for products with less than 1000 candidates. Note that the optimum number of clusters for all of them is almost the same

Table 2 Clustering properties of the design spaces for each product

ID	Product name	m	Number of candidates	Approximate elbow point location (cluster count)	Complete design space
A	Common Alternator	-0.204	210	40 to 50	Yes
B	Troy Bilt Leaf blower	-0.151	480	25 to 35	Yes
C	Hair dryer	-0.117	768	16 to 20	No
D	Stanley Staple Gun	-0.132	928	15 to 25	Yes
E	VW Bug Carburettor	-0.141	1000	15 to 25	No
F	Hydraulic Jack	-0.138	1000	20 to 30	No
G	Lycomingt53 Turbine	-0.120	1000	20 to 30	No
H	Portable Air Compressor	-0.115	1000	20 to 30	No
I	Proctor Toaster	-0.138	1000	20 to 30	No
J	Squirt Gun	-0.138	1000	20 to 30	No

This indicates that the clustering error decreases at a high rate even as the number of clusters increase which means that the solutions are equally spaced and an efficient clustering can only be obtained with a large number of clusters. On the other hand, a high value of *m* indicates a sharp curve or a setting that after a certain number of clusters the organization of the design space does not improve. A high value of *m* also means that the candidates in the design space are organized into clumps by themselves and that a clustering algorithm is effective in identifying these clumps. For the products with large number of candidates, the value of m remains in the range 0.12 to 0.14. Compared to the case of the common alternator and the presence of an elbow point this is a relatively higher value of *m*. This means that the algorithm has detected the presence of lumps in the design space. In the case of the Common Alternator and the Troy Bilt Leaf blower, the lower values of m indicate that the designs are more uniformly distributed.

Conclusion

This paper presents ongoing research to automate the conceptual design of common electromechanical products. Through studies ten products, a set of 91 graph grammar rules are extracted automatically for use in generating new designs. Based on the characteristic CFGs created in this work and their corresponding DSMs, it can be seen that a global K-medoid algorithm can effectively organize the space into 20 to 30 clusters.

The number of candidates generated is indicative of the width of the search tree which is related to the number of options at each level of the tree. A lower number of candidates indicate a small set of rules used to create them, indicating that designs are more similar which is the reason for the inefficiency of the clustering algorithms on this design space. Clustering is difficult in such situation when the variety of rules used in the process is also low. Using a similar argument, it can be said that larger design space will have a larger variety of solutions and there will definitely exist designs that are very different from each other. Thus we will be able to differentiate the designs present in these design spaces better. This shows that clusters of designs exist and it is possible to identify them. Thus, if we have a large unmanageable design space, clustering appears to be an efficient approach in organizing it. It can be used to efficiently capture the variety in the designs generated by selecting candidates from different clusters that are topologically different from each other. This can assist an interactive tool similar to the interface present in [23] which uses a simple ranking scale where the knowledge of a human engineering designer is leveraged.

The use of clustering to compare and group divergent graphs such as these CFGs appear to be the new work on several levels. Ongoing research is investigating whether additional changes will need to be made to the global K-medoid algorithm in order to improve the accuracy and subsequently use this method to improve the efficiency in searching for the best design.

References

1. Kurtoglu, T., Swantner, A., Campbell, M.I.: Automating the Conceptual Design Process: From Black-box to Component Selection. In: Design Computing and Cognition 2008, pp. 553–572 (2008)
2. Pahl, G., Beitz, W.: Engineering Design: A Systematic Approach. Springer, London (1999)
3. Qian, L., Gero, J.S.: Function-behavior-structure paths and their role in analogy-based design. AI EDAM 10, 289–312 (1996)
4. Hundal, M.: A Systematic Method for Developing Function Structures, Solutions and Concept Variants. Mechanism and Machine Theory 25(3), 243–256 (1990)
5. Ward, A.C., Seering, W.P.: The performance of a mechanical design compiler. ASME, Design Engineering 17, 89–97 (1989)
6. Bracewell, R.H., Sharpe, J.E.E.: Functional Description Used in Computer Support for Qualitative Scheme Generation- Schemebuilder. Artificial Intelligence for Engineering Design, Analysis and Manufacturing 10(4), 333–345 (1996)
7. Chakrabarti, A., Bligh, T.: An Approach to Functional Synthesis of Mechanical Design Concepts: Theory, Applications and Emerging Research Issues. AI EDAM 10, 313–331 (1996)
8. Campbell, M., Cagan, J., Kotovsky, K.: Agent-based Synthesis of Electro-Mechanical Design Configurations. Journal of Mechanical Design 122(1), 61–69 (2000)
9. Rozenberg, G.: Handbook of Graph Grammars and Computing by Graph Transformation. World Scientific Publishing Company, Singapore (1997)
10. Sridharan, P., Campbell, M.I.: A Study on the Grammatical Construction of Function Structure. Artificial Intelligence for Engineering Design, Analysis and Manufacturing 19(3), 139–160 (2005)
11. Agarwal, M., Cagan, J.: A Blend of Different Tastes: The Language of Coffee Makers. Environment and Planning B: Planning and Design 25(2), 205–226 (1998)
12. Shea, K., Cagan, J., Fenves, S.J.: A Shape Annealing Approach to Optimal Truss Design with Dynamic Grouping of Members. ASME Journal of Mechanical Design 119(3), 388–394 (1997)

13. Patel, J., Campbell, M.I.: An Approach to Automate and Optimize Concept Generation of Sheet Metal Parts by Topological and Parametric Decoupling. Journal Of Mechanical Design (2010)
14. Starling, A.C., Shea, K.: A Grammatical Approach to Computational Generation of Mechanical Clock Designs. In: Proceedings of ICED 2003 International Conference on Engineering Design, Stockholm, Sweden (2003)
15. Swantner, A., Campbell, M.I.: Automated Synthesis and Optimization of Gear Train Topologies. In: Proceedings Of The ASME 2009 International Design Engineering Technical Conferences IDETC/CIE 2009, vol. DETC2009/86780. ASME, San Diego (2009)
16. Starling, A.C., Shea, K.: Virtual Synthesizers for Mechanical Gear Systems. In: Proceedings of ICED 2005 International Conference on Engineering Design, Melbourne, Australia (2005)
17. Hartigan, J.A.: Clustering Algorithms. John Wiley & Sons Inc., Chichester (1975)
18. `http://www.graphsynth.com`
19. `http://www.designfiles.org/packages/VOICED/Assemblies`
20. Yassine, A.: An Introduction to Modeling and Analyzing Complex Product Development Processes Using the Design Structure Matrix (DSM) Method. Quaderni di Management (Italian Management Review), No.9 (2004)
21. Browning, T.R.: Applying the design structure matrix to system decomposition and integration problems: a review and new directions. IEEE Transactions on Engineering management 48(3), 292–306 (2001)
22. Kurtoglu, T., Campbell, M.I.: A component taxonomy as a framework for computational design synthesis. Journal Of Computing And Information Science In Engineering 8(4), 1–10 (2008)
23. Campbell, M.I., Rai, R., Kurtoglu, T.: A stochastic graph grammar algorithm for interactive search. In: Proceedings of ASME 2009 International Design Engineering Technical Conferences & Computers and Information in Engineering Conference IDETC / CIE 2009 San Diego, California, USA (2009)
24. `http://www.ams.org/mathscinet-getitem?mr=0035935`
25. Likas, A., Vlassis, N., Verbeek, J.J.: The global k-means clustering algorithm. Pattern Recognition 36(2), 451–461 (2003)
26. Kurtoglu, T., Campbell, M.I.: Automated synthesis of electromechanical design configurations from empirical analysis of function to form mapping. Journal of Engineering Design 19(6) (2008)
27. Campbell, M.I.: A Graph Grammar Methodology for Generative Systems. University of Texas, Austin (2009)
28. Bishop, B., Nazmul, T., Campbell, M.: A Proposed Extensible Formalism and Initial Development for Representing Electromechanical Design Architecture and Fabrication. In: Proc. DESIGN 2010, The Design Society (2010)
29. Shigley, J., Mischke, C.: Mechanical Engineering Design. McGraw-Hill Science/Engineering/Math., New York (2004)

30. Caelli, T., Kosinov, S.: An Eigenspace Projection Clustering Method for Inexact Graph Matching. IEEE transactions on pattern analysis and machine intelligence, 515–519 (2004)
31. Campbell, M.I., Nakhjavani, O.B.: A Deterministic Global Optimization Method for Multimodal Spaces. Journal of Global Optimization (accepted, 2010)
32. Pine, B.J., Davis, S.: Mass customization: The new frontier in business competition. Harvard Business School Pr., Boston (1999)
33. Brown, K.N., Cagan, J.: Optimized Process Planning by Generative Simulated Annealing. Artificial Intelligence in Engineering Design, Analysis and Manufacturing 11, 219–235 (1997)
34. Schmidt, L., Cagan, J.: Recursive Annealing: A Computational Model for Machine Design. Research in Engineering Design 7(2), 102–125 (1995)

USING DESIGN COGNITION

Imaging the designing brain: A neurocognitive exploration of design thinking
Katerina Alexiou, Theodore Zamenopoulos and Sam Gilbert

A computational design system with cognitive features based on multi-objective evolutionary search with fuzzy information processing
Michael S Bitterman

Narrative bridging
Katarina Borg Gyllenback and Magnus Boman

Generic non-technical procedures in design problem solving: Is there any benefit to the clarification of task requirements?
Constance Winkelmann and Winfried Hacker

Virtual impression networks for capturing deep impressions
Toshiharu Taura, Eiko Yamamoto, Mohd Yusof Nor Fasiha and Yukari Nagai

Imaging the Designing Brain: A Neurocognitive Exploration of Design Thinking

Katerina Alexiou[1], Theodore Zamenopoulos[1], and Sam Gilbert[2]
[1]*The Open University, UK*
[2]*University College London, UK*

The paper presents a functional magnetic imaging study (fMRI) aimed at exploring the neurological basis of design thinking. The study carried out brain scans of volunteers while performing design and problem solving tasks. The findings suggest that (ill-structured) design thinking differs from well-structured problem solving in terms of overall levels of brain activity, but also in terms of patterns of functional interactions between brain regions. The paper introduces the methodology and the developed experimental framework, presents the findings, and discusses the potential role and contribution of brain imaging in design research.

Introduction

Design thinking is a fundamental human ability and an important vehicle for innovation and change in society. In the most general sense, design is perceived as a high level cognitive function responsible for our ability to construct or change our environment, in order to achieve some desire, need, idea or purpose. The cognitive nature of design has preoccupied researchers since the 60s, for example [1]. In the literature, design is perceived as a special kind of cognitive activity that involves dealing with open-ended, ill-structured problems, which do not have a single, optimal solution and require subjective interpretation and evaluation [2], [3], [4].

Although design is customarily taken to be a high level cognitive ability, to date there is very little research that provides evidence about the nature of design cognition from a biological or neurological perspective [5] and [6]. However, with new techniques for imaging brain activity becoming

J.S. Gero (ed.): *Design Computing and Cognition'10*, pp. 489–504.

more widely available, this area of research presents a significant opportunity for exploration.

Contemporary cognitive science generally considers that (at a smaller or greater extent) the brain has a modular organization, meaning that it is "structurally and functionally organized into discrete units or 'modules' and that these components interact to produce mental activities" ([7] page 947). Cognitive neuroscience uses various methods, including behavioral tests and brain imaging techniques to investigate the structure and organization of the brain that supports different cognitive functions. Cognitive neuroscience can thus help explore how design thinking is realised in the brain: what are the mental processes and cognitive functions that support design activity; and how design thinking is realised in terms of connectivity and interactions between neural circuits. In this sense, cognitive neuroscience research can offer insights to support the development of a theory of design cognition, as well as the development of computational design models.

Functional Magnetic Resonance Imaging or fMRI, which is the brain imaging technique used in this study, was developed at the beginning of the 1990s [8]. In contrast to typical brain MRI which uses magnetic and radio waves in order to visualize the 'structure' (or form) of the brain, fMRI captures changes in blood oxygenation which are associated with neural activation, thus aiming to capture the 'function' of the brain. The fMRI technique is non-invasive and has particularly good spatial resolution (picking up activity at the level of voxels of around 2-4 millimetres). The use of fMRI in cognitive science is one of the more rapidly growing areas of research focussing on the identification of brain areas that are specifically associated with different cognitive functions [9].

Certainly, localization of cognitive functions is not straightforward; it is often the case that a number of spatially distributed areas in the brain work together during a cognitive task, and so determining the interaction between different regions becomes of critical importance. Additionally, it is possible that the same cognitive process may be performed by recruiting different networks of neurons; and so it may not be possible to discover a unique association between certain functions and structures in the brain. Nonetheless, fMRI research is particularly well-suited to the investigation of the spatial organization of brain processes supporting cognitive functions and has already contributed greatly to the understanding of the neurological basis of cognitive abilities.

The paper reports a recent study, in which we used fMRI in order to identify brain areas and cognitive functions associated with design thinking, and examine the potential methodological role of brain imaging in design research [10], [11].

Research Questions and Objectives

What makes design thinking special?: This is one of the most important questions for design research, but also design education and design practice in general. The identification of the peculiarity of design as a cognitive process is important for the establishment of design as a discipline, a distinct activity, which can be taught and learned in design schools.

Perhaps the most common characterization of design as a distinct type or mode of thinking derives by reference to the nature of design 'problems' or tasks. Simon [12] conceptualized design as a process involving 'ill-structured' or 'ill-defined' problems where the goal state is unknown but also the criteria for deciding when the problem is solved are not well specified. Similarly, Rittel and Webber [2] characterized design problems as 'wicked'. Most contemporary literature agrees with those characterizations of design problems as open-ended and indeterminate. In response to such problems, design thinking requires a hermeneutic act; it necessitates the use of subjective interpretations and value judgments. It also requires a reflective practice of finding and framing the problem together with its solution [13]. To put it differently, design requires the formulation of an interpretation or vision about a design problem/task, together with a plan, a solution, that will satisfy this vision. These characterizations have been derived mainly from observations of designers at work, empirical studies of individual design processes and even personal experiences of design thinking in practice, often in association with general theories of cognition. Here we look for evidence in the brain to support the characterization of design as a distinct type of cognitive function.

Previous evidence about the brain areas involved in design or ill-structured problem solving indicates the involvement of the prefrontal cortex. Goel and Grafman [5] studied an architect with a right dorsolateral prefrontal cortex lesion on an architectural planning task, involving designing a new office space. Despite good performance on a range of other tests (Stroop test, Wisconsin Card Sorting Test, Tower of London, verbal fluency), the patient performed poorly on the design task, compared with an age- and education-matched control participant (also an architect). In the present study, we seek to investigate ill-structured design cognition, following on from these results using functional magnetic resonance imaging (fMRI).

In particular, we wish to explore two issues. The first is to identify brain regions that support design thinking, and the second is to identify patterns of functional interactions between brain regions that characterize design activity. Results can contribute both to the formation of psychological theories of the mental processes that contribute to design abilities, and to the development of neuroscience accounts of how these processes relate to the function of specific brain areas. In effect, there is a two-way exchange between design research and cognitive science research. On the one hand, existing neurophysiological knowledge about the specialized function of certain brain areas, may help us unpick certain characteristics of design cognition and inform design theory. That is, knowing which areas support which cognitive function (e.g. visual or spatial thinking), and knowing which areas support design activity (and how they interact), we can construe a more detailed understanding of the mental processes involved in design thinking. On the other hand, by identifying the 'signature' of design thinking in the brain we can develop a better understanding of the role and possible involvement of design thinking in other tasks or cognitive activities.

Methods

The use of neuroimaging for the study of ill-structured tasks (like design) is a novel and underdeveloped area. This can probably be traced to the fact that neuroimaging studies require strict experimental control over participants' behaviour, which is notoriously difficult to achieve with ill-structured tasks where participants need to structure their behaviour themselves.

Experimental Set-Up and Tasks

The lack of experimental control in ill-structured tasks creates at least two methodological problems. The first methodological problem relates to the difficulty of finding a suitable control condition to match to the ill-structured task in terms of basic input/output operations, because there may be relatively little control over the sensorimotor processing involved in the ill-structured task. In order to deal with this potential problem, we made efforts in the present study to match the experimental and the control conditions as closely as possible in terms of visual input and motor output.

More specifically, for this study we had to develop an experimental setting that would allow us to compare design with another closely related cognitive function, measure the accompanying brain activity, and correlate differences in brain activity with differences in cognitive activity. To do this we devised an experiment where subjects were asked to perform two types of tasks while in the fMRI scanner: one that corresponds to well-structured problem-solving and one that corresponds to ill-structured design. As discussed in the introduction, design is most commonly considered as an ill-structured task, which can be compared to well-structured problem solving. Although there is some ambiguity as to whether design is a special case of problem-solving or a completely distinct mode of thinking, the distinguishing characteristics of design are more or less generally agreed. In problem-solving theory, the problem space is a representation of a set of possible states, a set of 'legal' operations, as well as an evaluation function or stopping criteria for the problem-solving task [14], [15]. The solution space incorporates all those states that achieve the requirements expressed by the problem space. In same cases, the solution space is thought to constitute the set of all possible entities or states upon which an evaluation function is applied. In well-defined problems the task environment determines a set of legal operations over possible states, as well as the evaluation criteria that effectively determine when a solution has been identified. In ill-defined problems the task environment does not effectively determine when a solution is found. According to this view, design problems are ill-defined problems, in the sense that the means (i.e. the representation of the problem space and the possible operations over the problem space), as well as the ends (i.e. the evaluation function or the stopping criteria) are not given in the task environment but are part of the design process [12], [16]. Other researchers prefer to talk about the mutual influence between problem and solution in design tasks: while problem-solving supposes the existence of a defined problem that circumscribes the solution, designing involves defining the problem together with the solution [13], [17]. For this reason we took the distinction between design and (well-defined) problem-solving as the basis for our investigation.

The unique difference between the two types of tasks as defined here is that the design task requires not only generation of solutions but also interpretation of the problem requirements and definition of the criteria for evaluating the solution, Table 1. It is important to note that the distinction between well-defined problem solving tasks and design tasks has a methodological role in this study: it allows us to identify whether the two tasks are accompanied with different patterns of brain activation and therefore associate these differences with differences in cognitive functions. However, it is not necessary to assume a strict separation between design and problem solving tasks to interpret the present results. Even if the two types of task are considered to vary along a continuum, with the design tasks being relatively ill-defined in comparison with the problem-solving tasks, a cognitive subtraction between the two will reveal brain areas more strongly engaged in solving ill-defined design problems.

Table 1 The table summarizes how this study sees the definition and distinction between well-structured problem solving tasks and design tasks for the purpose of the experimentation. The distinction is made based on terms that appear in problem solving literature [12], [18]. The results of this study should be interpreted in the context of this definition.

Well-structured problem solving tasks	Ill-structured or open-ended problem solving tasks	Design tasks
The problem statement contains the evaluation criteria that determine when a solution is found.	The problem statement contains evaluation criteria that cannot determine when a solution is found.	The problem statement requires constructing an interpretation of the evaluation criteria that determine when a solution is found.
There is a unique family of correct solutions. (there is a unique space of possible solutions and legal operations)	There is no unique family of correct solutions or no solutions at all. (there is no unique space of possible solutions and legal operations)	The task requires the construction of solutions that will satisfy the constructed interpretation of evaluation criteria.

A second methodological problem caused by the lack of experimental control over ill-structured tasks is that such tasks may potentially involve a wide range of cognitive processes. With little experimental control over the time at which particular processes are engaged, it is difficult to link brain activity with specific processes. In order to address this difficulty, we split our tasks into study and performance phases, so that we could investigate differences between ill-structured and well-structured conditions specifically associated with thinking phases of problem structuring and solution generation, rather than with the phase of executing solutions, or simply performing the tasks.

Figure 1 shows examples of the design and problem solving tasks used in the experiment. Note that the particular tasks shown in the figures are essentially spatial in nature and very close to the type of task that has been employed to empirically study design cognition since the 60's [1]. However, the set of tasks used in the experiment were designed so as to equally include other more visual or abstract reasoning tasks (e.g. graphic design, reasoning with abstract shapes etc). Note that the two types of tasks required the same amount of time to study and solve, as well as a similar number and type of operations (such as moving and rotating objects) to perform.

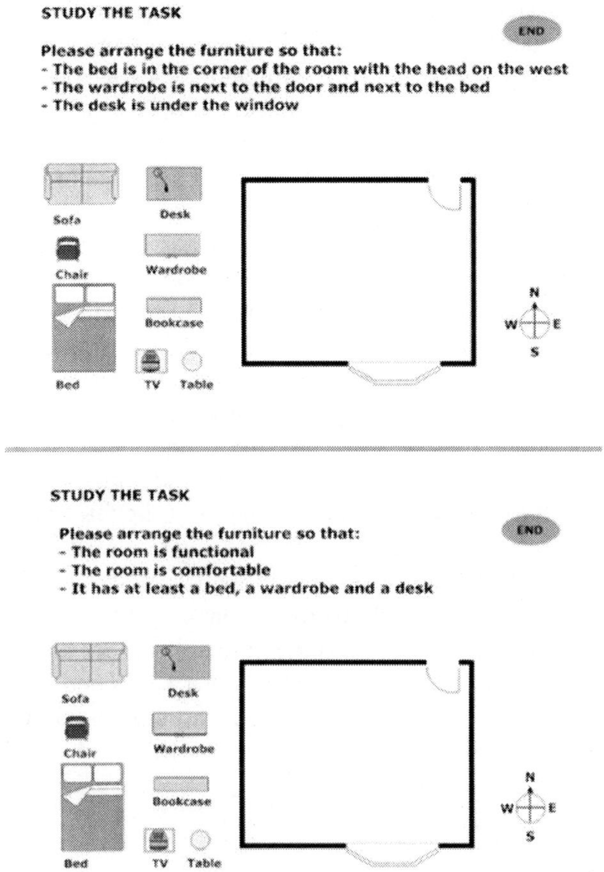

Fig. 1. An example of a problem-solving task (top) and its matched design task (bottom), which were used in the experiment.

To evaluate the appropriateness of the tasks chosen for the fMRI study, and ensure that the level of difficulty and time given was apposite, we also conducted semi-structured interviews after the end of the scanning sessions to elicit participants' views. Details and results from the participants' evaluation are discussed in more detail in the next sections.

Procedures

The fMRI study involved 18 participants, 11 female and 7 male, aged 27-60. All participants had some experience and familiarity with design, and 10 of them had formal training in a design discipline (architecture, multimedia or graphic design, interior design, product design, art etc). Data from one participant were discarded due to data quality problems. All the volunteers provided written informed consent before participating. The study was carried out in accordance with an ethics approval granted by the Ethics Committee of the Open University and following guidelines of the British Psychological Society and the Data Protection Act 1998.

Imaging was performed with a Siemens TIM Avanto 1.5 Tesla MRI scanner. A head coil was placed on the top of the head of each participant. A mirror was attached to the head coil, allowing participants to view the stimuli projected clearly onto a screen hanging outside the magnet and within their visual field. Headphones were used to reduce the noise made by the scanner while in operation. There were 8 design and 8 problem-solving tasks presented to participants in alternate order, covering different design domains. In both cases, participants were first presented with a study phase in which they saw a collection of items next to a blank space and instructions describing each task. The study phase lasted for 30s. Following the study phase, a performance phase of 50s commenced. The time was the same for each type of task (design or well-defined problem solving). The phase was indicated by an instruction at the top of the screen saying ‘Study the task’ or ‘Perform the task’. The participants used a trackball mouse to click-and-drag objects displayed in order to fulfil the given instructions. After each task, there was a 15 second rest period (that was used as baseline condition), in which participants viewed a fixation cross, until the next task.

Results

Semi-Structured Interviews

After completing the experiment in the scanner, the participants were asked to reflect on the tasks and their own cognitive process. It is important to note that the participants were informed about the general aim of investigating design and problem-solving cognition, but were not informed in advance about the hypothesised difference between design and problem-solving. So in the interview participants were asked whether they identified the existence of different types of tasks, and were invited to express their own perception of any difference.

All but three participants identified that there were two groups of tasks. According to the participants’ own words, one group contained tasks which were “more logical”, “more prescribed”, or “more objective”. In these tasks “you had to follow the instructions”, “do what you were told”, “understand the rules and obey them”. The tasks “were right or wrong” contained “clear instructions” and had “a finite answer”. The other group contained tasks that were more "open-ended", "free-style" or "subjective". In these tasks "you had to use your own interpretation", "think about more options, or more implications" "take control of what you are doing" and "decide how you interpret, how you want to create". The tasks "were more subjective, you couldn't say there was a right or wrong answer", they were "open to interpretation" and required "qualitative judgements". As discussed, the aim of our experimental design was to have two distinct groups of tasks: the first was meant to include tasks for which the criteria for deciding when a solution is found would be given, as well as a definition of legal moves leading to the solution. The second was meant to consist of tasks which would be open ended, and would require interpretation and evaluation of the criteria for deciding what constitutes a solution. The participants' observations confirm that our experimental design was successful in that respect.

Behavioral Analysis

Videos of participants' behaviour in each performance phase were analysed, and the following behavioural measures were recorded: 1) time until first movement of the trackball; 2) time until first object was clicked; 3) total number of clicks (excluding clicking on the same object twice in a row); 4) total number of revisits, i.e. returns to objects that had already been clicked previously. None of these measures differed significantly between design and problem-solving tasks, suggesting that the two types of task were well matched on initiation time following the study phase, and total movement complexity. In other words, both tasks were similar in terms of level of complexity.

fMRI Analysis

The data obtained from the experiment were studied using SPM8, a statistical package developed by the Wellcome Trust Centre for Neuroimaging in UCL for the analysis of brain imaging data sequences. The software works within Matlab (Mathworks, Inc). SPM involves procedures so that the images obtained from the MRI scanner are realigned in order to compensate head movements, and spatially normalised into a standard space, a template brain (the Montreal Neurological Institute MNI template), in order to be able to compare activation between participants. The data are also spatially smoothed in order to improve registration between participants and increase statistical reliability. For more details about SPM see [19].

The analysis was carried out on the basis of a comparison between different types of activity (or phases) defined in the experiment: studying (S), performing (P), studying design tasks (DS), studying problem-solving tasks (PS), performing design tasks (DP) and performing problem-solving tasks (PP). The software allows looking at the brain activation of participants at an aggregate level, and making comparisons so as to identify whether specific areas are more activated during particular phases.

First we made direct comparisons between the study and performance phases, collapsing over the Problem Solving and Design tasks. The results show a clear pattern of activation differentiating the two phases (Figure 2). The study phase was associated with greater activity in a predominantly right-lateralized network including right occipital cortex, right lateral temporal cortex, right intraparietal sulcus, right lateral PFC and bilateral ventromedial PFC. The performance phase was associated with widespread bilateral activation in motor and premotor cortices, inferior parietal cortex, medial occipital cortex, cerebellum, and thalamus. It is known from neuroscientific research that motor and premotor areas are associated with movement and planning of movement, whereas the cerebellum is associated with the integration of sensory perception, coordination and control of movement. Activation in the prefrontal and temporal cortex is on the contrary usually associated with high-level cognitive processes. The results reflect the experimental separation of the tasks in two distinct phases, and confirm the hypothesis that during the study phase the participants were primarily involved in thinking about problems and solutions, while in the performance phase they were primarily engaged in carrying out their solutions by using the mouse.

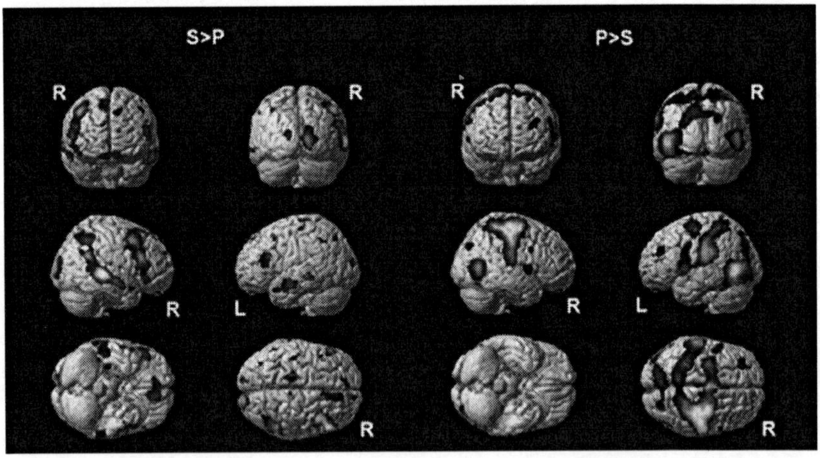

Fig. 2. 3D brain images produced from SPM showing activation when comparing studying versus performing (left), and performing versus studying (right) ($p<0.001$) collapsed over design and problem-solving. The top row shows anterior and posterior views, the middle row shows lateral views and the bottom row shows inferior and superior views of the brain.

Having compared the study and performance phases we moved on to consider the comparison between the phases of studying design (DS) and studying problem-solving (PS) tasks. Table 2 shows the areas that are activated in DS versus PS.

Table 2 Regions showing significant difference in activation when comparing the phase of studying design versus studying problem-solving ($p < 0.001$). Regions are designated using MNI (x, y, z) coordinates. Results are shown for $Z > 3.5$.

Region	Hemisphere	x	y	z	Z score
Anterior Cingulate Gyrus	L	-14	6	38	4.15
	R	14	22	38	3.33
Middle Temporal Gyrus	R	44	0	-22	3.96
Middle Frontal Gyrus	R	24	22	38	3.57
Dorsolateral Prefrontal Cortex	R	50	30	36	3.53

We see that there is statistically significant accompanying activation in the anterior cingulate cortex, middle temporal gyrus, and middle frontal gyrus. Let us examine these results in more detail.

The *dorsolateral prefrontal cortex* (DLPFC) is generally thought to be involved in executive function, working memory and directed attention [20]. Research shows that damage in this area may result in impaired executive function. We have already discussed that dorsolateral prefrontal cortex was also specifically found to be involved in ill-structured problem solving and particularly the ability to perform lateral transformations or set-shifts necessary for solving ill-structured problems [5]. The *anterior cingulate cortex* (ACC) is also part of the PFC and like the dorsolateral prefrontal cortex is generally thought to take part in executive function, particularly in supporting the coordination and modulation of information processing in other brain areas. It is also generally acknowledged that the ACC is associated with cognitive as well as emotional (affective) functions which are linked structurally to the dorsal and rostral parts of the cingulate cortex respectively. What is particularly relevant to our study is that dorsal ACC and areas of the lateral prefrontal cortex work together during tasks that involve high levels of cognitive effort. The exact role played by each area however is an open question. Perhaps the most general conjecture is that ACC mediates attention and selection of appropriate responses or behaviors, while the lateral PFC is engaged in the generation and maintenance of schemata (goals and means) for responding to novel tasks. It has also been suggested that ACC plays an evaluative role, being part of a network of cells that partake in evaluation of motivation, anticipation of tasks and events, error detection and encoding of reward values. For more details on this discussion see [21], [22], [23] and [24]. We also find heightened activation in *temporal gyrus area*. The temporal lobe is associated with language and semantic processing, multi-sensory integration, as well as memory encoding and retrieval. Finally, heightened activation in DS>PS was also found in the *medial frontal gyrus*. This area includes the frontal eye fields, a region associated with voluntary eye saccades and gaze control. Although it is difficult to ascertain whether the area found in the study is indeed located in the frontal eye fields, the heightened activation in studying design versus studying problem-solving tasks may be due to increased demand for examining, comparing and attending to various features of the stimuli.

Of these regions only dorsolateral prefrontal cortex was also activated in the study versus perform comparison and as this area was our a-priori region of interest (following Goel and Grafman [5]), we went on to investigate whether this region showed functional connectivity with other brain regions that differed between the two conditions (design and problem-solving). We investigated functional connectivity using as a seed co-ordinate the region defined by the contrast of study versus perform phases (orthogonal to the design / problem solving distinction). An exploratory analysis at an uncorrected threshold of $p < .001$ with 5 voxel minimum extent revealed widespread activation for the positive contrast (i.e. greater functional connectivity with right dorsolateral PFC during design study compared with problem-solving study phases). A total of 26 clusters were activated in this contrast, yielding a set-level probability of $p<.001$ [25]. This reveals that there was significantly greater activity associated with this contrast than would be expected by chance. Seeing as the voxel-level analysis did not reveal any significant activations at a corrected threshold, results from specific regions are preliminary. However, two brain regions showing increased coupling with right DLPFC were of particular theoretical interest (Figure 3).

The region showing the strongest modulation of coupling with right DLPFC, depending on problem type, was the *precuneus*. This area has previously been suggested to support visual imagery [26]. We might therefore speculate that during the study phases of design tasks, where participants had to generate potential solutions without yet interacting with the visual display, participants engaged strongly in visual imagery, mediated by interactions between right DLPFC and precuneus. A second region showing greater coupling with right DLPFC during the study phases of design than problem-solving tasks was *left frontal pole*. This region has been particularly implicated in dealing with ill-structured situations [27] and with attending to self-generated, internally-represented information [28], [29] and [30].

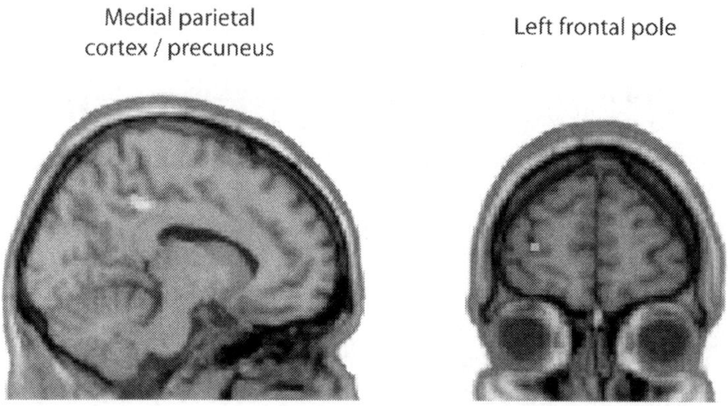

Fig. 3. Two regions showing significantly greater effective connectivity with right dorsolateral prefrontal cortex during design versus problem-solving tasks. The illustration on the left is a saggital slice at x=-10 showing the activated area in the precuneus; the right illustration is a coronal slice at y=58 showing the activated area in left frontal pole.

Discussion and Conclusions

Although this is only a limited study of design cognition, the results show that research in cognitive neuroscience and particularly neuroimaging studies may offer interesting insights into the nature of design tasks and design thinking.

The findings suggest that design and problem-solving involve distinct cognitive functions associated with distinct brain networks. The discovered activation in the right dorsolateral prefrontal cortex for design versus problem solving is consistent with previous studies focussing on features of design and problem solving such as insight and the ability to perform lateral transformations and set-shifts. Additionally, the results are consistent with the view that design cognition essentially also involves evaluation and modulation of alternative goal states, which may be supported by the anterior cingulate cortex. The experimental setting suggests that the activation of these two areas in design must correspond to the fact that the task environment of design is open-ended and affords different interpretations about what is the task at hand or how this task should be evaluated. In this environment, the brain needs to develop interpretations and visions about the task, but also to devise plans that satisfy this vision.

We therefore suggest that the ACC worked together with the DLPFC in order not only to construct new schemes of actions in response to a problem, but also in order to construct representations and semantics within which these actions are defined. So this observation leads us to the idea that as different areas of the brain spontaneously react to this open-ended environment, a number of different interpretations and possible responses are formulated in the brain leading to representational conflicts. It is these conflicts that the ACC and DLPFC need to address.

Our findings also suggest that compared to problem-solving, studying design tasks recruits a more extensive network of brain regions including areas involved in visual imagery, semantic processing and multi-sensory integration.

Based on these results we can construct a model of design thinking as a characteristic phenomenon, which involves the coordination of two layers of brain activity, one that incorporates bottom-up processing of information and one that incorporates top-down executive processing. The first layer consists of spontaneous brain activity responsible for constructing an emotional and cognitive representation of the task environment. This activity is thought to be realized mainly in the temporal, occipital and parietal (TOP) regions of the brain, including areas for representation of somatosensory inputs, spatial and visual representations as well as linguistic representations. Brain activity at this level is spontaneous in the sense that it constitutes a direct response to the task environment. The second layer consists of supervisory brain activity responsible for detecting and monitoring conflicts in different brain areas (conflicting representations) and constructing executive schemes of action. This activity is mainly realized in the PFC.

Design thinking in the brain initiates with the appearance of incomplete representations and conflicts among different representations of the task environment; or conflicts between internal and external representations of the task environment. These inconsistencies or conflicts are formed because design tasks afford different interpretations and visions about what is an appropriate response or 'solution'. The activation of the anterior cingulate and dorsolateral areas of the prefrontal cortex which characterises design thinking, in effect signifies a process of coordinating different representations and involves the formulation and reformulation of internal representations and appropriate courses of action. Further studies will of course be required to evaluate the precise role of these two brain regions in design cognition and support this hypothesis.

From a methodological perspective, the findings of the study indicate that ill-structured tasks may be fruitfully examined using fMRI as well as neuropsychological approaches. Future studies will be required to investigate in greater detail the cognitive processes, as well as the brain regions and networks involved in design thinking, focusing for example on problem structuring, evaluation, or hypothesis generation.

References

1. Eastman, C.M.: Cognitive processes and ill-defined problems: a case study from design. In: Proc. of the First Joint International Conference on Artificial Intelligence, Washington, DC (1969)
2. Rittel, H.W.J., Webber, M.M.: Planning problems are wicked problems. In: Cross, N. (ed.) Developments in Design Methodology, pp. 135–144. John Wiley & Sons, New York (1984)
3. Buchanan, R.: Wicked problems in design thinking. Design Studies 8, 5–22 (1992)
4. Lawson, B.: How Designers Think: The Design Process Demystified. Architectural Press, Oxford (1997)
5. Goel, V., Grafman, J.: Role of the right prefrontal cortex in ill-structured planning. Cognitive Neuropsychology 17, 415–436 (2000)
6. Vartanian, O., Goel, V.: Neural correlates of creative cognition. In: Martindale, C., Locher, P., Petrov, V.M. (eds.) Evolutionary and Neurocognitive Approaches to the Arts, pp. 195–207. Baywood Publishing, Amityville (2005)
7. Gazzaniga, M.S.: Organization of the human brain. Science 245(4921), 947–952 (1989)
8. Ogawa, S., Lee, T.M., Kay, A.R., Tank, D.W.: Brain magnetic resonance imaging with contrast dependent on blood oxygenation. PNAS 87(24), 9868–9872 (1990)
9. Cabeza, R., Nyberg, L.: Imaging Cognition II: an empirical review of 275 PET and fMRI studies. Journal of Cognitive Neuroscience 12(1), 1–47 (2000)
10. Alexiou, K., Zamenopoulos, T., Johnson, J., Gilbert, S.: Exploring the neurological basis of design cognition using brain imaging: some preliminary results. Design Studies 30(6), 623–647 (2009)
11. Gilbert, S., Zamenopoulos, T., Alexiou, K., Johnson, J.: Involvement of right dorsolateral prefrontal cortex in ill-structured design cognition: An fMRI study. Brain Research 1312, 79–88 (2010)
12. Simon, H.A.: The structure of ill-structured problems. Artificial Intelligence 4(4), 181–201 (1973)
13. Dorst, K., Cross, N.: Creativity in the design process: Co-evolution of problem-solution. Design Studies 22, 425–437 (2001)
14. Ernst, G.W., Newell, A.: GPS: A Case Study in Generality and Problem Solving. Academic Press, Inc., New York (1969)
15. Newell, A., Simon, H.A.: Human Problem Solving. Prentice Hall, Englewood Cliffs (1972)
16. Goel, V., Pirolli, P.: The structure of design problem spaces. Cognitive Science 16(3), 395–429 (1992)
17. Dorst, K., Dijkhuis, J.: Comparing paradigms for describing design activity. Design Studies 16(2), 261–274 (1995)

18. Schraw, G., Dunkle, M.E., Bendixen, L.D.: Cognitive processes in well-defined and ill-defined problem solving. Applied Cognitive Psychology 9(6), 523–538 (1995)
19. Friston, K.J., Ashburner, J.T., Kiebel, S.J., Nichols, T.E., Penny, W.D.: Statistical Parametric Mapping: the Analysis of Functional Brain Images. Academic Press, London (2007)
20. Miller, E.K., Cohen, J.D.: An integrative theory of prefrontal cortex function. Annual Review of Neuroscience 24, 167–202 (2001)
21. Bush, G., Luu, P., Posner, M.I.: Cognitive and emotional influences in anterior cingulate cortex. Trends in Cognitive Sciences 4(6), 215–222 (2000)
22. Milham, M.P., Banich, M.T., Webb, A., Barad, V., Cohen, N.J., Wszalek, T., Kramer, A.F.: The relative involvement of anterior cingulate and prefrontal cortex in attentional control depends on nature of conflict. Cognitive Brain Research 12(3), 467–473 (2001)
23. Botvinick, M.M., Cohen, J.D., Carter, C.S.: Conflict monitoring and anterior cingulate cortex: An update. Trends in Cognitive Sciences 8(12), 539–546 (2004)
24. Carter, C.S., Van Veen, V.: Anterior cingulate cortex and conflict detection: an update of theory and data. Cognitive, Affective and Behavioral Neuroscience 7(4), 367–379 (2007)
25. Friston, K.J., Holmes, A., Poline, J.B., Price, C.J., Frith, C.D.: Detecting activations in PET and fMRI: Levels of inference and power. Neuroimage 4, 223–235 (1996)
26. Fletcher, P.C., Frith, C.D., Baker, S.C., Shallice, T., Frackowiak, R.S., Dolan, R.J.: The mind's eye–precuneus activation in memory-related imagery. Neuroimage 2, 195–200 (1995)
27. Burgess, P.W., Alderman, N., Volle, E., Benoit, R.G., Gilbert, S.J.: Mesulam's frontal lobe mystery re-examined. Restorative Neurology and Neuroscience (in press)
28. Burgess, P.W., Dumontheil, I., Gilbert, S.J.: The gateway hypothesis of rostral prefrontal cortex (area 10) function. Trends in Cognitive Sciences 11, 290–298 (2007)
29. Christoff, K., Ream, J.M., Geddes, L.P., Gabrieli, J.D.: Evaluating self-generated information: anterior prefrontal contributions to human cognition. Behav. Neurosci. 117, 1161–1168 (2003)
30. Gilbert, S.J., Frith, C.D., Burgess, P.W.: Involvement of rostral prefrontal cortex in selection between stimulus-oriented and stimulus-independent thought. European J. of Neuroscience 21, 1423–1431 (2005)

A Computational Design System with Cognitive Features Based on Multi-objective Evolutionary Search with Fuzzy Information Processing

Michael S. Bittermann
Delft University of Technology, The Netherlands

A system for architectural design is presented, which is based on combining a multi-objective evolutionary algorithm with a fuzzy information processing system. The aim of the system is to identify optimal solutions for multiple criteria that involve linguistic concepts, and to systematically identify a most suitable solution among the alternatives. The system possesses cognitive features, where cognition is defined as final decision-making based not exclusively on optimization outcomes, but also on some higher-order aspects, which do not play role in the pure optimization process. That is, the machine is able to distinguish among the equivalently valid solution alternatives it generated, where the distinction is based on second order preferences that were not pin-pointed by the designer prior to the computational design process. This is accomplished through integrating fuzzy information processing into the multi-objective evolutionary search, so that second-order information can be inductively obtained from the search process. The machine cognition is exemplified by means of a design example, where a number of objects are optimally placed according to a number of architectural criteria.

Introduction

Design is complex. This is because it involves conflicting goals that are often vague. For example a design is demanded to be functional, look appealing and have moderate costs. The vagueness of these objectives makes it problematic to precisely compare alternative solutions during design, and the conflicting nature of the objectives make it problematic to

J.S. Gero (ed.): *Design Computing and Cognition'10*, pp. 505–524.

reach optimality. Another source of complexity is that the amount of possible solutions is excessively large in general. This is due to the combinatorial explosion in the parameter domain, making it difficult to ensure a designer does not miss a superior solution during the design process. A third source of complexity is that prior to the design it is generally not clear how important goals are relative to each other. For example it is difficult to specify exactly how important the functionality of a design should be taken compared to cost aspects prior to knowing what the implication of such a commitment is. That is, before finding most suitable solutions for the goals, and thereby becoming aware of the nature of the inevitable trade-offs occurring in the task at hand, it is premature to commit to a relative importance among high-level goals.

The complexity makes it difficult to accomplish scientific means for design enhancement, i.e. providing designers with means that ensure to some extend the suitability of their designs for an intended purpose. There have been a number of works addressing this issue. The first ones were using methods of classical artificial intelligence [1-3]. These methods are based a rigid inference mechanism. Namely, the identification of a deficiency in a design has a set of actions associated for improving the deficiency that is defined a-priori. For example in [3], when it is detected that a room has insufficient openness, the computations move or remove elements blocking the view. This predefined association of goal and action is clearly a problematic approach when *openness* for instance is not the only objective at hand, but other factors play a role as well, such as costs, functionality etc. Also the approach is challenged when the action to take involves a number of parameters to decide upon in combination, e.g. determining the most suitable location to move the object blocking the view to.

A promising approach in this respect is the emerging information processing paradigm known as computational intelligence (CI), also referred to as soft computing [4]. CI methodologies are superior over the classical AI methodologies in particular with respect to dealing with vagueness of objectives and combinatorial explosion in the parameter domain. Among the CI methodologies, in particular evolutionary algorithms became popular in the last decade for identifying optimal solutions for different aspects in design problems [5-9]. A difference of evolutionary approaches compared to the classical AI approach lies in the fact that the former does not involve a-priori defined solutions for preconceived conditions. Instead the solutions are the result of a stochastic process involving an evaluation-generation loop, where the evaluation and the generation are *separate* sub-processes, i.e. they are not associated rigidly and a-priori as it is the case in classic AI approaches. Among the evolutionary algorithms the class known as multi-objective evolutionary algorithms (MOEA) is particularly promising for design, as it is able to deal with the multiplicity of conflicting criteria that is common in design. However, it is important to note that the ability of MOEAs to handle multiple objective dimensions is generally limited to about four or five [10]. This issue will be further explained in section three. Design usually involves much more than five requirements, while some of the requirements may be vague in nature. Therefore it is clear that another approach to represent the objectives is needed compared to conventional applications of MOEAs.

This paper presents a novel system for performance-based computational design that addresses the complexity issues mentioned above using a combination of two CI methodologies. One bottleneck addressed concerns MOEAs ability to deal with many objectives at the same time. The strategy used in the present work is to reduce the amount of objective dimensions. This is achieved by establishing a neuro-fuzzy model in which a number of elemental requirements are aggregated forming fewer, more complex and more abstract concepts. In this way the model mimics the human-like comprehension of the design objectives. It is noted that aspects of the approach have been presented in an earlier paper [11], where the focus was on enhancing the effectiveness of multi-objective evolutionary algorithm. This paper elucidates the role of the cognitive approach in design applications. In particular attention is paid to the role of fuzzy information processing in a computational design process, and to the implications of the machine cognition concept on alleviating the decision making.

The paper is structured as follows. In section two fuzzy modeling for design evaluation is described. In section three multi-objective evolutionary search is described. In section four the two components are combined yielding an intelligent system for performance-based design exhibiting cognitive features. This is followed by conclusions.

Evaluating Design Performance

During design we need to estimate the suitability of a solution for our intended purpose. This means beyond observing the direct physical features of a solution, they need to be interpreted with respect to the goals pursued. For example, designing a space it may be desirable that the space is *large* or it is *nearby* another space. Clearly these requirements have to do with the size of the space, and the distance among spaces respectively, which are physical properties of the design. However, it is clearly noted that largeness is a concept, i.e. it does not correspond immediately to a physical measurement, but it is an abstract feature of an object. It is also noted that there is generally no sharp boundary from on which one may attribute such a linguistic feature to an object. For instance there is generally no specific size of a room from on which it is to be considered large, and below which it is not large. Many design requirements have this character, i.e. they do not pin-point a single acceptable parameter value for a solution, but a range of values that are more or less satisfactory. This is essentially because design involves conflicting requirements, such as spaciousness versus low cost. Therefore many requirements are bound to be merely partially fulfilled.

Such requirements characterized as *soft*, and they can be modelled using fuzzy sets and fuzzy logic from the soft computing paradigm [12]. A fuzzy set is characterized via a function *termed fuzzy membership function*, which is an expression of some domain knowledge. Through a fuzzy set an object is associated to the set by means of a membership degree μ. Two examples of fuzzy sets are shown in Figure 1.

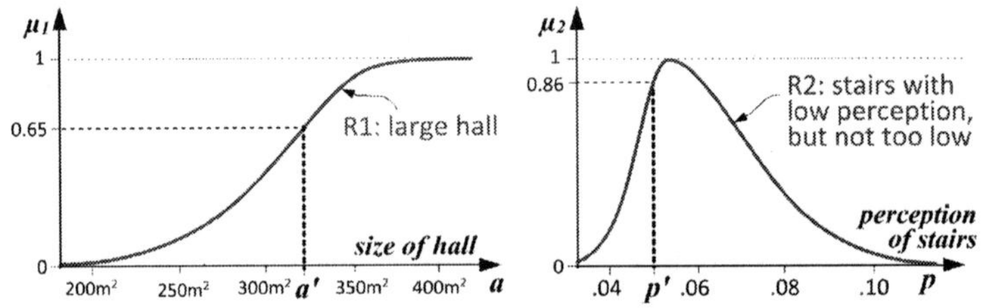

Fig. 1. Two fuzzy sets expressing two elemental design requirements

By means of fuzzy membership functions a physical property of a design, such as size, can be interpreted as a degree of satisfaction of an elemental requirement. The degree of satisfaction is represented by the membership degree.

The requirements considered here are relatively simple, whereas the ultimate requirement for a design - namely a high design performance - is complex and abstract. Namely the latter one is determined by the simultaneous satisfaction of a number of elemental requirements.

In this work the performance is computed using a fuzzy neural tree [13]. It is particularly suitable to deal with the complex linguistic concepts like design performance. A neural tree is composed of one or several model output units, referred to as *root nodes* that are connected to input units termed *terminal nodes*, and the connections are via logic processors termed *internal nodes*. An example of a fuzzy neural tree is shown in Figure 2.

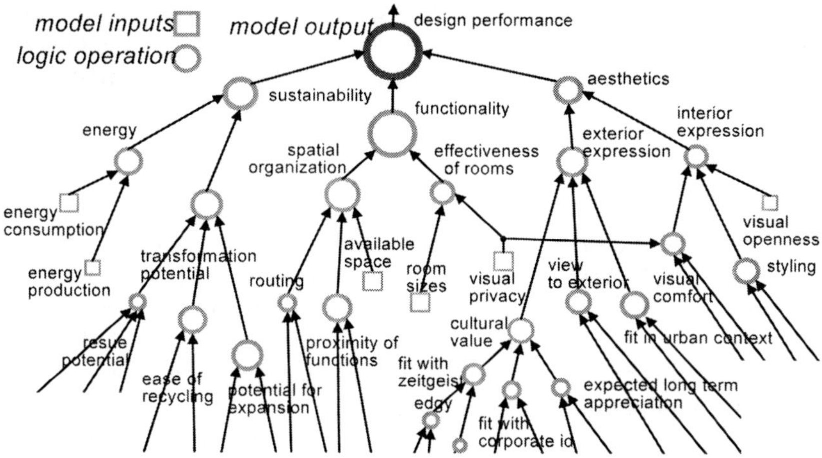

Fig. 2. The structure of a fuzzy neural tree model for performance evaluation

The neural tree is used for performance evaluation by structuring the relations among the aspects of performance. The root node takes the meaning of *high design performance* and the inner nodes one level below are the aspects of the performance. The meaning of each of these aspects may vary from design project to project and it is determined by experts. The model inputs are shown by means of squares in Figures 2 and 3, and they are fuzzy sets, such as those given in Figure 1.

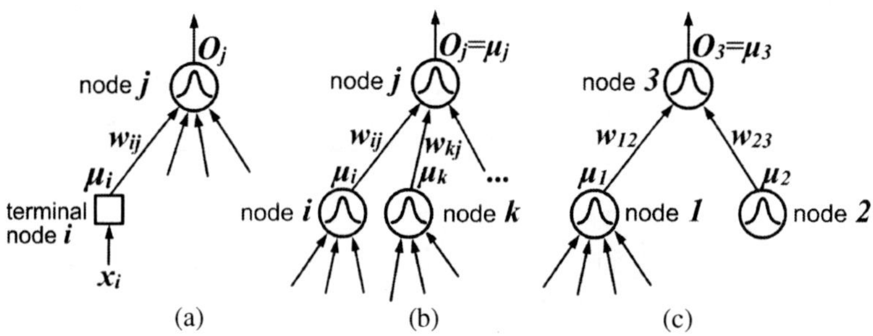

Fig. 3. Different type of node connections in the neuro-fuzzy model in Figure 2

The detailed structure of the nodal connections with respect to the different connection types is shown in Figure 3, where the output of i-th node is denoted μ_i and it is introduced to another node j. The weights w_{ij} are given by domain experts, expressing the relative significance of the node i as a component of node j.

The centres of the basis functions are set to be the same as the weights of the connections arriving at that node. Therefore, for a *terminal node* connected to an *inner node*, the inner node output denoted by O_j, is obtained by [13].

$$O_j = \exp\left(-\frac{1}{2}\sum_i^n \left[\frac{(\mu_i - 1)}{\sigma_j / w_{ij}}\right]^2\right) \tag{1}$$

where j is the number of the node; i denotes consecutive numbers associated to each input of the inner node; n denotes the highest number of the inputs arriving at node j; w_i denotes the degree of membership being the output of the *i-th* terminal node; w_{ij} is the weight associated with the connection between the i-th terminal node and the inner node j; and σ_j denotes the width of the Gaussian of node j.

It is noted that the inputs to an inner node are *fuzzified* before the AND operation takes place [11]. This is shown in Figure 4a. It is also noted that the model requires establishing the width parameter σ_j at every node. This is accomplished by means of imposing a consistency condition on the model [13].

This condition is to ensure that when all inputs take a certain value, then the model output yields this very same value, i.e. $\mu_1 = \mu_2 \approx O_j$ This is illustrated in Figure 4b by means of linear approximation to the Gaussian. The consistency is ensured by means of gradient adaptive optimization, identifying optimal σ_j values for each node.

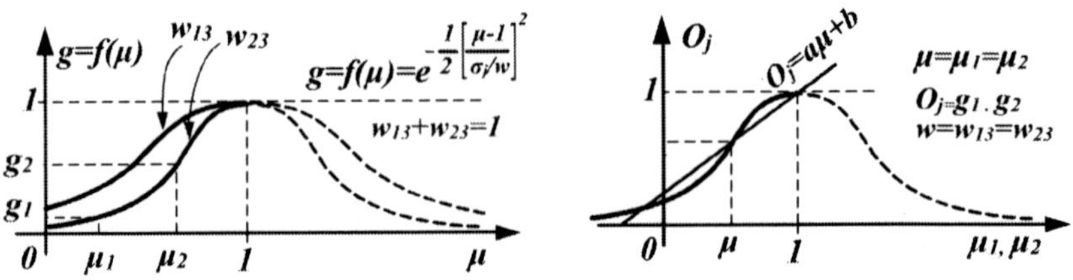

Fig. 4. Fuzzification of an input at an inner node (a); linear approximation to Gaussian function at AND operation (b)

It is emphasized that the fuzzy logic operation performed at each node is an AND operation among the input components μ_i coming to the node. This entails for instance that in case all elemental requirements are highly fulfilled, then the design performance is high as well. In the same way, for any other pattern of satisfaction on the elemental level, the performance is computed and obtained at the root node output. The fuzzy neural tree can be seen as a means to aggregate elemental requirements yielding fewer requirement items at higher levels of generalization compared to the lower level requirements. This is seen from Figure 5.

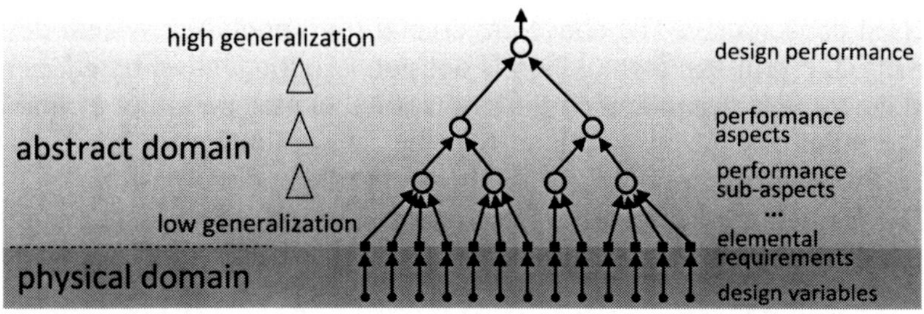

Fig. 5. Degrees of generalization in the neuro-fuzzy performance evaluation

Multi-objective Evolutionary Search with a Relaxed Dominance Concept

In design generally multiple objectives are subject to simultaneous satisfaction. Such objectives are for example high functionality and low cost. To deal with multi-objectivity, evolutionary algorithms with genetic operators are effective in defining the search direction for rapid and effective convergence [14]. Basically, in a multi-objective case the search direction is not one but may be many, so that during the search a single preferred direction cannot be identified and even this is not desirable. In the evolutionary computation case a population of candidate solutions can easily hint about the desired directions of the search and provoke the emergence of candidate solutions during the search process that are more suitable for the ultimate goal. Next to the principles of genetic algorithm-directed optimization, in multi-objective (MO) algorithms, in many cases the use of Pareto ranking is a fundamental selection method. Its effectiveness is clearly demonstrated for a moderate number of objectives, which are subject to optimization simultaneously [15]. Pareto ranking refers to comparing solutions of a population regarding their degree of being non-dominated by other solutions in the population. The evolutionary search using Pareto ranking converges to a set of solutions that lie on a surface in the multidimensional objective space. On this surface, the solutions are termed Pareto optimal solutions. They are different, both in terms of the solution parameters and their associated features. But they are assumed to be equivalently valid as there are no other solutions outperforming them for every objective dimension at the same time, i.e. the solutions on the Pareto surface are all *non-dominated*. Selection of one of the solutions among these is based on some higher-order preferences, which require further insight into the problem at hand. This is necessary in order to make more refined decisions before selecting any solution represented along the Pareto surface. From the cognitive viewpoint, this means among the solutions available for the task, one is selected consciously. The above are crucial for a cognitive system design. Namely, the problem formulation is not purely optimization-based, but the final outcome is dependent on the availability and the nature of availability of the solutions. Even solutions may be sub-optimal as a trade-off for diversity, when cognition plays an important role in decision-making.

The formation of the Pareto front is based on objective functions of the weighted N objectives which are of the form

$$F_i(\boldsymbol{x}) = f_i(\boldsymbol{x}) + \sum_{j=1, j\neq i} a_{ji} f_j(\boldsymbol{x}), i = 1,2,\ldots,N\ ) \tag{2}$$

where $F_i(x)$ is the new objective function; a_{ij} is the designated amount of gain in the j-th objective function for a loss of one unit in the i-th objective function. Therefore the sign of a_{ij} is always negative. The above set of equations require fixing the matrix a, which has all ones as diagonal elements. For the Pareto front we assume that, a solution parameter vector $\boldsymbol{x}_1$ dominates another solution $\boldsymbol{x}_2$ if $F(\boldsymbol{x}_1)\geq F(\boldsymbol{x}_2)$ for all objectives, and a contingent equality is not valid for at least one objective.

Applying the Pareto concept in its conventional strict form has drawbacks when the problem involves more than four or five objectives. The cause of this issue is the *greediness* of the conventional Pareto ranking. Namely with many objectives most solutions of the population will be considered non-dominated, although the search process is still at a premature stage. This means the search has little information to distinguish among solutions, so that the selection pressure pushing the population into the desirable region is too low. This means the algorithm prematurely eliminates potential solutions from the population, exhausting the 'creative' potential inherent to the population. As a result the search arrives at an inferior Pareto front, and with aggregation of solutions along this front. This is a well known topical issue in the area of evolutionary multi-objective optimization [10, 16, 17]. For the *greedy* application of the MO algorithm, one uses the orthogonal contour lines at the point $\boldsymbol{P}$ as shown in Figure 6. In this Figure the point $\boldsymbol{P}$ denotes one of the individuals among the population in the context of genetic algorithm (GA) based evolutionary search. In the greedy search many potential favourable solutions are prematurely excluded from the search process. This is because each solution in the population is represented by the point $\boldsymbol{P}$ and the dominance is measured in relation to the number of solutions falling into the *search domain* within the angle $\theta=\pi/2$. To avoid the premature elimination of the potential solutions, a relaxed dominance concept is implemented where the angle θ can be considered as the *angle for tolerance* provided $\theta>\pi/2$. The resulting Pareto front corresponds to a non-orthogonal *search domain* as shown in Figure 6.

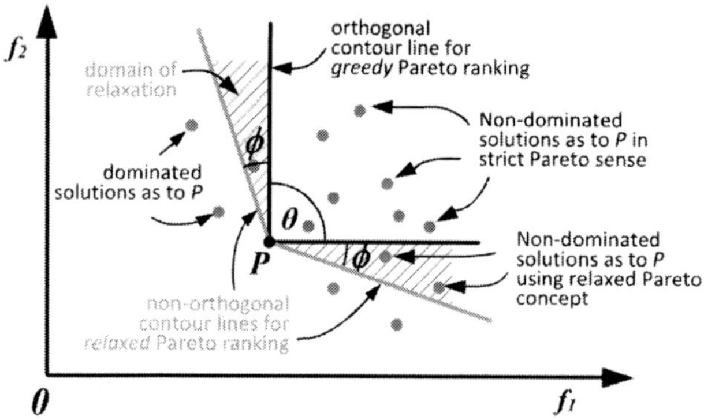

Fig. 6. Contour lines defining the search areas

The wider the angle beyond π/2 the more tolerant the search process and vice versa. For $\theta<\pi/2$, θ becomes the *angle for greediness*. Domains of relaxations are also indicated in Figure 6. In the greedy case the solutions are expected to be more effective but to be aggregated. In the latter case, the solutions are expected to be more diversified but less effective. In both cases, the fitness of the solutions can be ranked by the fitness function

$$\boldsymbol{R}_{fit} = \frac{1}{N(\theta)+n} \tag{3}$$

where n is the number of potential solutions falling into the *search domain*. Although $N(\theta)$ can be on-line modified during the search, it is expectedly constant once θ is determined. However, without the analysis of the functionality of $N(\theta)$ it is difficult to establish such a function through experiments.

To obtain n in Eq. (3), for each solution point - say $\boldsymbol{P}$ in Figure 6 - the point is temporarily considered to be a reference point as origin, and all the other solution points in the orthogonal coordinate system are converted to the non-orthogonal system coordinates. This is accomplished through a matrix operation [11]. The importance of this coordinate transformation becomes dramatic especially in higher dimensions. In such cases the spatial distribution of domains of relaxation becomes complex and thereby difficult to implement. Namely, in multidimensional space the volume of a relaxation domain is difficult to imagine, and more importantly it is difficult to identify the population in such domains. Therefore the approach through the coordinate transformation is a systematic and elegant approach. In this way the bottleneck of conventional Pareto ranking dealing with many objectives is alleviated to some extend, so that the evolutionary paradigm becomes more apt for applications in architectural design usually containing a great many requirements. It is noted that the relaxed Pareto approach requires significantly less computations to rank the population compared to the existing strategies aiming to avoid aggregation on the Pareto front, such as *niched* Pareto ranking [17].

Multi-objective Evolutionary Algorithm & Fuzzy Neural Tree = A Computational Design System with Cognitive Features

Next to relaxing the greediness of multi-objective evolutionary algorithm (MOEA), a second measure that alleviates the limitations of evolutionary search for design applications is to couple the algorithm with the fuzzy model in section 2. This yields the system shown in Figure 7. In the system the fuzzy model is the playing the role of fitness function in the MOEA. When we first consider a design containing many elemental requirements, on can consider that within the fuzzy model the amount of objectives is reduced by aggregating some requirements forming fewer, more complex concepts, which become the objectives subject to simultaneous satisfaction. Examples of such complex objectives are *functionality* and *sustainability*. Using the fuzzy information processing in this way entails that the search process makes use of human-like reasoning during its strive for optimality.

From Figure 7 we note that the computational design system starts its processing by generating a population of random solutions within the boundaries put forward by the designer in advance. Then several properties of these solutions are measured, such as sizes, distances, and perceptual properties. These are interpreted with respect to the elemental design requirements at the input layer of a fuzzy neural tree. This information is propagated through the tree yielding the degree of satisfaction of the solution at the penultimate level right below the root node. That is the evaluation using the neural tree is able to express the features of a solution in a few abstract, linguistic terms. For example it provides the performance regarding functionality, perception and cost effectiveness. These outputs are then used to compare the randomly generated solutions regarding their respective non-dominance using the relaxed Pareto concept.

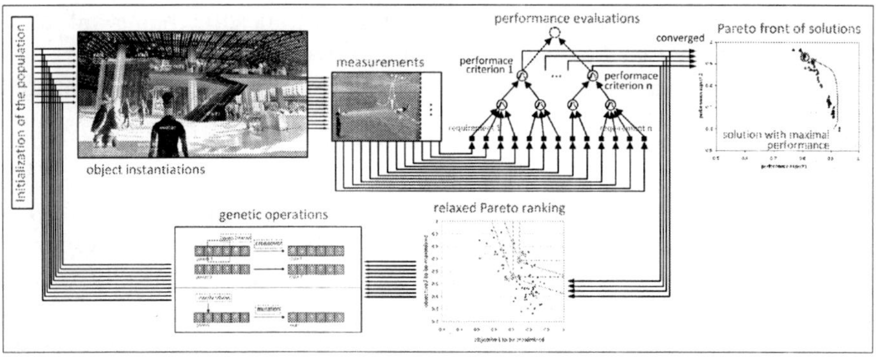

Fig. 7. Cognitive system based on MOEA and fuzzy neural tree

Relatively non-dominated solutions are then favoured for reproduction and the genetic operations, so that the next generation is more likely to contain non-dominated solutions. This generation-instantiation-evaluation-loop is executed for a number of generations, finally resulting in a set of Pareto optimal solutions. A designer or decision-maker is then able to compare these solutions in order to select his favourite design among the apparently equally valid solutions. In case the favourite solution is completely satisfying the designer's preferences, the design solution is found. Otherwise the designer may change the criteria of the computational design process and re-run the algorithm. This process iterates as shown in Figure 8, where the box containing the term *IDO (Intelligent Design Objects)* represents the system shown in Figure 7.

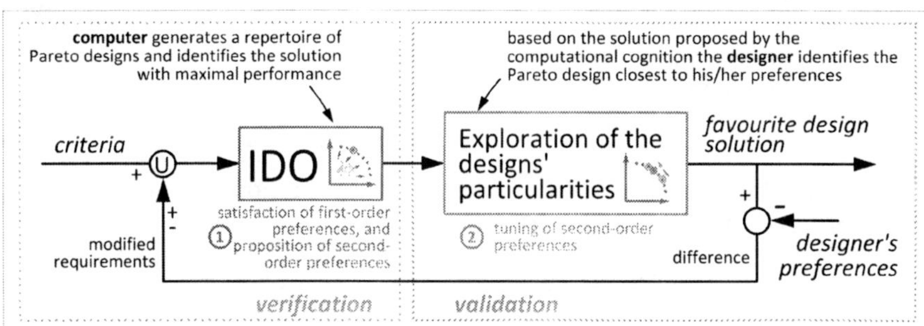

Fig. 8. Cognitive design approach

The Cognitive Features of the System

In contrast to conventional multi-objective optimization, due to the special fitness evaluation in this work involving a fuzzy model, the solutions on the Pareto front are not completely equivalent despite being all non-dominated. They may be distinguished as follows.

At the root node of the neural tree, the performance score is computed by the de-fuzzification process given by

$$w_1(1)f_1 + w_2(1)f_2 + ... + w_n(1)f_n = p, \tag{4}$$

where $w_1+w_2+...+w_n=1$; f_1-f_n are the outputs at the penultimate nodes; p is the design performance, which is requested to be maximum. The vector $\boldsymbol{w}$ containing the weights on the penultimate level is termed priority vector. The node outputs f_1 - f_n can be considered as the *design feature vector* $\boldsymbol{f}$. It is important to note that a certain feature vector $\boldsymbol{f}$ will yield the greatest performance p at the model output if the weights w_1-w_n define the same direction as that of the feature vector. In order to consistently compare among different solutions' feature vectors with their different magnitudes, it is necessary to obtain the unit vector $\boldsymbol{u}$ along every feature vector. This unit vector is given by

$$u_1 = \frac{f_1}{\sqrt{f_1^2 + f_2^2 + ... + f_n^2}};\quad u_2 = \frac{f_2}{\sqrt{f_1^2 + f_2^2 + ... + f_n^2}};\quad ...;\quad u_n = \frac{f_n}{\sqrt{f_1^2 + f_2^2 + ... + f_n^2}} \tag{5}$$

In order to meet the condition imposed by the de-fuzzification process in Eq. (4), namely that the weight components sum up to unity, it is necessary to normalize the components u_1, ..., u_n of the unit vector. Explicitly this is given by

$$u'_1 = \frac{u_1}{u_1 + u_2 ... + u_n} = \frac{\dfrac{f_1}{\sqrt{f_1^2 + f_2^2 + ... + f_n^2}}}{\dfrac{f_1 + f_2 + ... f_n}{\sqrt{f_1^2 + f_2^2 + ... + f_n^2}}} = \frac{f_1}{f_1 + f_2 + ... + f_n};$$

$$... \tag{6}$$

$$u'_n = \frac{u_n}{u_1 + u_2 ... + u_n} = \frac{\dfrac{f_n}{\sqrt{f_1^2 + f_2^2 + ... + f_n^2}}}{\dfrac{f_1 + f_2 + ... f_n}{\sqrt{f_1^2 + f_2^2 + ... + f_n^2}}} = \frac{f_n}{f_1 + f_2 + ... + f_n}.$$

Equating the normalized components to the components of a priority vector $\boldsymbol{w}_{max}$ yields

$$w_{1,\max} = \frac{f_1}{f_1 + f_2 + ... + f_n};\quad w_{2,\max} = \frac{f_2}{f_1 + f_2 + ... + f_n};\quad ...;\ w_{3,\max} = \frac{f_n}{f_1 + f_2 + ... + f_n} \tag{7}$$

The steps in Eq. 5-7 are illustrated in Figure 9 for a case with two objectives. Due to Eq. 7 the performance given by Eq. 4 becomes the maximal performance p_{max}

$$p_{\max} = \frac{f_1^2 + f_2^2 + ... + f_n^2}{f_1 + f_2 + ... + f_n} \tag{8}$$

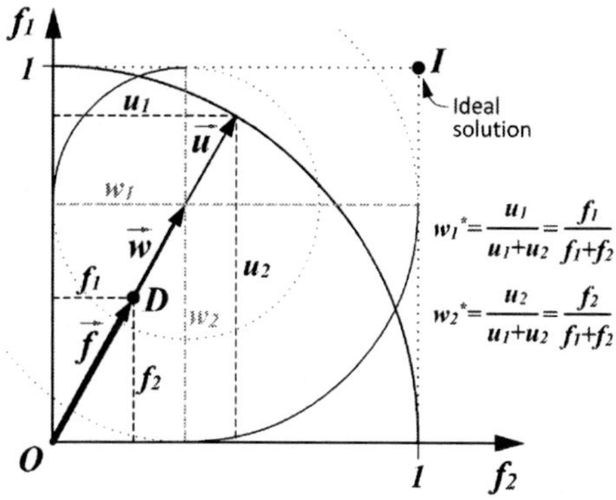

Fig. 9. Feature vector ***f***, unit vector ***u***, and priority vector ***w*** in objective space

Every solution on the Pareto front has an associated p_{max} value that characterizes it. This value gives the maximum design performance the solution attains, when there is no a-priori preference regarding the objectives. The solutions can be compared regarding their p_{max} value, and the solution with the highest value is a preferable choice among the Pareto solutions. This solution has a characteristic priority vector $\boldsymbol{w}^*$. It is noted that the computation used to identify p_{max} is elegant in the sense that it does not require, for instance, solving a linear programming problem to maximize p for a given feature vector $\boldsymbol{f}$.

This vector $\boldsymbol{w}^*$ implies that the computer advises the decision maker which goal he should take as more or less important in the present task, while this information was not known prior to the search process. This means the machine performs an act beyond mere optimization through intelligent information processing. Namely it is an act of cognition, yielding information about second-order aspects that were not included in the criteria given by the human decision maker. The artificial cognition alleviates decision-making in the sense that the designer need not explore the entire Pareto front, but has information on proficient areas along the front.

Let us consider the possibility that we obtain solutions having identical p_{max} values while their components are different. For example a solution with $\boldsymbol{f}$=(0.9, 0.1, 0.1) yields the same p_{max} value as a second solution (0.1, 0.1, 0.9). For clarifying this issue it is noted that Eq. (8) can be transformed in such a way that it describes a spherical surface in the multidimensional objective space, on the surface of which p_{max} is constant. In a two objective case the sphere becomes a circle given by [18]

$$f_1^2 + f_2^2 - pf_1 + pf_2 \equiv (x-x_1)^2 + (y-y_1)^2 - R^2, \text{ where} \quad (9)$$

$$x_1 = p/2$$
$$y_1 = p/2 \qquad (10)$$
$$R = p/\sqrt{2}$$

The circle of constant maximal performance is shown in Figure 10. It is noted that due to the Pareto dominance criterion a Pareto front has a flatter curvature compared to the circle of constant maximal performance. Therefore the greatest p_{max} values are generally to be found at the extremities of the front, i.e. close to the axes. From Figure 10b is noted that for Pareto fronts that are asymmetrically shaped w.r.t. the line passing through the origin and (x_l, y_l) *where* $x1=y1$ merely a single maximal p_{max} value may exist in the population. In the latter case provision of this solution clearly alleviates the decision making. From figure 10a we note that, tn case of symmetry of the front providing the p_{max} value to a decision maker alleviates the decision making as the two solutions will be very different regarding their respective features, so that the decision may be easier to make compared to the case p_{max} were unknown.

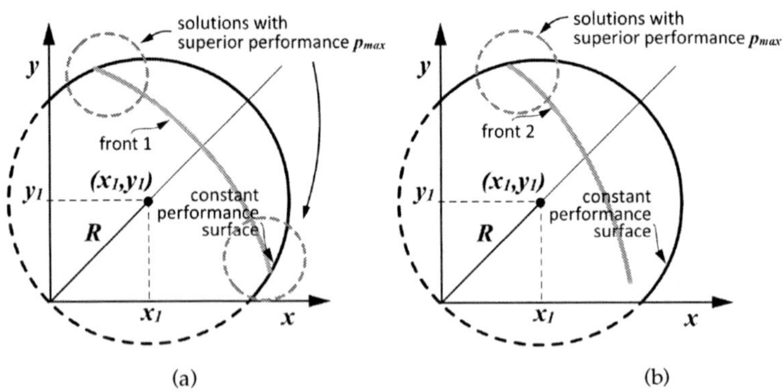

Fig. 10. Constant performance surface and Pareto front

It is also noted that a decision maker's preferences are also important with respect to determining to what extend an extremity is to be exercised.

Application

The design task concerns the design of an interior space. The space is based on the main hall of the World Trade Centre in Rotterdam in the Netherlands. The perception of a virtual observer plays a role in the design, as the task involves a number of perception-based requirements. An example is that the stairs should not be very noticeable from the entrance of the space, as seen from Figure 11(b). The perception computation yielding x_{12} in Figure 11(b) is accomplished using a probabilistic perception model [19]. Another example is that the building core should be positioned in such a way that the entrance hall is spacious, while the elevators should be easily perceived at the same time. This is seen from figure 11a.

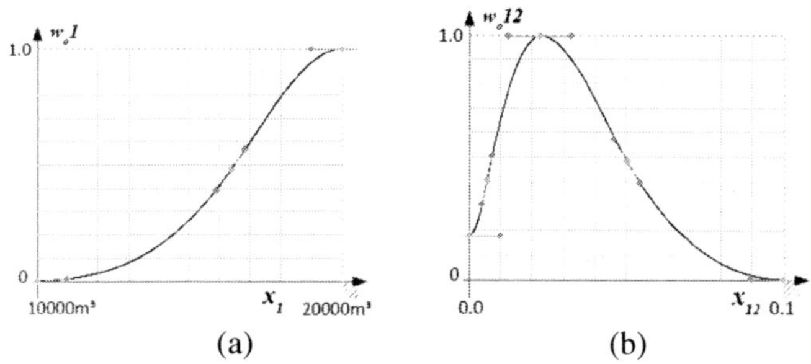

Fig. 11. Two requirements subject to satisfaction, concerning (a) spaciousness of the entrance hall; (b) perception of the stairs

The task is to optimally place the design objects satisfying a number of perception and functionality requirements. The objects are a vertical building core hosting the elevators, a mezzanine, stairs, and two vertical ducts. The goals are maximizing the performance of every design object forming the scene, as seen from the fuzzy neural tree structure in Figure 12. From the structure of the model it is seen that the amount of objectives to be maximized is four, namely the outputs of nodes 4-7, whereas the elemental requirements total an amount of 12.

The resulting Pareto optimal solutions are shown in Figures 13, and 14. Figure 13 shows the results using the greedy Pareto ranking approach, whereas Figure 14 shows the relaxed Pareto ranking approach. It is noted that the objective space has four dimensions, one for the performance of every design object. The representation is obtained by first categorizing the solutions as to which of the four quadrants in the two-dimensional objective space formed by the building core and mezzanine performance they belong, and then representing in each quadrant a coordinate system showing the stairs and ducts performance in this very quadrant. This way four dimensions are represented on the two-dimensional page. Comparing Figure 13 with Figure 14 we note that the relaxed approach is superior compared to the greedy case in Figure 13, as it yields a Pareto front having solutions with a higher performance p_{max}.

Two Pareto optimal designs are shown in Figure 15 and 16 for comparison. The maximal performance score as well as the performance feature vector for these solutions is shown in Table 1. From the table it is seen that design D2 outperforms design D4 with respect to the maximal performance p_{max} obtained using Eq. 8. It is also noted that the performance of D4 as to its features varies less compared to D2. The fact that D2 has a greater p_{max} confirms the theoretical expectation illustrated by Figure 10 that solutions with more extreme features generally have a greater maximal performance compared to solutions with little extremity.

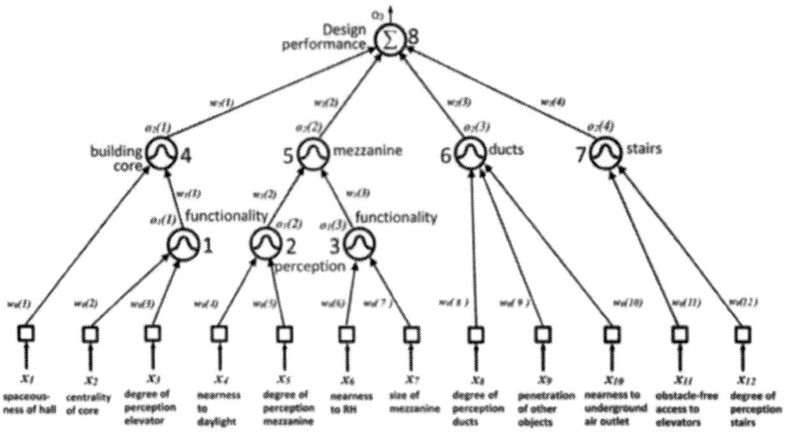

Fig. 12. Neural tree structure for the performance evaluation

The greatest absolute difference among D2 and D4 is the performance of the mezzanine. In D2 the mezzanine is located closer to associated functions, and this turns out to be more important compared to the fact that D4 yields more daylight on the mezzanine. Therefore D2 scores higher that D4 regarding the mezzanine. Additionally D2 slightly outperforms D4 regarding the performance of the ducts. This is because the ducts do not penetrate the mezzanine in D2, whereas in D4 they do. The latter is undesirable, as given by the requirements in D4. Regarding the building core D2 is inferior to D4, which is because the spaciousness in D4 is greater and also the elevators are located more centrally. Regarding the stairs' performance, the difference among D2 and D4 is negligible.

The latter exemplifies the fact that an objective may be reached in different ways, i.e. solutions that are quite different regarding their physical parameters may yield similar scores as to a certain goal. In the present case the greater distance to the stairs in D2 compared to D4 is compensated by the fact that the stairs is oriented sideways in D2, so that the final perception degree is almost the same. It is noted that D2 is the solution with the greatest maximal performance p_{max}, so that from an unbiased viewpoint it is the most suitable solution among the Pareto optimal ones. This solution is most appealing to be selected for construction.

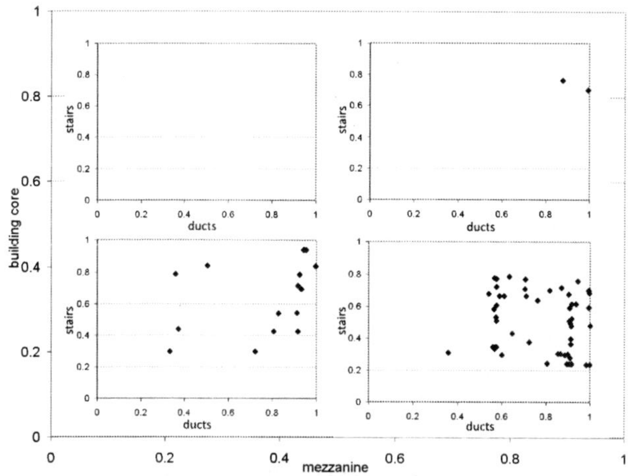

Fig. 13. Pareto optimal designs with respect to the four objective dimensions using greedy Pareto ranking

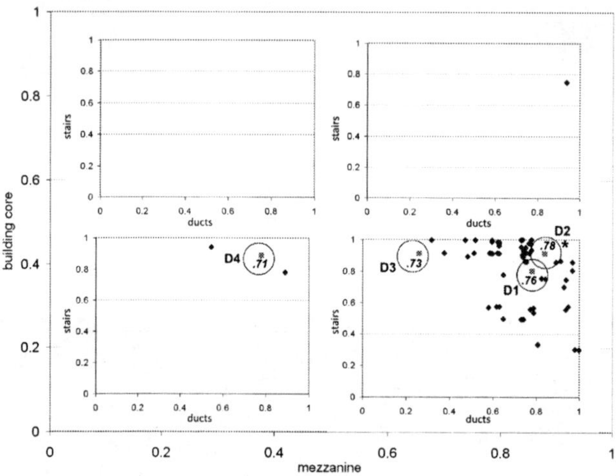

Fig. 14. Pareto optimal designs with respect to the four objective dimensions using relaxed Pareto ranking

This result is an act of machine cognition, as it reveals that pursuing maximal performance in the present task the stairs and ducts are more important compared to the building core from an unbiased viewpoint. This information was not known prior to the execution of the computational design process. It is interesting to note that the solution that was chosen by a human architect in a conventional design process without computational support was also similar to solution 2.

Table 1 Performance of design D2 versus D4

	core	mezzanine	ducts	stairs	p_{max}
D2	0.27	0.73	0.83	0.93	0.78
D4	0.48	0.49	0.78	0.89	0.71

This indicates that the architect must have had similar requirements in mind, and followed a similar reasoning as involved in the computational process. The benefit of the computational approach is that it ensures identification of most suitable solutions, their unbiased comparison, and precise information on their respective trade-off as to the abstract objectives. This is difficult to obtain using conventional means.

Fig. 15. Pareto-optimal design *D2* in Figure 14

Fig. 16. Pareto-optimal design *D4* in Figure 14

Conclusion

A novel computational system for architectural design is presented. It generates designs that satisfy multiple criteria put forward by decision makers, such as architects.

The contribution of the present work is twofold. First, the use of the fuzzy information processing enables multi-objective evolutionary algorithm (MOEA) to deal with the many, soft requirements characterizing design tasks. Secondly it enables the machine cognition. Namely the fuzzy model entails that the objective space takes on a particular shape, being a unit hypercube. This way, different properties of a design are subject to consideration on a common ground, which allows distinction among the Pareto solutions regarding their unbiased maximal performance being an act of machine cognition. It is noted that cognition is understood in this work as the faculty to take a decision based on an awareness of, in some sense, equivalently valid solution alternatives fulfilling first-order objectives. The cognitive act is selection among the alternatives based on second-order preferences that were not included in the process yielding these solutions. Such second-order preferences are due to the particularities among the alternative solutions, which cannot be foreseen prior to designing.

Machine cognition is a feature of scientific and practical interest. From the scientific viewpoint the introduction of human-like reasoning into an evolutionary search process increases the applicability of evolutionary algorithms in particular for design problems. Thereby it may contribute to the growing use of MOEAs in this domain. From the practical viewpoint it is an innovative direction for alleviating decision making tasks. A designer gets to know - from an unbiased viewpoint - which trade-offs inherent to the design task are worthwhile to accept compared to others. The attribute *unbiased* refers that no objective is favored over another one a-priori. This way the decision maker has an indication which regions on the Pareto front are outstanding, and to what extend this is the case. This information is challenging to obtain when conventional means are used.

References

1. Eastman, C.M.: Automated space planning. Artificial Intelligence 4, 41–64 (1973)
2. Flemming, U., Woodbury, R.: Software environment to support early phases in building design (SEED): Overview. J. of Architectural Engineering 1, 147–152 (1995)

3. Koile, K.: An intelligent assistant for conceptual design. In: Gero, J.S. (ed.) Design Computing and Cognition 2004, pp. 3–22. Kluwer Academic Publishers, Massachusetts Institute of Technology, Boston, USA (2004)
4. Jang, J.S.R., Sun, C.T., Mizutani, E.: Neuro-Fuzzy and Soft Computing. Prentice Hall, Englewood Cliffs (1997)
5. Gero, J., Kazakov, V.: Evolving design genes in space layout problems. Artificial Intelligence in Engineering 12, 163–176 (1998)
6. Damsky, J., Gero, J.: An evolutionary approach to generating constraint-based space layout topologies. In: Junge, R. (ed.) CAAD Futures 1997, pp. 855–874. Kluwer Academic Publishing, Dordrecht (1997)
7. Jo, J., Gero, J.: Space layout planning using an evolutionary approach. Artificial Intelligence in Engineering 12, 163–176 (1998)
8. Mawdesley, M.J., Al-jibouri, S.H., Yang, H.: Genetic Algorithms for Construction Site Layout in Project Planning. J. Constr. Engrg. and Mgmt. 128, 418–426 (2002)
9. Caldas, L.: GENE_ARCH: An evolution-based generative design system for sustainable architecture. In: Smith, I.F.C. (ed.) EG-ICE 2006. LNCS (LNAI), vol. 4200, pp. 109–118. Springer, Heidelberg (2006)
10. Coello, C.A.C., Veldhuizen, D.A., Lamont, G.B.: Evolutionary Algorithms for Solving Multiobjective Problems. Kluwer Academic Publishers, Boston (2003)
11. Bittermann, M.S., Ciftcioglu, O.: A cognitive system based on fuzzy information processing and multi-objective evolutionary algorithm. In: IEEE Conference on Evolutionary Computation. IEEE, Trondheim (2009)
12. Zadeh, L.A.: Fuzzy logic, neural networks and soft computing. Communications of the ACM 37, 77–84 (1994)
13. Ciftcioglu, O., Bittermann, M.S., Sariyildiz, I.S.: Building performance analysis supported by GA. In: Tan, K.C., Xu, J.X. (eds.) 2007 IEEE Congress on Evolutionary Computation, pp. 489–495. IEEE, Singapore (2007)
14. Deb, K.: Multiobjective Optimization using Evolutionary Algorithms. John Wiley & Sons, Chichester (2001)
15. Tan, K.C., Xu, J.-X. (eds.): Proc. 2007 IEEE Congress on Evolutionary Computation. IEEE Cong., Singapore (2007)
16. Hughes, E.J.: Evolutionary many-objective optimisation: many once or one many? In: IEEE Congress on Evolutionary Computation CEC 2005, pp. 222–227. IEEE Service Center, Edinburgh (2005)
17. Horn, J., Nafploitis, N., Goldberg, D.E.: A niched Pareto genetic algorithm for multiobjective optimization. In: Michaelwicz, Z. (ed.) First IEEE Conf. on Evolutionary Computation, pp. 82–87. IEEE Press, Los Alamitos (1994)
18. Ciftcioglu, Ö., Bittermann, M.S.: Adaptive formation of Pareto front in evolutionary multi-objective optimization. In: Santos, W.P.d. (ed.), Evolutionary Computation. In-Tech, Vienna, pp. 417–444 (2009)
19. Bittermann, M.S., Sariyildiz, I.S., Ciftcioglu, Ö.: Visual perception in design and robotics. Integrated Computer-Aided Engineering 14, 73–91 (2007)

Narrative Bridging

Katarina Borg Gyllenbäck[1] and Magnus Boman[2]
[1]*Stockholm University, Sweden*
[2]*KTH/ICT/SCS and SICS, Sweden*

In the design of interactive media, various forms of intuitive practice come into play. It might prove tempting to use templates and strong narrative structures instead of developing the narrative directly for interactive media. This leads towards computer implementation too swiftly. The narrative bridging method focuses on the initial design phase, in which the conceptual modeling takes place. The purpose is to provide designers with a non-intrusive method that supports the design process without interfering with its creative elements. The method supports the sentient construction of digital games with a narrative, with the ultimate goal of enhancing the player's experience. A prototype test served as a first evaluation, and two games from that test are showcased here for the purpose of illustrating the hands-on use of narrative bridging. The test demonstrated that the method could aid time-constrained design, and in the process detect inconsistencies that could prevent the design team from making improvements. The method also provided teams with a shared vocabulary and outlook.

Introduction

The canonical story format and dramaturgy from the film medium have dominated game development and analysis for a long time. The benefits of this approach have been realized at the expense of learning precisely how to apply the narrative to interactive media. There is nothing inherently wrong in using strong structures or templates. The main problem resides in being unaware of these structures or categories and in applying them too quickly. This can lead to the creation of stereotypes, lack of depth, and to not exploring what lies behind the strong structures [1], [2]. In particular, the aesthetics of digital games have been studied [3], [4], and [5]. The so-called structural perspective (which includes canonical story formats, and

J.S. Gero (ed.): *Design Computing and Cognition'10*, pp. 525–544.

structures, as used in other media) has lead to intense discussions about the terms story and narrative. A story represents a fixed event that can be referred to, and a story can be retold. Problems usually occur when using the term to allude to the process of construction, i.e. to the practise of cognition-based construction of causal, spatial and temporal links which include the interpretation of the receiver (i.e., the person listening to the story). This process can in fact be explained by other terms: the *syuzhet* (plotting) and *fabula* (interpretation as made by the receiver) [6], and adapted to the media at hand. A reason for seeking to understand and take seriously the narrative process is that within game studies and design, various forms of tacit knowledge and intuitive practice are employed. While everyone knows a good story by heart, to iterate its narrative components is a different matter. Preconceptions can, in the case of complex game worlds, push an idea towards manuscript, systematization, graphics, and computational issues too quickly. This might result in increased costs, due to the difficulties associated with the detection and late adjustment of inconsistencies. Moreover, making the most of the capacity of the media involved may prove difficult.

In game design, the use of genre is popular because it helps and engages large teams to focus on a joint target [7]. Genre is also used as a contract of sorts between the producer and the publisher, and as a bridge between the audience and their expectations [8]. Deciding upon a genre early on frames the possibilities for how to construct an idea and to see its full potential with respect to conventions, structures, familiarity, expectations, and interpretation [9] by classifying the idea as, e.g., adventure game, first person shooter, or crime story. Although genre can be an excellent guide for communication, it may produce a bound on creativity if adopted too early; the contract must not be breached. We take a constructive approach to the problem of making the narrative more prominent in the creation of interactive media, in order to provide the process of design with a possibility to model a game with a narrative. *Narrative Bridging* is a method that supports the design process with a narrative, throughout creation, organization, control, and generation of information for interactive media. The method is not restricted to digital games, but can also be used to guide the design of interactive installations, narrative characters (agents), and pervasive games. The primary purpose of the method is to aid the initial design phase, in which the conceptual modeling [10] takes place. This is a tiny part of the process, but perhaps the most critical as it is here that all the components are set that constrain the end product.

A secondary purpose is to provide researchers, students, and the game design business with a means to a controlled use of narratives for interactive media. Narrative bridging has its origin in a situation where we found ourselves short of helpful tools (cf. [11], [12]) for analyzing and explaining narrative processes, in creation of a meaning within the interactive media that support a design process involving a narrative [13]. Grünvogel [14] considers the drawbacks of formal models to be that they force designers into a standardized workflow that they need to learn; one that is not integrated in language. A method should instead give designers insights into their practise, and to allow for them to decide when to use it [15]. Upon realizing that our method could be useful to others, helping how to communicate about narrative in design, and providing designers with a possibility to consciously work with narrative process, work began to offer the method to others. Narrative bridging uses the strong cognitive impact that the narrative has on a user by assigning a meaning to an experience. The will to interpret is so strong that even when a receiver is not given any information at all, the need to create meaning will still take place via patterns [1], [16], and [17]. Therefore, narrative bridging builds on cognitive vehicles that bring logic and rules to a game world, for a receiver to take action upon. Balancing an interactive media with a narrative is about creating meaning so that the receiver can create patterns and strategies for choices.

The following section describes the key elements of narrative bridging, at a high level of abstraction. The practical use of the method is made more concrete in the third section, and then initial empirical tests are described. In the penultimate section, we present lessons learned from the tests, before concluding and describing our next steps in the final section.

The Key Elements of Narrative Bridging

In order to achieve meaningful play, the elements of character, world and action form a base for causal interplay, given a premise and a goal. The character represents the user's position for interaction and the perspective from which the construction is set. Meaningful play occurs when the causes and effects of possible events and user actions are consistent with the set premise, and support the goal.

Media

As narrative is not media-specific [6], it is important to determine which media the narrative construction could be applied to. Narrative bridging works on the supposition that the interactive medium has a user and that there is a piece of software that embodies the user experience. The choice to take a user's perspective is derived from the mimetic theory of narration: how an object is perceived and presented to the beholder. Interactivity can be seen as an exchange of input and output between user and digital artifact, presenting a space of possibilities that represents the user's options to influence the *diegesis*: the fictional world and how it is presented to the user [18]. The most important elements will be expressed here as *character*, *world* and *action* (cf. [1]), and the method also captures relevant aspects of game play and mechanics. The game play is the interactivity shaped by what the space offers the player while the mechanics, in short, provide the visuals and audio supporting the narrative and the game play.

Premise

A *premise* is a realization of an idea with an identified situated subject and predicate. The fabula consists of the bits of information linked together and read by the receiver (to ultimately sum up to fit the overall goal of the receiver) [6]. The minimum of information needed to construct a fabula is a premise and a causal relationship referred to here as *cause-and-effect*. The premise can be anywhere on the scale from a piece of an event, a beat, or a sequence, to a whole concept. It initiates a process by which causes-and-effects are plotted towards the premise, and it includes the setting of time, place, the main characters, as well as the goal and the activities that propel the story [19]. The term premise should not be confused with the term genre, as the latter is not an idea, but a classifier of ideas.

Goal

To set a goal means to decide what the receiver shall experience. Fullerton [19] refers to the goal as the experience the designer likes the player to have, and stresses the importance of keeping the mimetic perspective in mind throughout every single stage of the development. Within digital games, it is important to remember to set the goal so that the player will be able to carry out activities, create mental patterns, and devise strategies for choices and even empathy, if well established. It is in turn important to establish the diegetic world and its logic, rules and relations, so that the player can read the conditions for action and see the consequences of it.

Syuzhet

The diegetic world does not generate itself. The role of the syuzhet (i.e., the creation of the plot) provides three principles for this process: narrative logic, time, and space [6]. These principles are used in the plotting of causes and effects, given by the setting of the premise, goal and media (cf. [18], [19], [20], [21], and [22]). The suyzhet creates consistency, allowing the diegetic world to make sense, and creating a relational vehicle between the elements in the system. Another system, connected to the syuzhet, is the style. The style is the means by which sound, audio, text, graphics and related techniques support the syuzhet. In the *narrative logic*, events are arranged and the relations between events are established. In games, this governs how the player learns about the world. The syuzhet can cue events in any sequence, and these can occur in any time span and frequency through repetition. The syuzhet can block the receiver's construction of the fabula, as in genres like crime stories in which the fabula is known to the receiver, or in those genres where one has to stay in a predefined cueing of the syuzhet because this is what the receiver expects. In an emergent system, for example, a frequency or value might increase to assure that the player comes across a certain piece of information in the open space. If a canonical structure is used, however, the receiver might experience restraint as a side-effect. The syuzhet also helps to create a spatial environment by describing surroundings, positions, and paths, but it can also hinder comprehension delaying, puzzling, and even fooling with the receiver's construction of the fabula. This principle is highly relevant to games because of how the player experiences how to move about in the world, what to avoid, how to access the world, what to control, etc.

The Process of Using Narrative Bridging

The method is divided into three non-consecutive and iterative phases. In Phase 1, we set the premise and the goal. To set the premise at this stage is to state a simple one-liner about what we are trying to construct, by setting the causes and effects. To set the goal means to explain what the receiver should experience, again in only one line. It is important to keep both the premise and the goal in mind throughout the whole process.

Phase 1

Creation of the diegetic world can be described as follows, see Figure 1.

- Fill in the information you have that fits into the categories: Character, World and Action.
- Look at the causes and effects between the values you have set to the Character and World (CW). Continue this for WA and CA by asking who, what, when, where, and most of all, why?

- Compare how the three relate to each other in CWA to match the premise and the goal.
- Iterate, adding the new information generated by the process to the elements of information, until you consider the diegetic world to be complete enough to match the premise and the goal, without any loose ends.

  Move the information to Phase 2 to generate, distribute, organize, and control new and detailed information.

Fig. 1. An example of a diegetic world and its elements of information. Screenshots from World of Warcraft: courtesy Blizzard Entertainment.

Phase 2

One must decide where to start, where to go; and what, when, and how to meet the world, and its objects and inhabitants. Here one can test if the laws, rules, and relations constructed in the diegetic world (Phase 1) are developed in such a way that they create clear conditions and consequences for the receiver. The working process is as follows, see Figure 2.

- With the premise and goal in mind throughout, bring in the information from Phase 1 and focus on how the narrative is experienced from a user perspective. Move through the elements Narration, Spatiality, and Interactivity (NS, SI, NI, and NSI) and see what is generated asking:

Where? Think about the spatiality in the environment and where the information about surroundings, positions, and paths are. What information would you like to delay, obstruct, or puzzle the receiver with, and what shall be pushed forward as the first information for the receiver to encounter?

When? Think about the time and cueing of the information: when shall events occur, in what frequency, and in how many repetitions?

How? Think about how the information should be forwarded to the receiver. How will the receiver get to know and learn about the world and its inhabitants? Events can be constructed as linear or non-linear by blocking or complicating the construction of relations. How do these blocks and complications appear?

- Loop back into Phase 1 and add any new information to the elements of the diegetic world. This process will in turn generate more information that is fed back into Phase 2. Repeat until the information matches the desired premise and goal. If this proves impossible, check if the premise or goal has to be changed.
- Find out what kind of causal chains create conditions and consequences that motivate the receiver to create patterns and strategies for choices and to take action upon them. This is done in the NSI interplay. Move the new values to Phase 3, to set up a system for game play.

Phase 3

In order to get the complete picture of narrative bridging, the components are organized, distributed, and balanced. The information retrieved from the first and second phase is here used to shape the digital game, see Figure 3. In Phase 3, they will present a pattern of activities in the world for governing game play, by the derived interactivity, modulo the diegetic world.

If one cannot set the game play in Phase 3, one needs to move back to see what has been missed in the diegetic world. The organization of information will generate the possibility to enhance the narrative, the game play, and the experience for the player by setting the *style*: the detailed information that governs sound, music, and graphics, and the *mechanics*: the detailed information that governs the user interface (the health bars, hit points, weaponry, etc.).

Fig. 2. An example of the interplay between the elements of narrative, interactivity, and spatiality. Screenshot images courtesy Blizzard Entertainment.

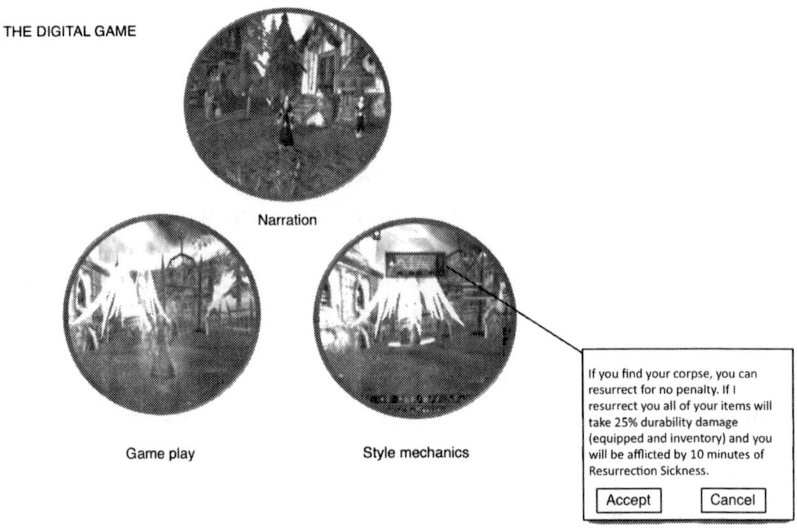

Fig. 3. The specifics of dying in the diegetic world, and how it is handled in the user interface belonging to the style mechanics of World of Warcraft. Screenshot images courtesy Blizzard Entertainment.

Prototype Testing of the Method

Finding test subjects from within the game design industry for a prototype study was deemed highly unlikely. The plan was instead to find a course in game design, and to give a lecture and hold a workshop within that course. A Swedish university was about to start a course in advanced design for their game design undergraduate students (in their third year), and were set on trying out different methods for rapid prototyping, so we seized this opportunity for putting together a test group, in the Autumn of 2008. We also provided students with a chance to reflect upon narrative processes before the method was presented to them. The intent was to develop interplay between narration and game play, and to get some understanding across about how these systems are interrelated, Figure 4.

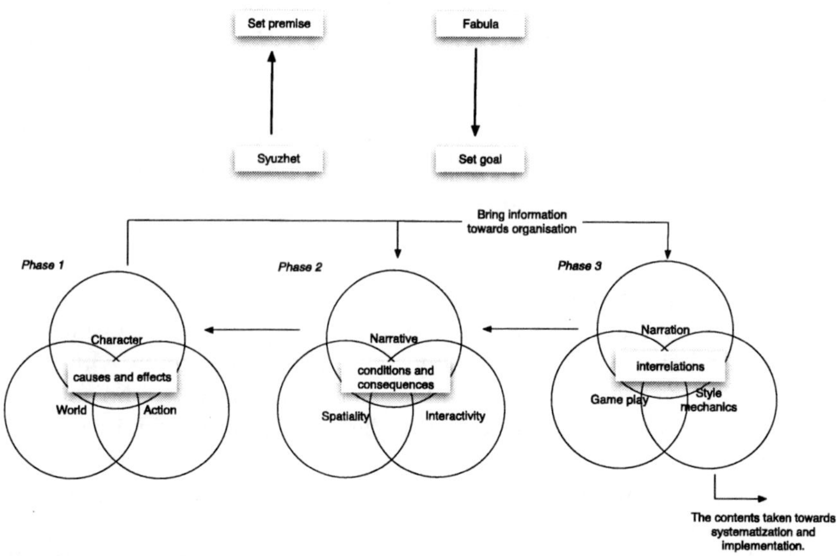

Fig. 4. An overview of the method and the iterative process

An introductory seminar and instruction were provided. The students, in groups of three, were then asked to hand in a scripted game idea, and had one week to complete the task, given only the theme 'narrative'. They were asked to give a detailed description of a certain part of their game, showing how narrative and game play interacted there. The students were aware of the fact that the method was new and part of a research project. Six game ideas were handed in after the first week, and a questionnaire was subsequently distributed to determine the subjects' previous knowledge of narrative. One female and 19 male students participated, between 22 and 29 years old. Details on two game ideas will be given below, for the purpose of illustrating test use of the method. Details on all game ideas and the responses to the questionnaire are available in full [23].

The Masquerade Game

The premise was to make a historical game about the assassination of king Gustav III at the 1792 masquerade in Stockholm. The goal for the player was to socially interact (bow, salute, flirt, etc.) to find out who was guilty of murder. In the real drama, Anckarström was arrested and executed for the crime. It is not clear how many were involved or who the brain behind the murder plan was. Inspired by that unsolved murder mystery, the player had to find out through socializing with the royal court what happened before the murder, and find the assassin before it was too late. The pitch was well formulated and the descriptive pictures gave a good idea about the world. The group used keywords to express the world: Russian war, Gustav III, party, gossip, costumes, etc. The platform for the game was a console with accelerometer for Playstation 3 or Wii, to support the game play of social interactions using different kinds of greetings. The surroundings were the castle, filled with well-known people from the political scene, artists, and the royal court. The drama took place in a traditional environment with a ballroom, dinner room, park, theatre, etc. The group set up a narrative structure that constrained the game to 24 hours. The player would wake up in the morning (or night), told by a butler that the player was expected to meet X. The player's position was presented dramatically by addressing the reader of the pitch with 'you': "Dazed, you are led through corridors to finally meet X who tells you about a conspiracy against the king and they expect you to find out who is behind it". The player would gradually get a picture of the conspiracy and be given a chance to change the outcome, should the player find out who was guilty. In this setting, the player would also slowly understand what social class he or she belonged to. The group explained how the perspective would change through three acts. The player started with a first-person perspective to end up in a third-person perspective: a way for the player to grow with the character.

Premise/Syuzhet: The murder of Gustav III.
Goal/Fabula: Socializing to find out who murdered Gustav III.

In Phase 1, the group developed the rules, relations, and logic of the diegetic world by choosing action to take place in the castle and at the masquerade, Figure 5. They thought about all participants that could be present in the world and how they could all socialize. They used gestures as the player's means to find out about who was behind the murder. When transferring the information developed in Phase 1 to Phase 2 and finding out where, how, and why the action took place, the group faced problems.

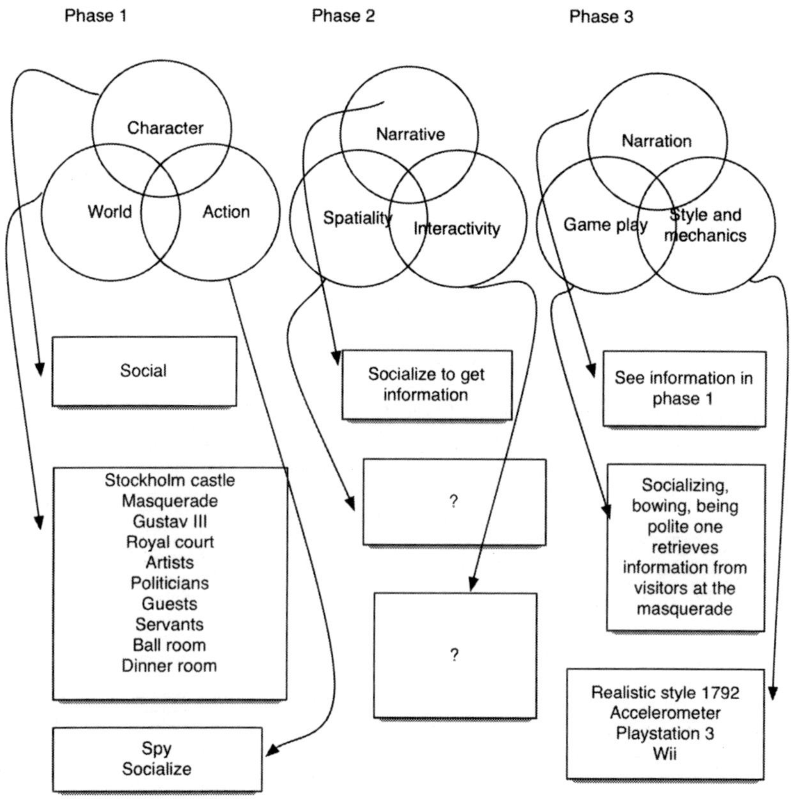

Fig. 5. First iteration of the masquerade game.

When looking at the causes and effects for the player's spatiality and interactivity, a question arose: Who was the player? Uncertainties regarding how the player moved about, and where and when to find information about the assassination, proved hard to deal with. Given the player's goal (fabula), how should interactivity be governed? To where could the player walk? Could the player be near the king? In what ways did other people pay attention to the player? Did the player's character have a strong position or should it be set as neutral (as in the case of a visitor from the future just falling into a historical documentary)? For Phase 3, the group had well thought through the socializing game play, and how to support it with different technical devices, such as an accelerometer, for the platform Playstation 3 or Wii. The group also presented a scripted structure for where the drama should start, the duration of it, and its acts as well as actions.

In a second iteration, the group presented a solution to their problem of what the player's position should be. The group had produced a strong structure with three acts already, and an idea about how to begin the game, but needed to elaborate on the idea of what the spatiality and interactivity should be like, Figure 6.

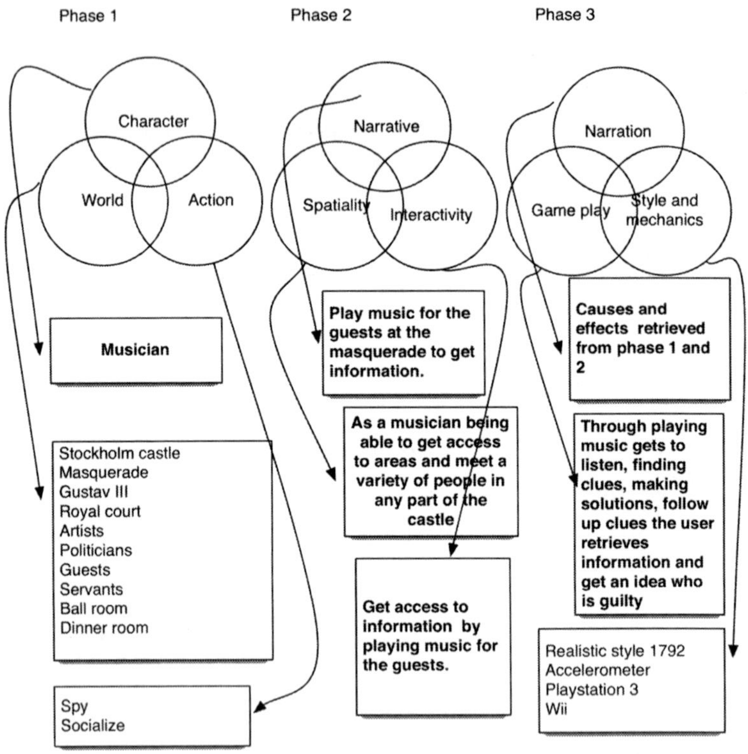

Fig. 6. Second iteration of the masquerade game.

Initially the group wanted the player to slowly grow with the character but this caused problems. The group's choice of a musician allowed them to keep the original premise and goal, to go deeper into the construction of the spatiality and retrieving interactivity in Phase 2, and to investigate what spying for game play could look like. The group chose to narrow down the spatiality to a dinner room, in which they gathered eleven people. Game play now included to find out who would attend the dinner, and to see how the musician could manipulate dinner guests in order to retrieve information. This revealed more details relevant to game play, e.g., by having one missing guest at the table, and letting the player hear details about the missing guest. By letting the musician choose how to follow up on this information, new traces of information about the murderer(s) would be revealed. The group felt the method pointed out where the idea lacked flow, and helped them to find the key problem and fix it. The group regretted that time was short, not least because they were simultaneously learning more about narrative and about using the method.

The Parasite Game

The title for this game was 'Life begins as an egg'. It was a single-player game, and the premise was about a parasite that takes over a body for the purpose of its own survival. The goal was to infect and affect the body. To grow, the egg needed to infect different parts of the body, in turn affecting the carrier in various ways. The group said they wanted to create two narratives: the life of the parasite, and the carrier's life, as affected by the parasite. The game play was to create different reactions, such as stress, love, and happiness, for different parts of the body of the carrier. The game had two levels, each divided into two parts. The first level and first part was for the egg to grow and to take over part of the body that the parasite made into its headquarters. In the second part, the parasite sent out 'commando parasites', equipped with different skills and armor. The equipment was bought with money earned via 'parasite points' that one received taking over certain body parts. To reach the second level, one had to gain control over the body parts and then from there control the carrier's actions and reactions. The first level had a first- and a third-person's perspective. The player should be able to move in a three-dimensional way such as up and down, sideways, etc. in an open space. Some areas were closed out by enemies, while others were the immune defenses represented by white corpuscles. All levels had control points that were difficult to find. This offered exploration, as well as danger, and some points should have stronger defenses than others. The second part was seen from a third-person perspective, controlling the carrier and having access to his or her life, and understanding what the carrier was about to do in life. At their oral presentation, the group also talked more about their contemplated carrier. It was a man in the prime of life, having a date with a girl, with whom the parasite complicated interaction. The group presented pictures from games like Spore that showed a germ's life from being a seed to growing into a full life form. They also showed a picture of a marriage from Sims, a picture of person with a cold, and pictures of spacecrafts shooting at each other.

Premise/Syuzhet: A parasite that tries to take over a body, to survive.
Goal/Fabula: Infect, control, and defend to survive.

The diegetic world was developed by taking a parasite's world view, by focusing on its survival and on the completion of different tasks in the game. The premise and goal were clearly defined as a parasite that wanted to take over the world for its survival and for the player to infect and assume control of a body. The logic, relations, and rules were balanced and created a clear picture of the diegetic world, Figure 7.

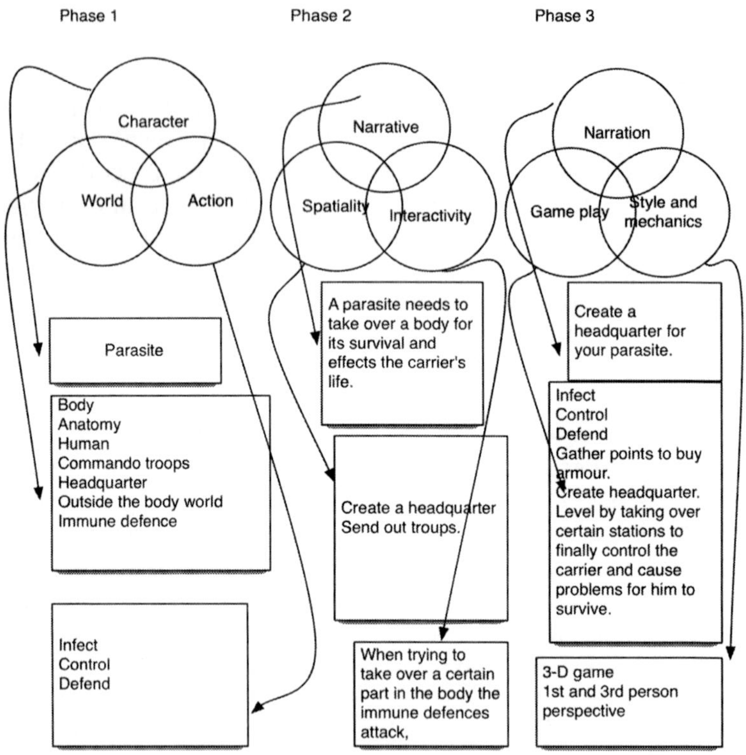

Fig. 7. First iteration of the parasite game.

When moving information from Phase 1 to Phase 2, the two narratives the group had in mind were not described. The group referred to a parasite's relation to the carrier and, through visiting certain parts of the body, how it should affect the life of the carrier. One thing the group had done was to define the setting of the parasite's headquarters, but the directions and the targets and their relations to the carrier had to be elaborated on, if that was the game they would like to create (the premise). By defining the two worlds (the carrier and the parasite) in Phase 1 and seeing how the systems related, the group could see how the causes and effects created conditions and consequences for game play. Seen as targets, the relations between the parasite and the carrier were not clear, and the group had to move back to Phase 1 to divide the worlds. By running this new information through another transition from Phase 1 to Phase 2, it got easier to decide where the parasite was born, what the first target was, what had to be defended, what the next target was, when the target needed the troops, which the different effects on the carrier were, and what the parasite should do or use to gain maximum effects on the carrier. Even if the group would stay with a simple graphic or a simple narrative (not involving a carrier), they would still need to decide upon these things in order to fix the targets for the parasite.

What happened after the first iteration was that the parasite's goal to survive and take over the body got an extra layer and motive for the player, Figure 8. From the beginning, it was about a parasite's biological concern to take over a body and to affect a person's life. The group now developed the carrier's world and made him a rude person at the top of his career, and that they wanted to see him fail in (i.e., to learn a lesson). What the group also did by elaborating the two worlds of the parasite and the carrier was to give depth to the game play. The group could have chosen to work with game play based on a system where the heart and brain would have been the hot targets for attacks. Instead, they related the targets to what the carrier was about to do, in order to let the parasite attack more subtle systems in the personal life or career of the carrier. The group had too little time to develop a full system of play, but they did see new possibilities like making the carrier sneeze, fart, have involuntary reflexes, etc. Even though this information was found early on, its importance became evident to the group only in the second iteration.

Results

Looking at the key elements of narrative bridging—the media, premise, goal, and syuzhet—the interplay in the diegetic world can have a strong impact on game design. In our prototype tests, the syuzhet was supported by defining the media, which in turn enabled control of the information required to reach a good outcome; the latter being defined through the premise and goal. In all of the game ideas produced in the test period, six in total [23], the groups moved back to develop the diegetic world to create stronger motives, restricting their narrated objects and attributes to create a stronger game play.

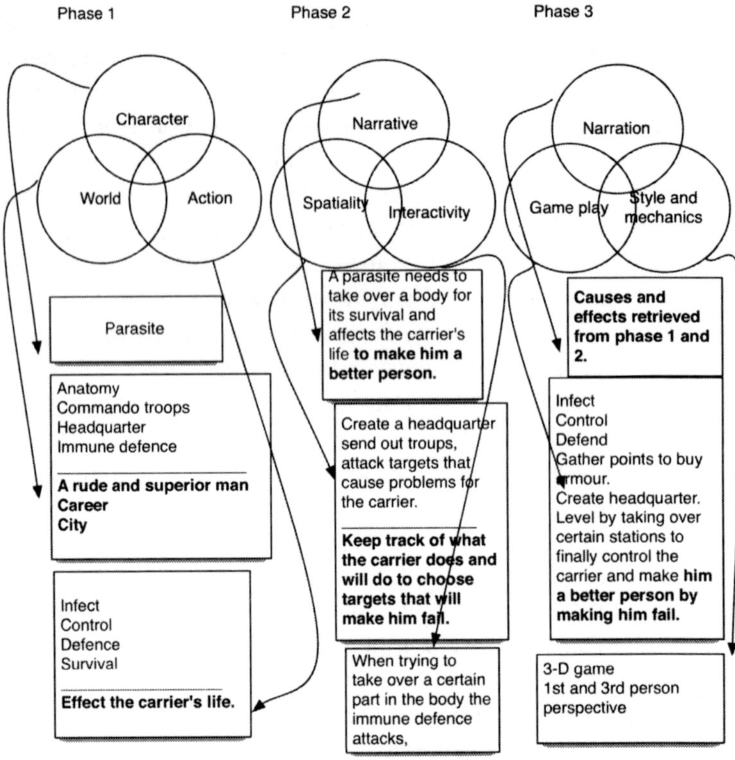

Fig. 8. Second iteration of the parasite game.

The groups here used the means, conditions, and consequences from the choices that the user could make and gave them a deeper meaning. This says something about how the narrative practice can manipulate cognitive patterns to create meaning, and about how the construction of the method supports this process. In general, the proponents of ideas that did not exploit the diegetic world faced problems when trying to plot the information, using the syuzhet principles, in establishing the spatiality. This in turn affected the interactivity, as ideas did not use walkthroughs or maps, and did not convey meaning. The groups using templates with an already familiar structure, such as the historical drama in the Masquerade game, or the human body (with its heart, brain, limbs, etc.) in the Parasite game, got quick access to both the spatiality and the interactivity. What made them move back to model the diegetic world was that the character had to be developed in order for the player to see what conditions and consequences the diegetic world produced. In the Masquerade game, an inconsistency was detected in the second phase, and the element of spatiality as a condition and consequence of the activity could not be created, as in answering the question "How do I approach the king?", for example. This led the group back to the question of how to either elaborate on the character, or to revise their premise or goal, Figure 9. In the Parasite game, an inconsistency was detected in the second phase too, and in this case the narrative was made more complex so that actions affected not only a body, but also the cognitive features of the carrier, Figure 10.

All groups naturally felt time was too short for handling the emerging complexity of their systems. All the inconsistencies detected would probably have been found, sooner or later, time permitting. But the fact remains that the method gave an immediate response to the idea in a time-constrained design process, thus enabling early detection of anomalies. By prompting a division of the systems and their elements, the method supported teams in presenting and understanding their own design as well as the complexity of the media. The three phases provided teams with a vocabulary so that they could follow each other's reasoning, track inconsistencies and resolve these. By using the graphics prescribed by the method, the idea could also be drawn on a wyteboard, for others to inspect.

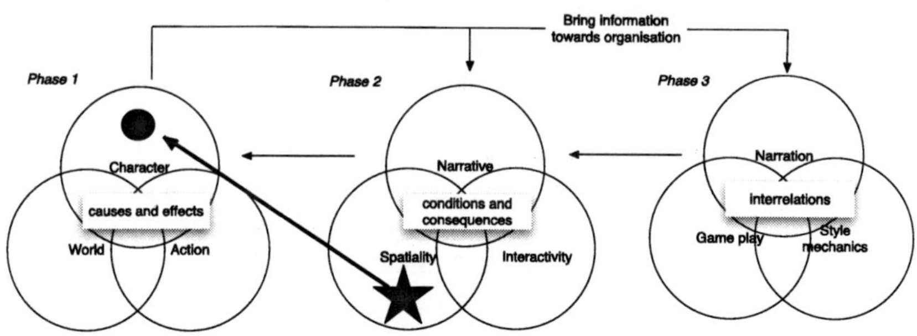

Fig. 9. Detection of inconsistency in the Masquerade game

Without explicit intrusion, the method provided groups with hints on where to go, and directed by the premise and goal, the groups could then take design decisions from there. Rapid prototyping was supported directly through the control of causal chains that the syuzhet principles offered. Asking why, where, when, and how is nothing new, and neither is dividing and defining the media-specific attributes.

What is new is the division into three phases, where the first phase offers a possibility to overview the logic and relations. The definition of media in the second phase then checks this overview and contributes new information to the work, such as maps, elaborated characters, and walkthroughs.

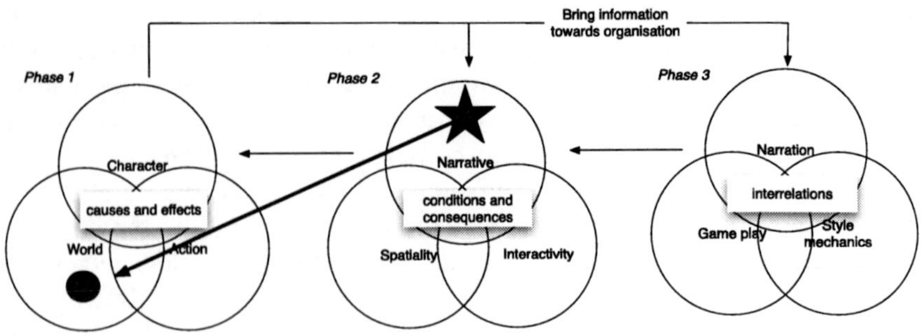

Fig. 10. Detection of inconsistency in the Parasite game.

For instance, the method could force the Masquerade game to go into depth by modeling the social interaction between the king and the player's avatar. If the king were not to like the music, for example, the troubadour would have to find a new song quickly or risk being thrown out. This would introduce values into a database of songs, to provide a balance to the game play and its mechanics around the king's taste of music and his moods. Recalling the musician's goal of finding the murderer in time, choosing the wrong music would result in a causal chain leading to a failed goal. The narrative process defining the king's taste in music could have been put into text, or in a cut scene (influenced by the canonical structure from film), but narrative bridging yields enough control so as not to require such information handling. This explains why one group said it helped them avoid 'cut scenes' and 'text', and instead render the story material within the game [23]. This is evidence of how the method supports the motto: "Do not show, involve", meaning to let the player take part of the narrated world by interacting with it instead of reading about it.

Conclusions and Further Work

The three phases of narrative bridging provide a backdrop against which teams may detect inconsistencies in design for interactive media, as well as find ways of resolving those inconsistencies. By using the graphical depiction of design as devised by the method, ideas can be sketched more easily, and the graphics and terminology together form an ontological basis for the design teamwork. Grünvogel criticized formal models for forcing designers into standardized workflows [14], but our initial empirical tests suggest that narrative bridging mimics the creative processes in design in a non-intrusive way. Narrative bridging also stresses the narrative as a process to be treated as at least equal to game play and mechanics when it comes to balancing logic and relations to create meaning. It renders the syuzhet, as a system, superior to style. In the prototype tests, all groups revisited their first description of the diegetic world. This was triggered by not being able to answer what someone in the game at hand encountered and where. The method then called for new information to be created in the diegetic world, sending groups back to Phase 1 before scrutinizing their game design again in Phase 2; producing walkthroughs and maps, as well as conditions for interactivity. Another important issue brought forth in the prototype test was the use of templates and strong structures, including game-specific structures, like labyrinths and puzzles. At first sight, those strong structures provided an illusion of being worked out when presented, but in taking a closer look via narrative bridging, most of those illusions burst.

It must be pointed out that the test subjects constituted a homogenous group of students (without a control group), and that more empirical work is needed. In preparation for such work, the instructions on how to use the method have been streamlined. An unexplored research strand is to go deeper into the user's means to character creation, looking at avatars, for instance. In the prototype testing, no group created avatars. Narrative bridging can support character creation, however, and future tests will provide design teams with such assignments. Future studies will also go deeper into the method's interrelating elements, where the causes and effects, and the conditions and consequences, are constructed. A future goal is to try and map the motivational junctions in a construction by simulating and analyzing a variety of games, as well as constructing new games to take the research beyond the strong market forces which govern industry output. Not only film influences the development of digital games. Engineer-driven cultures and software engineering, e.g., agile development, also affect the conceptualization of digital games, using concepts like genre, target groups, and fun factor [8]. Finally, since the empirical tests reported on here were completed, several validation efforts have been made within the realm of game dramaturgy. The method was used in three consultations for game development companies, two for personal computer games and one for a mobile phone game. No formal validation of the results of these consultations have yet been made, but these experiences provided several clues as to how the method could be employed within a competitive commercial game design environment.

References

1. Ryan, M.-L.: Beyond myth and metaphor: the case of narrative in digital media. Game Studies 1(1) (2001)
2. Ryan, M.-L.: Avatars of story. University of Minnesota Press (2006)

3. Aarseth, E.: Playing research: methodological approaches to game analysis. In: Digital Arts and Culture Conference, Melbourne, Australia (2003)
4. Aarseth, E.: Genre trouble: narrativism and the art of simulation. In: Harrigan, P., Wardrip-Fruin, N. (eds.) First Person: new media as story, performance and game, pp. 45–55. MIT Press, Cambridge (2004)
5. Frasca, G.: Ludology meets narratology: similitudes and differences between (video) games and narrative. Parnasso 3, 365–371 (1999) In Finnish, English version available at http://www.ludology.org
6. Bordwell, D.: Narration in the fiction film. Methuen (1985)
7. Frow, J.: Genre. Taylor & Francis Ltd., Abington (2005)
8. Chandler, M.H.: The game production handbook, 2nd edn. Jones & Bartlett Publishers (2008)
9. Chandler, D.: An Introduction to Genre Theory (2000), http://www.aber.ac.uk/media/Documents/intgenre/chandler_genre_theory.pdf (Last accessed May 2010)
10. Boman, M., Johannesson, P., Bubenko, J.: Conceptual modelling. Prentice-Hall, Englewood Cliffs (1997)
11. Hunicke, R., Leblanc, M., Zubek, R.: MDA: A formal approach to game design and game research. In: Proceedings of the challenges in game AI Workshop, AAAI 2004. AAAI Press, Menlo Park (2004)
12. Church, D.: A language without borders. Game Developer (August 1999)
13. Jarvinen, A.: Understanding video games as emotional experiences in The video game theory reader 2, ch. 5, Routledge (2008)
14. Grünvogel, S.M.: Formal models and game design. Game Studies 5 (2005)
15. Löwgren, J., Stoltenman, E.: Thoughtful interaction design. MIT Press, Cambridge (2004)
16. Mateas, M., Sengers, P.: Narrative intelligence. In: Mateas, M., Sengers, P. (eds.) Narrative Intelligence, John Benjamins Publ Co, pp. 1–11. John Benjamins, Amsterdam (2000)
17. Gärdenfors, P.: How homo became sapiens—on the evolution of thinking. Oxford University Press, Oxford (2004)
18. Salen, K., Zimmerman, E.: Rules of Play—Game Design Fundamentals. MIT Press, Cambridge (2003)
19. Fullerton, T.: Game design workshop—a playcentric approach to creating innovative games. Morgan Kaufmann, San Francisco (2008)
20. Crawford, C.: On game design. New Riders Publishing, Indianapolis (2003)
21. Crawford, C.: On Interactive Storytelling. New Riders Publishing, Indianapolis (2005)
22. Juul, J.: Half-real—video games between real rules and fictional worlds. MIT Press, Cambridge (2005)
23. Borg Gyllenbäck, K.: Narrative bridging—a specification of a modelling method for game design. M.Sc., thesis, Stockholm university (2009)

Generic Non-technical Procedures in Design Problem Solving: Is There Any Benefit to the Clarification of Task Requirements?

Constance Winkelmann and Winfried Hacker
Technische Universität Dresden, Germany

The quasi-experimental field study with 174 advanced engineering students analysed the possibilities to assist the requirements analysis when solving design problems. Technical check lists are in common practice for assisting the requirement analysis. We wondered if a generic question answering system (GQAS) aiming at the 'semantic relationships' would offer an additional benefit to the exhaustive identification of the requirements of a design task when a technical check list were offered at the same time. Therefore, two groups of students of mechanical engineering were asked to develop a list of requirements for the design of a machine collecting windfall. Whereas one group was offered a technical check list together with the generic question answering system, the other group was only offered the technical check list. The entire number of identified applicable requirements is significantly higher for the group with the additional GQAS. The additional benefit of answering generic interrogative questions holds for the majority of the individual categories of technical requirements, e.g. the weight of the device to be designed, manufacturing costs or recycling. The benefit is explained hypothetically with the proven stimulation of meta-cognitive processes by means of systems of interrogative questions. Practically, the consideration of generic procedures of problem solving in engineering design education may be proposed.

Introduction

Experts for engineering design among other things differ from less efficient designers by a more sophisticated analysis of the requirements of their design tasks [1], [2], and [3]. They identify the needs of clients in

J.S. Gero (ed.): *Design Computing and Cognition'10*, pp. 545–558.

detail, try to determine the requirements as completely as possible, including implicit ones, distinguish between mandatory and desirable requirements, look for contradictions between requirements, and for remaining degrees of freedom for design solutions.

Moreover, in their ongoing design process the experts systematically specify their thorough analyses of the requirements, iteratively. The Munich Model of the Design Procedure [4] recommends exactly this feedback-circle between the evaluation of intermediate results of design and the specifying re-analyses of task requirements.

Text books for Engineering Design offer technical check lists meant to assist this weighty analysis of task requirements [4] or [5].

Experimental research in generic or non-technical skills had shown that quite another type of check lists might assist the clarification of the design problem as well as the identification of task requirements. The thorough application of a system of generic non-technical questions may assist the requirement analysis if applied mandatorily and completely [6], [7], [8], and [9]. This is a system of interrogative questions which stimulate reflection on the problem solving procedure. Answering these questions requires dealing with a set of *semantic relationships*, particularly causal, final, conditional, and instrumental relations. The set of these relationships exhaustively characterizes the possible relations within any system [10] and [11].

So far, however, the system of the non-technical generic interrogative questions was applied separately, but not in combination with the mentioned specific technical check lists assisting the identification of task requirements. Thus, the question arises, whether a generic questioning system might contribute an additional benefit, even if applied in combination with a technical check list. Answering this question is the novel contribution of this study. More generally, this question is a part of the problem, whether there is an additional benefit in design problem solving by non-technical or generic skills or strategies. The approach of "Rethinking engineering education" [12] stresses generic skills and strategies such as scheduled problem solving techniques, communication skills or strategies of efficient teamwork. A generic question answering system assisting the analysis of task requirements in problem solving is one of these strategies.

For other professions with complex problem-solving tasks, especially pilots, anaesthetists, and operators of high-risk technologies, the additional benefit of non-technical strategies was verified and has become an essential part of their professional education and training [13] and [14].

Aims

The following questions result from the outlined issue: Does a generic question answering system at all offer an additional benefit for the exhaustive identification of the requirements of a design task if a technical check list is offered at the same time?

We expect that participants who are offered an additional GQAS along with a technical check list will identify significantly more applicable task requirements than those without the GQAS.

Moreover, we are interested in hints on the participants` perception of the additional GQAS.

Our questions are: Do the participants perceive a benefit from the entire GQAS? Which of the questions are perceived as especially helpful? Which improvements of the GQAS are proposed by the participants?

Method

Sample

174 engineering students of a University of Technology volunteered in the investigation (age 19-33 years, mean and standard deviation 22.4 ± 1.4 years; 88% male). The investigation took place during the final period of the students` course in methods of engineering design.

Task

Each participant was asked to prepare an individual list of requirements to be considered in the design of an all-terrain machine collecting windfall. For the task, see the appendix.

Independent Variable, Experimental Design and Procedure

The investigation applied a quasi-experimental design with two randomized groups of engineering students (group A, $n_a = 113$; group B, $n_b = 61$). The participants of both groups were offered the technical check list by [5] with the main categories of technical requirement analyses. Additionally, the participants of group B were offered the GQAS, which is thought to stimulate the decomposition of the entire design order, to identify its explicit as well as implicit requirements, and to deduce further requirements that may result from relationships between the requirements.

As a precondition we tested experimentally the impact of the GQAS separately: In a previous study [8] with other students the group with the GQAS (but no further tool) achieved better results than a control group without any tool.

Concerning the obligation of the treatment in this pilot-study, a voluntary application of the tools as to their sequence and the completeness of coping with them took place.

The participants received an oral instruction and the written task with the relevant tools. Their working time was limited to 45 minutes for both groups of students. All of them spent these 45 minutes processing their task.

Dependent Variables and Instruments

The main dependent variable is the number of the applicable requirements identified by each participant. For reasons of data analysis, a group of experienced engineering designers of the Chair of Engineering Design at TU Dresden developed a reference-list of applicable requirements for the task (see appendix). The requirements listed by each participant were compared with this reference-list. This evaluation was carried out independently by two raters. The interrater-reliability was satisfactory (κ = 0.69 to 0.75 for the different kinds of requirements).

Moreover, the participants of group B answered the questions of a semi-standardized questionnaire on the benefit of the GQAS and were asked which individual questions they perceived as being highly helpful and which as less helpful in the requirement analysis (see appendix). The questions correspond with learning goals for the analysis of task requirements which were determined by a group of experienced co-workers of the Chair of Engineering Design.

Control variables. Beyond age and gender, we controlled for aspects of the participants' knowledge in non-technical strategies in engineering design. Contrasting with generic strategies, technical ones were taught and trained in the ongoing lectures and training sessions; therefore, severe differences between the groups were not expected here.

Moreover, we controlled for relevant dimensions of Action Style of the participants (planning attitude; flexibility and persistence of goal attainment; analysed by a questionnaire of [15]).

A possible impact of dimensions of Action Style on problem solving is reported by [15].

Results

Check of Randomization of the Intervention Groups

There are no significant differences between the two intervention groups according to age (t (172) = 0.16, $p > .05$), gender (χ^2 (1, N = 174) = 0.03, $p > .05$), relevant knowledge (t (87) = 4.31, $p > .05$), and the dimensions of Action Style, i.e. planning attitude (U-test (Mann & Whitney), z = -0.71, $p > .05$; [22]) and flexibility of goal attainment (z = -1.08, $p < .05$). However, the participants of group B are significantly more persistent in goal attainment on average (z = -2.95, $p < .05$). Thus, an objection is possible: More persistent goal attainment of the subjects may result in a higher number of identified requirements. Therefore, we checked the relation between persistence and the total of identified task requirements. There is no significant correlation (r = 0.18, p > .05); an impact of persistence on the number of identified requirements consequently might be marginal at best.

Thus, the randomization check confirmed that the intervention groups do not differ in hypothetically influential factors to a substantial extent.

Performance Benefit of the Additional Offer of the GQAS

The entire number of identified applicable requirements is significantly higher for the intervention group B with the additional GQAS, even if persistence is considered (t (172) = 5.51, p < .05).

Considering the 15 individual categories of requirements, Table 1, the results of this group are significantly higher for 10 of the 15 categories. Thus, the additional GQAS improved the requirement analysis in the majority of the categories of technical requirements.

Reported Benefit of the Additional Offer of the GQAS

Half of the participants of intervention group B (51%) reported being able to cope well with the identification of the requirements of the device whereas in group A this was reported only by a forth (28%) of the group.

The difference (D) is statistically significant ($D_{observed} = 23\% > D_{0.05,\ one\text{-}sided} = 16.4\%$).

Table 1 Number of identified requirements of group A (with technical check list) and B (check list and generic question answering system) and significance of differences

Requirements	Intervention group A M ± SE	Intervention group B M ± SE	T	*p*
Total number	12.61 ± 0.59	17.92 ± 0.73	5.51	< .05
1. Geometry	1.04 ± 0.09	0.70 ± 0.09	-2.29	< .05
2. Kinematics	1.66 ± 0.15	1.54 ± 0.16	-0.54	> .05
3. Weight	0.34 ± 0.06	0.79 ± 0.09	4.59	< .05
4. Energy	0.71 ± 0.08	1.13 ± 0.13	2.99	< .05
5. Function	3.66 ± 0.19	4.69 ± 0.32	2.87	< .05
6. Material	0.75 ± 0.09	1.74 ± 0.18	4.98	< .05
7. Safety	0.60 ± 0.08	0.46 ± 0.09	-1.09	> .05
8. Ergonomics	1.49 ± 0.13	2.87 ± 0.21	5.86	< .05
9. Control	0.21 ± 0.05	0.23 ± 0.06	0.19	> .05
10. Assembling	0.19 ± 0.05	0.18 ± 0.06	-0.09	> .05
11. Transportation	0.27 ± 0.05	0.66 ± 0.10	3.87	< .05
12. Maintenance	0.62 ± 0.08	0.95 ± 0.11	2.52	< .05
13. Costs	0.86 ± 0.08	1.41 ± 0.13	3.70	< .05
14. Recycling	0.12 ± 0.03	0.25 ± 0.06	1.97	< .05
15. Design	0.12 ± 0.03	0.33 ± 0.07	2.91	< .05

Note: The independence of the requirements was proven. Therefore, an α-adjustment is not necessary.

Consequently, our expectation concerning an additional performance benefit of a non-technical question answering strategy aiming at semantic relationships is verified for the given conditions.

Half of the students of group B (51%) perceived the GQAS as being very helpful respectively helpful, the other half (49%) not much helpful for coping with the task. These ambiguous findings might correspond with the perceived different impact of some questions:

On the one hand, most of the participants made statements on the benefit of the questions stimulating the identification of implicit requirements, Table 2. The majority of them perceived these questions as being highly helpful. If one adds the statements on a related question concerning requirements following from other ones, about 60% of the participants mentioned these aspects in their statements, and the majority (77%) perceived them as being highly helpful.

On the other hand, at least a quarter of the participants criticized a couple of other questions of the GQAS as to their redundancy. These are the questions 3, 6, and 7 concerning aspects of the relationships between requirements. They were perceived as being less helpful. Accordingly, suggestions for improvements to the GQAS stress the reduction of this supposed redundancy, and, moreover, the vague wording of these questions.

Table 2 Statements of the participants (group B; n = 61) concerning the support by the question answering system

	Share of participants who commented the questions		**Out of these commented "highly helpful"**		**Out of these commented "less helpful"**	
Questions	absolute	%	absolute	%	absolute	%
1. explicit requirements	10	16	9	90	1	10
2. implicit requirements	26	43	22	85	4	15
3. relationship between requirements	13	22	4	31	9	69
4. requirements resulting from relationships	10	16	6	60	4	40
5. translation into technical requirements	7	12	4	57	3	43
6. requirements influencing others	15	25	3	20	12	80
7. type of influence of requirements	19	31	5	26	14	74
8. identical requirements listed	14	23	10	71	4	29
9. clustering of requirements	6	10	4	67	2	33
10. indispensable requirements versus desirable ones	11	18	7	64	4	36
11. importance of requirements	7	11	6	86	1	14
12. black-box diagram of functions	2	3	0	0	2	100

Besides, the sophisticated statements made by nearly a quarter of the participants of group B confirm the actual application of the GQAS.

All in all: participants with the additional GQAS reported significantly more frequently being able to cope well with the requirements analysis. Questions of the GQAS aiming at implicit requirements and at requirements resulting from other ones are reported to be most helpful.

Conclusions

The quasi-experimental field study involving 174 advanced engineering students analysed the possibilities to assist the requirements analysis. This work step is decisive in the entire process of engineering design: overlooked requirements are not considered in the design process at all or are not considered in decisions between alternative solution principles.

Technical check lists are common practice for assisting the requirements analysis [4], and [5]. A further approach for this reason may be the application of a non-technical GQAS, aiming at the semantic relations (e.g. causal, conditional or final) within any system and thus initiating reflective thinking on the completeness of the analysis and the significance of the requirements found. Although the efficiency of the isolated GQAS was repeatedly verified [8] and [9], it has so far remained unknown whether a GQAS contributes additional benefits if applied together with a proven technical check list as offered in engineering textbooks [4], and [5].

Therefore, in this experiment, the participants were asked to list all requirements for the design of a machine that collects windfall and for doing so, one group of participants was offered the technical check list together with the GQAS to assist the task, the other group received the technical check list only.

The total number of identified applicable requirements is significantly higher for the group with the additional GQAS. The additional benefit of answering generic interrogative questions holds for the majority of the individual categories of technical requirements, e.g. the weight of the device to be designed, the costs of manufacturing or the materials used.

Correspondingly, the participants` perceived mastering of the task is consistent with this performance difference: half of the group of participants who received both instruments reported they felt able to tackle the task well. In contrast, only a quarter of the group with the technical check list only reported so. Most of the participants remarked on the helpful benefit of questions considering the identification of implicit requirements and requirements following from other ones. Implicit requirements are those not explicitly addressed in the design order.

The results are of theoretical and practical impact. First, the results confirm that a GQAS improves the identification of the requirements of engineering design orders as already shown for the application of the GQAS without a further technical check list. The validity of these former results [8], [10], and [11] could be verified.

The essential new result is the significant benefit of a GQAS if applied in addition to a proven technical check list. This benefit is not due to a longer working time.

An *explanation* of the additional benefit of a GQAS without increased processing time may hypothetically consist in a qualitative change of the process of the identification of requirements: as shown in a previous experimental study – however applying the GQAS only [8] – answering a system of interrogative questions requires the processing of semantic relations [11] and as a result forces to complete and deepen the analysis of the task. This last argument is confirmed in this study by the mentioned perceived benefit just of questions aiming at implicit requirements and at those following from other ones.

A *limitation* of our study is the lack of a qualitative analysis of possible differences within the processes of the identification of requirements with versus without the GQAS. Thus, final evidence of an explanation may not be offered so far.

Suggestions for further research: the impact of the GQAS should be tested for different kinds of design orders. In addition, the voluntary application of the tools is to be replaced by a mandatory and sufficiently checked one.

Nevertheless, from a *practical point of view* a consideration of generic questioning techniques – as instances of generic skills and strategies – in rethinking engineering education [12] may be useful.

References

1. Wallace, K., Ahmed, S.: How Engineering Designers Obtain Information. In: Lindemann, U. (ed.) Human Behaviour in Design. Individuals, Teams, Tools, pp. 184–194. Springer, Berlin (2003)
2. Müller, A.: Iterative Zielklärung und Handlungsplanung als Faktoren erfolgreichen Gruppenhandelns bei der Lösung komplexer Probleme: Eine handlungstheoretische Betrachtung des Konstruierens in Gruppen. Mensch & Buch Verlag, Berlin (2007)
3. Görner, R.: Zur psychologischen Analyse von Konstrukteur- und Entwurfstätigkeiten. In: Bergmann B., Richter P. (Hrsg.) Die Handlungsregulationstheorie. Von der Praxis einer Theorie. Hogrefe, Göttingen, pp. 233–241 (1994)

4. Lindemann, U.: Methodische Entwicklung technischer Produkte: Methoden flexibel und situationsgerecht anwenden, 3. korrigierte Aufl. Springer, Heidelberg (2007)
5. Pahl, G., Beitz, W., Feldhusen, J., Grote, K.: Konstruktionslehre. Methoden und Anwendung, 4. neubearb. Aufl. Springer, Heidelberg (2004)
6. Wetzstein, A., Hacker, W.: Reflective Verbalization Improves Solutions – The Effects of Question-based Reflection in Design Problem Solving. Applied Cognitive Psychology 18, 145–156 (2004)
7. West, M.A.: Managment of Creativity and Innovation in Organizations. In: Smelser, N.J., Baltes, P.B. (eds.) International Encyclopedia of the Social & Behavioral Sciences, vol. 5, pp. 2895–2900. Elsevier, Amsterdam (2001)
8. Winkelmann, C., Hacker, W.: Erklärungsansätze für die Wirkung einer Frage-Antwort-Technik zur Unterstützung beim Design Problem Solving. Zeitschrift für Psychologie 214(2), 73–86 (2006)
9. Winkelmann, C., Hacker, W.: Unterstützungsmöglichkeiten der Produktentwicklung: Welche Veränderungen am Ergebnis löst das fragengestützte Nachdenken über eigene Lösungen aus? Zeitschrift für Arbeitswissenschaft 61, 11–21 (2007)
10. Daudelin, M.: Learning from experience through reflection. Organizational Dynamics 24, 36–48 (1996)
11. Krause, W.: Denken und Gedächtnis aus naturwissenschaftlicher Sicht. Hogrefe, Göttingen (2000)
12. Crawley, E.F., Malmqvist, J., Östlund, S., Brodeur, D.R.: Rethinking Engineering Education. The CDIO Approach. Springer Science + Business, Heidelberg (2007)
13. Flechter, G., Flin, R., McGeorge, P., et al.: Rating non-technical skills: Developing a behavioral marker system for use in aneasthesia cognition. Technology and Work 6, 165–171 (2004)
14. Yule, S., Paterson-Brown, S., Maran, N.: Non-technical skills for surgeons in the operating room: A review of the literature. Surgery 139(2), 140–149 (2006)
15. Heisig, B.: Planen und Selbstregulation: Struktur und Eigenständigkeit der Konstrukte sowie interindividuelle Differenzen. Peter Lang, Frankfurt am Main (1995)
16. Winkelmann, C., Hacker, W.: Question-answering-technique to support freshman and senior engineers in processes of engineering design. International Journal of Technology and Design Education (2009), http://springerlink.com/content/g70n16075001x4u8/?p=1ace059a5f514205b6f9bcdca2e5c4f9&pi=2 (Last accessed May 2010)

Appendix

Task Windfall Collector

Collecting windfall from fruit trees is an exhausting and time-consuming task. If the pieces of fruit remain laying on the soil for a longer period of time, unpleasant odour and dangerous situations may occur.

Your task is to design a machine that collects the fruit easily and swiftly. The fruits should remain usable as much as possible. The machine is to be designed for use in small businesses, e.g. small orchards. List all requirements that should be taken into consideration for the design (please use backside as well).

Check List Concerning the Completeness of the Requirements Specification

Check lists help to test and correct e.g. the clarification of the requirements and functions.

The reason is: An incomplete clarification may result in incomplete solutions.

This list of requirements is necessary
- to affiliate and deliberate ideas on solutions
- for the final evaluation of the various solution alternatives and the selection of the best one.

A. Check the completeness of gathering the requirements

1. What are the explicit requirements of the order concerning the artifact and its functions?
2. What requirements are provided implicitly in the order, in particular
   - in relation to market trends?
   - regarding the product and the purchased parts (benchmarking)
   - in relation to the users' interests?
   - in terms of standard solutions for modules or components?
   - in relation to the product life cycle (i.e., manufacturing, mounting, transport, maintenance, disassembling/ reuse)?
   - in relation to occupational health and safety and other legal requirements and international or national standards?
3. Which relationships (dependencies, linkages, contradictions) do exist between requirements?
4. Are there any modifications or new demands resulting from relationships between requirements?

5. How can colloquial or unspecified requirements (for example "cheaper than competing products", "muted" etc.) be "translated" in and fulfilled by the technical requirements?
6. What technical requirements influence other ones and which are influenced by others?
7. Which technical requirements
   - support each other ,
   - contradict each other,
   - are mutually exclusive?
8. Which requirements – different in wording – are identical?
   Sort these out!
9. How can the technical requirements be grouped, e.g. by spatial, energetic or informational functions?

B. Clarify the importance of the requirements, determine priorities

10. Which requirements are indispensable and which are desirable only?
11. How important are the desirable requirements?
    (Weigh them (e.g. "high" (e.g. since essential during the entire life cycle of the product); "low" (e.g. since due to additional comfort))
    Create and exhaustive list of the clustered and weighted technical requirements and determine requirement priorities in abstraction from details!
12. What total functions are to be fulfilled? Visualize them as a black-box representation!

Reference-List of Applicable Requirements

	List of requirements: Windfall collector	
1.	Geometry:	
1.	Consideration of the distance between the trees	]
2.	Consideration of the passage height	I
2.	Kinematics:	
1.	only forward	]
2.	steerable, reversible (turn over) at the place	]
3.	possible reverse	]
4.	all-terrain (grass, mole hills, holes ...)	]
5.	walking speed at maximum, continuously variable or different speed levels	]
10.	Weight:	
1.	empty max. 45 kg	]
2.	with fruit max. 90 kg	]
10.	Energy:	
1.	combustion engine + tank or electric motor + accumulator	]
2.	8h or 2x4h duration of use per day	]
3.	ambient air temperature while operation 0° … 50 °C	]
10.	Material:	
1.	materials against aggressive fruit acids, moisture/ juice resistant	]
2.	UV-resistant materials and food safe (depending on the installation location of the machine)	]
3.	weatherproof, suitable for outdoor use	
4.	reservoirs easily reversible/ emptying, possibly standard crate	]
5.	not/ or hardly fruit damaging	]
		]
10.	Safety:	
1.	drive capsule, covering moving parts	]
2.	if exhaust routing: no nuisance to the operator	]
10.	Ergonomics:	
1.	huge wheels, adjustable handles / mirrors	]
2.	noise isolation (max. 80 dBA noise development)	]
3.	simple operating and control elements	]
10.	Control:	
1.	motor proven by TÜV/ GS	]
2.	own CE statement	]
10.	Construction:	
1.	completely delivery, possibly only pre-assembled, handles/ steering/ wheels possibly to be mount by the customer	]
10.	Transportation:	
1.	assembled in a large wagon/ vans	]
2.	one person sufficient for transportation	]

cont…List of requirements: Windfall collector		
11.	Maintenance:	
1.	cleaning with water/ hose and broom street	]
2.	no wearing parts, if easily replaceable	]
3.	no contact between operating materials/ lubricant and fruits	]
12.	Costs:	
1.	budget-priced (max. 1.000 €)	]
13.	Recycling:	
1.	no / little Composites; enable separation of waste disposal	]

Questionnaire on the Benefit of the GQAS

Interview with Intervention Group B

To guarantee the anonymity of the interview, we ask you to fill out the following code:

Letter of first name of mother	
Letter of first name of father	
Birthday of mother (two-digit; e.g. 25th of March 1949: 2 5)	

Please fill in the number of your group:

1. How well were you able to cope with the identification of the requirements?

☐ very good ☐ good ☐ middle ☐ not so good ☐ not good

2. Did you have any difficulties in generating the list of requirements? Why?

3. How useful did you find the check list when generating the requirements list, and why?

☐ very usefully ☐ usefully ☐ middle ☐ less usefully ☐ not usefully ☐ obstructively

4. How were you able to cope with the check list? (in terms of understanding of the questions, complexity of the check list, etc.)
5. In your opinion, which questions were very helpful? Why?
6. In your opinion, which questions were less helpful? Why?
7. What was missing?

Virtual Impression Networks for Capturing Deep Impressions

Toshiharu Taura[1], Eiko Yamamoto[1], Mohd Yusof Nor Fasiha[1], and Yukari Nagai[2]
[1]*Kobe University, Japan*
[2]*Japan Advanced Institute of Science and Technology, Japan*

In this study, we focus on deep impressions, which are defined as the impressions that are related to deep feelings towards a product and lie under surface impressions. In order to capture the nature of deep impressions, we developed a method for constructing "virtual impression networks," which involve the notions of "structure" and "inexplicit impressions", using a semantic network. This paper, in particular, aims at understanding the manner in which people form impressions of preference. Our results indicated that it is possible to explain the difference between feelings of "like" and "dislike" using several indicators in the network theory of virtual impression networks. The process of forming the impressions of "like" is shown to differ from that of "dislike" at a deep impression level.

Introduction

When designing products, designers need to create products that are congruent with the emotional feelings of users; in other words, the products should be preferred by most people [1]. This study assumes that there are some sorts of emotional "impressions", which a user derives from a product that is expected to affect user preference. We therefore focus on impressions that may underlie "surface impressions" that a user ordinarily holds when viewing a product, which we refer to as "deep impressions." An understanding of such impressions will result in a better understanding of the preferences an individual holds towards a product. This will also furnish designers with practical ideas for designing better products since a user's actual preference for a product cannot be evaluated based purely on

J.S. Gero (ed.): *Design Computing and Cognition'10*, pp. 559–578.

surface impressions. In particular, this study will focus on the difference between the feelings of "like" and "dislike," which assumes that the deep impression of "like" differs from that of "dislike." In approaching this issue, we attempt to validate the method developed in our study for the identification of deep impressions by using two categories of target objects: artifacts and natural objects.

Surface and Deep Impressions

Several methods have been proposed as a way of enabling designers to understand the impressions of user's, with the most general one being the Semantic Differential (SD) Method [2]. This method has been applied to several products in various fields. It quantitatively measures a user's impression of products and therefore solves the difficultly of expressing the user's impression of a designed product by using words and scales provided in answer form. This method, however, requires that the evaluation items are determined beforehand. The items concerned involve pairs of antonymic adjectives or nouns, for example, bright and dark, and the scales range from 1 to 5 or 7. The SD method persistently measures the difference in impressions of some products made by individuals, and the results cannot be evaluated without the products or the people that were compared. Thus, it seems that impressions undergo a shallow analysis and we call such impressions "surface impressions." Although some methods are proposed on the basis of fuzzy theories such as rough sets [3], which are expected to overcome these problems, it is difficult to capture deeper impressions.

On the other hand, we assume that "like" and "dislike" impressions are different at the level of basic cognitive process; a different cognitive process occurs when people "like" or "dislike" an object and this difference could not be captured by employing the SD Method. We view deep impressions as being closely related to this difference.

Preferences and Deep Impressions

Researchers in psychology have discussed the idea of feelings or preferences with regard to whether they require extensive cognitive processing. A widely-argued paper specifically suggested that affect (responses of liking or disliking a stimulus) is so fundamental to an organism that it occurs non-consciously and without cognition [4]. Of course, some form of recognition must occur, but it must be primitive or minimal, not at a conscious level. The conditioning of affective responses is known as Evaluative Conditioning (EC). A study examined the role of awareness in EC and suggested that awareness is in no way influenced by

EC [5]; however, the results are still refutable since the evidences are unequivocal [6].

We consider that certain "deep factors" may function in tandem with affective processing and result in the development of preferences. In our study, we construct a methodology for capturing deep impressions. Specifically, we capture the nature of human preference, which is not possible when analyzing surface impressions.

Viewpoint of This Study

In order to capture deep impressions, we focus on the "structure of impressions" and "inexplicit impressions."

Structure of Impressions

Aristotle said, "The whole is more than the sum of its parts." This concept is well-known in various sciences for its complexity and is called "holism." The existence of "something" appears as a whole.

In an attempt to capture the "something," we focus on constructing a structure based on each impression and assume that the structure can play the role of the whole impression that leads to deep impressions.

In the field of design studies, some researchers have focused on the notion of a structure in order to capture the user's impression. Georgiev et al. proposed a method of design by focusing on the structured meanings that are constructed among multiple impression words [7]. Their method is based on the analysis of logo designs using semantics analysis. They analyzed the meaning of one prototype design by the sum of relatedness between the impression word and an associated word. In fact, their analysis has led to the development of another design evaluation method that focuses on the depth of emotional impressions.

These studies indicate that the structure of impressions may be related to deep impressions.

Inexplicit Impressions

In Cognitive Psychology, arguments about implicit cognition have become active since the 1970s [8]. No obvious borderline separates implicit processes from explicit knowledge: but a border zone referred to as "inexplicit" cognition has been discussed, e.g. language that derives from both explicit and inexplicit cognitions [9]. Thus, inexplicit language has been employed as a means of investigating the nature of language [10].

From the perspective of language, we consider that human beings are unable to express every impression explicitly. However, even though it may be difficult to express this exactly by using words, humans might be capable of recognizing the expressible feelings, which can be implied through describing impressions.

In this study, we focus on "inexplicit impressions," which are impressions that are not expressed explicitly by human beings. Here, "inexplicit" is understood to be that which is not explicitly recognized or verbalized, but can appear if there is any circumstance [11]. Notably, inexplicit impressions could exist in the feelings and are implied underneath explicit impressions. Inexplicit impressions may thereby be related to deep impressions.

Purpose and Method

The purpose of this study is to construct a methodology to capture deep impressions. In order to accomplish this, we developed a method for constructing "virtual impression networks," which involve the notions of "structure" and "inexplicit impressions" using a semantic network.

We previously have conducted a preliminary experiment [12] and found interesting results. In order to refine and confirm the method developed, as well as to accumulate a wider range of data, the present study was therefore carried out involving an experiment with extended target objects.

Virtual Impression Network

In constructing a virtual impression network, we use words that express the impressions held by a user who is required to describe his/her impressions of a product using certain words. These words are called "explicit impression words." A semantic network (explained in the next section) is used to trace a virtual chain of nodes (representing the meaning of words), that is, a path from an explicit impression word to another. We assumed the nodes that appear in the paths to be inexplicit impressions. We constructed a network in which the traced paths are considered a representation of a virtual impression network. Thus, a virtual impression network consists of two types of nodes—explicit impressions and inexplicit impressions. We extracted the structure of the virtual impression network and analyzed it using network theory, which is expected to involve the notion of "structure" of impressions.

Figure 1 shows the construction of the virtual impression network. This modeling method consists of the following three steps:

Step 1: Tracing paths linking every two explicit impression words (first diagram in Figure 1). Here, a node represents the meaning of a word, and a path is a set of direct links joining two word meanings. The words expressing the meanings that are found along each path are regarded as inexplicit impression words.

Step 2: Drawing a network with the explicit and inexplicit impression words as nodes, and the links as the traced paths (second diagram in Figure 1).

Step 3: Extracting the structure of a virtual impression network (third diagram in Figure 1).

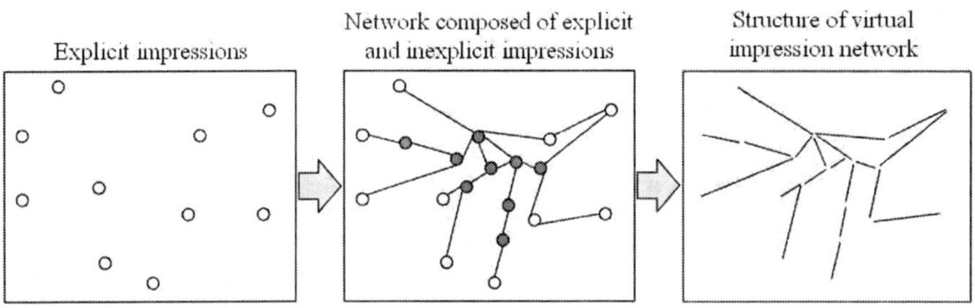

Fig. 1. Flow of modeling

Semantic Network

Semantic networks have structures composed of the semantic relations between words, such as the hypernym-hyponym and associated relations. Semantic networks are useful in searching for links between words via virtual chain processes between two words, Figure 2.

In order to obtain a better understanding of the nature of design creativity, a virtual thinking network has been constructed using a semantic network, and the structure of the virtual thinking network was found to correlate with the evaluated creativity of the actual design idea [14]. In this study, we apply them to the virtual modeling of the impressions.

The circles represent nodes in the semantic network. The white circles are nodes of explicit impression words and the black ones are nodes appearing on the path between the nodes of the explicit impression words, and the arrows are the links comprising the paths.

In semantic networks, the word meanings are nodes. Therefore, we tried to trace the shortest path for any pair of meanings of the explicit impression words and extracted words expressing the meanings in the path between the explicit impression words as inexplicit impression words. The number of links in each path was used as the length criterion for the path. When a pair had two or more shortest paths, we selected the path with the most popular word meanings on the semantic network as the shortest path.

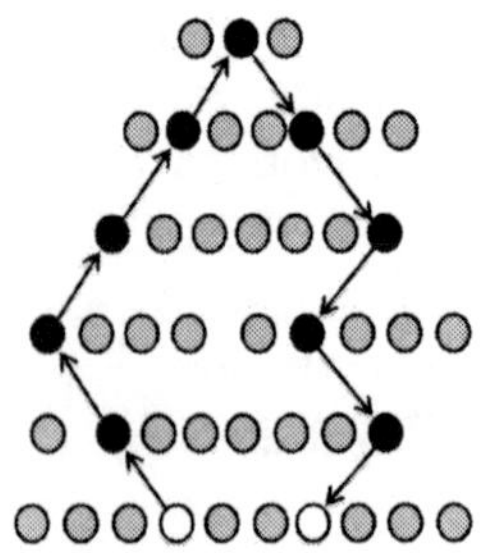

Fig. 2. Image of tracing a path linking explicit impression words in a semantic network

Structure Analysis

Using network theory, we analyzed the structure of the virtual impression network in order to identify clues that might capture deep impressions. We prepared a graph consisting of a set of nodes and a set of links joining the nodes. Although a wide range of statistical criteria exist in network theory, in this study we used a few important criteria to characterize the virtual impression network, Table 1. We chose the same network statistical criteria as those in the study of Steyvers and Tennenbaum [15]. They used the criteria to examine whether semantic networks have a structure that is necessarily different from that of other complex natural networks. Since we are using a semantic network, and the network of the user's impressions is part of a complex natural network, we decided to use the same criteria.

On the basis of the assumptions stated below, we applied these criteria for the purpose of capturing deep impressions.

- n, $\langle k \rangle$, and *Density* can indicate the expansion of the impressions.
- C, L, and D can indicate the complexity of the impressions.

The extension of the impressions is thought to be related to deep impressions since the product that resonates with deep feelings is expected to induce the user to associate them with many impressions. Further, human knowledge is like a complex network composed of pieces of knowledge and the relationship between each piece and its complexity appears in an individual's feeling space. Therefore, we consider that the complexity of a virtual impression network can be regarded as a clue to capturing deep impressions.

Table 1 summarizes the definitions of the network theory criteria used in this study. The number of nodes n denotes the number of nodes, which are expressed as words, appearing in each network. The number of links that is joined to a node is called the *degrees*, and the average degree $<k>$ is the average number of links joining a node in the network.

Table 1 Definitions of terms in network theory

Term	Definition
n	Number of nodes
$<k>$	Average degree (degree = number of links)
C	Clustering coefficient
L	Average length of the shortest path between every pair of nodes
D	Diameter of the network (the longest path among the shortest paths)
Density	Sparseness of the network (the percentage of how a node is joined to other nodes)

Two joined nodes are said to be *neighbors*. The probability that the neighbors of an arbitrary node are neighbors to each other is defined by the clustering coefficient C. In terms of network topology, a high probability signifies the existence of “shortcuts” or “triangles” in the network. The presence of shortcuts or triangles is common in complex networks. In other words, C indicates the complexity of the network. Figure 3 is an illustration of the calculation of C in this study. We calculate C by taking the average of C_i:

$$C_i = T_i \Big/ \binom{k_i}{2} = 2T_i / k_i(k_i - 1) \tag{1}$$

where T_i denotes the number of links between the neighbors of node i, and $k_i(k_i - 1)/2$ denotes the number of links that would be expected to occur between the neighbors of node i if they formed a fully joined subgraph. The character L denotes the average number of links of the shortest (or geodesic) paths between every pair of nodes over the entire network, that is, the shortest path among the sets of links comprising the paths between the nodes and the diameter of the network, and D denotes the longest path in the set of the shortest paths.

The *Density* of a network indicates the sparseness of the network and is calculated by dividing $<k>$ by size n of the network; thus, the network has a high sparseness when the *Density* is low.

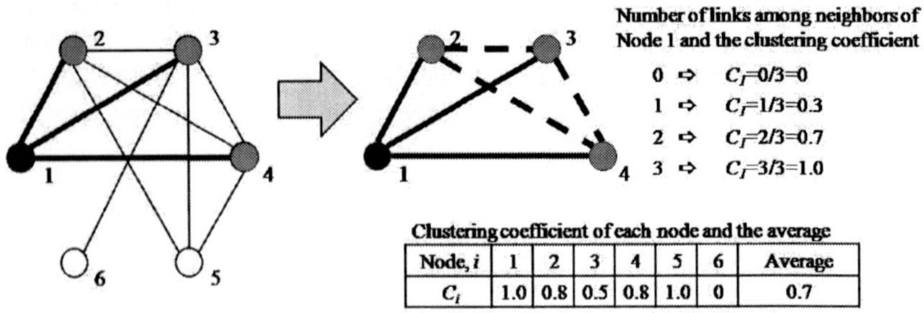

Clustering coefficient of each node and the average

Node, i	1	2	3	4	5	6	Average
C_i	1.0	0.8	0.5	0.8	1.0	0	0.7

Fig. 3. Example of the calculation of the clustering coefficient

Experiment

An experiment extended from our preliminary experiment [12] was conducted in this study. Thirteen adult graduate students participated in this experiment.

We used two categories of target objects: artifacts and natural objects. Six types of cups were used as artifacts and eight types of animals as natural objects. Cups were selected as the target for the artifact because we think that people's preference for cups could simply be influenced by the material, design, etc. Animals were selected as the target for natural objects because we think that people simply have their own preference for animals.

The target objects were selected in the preparatory experiment. In the preparatory experiment, first, we prepared pictures of 30 different types of cups and 24 different types of animals. Then, the pictures were shown to 15 students who were required to state their preference for cups/animals. We selected the target objects according to the following categories.

<Category A> Two cups and two animals that were mostly liked

<Category B> Two cups and two animals that were mostly disliked

<Category C> Two cups and two animals that were neither liked nor disliked

<Category D> Two animals that some subjects evaluated to like them very much but some different subjects evaluated to dislike them very much (which was found not to be the case for cups).

Method

The method for the main experiment consisted of two tasks. In the first task (description of impression), the subjects were shown a picture of each target object and were asked to describe their impression of the object using a number of Japanese words. These words were processed as explicit impression words. The subjects were given two minutes for each object in this task.

In the second task (indication of boundary), the subjects were asked to rank all the objects in both categories (artifacts and natural objects) according to their preference and draw a boundary of their likes and dislikes.

Results

Figure 4 shows a picture of one cup that was used in the experiment. The number of explicit impression words described by one subject for the cup shown in the figure is 21. Table 2 shows the explicit impression words. Table 3 shows the subject's preference rank for the cup.

Fig. 4. Picture of one particular cup that was shown to the subjects

Analysis

In order to appropriately use the semantic network (WordNet 3.0 [13]), we first conducted two preprocesses (explained in the next section). We then constructed a virtual impression network for each object with regard to each subject while looking at each object according to the network construction method explained previously. We then visualized the networks using an analysis tool, Pajek [16], which was also used to analyze the structural characteristics of the network. Finally, we categorized the virtual impression networks into two groups—the "like" networks and the "dislike" networks—and examined the difference in the structural characteristics between the two groups. The "like" network signifies the virtual impression network for each subject of the objects that were evaluated as "like" by the subject; the "dislike" network signifies the virtual impression network for each subject of the objects that were evaluated as “dislike” by the subject.

Table 2 Example of impression words given by one particular subject regarding the cup in Figure 4

Impression words		
ribbon	tea	gift
present	green tea	shop
coffee	functionality	restaurant
lemon	drink	department store
black tea	health	syndicate
milk	tea	retailing
herbal tea	manufacturer	baseball team

Table 3 Example of preference rank given by one particular subject regarding the cup in Figure 4. The line shows the boundary of the subject’s likes and dislikes

Preference	Cup	
1	Cup 5	likes
2	Cup 6	
3	Cup 4	
4	Cup 2	dislikes
5	Cup 1	
6	Cup 3	

Preprocess

WordNet 3.0 [13] is a vast electronic lexical database that contains information regarding the manner in which human beings process language and concepts. Currently, the database comprises over 150,000 words. Words are organized in hierarchies and are interconnected by various semantic relations such as synonyms, hypernyms-hyponyms, and meronyms. The advantage of using WordNet as a semantic network is that it provides a practical means of finding links between words.

In order to appropriately use WordNet, we conducted two preprocesses. Since the WordNet database was developed in English, we initially translated the impression words collected from the experiment (in Japanese) into English. In this process, we confirmed that the meanings were consistent with the original meanings. Further, in WordNet, links are presented only between words belonging to the same POS (part of speech; for example, noun-noun). Thus, to enable the search for links between words, we replaced all the words other than the noun with their corresponding nouns.

In these preprocesses, some noise and bias must have appeared due to the difference between the languages. The analysis by comparison would, however, have invalidated the noise effect as well as the bias because this comparison was conducted using translated explicit impression words which were expected to involve the same degree of noise and bias.

Analysis of the Structure of the Virtual Impression Network

We constructed virtual impression networks for each object of each subject. In total, 78 virtual impression networks were constructed for cups—45 were "likes" and 33 were "dislikes," and 104 virtual impression networks were constructed for animals—64 were "likes" and 40 were "dislikes." We extracted the structural characteristics of the networks using the criteria listed in Table 1 and examined the structural differences of virtual impression networks among the following.

1. All the "like" and "dislike" networks
2. "Like" and "dislike" networks by objects (the comparison of the structure of virtual impression networks of subjects who evaluated a particular object as "like" or "dislike"
3. "Like" and "dislike" networks by subjects (the comparison of the structure of virtual impression networks of a particular subject regarding the objects that he/she evaluated as "like" or "dislike")

Regarding animals, the structural differences in the networks for two animals (Animal 4 from Category B and Animal 6 from Category D) could not be calculated, as all the 13 subjects liked (or did not like) the animals. Furthermore, the significance level of the differences could not be determined for three animals (Animal 1 and Animal 2 from Category A, and Animal 5 from Category D) as only one subject liked (or did not like) them.

In this analysis, we focus on and discuss data that produced significant results.

Table 4 shows the means of the values of the network criteria for cups. We found that the average for the clustering coefficient C of the "like" virtual impression networks for Cup 4 was significantly higher ($t(11) = 3.614$, $p = 0.004$) compared to the average for those of the "dislike" networks. As C can indicate complexity with regard to explicit and/or inexplicit impressions, this significant difference implies that the complexity of the explicit and/or inexplicit impressions may also be what differentiates human preference.

Table 4 Means of the values of the network criteria for cups

Cup		*n*	<*k*>	*C*	*L*	*D*	*Density*
Cup 1  Category B	Likes	54.143	2.129	.046	6.769	16.571	.046
	Dislikes	71.000	2.249	.062	6.808	17.500	.040
	Significance Level	.329	.179	.576	.931	.586	.574
Cup 2 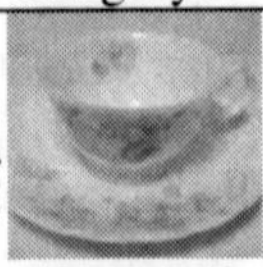 Category B	Likes	56.286	2.067	.027	7.208	17.000	.053
	Dislikes	70.333	2.075	.025	7.936	19.000	.034
	Significance Level	.404	.920	.919	.258	.249	.318
Cup 3  Category A	Likes	51.091	2.114	.051	6.727	16.636	.056
	Dislikes	51.000	2.203	.070	6.549	16.500	.064
	Significance Level	.997	.400	.554	.867	.961	.792
Cup 4 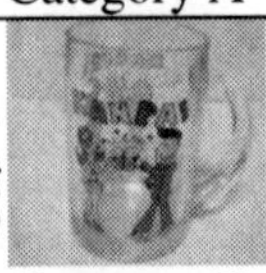 Category C	Likes	71.667	2.323	.105	7.765	19.000	.038
	Dislikes	60.800	2.133	.030	7.243	18.200	.043
	Significance Level	.537	.061	.004**	.384	.586	.723
Cup 5  Category A	Likes	59.000	2.088	.038	7.586	18.667	.047
	Dislikes	58.714	2.108	.039	7.452	18.429	.042
	Significance Level	.984	734	.978	.825	.940	.702
Cup 6  Category C	Likes	63.545	2.177	.035	7.250	17.727	.039
	Dislikes	24.000	0.955	.014	5.779	14.500	.082
	Significance Level	.031*	.031*	.317	.056	.101	.004**
All Networks	Likes	57.844	2.135	.045	7.120	17.378	.047
	Dislikes	61.121	2.132	.038	7.204	17.939	.044
	Significance Level	.589	.913	.477	.731	.476	.581

**1% level of significance by the standard (both sides); *5% level of significance by the standard (both sides)

With regard to Cup 6, we found significant differences between the means of the number of nodes *n* ($t(11) = 2.480$, $p = 0.031$), average degree <*k*> ($t(11) = 2.471$, $p = 0.031$), and *Density* ($t(11) = -3.607$, $p = 0.004$). All the criteria indicate that many explicit and/or inexplicit impressions are associated with one another; impressions of the subjects who like Cup 6 are expanded.

However, no significant difference was found for the other cups.

The structural difference in the network criteria easily identified for cups that were not commonly specified as liked or disliked (Category C). This suggests that when a person likes or does not like an object that is not commonly specified as liked or disliked, the process by which impressions are realized would be unique at the level of deep impressions; thus, the difference could be identified.

No significant difference was found in the analysis of subjects or in the analysis of all the networks of cups.

Table 5 shows the means of the values of the network criteria for animals. When comparing the structural differences of animals, we found significant differences in the average shortest path length L and diameter D for Animal 3 from Category B. The means of L and D of the "dislike" networks are significantly higher ($t(11) = -2.253$, $p = 0.046$ and $t(11) = -2.255$, $p = 0.046$, respectively) compared to the means of those of the "like" networks. Low values of L and D may indicate the high level of complexity of the impressions. Animal 3 was identified as a disliked animal in the preparatory experiment. Hence, the significant difference found in the criteria showing complexity (higher level of complexity in "like" networks) suggests that the feeling space for a subject who likes the commonly disliked animal is rather complex; in other words, having a high level of interest or knowledge for an animal. The complexity can be evidenced at the level of deep impressions.

No significant difference was found for other animals and also in the analysis of subjects. However, in the analysis of all the networks, we found significant differences in the clustering coefficient C, $t(102) = 2.121$, $p = 0.036$, where the clustering coefficient C of "like" networks were found to be higher than those of the "dislike" networks. The result also suggests the complexity of the "like" virtual impression networks.

Figures 6, 7, and 8 represent the virtual impression networks of Cup 4, Cup 6, and Animal 5, respectively, which were drawn using Pajek [16]. Both networks in the figures virtually represent the impressions of (a) a "like" network and (b) a "dislike" network. The figures illustrate the larger expansion of the impressions and the higher complexity of the structure in the "like" virtual impression networks compared to the "dislike" virtual impression networks.

Table 5 Means of the values of the network criteria for animals

Animal		*n*	<k>	*C*	*L*	*D*	*Density*
Animal 1  Category A	Likes	121.330	2.015	.011	7.945	19.500	.021
	Dislikes	220.000	2.056	.012	8.870	20.000	.009
	Significance Level	-	-	-	-	-	-
Animal 2  Category A	Likes	112.830	1.992	.006	7.630	18.000	.023
	Dislikes	128.000	2.000	.000	8.999	21.000	.016
	Significance Level	-	-	-	-	-	-
Animal 3  Category B	Likes	96.000	1.967	.000	6.583	15.333	.027
	Dislikes	135.200	2.005	.006	8.341	20.200	.018
	Significance Level	.345	.402	.320	.046*	.046*	.256
Animal 5  Category D	Likes	127.500	2.109	.042	7.547	19.417	.024
	Dislikes	214.000	2.078	.013	8.398	21.000	.010
	Significance Level	-	-	-	-	-	-
Animal 7  Category C	Likes	142.000	2.034	.006	7.822	17.875	.016
	Dislikes	119.200	1.992	.004	7.148	17.800	.025
	Significance Level	.523	.364	.726	.283	.960	.170
Animal 8  Category C	Likes	72.500	1.991	.027	6.941	17.000	.034
	Dislikes	132.890	2.000	.006	7.726	18.222	.020
	Significance Level	.120	.840	.279	.336	.565	.191
Animal 4  Category B	The difference could not be calculated as all 13 subjects did not like Animal 4	Animal 6  Category D	The difference could not be calculated as all 13 subjects liked Animal 6				
All Networks	Likes	120.940	2.027	.017	7.571	18.547	.023
	Dislikes	129.900	2.004	.007	7.761	18.575	.022
	Significance Level	.461	.231	.036*	.493	.969	.726

- **1% level of significance by the standard (both sides); *5% level of significance by the standard (both sides)
- For animals 1, 2, and 5, the significance level could not be calculated as only one subject liked (or did not like) the animals

The expansion of impressions is demonstrated by the high number of nodes n and average degree $<k>$, and the low *Density* in the "like" network. In addition, there exist many shortcuts or triangles (that illustrate the value of the clustering coefficient C) in the "like" network; in other words, a higher complexity is seen in the "like" network than in the "dislike" one. The existence of shortcuts or triangles in the "like" networks also explains that average shortest path degree L is short and the diameter D is small.

In addition, we observed the words having a value of clustering coefficient C_i in those composing the example networks in Figures 6, 7, and 8 (Table 6). The words, in general, are abstract words, and are located in the center of the networks. However, in the "like" networks, less abstract words (e.g. *drug*, *liquid*, *fluid*, *constituent*, *relation*, *location*, *communicator* and *treater*) also have a value of C_i (C_i>0). None of the words are explicit impression words; however, those are located close to the explicit impressions words in the network. This suggests that less abstract words also affect the complexity of the networks; the "like" networks are more unique than the "dislike" networks. Furthermore, the clustering coefficient for the less abstract words implies that the cognitive process generating impressions is more focused when people have good impressions of an object.

We constructed the virtual impression networks by tracing the shortest path between each pair of explicit impression words. This was to demonstrate how words interact in human language, where two arbitrary words are often found to be relatively close to each other [17]. Figure 5 explains the relation between this step and the clustering coefficient. In Figure 5(a), the shortest path from Node 1 to Node 2, Node 3 and Node 4 is one, two and three, respectively. In this case, the clustering coefficient C is zero. In Figure 5(b), the length of each the shortest paths between any two nodes is one. In this case, the nodes are fully connected, that is, the clustering coefficient is one (refer to the example in the calculation of Figure 3). Thus, the clustering coefficient would be affected by the shortest paths in the network. In our results, we found that the clustering coefficient of the "like" networks, are on average, higher compared to the "dislike" networks. This indicates that the words in the "like" networks are more interconnected or closer to each other.

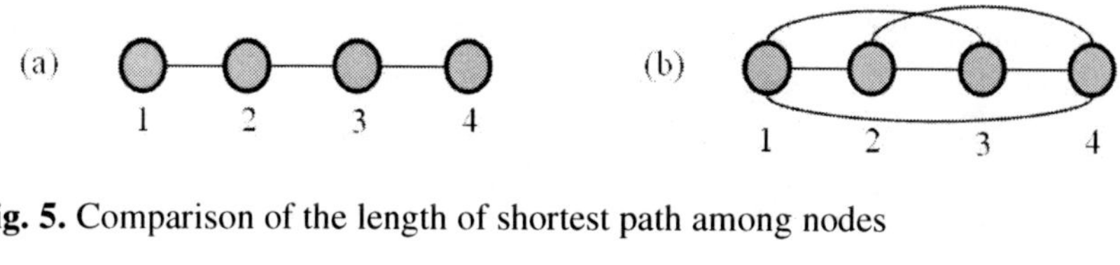

Fig. 5. Comparison of the length of shortest path among nodes

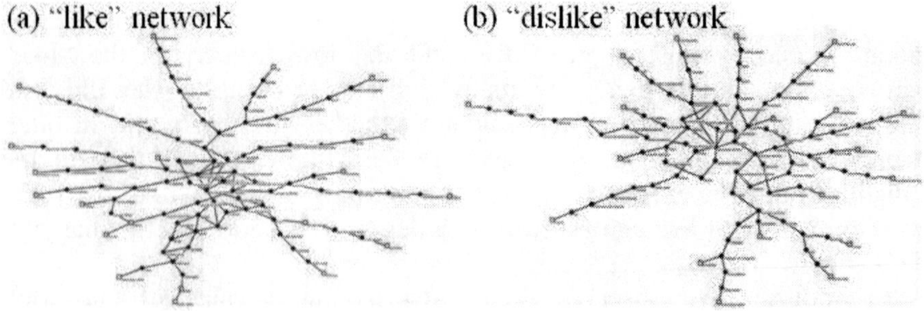

Fig. 6. Examples of virtual impression networks for Cup 4; (a) is a "like" network and (b) is a "dislike" network. □ is the explicit impression node, and ● is the node appearing in the path between the explicit impression nodes.

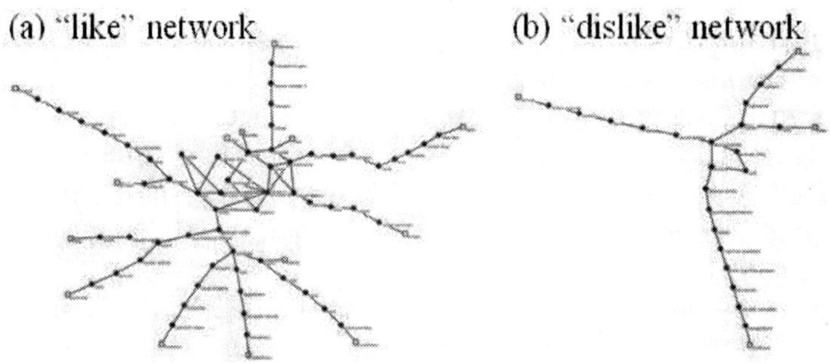

Fig. 7. Examples of the virtual impression networks for Cup 6; (a) is a "like" network and (b) is a "dislike" network. □ is the explicit impression node, and ● is the node appearing in the path between the explicit impression nodes.

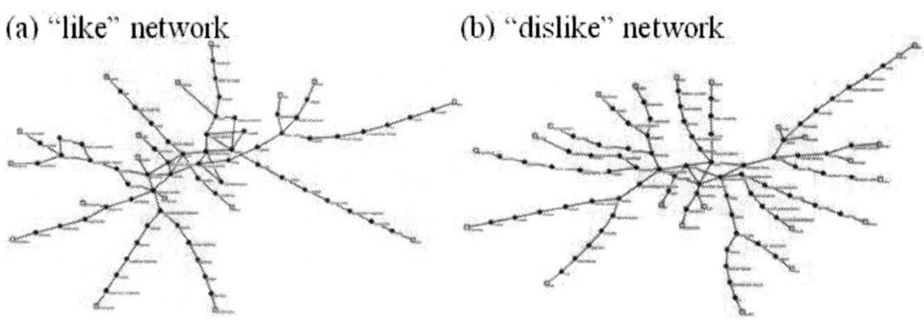

Fig. 8. Examples of virtual impression networks for Animal 5; (a) is a "like" network and (b) is a "dislike" network. □ is the explicit impression node, and ● is the node appearing in the path between the explicit impression nodes.

Table 6 Example of words having clustering coefficient C_i value (C_i>0) in the networks in Figures 6, 7, and 8

	(a) "like" network		**(b) "dislike" network**	
Figure 6	causal agency	soul	physical entity	matter
(Cup 4)	animate thing	being	chemical compound	stuff
	physical entity	event	chemical substance	
	inebriant	drug		
	street drug	liquid		
	biological process	fluid		
	substance	matter		
	abstract entity	agent		
	psychological feature	entity		
	physical process	potable		
	nutrient			
Figure 7	physical entity	matter	(the value of clustering	
(Cup 6)	substance	entity	coefficient for all words is 0)	
	physical object	unit		
	causal agency	article		
	constituent	relation		
Figure 8	physical object	location	physical object	unit
(Animal 5)	physical entity	being	physical entity	entity
	causal agency	treater	abstract entity	location
	animate thing	unit		
	communicator	entity		
	abstract entity			

Comparison with the Preliminary Experiment

In order to capture the nature of deep impressions, we developed a method for constructing "virtual impression networks" and conducted an experiment to identify the applicability of the method. Here, we discuss the results of this experiment by comparing them with those from our preliminary experiment [12].

In the preliminary experiment, subjects were asked to describe their impressions by using words grouped in categories of nouns, adjectives, and verbs by looking at a picture of each object. The subjects were also required to indicate their boundaries of "like" and "dislike" for the objects. Ten adult Japanese graduate students participated in the previous experiment, and six different cups were used. Sixty virtual impression networks were constructed—34 for the "like" and 26 for the "dislike" virtual impression networks. The results indicated significant differences in two of the network criteria—the average degree $<k>$, $t(58) = 2.037$, $p = 0.046$, and the clustering coefficient C, $t(58) = 2.262$, $p = 0.027$.

In the experiment of the present study, we used six types of cups that differed from the preliminary experiment and eight types of animals. The subjects comprised thirteen Japanese adults. Compared to the results in the preliminary experiment, in addition to the average degree $<k>$ and clustering coefficient C, the other structural criteria in the "like" and "dislike" networks were also found to be significantly different. Not all analyses yielded significant differences; only two types of cups, one type of animal, and all the networks of animals showed a significant difference.

The results show a similar tendency for two of the criteria; the structure of the "like" virtual impression networks have a higher average degree $<k>$ and clustering coefficient C than the "dislike" networks. Hence, we assume that the methodology proposed when analyzing the structure of the virtual impression network is appropriate for understanding the nature of deep impressions.

Moreover, from some example networks in the preliminary experiment, we observed that explicit impression words in the "like" networks were noticeably more related to each other. In the experiment of this study, by also observing some example networks, we noticed that less abstract words are included in the words with a value of clustering coefficient C_i. In other words, the triangles or shortcuts appeared in the less abstract words, that is, the explicit impression words are less separated in meaning. This suggests that the observations from both experiments indicate the same idea—when people have a good impression of a product, they tend to think in a certain direction at a deeper level; the thought process initiating the impressions are more focused and unique. On the other hand, Norman mentioned that attractive objects make people feel good, which in turn expands their thought processes, thereby enabling them to be more creative and imaginative; when people are anxious their thought processes tend to become narrow, concentrating upon aspects directly relevant to a problem [1]. Our observation in the present study seems to imply the same meaning as this mention.

Conclusion

In order to capture the nature of deep impressions, we developed a method for constructing "virtual impression networks," which involves the notions of "structure" and "inexplicit impressions" using a semantic network and confirmed its applicability by using two categories of target objects: artifacts and natural objects. We were particularly interested in understanding the manner in which people form impressions, whether "like" or "dislike". Our results showed that it is possible to explain the difference between the feelings of "like" and "dislike" using several criteria from network theory that were applied to the virtual impression networks. This result suggests the effectiveness of the proposed method, suggesting that the impression process of "like" is different from that of "dislike" at the level of deep impressions. This method of analyzing deep impressions with regard to products is anticipated to be useful for analyzing the direction of brand leaders or for evaluating product prototypes at an early stage in the design process.

This, however, also involves much conjecture. Certainly, there are other factors that affect human impression formation, such as cultural differences. In order to confirm the validity of this method for future studies, it will be necessary to examine the structure of virtual impression networks in relation to other racial groups, especially native English speakers. We will therefore attempt to refine our experiment and find a concretely applicable methodology. We will then attempt to extract clues in order to facilitate the design of "good products."

In addition, in the experiment, we used pictures of the targeted objects, and found a significant difference in some network criteria for several objects. In the future, it will be necessary to employ actual objects for this purpose.

Acknowledgements

The authors would like to thank Miss Yuiko Sakayama for conducting the experiment and Mr. Masanori Yoshizuka for constructing the networks. Their assistance in the analysis is also greatly appreciated. The authors would also like to express their gratitude to Tennoji Zoo (http://www.jazga.or.jp/tennoji/) for allowing them to use the information from their website. This work was supported by KAKENHI (21700240).

References

1. Norman, D.A.: Why we hate (or love) everyday things. Basic Books, New York (2004)
2. Osgood, C.E., Suci, G.J., Tennenbaum, P.H.: The measurement of meaning. University of Illionis Press, Urbana (1957)
3. Pawlak, Z.: Rough sets. International Journal of Computer and Information Sciences 11(5), 341–356 (1982)
4. Zajonc, R.B.: Feeling and thinking: preferences need no inferences. American Psychologist 35, 151–175 (1980)
5. Baeyens, F., Eelen, P., Van den Bergh, O.: Contingency awareness in evaluative conditioning: a case for unaware affective-evaluative learning. Cognition and Emotion 4, 3–18 (1990)

6. Field, A.P.: I like it, but I'm not sure why: Can evaluative conditioning occur without conscious awareness? Consciousness and Cognition 9, 13–36 (2000)
7. Georgiev, G.V., Taura, T., Chakrabarty, A., Nagai, Y.: Method of design through structuring of meanings. In: Proceedings of the International Design Engineering Technical Conference and Computers and Information in Engineering Conference (IDETC/CIE) - ASME (DETC 2008-49500) (2008)
8. Reingold, E., Colleen, R.: Implicit cognition, in Encyclopedia of Cognitive Science, pp. 481–485. Nature publishing group (2003)
9. Searle, J.: Indirect speech acts. In: Cole, P., Morgan, J.L. (eds.) Syntax and semantics. Speech acts, vol. 3, pp. 59–82. Academic Press, London (1975)
10. Higginbotham, J.: On linguistics in philosophy, and philosophy in linguistics. Linguistics and Philosophy 25, 573–584 (2002)
11. Grice, P.: Logics and conversation. In: Cole, P., Morgan, J.L. (eds.) Syntax and semantics. Speech acts, vol. 3, pp. 45–58. Academic Press, London (1975)
12. Fasiha, M.Y.N., Sakayama, Y., Yamamoto, E., Taura, T., Nagai, Y.: Understanding the Nature of Deep Impressions by analyzing the Structure of Virtual Impression Networks. In: Proceedings of International Design Conference - DESIGN 2010 (2010)
13. Felbaum, C.: WordNet: an electronic lexical database. MIT Press, Cambridge (1998)
14. Yamamoto, E., Goka, M., Yusof, N.F.M., Taura, T., Nagai, Y.: Virtual modeling of concept generation process for understanding and enhancing the nature of design creativity. In: Proceedings of the 17th International Conference of Engineering Design, Stanford University, San Frasncisco (2009)
15. Steyvers, M., Tennenbaum, J.B.: The large-scale structure of semantic networks: Statistical analyses and a model of semantic growth. Cognitive Science 29, 41–78 (2005)
16. Pajek—Program for large network analysis, `http://vlado.fmf.uni-lj.si/pub/networks/pajek` (Last accessed May 2010)
17. Cancho, R., Solé, R.V.: The small world of human language. Proceedings of the Royal Society of London B 268, 2261–2265 (2001)

COLLABORATIVE/COLLECTIVE DESIGN

Scaling up: From individual design to collaborative design to collective design
Mary Lou Maher, Mercèdes Paulini and Paul Murty

Building better design teams: Enhancing group affinity to aid collaborative deisgn
Michael A Oren and Stephen B Gilbert

Measuring cognitive design activity changes during an industry team brainstorming session
Jeff WT Kan, John S Gero and Hsien-Hui Tang

Scaling Up: From Individual Design to Collaborative Design to Collective Design

Mary Lou Maher, Mercedes Paulini, and Paul Murty
The University of Sydney, Australia

This paper presents a conceptual space for collective design to facilitate development of design environments that encourage large-scale participation in the next generation of challenging design tasks. Developing successful collective design starts by understanding how individual and collaborative design are supported with computing technology and then goes beyond collaborative design to structure and organize the design tasks so that people are motivated to participate. The analysis in this paper develops and illustrates several categories of motivation to be considered when implementing an environment for collective design.

Introduction

We are facing design challenges on a much larger scale as we become an increasingly global and technological society. Our design solutions not only need to respond to the needs and desires that may be included in a specific design brief, but they also need to be environmentally sustainable, attractive to multiple cultural groups, adaptable as technology changes, and intuitive to potential users. In *Cradle to Cradle*, McDonough and Braungart [1] argue that design for environmental sustainability has lost its way by focusing on reducing the impact of our designs on the environment, and advocate designing for reuse of natural resources when a product is no longer required. Tim Brown from IDEO proposes that designers cannot meet all of these challenges alone[1]. Both of these accounts, and a growing number of others, propose that we need to rethink design and extend the capability and responsibility of design to all people.

[1] http://www.ted.com/talks/tim_brown_urges_designers_to_think_big.html

J.S. Gero (ed.): *Design Computing and Cognition'10*, pp. 581–599.

Many innovative World Wide Web developers of applications, including Amazon, Google, Second Life and Wikipedia, have successfully implemented novel uses of the internet for large scale communication and collaboration. These developments offer us the opportunity to reconsider designing as a vital role of collective intelligence. There are many examples of the collective construction of knowledge and collective problem solving on the WWW, including examples of collective creativity. Collective design can facilitate a more inclusive design process by designers and non-design specialists by motivating the broader community to participate in design thinking.

When we describe and study design cognition, we consider the characteristics of the designer and design processes with a focus on the individual's cognitive processes when responding to an ill-defined problem. We value creativity and ingenuity and study highly creative individuals in order to better understand and encourage these processes in others. When we study collaborative design, we consider the ways in which multiple perspectives from a group of designers with different backgrounds can be brought together to create a synergistic solution. We value the emergence of solutions that could not be seen by any individual. Recognizing that creativity takes place in a community, studies show how computer support can enable the influence of the community on individual creativity, referred to as collective creativity by Nakakoji, Yamamoto, and Ohira [2]. Recently, the use of the internet for encouraging collective intelligence on a large scale allows us to go beyond studying individuals or teams of designers. We now have the opportunity to encourage and study large scale participation from individuals that may or may not be qualified as designers with the potential for a very large number of contributions to produce results that go beyond the capability of a more carefully constructed team of designers.

This paper presents a conceptual space for large-scale collective design. A conceptual space defines the dimensions along which a class of artifacts is described. When we encounter a novel way of designing, such as collective design, our conceptual space is expanded. The precedent for this is the change in our conceptual space for computer-supported design when digital communication technologies were introduced to the design process, enabling computer-supported collaborative design. The conceptual space for computer-supported design started with an articulation of technologies that facilitate various digital representations of the design artifact that support synthesis, analysis, evaluation, prototyping, and other design processes. This conceptual space expanded to include technologies for communication and sharing models in order to support collaborative design. The conceptual space for collective design adds another dimension to include principles associated with motivation: requiring incentives and structures that motivate selected designers and others to participate in collective design.

A Conceptual Space for Collective Design

Collective design is a phenomenon that is possible because we can easily communicate and share ideas, digital models, files, etc. on the internet. Computational support for design started with support for a single designer, primarily by providing a digital model of the design description as the basis for visual feedback and analysis. The early systems, known as CAD, were developed to assist in developing design drawings, and then design models. Successful collective design should build on the developments in and studies of computing technology that have been the basis for successful individual design and collaborative design.

Use of CAD for design is the norm now and the technology has developed to incorporate an extensive range of modeling and virtualization capabilities. In parallel with these developments, groupware for computer-supported collaborative work has developed and access to the internet has become more common. Computational support for design has been extended to support communication via email and collaborative portals. CAD systems have also been extended to support versions and sharing among a distributed team of designers. Research in virtual design studios and computer supported collaborative design has led to new tools and studies of teams of designers using computational systems. Studies of computer-supported collaborative work for creativity, such as Farooq, Carro, and Ganoe [3], show how collaborative tools can better support the creativity of small groups by improving the awareness of ideas generated by members of the group.

The development of social networks based on easy to use interfaces and the emphasis on communication and contribution made possible by Web 2.0 technologies has enabled many passive internet users to become active participants: engaging in discussion forums, creating social networks, taking part in opinion polls and building online communities and portals of knowledge. These developments provide opportunities for designing to be shared among large numbers of people, extending beyond the designated design team; opportunities for collective intelligence and therefore, collective design.

The term collective intelligence is commonly used to characterize the phenomenon of large numbers of people contributing to a single project and exhibiting intelligent behavior. The phenomenon is not new but it is being defined and redefined as new *variations on the theme* are emerging on the Internet at an increasing rate. In general collective intelligence can be described along a continuum: from aggregating the knowledge or contributions of individuals, a kind of collected intelligence, through to collaboration among individuals with the goal of producing a single, possibly complex output as a kind of collective intelligence. Rather than thinking of collected intelligence and collective intelligence as two separate entities, we can view them as two ends of a continuum, as illustrated in Figure 1, where the degree of direct interaction between individuals and their contributions differs. Systems may lie anywhere along this continuum as they incorporate more or less collaboration.

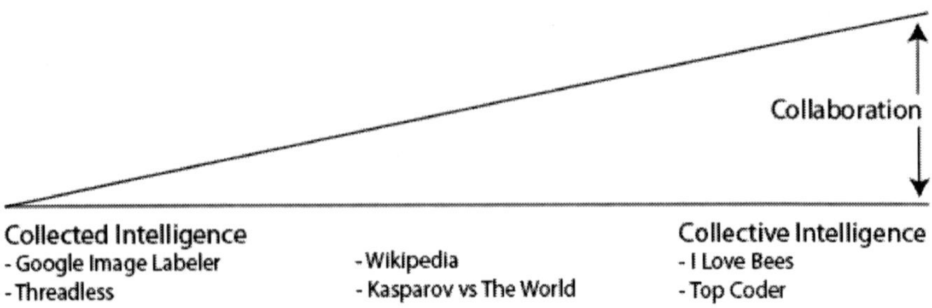

Fig. 1. The Group intelligence continuum: collected intelligence to collective intelligence

Collected intelligence, on the left side of the continuum in Figure 1, describes systems in which an individual contributes to a discrete task. The solution or outcome for each task is not synthesized with other solutions and therefore stands alone. The Image Labeler is an example of collected intelligence where each person contributes one or more labels to an image, but the labels need not be synthesized to a single coherent description of the image. The underlying principle behind collected intelligence lies with individuals providing the system with a single data item based on their own interpretation of the solution to a given problem. On the right side of Figure 1, collective intelligence involves both collaboration and synthesis: individuals collaborate in the production of the solutions and individual solutions are synthesized for a synergistic solution. I Love Bees is a good example of collective intelligence where many individuals worked together to find and share more clues and to find the answer to the mystery.

Crowdsourcing is a term used to describe situations where the key to success lies in large numbers of individuals providing input at many stages of the process: In Threadless the individuals contribute the designs and vote for the best. In TopCoder the individuals contribute the code and vote for the best. Crowdsourcing can be used to achieve collected or collective intelligence. The significant feature of crowdsourcing is that the large numbers of people contributing are self-selected rather than preselected based on qualifications. According to Howe [4], crowdsourcing works when the number of participants is very large so that the small percentage of good ideas is a large number when collecting ideas or solutions, and the popularity of a solution is identified when voting.

As people and computers begin to work synergistically within systems, it becomes important to recognize the interaction between them and the role of collaboration in forming a collective intelligence. The concept that through interaction the whole can produce something greater than the sum of its parts, is a key idea in understanding collective intelligence.

One can conceptualize systems that enable collected and collective intelligence in design by looking at the progression of computer support for individual design through collaborative design to collective design as illustrated in Figure 2.

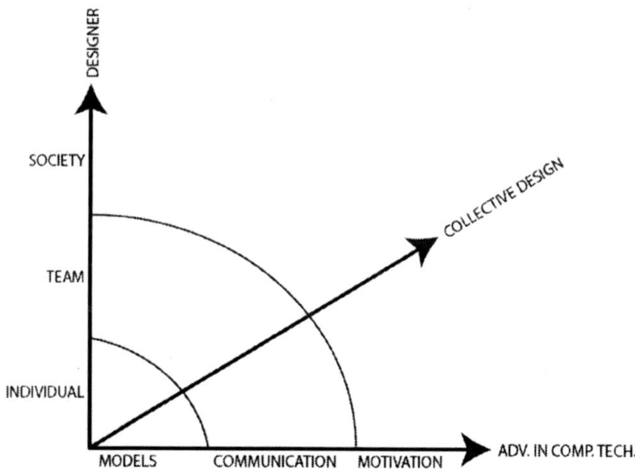

Fig. 2. Elements of collective intelligence in design

The vertical axis represents the designer dimension. For the individual, the primary computational support for design is the digital model. For a team of designers, computer support for communication is a necessary component of successful collaborative design. To engage the broader population or society in design, motivation becomes critical. Popular participation is fundamental to the development of the unique social chemistry that precipitates collective intelligence in design. The role that advances in computing technology play in enabling individual, group and collective design is represented by the horizontal axis of Figure 2.

The conceptual space for collective design is illustrated in Figure 3. The three axes for defining the space are: Representation, Communication, and Motivation. Representation refers to the digital models and files that support visualization, analysis, synthesis, etc. The representation can be text, sketches, 2D models, 3D models, etc. Communication refers to the ways in which people can communicate during the design process, for example via blogs and email, and can be characterized as synchronous or asynchronous and as direct or indirect. Motivation refers to the principles of motivation and the way the participation in the design process is structured. The three axes are elaborated in the next sections.

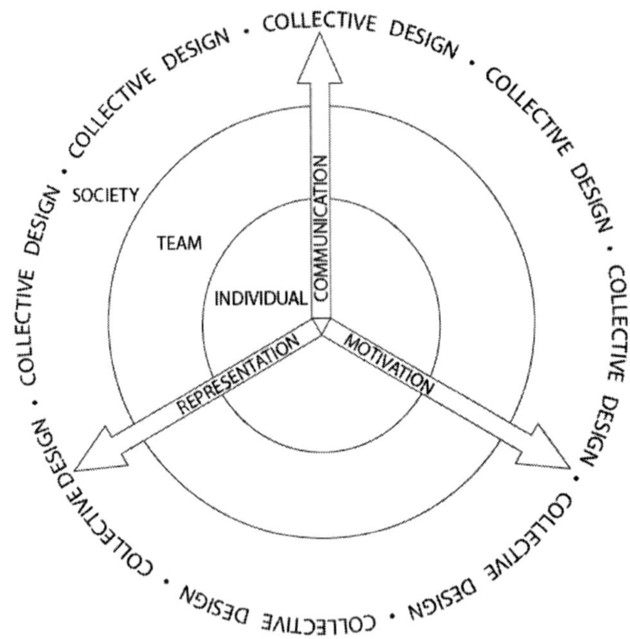

Fig. 3. Conceptual space for collective design

The Representation Dimension

We use two categories to describe the representation dimension:

- *type* such as images, sketches, databases, audio, 2D drawings, 3D models;
- *content* such as design problem, solution, and constraints.

A shared representation is required for collective intelligence, with or without collaboration. A (near) real-time external representation acting as shared memory is described in [5], considering the role of representation in reference to theories in philosophy and psychology. Halpin [5] asserts that the wide uptake of socially-generated content provides a community with the ability to influence each other for their greater collective success, and that Web 2.0 is a powerful facilitator for this. Since the individual has a limited and finite memory, they are able to record their thoughts onto the external environment of Web 2.0 and bring about social and collaborative creation and sharing of content. This is possible through intuitive interfaces, social networking tools and shared documents. Gruber [6] suggests in his paper on collective knowledge systems, that when the social web (Web 2.0) is combined with the semantic web, then collective intelligence is unlocked.

The first category for shared representation is the *type* of representation. Designers externalize their design ideas on the computer predominantly as structured and unstructured text, sketches, images, 2D/3D models, and more recently in databases. When one designer works on a problem alone, computer support for creating, editing, and sharing the representation relies on applications such as CAD, image processing software, and more recently 3D virtual worlds. This dimension of the conceptual space allows us to characterize the principles and alternatives for an external representation. Gul and Maher [7] studied how the type of external representation influences design cognition and the collaborative design process, showing that sketches facilitate more conceptual thinking than 3D models.

The second category of the shared representation is the *content*. In collaborative design, the participants share a description of the design problem, versions and components of the design solution, various constraints derived from domain knowledge, etc. Since designing involves creating new solutions to satisfy requirements and constraints, the shared representation is not static but is modified by the participants. Maher and Tang [8] describe the adaptation of the problem and solution spaces as a co-evolutionary process.

Levy [9] defines the role of a shared object in organizing efforts, such as the ball in a soccer game to coordinate movements. Heylighen [10] proposes a collective mental map, defined as an external, shared cognitive system. A collective mental map acts as an external memory accessed and contributed to by the collective. It represents problem states, actions and preferences for actions. In addition to components, relationships are relevant in a shared representation. The EWall project [11] is an example of a shared representation supporting brainstorming, decision-making, and problem solving. EWall supports individual and collective sense making activities by relating pieces of information in order to develop an understanding of a particular situation. The focus is on the explicit and implicit relations, shown in a spatial arrangement for collaborative use.

The Communication Dimension

We use four categories to characterize this dimension:

1. *mode*: synchronous and asynchronous;
2. *type*: direct in which a person communicates to one or more others or indirect in which a change is made to the shared representation;
3. *content* such as design process communication, participant attribution for contributions, design ideas and suggestions, design critique, and social communication;
4. *structure* of communications network, such as random or scale-free.

The *mode* of communication depends on whether the participants are present at the same time. Synchronous communication, requiring that participants be present at the same time, is supported by a chat window or by voice over IP. Asynchronous communication, where participants need not be aware of each other's presence and can contribute at different times, is supported by blogs, wikis, email, discussion forum, or documents.

The *type* of communication can be direct or indirect. Direct communication occurs when one participant sends or posts a message to one or more other participants with the intention of communicating about the problem. Indirect communication occurs when one participant makes a change to the shared representation that can then be seen by other participants. Heylighen [10] illustrates indirect communication with an example from nature: the construction of termite mounds, where the physical environment acts as the shared medium for collective knowledge. As each termite follows simple rules governing where to deposit mud (to place mud where the most mud is), the muddy towers provide a physical encoding of their collective efforts, a stigmertic signal available for all individuals to interpret. Wikipedia is an example of collective intelligence that occurs with both direct and indirect communication. Individuals can edit Wikipedia articles and thereby engage in indirect communication, and an individual can post a notice on the discussion forum and engage in direct communication.

The *content* of the communication is either a contribution to the shared representation of the problem, solution, or relevant domain knowledge, or is about the process. In collective design, communication about the process takes the form of design task or resource allocation, suggestions, critique, evaluation, and social communication. Design cognition studies using a protocol analysis have contributed to our understanding of the content of design communication by developing coding schemes that characterize this content. For example, Kim and Maher [12] develop a communication coding scheme, to compare collaborative design, using a keyboard and mouse, with collaborative design using tangible interaction technologies.

The *structure* of the communications network is an emergent property in collective design. Individuals are not equally connected to other individuals in a social network. Some people are highly connected with others, while some only possess a few connections, Figure 4.

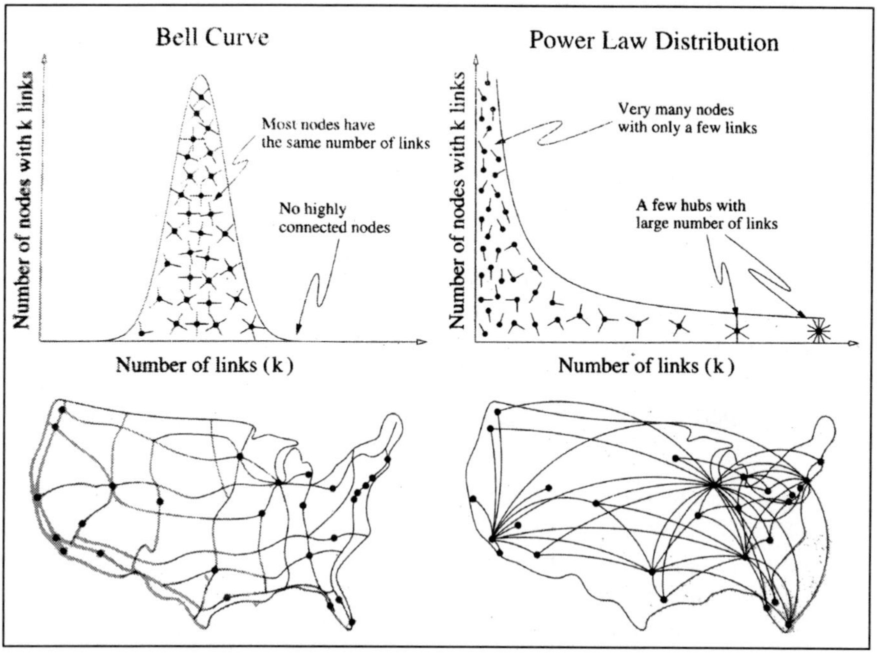

Fig. 4. Random and scale-free networks from Barabasi [13]

Although not all people will know each other in a large network, any two people can usually be connected by only a few links – links which usually pass through well-connected hubs. The structure of a social communications network is governed by the strength of the relationship between individuals. In a design team, the connections between individuals is usually evenly distributed as the team size is limited and the structure is formally defined, leading to a bell curve link distribution where most nodes have the same number of links, as shown in the left part of Figure 4.

Design communications occurring within an open community are not formally defined and links within them are more likely to form a power-law distribution, as shown in the right part of Figure 4.

The Motivation Dimension

For a conceptual space for collective design, we have developed the following categories of motivation. These have been drawn from categories of motivation identified in collective intelligence, open source software, and social psychology literature, briefly described after the list.

- *Ideology* – participation for the purpose of contributing to a larger cause.
- *Challenge* – participation that provides a sense of personal achievement through acquiring additional knowledge or skill.
- *Career*– participation that may lead to an advance in the individual's career.
- *Social* – desire to have a shared experience with one or more individuals.
- *Fun* – participation for the purpose of entertainment, enjoyment, excitement, relief from other experiences, or simply furnishing or structuring the passage of time.
- *Reward* – participation to receive tangible rewards includes money, points in a game, a gift or voucher.
- *Recognition* – participation in order to receive private or public acknowledgement.
- *Duty* – participation in response to a wish or command expressed personally.

A key dimension of the conceptual space that describes collective design is motivation: that is, the technologies and organizing principles that attract people to participate. Understanding the range of motivations is an essential dimension of collective design. It leads to guidelines for achieving participation from both designers, who may be involved because it is part of their job, and society at large, who may be volunteering their effort. Motivation theories have been developed from a range of perspectives: from Darwin's evolutionary theory contributing a biological basis for human motivation to intrinsic motivation as described in Maslow's hierarchy of needs which spans from the purely physiological to self-actualization. Merrick and Maher [14] provide an overview of motivation theories and their relevance to computational models of motivation as the basis for implementing a curious agent. Here we focus on studies of motivation as related to volunteer activities.

Malone et al [15] present an analysis of mechanisms that induce mass-individual participation in several computer-enabled collective intelligence systems. In this study the range and instances of four parameters, or "building blocks" of a collective intelligence task, are framed as question pairs. *Who* is performing the task? *Why* are they doing it? *What* is being accomplished? *How* is it being done? Malone et al. identify three personal motivations, associated with the question, *Why are they doing it?* as money, love, and glory. The categories, money, love and glory, are useful generalizations, and are embedded in our categories: money is what we more generally refer to as reward, love is what we more generally refer to as social, and glory is what we more generally refer to as recognition.

Nov [16] identified 8 categories of motivation in a survey of people that contribute to Wikipedia, starting with 6 categories of motivation associated with volunteering defined by Clary [17]: values, understanding, enhancement, protective, career, and social. Nov's additional categories for understanding motivation in Wikipedia are fun and ideology, which are also used in research on motivation to contribute to open software development. Nov's survey found that the top motivations were fun and ideology. Our categories of motivation are more similar to Nov's categories, but we also incorporate categories that describe the motivation of a selected design team and non-designers whose participation may be entirely informal and voluntary.

Mapping Collective Intelligence to the Conceptual Space for Collective Design

In order to better understand collective design, we review 6 successful examples of Internet applications that engender varying combinations of collected intelligence and collective intelligence and identify how they map onto our conceptual space for collective design. This process allows us to explore the conceptual space and develop principles for collective design.

Threadless[2] *– Crowdsourcing*

Threadless is a web site that encourages individuals to submit T-shirt designs. Every week the Threadless community votes for the best designs to go into production. The winning entrants receive a one-off monetary reward as well as a percentage of sales. Threadless has a thriving community actively engaged on a forum.

- *Representation.* A textual description of the competition serves as representation. All current and archived T-shirt designs provide precedents.
- *Communication.* If the artist works independently, no communication occurs. Should the artist engage in the online community forum, collaboration on a design may occur, resulting in direct, asynchronous communication.

[2] www.threadless.com

- *Motivation.* Primary motivators may include: the challenge of having their design selected for production; recognition arising from their username being associated with that design and promoted; the financial reward. Secondary motivations may include: the social aspect of communicating with like-minded people; fun – participating in something that may be a hobby or performed beside their primary source of income; career – the leverage their status provides to potential employers within the field (or a related field).

Google Image Labeler[3] - Collected Intelligence

Google's Image Labeler presents a game-like scenario, to add tags to images, inviting users to work at categorizing online pictures in order to improve Google's search engine in exchange for points and gifts. Keywords from multiple sessions, of an image are compared and frequently occurring terms are allocated to the image permanently.

- *Representation.* The type of shared representation is the image and the content is the problem description. The shared representation for each problem is a simple structure and the contribution to the collected intelligence comprises one or more labels.
- *Communication.* The type of communication is indirect and therefore the mode is asynchronous. The content of the communication is a contribution to the solution, that is, the image label. The structure of the network is evenly distributed since the individuals do not communicate with each other directly.
- *Motivation.* The Google Image Labeler is structured as a game, where individuals are motivated to play in order to win and achieve recognition. Individuals are recognized when they become "today's top pairs" and when they become one of the top 5 "all-time contributors". The relevant categories of motivation are fun, recognition, and reward.

Wikipedia[4] – Hybrid Collected/Collective Intelligence

Wikipedia is often cited as an example of a kind of collective intelligence where many individuals work together to create a vast and socially constructed knowledge base. Any one individual contributes to only a few specific articles of interest, adding their knowledge to them. *Collective* intelligence is the phenomenon that lies behind the creation of each article, but Wikipedia itself, or rather the collection of articles that make up

---

[3] For more information see: images.google.com/imagelabeler

[4] For more information see: en.wikipedia.org/wiki/Wikipedia:About

Wikipedia, is *collected* intelligence. This is why we place this example at the middle of the continuum, in Figure 1.

- *Representation.* The type of shared representation comprises images, text, and links. The content of the shared representation is the "solution" or the shared knowledge on specific topics.
- *Communication.* Communication in Wikipedia is either direct, where participants can contribute to a discussion forum, or indirect, where participants can edit an article. The mode of communication is asynchronous so participants are not aware of the presence of others that may also be editing an article. The content of the communication is either the "solution" or knowledge in an article or comments about the changes to the article. The emergent structure of the social network of Wikipedia has been studied by many with varying conclusions.
- *Motivation.* The motivations of individuals that contribute to Wikipedia have also been studied by many. Based on Nov [16], the motivations that map onto our categories include: ideology, challenge, career, social, fun, recognition, and duty. The only motivation in our list that is not included is reward since there is no tangible reward for contributing to Wikipedia.

Kasparov vs The World[5] – Hybrid Collected/Collective Intelligence

A 1999 game played over the Internet by Gary Kasparov, the (now former) reigning world chess champion, against Team World, which comprised five consulting chess champions, chess clubs distributed internationally, any person with an internet connection wishing to participate, and strong chess analysis software. The combination of discussion over a forum and the played move being voted for by plurality, suggests both the collaborative aspect of *collective* intelligence and the wisdom-of-crowds-aggregation of *collected* intelligence. Through their combined effort, a novel move was played by Team World; one never made before in a recorded game.

- *Representation.* The type of the shared representation comprises image and text and the content is the description of the problem and solution, that is, the chessboard, the rules of chess, and a decision tree.
- *Communication.* Both types of communication are supported: direct and indirect. The mode of communication is asynchronous. An MSN bulletin board was the platform on which online communication primarily took place. The content of communication included ideas and suggestions, critique and voting. The network structure was scale-free,

[5] For more information see [18].

with certain individuals communicating more frequently or influencing others more heavily.

- *Motivation.* Several categories of motivation apply to this example: fun, reward, recognition, challenge, career, social, and duty. Individuals were playing for points, to see if they could win against the world's master chess champion. Participants collaborated and were intellectually challenged. Participants became part of a global community and chess clubs also became involved.

I Love Bees[6] – Collective Intelligence

I Love Bees is a detective game that was played by over 600,000 participants, most of whom were avid fans of an earlier game, Halo, and were eager to learn more about the sequel to their game. Abstract clues were provided across a variety of media, including a "corrupted" web site. Users were not given any explicit instruction, although the game's designers intended the output to be a narrative providing the back-story to the Halo 2 game. Levels of collaboration were extremely high, with information amassed and elaborated on by many players as the narrative structure evolved.

- *Representation.* The type of shared external representation is text and the content is both the clues provided by the administrators as well as content created by the users. Ideas and theories about the mystery also became a shared representation. The problem was not given, but discovered by the players, as was the solution. The organizers provided constraints, in the form of raw data.
- *Communication.* Direct communication occurred over multiple channels, such as email, message board, phone etc. The I Love Bees project is a good example of how an ongoing conversation developed between players to enable a collective intelligence. The conversation emerged spontaneously from the use of existing tools they had readily available. Linking multiple platforms allows user interactions with data to be more fluid as information can be interrelated, manipulated and analyzed across a variety of tools and representations. Encouraging user deployment across a spectrum of situations/contexts enhances the accessibility and ubiquity of a system, and acts to maximize user involvement. Communication was direct and mostly asynchronous, although on occasion (such as during live chat) it was synchronous. Indirect communication took place by tweaking the original representations (raw data).

[6] For more information see [19].

- *Motivation.* The participants were motivated by the challenge of the mystery, as well as for the social, fun, and recognition aspects of the activity.

TopCoder*[7] *– Collective Intelligence – Crowdsourcing

Top coder is inspired by the open source software movement adding a modern crowdsourcing approach. Coding projects are broken up into discrete elements, which are made available to anyone who wants to contribute. The submissions are checked for correctness and the community can then vote on the best code for each element. The best pieces of code are selected and the community is again challenged to synthesis the elements to a larger working program.

- *Representation.* A textual description of the code is provided. Inputs and outputs are identified.
- *Communication.* If the individual codes the task alone there is no communication. If collaboration is involved it could be indirect asynchronous (the code itself is modified by many) or direct asynchronous (ie. contributors discuss the coding over email) or direct synchronous (ie. they discuss the coding over chat).
- *Motivation.* Top Coder appears to invoke every motivation except duty. This is partly a legacy of the open source software movement's ideology of free software designed by the public, for the public, that was 'better' than the alternatives offered by industry. Individuals may contribute due to: the challenge of writing the code; the social aspect addressed by the community; or to enhance their career. Even though there is a financial reward if the code is selected, members don't participate simply because they *expect* to benefit financially.

Principles for Collective Design

The analysis of the examples described in section 3 are summarized in Table 1.

The table provides a basis for developing principles for collective design by considering the extremes in the continuum: collected design intelligence to collective design intelligence. Collected design intelligence, that follows from successful examples of collected intelligence, should be structured around encouraging and enabling large scale participation in design tasks, such as identifying novel and useful features of design

[7] http://www.topcoder.com/

alternatives, and identifying labels for new design ideas. Collective design intelligence should be structured around encouraging and enabling large-scale participation in design tasks, such as brainstorming, concept analysis, ideation and competitive design solutions. Key questions and answers are:

What is the function of the shared representation? A shared representation has multiple vital roles. The type and content provide the basis for defining what participants can do and for motivating people to participate. A collected intelligence task like the Image Labeler has a very simple shared representation and a collective intelligence task has multiple types of representation and requires skill to navigate and manipulate the shared representation. A principle for collected intelligence in design is to keep the shared representation simple and modular. A principle for collective intelligence in design is to develop an adaptive and dynamic shared representation that allows the individuals to express themselves through the shared representation.

How do people communicate? The successful examples of collective intelligence do not provide a recipe for an ideal type and mode of communication. The I Love Bees example shows that people will create their own way of communicating if they are highly motivated to participate. However, in collective design, facilitating communication across a range of types and modes will make it easier for participants to join and interact. There are many studies and lessons learned from computer supported collaborative work and social networks for principles of effective communication.

Table 1 Analysis of Successful Examples of Collective Intelligence

	Google Image Labeller	Threadless	Wikipedia	Kasparov vs The World	I Love Bees	Top Coder
Representation						
type I = image(s) T= text 3D = 3D model DB = database	I	T	I+T	I+T	T	T
content P= problem description S= solution, knowledge, world model	P	P	S	P+S	P+S	P+S

Communication						
mode S= synchronous A= asynchronous	S	A	A	A	A+S	A+S
type D= direct I = indirect	I	D	D+I	D+I	D	I
Content C= comments E= entries V= votes B=broad range	0	C+V	C+E	C+V	B	B
structure 0= no structure S= scale free M= multiple E= emergent	0	S	S	E	M+E	S+E
Motivation						
Ideology	X		X		X	X
Challenge		X	X	X	X	X
Career		X	X	X	X	X
Social		X	X	X	X	X
Fun	X	X	X	X	X	X
Reward		X		X	X	X
Recognition		X	X	X	X	X
Duty			X	X	X	

In addition to computer supported communication, there are examples of machine learning systems that can aggregate and find patterns in communication data and human activity, such as recommender systems, that will enhance indirect communication in collective design.

Why do people participate? While some of the participants in collective design will be motivated by duty and career, the non-specialist designers need to be motivated by the categories typically associated with volunteer activities. These categories include fun, challenge, ideology, social, reward, and recognition. In developing collective design, a mapping from these categories of motivation to organizing principles in the way the design tasks are presented and structured is essential. The most popular motivation for Wikipedians is fun, suggesting that a game-like environment is a good starting point. A game-like environment addresses the motivation categories; fun, social, reward, and challenge. A design task that promises to make the world a better place addresses the motivation categories; ideology and recognition.

In summary, this paper presents a conceptual space for collective design that leads to design environments that encourage large scale participation in the next generation of challenging design tasks. Developing successful collective design starts by understanding how individual and collaborative design is supported with computing technology and then going beyond collaborative design to structure and organize the design tasks so that people are motivated to participate. The analysis in this paper develops and illustrates several categories of motivation to be considered when implementing an environment for collective design.

Conclusions

Collective design is a phenomenon that can occur when large numbers of motivated professional and amateur designers contribute to a collective intelligence that emerges from their mutual communication, collaboration, and competition. This paper proposes that design outcomes from collective processes can be greater than can be achieved by a preselected team of designers, participating in a collaborative process. The successful examples in design domains "crowdsource" individual designs from very large crowds where the individual benefits from participation in a community. Beyond this, collective design can draw on contributions from large numbers of human and computer agents to complex design problems. Collective design is possible because the internet facilitates participation from individuals who are not preselected, but are motivated to participate for personal reasons that go beyond financial reward. This paper articulates several kinds of motivation in successful collective intelligence as part of a framework for understanding collective design and serves as a basis for designing systems that enable large-scale collective design.

References

1. McDonough, W., Braungart, M.: Cradle to Cradle. North Point Press (2002)
2. Nakakoji, K., Yamamoto, Y., Ohira, M.: A Framework that Supports Collective Creativity in Design using Visual Images. In: Creativity and Cognition 1999, pp. 166–173. ACM Press, New York (1999)
3. Farooq, U., Carroll, J.M., Ganoe, C.H.: Supporting creativity in distributed scientific communities. In: ACM SIGGROUP conference on Supporting Group Work, pp. 217–226 (2005)
4. Howe, J.: Crowdsourcing: Why the Power of the Crowd is Driving the Future of Business, vol. 336. Three Rivers Press (2009)
5. Halpin, H.: Foundations of a philosophy of collective intelligence. In: Guerin and Vasconcelos (eds) AISB 2008 Convention: Communication, Interaction and Social Intelligence, pp. 12–19. University of Aberdeen, UK (2008)

6. Gruber, T.: Collective knowledge systems: Where the social web meets the semantic web. Web Semantics: Science, Services and Agents on the World Wide Web 6, 4–13 (2008)
7. Gul, L.F., Maher, M.L.: Co-Creating external design representations: comparing face-to-face sketching to designing in virtual environments. CoDesign International Journal of CoCreation in Design and the Arts 5(2), 117–138 (2009)
8. Maher, M., Tang, H.: Co-evolution as a computational and cognitive model of design. Research in Engineering Design 14, 47–63 (2003)
9. Levy, P.: Collective Intelligence: Mankind's Emerging World in Cyberspace, Plenum, New York (1997) (Translated from the French by R. Bononno)
10. Heylighen, F.: Collective intelligence and its implementation on the web: Algorithms to develop a collective mental map. Computational & Mathematical Organization Theory 5(3), 253–280 (1999)
11. Keel, P.E.: EWall: A visual analytics environment for collaborative sense-making. Information Visualization 6(1), 48–63 (2007)
12. Kim, M.J., Maher, M.L.: The Impact of Tangible User Interfaces on Spatial Cognition During Collaborative Design. Design Studies 29(3), 222–253 (2008)
13. Barabási, A.-L.: Linked: How Everything is Connected to Everything else and what it means for Business. In: Science and Everyday Life, Plume, NY, vol. 71 (2003)
14. Merrick, K.E., Maher, M.L.: Motivated Reinforcement Learning: Curious Characters for Multiuser Games, vol. 206. Springer, Heidelberg (2009)
15. Malone, T.W., Laubacher, R., Dellarocas, C.: Harnessing crowds: Mapping the genome of collective intelligence, MIT Sloan School Working Paper 4732-09 (2009), `http://ssrn.com/abstract=1381502` (Last accessed January 2010)
16. Nov, O.: What Motivates Wikipedians? Communications of the ACM 50(11), 60–64 (2007)
17. Clary, E., Snyder, M., Ridge, R., Copeland, J., Stukas, A., Haugen, J., Miene, P.: Understanding and assessing the motivations of volunteers: A functional approach. J. Personality and Social Psychology 74, 1516–1530 (1998)
18. MSN World Chess Champion Garry Kasparov Defeats World Team in Kasperov vs The World on MSN.com (1999), `http://www.microsoft.com/presspass/press/1999/oct99/kasparovwinspr.mspx` (Last accessed March 2010)
19. McGonigal, J.: Why I love bees: A case style in collective intelligence gaming. In: Salen, K. (ed.) The Ecology of Games: Connecting Youth, Games, and Learning, Cambridge, MA, pp. 199–228 (2008)

Building Better Design Teams: Enhancing Group Affinity to Aid Collaborative Design

Michael A. Oren and Stephen B. Gilbert
Iowa State University, USA

This paper discusses ConvoCons, a novel system of conversational icons intended to encourage affinity between collaborators unobtrusively. Using a reification of Bonnie Nardi's framework for social connection and affinity, ConvoCons overlay an existing application and display a varying media that can encourage collaborating partners to begin developing affinity through informal conversations. This research explores whether dyads working on a collaborative multitouch application with ConvoCons develop more affinity than dyads that do not while solving simple design problems and a freeform design task. Results indicate that after an average of 23.25 minutes, affinity, defined as a function of conversational and behavioral cues, was 40% higher ($p < 0.001$) in the ConvoCons group than in the control group. This research offers a framework for evaluating affinity within groups and a foundation for exploring software-based methods of improving the effectiveness of collaboration within design teams.

Introduction

Imagine a new employee joins a team of designers to create an interface for an application. The design team typically works in pairs, and her partner for this project is somebody she has never met. The designers share a multitouch device to place the GUI components and create the interface of their product. There is some awkwardness and formality as they work. Having never worked together, they are strangers trying to create a shared vision of a design. Suddenly, a message pops up on the screen; the new employee reads it—some sort of a question. She looks across the screen and her partner has one as well, so she reads hers aloud. The other employee reads hers, and they realize it is a riddle. They begin talking

J.S. Gero (ed.): *Design Computing and Cognition'10*, pp. 601–620.

about the riddle and about their past work experience. They continue to work, and when a work-related question arises about a design idea, the partners now have no problem asking each other questions; the awkwardness has been removed.

The story above illustrates the social awkwardness that can occur when working with a new partner for the first time. Typically one has little idea of what to expect from a partner and unless there has been an introduction before, a person may find him- or herself hesitant to start a conversation as no affinity has been developed. When individuals work together for the first time they lack knowledge of one another's reputations and other relational elements typically useful for successful cooperation [3]. Strangers cooperating for the first time without a shared connection to facilitate introductions and establish common ground may at first struggle to establish a level of affinity needed for productive cooperation [8][23]. Individuals seek affinity as a means to fill a need for interpersonal relationships and established affinity is necessary for sustained cooperative relationships [15][34].

This research describes the evaluation of a user interface technique developed to more quickly build affinity and effective collaboration strategies among strangers by promoting incidental conversations. This system of conversation-starting icons, called ConvoCons, offers conversation starters to encourage an informal discourse between new partners that Nardi identified as a central component of group affinity [23]. An example screenshot of an application with overlaid ConvoCons can be seen in Figure 1; in this case, the ConvoCons are text within circles that are oriented towards two partners facing each other and collaborating on a tangram puzzle.

ConvoCons are part of an ongoing research effort to explore means of using interfaces to promote constructive collaborative strategies among groups of individuals using computers to facilitate their work, with a particular emphasis on collaboration involving creativity and design [24].

In this paper, we provide an analysis of ConvoCons used in a multitouch tangram application and their effectiveness in encouraging affinity between dyads, a pair of individuals treated as one unit. Tangrams is a Chinese puzzle game consisting of seven geometric shapes (five triangles, a square, and a parallelogram) that is used to create a variety of shapes both freeform (creative) as well as filling in a pattern (problem solving). It is important to note that ConvoCons appear concurrently with the tangram puzzles, but are semi-transparent and serve as passive interface elements (they do not recognize user input), thus users are free to attend to or ignore the ConvoCons without adversely effecting their ability to get work done.

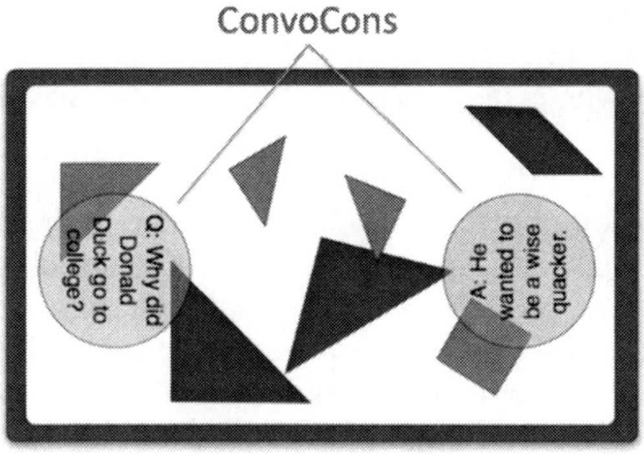

Fig. 1. A sample joke ConvoCon, the participant on the left would have a privileged view of the question while the one on the right would have a privileged view of the answer.

The first research question (Q1) is "Does the presence of ConvoCons lead to increased incidental conversations?" In order to answer this question, we defined incidental conversation to be dialogue unrelated to the tangram task at hand and looked at the amount of such dialogue between dyad members that worked on tangrams with ConvoCons vs. without Convocons. Our secondary research question, (Q2) is, "Do ConvoCons lead to increased affinity between participants?" For this research question, we operationalized a definition of affinity based on two components: conversational affinity and behavioral affinity. Total affinity is based on a percentage of the interactions that demonstrated affinity vs. those that did not.

Background and Context

The system and research model for ConvoCons are an applied reification of Nardi's observations suggesting that affinity plays a central role in the process of creating and sustaining connections necessary for productive collaboration. While the majority of the dimensions of affinity that Nardi observed were physical in nature, we chose to focus on the aspect of incidental communication, i.e., conversations outside of productive work such as commenting on the weather. Nardi suggests that this informal discourse leads to connections that aid in collaboration critical to productive collaborative strategies [23]. We hypothesize that these affinity bonds, promoted through discussing ConvoCons, lead to the critically important state of social cohesion [19]. By conducting this experiment with two people sharing a small (15.4") multitouch device, two of the other activities that promote affinity are added to the work context: human touch (the occasional brush of the hand) and a shared experience in a common space (where the common space is both physical and virtual) [23].

The importance of affinity for effective collaboration can also be seen in Schmid's discussion of affinity as the cornerstone to the development and use of social capital [30]. Specifically, ConvoCons seek to improve what Schmid refers to as positive affinity, which connects individuals to one another through a build up of social capital and, as a result, reduces the free rider problem. While this study does not examine the effect of affinity overtime, assuming the validity of Schmid's theory, the ConvoCon system is designed to accelerate this accumulation of affinity and decrease the time needed to build social capital. Kellogg and Erickson [18] suggest that social translucence, the idea that user activity needs to be apparent to other users, is a key to effective collaboration. ConvoCons are designed to increase social translucence in that by building affinity between partners, group members will be able to understand their partners' cues better in order to collaborate through turn taking and directing work with their partners. In addition, Convertino *et al.* [8] suggest that in order for group members to successfully collaborate, they must develop converging measures, which is the idea that they have a common ground or shared representation of the task. Both of these concepts are components of the third dimension of affinity identified by Nardi, that of a shared experience within a shared space [23].

However, the creation of common ground has been shown to cause problems within groups when individuals focus on the elements they share and never move beyond that to share expert knowledge needed to solve a problem. Larson described this tendency in his study of doctors working on collaborative diagnosis where each doctor had been shown a different piece of the medical problem and a successful diagnosis could only occur when information was shared [21]. Analogously, earlier ConvoCons prototypes explored the use of a centralized, shared conversation starter, but designs that provided each participant with separate, privileged information, Figure 1, were more likely to prompt participant conversation. Participants asked about one another's pieces of the question and answer [26].

The ConvoCons Approach

ConvoCons were developed based on an initial collaboration study that suggested that the ambiguity in a somewhat confusing user interface served as a means of creating converging measures through users' discussion of the interface. However, given the cost of encouraging poor interface design to promote collaboration, the ConvoCon system was developed to serve the same role without the cost to general usability of the system [24]. Related research by Clear and Daniels has explored the use of icebreakers to encourage better collaboration techniques between distant learners [6]. In addition, Fisher and Tucker have used online games as a means of providing an out-of-classroom means for online students to gain affinity with one another [10]. However, unlike these previous approaches, ConvoCons are built into the tasks and collaborators are free to attend to them or not; there is no structured ice-breaking mechanism or time required outside of the task.

Also, while ConvoCons are designed to encourage incidental conversations and affinity, the goal is not simply to connect people, but rather to encourage stronger collaborative working behavior. In our research we have often observed members of dyads who, although working on the same problem within the same virtual and physical space (on the same device), failed to acknowledge or utilize their partners, instead tackling the problems separately, avoiding interaction with their partners when possible. Previous research by Rogers looked into the use of shared displays to serve as an icebreaker to promote and track conversations within a social setting; in this research we use the shared display as a work tool (rather than an icebreaker) although Rogers' and our own work share the goal of using technology to bring people closer together [28]. Finally, while the underlying goal of Karahalios' social catalysts is similar [17], in calling for designers to consider interfaces as a means of promoting social connections, her work focuses primarily on aiding individuals in finding collaborators. In contrast, our work assumes that group members are already paired by a work assignment. ConvoCons are intended to ease new partners' transition into working with higher affinity on a shared goal.

ConvoCons System Architecture

The ConvoCons system is designed to be overlaid on any Java application and can be used with several simultaneous client applications, e.g. two people using ConvoCon-enabled applications at different sites. By allowing this level of adaptability of ConvoCons, we are able to test a variety of configurations to test their effectiveness in encouraging collaboration within a wide range of applications and environments. The touch interface is not required for ConvoCons; it is helpful in this context for affording an easy approach to simultaneous use of an application by two co-located users. The touch-based gesture recognition system is built using Sparsh-UI, an open-source API for platform independent touch-based applications [26,32]. While this study used visual ConvoCons containing text, the ConvoCon architecture supports media including auditory signals, videos, and images.

Methods

Thirty-six participants were recruited from the Iowa State University psychology department participant pool and paired into 18 dyads (the dyad will be the unit of analysis). Each dyad was then randomly assigned to be in either the experimental group (ConvoCon enabled tangrams; n=9) or control group (plain tangrams, n=9). Participants in all dyads had no previous relationship beyond "seeing each other around" (n=1 dyad) although the majority (n=17) had never met before. Participants were instructed to arrive and wait at different entrances to the research lab to prevent interaction before the start of the study.

Dyads were instructed to sit across the table from one another to allow co-located collaboration with the multitouch device, a Stantum SMK 15.4, placed length-wise between them (see Fig. 2) [29,4,31]. All dyads then followed the procedure below (see Fig. 3). Dyads were first given a brief description of the technology and told they would have five minutes to play with the interface and teach themselves how to use it. After dyads completed the five minutes of playtime, dyads were then given the first pattern to create with the tangram pieces. Dyads worked on the pattern until completion, before being given the next pattern for a total of three patterns. Upon completion of the patterns, the dyads were then given up to five minutes to create any new pattern of their choice. All interactions were video recorded, and the software logged user inputs.

The dyads assigned to the ConvoCon group were exposed to ConvoCon riddles and jokes upon the first touch of the multitouch interface. One participant was given a privileged view of the riddle while the other was given a privileged view of the answer (see Fig. 2). Each ConvoCon remained visible for thirty seconds, followed one minute later by another ConvoCon. ConvoCons did not affect the interaction with the tangrams application; they did not block access to the application, nor could users control them. There were a total of ten ConvoCons displayed to dyads over a fifteen minute time period (see Fig. 3). The ConvoCon group had a mean completion time of 14.75 minutes (SD=5.75); this resulted in most participants completing part of Task 2 as well as all of Task 3 and the freeform task without ConvoCons. Turning ConvoCons off midway was intended to allow the researchers to observe whether or not the effects of ConvoCons would be sustained throughout the working session.

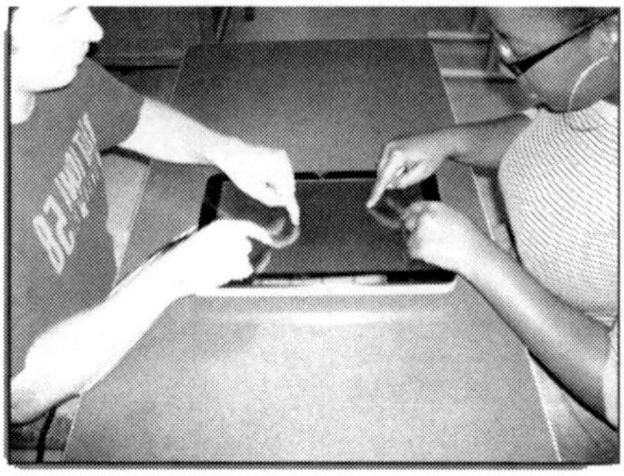

Fig. 2. The multitouch device between two participants.

Finally, we administered an exit survey to all participants based on a modified version of the survey Convertino developed to assess the similar concept of common ground development [7]. Using this survey, consisting primarily of five-point Likert scale questions, we compared the control and experimental groups to determine how well they felt their group worked together and the agreement that was reached within the group. Dyads in the ConvoCon group also participated in a brief, unstructured interview in order to obtain their input on ConvoCons to obtain qualitative feedback and aid future design of the system.

Our choice of using visual versions of ConvoCons was taken due to research indicating that visual background noise has less of an adverse effect on performance compared to auditory background noise [9]. Since ConvoCons are not directly related to the task at hand in this experiment they may, at times of particularly difficult work, be viewed as background noise. The choice of using text-based ConvoCons for the initial study was due to the two-part reasoning that they were the simplest to implement and that it was easier to detect whether participants were attending to the ConvoCons since reading text requires some cognitive load and, with a partner present, is often spoken.

Through an iterative process we discussed in an earlier paper, we settled on riddles and jokes over news headlines and trivia or facts about tangram puzzles. Participants lacked the contextual information to discuss news headlines and trivia, while facts about tangram puzzles afforded little discussion [24].

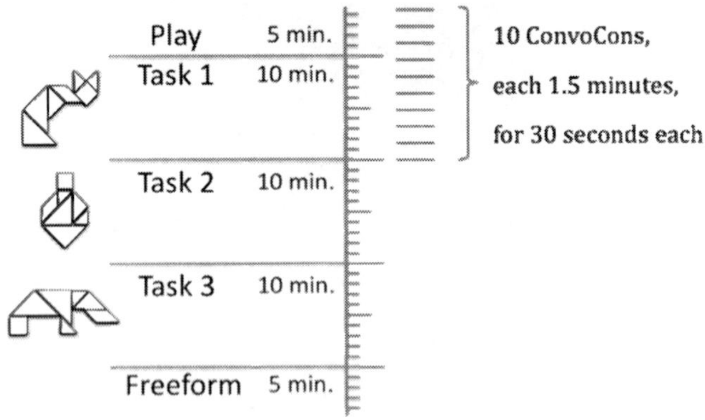

Fig. 3. Sample timeline of the procedure (task times varied based on completion time), the 10 minutes per puzzle is just an example, some took less time and some took more.

The jokes (obtained from children's collections) were chosen as a way of lightening the mood, which Goffman's study of role distances has shown to be an effective means of initiating new members into a group and allowing senior members a break from the stress of the roles they play [11]. Riddles alternated with jokes in the same order for all ConvoCon dyads. There was no observable difference between dyads' attentiveness to riddle or joke-based ConvoCons, although in informal observation, the joke-based ConvoCons did appear to be more effective in creating affinity bonds within a dyad, particularly in stimulating discussion after the fact. The riddle-based ConvoCons often had more discussion while they were still present, however, as participants sometimes tried to solve them before reading the answer or would comment on how the answer "made sense."

Framework for Measuring Affinity

To measure whether the ConvoCons system increases affinity, a measurable definition is required. Nardi [23] defines affinity as a "feeling of connection between people." The issue of empirically measuring affinity is similar to the problem Goudy observed with "rapport" where there are multiple, sometimes conflicting, definitions and limited clearly defined metrics for measurement [12]. With this problem in mind, we adapted Nardi's definition and framework and narrowed it in the context of our multitouch environment to the "convergence of thoughts, actions, or ideas" and made the following operationalized assumptions for measurement purposes within a multitouch collaborative context.

Seven tag categories were established partially based on observations of over forty dyads performing collaborative work on multitouch devices, and partially based on common notions of affinity, such as socially appropriate conversational distance [13,27].

Peshikin has suggested that qualitative methods are the most useful means of observing social interactions [25]. To quantify the affinity that we observed within dyads, we used an approach based on Anfara et al. in their discussion of making qualitative data gathering techniques transparent [1]. Videos of dyads were tagged with codes according to the behavior observed. The coding tags were derived from the below seven categories of affinity, Table 1.

Coding for Affinity

Using the video recordings of the hand movements and voices of dyads, we divided the videos up between each task (Play/Training, Task 1, Task 2, Task 3, and Freeform/Creative). The researchers then classified each five-second block of the video based on two overall constructs: the type of **behavior** (9 codes) and type of **conversation** (16 different codes). See Table 1. For conversations, the codes were then grouped into four larger categories: ConvoCon-related (e.g. reading ConvoCon text to each other, discussing ConvoCons, trying to solve ConvoCon riddles, etc.), *ConvoCon-indirect* (laughter within 1 minute of ConvoCon appearance and non-work talk within 1 minute of ConvoCon appearance), *non-ConvoCon affinity* (talking about year in school, major, directing partner, etc.), and *low/no affinity conversations about task* (e.g. getting unstuck, teaching their partner how to perform a system action, etc.) These four conversation categories were further grouped into affinity-related and low/no affinity. Participants' behaviors were coded as *affinity-related* (e.g. close proximity of hands, turn taking where one places and the other adjusts, etc.) or *low/no affinity* (e.g. hand avoidance, independent work where one partner is working on one section of the pattern while the other is working on another without any shared vision).

Each five-second block of video received one tag related to dyad behavior and one related to dyad conversation. The total number of affinity related blocks were then calculated and divided by the total number of blocks for each task. The overall affinity score is based on two parts: the proportion of affinity conversation and the proportion of affinity work (all affinity blocks / all blocks that exhibited some conversation or behavior). The proportion of affinity was then compared for each task between the experimental and control group through a Student's T-Test (see Table 2 for an illustration of the proportion calculation).

Table 1 The tags used to code the videos

- **Types of Conversation**
  - Affinity – Directly Tied to ConvoCons
    - Riddle Solving
    - Both Reading
    - Laughing (ConvoCon)
    - Talking (ConvoCon)
  - Affinity – Indirectly Tied to ConvoCons
    - Talking (within 1 min. of ConvoCon)
    - Laughing (within 1min. of ConvoCon)
  - Affinity – non-ConvoCon (not tied to ConvoCons)
    - Playful Conversation
    - Conversation About Partner
    - Planning Solution (not fixing)
    - Discussing Freeform
    - Directing Partner
    - Affirmation, gratitude, etc.
  - Low/no Affinity
    - Getting 'unstuck'
    - Teaching
    - Other Talking
    - Work related (w/i 1 min. of ConvoCon)
- **Types of Behavior**
  - Low/no Affinity
    - Independent
    - Turn Taking (independent)
    - Avoidance (hands)
    - Grabbing (taking pieces from other's 'personal space')
  - Affinity Related
    - Turn Taking (one places, other adjusts)
    - Directing-Following
    - Close Proximity (hands)
    - Shared Plan
    - Building--adding on to other's (Creative/freeform only)
- No Talking or Action

A total of 5,149 blocks were given conversational and behavioral codes by a single coder, one of the researchers. In order to ensure the coding method was valid, two videos (674 blocks) were randomly selected and the category codes for behavior and conversation were compared between the researcher and a second coder using percent agreement and Cohen's

Kappa (calculated in SPSS version 18). The second coder was an undergraduate who was not informed about the purpose of the experiment. She was trained for approximately an hour. The second coder was also provided with a 1-3 sentence description of each code but was not provided a specific video example of the code. She then completed one practice video getting feedback from the researcher after each task had been coded, which were only checked to ensure that she understood the process—particularly that each block should have only one conversation and one behavioral code. After completing the practice video she then tagged the two videos used for the calculation of interrater reliability. For the behavior category codes across both videos, there was a 90% agreement between coders with a Cohen's Kappa of 0.612. For the conversational category codes across both videos there was a 90.7% agreement with a Cohen's Kappa of *k*-0.708. Both of these Kappa scores fall into the range of scores that Landis and Koch referred to as "substantial agreement" [20].

Table 2 For a 4-block task (20 seconds), conversational affinity is 25% (1=affinity, 0=low/no affinity, blank=no talking). Behavioral affinity is 75%. The overall affinity is 67% (all affinity blocks / all blocks that exhibited some conversation or behavior).

Conversation	1			0
Behavior	0	1	1	1
	5 sec	10 sec	15 sec	20 sec

Results

As this was an initial study intended to verify the feasibility of ConvoCons as an interface technique to encourage increased affinity in collaborative work, we have only analyzed the data using basic statistical methods and have not identified any other variables of interest at this time. In reading these graphs it should be noted that ConvoCons typically stopped appearing between the end of task 1 and the middle of task 2. In addition, the puzzles used for each task, presented in a consistent order, were intended to go from simplest to solve to hardest to solve. All Student T-Tests were conducted with an $\alpha=0.05$.

Exit Survey

The control group (n=18; 9 dyads) had a mean age of 20 (SD=2.09) with 12 males and 6 females. All but one participant in the control group indicated that they had used a multitouch device (such as an iPhone) at least once with a mean response on a 5-point Likert scale of 3.0 ("A Few Hours") and a median response of 2 ("Tried it Once"). One participated self-reported as "Life of a Party" on a 5-point Likert scale of sociability, with a mean rating of 2.89 and a median score of 2 ("Prefer tight groups").

The experimental group (n=18; 9 dyads) had a mean age of 21 (SD=3.21) with 9 males and 9 females. Five participants in the experimental group indicated they had never used any form of multitouch device, with a mean score of 2.53 and a median score of 1 ("None") and 3 ("A Few Hours"). One participated self-reported as "Life of a Party" on a 5-point Likert scale of sociability, with a mean rating of 3.00 and a median score of 2 ("Prefer tight groups").

Analysis of survey results was conducted both through an analysis of individuals within each group as well as by grouping data into dyads where agreement at an appropriate level was scored a "1" and disagreement with one partner providing a score of "neutral" or lower scored as "0". There were no statistically significant differences between groups on the questions intended to assess participant's feelings of affinity toward their partners.

Completion Time – Log Data

Including the play time and the freeform task, there was no significant difference between the experimental group (mean=23.25 minutes, SD=7) and the control group (mean=23.25 minutes, SD=6.5). While groups were told they had five minutes to "play" with the system and learn the controls, the ConvoCon group was more likely to utilize the full play time while the control group often reached a point where both partners would awkwardly stare at their feet, the screen and away from each other before asking to move on to the puzzle, resulting in a statistically significant difference between the groups (p=0.008). Since one concern we had was that ConvoCons and the incidental conversations might distract groups from the work at hand, we calculated the mean completion time just for the three puzzles to look at just the effects of ConvoCons on work efficiency. There was no significant difference in time spent on the three puzzle tasks between the experimental group (mean=14.75 minutes; SD=4.75) and the control group (mean=16.25 minutes, SD=6.5).

Quantitative Evaluation of Video Data

Q1 of this study, whether ConvoCons produce more incidental conversations was part of our score for conversational affinity with the means and standard deviations seen in Table 3. It should be noted that for frequency of incidental conversations, we did not count all conversational labels that we classified as signs of affinity—only the tags that were not related to work were counted (e.g. "playful conversations" and "talking about partner"). There was a significant difference between the frequency of incidental conversations between groups for the playtime, task 1, task 2, and task 3 (p=0.001, p=0.002, p=0.01, and p=0.021 respectively). However, there was not a significant difference between groups in the frequency of incidental conversations during the freeform task (p=0.11). Overall, there was a significant difference between groups (p<0.001) thus supporting the idea that ConvoCons increase the frequency of incidental conversations.

Table 3 Means and standard deviations of incidental conversations over all tasks indicate a higher frequency in the experimental (ConvoCons) group.

	Play	Task 1	Task 2	Task 3	Freeform	Overall
ConvoCons	15.78	8.22	5.44	7.67	2.67	7.96
SD	7.68	6.96	4.33	9.82	3.00	2.64
Control	3.11	0.44	1.22	0.33	1.22	1.27
SD	7.20	0.88	2.39	0.71	1.72	2.35

Table 4 corresponds directly with a portion of Q2 of this study: whether or not the use of ConvoCons leads to increased affinity. As expected from the literature on ice breakers that providing a shared framework for conversation allowed participants to begin incidental conversations at an early stage, resulting in a 20% increase in conversational affinity which was statistically significant at p=0.006. Furthermore, a statistically significant difference between the experimental and control groups were maintained throughout the duration of the study (task 1, p=0.025; task 2, p<0.001; task 3, p=0.004; and freeform, p<0.001). While the researchers expected a significant increase in conversational affinity for the freeform task in both groups, the control group only saw a 4% increase compared to the experimental groups 26% increase in conversational affinity. This difference came from the control group discussing the freeform task less, with conversation centering on the general shape that would be made and very little planning and coordination of the task compared with the experimental group. In fact, it was not uncommon to observe one individual in the control group take control of the pattern rather than sharing the work and design with his or her partner. Taking the mean across all tasks (not seen in the graph), there was a significant difference (p<0.001) between the experimental group with a mean of 32.4% affinity

(SD=10.2%) and the control group with a mean of 10.6% affinity (SD=9.3%).

Table 4 ConvoCons serve as an early conversation starter for groups and the conversational affinity increases steadily.

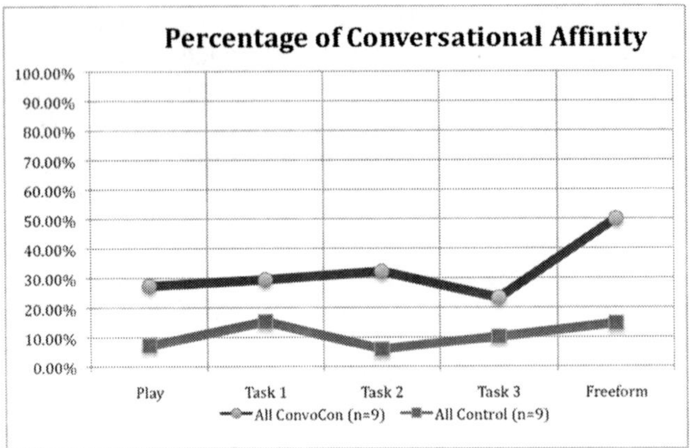

Table 5 also corresponds to Q2, whether or not the use of ConvoCons leads to increased affinity. To the researchers, this is the more important question since our ultimate goal for ConvoCons is to get individuals to work with one another in a collaborative manner. As expected with groups that are working together for the first time, the level of behavioral affinity for both the experimental and control groups starts out with a nonsignficant difference (p=0.456). Once the first pattern is given and the individuals start trying to complete a shared puzzle, the proportion of behavioral affinity goes up for both the control and experimental group, although the increase for the experimental group is larger with a marginally significant difference between groups (p=0.069). Task 2 and task 3 see a minor, but steady increase in the percentage of behavioral affinity for the experimental group with a statistically significant difference compared to the control group (p=0.024 and p=0.013, respectively). In the final, freeform task the researchers expected both control and experimental groups to have a rapid increase in behavioral affinity as they work to realize a shared vision for a new pattern; however, while both groups did see a jump in behavioral affinity, the experimental group saw a much larger increase from 29.6% in task 3 to 60.6% compared with a 10.6% to 21.8% for the control group. This difference between the behavioral affinity and the control group for the freeform task was significant (p=0.002). Taking the mean score across all tasks results in a significant difference (p=0.004) between the experimental group with a mean of 30.6% affinity (SD=16.9%) and the control group with a mean of 12.3% affinity (SD=5.9%).

Table 5 Behavioral affinity begins the same for both groups; however, when work on the puzzles begins there is sharp rise followed by a steady increase. Both groups see an expected jump in behavioral affinity in the freeform task; however, the experimental group ends nearly 40% higher.

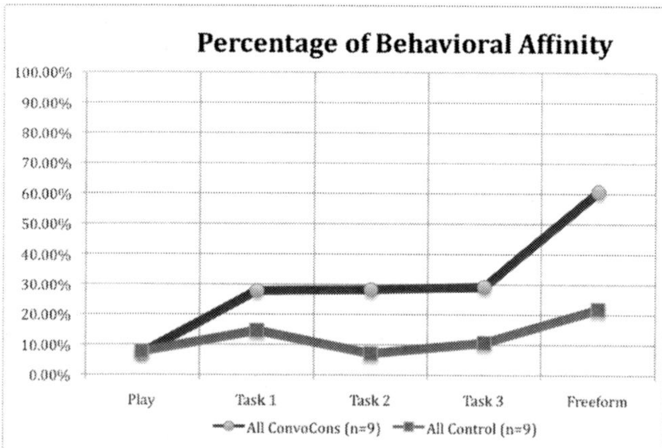

Exit Interviews

During the interviews of the nine ConvoCon dyads, four dyads indicated no signs of affinity as they answered questions. Out of the four dyads that exhibited no signs of affinity, two of them exhibited a disconnect regarding their feelings toward ConvoCons where one person found them interesting/funny and the other had no opinion. Three of the nine dyads thought the ConvoCons were irritating or distracting and all three of these dyads expressed affinity or high affinity during the interviews. Only one of nine dyads had both members that enjoyed the ConvoCons and this dyad showed signs of high affinity.

Out of the eight dyads asked about the ConvoCons they remembered, six remembered three ConvoCons (a total of ten were displayed during their working time). One dyad remembered one ConvoCon and one remembered two ConvoCons. No relationship exists between the number of ConvoCons remembered and a group's feelings toward ConvoCons or their apparent affinity during the interview.

All dyads, regardless of their feelings toward ConvoCons, indicated that after some time they began ignoring the ConvoCons and focusing more on the tasks. This was expected and part of our reasoning behind stopping ConvoCons after a period of time had elapsed. Further studies may indicate what the "sweet spot" is for displaying ConvoCons long enough to get a dyad conversing but stopping them before the dyads decide to ignore the ConvoCons.

Four of the nine dyads indicated that they felt the ConvoCons were somehow related to the task. Of these four dyads, two showed no signs of affinity within the interview and two of them showed some signs of mild affinity. In addition, five dyads expressed feelings of pressure to complete the puzzles as part of the reason they began ignoring the ConvoCons—this was despite the fact that the groups were told that completion time was unimportant and that they would be given as much time as they needed or desired to complete the puzzles.

Finally, three of the nine dyads specifically mentioned that they felt the ConvoCons had an influence in getting them to begin conversations. However, one of these three dyads indicated no affinity during the interview.

Discussion

The 9% drop in conversational affinity for the experimental group from task 2 and 3 (mirrored in the overall affinity) is likely due to ConvoCons no longer appearing after task 2. Despite this drop, the freeform task saw the experimental group increasing conversational affinity to 45% higher than that of the control group.

One thing we noted, was the apparent unreliability of the survey data—while multiple observers saw groups where a single individual performed almost all work, however, the survey data reported that both participants felt they worked equally and the final product equally represented one another's goals. This surprised us because previous studies that have examined similar concepts of rapport and development of common ground have relied almost exclusively on survey data [8,5]. To us, this disparity between the survey data and the empirical observations suggests a need to explore new methods for assessing group work that has been echoed by other researchers who study improving group work [2].

Given the power of authority and the tendency to conform to assigned roles demonstrated by Milgram and Zimbardo [22,35], participants may have been strongly focused on the puzzle tasks by 1) hearing our experimenter conduct training on the tasks and 2) knowing that they would receive a departmental research credit for participating in the study. The playtime (task 1) was designed to lessen these influences.

One of the key findings from the interviews is that groups that showed affinity during the interviews were likely to agree on their feelings about ConvoCons with three of nine dyads agreeing that they had negative opinions toward ConvoCons. These findings may suggest that the key to ConvoCons building affinity may not be the content but rather the creation of a shared experience outside of the work although further research would need to confirm this. Such a finding would be consistent with earlier work where we discovered that an ambiguous interface could lead to increased affinity within groups [24].

The lack of difference in completion time between groups provides support that ConvoCons do not increase the time groups take to complete work even though they produce more incidental conversations. These results are encouraging to us because they indicate that it is possible to use the interface to increase conversational and behavioral affinity without adversely affecting the efficiency of work.

Conclusion

To return to our initial research question of "Does the presence of ConvoCons lead to increased incidental conversations," the data presented in this study suggests that they do, in fact, promote incidental conversations. Furthermore, the increase in incidental conversations does not appear to come at the cost of efficiency as measured by completion time.

At this point, our research focuses on affinity creation and does not look at the length of affinity bonds created nor does it explore whether or not affinity creation through our system promotes cooperation in a competitive environment; it simply seeks to explore a low-cost method of promoting affinity within a co-located dyad where neither partner has previous knowledge of the other. Some limitations of the current results include the possibility that the system will not work in a competitive team climate, it may not scale to larger teams without modification, and part of the benefit may be task specific. Future research will seek to answer these larger problems, but in this research we are seeking to establish a foundational framework for the design of interfaces to encourage specific collaborative behavior within working groups.

Regarding the secondary research question, "Do ConvoCons lead to increased affinity between participants," based on this study it appears that the incidental conversations promoted by ConvoCons are effective in producing a greater level of behavioral affinity, reifying Nardi and Whittaker's framework for affinity as a central element to collaboration. These results may also suggest that Schmid's theory of the role of affinity within the buildup of social capital for the reduction of free riders and increased motivation may be realized through ConvoCons, although further studies would be needed to explore free riders within larger groups rather than simple dyads. This effect may be enhanced through the use of privileged, as opposed to shared, information within the ConvoCon display.

Acknowledgements

We thank Prasad Ramanahally and Jay Roltgen for their assistance in creating the drivers and software used for this study. In addition, we thank Joanne Marshall for her advice and help in exploring qualitative data. This research was performed with support from the Air Force Research Lab.

References

1. Anfara, V., Brown, K., Mangione, T.: Qualitative Analysis on Stage: Making the Research Process More Public. Educational Researcher 31(7), 28–38 (2002)
2. Bolia, R., Nelson, W., Summers, S., Arnold, R., Atkinson, J., Taylor, R., Cottrell, R., Crooks, C.: Collaborative Decision Making in Network-Centric Military Operations. In: Proc. of the Human Factors and Ergonomics Society (2006)
3. Bolton, G., Katok, E., Ockenfels, A.: Cooperation among strangers with limited information about reputation. Journal of Public Economics 89, 1457–1468 (2005)
4. Buxton, W., Hill, R., Rowley, P.: Issues and techniques in touch-sensitive tablet input. In: Proc. of SIGGRAPH 1985, pp. 215–224 (1985)
5. Carey, J., Hamilton, D., Shanklin, G.: Development of an Instrument to Measure Rapport Between College Roommates. Journal of College Student Personnel 27(3), 269–273 (1986)
6. Clear, T., Daniels, M.: A Cyber-Icebreaker for an Effective Virtual Group? In: Proc. of the Conference on Innovation and Technology in Computer Science Education (2001)
7. Convertino, G., Ganoe, C., Schafer, W., Yost, B., Carroll, M.: A multipleview approach to support common ground in distributed and synchronous geo-collaboration. In: Proc. CMV 2005, pp. 121–132 (2005)
8. Convertino, G., Mentis, H., Rosson, M., Carroll, J., Slavkovic, A., Ganoe, C.: Articulating common ground in cooperative work: content and process. In: Proc. of CHI 2008, pp. 1637–1646 (2008)
9. Ephrem, A., Brungart, D., Parisi, J.: Effects of Background Noise on Communication in a Collaborative Team Environment. In: Proc. of the Human Factors and Ergonomics Society (2006)

10. Fisher, M., Tucker, D.: Games Online: Social Icebreakers That Orient Students To Synchronous Protocol and Team Formation. Journal of Educational Technology Systems 32(4), 419–428 (2003)
11. Goffman, E.: Role-Distance. In: Brisset, D., Edgley, C. (eds.) Life as Theater: A dramaturgical sourcebook, pp. 123–132. Aldine Publishing Company, Chicago (1961)
12. Goudy, W., Potter, R.: Interview Rapport: Demise of a Concept. The Public Opinion Quarterly 39(4), 529–543 (1976)
13. Hall, E.T.: The Hidden Dimension. Anchor Books (1966)
14. Hollan, J., Hutchins, E., Kirsh, D.: Distributed cognition: toward a new foundation for human-computer interaction research. TOCHI (2000)
15. Honeycutt, J., Patterson, J.: Affinity Strategies in Relationships: The role of gender and imagined interactions in maintaining college roommates. Personal Relationships 4, 35–46 (1997)
16. Kafai, Y., Resnick, M.: Constructionism in Practice. Lawrence Erlbaum Associates, Mahwah (1996)
17. Karahalios, K.: Social Catalysts: enhancing communication in mediated spaces. Doctoral dissertation (2004)
18. Kellogg, W., Erickson, T.: Social Translucence, Collective Awareness, and the Emergence of Place. In: Proc. of CSCW (2002)
19. King, J., Star, S.: Conceptual Foundations for the Development of Organizational Decision Support Systems. In: Proc. of the Hawaii International Conference on Systems Science, pp. 143–151 (1990)
20. Landis, J.R., Koch, G.G.: The measurement of observer agreement for categorical data. Biometrics 33, 159–174 (1977)
21. Larson Jr, J.R., Cristensen, C., Franz, T.M., Abbott, A.S.: Diagnosing Groups: The pooling, management, and impact of shared and unshared case information in team-based medical decision making. Journal of Personality and Social Psychology 75(1), 93–108 (1998)
22. Milgram, S.: Obedience to Authority. Tavistock Publication, Londong (1974)
23. Nardi, B.: Beyond Bandwidth: Dimension of Connection in Interpersonal Communication. In: Proc. of CSCW 2005, pp. 91–129 (2005)
24. Oren, M., Gilbert, S.: ConvoCons: Encouraging Affinity on Multitouch Interfaces. In: Proc. of HCI International 2009 (2009)
25. Pehsikin, A.: The Goodnes of Qualitative Research. Educational Researcher 22(2), 23–29 (1993)
26. Ramanahally, P., Gilbert, S., Anagnost, C., Niedzielski, T., Velázquez, D.: Creating a Collaborative Multitouch Computer Aided Design Program. In: Proc. of WinVR 2009 (2009)
27. Richmond, V., McCroskey, J.: Immediacy - Nonverbal Behavior in Interpersonal Relations. Allyn & Bacon, Boston (1995)

28. Rogers, Y., Brignull, H.: Subtle ice-breaking: encouraging socializing and interaction around a large public display. In: Proceedings of Public, community and situated displays: Design, use and interaction around shared information displays, CSCW 2002 Workshop, New Orleans, Louisiana, USA (2002)
29. Russell, D., Sue, A.: Secrets to Success and Fatal Flaws: The Design of Large-Display Groupware. IEEE Computer Graphics and Application (2006)
30. Schmid, A.: Affinity as social capital: its role in development. Jounral of Socio-Economics 29(2), 159–171 (2000)
31. Scott, S., Grant, K., Mandryk, R.: System guidelines for co-located, collaborative work on a tabletop display. In: Proc. of the Eighth European Conference on Computer-Supported Cooperative Work (2003)
32. Sparsh UI (2009), `http://code.google.com/p/sparsh-ui` (accessed September 2009)
33. Tuddenham, P., Robinson, P.: Distributed Tabletops: Supporting Remote and Mixed-Presence Tabletop Collaboration. In: Proc. of IEEE International Workshop on Horizontal Interactive Human- Computer Systems 2007 (2007)
34. Whittaker, S.: Theories and Models in Mediated Communication. In: Graesser, A. (ed.) The Handbook of Discourse Processes, Hillsdale, Lawrence Erlbaum, Cambridge, NJ (2003)
35. Zimbardo, P., Maslach, C., Haney, C.: Reflections on the Stanford Prison Experiment: Genesis, Transformation, & Consequences. In: Blass, T. (ed.) Obedience to Authority: Current Perspectives on the Milgram Paradigm, pp. 193–238. Lawrence Erlbaum Associates, Mahwah (2000)

Measuring Cognitive Design Activity Changes during an Industry Team Brainstorming Session

Jeff W.T. Kan[1,2], John S. Gero[2], and Hsien-Hui Tang[3]
[1]*Taylor's University College, Malaysia*
[2]*Krasnow Institute for Advanced Study, USA*
[3]*National Taiwan University of Science and Technology, Taiwan*

This paper presents the results of using an ontologically-based method of measuring cognitive design issues and design processes on an in-situ team brainstorming session to study the changes in cognitive design issues and design processes at the beginning, middle and end of the session. Detailed results of the distributions of issues and processes are presented.

Introduction

This paper uses an ontological view of understanding design cognition. The ontological view consists of issues and processes. It is assumed that designers use different cognitive resources for different design issues and different design processes, and that they use different cognitive resources to handle the states before and after any design processes. This ontological view is used to analyse a protocol of a design session of a team of designers from industry brainstorming.

Quantifying Design Processes

In order to establish a common ground to study design activities, an established ontology from the literature is used as an overarching principle to guide the protocol study.

J.S. Gero (ed.): *Design Computing and Cognition'10*, pp. 621–640.

The FBS Ontology

The research commenced with the following statement about designing:

"The meta-goal of design is to transform requirements, more generally termed functions which embody the expectations of the purposes of the resulting artefact, into design descriptions. The result of the activity of designing is a design description." [1]

This view centers design around the creation of artefacts, whether physical or virtual. Anything that is not related to the resulting artefacts is not considered within this framework. For example, supporting activities such as planning and scheduling are not included. People can spend all their time planning and scheduling without producing any design description.

The FBS design ontology [1], as a formal model, models designing in terms of three fundamental classes of issues: function, behavior, and structure; along with two external classes: design descriptions and requirements. In this view the goal of designing is to transform a set of functions into a set of design descriptions. The function (F) of a designed object is defined as its teleology; the behavior (B) of that object is either its expected behavior (Be) or behavior derived from the structure (Bs), where structure (S) are the elements and their relationships that go to make up the artefact. A design description cannot be transformed directly from the functions, which undergo a series of processes among the FBS issues. Figure 1 shows the relationship among these issues with the resulting processes that link the issues.

Formulation (F→Be) is the transformation of the function issues of a design into issues of expected behavior. Synthesis (Be→S), is the transformation of the expected behavior issues (Be) into structure issues of the artefact that aim to satisfy the requirements. Analysis (S→Bs) is the derivation of "actual" behavior issues from the synthesized structure (S). Evaluation (Bs←→Be), is the comparison of the actual behavior (Bs) with the expected behavior (Be) to decide whether the artifact is to be accepted. Documentation (S→D), is the production of any design description from structure issues of the designed artefact.

Traditional models of designing reiterate the analysis -synthesis - evaluation processes until a satisfactory design is produced. In the FBS ontology, Figure 1, three types of reformulations is introduced to expand the design state space so as to capture the innovative and creative aspect of designing, which have not been well articulated in most models because they have not been adequately understood.

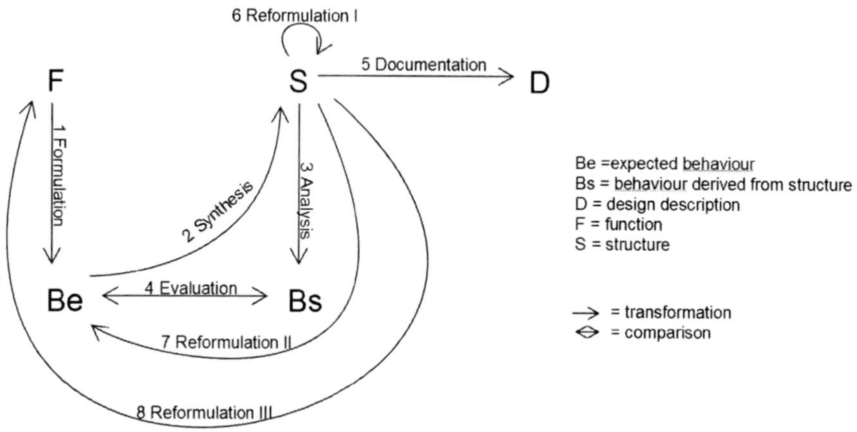

Fig. 1. The FBS ontology of designing

Reformulation type I (S→S'), addresses changes in the design state space in terms of structure issues. Reformulation type II (S→Be'), addresses changes in design state space in terms of behavior issues. A review of synthesized structure may lead to the addition of expected behavior variables. Reformulation type III (S→F'), addresses changes in design state space in terms of function issues.

The Brainstorming Session

Data for a brainstorming design session was obtained from the 7th Design Thinking Research Symposium [2]. The source data was a video of design meetings taking place in a product design practice. The data is made up of a 4-camera video recording, Figure 2, and the transcripts of the voice communication. The team consisted of a business consultant, who acted as the moderator (Allan), three mechanical engineers (Jack, Chad and Todd), an electronics business consultant (Tommy), an ergonomicist (Sandra), and an industrial design student (Rodney). They were all from the same company and the student, Rodney, was on an internship with the company.

In this brainstorming session, the team was asked to provide ideas for solving technical issues of a working demonstrator of a thermal printing pen. The two main issues were: 1) keeping the print head in contact with the paper surface and optimum angle with the media despite wobbly arm moment; and 2) protecting the print head from abusive use and overheating. Observing the protocol, it can be divided into two episodes corresponding to these two primary concerns. The video and the transcript of the utterances formed the basis of a think-aloud protocol [3], [4].

The remainder of the paper commences with a brief qualitative description of the design session and is followed by sections on quantitative measurements of the cognition of the design activity. The quantitative sections commence with the word counts and turn taking of the team members. This is followed by a more detailed analysis based on the FBS ontology issues and processes coding scheme.

Qualitative Observations

The whole session lasted about one hour and thirty-five minutes. The session can be divided into two episodes; the first one concerned the problem of keeping the print head in contact and at the optimum angle to the media, despite wobbly arm moments. The second episode dealt with protecting the print head from abusive use and overheating. In the first episode, participants were asked to generate ideas from available products that follow a contour. Several products were mentioned, such as a sledge, snowboard, wind surfboard, shaver, snowmobile, train, and slicer. Other concepts such as wheels, spirit level, feedback to user, and laser leveler were also discussed. Loosely related to those analogies, a few shapes, such as a mouse-type pen, were proposed. Besides product behavior, user behavior was also considered.

In the second episode, ways of protecting the print head were discussed. A sheath protecting the print head was proposed, the idea of a viscous damper such as leaky syringe was discussed. Other ideas like spring loaded cap, dead man's handle and a dock or cradle, that provides cleaning and charging when the pen is not in use were also discussed. Sandra left about thirty minutes before the end of the session.

Quantitative Observations

This section quantifies some of the qualitative observations in terms of word count and turn taking.

Word Count and Turns Variation during the Session

This section commences with the total number of words and turns in five minutes intervals across the session, Table 1, published by Gero and Kan [5]. Figure 2 is the corresponding trends of word count and turns.

Table 1 Total number of words and turns in five minutes intervals [4]

Interval	**1**	**2**	**3**	**4**	**5**	**6**	**7**	**8**	**9**	**10**
Total no. words	907	863	828	761	830	700	804	800	758	845
Total no. turns	21	61	78	63	74	45	68	77	84	94
Interval	**11**	**12**	**13**	**14**	**15**	**16**	**17**	**18**	**19**	
Total no. words	786	717	766	782	684	717	700	738	810	
Total no. turns	88	91	84	93	59	46	63	94	109	

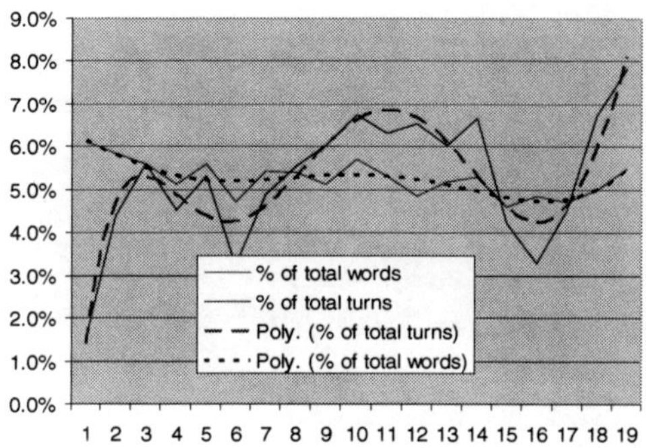

Fig. 2. The percentages of total words and total turns in five minutes intervals for the entire team, "Poly" is the polynomial line of best fit [5].

Looking at the team as a whole their percentages of interactions increased but their number of words remained fairly constant. This may indicate they have learned by producing a shared mental model [6], which is implied by the increased interaction with each interaction requiring fewer words to communicate.

Word Count and Turns of Groups

The group of mechanical engineers was the biggest group with the same background. Since they belong to the same design profession, it is expected that they will share similar mental models, hence their conversation may display a common pattern. However, neither the curves in Figure 3, nor a statistical analysis, suggest any correlating patterns.

Todd was the most active participant in this group based on the word count. Figure 4 shows that the word count curves of Todd and Tommy are of similar shape from the 2nd interval until the 17th interval; they seemed to form a cross-discipline sub-team.

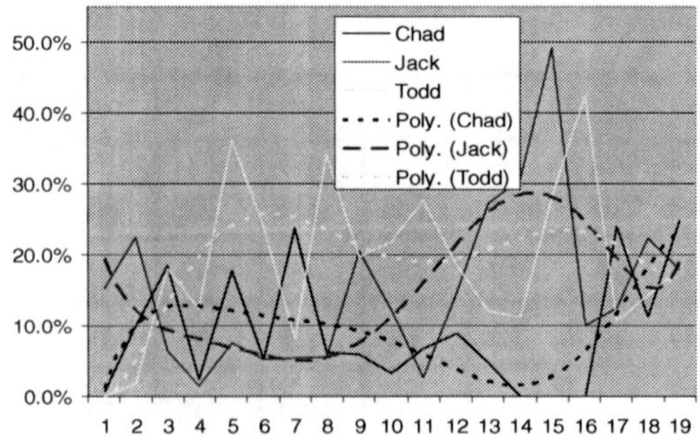

Fig. 3. The percentages of word count of the three mechanical engineers in a five minutes interval [5]

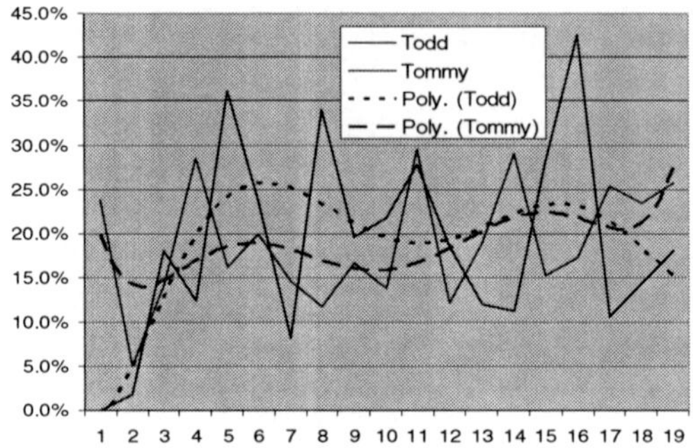

Fig. 4. The percentages of word count of Todd and Tommy [5]

FBS Segmenting and Coding

The brainstorming protocol is segmented and coded strictly based on these six categories of issues according to the following rule: one segment per code/ one core per segment. This deals with the problem of how many codes should there be to a segment. Those utterances that do not fall into these categories were not coded.

Results

This section presents the statistical results of the FBS coding that quantifies the protocol into issues in terms of FBS. Two remotely-located independent coders segmented and coded the sessions and arbitrated through internet telephony. The inter-coder agreement was over 80% and each coder's agreements with the arbitrated set was over 88%. Table 2 shows examples of each code.

Table 2 Examples of coding

R Requirements	"quite important is it's about the thermal-incli-inclis ( ) pen"
F Function	"I mean it only moves in one axis that's the standard plain thermal paper err and then it can draw"
Be Expected Behavior	"so there needs to be this contact maintained"
Bs Behavior from Structure	" I mean it only moves in one axis "
S Structure	"a sledge or a snowboard a skis or snowboard"
D Design Description	(write: sledge)
Not Coded	"yeah, we'll come to that in a minute"

Issue Distribution of the Session

After arbitrating there were 1,280 segments that contained FBS issues. Table 3 summarizes the counts and percentages of each issue.

Table 3 Count and Percentage of Issues

	F	Be	Bs	S	D
Count	47	275	369	512	77
Percentage	4.0	21.0	29.0	40.0	6.0

The percentages of FBS issues reflect the nature of the session – they were mainly borrowing behavior and structure of other objects. The behavior issues occupied half of the counts followed by structure. The low count of D was because only shared documentations were coded. Not withstanding the nature of the session, functional aspects were discussed.

The percentages in Table 2 only give an account of the whole session, however according to Asimow elementary model [7], design can be characterized by a series of cycles through analysis of the problem, synthesis of a solution, and evaluation of the solution. If Asimow's model is mapped onto this ontology, the analysis of the problem will involve function issues; the synthesis of solution will involve structure issues and the evaluation of solution will involve behavior issues. To observe Asimow's cycles of designing the number of the FBS issues needs to be counted within a sequence.

In the next sub-section, the contribution of FBS issues over the session will be presented. Results are also presented that differentiate the contribution of individuals and sum them for the whole team.

Distribution and Variations of FBS Issues

In order to obtain a more fine-grained understanding of how the issues are distributed, a window of 128 segments is taken and moved segment by segment from the beginning to the end of the protocol. This produces an averaging of that issue over those segments. With this 128 segments sliding window the number of FBS issues produced by an individual is counted and presented in Figures 5 to 9 for issues F, Be, Bs, S and D respectively. The horizontal axes show segment numbers and the vertical axes show issue counts. In each of the graphs the top surface of the graph shows the overall behavior of the team, while the shading maps onto individual team members.

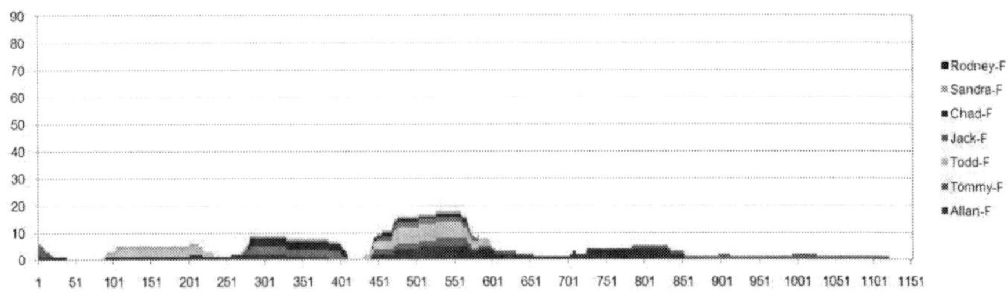

Fig. 5. F issue distribution of individuals with a 128 segment moving window

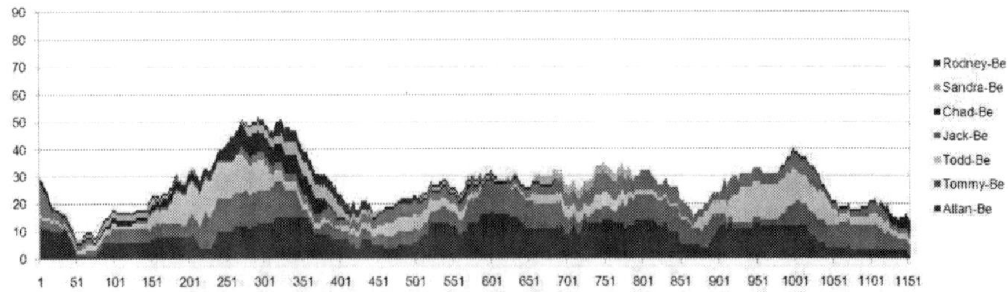

Fig. 6. Be issue distribution of individuals with a 128 segment moving window

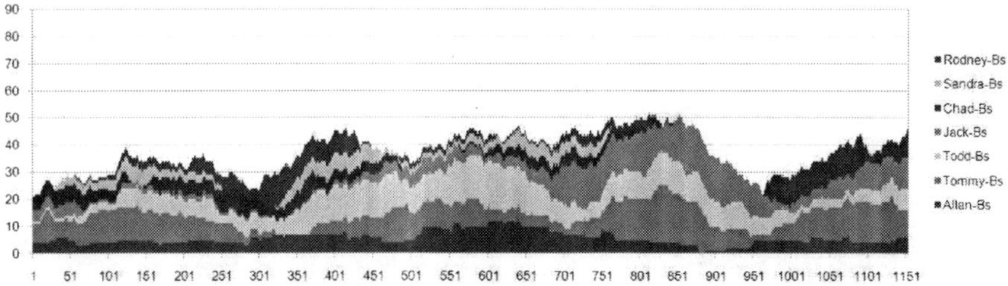

Fig. 7. Bs issue distribution of individuals with a 128 segment moving window

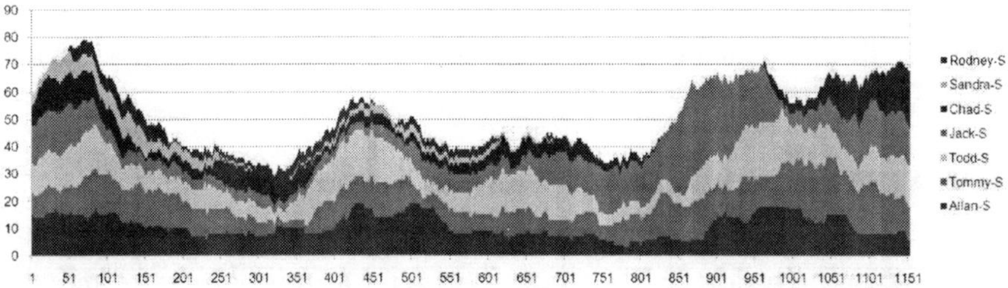

Fig. 8. S issue distribution of individuals with a 128 segment moving window

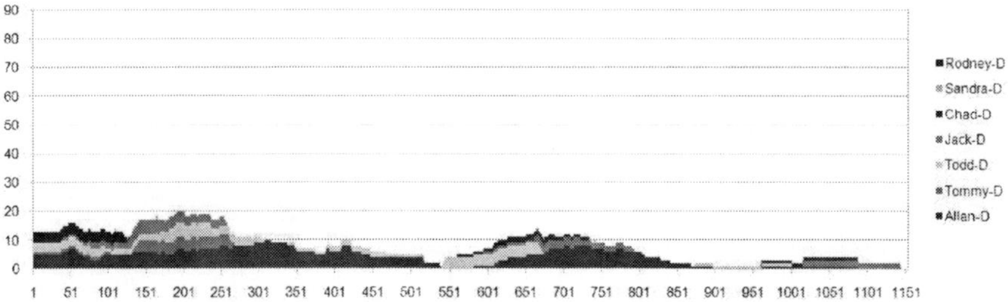

Fig. 9. D, issue, distribution of individuals with a 128 segment moving window

As expected in a brainstorming session the structure issue, S, is the dominant issue. Again as expected in a brainstorming session the issues of expected behavior, Be, and behavior from structure, Bs, are relatively low as team members had been advised to suspend judgement. One unexpected result is that the F issues concentrate towards the middle rather than at the beginning of the session as has been observed elsewhere [8].

From Figures 5 to 9, qualitatively it appears that there are changes in distributions across the design session. The significance of the changes is tested by counting the first, middle and last 300 segments of each session, giving us a total sample of 900 issues spread across all members of the team, Table 4 and Figure 10.

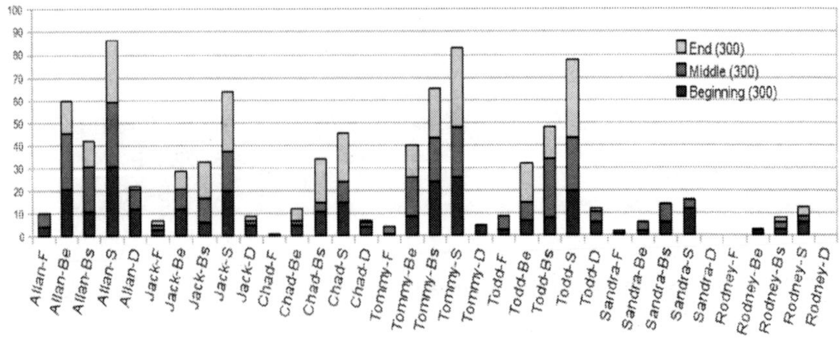

Fig. 10. The distribution of design issues of individuals within the first, middle and last 300 segments of the session.

Table 4 FBS issue variations of individual team members

		F	Be	Bs	S	D	Sum
Allan	First	4	21	11	31	12	79
	Middle	6	24	20	28	9	87
	Last	0	15	11	27	1	54
Chad	First	0	5	11	15	4	35
	Middle	1	2	4	9	2	18
	Last	0	5	19	21	1	46
Todd	First	3	7	8	20	6	44
	Middle	6	8	26	23	5	68
	Last	0	17	14	35	1	67
Rodney	First	0	1	3	6	0	10
	Middle	0	1	3	3	0	7
	Last	0	1	2	4	0	7
Jack	First	3	12	6	20	5	46
	Middle	2	9	11	17	2	41
	Last	2	8	16	27	2	55
Tommy	First	1	9	24	26	5	65
	Middle	3	17	19	22	0	61
	Last	0	14	22	35	0	71
Sandra	First	1	2	6	12	0	21
	Middle	1	4	8	4	0	17
	Last	0	0	0	0	0	0

Table 5 shows the results, for each individual member of the team, of two tailed paired-t tests carried out for testing the following:

- FBS issues distribution of the first 300 segments against the same issues in the middle 300 segments,

- FBS issues distribution of the first 300 segments against the same issues in the last 300 segments, and
- FBS issues distribution of the middle 300 segments against the same issues in the last 300 segments.

Table 5 Probabilities of paired-t test of individual's FBS issue count

	First and Middle (p)	Middle and Last (p)	First and Last (p)
Allan paired-t	0.51	0.01	0.05
Jack paired-t	0.55	0.25	0.56
Chad paired-t	0.08	0.17	0.35
Tommy paired-t	0.77	0.54	0.66
Todd paired-t	0.23	0.97	0.29
Sandra paired-t	0.69	N/A	N/A
Rodney paired-t	0.37	1.00	0.21

With paired-t test, the test statistic is t with n-1 degrees of freedom. If the p-value associated with t is low (< 0.05), there is evidence to reject the null hypothesis that the difference between the two observations is zero. Row 1 of Table 4 suggests, with a high probability, that Allens' distribution of FBS issues at the end of the session is different from the middle and start of the session. Other than for Allan, the variations of the distribution of FBS issues in the others are not conclusive.

Producing Design Processes from a Linkograph

Linkography was first introduced to protocol analysis by Goldschmidt [9] to assess the design productivity of designers. The design protocol is decomposed into small units Goldschmidt called "design moves". Goldschmidt defined a design move as "a step, an act, an operation, which transforms the design situation relative to the state in which it was prior to that move" [9]. A linkograph is then constructed by linking related moves. She states that the links are established by discerning, using domain knowledge and common sense, whether a move is connected to the previous moves.

Using the FBS coding scheme design processes are an automatic consequence of the generation of a linkograph as the two ends of a link have an FBS issue and the transition between those issues defines a design process. The 8 design processes in Figure 1 map onto the start and end issues of a link in the linkograph, Table 6.

Table 6 FBS Processes

Design Process	**Link**
Formulation	F>Be
Synthesis	Be>S
Analysis	S>Bs
Documentation	S>D
Evaluation	Be<>Bs
Reformulation I	S>S
Reformulation II	S>Be
Reformulation II	S>F

Figure 11 shows a linkograph connecting the design issues. Column 1 in this study is the participant, column 2 is the segment number and column 3 is the design issue of the protocol. The dots represent the segments and the links and the gray arrow lines represent the derived design processes. The four links represent four design processes.

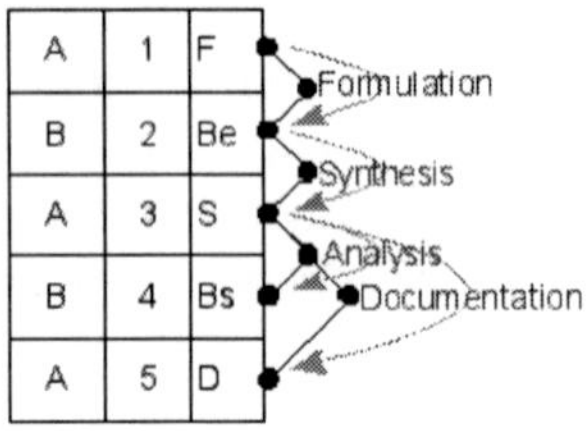

Fig. 11. Exemplar case showing the production of the design processes from the construction of the linkograph.

Segment 3 has two links, which indicate it spawns two processes: analysis (S > Bs) and documentation (S > D). Considering the participant who responded as the one who contributed to that design process, the link from segment 1 to segment 2 is a formulation process by participant B.

Team Design Processes

Figure 12 shows the means (average of the individuals) and the standard deviations of the design processes of the team over the entire design session.

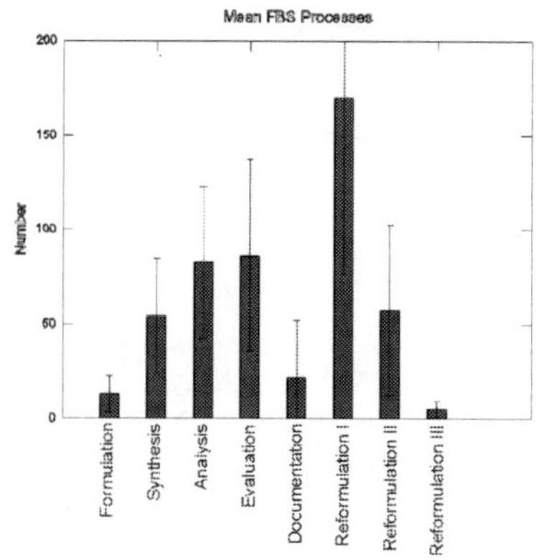

Fig. 12. The means and standard deviations of the distributions of design processes of the whole team

In order to compare if there are changes of the design processes the first, middle, and last 300 segments of the protocol were extracted and the total number of FBS processes in each group were counted, Figure 13.

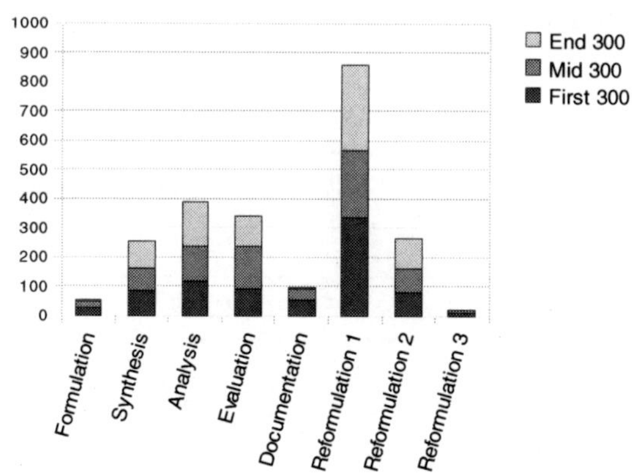

Fig. 13. The distribution of design processes counting 300 segments from the beginning, middle and end of the session

There is no formulation and reformulation type III at the end of the session which is in agreement with Asimow's model.

Comparing Design Process Distributions

Figure 14 shows and compares Allan's design processes over the three periods with those of Tommy and Todd.

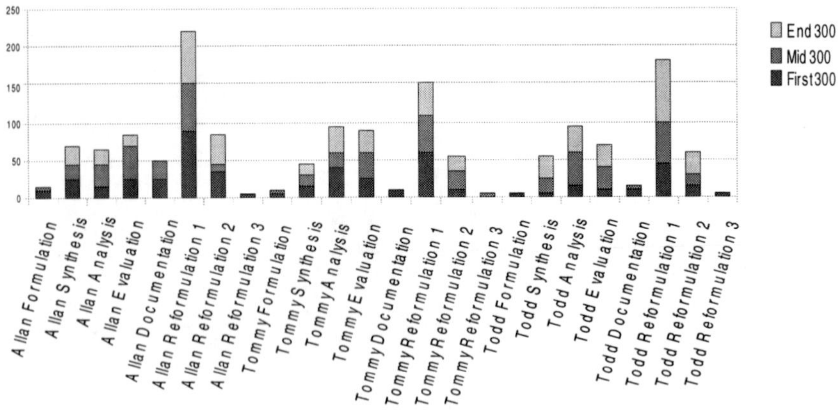

Fig. 14. The distribution of design processes of Allan, Tommy and Todd, measuring the first, middle and last 300 segments of the session

Table 7 shows the results of the paired-t test to determine if the distributions of individual's design processes change significantly over the course of the session.

Table 7 Probabilities of paired-t test of individual's design processes count

	First and Middle	Middle and Last	First and Last
Allan paired-T	0.78	0.42	0.10
Jack paired-T	0.61	0.95	0.62
Chad paired-T	0.05	0.11	0.36
Tommy paired-T	0.75	0.89	0.76
Todd paired-T	0.09	0.18	0.03
Sandra paired-T	0.27	N/A	N/A
Rodney paired-T	0.36	0.73	0.40

These results suggest that Chad's distribution of design processes changes from the start to the middle of the session. Within the first 300 segments, Chad's idea of "... a hot ball like a ball point pen" had not been considered by the group and received a very negative response by Todd "we've done that, we did that". Following that Chad's design processes contribution drop significantly, Figure 15. Further, they indicate that Todd's starting and ending distributions of design processes were different.

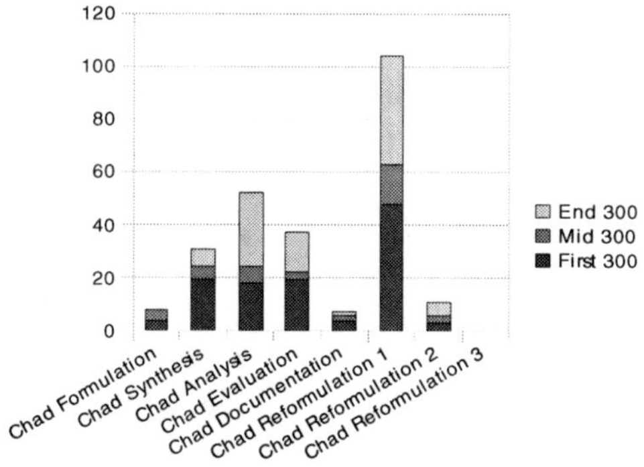

Fig. 15. Chad's change of distribution of design processes during the session

Interactions between Team Members Measured through Processes

The data in the linkograph allows for an examination of processes at a finer level of granularity. It is possible to examine the interactions between team members during design processes since each link in the linkograph has an individual at both ends. Each individual's processes can be measured using this information. Figures 16 and 17 show all the design processes of three members of the team: Allan and Tommy. The distribution of the processes is again in three groupings: the first 300, the middle 300 and the last 300 segments of the protocol. The label of the horizontal axis shows the people at the two ends of the link separated by ">". For example in Figure 16, in the subgraph titled Allan's Formulation, "Jack>Allan" means Jack raised an F issue and at some time later Allan responded to Jack with a Be issue. This indicates that the process "formulation" took place. Similarly in the subgraph of Figure 16 titled Allan's Reformulation I, "Todd>Allan" means Todd raised an S issue and some time later Allan responded with an S issue. This indicates that the process "reformulation I" took place. In both cases the processes were carried out by Allan.

A qualitative analysis of Figure 16 indicates that the sources of the issues of Allan's processes were primarily himself, irrespective of the process involved, with the exception of Reformulation I. In Reformulation I the interactions with issues raised by others outweighed Allan's interactions with his earlier issues. Further, it can be seen that in Reformulation I Allan's behavior changed from beginning to middle to end. It would be interesting to see the mean transit times for Allan. However, this is left for later analysis.

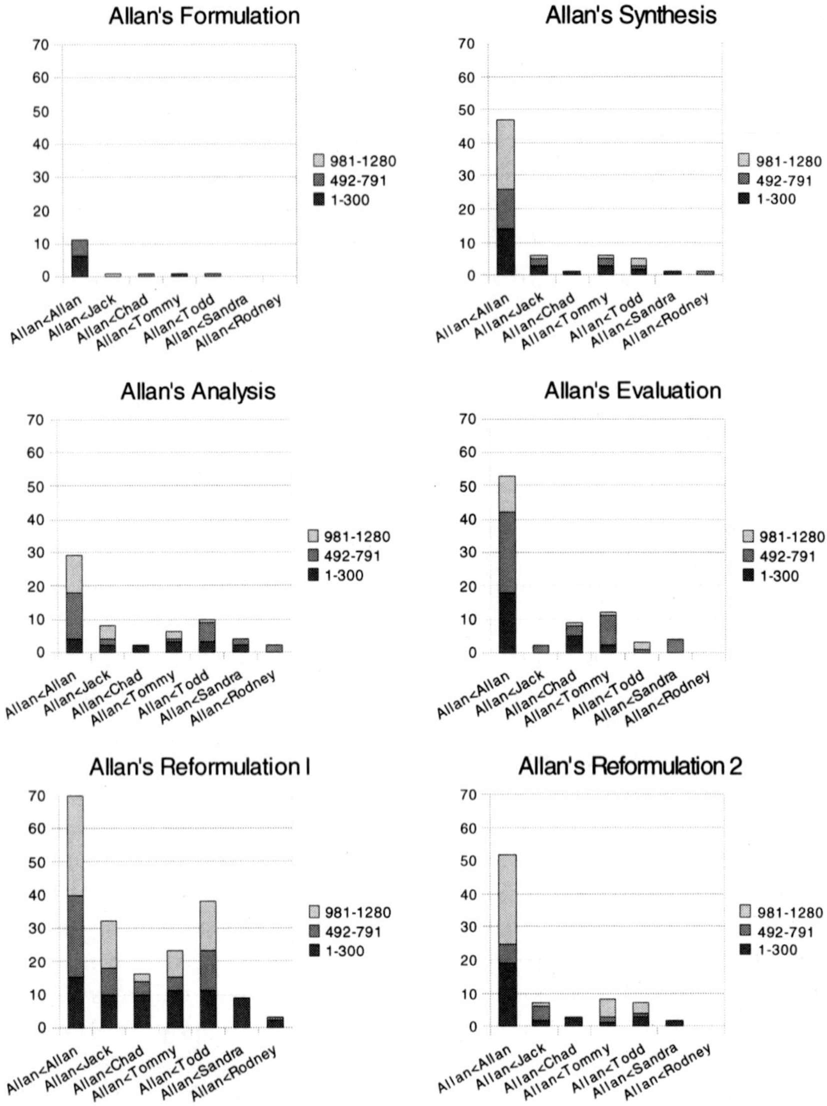

Fig. 16. The processes and their team member sources between Allan and other members of the team

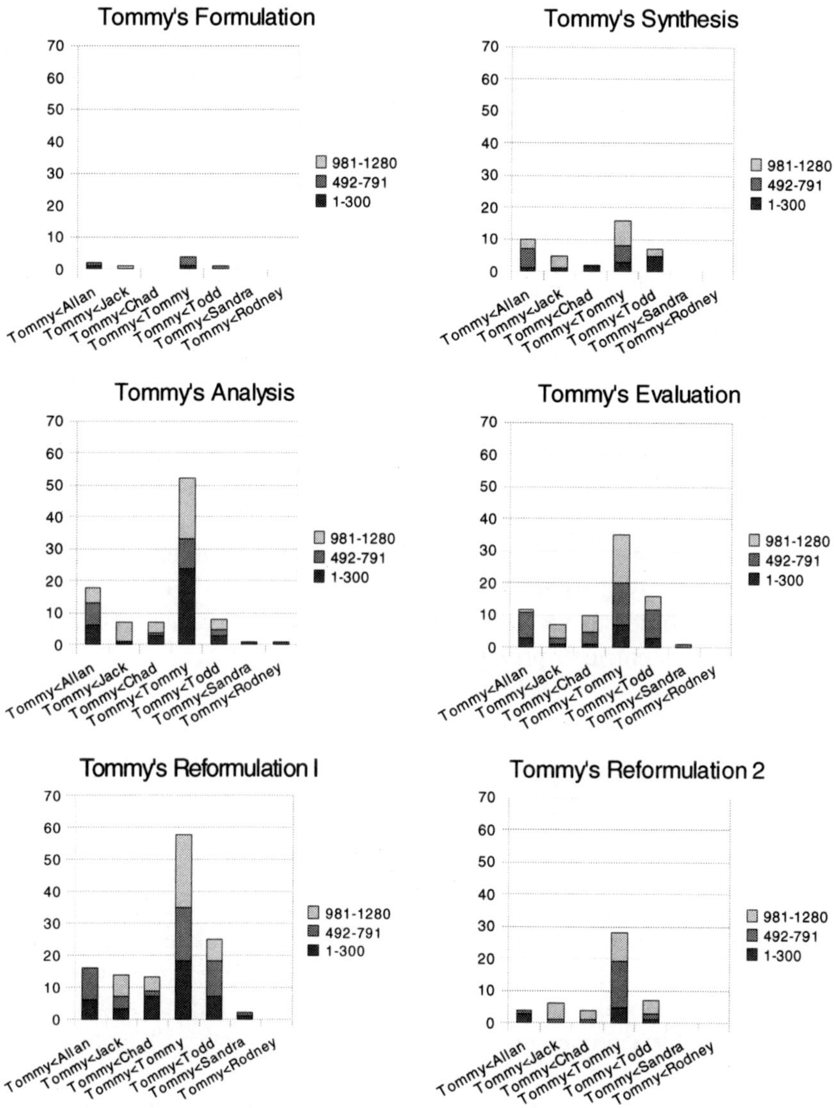

Fig. 17. The processes and their team member sources between Tommy and other members of the team

Tommy's process behavior was similar to Allan's in that he developed processes linked to issues that primarily he brought up. However, unlike Allan the distribution of sources of issues that are the bases of his processes are more varied.

These results indicating that self is a primary source of design processes is surprising as many believe that brainstorming or group process is one of the important sources for ideas. Minneman [10] even argued that design work emerges from the interactions of the group to establish, maintain, and develop a shared understanding. He suggested that designs are created through an interactive social process. However, in the study of communication and decision making, Hewes [11], based on socio-egocentric theory, claimed that the content of social interaction in small groups does not affect group outcomes; rather those non-interactive *inputs* factors are more important. The *inputs* are:

> "*shared variables that individual group members bring to a discussion, such as cognitive abilities and limitations, knowledge of the problem or how to solve problems individually or in a group, personality characteristics, motivations both individual and shared, economic resources, and power*" (Hewes, [12] p181).

The basis of Hewes' claim was from a cognitive resources point of view – when the task is difficult, individuals in the group will compromise social tasks in order to reduce cognitive load and increase efficiency in reasoning by egocentric or private speech.

Changes in Interaction during Design Session

In these results the changes that take place over the course of the design session as members of the team interact with each other can be observed and quantified.

Take as an example Tommy's interactions with other members of the team over the three periods presented in Figure 17. In the first 300 segments Tommy's analysis processes were primarily related to issues that he raised. However during the middle 300 segments almost half were raised by other members of the team. In the last 300 segments again almost half of his analysis processes were related to issues raised by other team members.

Tommy's evaluation process behavior changed from the middle 300 segments to the last 300 segments. In the middle 300 segments he interacted, in terms of issues raised, more with other members of the team than in the last 300 segments. The opposite behavior can be observed in Tommy's reformulation II processes. The behavior of each individual varied over the design session.

This range of different behaviors by individual team members changing during the design session implies a change in team behavior.

Conclusion

This paper has presented a course-grained set of results and then a fine-grained set of results derived from an ontologically-based coding scheme applied to a protocol of a team of designers in industry brainstorming.

Word counts and turntaking are one course-grained representation of the behavior of the team and its members. The results from word counts and turntaking showed that their interactions increased but their number of words remained fairly constant. This may indicate they have learned by producing a shared mental model, which is implied by the increased interaction with each interaction requiring fewer words to communicate.

The use of an ontologically-based coding scheme produced a number of increasingly fine-grained representations of the behavior of the team and its members based on issues and processes. The technique was able to isolate the issues and a statistical distribution of the team and its individual members in terms of issues was presented.

From a linkograph of the session, coupled with the issues at the segments at each end of a link, the design processes were derived. Formulation and reformulations dominated the session for the team and for the individuals in the team.

The design processes could be further articulated by examining not just the issues at each end of a link but also the individuals attached to those issues. This provided a very fine-grained representation of the process behavior of individuals by connecting them to each individual who raised the issue from which the process was generated. With this data it is possible to examine and measure the changes in behavior of individual team members and of the team as a whole across the design session.

Acknowledgements

This research is based upon work supported by the National Science Foundation under Grant No. 0750853. Any opinions, findings, and conclusions or recommendations expressed in this material are those of the author and do not necessarily reflect the views of the National Science Foundation.

References

1. Gero, J.S.: Design prototypes: A knowledge representation schema for design. AI Magazine 11(4), 26–36 (1990)
2. McDonnell, J., Lloyd, P. (eds.): DTRS7 Design Meeting Protocols: Workshop Proceedings, London (2007)

3. Van-Someren, M.W., Barnard, Y.F., Sandberg, J.A.: The Think Aloud Method: A Practical Guide to Modelling Cognitive Processes. Academic Press, London (1994)
4. Ericsson, K.A., Simon, H.A.: Protocol Analysis; Verbal Reports as Data. MIT Press, Cambridge (1993)
5. Gero, J.S., Kan, J.: Learning to collaborate during team designing: Some preliminary results from measurement-based tools. In: Chakrabarti, A. (ed.) Research into Design, pp. 560–567. Research Publications, India (2009)
6. Badke-Schaub, P., Lauche, K., Neumann, A., Ahmed, S.: Task – team - process: Assessment and anlaysis of the development of shared representations in an engineering team. In: McDonnell, J., Lloyd, P. (eds.) DTRS 7 Design Meeting Protocols Workshop Proceedings, London, pp. 97–109 (2007)
7. Asimow, M.: Introduction to Design. Prentice-Hall, New York (1962)
8. McDonnell, J., Lloyd, P. (eds.): About: Designing. Analysing Design Meetings. CRC Press, Boca Raton (2009)
9. Goldschmidt, G.: Linkography. In: Trappl, R. (ed.) Cyberbetics and System 1990, pp. 291–298. World Scientific, Singapore (1990)
10. Minneman, S.L.: The Social Construction of a Technical Reality: Empirical Studies of the Social Activity of Engineering Design Practice, PhD Thesis, Mechanical Engineering, Stanford University, Stanford (1991)
11. Hewes, D.E.: A socio-egocentric model of group decision-making. In: Hirokawa, R.Y., Poole, M.S. (eds.) Communication and group decision-making, pp. 265–291. Sage, Beberly Hills (1986)
12. Hewes, D.E.: Small group communication not ingluence dscision making: an amplication of socio-egocentric theory. In: Hirokawa, R.Y., Poole, M.S. (eds.) Communication and Group Decision Making, pp. 179–214. Sage, Thousand Oaks (1996)

DESIGN GENERATION

Interactive, visual 3D spatial grammars
Frank Hoisl and Kristina Shea

A graph grammar based scheme for generating and evaluating planar mechanisms
Pradeep Radhakrishnan and Matthew I Campbell

A case study of script-based techniques in urban planning
Anastasia Koltsova, Gerhard Schmitt, Patrik Schumacher, Tomoyuki Sudo, Shipra Narang and Lin Chen

Complex product form generation in industrial design: A bookshelf based on Voroni diagrams
Axel Nordin, Damien Motte, Andreas Hopf, Robert Bjärnemo and Claus-Christian Eckhardt

A computational concept generation technique for biologically-inspired, engineering design
Jacquelyn KS Nagel and Robert B Stone

Interactive, Visual 3D Spatial Grammars

Frank Hoisl and Kristina Shea
Technische Universität München, Germany

Since their introduction, shape or spatial grammars have been successfully used as a generative approach for creating alternative designs in different areas, e.g. visual arts, architecture or engineering. However, there are only a few three-dimensional spatial grammars that have been computationally implemented to date. Most are hard-coded, i.e. the vocabulary and rules cannot be changed without re-programming them, and only some provide limited rule parameter definition. This paper presents an approach for a basic 3D grammar interpreter that provides for the interactive, visual development and application of three-dimensional spatial grammar rules. It puts the creation and use of spatial grammars on a more general level and supports designers, who tend to think spatially, with facilitated definition and application of their own rules.

Introduction

More than 35 years ago Stiny and Gips [1] introduced shape grammars as a generative approach to shape design. Later, Stiny [2] further detailed this concept and since there has been a significant amount of research on shape or spatial grammars. Originally presented in painting, to date, they have been also successfully applied in other domains like architecture, industrial design, decorative arts and engineering [3], [4]. Most of the grammars that have been developed exist on paper and only a minority have been computationally implemented. However, many of the existing implementations are restricted to one specific design task or application. They usually do not provide for an easy way to change the existing grammar or to develop a completely new grammar, as they require coding of grammar rules in a textual form [5]. Therefore, at least some programming knowledge is needed. Practicing designers, however, tend to think spatially. They are used to working in a graphical environment and

J.S. Gero (ed.): *Design Computing and Cognition'10*, pp. 643–662.

are often not willing or do not know how to program. Instead, they want to focus on designing.

Recent research activity, especially in the area of 2D grammars, shows a growing interest towards creating more flexible shape grammar systems. These provide the grammar-user not only the possibility to apply rules but also to design or define and change rules (e.g. [6], [7]). However, none of the currently existing 3D implementations support the flexible development of rules from scratch by visually designing the geometric objects and their spatial relations. Especially in three-dimensional space, where spatial thinking is more demanding, it is easier to have a direct visualization of rules while they are designed instead of writing code first, compiling and executing it and then finally seeing whether it really created the intended geometric objects and spatial relations. Further, the definition of a new grammar "often tends to be a 'generate and test' cycle" [5]. This demands for proper user support not only for the development, but also for the application of rules since the grammar might be tested and modified several times before it fulfills the desired behavior.

The aim of this paper is to present an approach for a basic 3D grammar interpreter that includes aspects of the interactive, visual development and application of three-dimensional spatial grammar rules. It provides support for designing non-parametric and parametric rules as well as for matching the left hand side of a rule in an existing design and correctly applying it.

The paper is written from the viewpoint of providing design support to mechanical engineers. Therefore, comparisons to Computer-Aided Design (CAD) are drawn and the implementation is based on a three-dimensional solid CAD system. This reflects a statement by Gips [8] who pointed out the idea of developing a shape grammar plug-in for a traditional computer aided design program that would assist in creating a shape grammar, which in turn would help the practicing designer.

The paper starts with a background section introducing the shape grammar formalism, elucidating the meaning of interpreters and presenting relevant existing implementations. Next, it discusses the challenges involved in creating a general, 3D spatial grammar interpreter. The approach taken in this paper is then described including different concepts for developing non-parametric as well as parametric grammar rules, matching the left hand side of a rule to a current design and applying rules based on the calculation of the spatial relations between the involved geometric objects. A short section describing an early prototype implementation is followed by illustrative examples. The paper ends with a discussion of the benefits and limitations of the presented approach as well as future work.

Background

Spatial and shape grammars are generative systems that generate shapes by starting from an initial shape, which exists within a defined vocabulary of shapes, and applying defined shape rules iteratively to an evolving shape.

Formalism

A shape grammar is defined formally as $G = (S, L, R, I)$ where:

S is a finite set of shapes

L is a finite set of labels

R is a finite set of rules, and

I is the initial shape where $I \subset (S,L)^0$.

The set of labeled shapes, including the empty labeled shape, is $(S,L)^0$ and is also called the vocabulary. Shape rules are defined in the form $A \rightarrow B$, where A and B are both shapes in the vocabulary. To apply a shape rule to a given working shape C, first, the shape A in the left hand side (LHS) of a rule is detected in C. This shape matching process can make use of valid Euclidean transformations, t, e.g. translation and rotation, that are applied to the shape A, to find more possible matches of A in the working shape C. The transformed shape A then gets subtracted from C and the transformed shape B of the right hand side (RHS) of the rule is added, thus resulting in the new working shape C' where $C' = C - t(A) + t(B)$.

Strictly speaking, a shape grammar involves the use of a maximal shape representation, which can be broken down and re-represented in a large number of ways. For example, a line can be broken up into smaller line segments, called subshapes. This ability for re-representing shapes in a number of ways enables for wider matching of the shape A in the LHS of a rule to embedded subshapes in the working shape C. Spatial grammars, on the other hand, is the more general term and includes all kinds of grammars that define languages of shape, e.g. string grammars, set grammars, graph grammars and shape grammars [9].

Grammar Interpreters

Gips [8] defines a shape grammar interpreter in a general way as a "program for shape grammar generation". According to Chase [5] "the complete process of generating a design with a grammar involves two main stages": the development of the grammar and its application. A generalized interpreter that provides for facilitated use of grammars without programming has to support the tasks in both stages in an interactive, visual manner. These tasks include, for example, the design of the vocabulary and the rules, the selection of a rule to be applied, the determination of an object to apply the rule to, including calculating transformations, and the application of the rule.

Existing Spatial Grammar Implementations

Since the introduction of the detailed formalism by Stiny [2], few shape or spatial grammar systems have been computationally implemented, almost exclusively in academia as experimental prototypes or for educational purposes. An overview until 2002 can be found in [4]. Only during the last ten years, an increasing interest in implementing more general systems that, at the same time, provide interactive visual support for both the development and the application phase can be seen. The vast majority among them are two-dimensional systems. *GEdit* [10] allows for the interactive development of rules as well as their application, which includes subshape recognition and emergence. New shapes based on straight lines can be created in an external program and are converted to maximal line representation when they are imported. *Shaper2D*[1] [11] was implemented for educational purposes and provides for the visual manipulation of a rule by changing size, location or orientation of two shapes that are squares and/or triangles. A maximum of two rules are iteratively and fully automatically applied in real time according to the changes. Most recent research makes an effort to extend on interpreters that can additionally handle curvilinear 2D shapes. McCormack and Cagan [12] presented an interpreter that is able to match curve-based shapes including parametric shape recognition, whereas the strong focus is on the application phase. The rules are defined using a text-based interface. The *SubShapeDetector*[2] [6] allows for the import of hand-drawn sketches that are the basis for the interactive development and application of rules. It includes a pixel-based approach for the detection of subshapes enabling the use of curved shape grammars. The second version of this system, *SD2*[2] [13], additionally provides for direct computerized drawing of shapes within the software and the definition of an arbitrary number of rules. A system with a very similar range of functionality, but using a maximal line representation for subshape detection, was recently published by Trescak et al. [7][3].

---

[1] http://designmasala.com/portfolio/ui_shaper2d.html

[2] http://www.engineering.leeds.ac.uk/dssg/downloads/requestForm.php

[3] http://sginterpreter.sourceforge.net/

To date, only a few implementations for three-dimensional grammars exist. They are mainly designed to deal with very specific or restricted problems. Therefore the rules or rule schemata are fully pre-implemented using programming or scripting languages. *Genesis* is currently the only known commercially used implementation of a shape grammar and hence can be considered very mature. It was implemented at Boeing by Heisserman [14] based on the original version that was developed to generate alternative Queen Anne style houses [15]. It is designed to support the interactive application of rules for piping design in aircrafts. The geometric objects as well as the rules are developed by hard-coding upfront. Piazzalunga and Fitzhorn [16] used a commercial solid modeling kernel to develop an interpreter. The rules are defined using a programming language that makes most of the kernel's accessible capabilities. The application of these rules requires visual interaction whereas the user chooses a rule and an object to apply the rule to. Chau et al. [4] presented an approach that can handle curvilinear basic elements in 3D space. The shapes and rules are predefined in a data file and applied to generate wireframe-like shapes.

In addition to the three-dimensional systems described so far, a few implementations provide limited options to customize rules without programming. They are realized as pre-implemented rule schemata that the user can assign specific values to. That way the sizes of shapes and their spatial relations as well as the number of rule applications are defined before the actual application of a rule. The rule application itself is executed automatically. Wong and Cho [17] picked up the concept of *Shaper2D* (see above) and extended it to the use of three-dimensional blocks in a single rule. In the revised version of this system, *Shape Designer V.2*, the user can choose from several different predefined rule schemata and assign the required values writing a short command. The *3DShaper* [18] provides a dialog window for a pre-implemented rule schema to define the sizes and the spatial relationships of two blocks by typing in the required values. In doing so, one or two rules are specified which get immediately applied and saved in data files. A visual representation of the rules as well as of the generated design is only available after opening the created files in an external viewer.

Challenges

Using programming for the development of rules and their application provides high flexibility for the implementation of various geometry, or vocabulary, and especially expressing relations between or constraints on the geometric objects. However, as described in the introduction, programming in a design environment has several drawbacks. An approach that allows for the interactive visual design and application of grammar rules has to provide a set of standard commands whose functionality is on a higher level compared to programming, but visually usable via a user interface and therefore easier to work with. The interplay of the different single commands has to provide high flexibility to enable a generalized design and use of grammars.

In that regard specific challenges arise, especially in conjunction with a three-dimensional approach. Concerning the realization of an actual system, the underlying representation of the grammar has to be flexible enough to account for the following issues. The vocabulary, as the basis for the definition of a grammar, should be as unrestricted as possible to allow for the creation of a wide variety of geometric objects. For the definition of rules, these geometric objects have to be positioned in 3D space, which requires the robust handling of 3D transformation operations, at least for translation and rotation. For the later application of a rule, the location and orientation of the objects have to be defined in a way that enables the automatic matching of the LHS. To allow for an even wider variety of possible designs, not only the definition of non-parametric but also parametric rules should be possible along with their automatic application. Further, the number of geometric objects in a rule should not be restricted. The same applies to the number of rules that can be defined and applied, since the interaction of several rules generally leads to more alternative solutions.

Approach

The approach described in this paper is based on a set grammar formulation of spatial grammars. Generally speaking, the formalism follows the one used for shape grammars, but in a set grammar the generated designs are always parsed into the elements of the set from which they are formed and the grammar rules cannot be applied to subshapes [19]. While they are, to a certain extent, restricting for the generation of design alternatives, set grammars are very amenable to computer implementation [19]. The vocabulary used in this paper consists of a set of parameterized three-dimensional primitives, namely block, torus, wedge, cone, cylinder and sphere or ellipsoid. Although at first view this seems limited, it provides advantages for a 3D grammar system in dealing with objects. First, parameters are explicit for each object. Second, more than one object can be defined in a grammar rule to describe more complex geometry. These and the corresponding issues will be discussed in this section that is, in accordance to Chase's definition [5], subdivided in two main-sections: the development and the application of a grammar.

Development of Spatial Grammar Rules

This paper aims to develop an interactive visual approach to spatial grammar definition and application that is intuitive to designers. The defined rules are always visible right away. There is no code that has to be compiled or executed upfront to check whether the rule is geometrically and spatially what was intended.

Development of Non-parametric Rules

Spatial grammar rules are predominantly about geometric objects and their spatial relations. The fundamental aspects for the design of spatial grammar rules are therefore the creation and the positioning of geometric objects in 3D space.

As mentioned at the beginning of this chapter, the basis for the definition of geometric objects is a given set of parameterized 3D primitives, or vocabulary. An object is created by choosing one of the given primitives and assigning concrete values to its parameters. As is often the case in the design of solids in mechanical engineering CAD, this input is made numerically, so that exact values can be assigned to the parameters. The parameters describe the size as well as the location and orientation of the objects. In the process of designing a rule, the assigned values are allowed to be modified until the intended state of the rule is achieved. Once these values are set, they remain static in the later application of the rule.

Every primitive geometric object that is created has its own local coordinate system. This coordinate system, and therefore the attached object itself, can be translated and rotated in relation to the global coordinate system, Figure 1(a).

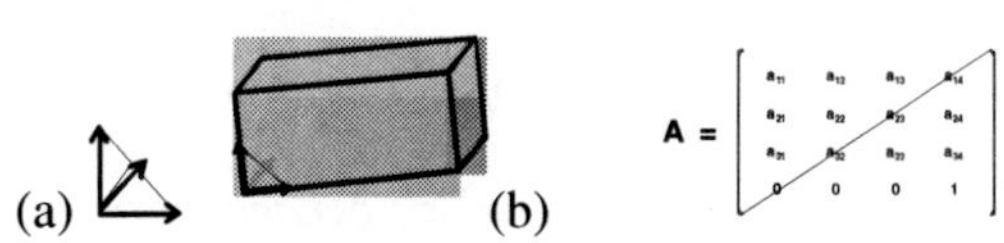

Fig. 1. (a) Global coordinate system and object with local coordinate system; (b) general transformation matrix

Through this, the location and orientation of the object is defined or changed in three-dimensional space. Mathematically, the information about the position is kept in a transformation matrix. As is common in 3D applications, this is a 4x4 matrix using homogeneous coordinates,

Figure 1(b), that enables the calculation of translations and rotations in one single matrix. Several transformations can be easily concatenated by multiplying the corresponding matrices. This is of use especially for the application of grammar rules as they require the substitution of the transformed LHS of a rule, *t(A)*, by the transformed RHS, *t(B)*.

Designing a grammar rule, i.e. the visual definition of the geometric objects as well as their positions in the LHS and RHS of a rule, makes the data about the objects' locations and orientations in the 3D space explicitly available. This is helpful to facilitate the detection of the transformations for the application of the rules and therefore to position the objects according to the spatial relations as defined in the rules themselves.

To ensure the correct definition of the spatial relations and easy access to the transformation information, a rule is designed in such a way that there is one reference object in its LHS. This reference object has to be located in the global origin and must not be rotated. It is the basis for the detection of the rule's LHS (see cp. 4.2.1). Due to the considerably higher complexity for automatically detecting a rule's LHS that consists of several objects, the approach is, for now, restricted to the usage of one single object in the LHS, which is therefore automatically the reference. All the other geometric objects in a rule are positioned in relation to this reference object and therefore in relation to the global origin. Thus, the relative spatial relations of the single objects are implicitly defined via the reference object. The transformation matrix of the reference object, T_{L1}, always equals the identity matrix I. For rule application (cf. cp. 4.2), the further transformation matrices of a rule with an arbitrary number of objects, n, in the RHS are denoted T_{Ri} for $i = 1, \ldots n$.

Figure 2 shows an example for a rule with several objects in the RHS and the reference object, L_1, in the global origin in the LHS.

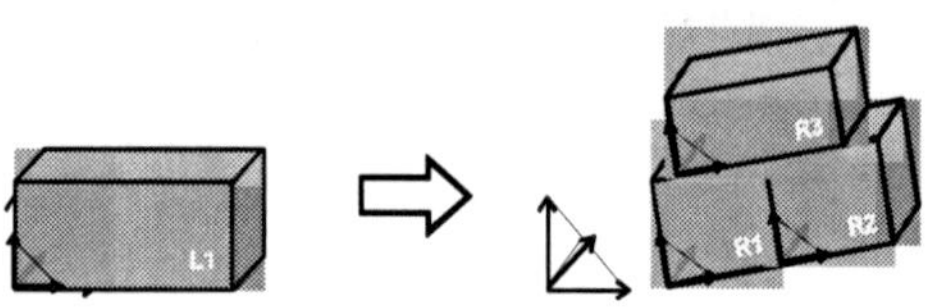

Fig. 2. Example for a non-parametric rule

Development of Parametric Rules – Relations between Parameters

In the previous section, the used parameterized primitives are assumed to be fully defined, i.e. specific values are assigned to *all* of their parameters. The outcome is rules that are based on fixed, fully determined objects. Stiny [20] describes parametric shape grammars where the rules are based on parameterized shapes and some or all of the parameter values are not predefined in the rule. Thus, a rule schema is defined that describes many different, but related, rules in one generalized rule. By assigning specific values to all parameters, one rule instantiation can be derived from this schema, which can be applied in the usual way to an existing design. This allows for a wider variety of possible designs generated by fewer rules.

On paper, this is a more manageable issue. However, a general computational implementation cannot be as easily provided. One of the crucial points is to automatically match a general parameterized shape in the LHS of a rule to an existing design. In this section, aspects for the development of parametric rules are elaborated and partially adapted from Stiny's original definition [20].

This rule type builds on the development of non-parametric rules, described in the previous paragraph. Even though the intention is to define a parametric rule, initially the geometric objects have to be fully determined, i.e. specific values have to be assigned to all of their parameters. This is because a visual grammar system, in comparison to a grammar on paper or a hard-coded grammar, would not be able to create and display objects that initially have one or more unspecified parameters. Once the initial, non-parametric state of an object is designed, one or more of its parameters can be "unlocked" – in the following denoted "free parameters" – to make it parametric. If a parameter is unlocked, by default it is completely unrestricted, i.e. any arbitrary value can be assigned to it.

However, there are two ways to constrain free parameters:

(1) The values that are allowed to be assigned to a free parameter can be restricted to a certain range. An example rule with restricted ranges of possible values for the length *l* of a box and its rotation angle around the y-axis is shown in Figure 3 (the global coordinate system is not shown in this case).

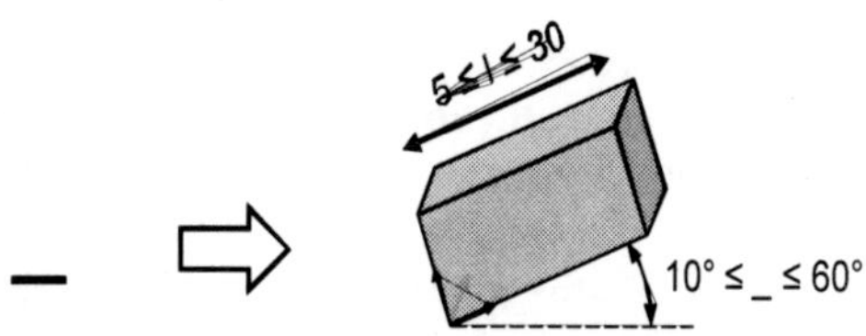

Fig. 3. Parametric rule with free parameters that are restricted to certain ranges

(2) Parametric relations can be defined between free parameters. As mentioned before, the parameters for every single object are not only the ones to define the geometric size ("size" parameters) like width, length, radius, etc., but also the ones to determine the location and orientation ("location" or "orientation" parameters). To make a free parameter dependent on another one, simple mathematical equations including one operator and one numeric operand can be defined. For example, in Figure 4,

the length l of the box is made dependent on the height h and is twice as long, $l = 2*h$, whereas h is restricted to a range between 10 and 30.

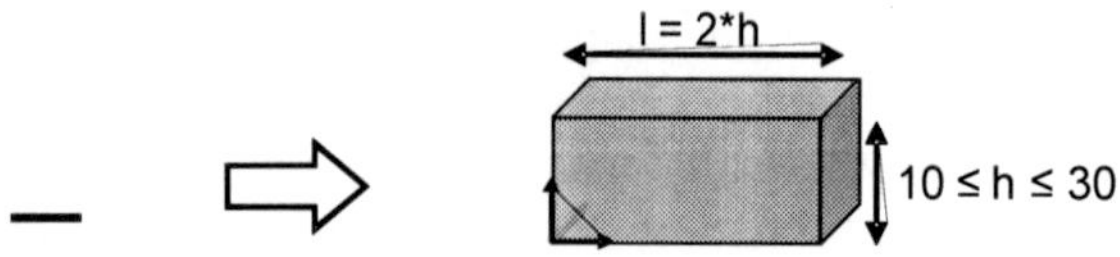

Fig. 4. Rule with parametric relation in the RHS

All the examples given so far considered objects in the RHS of a rule, but the described specification of free parameters can be analogously used for geometric objects in the LHS of a rule. The intention behind it, however, is slightly different in this case, as it is about enabling parametric matching of the objects. Once a parameter is "unlocked", by default it is completely unrestricted. This means that all objects with any arbitrary value for this free parameter can be detected as a match for the LHS. Assigning a range of possible values to a free parameter in the LHS restricts the matching of the object accordingly. Figure 5 shows an example, in which the parameter l_{L1} of the box in the LHS is completely unrestricted; additionally a parametric relation between the length l_{R1} in the RHS and l_{L1} is defined.

Parametric relations, as described above, can also be used to establish dependencies between free parameters in the LHS. This allows for a more restricted matching in the application of the rule. For example, a parametric relation between length and width of a box can be defined as $l = 3*w$, so that the matching is restricted to objects with a ratio of three between these two parameters. This concept can be further used for detecting scaled versions of an object, as the use of "scaling"- and "reflection"-transformations is currently not considered in this approach. For example, if the length of a box is completely unrestricted but its width and height are restricted by the parametric relations *width* = *length* and *height* = *length*, the matching will detect all cubes, no matter what size they are.

Fig. 5. Rule with completely unrestricted parameter in the LHS and parametric relation in the RHS

For the definition of free parameters in both the LHS and RHS of a rule, it is assumed for now that the user sets them in the correct order, i.e. in such a way that they can be evaluated correctly once the rule is applied. There is currently no rule validation. For example in the rule in Figure 4 the free parameter for the height has to be defined first, so that the length can be calculated correctly depending on the height. Further, cyclic relations between free parameters, e.g. *width* = *length*, *length* = *height* and *height* = *width*, are currently not checked.

Application of Spatial Grammar Rules

The application of grammar rules can be subdivided in three steps: (1) the determination of a rule to apply, (2) the determination of an object the rule is applied to and (3) the determination of a matching condition [5]. These can generally be done either fully manual, semi-automatic or fully automatic. In this approach, the selection of rules as well as the number of rule applications is assumed to be either done manually by the user or randomly by the system. The other two steps are supported automatically.

The semi-automatic application of rules in this approach is restricted to the scenario of "manually selecting a rule and automatically detecting all objects it can be applied to". The alternative scenario of "manually selecting an object and automatically finding all rules that can be applied to it" is not included for now.

Matching – Detection of the LHS of a Rule

Once a rule is selected, according to the grammar formalism (cf. cp. 2.1), the next step in the application is to detect the LHS of the rule in the current working design. Generally, the automatic matching of the LHS of a rule in a current design is a difficult problem, especially in three-dimensional systems. Existing 3D systems require manual intervention (e.g. [16]) or they circumvent the problem due to the nature of the provided rules, which only allow for adding in a new geometric object in relation to the last object that was inserted (e.g. [18]).

The aim of the presented approach is to automatically match the LHS of a rule in a current design. As it is restricted to the usage of basic primitives, it can benefit from the fact that for any primitive it is explicitly known which class it is derived from, e.g. box, cylinder, etc. Therefore, the first step in the automatic matching is to search the current working design for objects that are of the same primitive class as the object in the LHS. The further specification of the matching depends on how the parameters in the LHS of the rule were defined during the development phase. If the rule does not contain any free parameters, an exact search is performed, i.e. the identity of all size parameters of the object in the LHS and the objects that were detected is checked. If one or more free parameters were defined, it is additionally checked whether they are completely unrestricted, whether they are within the defined ranges or whether they can fulfill the assigned parametric relations (cf. cp. 4.1.2). For the special case that all size parameters of the object in the LHS are "unlocked" and completely unrestricted, all objects of the same primitive class are found, e.g. all boxes no matter which length, width or height. This is the most general parametric matching of the LHS in an existing design.

Once all valid objects in a design have been detected, the rule can be automatically applied to all of them, randomly to some of them or the user can manually select the ones it should be applied to.

Calculation of Positions and Sizes of the Objects in the RHS

Following the formalism, the rules are all realized as substitution rules, meaning the LHS is always fully subtracted from the working shape and replaced by the RHS. While this is perhaps not always computationally efficient, it is the most general way. The transformation information of the selected object(s) that match the LHS of the rule is explicitly available and so are the transformation matrices T_{Ri} of the objects in the RHS of the rule (cf. cp. 4.1.1).

As an example, Figure 6 shows a simple non-parametric rule with one single object on both sides, the current design (1) and the steps of the rule application. The transformation matrix L of the matched object in an existing design, corresponding to the LHS of the rule, is detected (2), the object is subtracted from the current working shape (3) and L is multiplied by the transformation T_{R1} so that the object in the RHS of the rule is added to the current design under the transformation $R = L \bullet T_{R1}$ (4). Extending this to the application of rules with n objects in the RHS requires the multiplication of L by the transformation matrices of all other objects in the RHS, i.e. $R_i = L \bullet T_{Ri}$ for $i = 1, 2, \ldots n$.

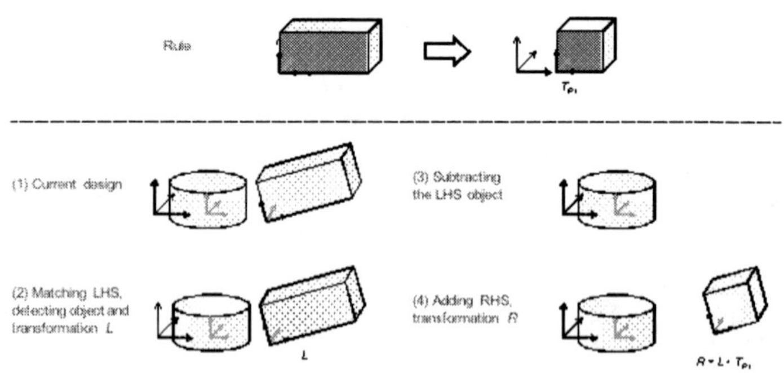

Fig. 6. Example for a simple grammar rule and the steps of its application

For parametric rules, besides the positioning of the objects in the RHS, in accordance to the LHS, also the defined relations are taken into consideration. After a valid object has been selected for rule application (cf. cp. 4.2.1), the parameter values of the objects in the LHS are all explicitly available. To adapt the geometry of the objects in the RHS in accordance to the defined parametric relations, first, values within the given ranges are assigned to all free parameters that are independent of any other parameters. This can be done either manually by the user or randomly in an automatic mode. Last, the remaining dependent parameters in the RHS are calculated in accordance to all given parameters in the LHS and the RHS. The order of the assignment is based on the order the parametric relations were defined (cf. cp. 4.1.2).

Implementation

A prototype software system of the approach described in this paper has been implemented. It is based on an open source 3D CAD system[4] that in turn is built on top of an open source geometric modeling kernel[5]. The approach for the interpreter is realized as a Python[6] module that is integrated at the startup of the system. For the grammar development phase, it provides two windows for the design of the geometric objects and their relations in the LHS and RHS of rules. Once defined, rules can be saved and later opened and edited if needed. For the definition of "unlocked" parameters it provides a dialog window that lists all the free parameters of the currently existing geometric objects in a rule. Completely unrestricted parameters, parameter ranges or parametric relations can be defined Saved rules can be loaded and applied to a current working design. The system therefore provides another dialog window to choose from manual, semi-automatic or automatic options for the application. Figure 7 shows a screenshot of the system during the design of a rule.

---

[4] http://sourceforge.net/apps/mediawiki/free-cad/

[5] http://www.opencascade.org/

[6] http://www.python.org/

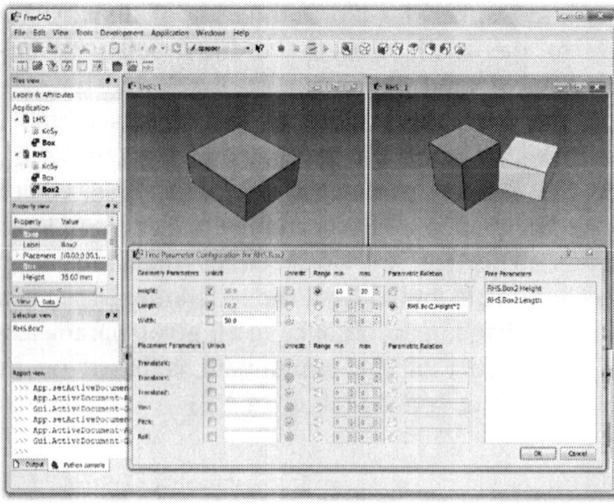

Fig. 7. Screenshot of the prototype software system

Examples

To evaluate the approach, several grammars were designed using the system. The first grammar that was developed is the "Sierpinski" grammar (Figure 8) that was previously presented as an implemented example by Piazzalunga and Fitzhorn [16]. The most basic version of this grammar consists of one parametric rule. It substitutes a box in the LHS by five smaller boxes in the RHS.

The size parameters of the cube in the LHS are specified as free parameters defining the length completely unrestricted and additionally using the parametric relations *width* = *length* and *height* = *length*, so that any scaled box that is a cube can be detected in the parametric matching. In the RHS there are several free parameters specified for the correct definition of the sizes and locations of the objects. They are all made dependent on the free parameters in the LHS. Width, length and height of the cubes are exactly half the size of the cube in the LHS.

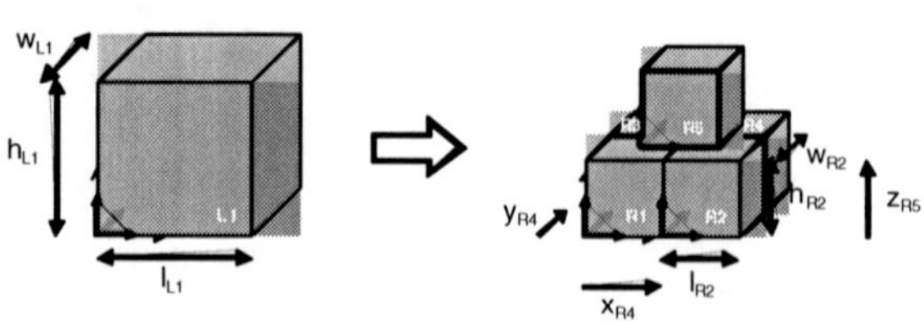

Fig. 8. Parametric "Sierpinski" grammar rule

The corresponding parametric relations are e.g. for block *R2* in the RHS: $l_{R2} = l_{L1} / 2$, $w_{R2} = w_{L1} / 2$ and $h_{R2} = h_{L1} / 2$. The parametric relations for the sizes of the other four blocks are defined in the same way. For their locations, which are dependent on the size of the block found in the current design in accordance to the LHS of the rule, different parametric relations for each single cube are set. While cube *R1* does not require any additional information (as it sits in the global origin just like the cube in the LHS), *R2* is moved along the x-axis by half the size of the cube in the LHS and *R3* in the same way but along the y-axis. Cube *R4* requires parametric relations for setting its location in both x- and y-direction: $x_{R4} = l_{L1} / 2$ and $y_{R4} = w_{L1} / 2$. Cube *R5* additionally needs to be moved in positive z-direction as it is "centered" on top of the first four cubes; the parametric relations are therefore: $x_{R5} = l_{L1} / 4$, $y_{R5} = w_{L1} / 4$ and $z_{R5} = h_{L1} / 2$.

Figure 9 shows four different solutions that are generated applying the rule fully automatically in four iterations. The current working design initially consists of one cube that is manually designed with a random edge length. A random selection out of all the detected valid objects to apply the rule to is performed every time the rule is applied.

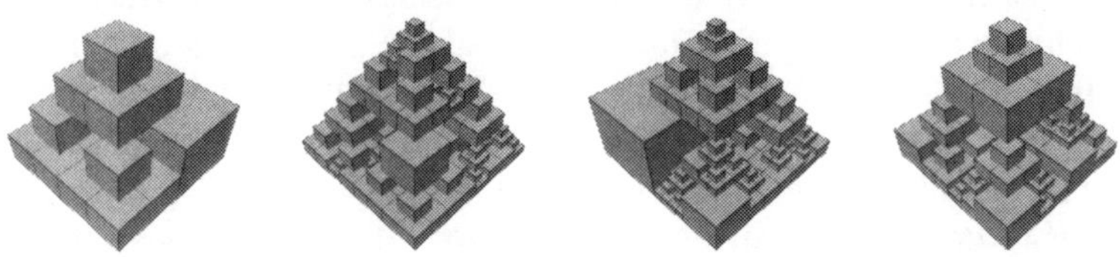

Fig. 9. Different design solutions generated applying the "Sierpinski" grammar

The rule used above is extended defining an additional free parameter that turns the box on top of the other boxes around the z-axis. The range of values that can be assigned to that parameter before applying the rule is restricted to angles between 0° and 60°. During rule application, the angle within this range is selected randomly. Figure 10 shows the adapted rule and one of the resulting designs after applying the rule five times.

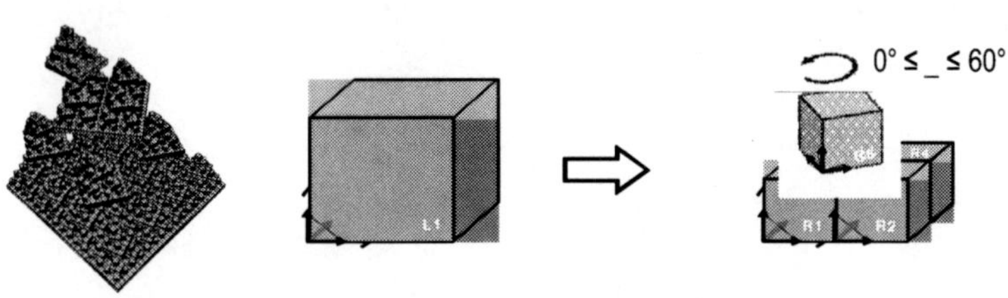

Fig. 10. Design created with the additionally introduced free parameter in the rule

The original "Sierpinski" grammar rule is now redesigned so that three out of the four lower boxes in the RHS are deleted (Figure 11). The remaining box is placed in the global origin and its length and width parameters are set to the length and width of the box in the LHS. The parametric relations are adapted accordingly: $l_{R1} = l_{L1}$, $w_{R1} = w_{L1}$ and $h_{R1} = h_{L1}/2$.

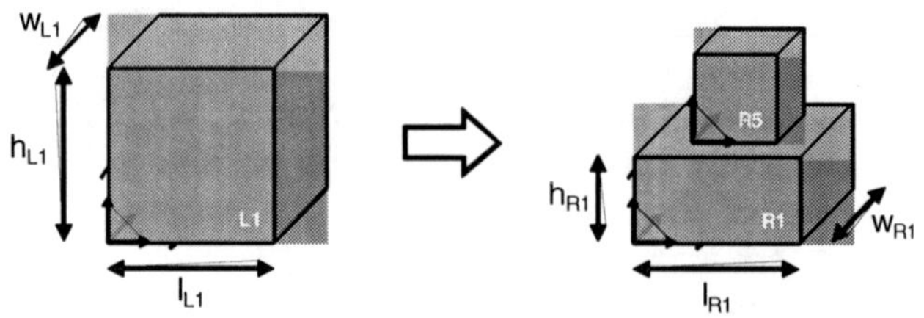

Fig. 11. Modified parametric "Sierpinski" grammar rule

The height relation stays the same and so do all the relations for the free parameters of the box that is located on top. Although, at first glance, the rule looks nearly the same, the resulting design is very different, Figure 12(a). This is not only due to the change of the number of objects and the sizes. It is also important to note that when applying the rule more than once, it now does not apply to all detected boxes. This is due to the parametric relations *width = length* and *height = length* that restrict the matching to cubes. In a further modification of the rule, these parametric relations are removed, so that not only cubes but any kind of box can be detected. The effect on the created design is shown in Figure 12 (b).

The generation of vehicle wheel rims as an engineering example for spatial grammars was shown by the authors earlier [21]. However, previously the rules were implemented as Python scripts. The rim rules are all non-parametric rules. This makes it relatively easy to visually design them and shows the benefit in comparison to the effort that was needed to write the scripts. Figure 13 shows the four rules for the generation of the spokes. The initial objects, felly and hub of the rim, Figure 14(a), are designed manually in this case using standard functionality of the underlying CAD system.

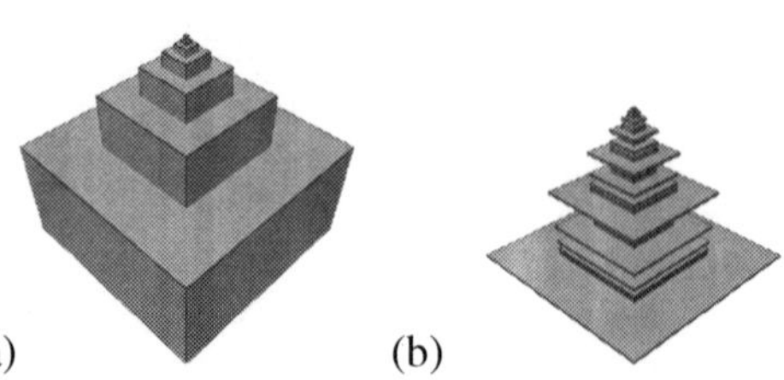

(a) (b)

Fig. 12. (a) and (b) designs generated after modification of the original rule

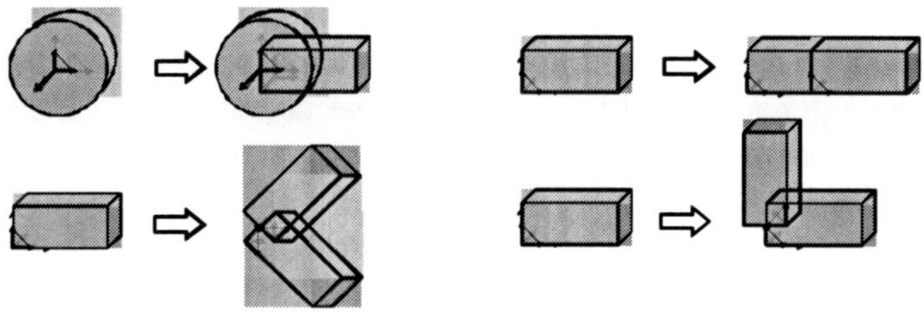

Fig. 13. Rules for the generation of the rim spokes

The rules are applied in a semi-automatic mode. Two examples for rims that were generated are shown in Figure 14. The one in the center is the design that was initially intended to be generated when the rules were defined. The rim on the right is one of the unexpected results.

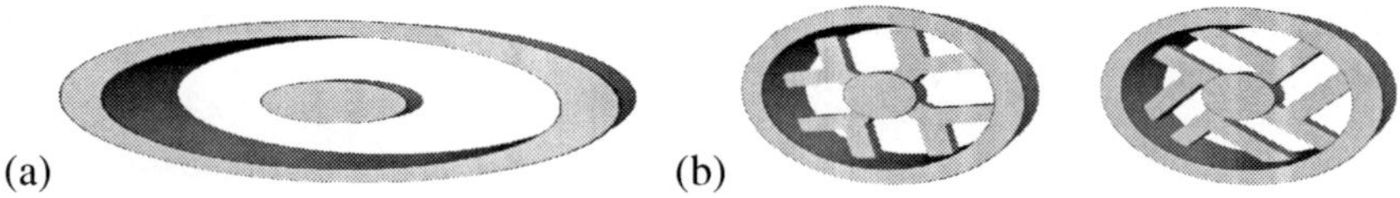

Fig. 14. (a) starting design; (b) two examples for generated rims

Discussion

For increased use and acceptance of spatial grammar systems, the development of interactive, visual grammar interpreters that are designer friendly is crucial. In an intuitive way, they allow designers to define their own shape rules and apply them interactively to generate different design alternatives.

The basic grammar interpreter presented in this paper comprises several aspects of interactive, visual development and application of three-dimensional spatial grammar rules. It provides several advantages over existing 3D implementations, specifically (1) the possibility to visually define 3D rules, both non-parametric and parametric, (2) automatically matching the LHS of a rule to a current design and (3) automatically applying rules including arbitrary rotations and translations in combination with adhering to parametric relations. Further, with regard to engineering design, this work is a step towards supporting engineers in formalizing their knowledge about shape design while they work in a familiar software environment, i.e. a CAD tool. While the approach is currently limited to the use of a set of geometric primitives as the vocabulary, creative and unexpected designs can still be generated.

Several extensions are now possible to increase the expressiveness of rules that can be defined and, thus, potential applications. First, the method needs to be extended to allow more than one object in the LHS of a rule. While this is easily realized for the development of a rule, it creates challenges for the automatic matching of the LHS in the current working shape due to possible interferences between combinations of objects and searching for two or more geometric objects with parametric relations defined among them. Second, enabling the use of Boolean operations in combination with parametric primitives would allow for the definition of more complex geometry. However, dealing with curved geometry, e.g. see [6] for a 2D approach, or freeform surfaces remains an open issue. Third, facilities for defining labels needs to be added since many existing grammars include them, for example, to guide the generation process or deal with symmetries in shapes, e.g. [18]. Finally, the addition of scaling and reflection transformations should be investigated as well as an extension of the approach to deal with cases where multiple transformations apply to an object.

Conclusion

Shape or spatial grammars have been successfully applied in different domains to describe languages of shapes and generate alternative designs. To date, only a few, limited three-dimensional spatial grammar implementations exist. This paper presents an approach for a basic 3D grammar interpreter based on a set of parameterized primitives. It includes the interactive, visual development of three-dimensional spatial grammar rules as well as their automatic application. For the rule development phase, this includes the creation and positioning of geometric objects in 3D space and the definition of non-parametric and parametric spatial grammar rules. For the rule application phase, automatic matching of the left hand side of a rule in a current working shape is carried out, with some restrictions, along with the calculation of the positions and sizes for the objects in the right hand side according to defined parametric relations. The approach is a step towards a more general implementation for the creation and use of 3D spatial grammars within a CAD environment and is expected to provide better support for designers. Future work includes the extension to an arbitrary number of objects in the left hand side of a rule, usage of Boolean operations to create more complex geometry in rules and the introduction of labels within the grammar rules.

Acknowledgements

This research is supported by the Deutsche Forschungsgemeinschaft through the SFB 768 "Zyklenmanagement von Innovationsprozessen".

References

1. Stiny, G., Gips, J.: Shape grammars and the generative specification of painting and sculpture. In: Information Processing, vol. 71, pp. 1460–1465. North-Holland Publishing Company, Amsterdam (1972)
2. Stiny, G.: Introduction to shape and shape grammars. Environment and Planning B: Planning and Design 73, 343–351 (1980)
3. Cagan, J.: Engineering Shape Grammars: Where We Have Been and Where We Are Going. In: Antonsson, E.K., Cagan, J. (eds.) Formal Engineering Design Synthesis, pp. 65–91. Cambridge University Press, Cambridge (2001)
4. Chau, H.H., Chen, X.J., McKay, A., de Pennington, A.: Evaluation of a 3D Shape Grammar Implementation. In: Design Computing and Cognition 2004, pp. 357–376. Kluwer Academic Publishers, Cambridge (2004)
5. Chase, S.C.: A model for user interaction in grammar-based design systems. Automation in Construction 11, 161–172 (2002)
6. Jowers, I., Prats, M., Lim, S., McKay, A., Garner, S., Chase, S.: Supporting Reinterpretation in Computer-Aided Conceptual Design. In: EUROGRAPHICS Workshop on Sketch-Based Interfaces and Modeling, Annecy, France, pp. 151–158 (2008)
7. Trescak, T., Esteva, M., Rodriguez, I.: General Shape Grammar Interpreter for Intelligent Designs Generations. In: Computer Graphics, Imaging and Visualization, CGIV 2009, Tianjin, China, vol. 6, pp. 235–240. IEEE Computer Society, Los Alamitos (2009)
8. Gips, J.: Computer implementation of shape grammars. In: NSF/MIT Workshop on Shape Computation, Cambridge, USA (1999)
9. Krishnamurti, R., Stouffs, R.: Spatial grammars: motivation, comparison, and new results. In: 5th International Conference on Computer-aided Architectural Design Futures, Pittsburgh, USA, pp. 57–74. North-Holland Publishing Co., Amsterdam (1993)
10. Tapia, M.: A visual implementation of a shape grammar system. Environment and Planning B: Planning and Design 261, 59–73 (1999)
11. McGill, M., Knight, T.: Designing Design-Mediating Software: The Development of Shaper2D. In: Proceedings of eCAADe 2004, Copenhagen, Denmark, pp. 119–127 (2004)
12. McCormack, J.P., Cagan, J.: Curve-based shape matching: Supporting designers' hierarchies through parametric shape recognition of arbitrary geometry. Environment and Planning B: Planning and Design 334, 523–540 (2006)

13. Jowers, I., Hogg, D., McKay, A., Chau, H., de Pennington, A.: Shape detection with vision: Implementing shape grammars in conceptual design. In: Research in Engineering Design (SpringerLink Online First: March 27, 2010)
14. Heisserman, J., Callahan, J.: Interactive grammatical design. In: AI in Design 1996, Workshop Notes on Grammatical Design, Stanford, CA (1996)
15. Heisserman, J.: Generative geometric design. Computer Graphics and Applications 142, 37–45 (1994)
16. Piazzalunga, U., Fitzhorn, P.: Note on a three-dimensional shape grammar interpreter. Environment and Planning B: Planning and Design 25, 11–30 (1998)
17. Wong, W.K., Cho, C.T.: A Computational Environment for Learning Basic Shape Grammars. In: International Conference on Computers in Education 2004, Melbourne, pp. 287–292 (2004)
18. Wang, Y., Duarte, J.: Automatic generation and fabrication of designs. Automation in Construction 113, 291–302 (2002)
19. Stiny, G.: Spatial relations and grammars. Environment and Planning B: Planning and Design 9, 313–314 (1982)
20. Stiny, G.: Kindergarten grammars: designing with Froebel's building gifts. Environment and Planning B: Planning and Design 74, 409–462 (1980)
21. Hoisl, F., Shea, K.: Exploring the Integration of Spatial Grammars and Open-Source CAD Systems. In: 17th International Conference on Engineering Design, Stanford University, California, USA, Design Society, vol. 6, pp. 6-427– 426-438 (2009)

A Graph Grammar Based Scheme for Generating and Evaluating Planar Mechanisms

Pradeep Radhakrishnan and Matthew I. Campbell
The University of Texas at Austin, USA

The paper focuses on representing and evaluating planar mechanisms designed using graph grammars. Graph grammars have been used to represent planar mechanisms but there are disadvantages in the methods presently available. This is due to the lack of information in understanding the details of a mechanism represented by the graph since the graphs do not include information about the type of joints and components such as revolute links, prismatic blocks, gears and cams. In order to overcome the drawbacks in the existing methods, a novel representation scheme has been developed. In this method, the authors represent a variety of mechanism types by the use of labels and x, y position information in the nodes. A set of sixteen grammar rules that construct different mechanisms from the basic seed is developed, which implicitly represents a tree of candidate solutions. The scheme is tested to determine its capability in capturing the entire set of feasible planar mechanisms of one degree of freedom. In addition to the representation, another important consideration is the need for an accurate and generalized evaluator for kinematic analysis of the mechanism which, given the lack of information, may not be possible with current design automation schemes. The graph grammar based analysis module is implemented in an existing object-oriented grammar framework and the results have found this to be superior to existing commercial packages.

Introduction – An Overview

The process of designing planar mechanisms usually begins with a clear sense of the mechanisms function which may be in the form of path through space [1]. Cognition of this process is complex since there are a number of variables involved. The typical design process involves

J.S. Gero (ed.): *Design Computing and Cognition'10*, pp. 663–679.

choosing a standard mechanism, customizing the mechanism by adding links, determining kinematic properties using analysis tools and then iterating to determine if the mechanism is able to satisfy the user requirements. In the process, the designer also determines the degrees of freedom, F, using Gruebler's criterion [2], which states

$$F = 3(n - 1) - 2 \times j_1 - j_2$$

where n is the number of links, j_1 refers to number of one degree-of-freedom joints (like pivots and sliding blocks) and j_2 refers to the number of two degree-of-freedom joints (like a pin-in-a-slot) in the mechanism. Since most mechanisms have only a single-degree-of-freedom (F = 1) and are comprised of links and joints, the equation (shown above) is compelling as it easily describes what is, and what is not a valid solution. Other than fairly simple geometry methods like three-position synthesis and its higher-order variations, there are no specific guidelines for the design of a mechanism – just this single equation. As a result the design process is time consuming and iterative. This paper attempts a systematic approach of designing planar mechanisms that augments the traditional mechanism design process.

The steps involved in the research is presented in Figure 1 which begins with the development of a new graph scheme for representing mechanisms followed by the creation of rules that generate the full language of mechanisms. In order to find an ideal solution for a particular problem, a method of generating candidate solution, evaluating their worth, and guiding the process to better solutions is also required. These three steps comprise a search process for finding mechanisms automatically and thereby reducing the design time of creating mechanism by hand.

Defining a representation scheme is one of the most important aspects of the process. Many of the challenges of the remaining tasks such as rules, generation and evaluation are dependent on the representation scheme. The scheme developed in this paper combines different elements commonly associated with planar mechanisms such as links, revolute joints, sliding blocks, pin-in-slots; and has the ability to integrate other elements such as gears, cam-followers and so on. The representation provides the language for the rules that are invoked to generate a mechanism. An important aspect considered at the time of representation and rule formulation was that of the tree-based design generation process wherein successive invocation of the rules generates the different states on the tree. The generation can thus be based on a tree-search technique such as breadth first search, depth first search or other similar techniques [3]. The rules were developed by reviewing traditional mechanism design literature [4-6] wherein links are added to existing structures thereby building a more complex structure. The representation and rules are explained in detail in the section on graph representation.

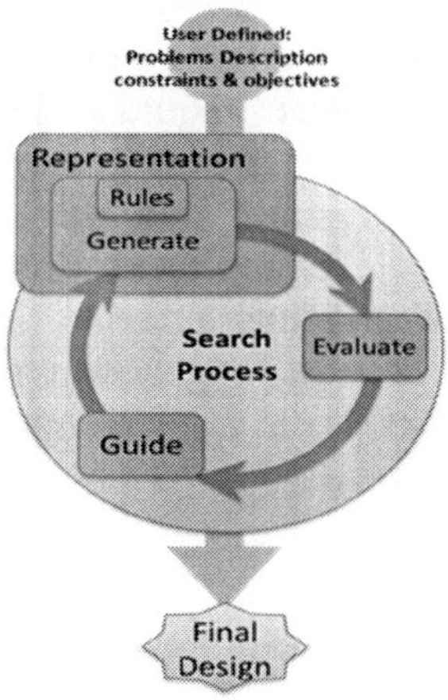

Fig. 1. The five step design automation process

Once the rules have been developed and mechanisms generated, it is important to evaluate the kinematic properties of the generated mechanism. The authors have developed an evaluator that utilizes the instantaneous center of rotation method [1] to determine the kinematics of mechanisms. Further details of the evaluator are available in the evaluation section. After the mechanism is kinematically analyzed, its motion is compared with user specifications and the resulting error (i.e. difference in paths) is stored as an objective function for further optimization. If the error cannot be effectively minimized, the particular design is discarded and another is chosen. Optimization is under active research and is presented only as an overarching principle in this paper.

Background

Automated design of planar mechanisms has been attempted by different researchers and [7-9] are some among them. One of the first graph grammar formalism of planar mechanisms was given by Freundenstein [10] which was further developed by Kota [11] and Tsai [12]. The Systematic method of Tsai was also used by Li and Schmidt [13] in the development of an epicyclic gear train grammar. But in many cases, more importantly with respect to graph grammar, Tsai's representation approach does not convey enough information about the mechanism. Moreover, the representations are not descriptive enough to encompass a variety of different elements within the same framework.

Planar mechanism synthesis requires integration with dynamic simulation tools. However, there are no generic and analytical evaluator modules that could be integrated into the graph grammar approach developed by the authors. Taking these drawbacks into account and building on recent graph grammar projects [14, 15], the authors have developed a more generic scheme for generation and evaluation. The graph transformation method has been implemented within the GraphSynth framework [16].

Graph Representation

The different elements of our planar mechanism representational language are shown using the four-bar mechanism example in Figure 2(a) which is converted to our graph-based schematic in Figure 2(b). The links are represented by rectangular nodes and the pivots are represented by circular nodes on the graph. Pivots and links are connected by arcs to form a complete mechanism.

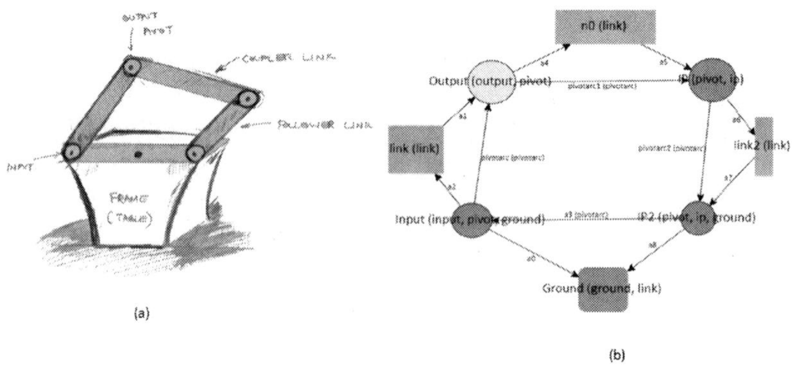

Fig. 2. A four bar mechanism (a) 2D schematic (b) Graph Representation

It may also be noted that there are labels associated with every node (links and pivots) and every arc. The labels indicate the function of the particular node or arc. For instance, a *ground* and *link* label for a node indicates that particular link is connected to ground (a reference frame). Similarly arcs with *pivotarc* label are used to connect adjacent pivots (which are required for manipulation in the evaluation module) and unlabeled arcs associate a pivot with its links that it joins. To illustrate the generality of the technique, a quick return mechanism is depicted in

Figure 3 and the highlight boxes indicate how a pin-in-slot and a sliding block are represented. A detailed list of the labels associated with nodes in the examples is given in Table 1.

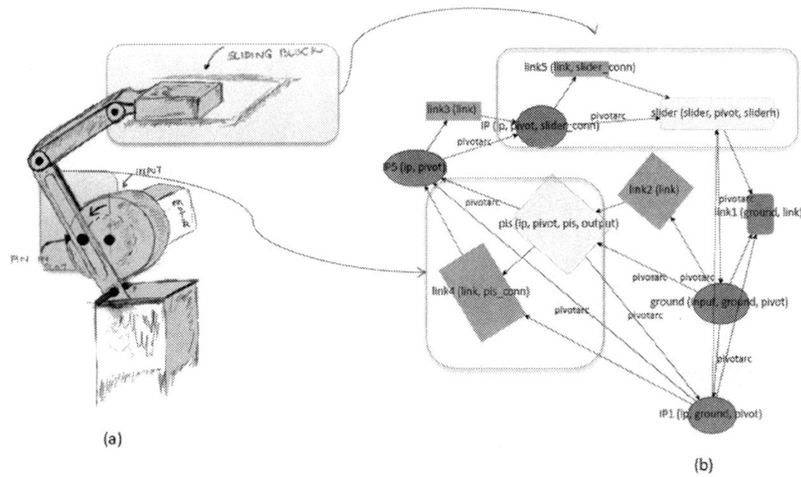

Fig. 3. Quick-return mechanism (pin-in-slot and sliding block are highlighted in box) (a) 2D schematic; (b) Graph representation

Table 1 Labels associated with nodes common to the three mechanisms

Node Name	Type	Labels
Input	pivot	input, link, ground
Output	pivot	output, pivot
IP (connected to binary links)	pivot	pivot, ip
IP (connected to pin-in-slot)	pivot	pivot, ip, pis
IP (connected to sliding block)	pivot	pivot, ip, slider_conn
Sliding Block	pivot	slider, pivot, sliderh (or sliderv)
Ground	link	ground, link
link (binary)	link	Link
link (pin-in-slot)	link	link, pis_conn
link (connected to sliding block)	link	link, slider_conn

From the examples it could be noted that the representation scheme developed by the authors is able to incorporate a wide variety of elements compared to other methodologies on planar mechanisms. This is due to the use of labels in the graph which clearly indicate the roles of the nodes and arcs. Labels are a common vehicle in graph theory to indicate the characteristics of a node, yet in previous efforts that utilize Tsai's systematic approach few or no labels are used and the physical description of the mechanism is determined only by the connectivity of nodes and arcs. Furthermore, that approach is limited to only revolute joints and cannot handle prismatic or cam-like joints. This necessitates greater bookmarking to indicate the type of mechanism or element the graph represents whereas the method proposed here eliminates that need.

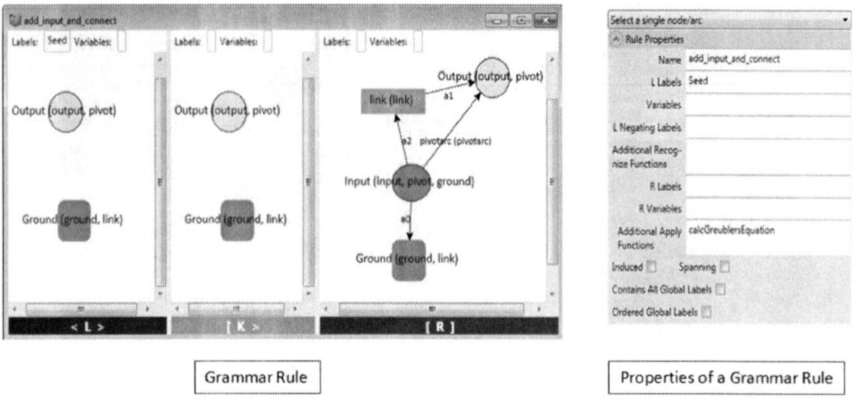

Fig. 4. A grammar rule shown with its properties window

Another unique feature is the position information (x and y coordinates) that permits manipulation of a given topology to optimize its dimensions to better meet user specifications. This is because planar mechanisms are spatial in nature. It is more realistic to include such information in the graph representation rather than simply nodes and arcs with no spatial information.

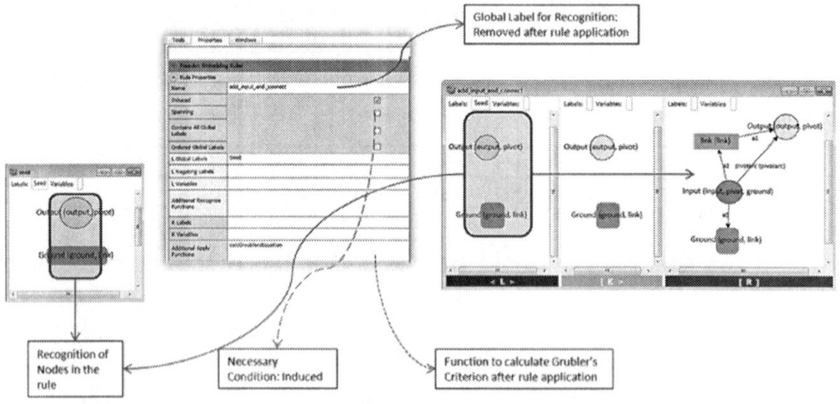

Fig. 5. Illustration of the rule recognition and application process

Rules for Design Generation

Now that the representation scheme is developed, rules are formulated to generate mechanism designs. A typical grammar rule developed is shown in Figure 4, wherein the left-hand side serves as the recognition section, the middle section consists of common elements between the left and right hand sides and the right-hand side is the desired outcome of applying the rule. In addition to nodes and arcs, rules can contain global labels that could be used to control the recognition and transformation of the candidate. Figure 5 shows an example of the rule recognition and application process along with associated properties. The rule in the figure is applied to the seed graph. The different properties for that particular rule are checked on the candidate to determine if it is satisfied. Some of the common Booleans that require being satisfied are "contains all global labels", "induced" and "spanning". It should be noted that not all rules require these Booleans to be satisfied. Further details on the variables are available in [17,18]. Once these Booleans are satisfied and the left-hand side matches with the particular candidate, the right-hand side transformation is applied to the candidate. There could be just one or multiple locations where the rule might be applicable on the particular candidate. After application of the rule, there is an additional function that computes the degree of freedom of the device based on Gruebler's equation. This is important since it aids in segregating mechanism designs depending on the degree of freedom.

The developed rules are grouped into a rule-set and called during the generation process. There are sixteen rules developed that result in full language of mechanisms. While it is possible to have more rules and more rule-sets, a small and manageable set is desirable and leads to a targeted space of solutions. For the automation of planar mechanisms illustrated by the authors, there is only one rule-set. Figure 6 show one rule that converts a link with three pivots into one with four pivots. In addition to the sixteen rules, five more rules have been created which transform mechanisms between common topological variations created using the original sixteen rules. For instance, a horizontal sliding block can be transformed to a vertical sliding block during optimization. The complete set of rules is available at http://www.graphsynth.com/mechsynth.

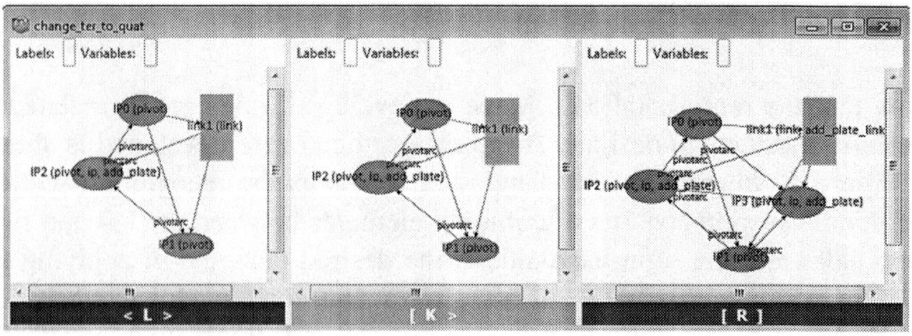

Fig. 6. A grammar rule that converts a 3-pivot link to a 4-pivot link

The generation process is initiated from a single starting point called the seed. The seed used by the authors for the design automation project is shown in Figure 7.

Fig. 7. Seed in the generation process

It consists of the ground link and an output pivot node that is used in determining the path trajectory provided by the user. The ground and output nodes are very essential in every candidate design and to the evaluation process. Some of the candidates that are generated using the rules are shown in Figures 8 and 9.

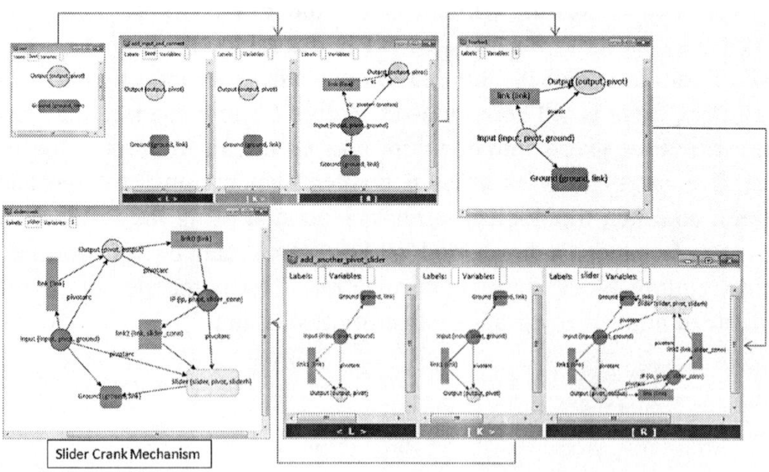

Fig. 8. Steps involved in the generation of a slider crank mechanism

Figure 8 demonstrates the creation of a slider crank mechanism [2] and Figure 9 shows a Watt–I mechanism [2] represented in our approach. In all the figures, the pivots are represented by elliptical nodes and links by rectangular nodes. The method is also capable of generating special linkages such as the double butterfly linkage [19]. The slider crank mechanism in Figure 8 is generated from two rules wherein the first rule connects the input link to the seed and the second adds the sliding block to the input link.

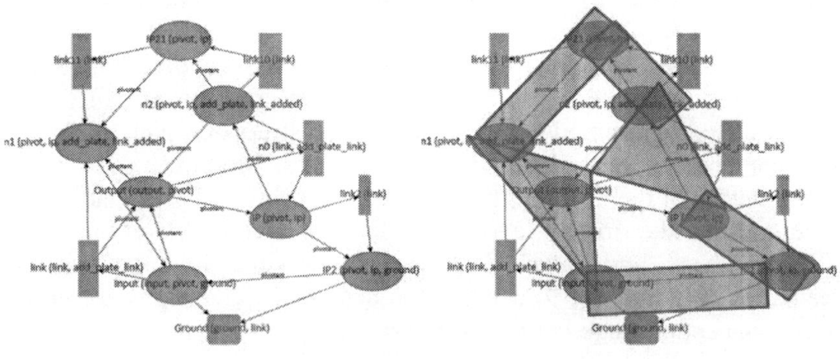

Fig. 9. Watt–I 6-bar mechanism

Evaluation

Evaluation is an important function since every generated mechanism design requires evaluation of its kinematic properties to conform to user specifications. Analytical kinematic analysis for four-bar mechanisms [1] and computational tools such as SAM [20], WATT [21], Working Model [22] and ADAMS [23] extend the functionality to *n*-bar mechanisms (*n*-may vary depending on the application). But none of the existing software tools other than WATT have a mechanism generator function and even in WATT, it is limited to fixed-topology six-bar mechanisms. In order to fulfill the requirement of integrating an analytical kinematic analysis, it was decided to adapt the instantaneous center of rotation method since 1) our focus is limited to single degree of freedom (DOF) systems, 2) the results are analytically accurate (no numerical approximations are involved), and 3) the algorithmic nature of the technique is amenable to computational programming. Despite these advantages, the method does not appear in the literature as an approach to computationally analyze planar mechanisms, and the authors are currently publishing this method in appropriate publications [24]. The inputs to the evaluation module are (a) candidate from the generation process and (b) user specifications such as angular velocity of input link and desired path. A flowchart for the kinematic analysis program is illustrated in Figure 10.

As the candidate is input to the evaluation module, the different parameters such as link lengths and positional information are recorded. Next, the instant centers are determined from where the velocities of the links and pivots are determined. Once velocities are determined, accelerations of the respective links and pivots are computed by forming equations and solving using a matrix reduction method. After velocities and accelerations are determined, new positions are determined using geometrical intersection techniques. The instantaneous method is innovatively programmed using different loop structures such as do-while loop that enables generalization of the process. Moreover the objective representation of the different links and pivots helps in easier manipulation in computing the instant centers, velocities and accelerations of the different pivots and links. The analysis module is concerned with kinematics only and not forces and torques since the designs are targeted for the conceptual stage in the design process.

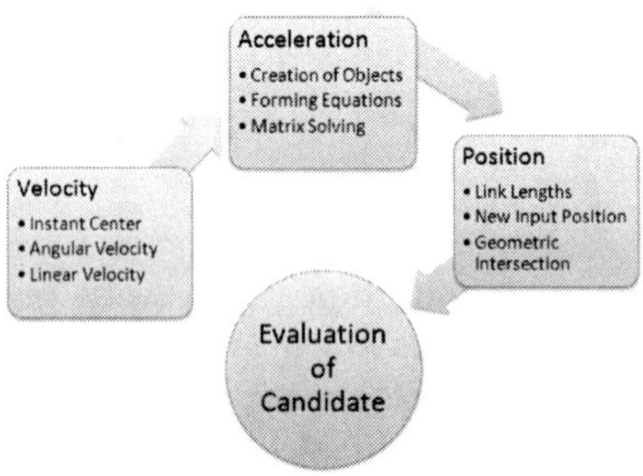

Fig. 10. Evaluation methodology adopted in the design automation process

The capabilities of the evaluator are shown in Figure 11, which displays the position, velocity and acceleration of the different pivots in the four-bar mechanism with a three-pivot coupler.

In order to test the accuracy of the computed values, a four-bar mechanism, Figure 11, is analyzed. This simple mechanism provides a good reference since analytical equations exist for the motion and thus the path can be determined with no numerical approximations. The baseline analytical result is compared with our method along with the results from

Working Model and SAM. Figure 12 (left side) shows how the ratio of the coupler link's length to the actual link length (a link is rigid, and should not change length throughout the simulation) changes with time. At the same time, Working model and our method were tested on a Watt-II mechanism and the variation in the link lengths between the two methods is shown on the right side of Figure 12. The figure clearly shows that our method is consistent with no variation in the link lengths. Though accuracy is important, equally important is computational time and our method was on the same order as the other commercial tools.

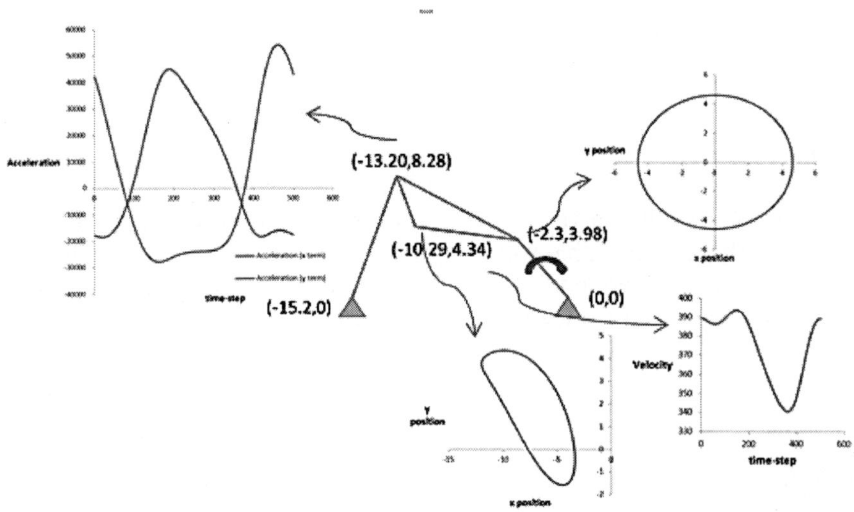

Fig. 11. Velocity and acceleration profile of a four-bar mechanism

Results: Rule Validity

In order to show that the 16 rules are valid, it is necessary to take a set theory approach. Consider the abstract set of all valid planar mechanisms, V. For the set of all candidates created through a rule-set, R; we gauge the validity of R by determining the amount of intersection with V, Figure 13(a). There are several approaches that could be taken to measure this. For example, one could directly take the intersection of V and R (V∩R) only if V is well understood. However, in many design problems the valid space, V, cannot be enumerated, but one can easily check when a solution is within the set.

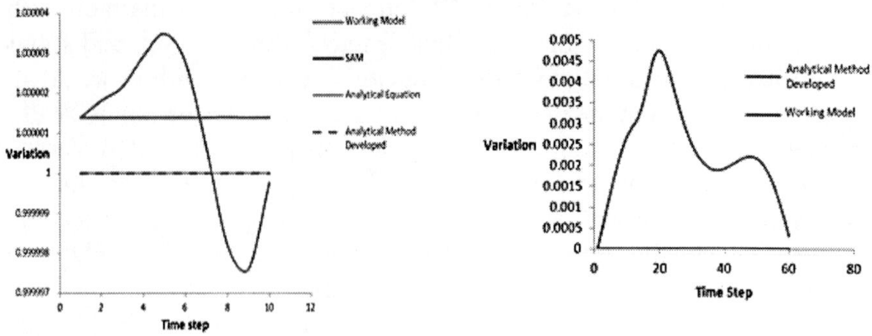

Fig. 12. Performance of the evaluation module with respect to commercial tools – Ratio of actual to desired link length variation

Our approach is to see if R encapsulates all the elements of V (V⊂ R; Figure 13(b) and then check whether there are any solutions in R that are not in V (R – V; Figure 13(c). If the latter is the null set (R – V = { }), then we prove that no infeasible solutions are found, and taken with the former, that the two spaces are equivalent. This is challenging in comparing graph languages that result from a set of grammar rules since graphs are difficult to compare [25] and the search tress are often large and intractable.

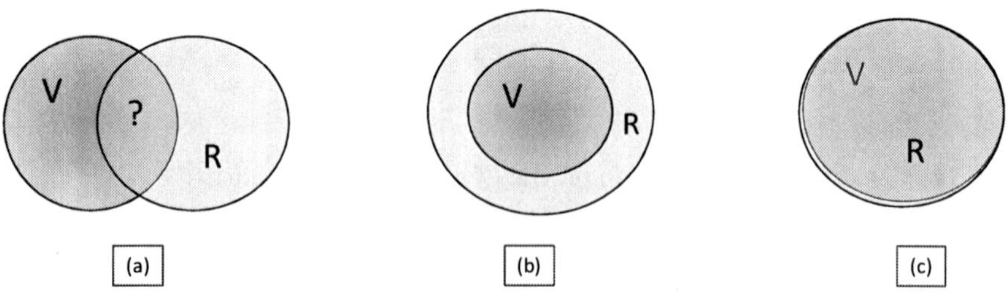

Fig. 13. Rule validation (a) the intersection of the valid set, V, with the set created by rules, R, is desired; (b) through a depth-limited depth-first search of the tree, it is found that R encompasses all of V; (c) additionally, no solutions are found in R that violate Gruebler's equation, hence the difference is minimal

A review of the literature [2] presents unique variations for different n-bar mechanisms of 1-DOF. Since the list is compiled considering only revolute joints without assignment of ground and input links, it is not appropriate to compare our capability on a one-to-one basis. Nevertheless, to test the validity of the rules we conducted a depth-first search of the tree (limited to a certain level) and found that the rules are able to generate all of the solutions shown in the literature. This is the first step at proving that R includes all of the solutions found in V for 1-DOF revolute joint mechanisms. In order to show that R does not include any infeasible solutions, each solution found in the tree-search is checked against the Gruebler's equation and none had a calculated degree-of-freedom with a value other than 1. Therefore, we are confident that our method is able to create a set R that completely encompasses the solutions found in V.

Discussion

The graph grammar based representation and evaluation schemes presented in this paper take advantage of the advances in object oriented programming in order to create a generalized and automated designer of planar mechanisms. The method presented is unique since it mimics the human way of iterative design wherein a chosen design is analyzed for its utility and then improved upon to suit the requirements of the designer. If the design does not conform to the specifications, it is discarded and another is chosen. Representing mechanism elements such as links and pivots as nodes and their connections as arcs presents an alternative approach when compared to existing methods like the Systematic Method [12]. The different labels associated with the links and pivots and their coordinate information are able to convey the type of mechanism represented by the graph without requiring any additional arbitration. Thereby any design generated can be easily visualized by the user who can continue with the post-processing activities using the generated design. The rules that are concerned with the design generation are few and are capable of generating varied designs as illustrated earlier. The rules also consist of several graph-centered Booleans that are used to specify when a rule is valid and this is crucial in carefully controlling the space of generated mechanisms. This space of generated candidates is compared to the other approaches in the literature that attempt to enumerate the space of planar mechanisms (as in the previous section). This is the first step towards proving the correctness of the rules and the utility of method.

In addition to gauging the completeness of the rules in creating the entire language of planar mechanisms, there are two additional concerns that the grammar presents: isomorphism and confluence. Isomorphism refers to the structural equivalence of topologies and Mruthyunjaya [26] presents a review of the different methodologies presented by different authors to overcome isomorphism. The sixteen rules created by the authors overcome some of the problems of structural equivalence through the rich sets of labels, the various recognitions Booleans, and the arc directions. Furthermore, finding isomorphically equivalent mechanisms does not necessarily mean the mechanisms have the same usable motion. Figure 14 shows a mechanism with the path followed by two pivots – one between input and coupler links and the other on the coupler link.

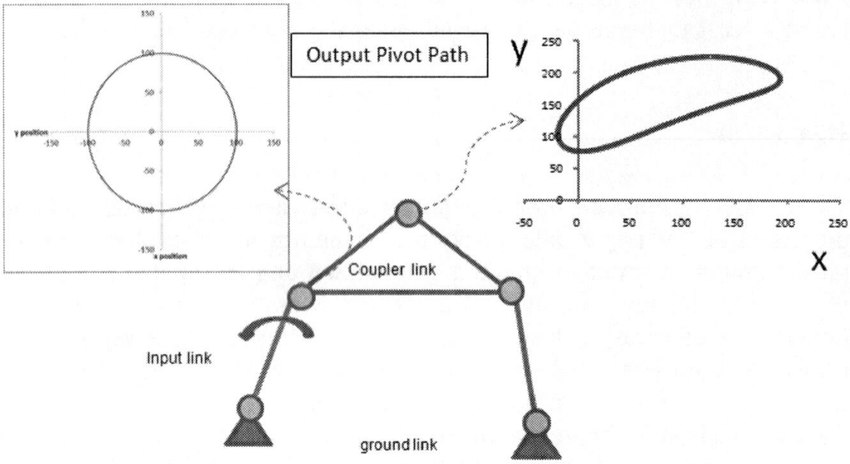

Fig. 14. Same topology but different output characteristic

If we consider the preferred output to be on the coupler link, then the path would be different from the mechanism whose output is located between the input and the coupler link. Therefore it is not desirable from a design perspective to prevent the creation of isomorphically equivalent mechanisms. The use in identifying isomorphically equivalent mechanisms resides only in categorizing and potentially simplifying the dynamic analysis. The kinematic analysis presented in the evaluation section and described in detail in [24] could be spared if a generic formulation of the kinematics could be prescribed for each unique isomorphism. However, this is, in itself, a large if not daunting research task. The dynamic analysis developed here is in fact very quick and accurate, and the savings in detecting isomorphically equivalent solutions would thus not be practical or significant.

The issue of confluence arises from the tree-based nature of the search space. Confluence between any two grammar rules means that the rules are operating on different sections of a host graph and as a result, they are independent of one another – meaning that they can be called in parallel, or in series (one before the other) and the resulting candidate will be the same. The reason this poses a challenge for efforts seeking to create a definitive language of solutions, is that different paths through the tree will arrive at the same solution. It is not always clear when such situations arise, and as a result the same solution (or topology) will be counted more than once. In the rules presented here, a four-bar mechanism may be found following multiple paths, Figure 15, some of which are at different levels of the tree. Knowledge of the recipe that creates these confluent states is important since it indicates the different ways of attaining a particular mechanism computationally. This information may be used to create better rules and is a separate area of research.

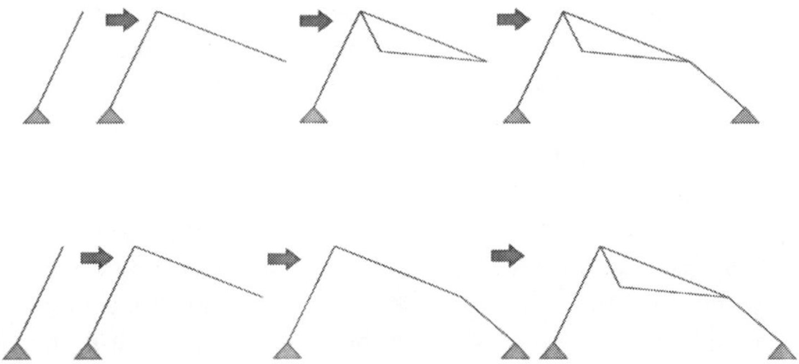

Fig. 15. Confluence – a four-bar mechanism shown with different design recipe

The evaluation module is very unique since it is based on the resulting candidates from the language prescribed here. The evaluation developed is generic and completely analytical and is equivalent or superior in accuracy to commercially available packages such as SAM and Working Model. The evaluation is also computationally efficient. While such claims seem "too good to be true", this is made possible by taking advantage of kinematic constraints not often leveraged in commercial software. This constraint limits our evaluation to single degree-of-freedom mechanisms containing four-bar chains. While the authors are currently working to include analysis of five-bar chains, such mechanisms are only theoretically derived and no practical device falls within this category. The creation of a valid representation (16 grammar rules) and the evaluation show promise for the future automated synthesis of planar mechanisms.

Conclusion

Replicating the human design and decision making process using computers is complex due to a variety of designs and variables associated with them. An attempt is made by the authors in this paper to address this issue in the automated design of planar mechanisms. Planar mechanisms are an interesting design problem to address since there is a complex (and previously unknown) set of rules that govern the creation of valid configuration. Yet, there is a simple equation which can be invoked to indicate whether a mechanism is a valid. One of the key innovations in the work is that the various elements of a planar mechanism have been represented in a completely different way compared to existing methods, where nodes represent both pivots and links and arcs simply connect between them. This graph scheme aids in the creation of graph-grammar rules, and replicate the typical manner of designing a mechanism. As the human way of designing is iterative, a tree of solutions is generated from which topologies of desired degree of freedom are evaluated to conform to user specifications. In addition to representation, there is also an evaluation module built-in that can operate on different topologies to determine the kinematics of the generated planar mechanism. The representation and evaluation schemes operate not only on rotary links but also on prismatic objects, which is another unique feature of the scheme. Search, optimization and guidance are still under active research and we hope that once they are integrated along with many other machine elements, the method would be able to provide new directions to automating mechanical design and understanding the cognitive process of mechanical design.

References

1. Waldron, K.J., Kinzel, G.L.: Kinematics, dynamics, and design of machinery. Wiley, Chichester (2004)
2. Norton, R.L.: Design of machinery. McGraw-Hill Professional, New York (2004)
3. Cormen, T.H.: Introduction to algorithms. MIT Press, Cambridge (2001)
4. Erdman, A.G., Sandor, G.N.: Mechanism design. Prentice-Hall, Englewood Cliffs (1997)
5. Myszka, D.H.: Machines and mechanisms. Prentice-Hall, Englewood Cliffs (2002)
6. Sclater, N., Chironis, N.P.: Mechanisms and mechanical devices sourcebook. McGraw-Hill, New York (2001)
7. Cabrera, J.A., Simon, A., Prado, M.: Optimal synthesis of mechanisms with genetic algorithms. Mechanism and Machine Theory 37, 1165–1177 (2002)
8. Martin, P.J., Russell, K., Sodhi, R.S.: On mechanism design optimization for motion generation. Mechanism and Machine Theory 42, 1251–1263 (2007)
9. Smaili, A.A., Diab, N.A., Atallah, N.A.: Optimum Synthesis of Mechanisms Using Tabu-Gradient Search Algorithm. J. of Mechanical Design 127, 917 (2005)
10. Freudenstein, F., Maki, E.R.: The creation of mechanisms according to kinematic structure and function. Environment and Planning B: Planning and Design 6, 375–391 (1979)

11. Kota, S., Chiou, S.J.: Conceptual design of mechanisms based on computational synthesis and simulation of kinematic building blocks. Research in Engineering Design 4, 75–87 (1992)
12. Tsai, L.: Mechanism design. CRC Press, Boca Raton (2001)
13. Li, X., Schmidt, L.: Grammar-Based Designer Assistance Tool for Epicyclic Gear Trains. Journal of Mechanical Design 126, 895 (2004)
14. Patel, J., Campbell, M.I.: Automated Synthesis of Sheet Metal Parts by Optimizing a Fabrication Based Graph Topology. In: 1st AIAA Multidisciplinary Design Optimization Specialist Conference, pp. 18–21 (2005)
15. Swantner, A., Campbell, M.: Automated Synthesis and Optimization of Gear Train Topologies. In: ASME Design Engineering Technical Conference. ASME, San Diego (2009)
16. Official Website for GraphSynth - UT Austin - Automated Design Lab `http://www.me.utexas.edu/~adl/graphsynth` (Last accessed May 2010)
17. Campbell, M.I., Nair, S., Patel, J.: A Unified Approach to Solving Graph Based Design Problems. In: 19th International Conference on Design Theory and Methodology; 1st International Conference on Micro- and Nanosystems; and 9th International Conference on Advanced Vehicle Tire Technologies, Parts A and B, Las Vegas, Nevada, USA, vol. 3, pp. 523–535 (2007)
18. Campbell, M.: A Graph Grammar Methodology for Generative Systems (2009)
19. Foster, D.E., Pennock, G.R.: A Graphical Method to Find the Secondary Instantaneous Centers of Zero Velocity for the Double Butterfly Linkage. Journal of Mechanical Design 125, 268 (2003)
20. ARTAS - Engineering Software
21. WATT Mechanism Suite, `http://www.heron-technologies.com/watt` (Last accessed May 2010)
22. Working Model 2D - Home, `http://www.design-simulation.com/WM2D/index.php` (Last accessed May 2010)
23. Adams - Overview, `http://www.mscsoftware.com/products/adams.cfm` (Last accessed May 2010)
24. Radhakrishnan, P., Campbell, M.: A completely analytical and implement kinematic analysis of planar mechanisms. Submitted to the Journal of Mechanisms and Robotics (to appear)
25. Arvind, V., Kurur, P.P.: Graph Isomorphism is in SPP. Information and Computation 204, 835–852 (2006)
26. Mruthyunjaya, T.S.: Kinematic structure of mechanisms revisited. Mechanism and Machine Theory 38, 279–320 (2003)

A Case Study of Script-Based Techniques in Urban Planning

Anastasia Koltsova[1], Gerhard Schmitt[1], Patrik Schumacher[2], Tomoyuki Sudo[2], Shipra Narang[2], and Lin Chen[2]
[1] *ETH Zurich, Switzerland*
[2] *AA School of Architecture, UK*

This paper introduces the use of parametric design tools in the domain of large-scale urban planning. The 253-hectare site in Moscow, Russia, was selected as the study case of the methodology. Through the exploration of a differentiated urban order functioning within a framework of overall coherence the site is interpreted as an informational data field. In the particular example, a data field represents a set of points distributed on the site. A manipulative set of input parameters is derived from the contextual forces around the site – such as 1950's socialist housing, the new urban Moscow City and the Moscow river - while output variables incorporate a distance value for each important point on the site to those of the surrounding elements. Using script-based techniques, values are translated into the urban and formal responses of building typology, height, connectivity and directionality. These data-holding elements then cumulatively outline the pattern and grain of the site. The use of multiple transformative building typologies creates an urban tapestry bringing about diversification within the urban field, whereas the negotiation and continual shift between singular and plural building masses becomes the premise of the architectural condition. Hence, the design process focuses on the formal resolution of the multiple juxtaposed patterns that emerge from the site. Diversity within the field is amplified at the component scale where dynamic building sub systems exhibit a spatial flexibility and changing image idea.

Introduction

The basics of parametric design approach can be found in the work of Otto F. with material machines, which Spuybroek L. describes in 'The

J.S. Gero (ed.): *Design Computing and Cognition'10*, pp. 681–700.

Structure of Vagueness' [1]. His "optimized path systems" illustrates how through multiple interactions between elements over time a system can reorganize. In 1962-1964 Archigram proposed an idea of a city that can reorganize and adapt to fit changing conditions in their 'Plug-In City' project and 'Computer City' project [2]. The 'Plug-In City' was a new type of city that had been designed and structured for change. It was a new solution for existing urban design problems such as the rapid city expansion, population growth etc. The 'Computer City' project proposed a model for a computational system that could sense requirements of the city and make the structure of the city respond to the requirements.

The rapid development of computer technologies brought new tools for design and simulation of responsive systems, which are available today in the form of time-based and code-based software such as Rhinoceros 3D, Maya, 3DS, Max, Digital Project. Many architectural offices use these tools to support the design process. Schumacher P. claims parametric architecture to be a new architectural style of today in the "Parametricism - A New Global Style for Architecture and Urban Design"[3]. The 'Kartal-Pendik Masterplan', Istanbul, Turkey, designed by Zaha Hadid Architects demonstrates the application of parametric design tools for urban design [4].

Diverse strategies for radical urban development and transformation have been investigated during the research. With investigations into associative parametric design systems we look at buildings as components that can be constructed from multiple sub-systems such as a system of internal subdivision (floors, walls), a structural system (primary skeleton), a navigation system (void with primary circulation/orientation), by associative relations. The crux of the use of parametric systems is to bring these tectonic sub-systems of the building, under the intention of the global urban differentiation. This means that the local architectural features will work towards the amplification for the urban vectors and thus facilitate orientation via an unprecedented level of overall associative integration. Architectural tectonics produces urban field effects and vice versa.

Significance

Embedding intelligence into the automatic scripting process would allow the creation of new types of urban spatial organizations and formations differentiated in response to contextual forces. The output geometry can be tailored to the diverse contingencies of various locations. This empowers the computational process by not only allowing it to manipulate the inputs for data generation but also trigger unprecedented responses. Once setup, the system can be used to catalogue multiple design variations in a short period of time. By altering the input parameters it is possible to use it at the architectural as well as at the urban scale.

Overview

This paper is structured as follows: the first part explains the development of computational tools; the next part illustrates the application of these tools at the urban scale; the last part shows the development of the computational system at the building level; the conclusion provides the image of the final master plan version and summarizes the research results and future work.

Computational Tools

A series of computational tools using Rhinoscript platform has been developed during the research [5]. This catalogue of computational tools uses distance as a driver to achieve differentiation in a given field. The base setup includes a grid of points and an influencing element. Every point in a point field can be described as a numerical value. For example, by converting each point into a value equivalent to its coordinates X, Y and Z, each point attains a "data" set that can be distinguished and classified within a data field. The design research focuses on the translation of local conditions of the site into data fields, which in turn become drivers for formal responses on the site allowing us to achieve differentiation within the framework of an overall coherent order.

The computational catalogue on Figures 1-4 demonstrates the principle of taking distance from an influencing element to generate a series of responses. The series originates from a basic point set outlining three conditions; completely random set of points, a grid of points and an equitable combination of the two, Figure 1. These effects are utilized in combination to create specific patterns on the field. The variety offered by the computational tools includes shifting the position of points on the grid, creating connections with its neighbouring points, Figure3, scaling of elements and rotation all in proportion with the distance value of each point from the influencing element, Figure 2. Voronoi cell packing is used as a method for space subdivision, Figure 4. The strength of the system emerges with a layering of multiple effects in correlation to one another.

These tools are applied to the site where the influencing elements become the immediate context of the site.

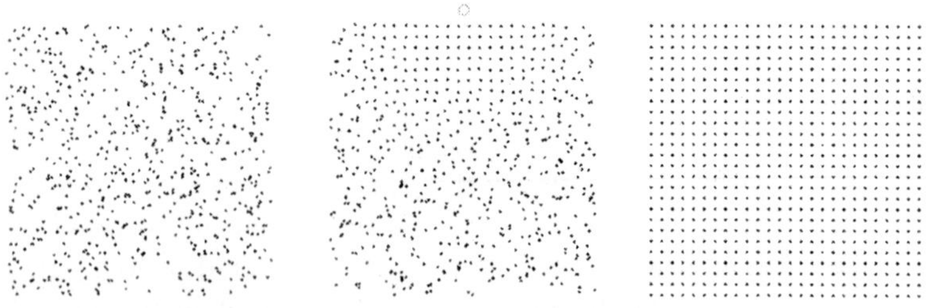

Fig. 1. Computational tools. Point set manipulation

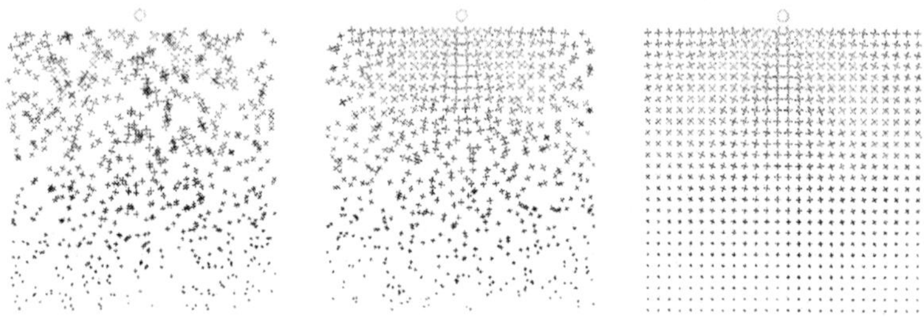

Fig. 2. Computational tools. Rotation and scaling of a component placed on points

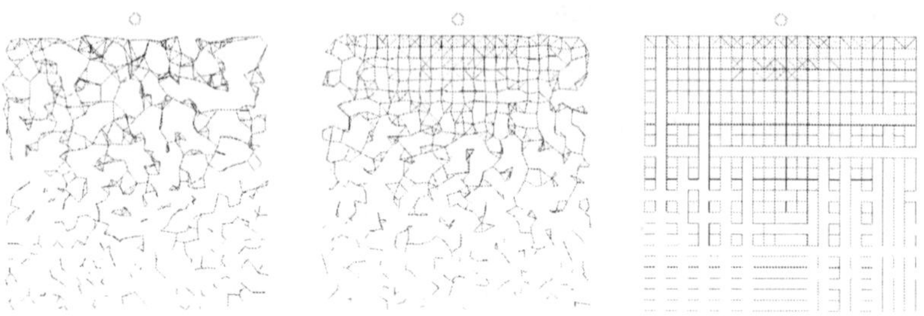

Fig. 3. Computational tools. Point connectivity pattern variation

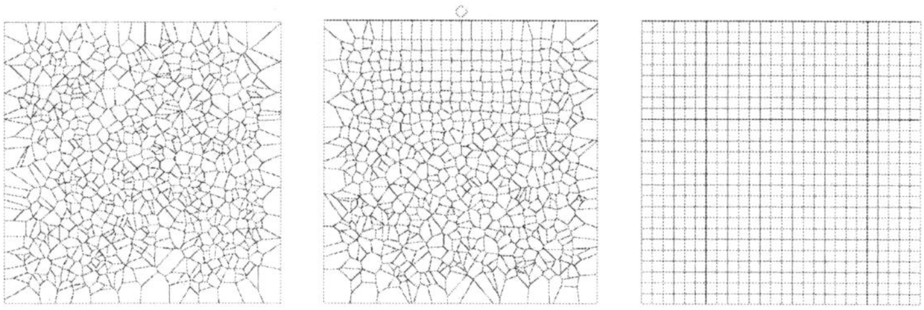

Fig. 4. Computational tools. Voronoi cell packing based on the point grid

Figure 5 demonstrates a first attempt to apply computational tools on the site. It shows a layering of computational tools in correlation to one another with the river as a primary influencing element.

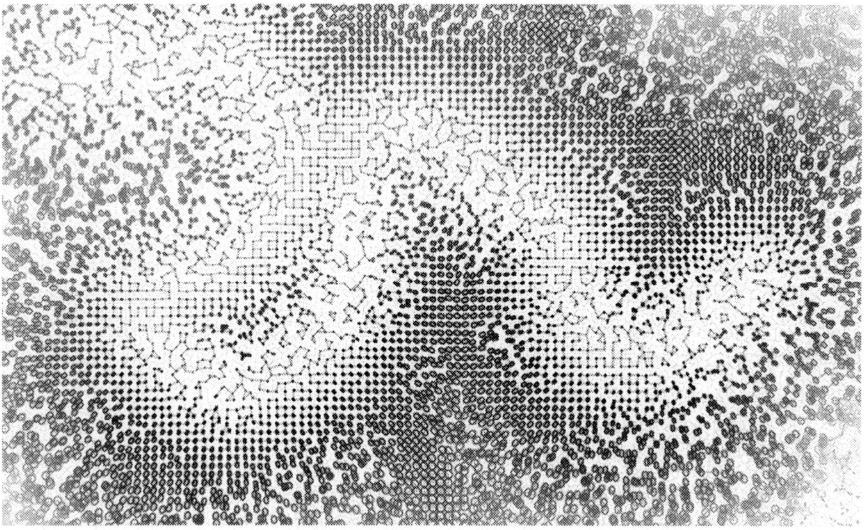

Fig. 5. Test of computational tools on the site

Application of Computational Tools at Urban Scale

In order to define the manipulative set of input parameters (influencing elements) for the system at urban scale it was crucial to analyze conditions around the site: building typologies, infrastructure, direction of urban growth and others.

The centre of Moscow is one of the most differentiated parts of the city. However the stamp of planned uniformity becomes evident as one moves towards the urban fringe. With the majority of the development being in the form of socialist housing blocks clustered around large industrial sites the urban fabric is a lot more simplistic as compared to the city centre. The project site is located on the threshold of these two distinct zones of the city. The 235-hectare industrial site, stretching between two river frontages, is now in the process of abandonment like many others around the city. This large hub of industries sustained itself on the housing blocks in the vicinity. The centrality of the site with respect to the socialist housing and a new business Moscow City district should be preserved and further enhanced.

The foundations of our proximetric tool rest on the features of the site that would eventually lend it character. Further the Moscow River, the socialist housing blocks and the Moscow City development are outlined as our primary influencing elements. These elements define the pattern of proliferation on site and act as initiators of subsequent layers of information. The data driven patterns of the field are seen as a series of informed layers that use similar influencing elements around the site but vary in magnitude and response (shifting of position, rotation and scaling). Ultimately this layering of information comes together to define an urban system that generates a complex entity.

In order to test out the computational tool system on the site this initial setup of a point grid that sketches out the current road network on site is created (Figure 6,a). The road network serves as a connector between primary influencing elements around the site such as housing blocks, Moscow River and Moscow city district. Therefore it dictates the initial condition for point distribution.

The series of differentiations that we aimed at begin by disturbing the original grid by shifting the position of points is shown on Figure 6(b). The distance value of each point from the closest point on the curve (outlining the original road layout) defineses the degree by which the point shifts from its original location. It is observed that the points closer to the road maintain their positions (min. shifting) whereas, as one moves away the pattern becomes more random (max. shifting).

Shifting focus from the road network to the river frontages on site; Figure 6(c) explores the concept of orientation on the site. The first effect observed here is rotation whereby every point is replaced by a component that rotates in a manner to orient towards the river whereas; the scaling of these components explains the degree of rotation they undergo. Since the river forms an important part of the site rotation of elements helps provide directionality to the master plan.

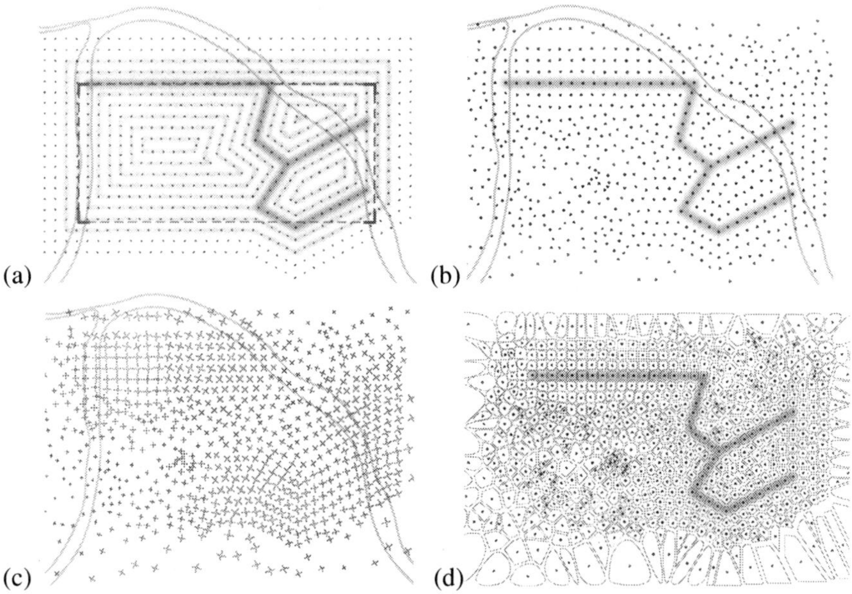

Fig. 6. (a) Initial setup of point grid; (b) Distorted point grid; (c) Scaling and rotation of components; (d) Voronoi cell packing

In order to negotiate the points into larger urban context we divide the site into cells based on the location of points. From the point set a packing of Voronoi cells is developed, where the density of the point reflects the size of each cell, Figure 6(d).

From the pattern of individual cells a system of connectivity is developed that establishes a relationship between the cells, Figure 7. In this case, each point connects with *n* number of nearest points according to the distance to the influencing elements around the site. The pattern is informed by the Moscow City development for that reason points become more connected towards the development. The number of connections is affected by the metro stations and the road network around the site. The number of connections decreases as one moves away from these infrastructure nodes. This map can be read as an accessibility mapping for the site.

Figure 8 explores the condition on site yielding a uniform connectivity across the site. However, this catalogue was developed with an aim to break the monotony; by introducing primary roads within the site, generating a more varied and diverse connectivity pattern.

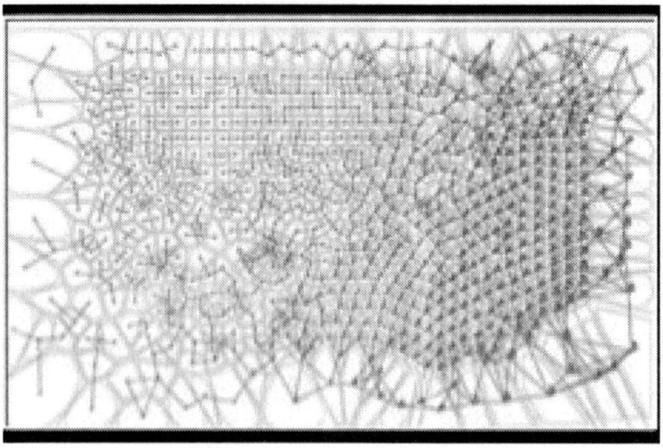

Fig. 7. Connectivity pattern establishes relation between Voronoi cells

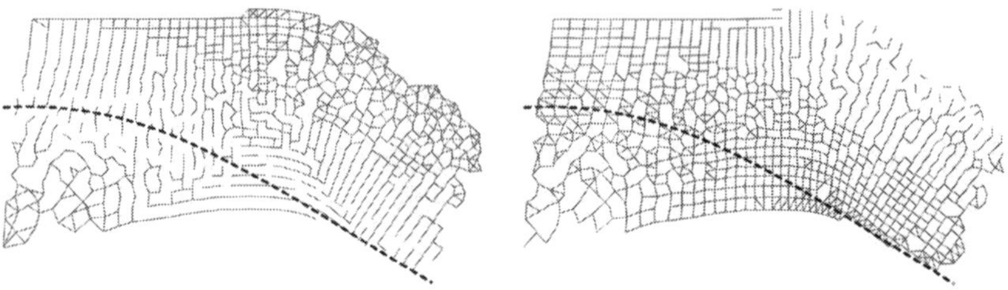

Fig. 8. Diversification of the connectivity pattern

We generate a data set that dictates the height information on the site, Figure 9. This data set is created by taking into consideration all the previous information sets and additional proximetric data from the influencing elements within the site such as proximity to the inner road network, Moscow City and housing blocks.

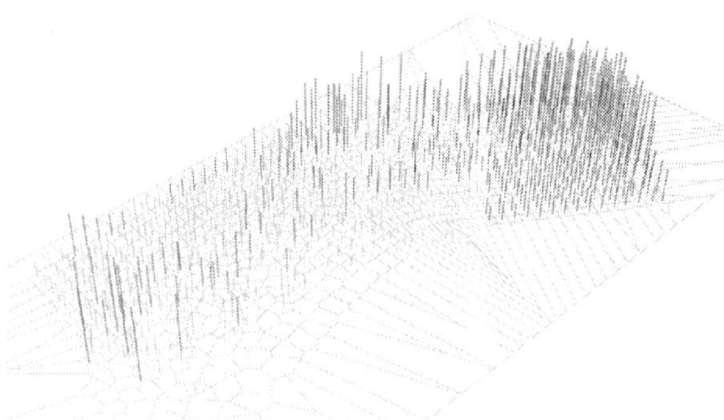

Fig. 9. Height information on the site

The studies described above suggest that the point set becomes a measuring tool whereas the cell gives this information a spatial connotation. Delving further into the anatomy of each cell it is observed that the cells differ from each other in terms of their area, number of sides and the degree of connectivity, Figure 10. Based on these elements a data set is derived for every cell leading to the formation of categories that define the different urban conditions such as open public spaces, tower typologies, low-rise typology and large landscaped areas.

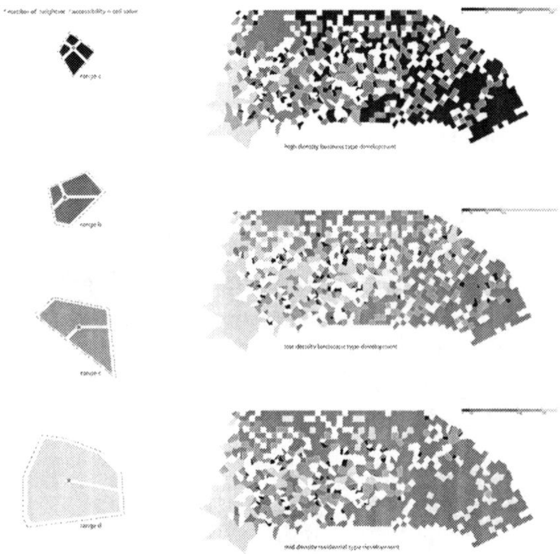

Fig. 10. Voronoi cell categories according to the cell area, number of sides and connection number

The cell so far exhibits a blank yet informative space, hence the focus now shifts to embedding formal information in the cell that reflects the original gradient of the Voronoi pattern. Since each pattern assumes its geometric information from the anatomy of the cell, variation becomes a virtue that is built in to the system. Figure 11 extends this informative drive to generate different pattern sets.

As the pattern spreads across the site it undergoes two sets of negotiations with the Voronoi cell geometry. The first is with the edges of the cells themselves that includes changes of curvature in proportion to the length of the side; the second is the occurrence of the central void in some cases. This space responds to proximity with the housing blocks around the site. Not only the occurrence of voids but also their proportion depends on the distance value the point holds from the influencing element.

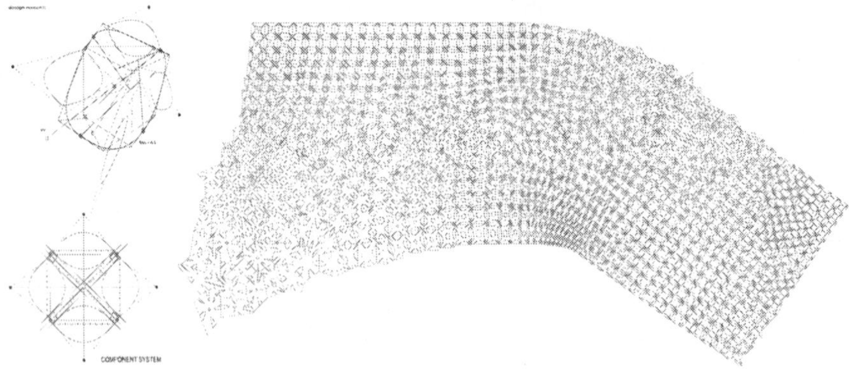

Fig. 11. Additional geometry embedded into the Voronoi cell

Applying the studies to the urban plan, versions of the typology within the broad framework of urban guidelines are created. These versions create a transition from the singular to the plural mass, Figure 12. Taking the housing to the north of the site as an influencing element, i.e. the building volume coalesces into singularity closer towards the housing blocks. From here it starts to disintegrate into individual elements that maintain a high degree of connectivity and eventually into single blocks.

Fig. 12. Building volume distribution on the site (in grey)

The patterns from the computational base are translated to an urban realm and are realized in the form of urban subsystems. The challenge faced is of keeping intact the sensibilities of the digital pattern and evoke a similar sense of dynamics in the urban form. Using numerous script based methods we generate different iterations for the master plan, Figure 13.

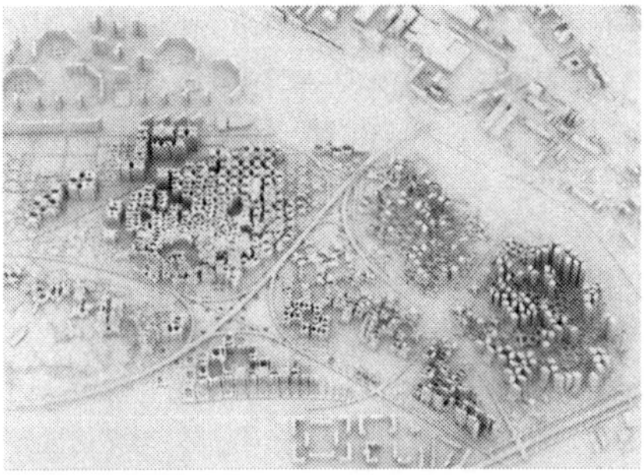

Fig. 13. Master plan iterations

Application of Computational Tools at Building Scale

The urban proposal is comprised of two parts, first are the systems that go into defining the urban fabric. They act as guidelines for the urbanism to develop and grow within. This structure is invisible without buildings as components that absorb and react to these guidelines. Component articulation therefore becomes the second system of the urban proposal. Our proposal comprises of a system of two types of components tower and low-rise. The difference between the two component systems exists in the logic behind them. The tower is seen as the more individual element whereas it is seated within a fabric of a connected low-rise typology. The ideology of the second component however is in its ability to connect with the elements beyond its boundaries. The objective is to blur out the boundary of the cell and to create a shared fabric.

Low-Rise typology

The low-rise typology acts as an architectural premise and a base for the tower. Hence, it is necessary for the surface to mediate from a singular to a plural building volume. This condition is met by dividing each cell into a multiple set of components that follow variant connectivity patterns. Connectivity patterns as have been mentioned earlier are a result of a selective network that generates patterns based on proximity. Hence, by overlaying a secondary layer of connectivity information, this time on the building components we generate a gradually coalescing mass. The architectural premise of this component lies in a hierarchical building up of spatial and circulatory logic.

The character that differentiates the low-rise typology from most of the Moscow's urban fabric is a system of internal voids. This particular catalogue, Figure 14, is targeted towards the structuring of voids within each cell with respect to the distance from a street as indicated by the splines. The objective is to explore a more dynamic set of solid void relationships within the field by changing the size of the voids with proximity to the street. In addition to this there is variation in the catalogue in terms of the void orienting towards the street. It underscores the need for the low-rise typology to start engaging with the street network that exists at the master plan level in order to derive its morphology.

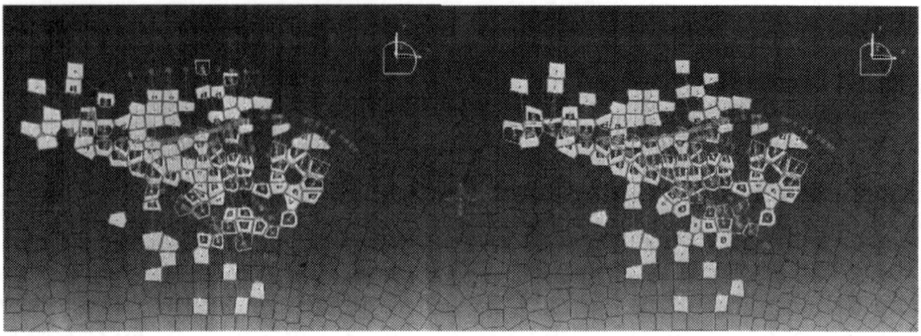

Fig. 14. Structuring of voids within cells

Supporting the previous studies this catalogue elaborates the transformation that each component undergoes according to the proximity from the influencing element, Figure 15. The parameters of control are the firstly the height of the component, secondly the voids start to orient towards the influencing element. Thirdly the degree of roundness and the amount of void increases as one moves away from the influencing element.

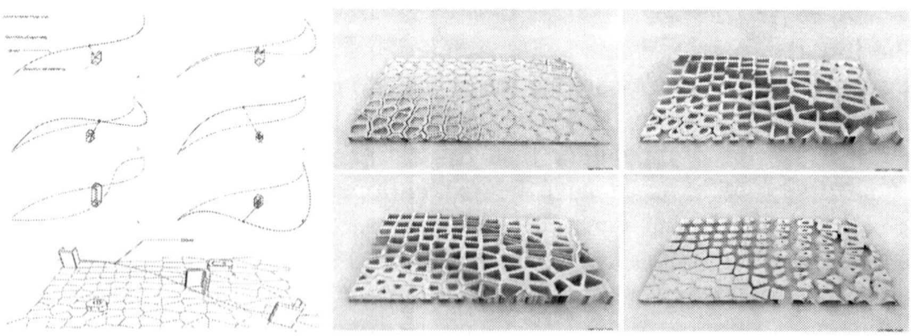

Fig. 15. Low-rise typology transformations

The transformative catalogue on Figure 15 studies how the continuous ground negotiates with the landscape, individual building forms and plural building volumes. In addition to these transformative qualities, a horizontal stacking of components is also studied. These are used as a mean to develop urban conditions that negotiate firstly with changing density patterns on the field and secondly help in integration of the network with the urban fabric

Tower

As compared to the low rise typology the tower is an individual element that adheres to the boundaries laid out by the cell. At the architectural scale the tower is seen as a correlated assembly of multiple tectonic elements. The first exploration within the tower is a volume-generating machine. This in turn provides us with a basic configuration of the tower typology consisting of a central core and four fragments. Within this system the following sub systems are embedded:

A. The core
B. Floor Plates
C. Structure for Segments

The "Volumetric machine" is a setup that is used to generate an articulated volume for the tower, Figure 16. It operates at four levels within one element; each level is composed of four control lines and a set of eight points that define the spatial structure of the tower .The four inner points define a central core whereas the outer points together with the core defines a plate within the tower. The setup lays down through the means of a model a computational setup that is taken further in order to achieve volumetric variations.

Potentially by manipulating these eight points at each level the scaling and rotating of both the core and the volume can be controlled while still maintaining their correlation to each other. It therefore, provides an opportunity to develop each volume with a specific ratio of solid and void.

By controlling the configuration of the tower at each level, different variations can be developed that fulfil certain programmatic requirements. The parameters that can be manipulated are firstly, the solid and void percentage within the building and secondly the location of the void within the envelope. The catalogue also shows that by tracing a curve through the centre point of the core a linear extrusion of the tower can be done away with (Figure 16). This spline represents the movement of void through the building envelope; while creating a footprint for the solid.

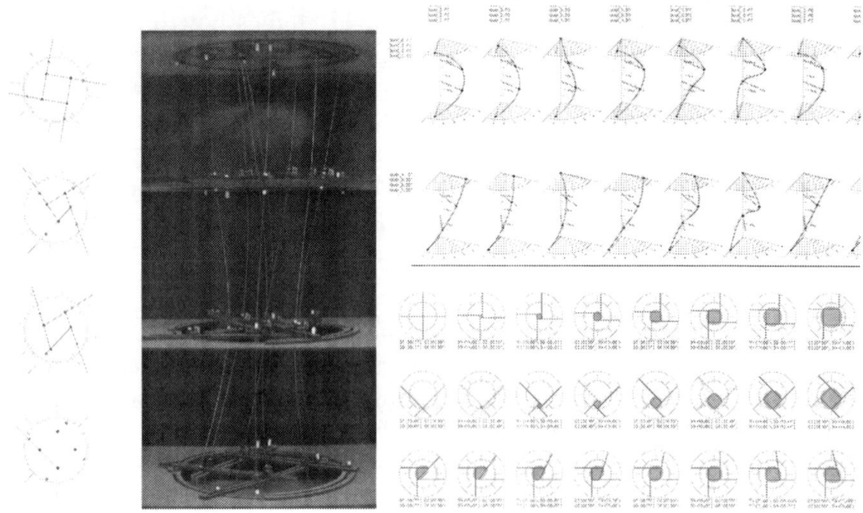

Fig. 16. Volumetric setup for tower component articulation

Assembling all these layers of information is a demonstration of the tower and how it develops, Figure 17. The diagram lays out the four data levels from which first the core develops and eventually the four fragments. This setup is further developed by embedding qualitative differences within the volume that can then be proliferated across the site. The parameters comprise of scaling and degree of rotation to emphasize and exaggerate the difference within the field.

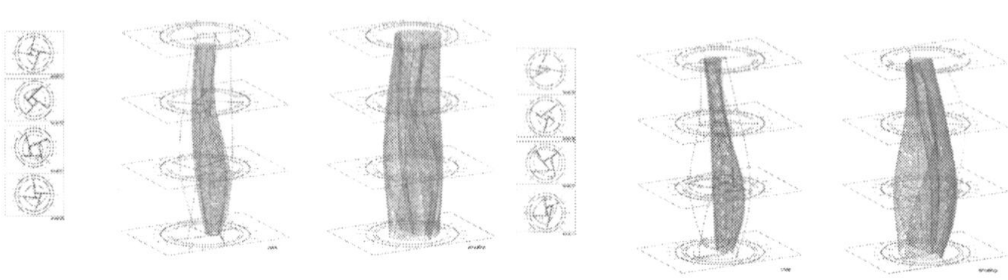

Fig. 17. Tower volume articulation

The data diagram below points out how the distance value from the river becomes a prime organizer of the tower typologies on site, Figure 18. The diagram conveys three types of variations firstly, the towers closest to the housing blocks and furthest away from the river exist as static elements i.e. they do not display any effects such as scaling or rotation. Secondly the towers that exist closest to the river exhibit the maximum degree of rotation. In addition to this the tower fragments start opening up to reveal the core within. The third is an intermediate stage that allows a lesser degree of rotation while no openings are created.

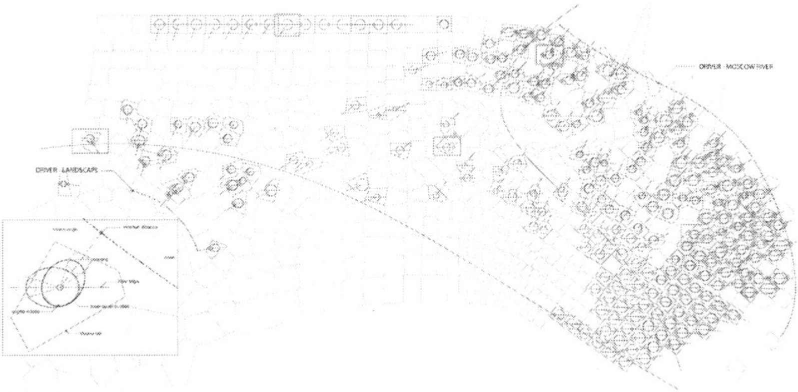

Fig. 18. Tower typology distribution on the site

The cell specific information that the previous diagram offered can be seen here visually with each map showing the extent of existence of each type, Figure 19. Since the information range that decides these typologies in itself is quite broad, subtle differences (owing to magnitude) occur within each section of the tower. In this stage a segregation of the towers closer to the river with the openings is created. The location of the opening travels from the bottom of the tower to the top according to the location on the site. In addition to this, upon close proximity of two towers there is a new cluster typology that emerges, by creating a bridge between the two.

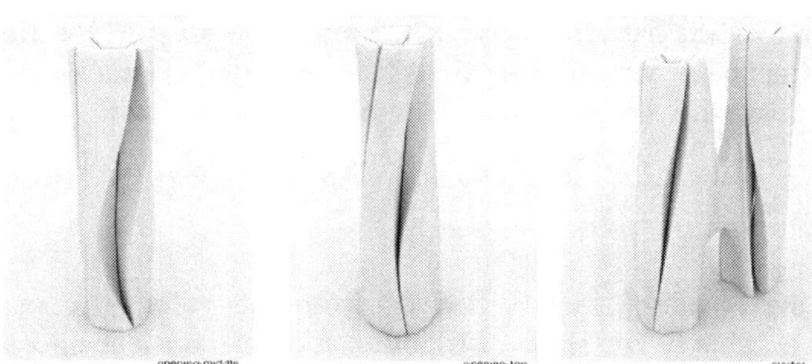

Fig. 19. Tower typological variations

Core

The core of the tower integrates within it the primary structural system and the circulation. In order to achieve a minimal structural form ANSYS is used to generate colour-mapped data correlating to the structural forces owing to the curvature of the form (Figure 20, a), [8]. This colour mapping informs the structural system in a densifying manner. A diagrid has been introduced as a structural configuration of the core. Hence the diagrid on the core is denser towards the base and so on in order to visually translate the forces within the form, Figures 15(a) and (b).

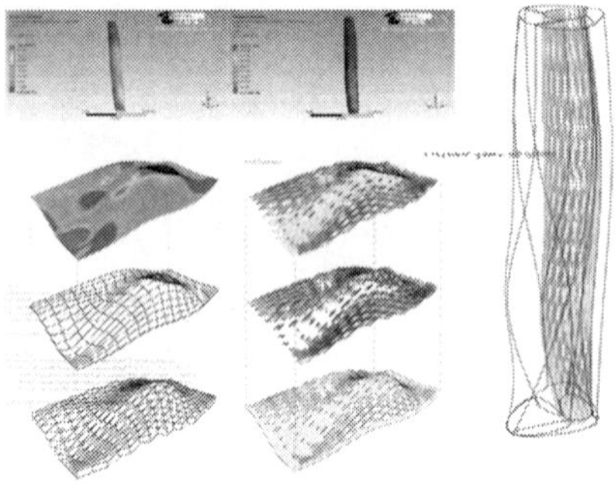

(a)

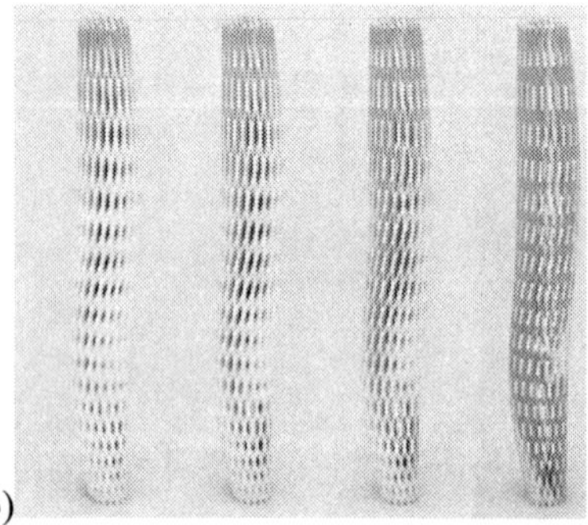

(b)

Fig. 20. (a) Densification of the grid according to stress analysis; (b) Diagrid variations according to stress distribution

Once we have the diagrid pattern, a thickness is given to the core according to spatial requirements; this triggers a process where the diagrid pattern is inset with components that manipulate this space creation. In addition to this the degree of openness of the component is driven by the distance data from the ANSYS mapping shown previously. Thus if the component is closer to the zone of maximum stress it will have smaller openings. Furthering the same ideology the diagrid is regularized to a proportion of six meters to give it the ability to snap to the floor plates. The parameters of variation are firstly the degree of openness and secondly the progressively increasing pattern of the diagrid with height, Figure 20(b).

Floors

Within the envelope that is derived from the volumetric machine we develop a strategy for floor distribution. For the vertical information we take the data from the volume itself, by taking the area of the plate at every level the height of the next plate is decided. Hence, if the area is smaller the floor-to-floor height is small and vice versa. The computational process further adapts itself to generate floor-to-floor heights that occur in multiples of 3 meters, Figure 21. In addition to this we also incorporate constraints like the minimum floor height being 3 meters whereas the maximum being 12 meters. This slicing of each of the four fragments generates different programmatic possibilities. Furthermore within each floor there is a possibility of insetting mezzanine floors.

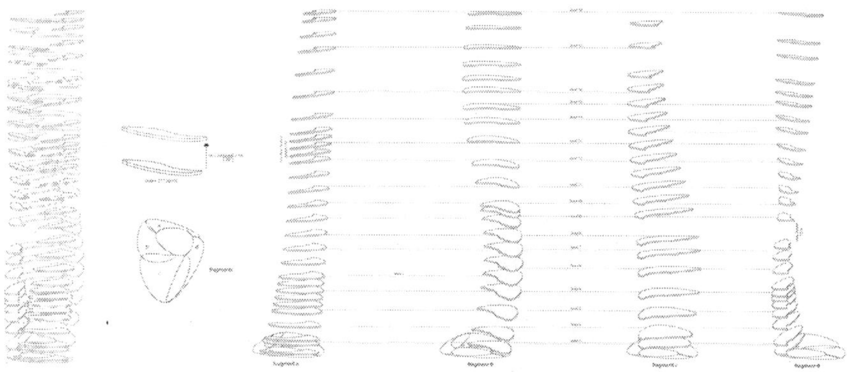

Fig. 21. Tower floor distribution

Structure

Each of the four fragments has a structural core that is meant to provide a system of vertical load transfer within the volume. Owing to the twisted nature of the volume the pivot point is located for the volume along which the rotation takes place Figure 22(a). This pivot point becomes the central core at every floor level from which secondary elements fan out to support the floor plate.

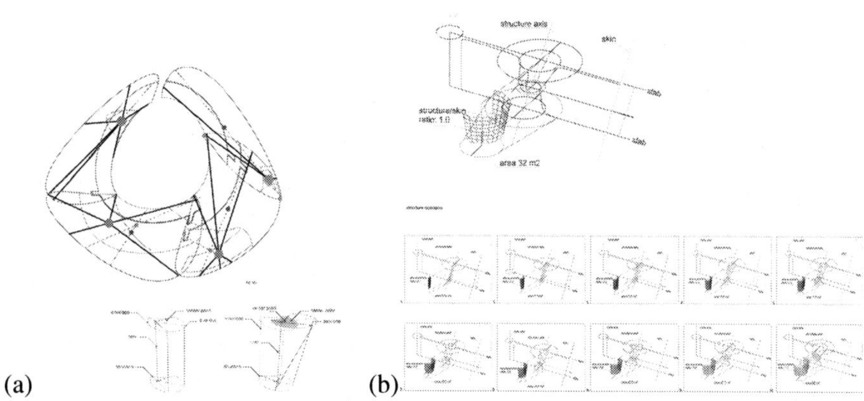

Fig. 22. (a) Footprint of the structural element on the floor plate; (b) Columne transformations

The structure undergoes transformations from one plate to another not only in its planometric configuration but also in the third dimension. The parameters that bring about the variation are the footprint of the structural element on the plate and the depth of the beams, Figure 23.

The sequential layout of the floor plans traces out the rotation of each of the fragments. Hence, even though the underlying logic behind the arrangement of each floor is the same each structural element varies in its configuration since its length is in continual progression, Figure 23(a).

The second parameter that is added is the length of the beam, if it exceeds a certain limit the column starts to bifurcate to provide additional support, Figure 23(b).

The formal expression of the column looks at how it becomes an extension of the floor and the ceiling. The junction where these two elements come together is further articulated in terms of the location of its level. This variation is affected by the area of the floor plate and the area of the column. Articulating the form of the column itself in a more three dimensional manner, the column becomes a two part assembly emerging from the floor and the ceiling respectively. The configuration of the column changes from being slender to stout depending upon the floor area. As the column tends to gain thickness and its area exceeds a certain limit, it starts to morph and open up creating space within, Figure 22(b).

(a)
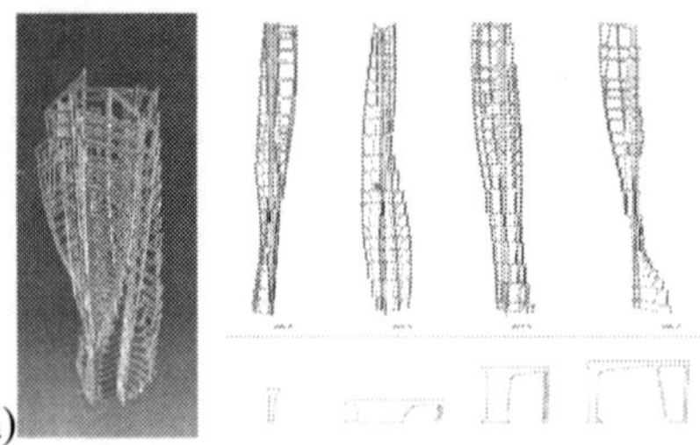
(b)
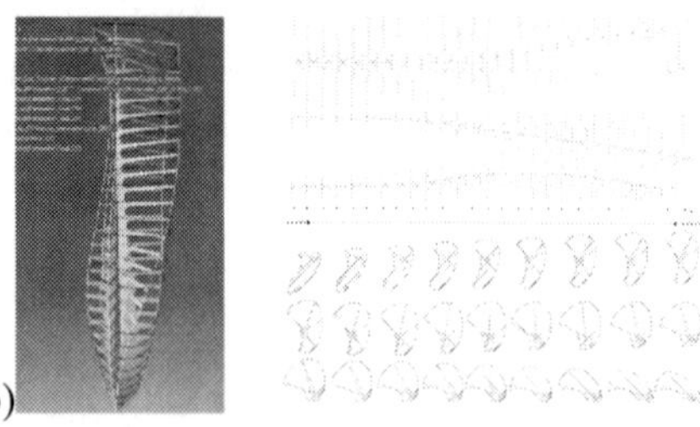

Fig. 23. (a) Beam extension; (b) Beam bifurcation

Conclusion and Future Work

The set of guidelines combined with the component information brings out a system of urbanism that is malleable and has the ability to adapt. The adaptability of the system is its greatest strength, based on its parametric setup. Parametrics plays an important role in the project not only in terms of creating a system responsive to the context conditions but also in its ability to produce a variegated result.

The final master plan proposal is an iteration that follows up from the prior versions and is certainly more developed than the earlier ones, Figure 24. However, the designer can strengthen this setup by more informative input and analyze it in Massive for human and vehicle traffic or in Ecotect for environmental effectiveness [6], [7]. The system developed here therefore has the potential and the ability to evolve continuously.

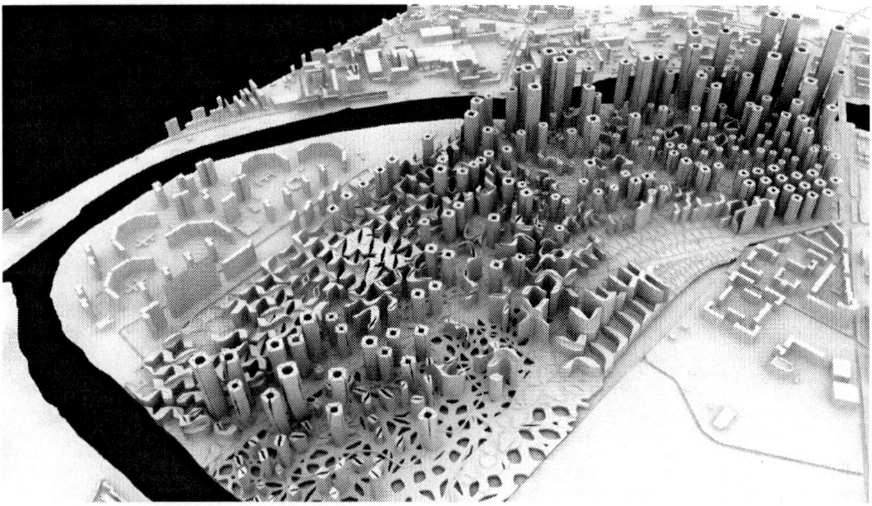

Fig. 24. Final master plan proposal

References

1. Spuybroek, L.: Nox, The Structure of Vagueness, transUrbanism, pp. 64–87. V2_Publishing/NAi Publishers, Rotterdam (2002)
2. Cook, P.: Plug-In City, Archigram, pp. 36–43. Studio Vista Publishers, London (1972)
3. Schumacher, P.: Parametricism - A New Global Style for Architecture and Urban Design. AD Architectural Design - Digital Cities 79(4) (2009)
4. Kartal-Pendik Masterplan project information, `http://www.arcspace.com` (Last accessed January 2010)
5. Rhinoceros NURBS Modeling for Windows official website, `http://www.rhino3d.com` (Last accessed January 2010)
6. Aschwanden, G., Haegler, S., Halatsch, J., Jeker, R., Schmitt, G., Van Gool, L.: Evaluation Of 3D City Models Using Automatic Placed Urban Agent. In: CONVR Conference Proceedings, University Of Sydney (2009)
7. Autodesk. Ecotect Analysis official website, `http://usa.autodesk.com` (Last accessed January 2010)
8. ANSYS official website, `http://www.ansys.com` (Last accessed January 2010)

9. Vanegas, C.A., Aliaga, D.G., Benes, B., Waddell, P.A.: Interactive Design of Urban Spaces Using Geometrical and Behavioral Modeling, SIGASIA, Purdue University. University of California, Berkley (2009)
10. Liaropoulous-Legendgre, G.: IJP: The Book of Surfaces. Architectural Association, London (2004)
11. Lee, C.C.M., Jacoby, S.: Typological Formations: Renewable Building Types and the City. Architectural Association, London (2007)
12. Hensel, M., Menges, A.: Morpho-Ecologies: Towards Heterogeneous Space in Architectural Space. Architectural Association, London (2007)

Complex Product Form Generation in Industrial Design: A Bookshelf Based on Voronoi Diagrams

Axel Nordin, Damien Motte, Andreas Hopf, Robert Bjärnemo, and Claus-Christian Eckhardt
Lund University, Sweden

Complex product form generation methods have rarely been used within the field of industrial design. The difficulty in their use is mainly linked to constraints – such as functionality, production and cost – that apply to most products. By coupling a mathematically described morphology to an optimisation system, it may be possible to generate a complex product form, compliant with engineering and production constraints. In this paper we apply this general approach to the designing of a bookshelf whose structure is based on Voronoi diagrams. The algorithm behind the developed application used here is based on a prior work submitted elsewhere [1], adapted to the bookshelf problem. This second example of product form generation, which includes specific constraints, confirms the relevance of the general approach.

The handling of complex morphologies is not straightforward. Consequently, an explorative study on that theme has been performed. A user interface has been developed that allows for designing a bookshelf based on Voronoi diagrams. The user interface was subsequently tested by peer designers. The results suggest that user attitudes diverge: one faction preferred maximum freedom of creation, that is, maximum control of the form creation process; the other faction wanted the application to generate a bookshelf based on their functional needs (e.g. adapt to the number and types of objects to be stored) and would ask for a "surprise me" effect for the final solution.

Introduction

Although complex – mathematical or nature-inspired – form generation methods have long been employed in the field of architecture [2, p. 137],

J.S. Gero (ed.): *Design Computing and Cognition'10*, pp. 701–720.

this has rarely been the case in industrial design. One barrier for such development in the latter discipline is the multitude of constraints linked to the form-giving of products; surfaces are often functional, the artefacts are produced in several exemplars – meaning that the product form must be modified to suit production systems; cost control is consequently important; finally, engineering constraints must also be respected. Another obstacle may be the lack of educational initiation in industrial design.

However, the situation is beginning to evolve; the ongoing digitalisation of the entire product design activity simplifies access to form generation tools whilst digital fabrication facilitates the production of physical prototypes. This digitalisation should allow for a much tighter integration of industrial design, engineering and production. Last but not least, one can sense an evolution of the by and large static relationship between the consumer and the product. There is an increasing desire to participate in the designing of products and the potential experiences consumers will share with them. As put forward by Friebe and Ramge [3], the upsurge of independent fashion labels, *crowdsourcing* initiatives or co-working spaces indicates the demand for consumer *empowerment*. This need for co-creation, implemented already in textile [4] but also in more advanced consumer goods businesses like sportswear, [5] and [6], goes well beyond mere material and colour choice – the future *prosumer* [7] desires control over form and function as well. Generative design can be one facilitator in such developments.

In a prior work [1], we have begun to study the use of complex forms in industrial design, taking into account functional, engineering and production constraints. The term complex is to be understood here in the sense that the form is virtually impossible to generate without computer aid. The present publication has two goals. First, we aim at partially validating the approach proposed in [1] by investigating another product type with different kinds of constraints and objectives. Second, the handling of complex forms is not straightforward. Most users (designers or consumers) cannot or do not want to manipulate directly the parameters linked to a morphology; in some cases this is even impossible. It is necessary to find alternative ways of controlling form that make sense for the user. We consequently reflect on the way the user can interact with these complex forms, and a user interface allowing for the development of the bookshelf based on Voronoi diagrams has been developed and tested.

Background

Expanding the Morphologic Repertoire in Design

The morphologic repertoire is the infinite repository of all two- and three-dimensional forms, structures and compositions thereof. Although no morphology has *a priori significance* – its adequacy pertaining only to the intended usage criteria – a designer's command of an extensive morphologic vocabulary and grammar enhances creative expressivity, which, in turn, is no end in itself, but essential for a designer's ability to rise to the present and future economic and ecologic challenges [8].

The prevalent *modus operandi* concerning the form-giving activity in industrial design is characterised by explorations that depend on the individual capability to mentally manipulate a solution space from which to select and express the intended result. In that sense a designer or team of designers is equivalent to an *auteur*, because the initial objective and resultant object are inextricably linked by a volitional act [9]. The morphologic repertoire, on which the form-giving activity is based, is by and large rooted in artistic experimentation (serendipity), cursory inspiration (mimicry) or canonical stipulation (methodology). Reliance on such rather traditional approaches is not problematic *per se*; a trained designer generally will come out with a *satisfying* solution to a brief.

However, individualist or formal aesthetic motivations preclude the creative potential of *generative* mathematical and natural morphologies that could be equally inspiring points of departure. Even more importantly, once these morphologies are coupled to algorithmic design processes, they provide access to *performative* and *emergent* qualities only found in dynamic systems [9]. Algorithmically controlled morphologies not only pave the way for the unimaginable, they also present methods to handle and adapt them to an intended purpose. In that sense, the form-giving activity is augmented or rather transmuted into a *form-finding* process – an almost literary *meta-design* activity concerned with the formulation of rules and constraints from which desired or unintended, but feasible, results emerge. Quite possibly, the self-conception of what a designer is and does will change considerably in the future; designers may eventually become *scriptwriters*, *moderators* or *curators* – or even redundant altogether?

Form Generation in the Larger Context

Apart from providing new creativity-enhancement tools, it is important to integrate them into the design context. As mentioned above, an industrial designer's activities in the product development process are intertwined with engineering and production preparation activities. Nevertheless, even in organisations were these different functions are well integrated, iterations are still unavoidable. Taking advantage of the digitalisation of engineering and production preparation activities, efforts have been made taking into account their different constraints early on in the development process (see e.g. [10]). In the latter context, the industrial design activity is still somewhat overlooked. By integrating critical engineering and production requirements in the design process, the likelihood for a designer to "get it right first time" – or at least to reduce the number of critical changes in the design– is higher.

If form is algorithmically generated and engineering and production constraints integrated in the process, partial or full transfer of the design activity to consumers becomes a concrete option – new business models will emerge as a result. Many businesses have already implemented mass-customisation to some degree, with the automotive industry being a precursor. That approach has been adopted in other industries like in the case of Threadless® (clothing) [4] or Innovate with Kraft® (food) [11]. The demand for bespoke products and services is increasing, even in markets where branding is important. "The world has changed. Consumers interact with brands on their own terms," says Trevor Edwards, Vice President, Brand and Category Management for Nike® [12]. Consequently, the NikeId® website and studios, [12] and [13], provide for high-level customisation of athletic footwear. Many consumers in saturated economies are decreasingly passive; it will be ever more important to hand over some control over the products-to-be they desire. Even if the traditional modes of product development remain dominant in the foreseeable future, the exploitation of niche markets becomes increasingly relevant and more profitable [14].

Handling complex forms requires the development of an adequate way for users to make sense of the creation process. The design environment is discussed in the section "User interaction".

Related Works on Generative Product Design Systems

Generative design systems that take into account functional and technical constraints (engineering and production) as well as aesthetic intent have existed for long in the field of architecture while such systems have rather been the object of isolated research studies in industrial design.

In industrial design, generative design has primarily been used for stylistic purposes. In the seminal work of Knight [15], a parametric shape grammar was developed for the generation of Hepplewhite-style chair-backs. Orsborn et al. [16] employed a shape grammar to define the boundaries between different automotive vehicle typologies. Recent works have focused on branding related issues. With the help of shape grammars, new designs based on the Buick® [17], Harley-Davidson®

[18], Coca-Cola® or Head & Shoulders® [19] brand were developed. Further research is undertaken towards rules that are linking form and brand (e.g. [20] for GA-based systems and [21] for shape grammars).

Some works are crossing the boundaries between engineering and industrial design, taking into account functional or technical constraints and aesthetics. Shea and Cagan [22] used a combination of shape grammar and simulated annealing for both functional and aesthetical purposes and applied it for truss structures (truss structures are commonly used for both heavy industrial applications and consumer products). Shape grammars are used to generate new designs while the simulated annealing technique directs the generation towards an optimum. The design objectives were functional (minimise weight, enclosure space and surface area), economic and aesthetic (minimise variations between lengths in order to get uniformity, make the proportions respect the golden ratio). Their model has been re-used in [23] (shape grammar and genetic algorithm, or GA) to develop stylistically consistent forms and has been applied to the design of a camera. The designs generated took into account the constraints linked to the spatial component configuration. A designer was in charge of the aesthetic evaluation, following the interactive genetic algorithm (IGA) paradigm. Ang et al. [23], also using shape grammars and GA, developed the Coca-Cola® bottle example of [19] and added functional considerations (the volume of the bottle), that were constrained to approach the classic Coca-Cola® bottle shape. Morel et al. [24], within the IGA paradigm, developed a set of chairs optimised for weight and stiffness. Finally, Wenli [25] developed a system that, through adaptive mechanisms, allows it to learn the designer's intent faster; that system was implemented as a plug-in for a CAD system and applied to boat hull design.

Approach

In the works presented above, shape grammar is the main technique used. In our study, we use a pre-determined computational geometry (namely, Voronoi diagrams) instead and optimise the form according to engineering and production constraints. The use of mathematical and natural morphologies has not been the object of much applied research in industrial design, but has been implemented for the development of several products and prototypes in industry (see [1] for a coarse typology of such products). An important observation is that, in many cases, plastics rapid prototyping is the fabrication system of choice. Restricting the application of an extended morphologic repertoire to rapid prototyping may not be sustainable in the long term as only a few types of products are suitable for this fabrication technology. Rapid prototyping is likely not to be the panacea. That is why we focused on "traditional" production systems such as laser cutting and CNC sheet metal bending. Defining and evaluating the aesthetics is up to the user through the IGA paradigm. This continuation of Nordin et al. [1] aims first at partially validating that paradigm by using a different product (bookshelf vs. table), another material (phenolic film – PF – coated veneer core plywood), and consequently a different manufacturing system (circular saw and strip-grinder) and different constraints (addition of a functional constraint). Second, the focus is on how new forms can be practically handled. The user (designer or consumer) may not necessarily be interested in a certain mathematical or natural morphology *per se*, but rather in its aesthetic potential. It is quite difficult to handle complex morphologies. Proper controls have to be defined and, to that end, a dedicated interface has been created and tested.

Short on the Voronoi Structure

A Voronoi (or Thiessen) structure is a simple 2D tessellation, as shown in Figure 1. Structures based on a Voronoi structure (or Voronoi diagram) are often found in very robust yet lightweight structures in nature. Apart from aesthetic aspects, a Voronoi-based bookshelf would consequently have a structure well suited for carrying heavy loads such as books whilst maintaining a low weight. A Voronoi structure can be described as follows: Let $p_1,.., p_n$ be a set of n distinct points in the plane; these points are called the Voronoi sites. For each site, the set of points that are closer to it than to any other site form a Voronoi cell. A Voronoi diagram is constituted of all such cells. An overview of a Voronoi diagrams' properties can be found in [26, chapter 7].

The Bookshelf

As in [1], the Voronoi structure is applied to a common type of furniture - a bookshelf. In the table case, the Voronoi structure was used as support for a glass tabletop, Figure 2. In case of the bookshelf, every Voronoi cell serves as a storage compartment.

The manufacturing process consists of cutting, gluing and assembling PF coated veneer core plywood parts. Each Voronoi cell is manufactured as an individually cut and glued compartment. The critical constraints related to functionality and production methods are described later.

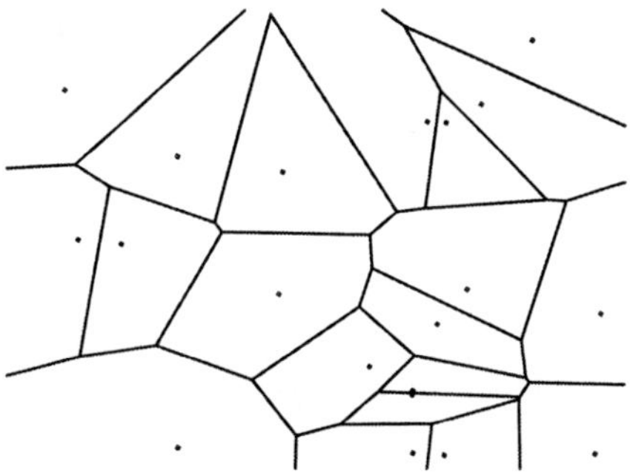

Fig. 1. Example of a Voronoi diagram

User Interaction

The initial question regarding functionality, interactivity and output of a generative design and optimisation application is whether the designer or consumer is willing to relinquish control to a certain degree. An *auteur* designer-personality may hold the view that algorithms *seemingly* restrict creativity; an experimentally open-minded design-personality may, in contrast, actively seek for emergent behaviour to find unexpected solutions. But it should be noted here that often, modern dance performances, music scores and contemporary architectures have been developed with help of algorithms – and the creativity of William Forsythe, Iannis Xenakis or Zaha Hadid has not been disputed. For consumers, control is important as it is not only the uniqueness of the product that matters, but also its personalisation [27].

However, in case the designer or consumer is attracted to generative design methodologies, the question then is in what way the degree of freedom is limited – and for what reason. The resultant output may either only remotely resemble the chosen start-up design, or turn out to be fairly predictable. One could therefore speak of *controlled serendipity*, wherein the number of constraints – whether aesthetic or functional – is the determining factor.

The sheer number of constraints – determining the degree of usability – must also be considered. Whereas a skilled designer may wish to *condition* a generative design and optimisation application with constraints beyond the aesthetic, e.g. complex functional and production constraints, an unskilled consumer is likely to not wanting to venture much beyond the aesthetic and overall dimensioning. Therefore, such application, if generalised for the widest possible range of input morphologies and output product typologies, needs a very customisable graphical user interface (GUI) in order to show and hide complexity depending on the task at hand.

An interface in accordance with such conception of the design work with respect to an expanded morphology has been produced and is presented next. The acceptance of that interface has then been explored by letting a group of peer designers use it and elaborate on their experiences. Their feedback is presented in the second part of this section.

Fig. 2. Example of table developed in [1]. The Voronoi structure was used as support for a glass tabletop and was optimised with respect to deformation and CNC bending constraints.

The Interface

In [1], users were able to define only the contour of a table and could not affect the layout of the Voronoi cells; in the case of the bookshelf it was decided to concede greater control to users – manipulating the compartmentalisation of the bookshelf for both aesthetic and functional purposes– and get feedback from a group of designers on that matter.

To enable users to control the appearance of the bookshelf, several strategies were discussed. It would have been possible to give them complete freedom to place the points of the Voronoi diagram. This, however, would have be a tedious task for someone not interested in controlling every aspect of the tessellation – and would be too complex for someone not familiar with the specific behaviour of Voronoi diagrams. Instead, it was decided to offer users the option of controlling the compartmentalisation via a set of parameters. The *staggering* and *randomness* parameters allow for the control of the generation of aesthetically interesting structures. *Staggering* controls the internal angles of the compartments so that the user can create cell-forms ranging from square to hexagonal. *Randomness* controls the randomisation of angles and compartment sizes.

Beside the functional parameters that determine the external dimensions of the bookshelf (*height*, *width* and *depth*), its usability is also very much dependent on compartment sizes and forms. To that end, it was decided to offer users three parameters ruling the distribution, size and form of the compartments, namely *growth*, *sparsity* and again *staggering* (the same parameter as above). *Growth* controls how the sizes of compartments are distributed to enable users to generate a bookshelf with small compartments on top and progressively larger compartments towards the floor. *Sparsity* controls the scale of the entire Voronoi diagram to allow users to uniformly scale the compartments up or down.

Additionally, three variables (*x-position*, *y-position* and *rotation*) enable the user to fine-tune the bookshelf structure. Users may want to avoid small compartments at the outer perimeter; rotating and/or shifting the entire structure horizontally and/or vertically can achieve this.

To reduce the complexity of handling nine variables, the creation process was divided into three steps – in the first step, the dimensions of the bookshelf are set; in the second step, its internal Voronoi structure is defined and in the third step, the structure can be fine-tuned.). The GUI is presented in Figure 3 with all steps shown at once. Finally, the bookshelf is optimised and presented to the user, Figure 4.

Interface Evaluation

To assess the application's usability and the selection of parameters, four professional industrial design peers tested the interface. They were asked to use the application to create and optimise one bookshelf per person and were questioned on the tool's usability and their opinion of the amount or lack of control over the creation process and final solution.

What became apparent from the usability testing was that the designers were split into two distinct factions – either wanting maximum freedom of creation, or wanting the application to generate bookshelves based on their needs.

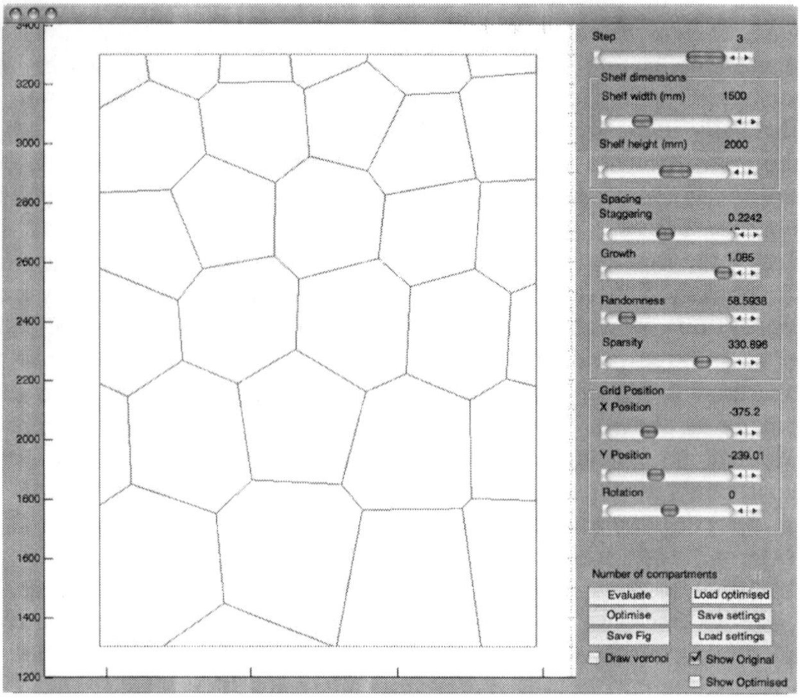

Fig. 3. The GUI for the bookshelf creation application (all steps shown).

The designers requiring maximum freedom wanted to be able to edit the Voronoi structure in detail by adding, removing and moving the Voronoi sites (points) as well as defining the bookshelf's contour. They felt deceived noticing that the algorithms had changed their structure after optimisation for production and they felt that their work had been taken away from them. They would have preferred to be able to get continuous visual feedback on the bookshelf' properties and adjust the structure themselves according to the feedback.

The other faction of designers requested a feature to input information on their particular requirements, such as the types of objects to be stored and/or the dimensions of a room – and then automatically receive a few optimised solutions fulfilling their needs to choose from.

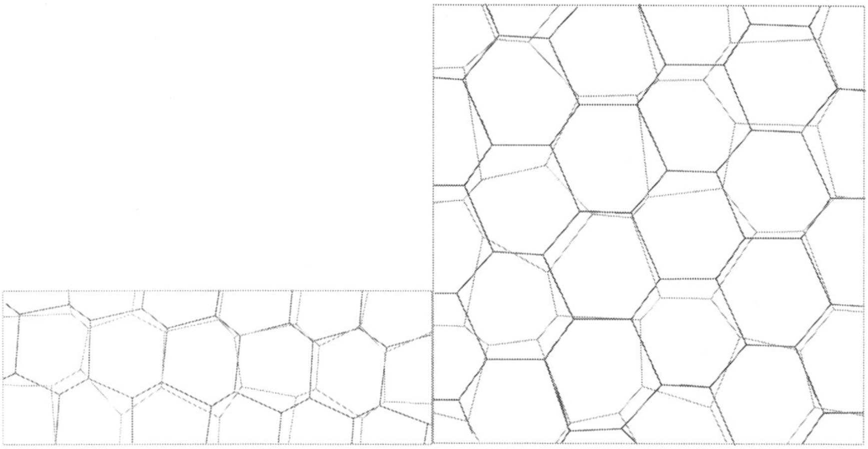

Fig. 4. The original user defined structure in black, optimised solutions in grey (in red, respectively green in the colour version)

The conclusion is that the application needs to accommodate both types of users; those requiring absolute control and those only interested in the bookshelf's functionality – something which could be achieved by either offering two modes of operation, or two entirely different applications. The goal of allowing users greater freedom in designing the structure was achieved, but evidently it will be necessary to concede to both types of users even more and/or different kinds of control.

Although at first feared to be a usability problem, the time users had to wait while the final solution was being optimised was not considered a major drawback in the application. In fact, some were willing to wait as long as one evening before getting the result – given that it would be satisfactory.

A photo-realistic rendering of the final solution was also requested – or at least an isometric representation including a depiction of the material thickness.

The General Search Algorithm

The algorithm for optimising the bookshelf's structure is essentially the same as in [1], with some adaptations for the present project. As described in the section above, the user "designs" the initial prototype that will be used to create the first generation of individuals. Only one final solution is presented to the user, who, if not satisfied, can re-launch the optimisation process an infinite number of times. The user can change the newly obtained bookshelf at each iteration step (subsequent testing by the designers showed that this was superfluous, see last section). Because the interaction is limited to choosing or not choosing the outcome of the optimisation, the principle of the algorithm is more of a “semi-interactive“ optimisation, than a usual IGA. The adapted algorithm is presented Figure 5.

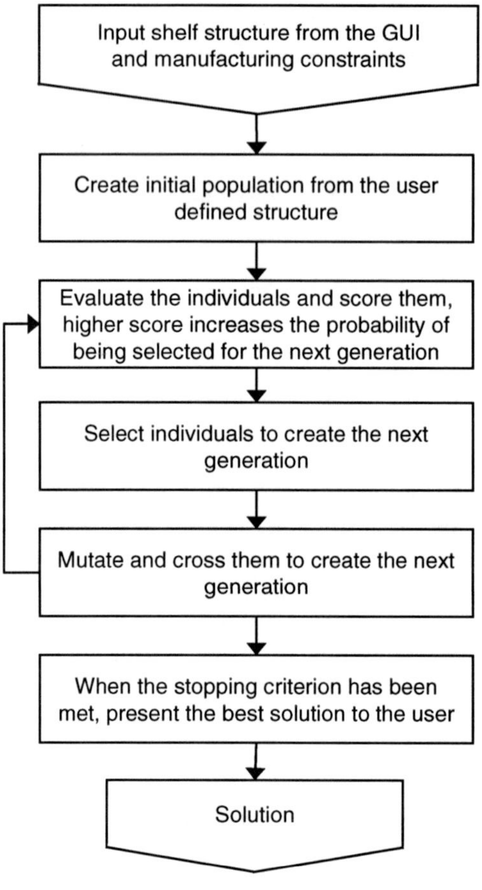

Fig. 5. Diagram of the general algorithm

The evaluation model for the multi-objective optimisation has also been modified to better suit the usability and production requirements of a bookshelf. In [1], a structural analysis was performed on the generated tables to ensure their stability – the most time-consuming evaluation step. Because a Voronoi structure is a robust tessellation and the material thickness can be over-dimensioned in the case of a bookshelf to eliminate the risk of structural failure, it was decided not to perform a structural analysis during the optimisation – but rather on the user's selected solution.

Simulations in [1] had also shown that one production constraint was more difficult to fulfil than others. These time-consuming tasks had resulted in a prioritisation of the resolution of the different constraints. In this case, such issues did not occur and the multi-objective optimisation model used is similar to [22], a weighted sum of both the constraints and the objective values.

The GA Characteristics

As in [1], the *representation* of the structure to be optimised consists of the coordinates of the Voronoi sites, $p_1,.., p_n$. The number of Voronoi sites, n, depends on the user input.

(1)
$$genome = \begin{bmatrix} x_{1,1} & x_{2,1} \\ \vdots & \vdots \\ x_{1,n} & x_{2,n} \end{bmatrix}$$

The Voronoi diagram is created from the coordinates of these sites and cut off along the edges of the user-defined contour of the bookshelf (see e.g. Figure 3). The resulting polygonal structure is used for evaluation (see below). The *mutation function* is identical to the one used in [1], and consists in displacing each coordinate, $x_{i,j}$ of the Voronoi sites in the representation by a random amount, $\delta_{i,j}$, varying between 0 and 10% of $x_{i,j}$. The maximum amount of change decreases linearly from the first generation to the last. The *crossover function* is also identical to the one used in [1], and consists in exchanging the coordinates of the Voronoi sites of two parents after a random mutation point to create two children.

As in [1] the *selection* is based upon ranking the individuals according to their fitness (the fitness function is described by Equation 5). An individual's probability for being selected is proportional to its ranking.

Duplicating the user-defined individual to fill the population creates the initial population; then the mutation function is applied to all individuals in the population apart from one, which remains unchanged.

The parameters of the search algorithm were chosen to give a feedback time of around one hour, which was deemed reasonable, as well as a high probability of receiving a solution that fulfilled all constraints. Therefore the stopping criterion was chosen to be the maximum number of generations, which was set to 300. The size of the population was set to 50 individuals. This proved to give feasible solutions to all user-defined structures during the testing. See Figure 7 for an example of a typical fitness curve.

Constraints, Objectives and Evaluation

For the table example of [1], it was required that the cells have walls longer than 30 mm and the internal angles are larger than 33° in order to be suited for production with a CNC sheet metal bending machine [28]. It was also necessary to take care of the stiffness and deformation (engineering constraint). In the case of the bookshelf, the properties of a strip-grinding machine constrain each PF coated veneer core plywood segment's geometry. The segments need to be attached to a support at a fixed distance from the strip. Given the minimal length needed to fix the parts to the support and the length from the support to the strip, which depends on the slipping angle, it was decided that the minimal segment length should be 100 mm (lmin) and the smallest angle 33° (min).

These requirements are expressed by the following function:

(2)
$$p_l = k_l \cdot \min(l)\ \textit{if}\ \min(l) \le l_{\min}$$
$$p_l = 100\,\textit{otherwise}$$
$$k_l = 100 / l_{\min}$$

where min(l) is the shortest cell wall found in an individual and

(3)
$$p_\alpha = k_\alpha \cdot \min(\alpha)\ \textit{if}\ \min(\alpha) \le \alpha_{\min}$$
$$p_\alpha = 100\ \textit{otherwise}$$
$$k_\alpha = 100 / \alpha_{\min}$$

where min( ) is the smallest angle measured in an individual.

The chosen plywood thickness is 9 mm, which is overdimensioned, meaning to be stable enough for whatever Voronoi network. Consequently, there was no engineering constraint. As a control, the final bookshelves presented in this paper were subsequently analysed (see the Results section), and showed no excessive deformation. A functional requirement for a bookshelf is how well books stack in the compartments. A critical factor for this is the angle between the lower walls in each compartment. A 90° angle is optimal for books to be stacked in a compartment, an angle $_0$ between 80° and 100° is considered acceptable. The individual is therefore scored after its percentage of cells that fulfil this specification.

(4)
$$p_\beta = k_\beta \cdot n(\beta) / n(cells)$$
$$k_\beta = 100$$

The objective function to be maximised is

(5) $$p = p_l + p_\alpha + p_\beta$$

which corresponds to the fitness of each individual. The constants k_l, k and k have been determined so that the maximum score for fulfilling each of the requirements is the same, in this case 100 points (in other words this is a sum for which all the requirements have the same weight).

Results

Sample bookshelves generated by the designers who tested the interface are presented in Figure 6, showing variations in terms of dimensions and compartments. Figure 7 shows the fitness curve of the evolution during the optimisation of one of the designer's bookshelves. The population converges towards the feasible solution space; using a population of 50 individuals during 300 generations was satisfactory.

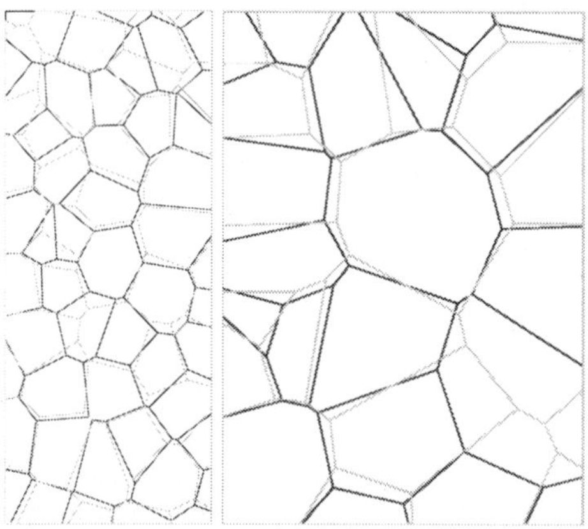

Fig. 6. Examples of bookshelves generated by the designers. The original user defined structure in black; optimised solutions in grey.

Two shelves are presented in more detail. Both (700x2000 mm and 2000x2000 mm, 9 mm plywood) were generated using the application and are shown in Figure 4 and Figure 8. Their optimisation took around 1 hour on a dual-core 2.2 GHz processor.

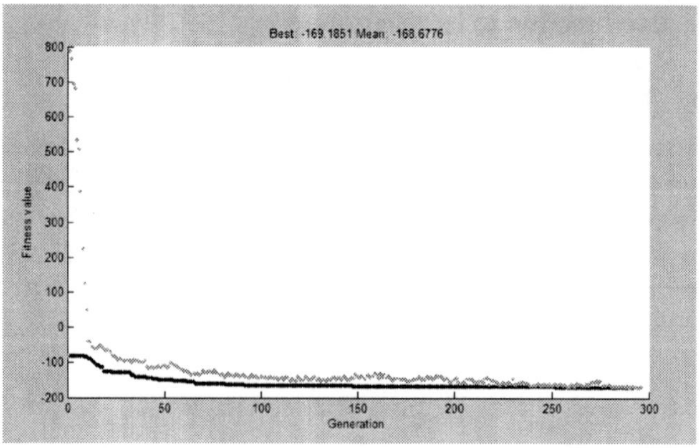

Fig. 7. Typical appearance of the fitness curve during the optimisation. In black: the fitness of the best individual, in grey: the mean fitness of the current population

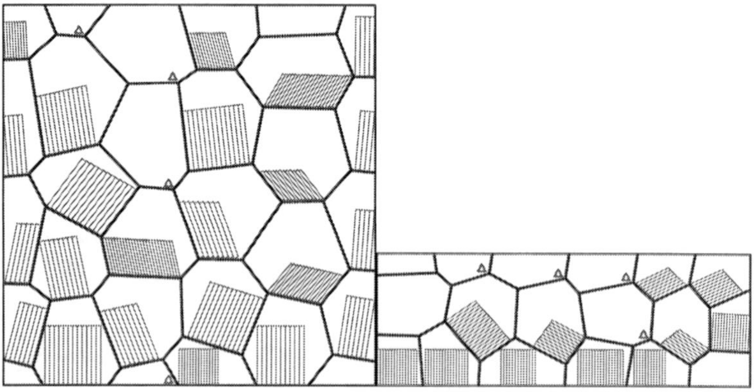

Fig. 8. Illustration of the final bookshelves with stacked books. The triangles indicate where the functional objective (equation 4) has not been met

To verify that these bookshelves were structurally sound and stable, they were tested with FEM-analysis. To simulate the weight of books, the bookshelves were subjected to a load of 10 kg in each compartment as well as the standard gravitational acceleration. The material used was 9 mm Low-density fibreboard (LDF-board, an engineered wood product that has properties similar to PF coated veneer core plywood, but is isotropic, which makes the analysis simpler and more conservative), with the modulus of elasticity being 8 GPa, Poisson's ratio being 0.3 and density being 500 kg/m^3 [29]. The analysis was done in ANSYS Workbench®.

The structural analysis indicated that the maximum deformation of either bookshelf never exceeded 0.4 mm. The results are shown in Figure 9 and Table 1.

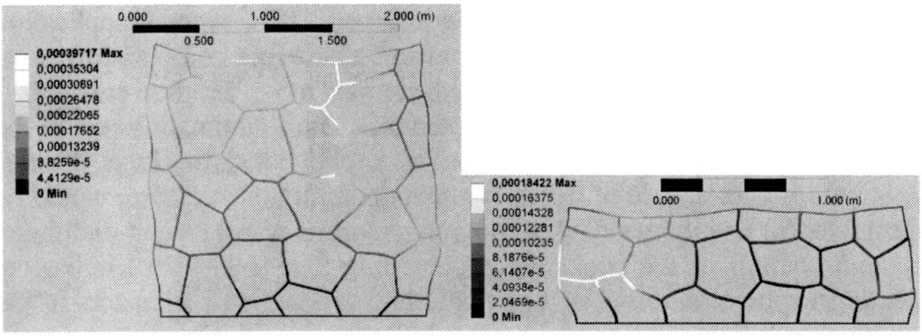

Fig. 9. Structural analysis of the two bookshelves using ANSYS Workbench®

Table 1 Results from the analysis in ANSYS Workbench®

Bookshelf	Volume(m^3)	Weight(kg)	Max. deformation(m)
700x2000	0.0538	26.911	1.84×10^{-4}
2000x2000	0.1225	61.259	3.97×10^{-4}

To verify the aesthetic qualities and their commercial potential, the two bookshelves were subsequently 3D-modelled with 9 mm board thickness; then photo-realistically rendered based on the output geometry of the application and subsequently shown to peer designers. The photo-realistic renderings are shown in Figure 10 and were very well received. To confirm the bookshelves' manufacturability, drawings have been produced from the output and two prototypes are being built.

Fig. 10. Photo-realistic renderings of two bookshelves

Conclusion and Further Research

In this study, a bookshelf generating system using computational geometry (Voronoi structure) – taking into account functional, engineering and production constraints – has been developed. This second application further validates our approach and confirms that using complex forms for designing artefacts has the potential to become a more common practice.

Higher levels of user control, feedback and automation could be implemented in the present and similar applications. In terms of user control, a higher degree of freedom in manipulating the structure would be desirable, for example adding, removing and moving points and defining a custom contour of the bookshelf. Concerning feedback, visual indicators relating to the monitoring of functional and production parameters in the creation and manipulation process could be integrated. Equally important would be to allow users to input individual requirements such as what types of objects are to be stored and in which quantity. Surprisingly, some designers actually desire to relinquish control and let the algorithm determine the design in the full for them. Automated production drawing or data generation would be a useful feature as well as an approximate estimate of cost dependent on material choice, but also weight and sizes of items to be packed and shipped. The interface has been tested by some designers, but the generation of the bookshelf could be equally controlled by consumers to provide them with *bespoke* products.

It is foreseeable that the continuation of this work may result in a generalised tool; that is to say an application that allows designers or consumers to chose from a much wider range of tessellations in order to generate an equally wider range of product types. It is well conceivable that such application will allow for any user-created or user-input tessellations but also integrate entirely new usability, functionality and production constraints for new types of products. This will be the object of future research.

References

1. Nordin, A., Hopf, A., Motte, D., Bjärnemo, R., Eckhardt, C.-C.: Using genetic algorithms and Voronoi diagrams in product design. J. of Computing and Information Science in Engineering - JCISE (2009) (submitted)
2. Shea, K.: Explorations in using an aperiodic spatial tiling as a design generator. In: 1st Design Computing and Cognition Conference DCC 2004, Cambridge, MA, pp. 137–156 (2004)
3. Friebe, H., Ramge, T.: The DIY Brand - The Rebellion of the Masses against Mass Production (In German. Original title: Marke Eigenbau: Der Aufstand der Massen gegen die Massenproduktion). Campus Verlag, Frankfurt (2008)

4. Lakhani, K.R., Kanji, Z.: Threadless: The Business of Community, Harvard Business School Multimedia/Video Case, pp. 608–707. Harvard Business School, Cambridge (2009)
5. Moser, K., Müller, M., Piller, F.T.: Transforming mass customisation from a marketing instrument to a sustainable business model at Adidas. International J. of Mass. Customisation 1, 463–479 (2006)
6. Bouché, N.: Keynote V: How could we create new emotional experiences with sensorial stimuli? In: 4th International Conference on Designing Pleasurable Products and Interfaces - DPPI (2009)
7. Toffler, A.: Future Shock, 2nd edn. Bantam Books, New York (1971)
8. Hopf, A.: Renaissance 2.0 - Expanding the morphologic repertoire in design. In: Hopf, A. (ed.) 23rd Cumulus Conference, Melbourne, September 23 (2009)
9. McCormack, J., Dorin, A., Innocent, T.: Generative design: a paradigm for design research, Futureground, Melbourne (2004)
10. Lin, B.-T., Chang, M.-R., Huang, H.-L., Liu, C.-Y.: Computer-aided structural design of drawing dies for stamping processes based on functional features. The International J of Advanced Manufacturing Technology 42, 1140–1152 (2009)
11. Foods, K.: Innovate with Kraft (2006), `http://brands.kraftfoods.com/innovatewithkraft/default.aspx` (Last accessed: January 2010)
12. Nike Biz Media, New NIKEiD studio opens at NikeTown London giving consumers a key to unlock the world of design (2007), available online: `http://www.nikebiz.com/media/pr/2007/11/01_nikeidlondon.html#` (Last accessed: January 2010)
13. NIKEiD (1999), available online: `http://www.nikeid.com` (Last accessed: January 2010)
14. Anderson, C.: The Long Tail: Why the Future of Business is Selling less of more. Hyperion, New York (2006)
15. Knight, T.W.: The generation of Hepplewhite-style chair-back designs. Environment and Planning B 7, 227–238 (1980)
16. Orsborn, S., Cagan, J., Pawlicki, R., Smith, R.C.: Creating cross-over vehicles: Defining and combining vehicle classes using shape grammars. Artificial Intelligence for Engineering Design, Analysis and Manufacturing - AI EDAM 20, 217–246 (2006)
17. McCormack, J.P., Cagan, J., Vogel, C.M.: Speaking the Buick language: Capturing, understanding, and exploring brand identity with shape grammars. Design Studies 25, 1–29 (2004)
18. Pugliese, M.J., Cagan, J.: Capturing a rebel: Modeling the Harley-Davidson brand through a motorcycle shape grammar. Research in Engineering Design 13, 139–156 (2002)

19. Chau, H.H., Chen, X., McKay, A., de Pennington, A.: Evaluation of a 3D shape grammar implementation. In: 1st Design Computing and Cognition Conference DCC 2004, Cambridge, MA, pp. 357–376 (2004)
20. Cluzel, F., Yannou, B.: Efficiency assessment of an evolutive design system of car contours. In: 17th International Conference on Engineering Design - ICED 2009, Stanford, CA (2009)
21. Orsborn, S., Cagan, J., Boatwright, P.: Automating the creation of shape grammar rules. In: 3rd Design Computing and Cognition Conference DCC 2008, Atlanta, pp. 3–22 (2008)
22. Shea, K., Cagan, J.: Languages and semantics of grammatical discrete structures. Artificial Intelligence for Engineering Design, Analysis and Manufacturing - AI EDAM 13, 241–251 (1999)
23. Ang, M.C., Chau, H.H., McKay, A., de Pennington, A.: Combining evolutionary algorithms and shape grammars to generate branded product design. In: 2nd Design Computing and Cognition Conference DCC 2006, Eindhoven, pp. 521–539 (2006)
24. Morel, P., Hamda, H., Schoenauer, M.: Computational chair design using genetic algorithms. Concept 71, 95–99 (2005)
25. Wenli, Z.: Adaptive interactive evolutionary computation for active intent-oriented design. In: 9th International Conference on Computer-Aided Industrial Design and Conceptual Design - CAIDCD (2008)
26. de Berg, M., Cheong, O., van Kreveld, M., Overmars, M.: Computational Geometry - Algorithms and Applications. Springer, Heidelberg (2008)
27. Franke, N., Piller, F.: Value creation by toolkits for user innovation and design: The case of the watch market. The J. of Product Innovation Management 21, 401–415 (2004)
28. Ferrum Lasercut GmbH, personal communication
29. Wood Handbook - Wood as an Engineering Material. Forest Products Laboratory, Madison (1999)

A Computational Concept Generation Technique for Biologically-Inspired, Engineering Design

J.K.S. Nagel and R.B. Stone
Oregon State University, USA

The natural world provides numerous cases for analogy and inspiration in engineering design. During the early stages of design, particularly during concept generation when several variants are created, nature can be used to inspire innovative solutions to a design problem. However, identifying and presenting the valuable knowledge from the biological domain to an engineering designer during concept generation is currently a somewhat disorganized process or requires extensive knowledge of a particular method. The proposed research aims to define and formalize the information identification and knowledge transfer processes, which will enable systematic development of biologically-inspired, engineering designs. The computational framework for discovering biological inspiration during function-based design activities is presented and discussed through an illustrative example.

Introduction

Engineering design is considered both an art and a science, which encourages the use of engineering principles, imagination and a designer's intuition to create novel engineering solutions. Nature is a powerful resource for engineering designers. The natural world provides numerous cases for analogy and inspiration in engineering design [1-4]. Biological organisms, phenomena and strategies, herein referred to as biological systems, are, in essence, living engineered systems. These living systems provide insight into sustainable and adaptable design and offer engineers billions of years of valuable experience, which can be used to inspire engineering innovation. Many engineering breakthroughs have occurred based on biological phenomena and it is evident that mimicking biological

J.S. Gero (ed.): *Design Computing and Cognition'10*, pp. 721–740.

systems or using them for inspiration has led to successful innovations (e.g., velcro, flapping wing micro air vehicles, synthetic muscles, self-cleaning glass, etc.).

Nature has influenced engineering and the engineering design process. While inspiration from nature can be taken at multiple stages in the engineering design process, it most notably occurs during concept generation. When inspiration in the form of analogies, metaphors and connections from multiple engineering domains and other sources (e.g., biological domain) are utilized for developing novel or creative solutions to a design problem. Concept generation methods and tools help stimulate designer creativity and encourage exploration of the solution space beyond an individual designer's knowledge and experience [5-13]. There are multiple approaches to concept generation for engineering design; however, most are not computational. Although in recent years, computation-based or automatic concept generation has gained importance in the engineering design research community and has taken many forms [14-20]. Identifying and presenting the valuable knowledge from the biological domain to an engineering designer during concept generation is currently a manual, in most cases, and somewhat disorganized, process. The proposed research aims to define and formalize the information identification and knowledge transfer processes, which will result in a systematic method for developing biologically-inspired, or biomimetic, designs.

This paper proposes and explores the third version of the established automatic concept generation software developed by researchers of the Design Engineering Lab. In order to facilitate concept generation for biologically-inspired, engineering design, two bodies of knowledge are required-successful engineered systems and biological systems-both indexed by engineering function. A Design Repository [21] containing descriptive product information serves as the engineered systems body of knowledge. Instead of creating a database containing functionally decomposed biological systems, similar to the design repository, an introductory biology textbook serves as the biological systems body of knowledge. To circumvent the terminology difference issue–indexed by natural language rather than engineering function–an engineering-to-biology thesaurus is utilized [22]. Structurally, the thesaurus acts as a set of correspondent terms to the functions and flows of the Functional Basis [23], which provides term mapping between the biological and engineering domains for the support of concept generation. Integrating biological system information with an established, computational method for concept generation enables designers to consider taking inspiration from biology without having to expend extra effort to learn a new method.

This paper begins by introducing the reader to related biologically-inspired design research and information retrieval in engineering design, followed by a section describing background research. Next, the proposed computational framework and algorithm for discovering biological inspiration during concept generation is presented and discussed. The paper ends with an illustrative example, conclusions and future work.

Related Work

With biology-inspired design emerging as its own field, engineering design research as turned to investigating methods and techniques for transferring biological knowledge to the engineering domain. Prominent research focuses on investigating individual aspects of the overall engineering design process involving biological systems for inspiration [24-29], and computational design techniques. The main goal of these research efforts is to create generalized methods, knowledge, and tools such that biomimetic design can be broadly practiced in engineering. Biology-inspired design "offers enormous potential for inspiring new capabilities for exciting future technologies" [30] and encourages engineering innovation [30, 31].

Work in the area of computational biologically-inspired design techniques involves the creation of databases, software and search methods. Chakrabarti et al. developed a software package entitled Idea-Inspire that searches a database of natural and complex artificial mechanical systems by chosen function, behavior, and structure terms [32]. Their database is comprised of natural and complex artificial mechanical systems and aims to inspire the designer during the design process [33]. Another database driven method is ontology driven bio-inspired design repository developed by Wilson et al. [34]. This ontology is encoded using description logics and uses subsumption, an inference mechanism, to precisely retrieve relevant biological strategies from the repository. Chiu and Shu have developed a method for identifying biological analogies by searching a biological corpus using functional keywords [35, 36]. A set of natural-language keywords is defined for each engineering keyword to yield better results during the analogy search. This method has successfully generated engineering solutions analogous to biological phenomena [37].

Work in the area of information retrieval in design related to this research involves the design of a hierarchal thesaurus, software and search methods. A general approach to design information retrieval was undertaken by Wood et al., which created a hierarchical thesaurus of component and system functional decompositions to capture design context [38]. Strategies for retrieval, similar to search heuristics, of issue based and component/function information were presented. Bouchard et al. developed a content-based information retrieval system named TRENDS [39]. This software aims at improving designers' access to web-based resources by helping them to find appropriate materials, to structure these materials in way that supports their design activities and identify design trends. The TRENDS system integrates flexible content-based image retrieval based on ontological referencing and clustering components through Conjoint Trends Analysis (CTA). Cheong et al. developed a set of search cases for determining sets of biologically meaningful keywords to engineering keywords [40]. Although the results are subjective, the process for retrieving the words is systematic and was successful in determining biologically meaningful words to several functions of the reconciled Functional Basis.

Supporting Design Tools

This section provides background information on the two computational tools and respective supporting design tools that are required to achieve the proposed computational framework for discovering biological inspiration. Researchers of the Design Engineering Lab developed each design tool described below–accessible at www.designengineeringlab.org.

Functional Basis Design Language

Functional representation through functional modeling has a long history of use in systematic design methods [7]. Stone et al. [41] created a well-defined modeling language comprised of function and flow sets with definitions and examples, entitled the Functional Basis. Hirtz, et al. [23] later reconciled the Functional Basis into its most current set of terms, with research efforts from the National Institute of Standards and Technology (NIST), two universities, and their industrial partners. In the Functional Basis lexicon, a function represents an action or transformation (verb) being carried out, and a flow represents the type (noun), material, signal or energy, passing through the functions of the system. There exist eight classes of functions and three classes of flows, both having an increase in specification at the secondary and tertiary levels. Both functions and flows have a set of correspondent terms that aid the designer in choosing correct Functional Basis terms during model creation. The complete function and flow lexicon can be found in [23]. Functional models for any product can be generated using this design language. Functional models reveal functional and flow dependencies and are used to capture design knowledge from existing products or define the dependencies for future products. Advantages to using a design language for modeling include repeatability, archival and transmittal of design information, comparison of functionality and product architecture development [6, 7, 41].

Concept Generation Software-MEMIC and Design Repository

Computational concept generation is an efficient way to generate several conceptual design variants. Also, it adds the benefit of providing lists of engineering components that may be used to solve a particular function. The Morphological Evaluation Machine and Interactive Conceptualizer (MEMIC) was created for use during the early stages of design to produce design solutions for an engineering design from a given functional model using knowledge of existing engineered products [15, 16]. The concept generator software MEMIC accepts an input functional model and uses functionality and compatibility information stored in the Design Repository to generate, filter, and rank full concept variants. The MEMIC algorithm utilizes the relationships contained in a function-component matrix (FCM) and the compatibility information contained in a design structure matrix (DSM), both of which are generated from the Design Repository contents [42]. MEMIC returns a listing of engineering component solutions for each function-flow pair of the input functional model. This allows a designer to easily choose between multiple solutions for a given function and interactively build a complete conceptual design.

The Design Repository housed at Oregon State University contains descriptive product information such as functionality, component physical parameters, manufacturing processes, failure, and component compatibility of over 113 consumer products. Each consumer product was decomposed and functionally modeled using the Functional Basis [23]. Each repository entry is designated as an artifact or assembly of artifacts, whether it performs a supporting function (secondary to the product's operation) and the class of the artifact when entered into the repository database. Additionally, several artifact attributes are captured and stored in a relational database where each record contains an artifact name, part number, and part family that can be used to catalog similar artifacts. Information about the actual function of an artifact is captured as a subfunction value.

Organized Search Tool

The organized search tool was originally developed for retrieving relevant biological systems that performed functions of interest. In this research we propose extending this search tool to better facilitate biologically-inspired design through combination with MEMIC. Specifically, the organized search tool is designed to work with non-engineering subject domain specific information. The majority of non-engineering domain texts are written in natural-language format, which prompted the investigation of using both a Functional Basis function and flow term when searching for solutions. Realizing how the topic of the text is treated increases the extensibility of the organized verb-noun search algorithm. This organized verb-noun combination search strategy provides two levels of results: (1) associated with verb only, of which the user can choose to utilize or ignore, and (2) the narrowed results associated with verb-noun. This search strategy requires the designer to first form an abstraction (e.g., functional model) of the unsolved problem using the Functional Basis lexicon. The verbs (functions) of the abstraction are input as keywords in the organized search tool to generate a list of matches, and subsequently a list of words that occur in proximity to the searched verb in those matches. The generated list contains mostly nouns, which can be thought of as flows (materials, energies and signals), synonymous with the correspondent words already provided in the Functional Basis flow set. The noun listing is then used in combination with the search verb results for a second, more detailed search to locate specific text excerpts that describe how the non-engineering domain systems perform the abstracted functionality with certain flows. The verb searches are constrained to the chosen corpus and the verb-noun searches are constrained to the extracted sentences that include the search verb.

This search strategy is embodied in an automated retrieval tool that allows an engineering designer to selectively choose which corpus or documents to search and to upload additional searchable information as it is made available. The user interface initially presents the designer with a function (verb) entry field and search options. Search options prompt the designer to choose from exact word, derivatives of the word, and partial word. Once the documents are searched for the function term the designer is presented with a flow (noun) listing for each searched document followed by a group of sentences that include the function and listed flows, as shown in Figure 1. If the designer does not want to search by verb-noun then the designer simply scrolls down to the group of sentences, which include the desired function. For this application, the non-engineering domain chosen for examples is biology.

The designer utilizing this organized search technique does not need an extensive background in the non-engineering domain but, rather, sufficient engineering background to abstract the unsolved problem to its most basic level utilizing the Functional Basis lexicon. The search tool typically yields more than one biological system for potential design inspiration.

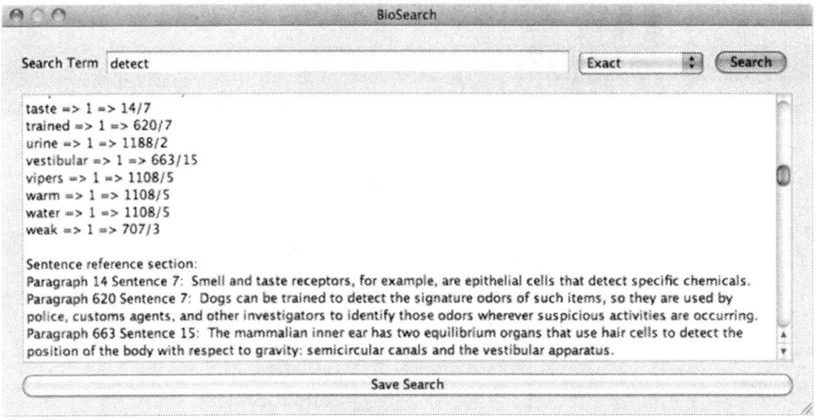

Fig. 1. Example output of the organized search tool

Engineering-to-Biology Thesaurus

The engineering-to-biology thesaurus [22] was developed to enhance the reconciled Functional Basis compiled by Hirtz et al. [23] to encourage collaboration, creation and discovery. The structure of the thesaurus (shown in Table 1) was molded to fit the knowledge and purpose of the authors; synonyms and related concepts to the Functional Basis are grouped at class, secondary and tertiary levels. It does not include an index nor does it include adjectives. Only verbs and nouns that are synonymous to terms of the Functional Basis are considered. The Functional Basis class level terms, however, do emulate the classes of a traditional thesaurus. Furthermore, the secondary and tertiary level Functional Basis terms emulate the categories of a traditional thesaurus. A tool such as the engineering-to-biology thesaurus increases the interaction between the users and the knowledge resource [43] by presenting the information as a look-up table. This simple format fosters one to make associations between the engineering and biological lexicons, thus, strengthening the designer's ability to utilize biological information. The thesaurus aids in many steps of the design process and it increases the probability of a creative or innovative design. Plausible applications of the thesaurus include design inspiration, comprehension of biological information, functional modeling, creative design and concept generation. Overall, the thesaurus provides a designer several opportunities for interfacing with biological information.

Table 1 Example function and flow terminology relationships

Functional Basis Terms			Biological Correspondents
Class	Secondary	Tertiary	
Material	Liquid		Acid, auxin, cytokinin, glycerol, pyruvate
	Solid	Object	Cilia, kidney, melatonin, nephron, xylem
		Composite	Enzyme, nucleotide, prokaryote, symplast
	Mixture	Solid-liquid	Cell, lipid, phytochrome, pigment, plastid
Energy	Chemical		Glucose, glycogen, mitochondria, sugar
Branch	Separate		Aneuploidy, bleaching, dialysis, meiosis
		Divide	Anaphase, cleavage, cytokinesis, metaphase
Connect	Couple		Bond, build, mate, phosphorylate
Control Magnitude	Regulate		Gate, electrophoresis, respire

Computational Concept Generation Technique

Automated concept generation methods promise engineers a faster realization of potential design solutions based upon previously known products and implementations. The technique described here requires the designer to input desired functionality and based on an algorithm several concept variants are presented to the designer. Functionality is a useful metric for defining a conceptual idea, as functional representation has been shown to reduce fixation of how a product or device would look and operate [6, 7].

The proposed computational concept generation technique requires the input of a functional model that abstractly represents a conceptual engineered solution. The functional model is then digitized and represented as a matrix of forward flows. Each function/flow pair of the model is then searched in the OSU Design Repository and the chosen biological corpus to identify solutions. The search algorithm parses the repository entries for the exact engineering function/flow pair, where as, the biological corpus is parsed repeatedly with the biological terms corresponding to the engineering terms per the engineering-to-biology thesaurus. Multiple solutions from both domains, to each function/flow pair, are returned and presented to the designer. The biological solutions are not indented for physical use, but are intended for spurring creative ideas, connections and designs that could be implemented in an engineered system. To make a biologically-inspired concept work, a leap is required from the designer to understand that component mapping is an activity that relates biological system attributes to engineering components. The computational technique proposed here assists with identifying biological solutions to engineering functions; however, to arrive at the final concept, the designer is required to identify principles within the engineering domain that support what the biological solution suggests.

Engineers have struggled with utilizing the vast amount of biological information available from the natural world around them. Often it is because there is a knowledge gap or terminology is difficult, and the time needed to learn and understand the biology is not feasible. Therefore, a computational technique that can identify and present valuable biological knowledge, indexed by engineering terms, to an engineering designer during concept generation would significantly increase the likelihood of biologically-inspired designs. The computational concept generation technique proposed here will promote biologically-inspired, engineering designs that partially (i.e., one or two components) to completely (i.e., entire design) mimic a biological system. This technique therefore lends itself more toward innovative design problems where novel solutions tend to dominate.

Computational concept generation provides the added advantage of limitless resources for inspiration. The technique described in this section can be extended by adding entries into the Design Repository and texts or documents into the biological corpus database. With this technique the well-known customer needs driven engineering design approach is utilized, which is an additional advantage.

Algorithm

Our proposed technique utilizes the Functional Basis, Design Repository, MEMIC, organized search tool and engineering-to-biology thesaurus to create, filter and inspire concept variants. The algorithm combines the research efforts that developed MEMIC and the organized search tool, but also adds recursive biological text search functionality using the engineering-to-biology thesaurus. There are two threads in this algorithm that execute simultaneously: (1) parse the Design Repository to find engineering solutions, and (2) parse chosen biological texts and documents with thesaurus terms to find biological solutions for engineering inspiration.

Thread 1: Parse Design Repository for Engineering Solutions

The proposed concept generation technique utilizes function-component relationships established through an FCM to compute a set of engineering components that solve the function/flow pairs of the input functional model. Next, the resultant set is filtered using component-component knowledge through a DSM. Each match is stored for display to the user. The resultant engineering components found in the repository and are compatible are displayed to the user as a list of potential solutions that have previously solved that function/flow pair.

Thread 2: Parse Biological Text for Biological Solutions Using Thesaurus Terms

First, the algorithm swaps the engineering function and flow terms for corresponding biological function and flow terms. The biological corpus is then searched for the biological function and all sentences containing the function are extracted for further processing. The algorithm then searches those sentences for any of the corresponding biological flow terms. Each match is stored for display to the user. When multiple biological function terms are present, the search is executed recursively until all corresponding biological functions have been searched. The resultant biological information is displayed to the designer as individual sentences containing the desired function/flow pairs, which are indicators of potential solutions from the biological domain.

Concept Generation Example

To illustrate the proposed computational concept generation technique, a smart flooring example is presented. The computational tool is utilized to search engineered and biological systems for solutions that can be implemented in the example product. When using the Functional Basis for product design there are a few basic steps needed before function-based concept generation can begin. First, one must define the customer needs and convert them into engineering terms [6, 7, 44]. Second, one must develop either a conceptual functional model of the desired new product or a functional model of an existing product to be redesigned using the Functional Basis lexicon. Examples can be found in [41, 42, 45]. With the functional model, the designer now has several function-flow pairs that represent the desired new product. These pairs are utilized by the computational concept generation technique to gather engineering and biological inspiration.

Smart Flooring

Consider the following scenario. A customer wants to create a security/surveillance product that looks like ordinary carpet, mats, rugs, etc. to detect intruders, a presence or movement. Requirements for the smart flooring include detection mechanism unseen by human eye, durability, and composed of common materials. Given these customer needs, the conceptual functional model in Figure 2 is created.

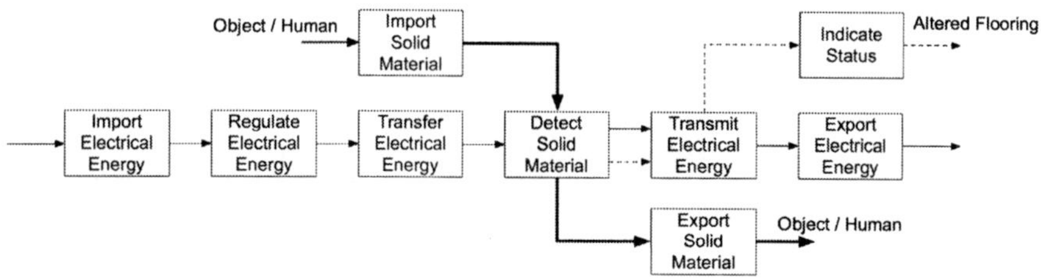

Fig. 2. Smart flooring conceptual functional model

The model of Figure 2 was created with a boundary of the flooring in place, energy is being supplied to the detection mechanism and when an object or human interacts with the flooring a signal is generated. Importation of human/object symbolizes interaction and exportation symbolizes the human/object exiting the boundary, which do not require solutions. In order to use the computational concept generation technique, an adjacency matrix, Table 2, representing the forward flows of the model in Figure 2 is created. The resulting engineering and biological solutions are shown in Table 3.

The computational concept generation technique resulted in a total of 31 engineering solutions and 60 biological text excerpts (in the form of individual sentences) for the seven function/flow pairs of Figure 2.

Table 2 Smart flooring adjacency matrix

	import solid material	import electrical energy	regulate electrical energy	transfer electrical energy	detect solid material	transmit electrical energy	indicate status signal	export electrical energy	export solid material
import solid material	0	0	0	0	**1**	0	0	0	0
import electrical energy	0	0	**1**	0	0	0	0	0	0
regulate electrical energy	0	0	0	**1**	0	0	0	0	0
transfer electrical energy	0	0	0	0	**1**	0	0	0	0
detect solid material	0	0	0	0	0	**1**	0	0	**1**
transmit electrical energy	0	0	0	0	0	0	**1**	**1**	0
indicate status signal	0	0	0	0	0	0	0	0	0
export electrical energy	0	0	0	0	0	0	0	0	0
export solid material	0	0	0	0	0	0	0	0	0

Table 3 (continued over next few pages) Smart flooring concept generation results – truncated, not all results presented

Function/ Flow	Engineering Solution	Biological Solution
Import/ electrical energy	Battery, circuit board, electric motor, electric wire, electric switch	1. The light energy absorbed by the antenna system is transferred from one pigment molecule to another as an electron. 2. To keep noncyclic electron flow going, both photosystems I and II must constantly be absorbing light, thereby boosting electrons to higher orbitals from which they may be captured by specific oxidizing agents. 3. Photosystem II absorbs photons, sending electrons from P680 to pheophytin-I-the first carrier in the redox chain-and causing P680 to become oxidized to P680+. 4. Electrons from the oxidation of water are passed to P680+, reducing it once again to P680, which can absorb more photons. 5. In sum, noncyclic electron flow uses a molecule of water, four photons (two each absorbed by photosystems I and II), one molecule each of NADP+ and ADP, and one Pi. 6. The attractive force that an atom exerts on electrons is its electronegativity. 7. Na+ ions would diffuse into the cell because of their higher concentration on the outside, and they would also be attracted into the cell by the negative membrane potential. 8. These positive charges electrostatically attract the negative phosphate groups on DNA. 9. Because opposite charges attract, the DNA moves toward the positive end of the field. 10. Leptin appears to be one important feedback signal in the regulation of food intake.
Regulate/ electrical energy	Actuation lever, capacitor, circuit board, automobile distributor, electric switch, heating element, transistor, transformer, thermostat, regulator, volume knob	1. Changes in the gated channels may perturb the resting potential. 2. An opposite change in the resting potential would occur if gated Cl- channels opened. 3. The inactivation gate remains closed for 1-2 milliseconds before it spontaneously opens again, thus explaining why the membrane has a refractory period (a period during which it cannot act) before it can fire another action potential. 4. When the inactivation gate finally opens, the activation gate is closed, and the membrane is poised to respond once again to a depolarizing stimulus by firing another action potential. 5. The binding of neurotransmitter to receptors at the motor end plate and the resultant opening of chemically gated ion channels perturb the resting potential of the postsynaptic membrane. 6. The structure of many plants is maintained by the pressure potential of their cells; if the pressure potential is lost, a plant wilts. 7. Negatively charged chloride ions and organic ions also move out with the potassium ions, maintaining electrical balance and contributing to the change in the solute potential of the guard cells. 8. This unloading serves two purposes: It helps maintain the gradient of solute potential and hence of pressure potential in the sieve tubes, and it promotes the buildup of sugars and starch to high concentrations in storage regions, such as developing fruits and seeds.
Transfer/ electrical energy	Battery, circuit board, electric wire, electric motor,	1. We have just noted proteins that function in blood clotting; others of interest include albumin, which is partly responsible for the osmotic potential in capillaries that prevents a massive loss of water from plasma to intercellular spaces; antibodies (the immunoglobulins); hormones; and various carrier molecules, such as transferrin, which carries iron from the gut to where it is stored or used. 2. Since the electronegativities of these elements are so different, any

Function/ Flow	Engineering Solution	Biological Solution
	electric socket, electric plate, electric switch, heating element, usb cable, light fixture, speaker	electrons involved in bonding will tend to be much nearer to the chlorine nucleus-so near, in fact, that there is a complete transfer of the electron from one element to the other. 3. Redox reactions transfer electrons and energy. 4. Another way of transferring energy is to transfer electrons. 5. A reaction in which one substance transfers one or more electrons to another substance is called an oxidation-reduction reaction, or redox reaction. 6. Thus, when a molecule loses hydrogen atoms, it becomes oxidized: Oxidation and reduction always occur together: As one material is oxidized, the electrons it loses are transferred to another material, reducing that material. 7. As we shall see, another carrier, FAD (flavin adenine dinucleotide), is also involved in transferring electrons during the metabolism of glucose. 8. The citric acid cycle is a cyclic series of reactions in which the acetate becomes completely oxidized, forming CO2 and transferring electrons (along with their hydrogen nuclei) to carrier molecules. 9. The transfer of electrons along the respiratory chain drives the active transport of hydrogen ions (protons) from the mitochondrial matrix into the space between the inner and outer mitochondrial membrane. 10. The light energy absorbed by the antenna system is transferred from one pigment molecule to another as an electron. 11. The high energy stored in the electrons of excited chlorophyll can be transferred to suitably oxidized nonpigment acceptor molecules.
Transmit/ electrical energy	Electrical wire, battery contacts, motor controller	1. These electrical changes generate action potentials, the language by which the nervous system processes and communicates information. 2. Ganglion cells communicate information about the light and dark contrasts that fall on different regions of their receptive fields. 3. Whether or not the sensory cell itself fires action potentials, ultimately the stimulus is transduced into action potentials and the intensity of the stimulus is encoded by the frequency of action potentials. 4. In the rest of this chapter we will learn how sensory systems gather and filter stimuli, transduce specific stimuli into action potentials, and transmit action potentials to the CNS. 5. Auditory systems use mechanoreceptors to transduce pressure waves into action potentials. 6. Earlier in this chapter, we saw how crayfish stretch receptors transduce physical force into action potentials. 7. Sitting on the basilar membrane is the organ of Corti, the apparatus that transduces pressure waves into action potentials in the auditory nerve, which in turn conveys information from the ear to the brain.
Detect/ solid material	Read head, line guide	1. Since both AT and GC pairs obey the base-pairing rules, how does the repair mechanism "know" whether the AC pair should be repaired by removing the C and replace it with T, for instance, or by removing the A and replacing it with G? The repair mechanism can detect the "wrong" base because a newly synthesized DNA strand is chemically modified some time after replication. 2. The cnidarian's nerve net merely detects food or danger and causes its tentacles and body to extend or retract. 3. Most sensory cells possess a membrane receptor protein that detects the stimulus and responds by altering the flow of ions across the plasma membrane. 4. The mammalian inner ear has two equilibrium organs that use hair cells to detect the position of the body with respect to gravity: semicircular canals and the vestibular apparatus. 5. These sensory cells enable the fish to detect weak electric fields, which

Function/ Flow	Engineering Solution	Biological Solution
		can help them locate prey. 6. This change is detected by the carotid and aortic stretch receptors, which stimulate corrective responses within two heartbeats. 7. Any objects in the environment, such as rocks, plants, or other fish, disrupt the electric fish's electric field, and the electroreceptors of the lateral line detect those disruptions. 8. Bats use echolocation, pit vipers sense infrared radiation from the warm bodies of their prey, and certain fishes detect electric fields created in the water by their prey. 9. In addition to genes for antibiotic resistance, several other marker genes are used to detect recombinant DNA in host cells. 10. Length of the night is one of several environmental cues detected by plants, or by individual parts such as leaves. 11. Animals whose eyes are on the sides of their heads have nonoverlapping fields of vision and, as a result, poor depth vision, but they can see predators creeping up from behind. 12. How does the sensory cell signal the intensity of a smell? It responds in a graded fashion to the concentration of odorant molecules: The more odorant molecules that bind to receptors, the more action potentials are generated and the greater the intensity of the perceived smell.
Indicate/ status signal	Light, tube, displacement gauge, LCD screen	1. The durability of pheromonal signals enables them to be used to mark trails, as ants do, or to indicate directionality, as in the case of the moth sex attractant. 2. To cause behavioral or physiological responses, a nervous system communicates these signals to effectors, such as muscles and glands. 3. The information from the signal that was originally at the plasma membrane is communicated to the nucleus. 4. A change in body color is a response that some animals use to camouflage themselves in a particular environment or to communicate with other animals. 5. The binding of a hormone to its cellular receptor protein, which causes the protein to change shape and provides the signal to initiate reactions within the cell. 6. Separation of the chromatids marks the beginning of anaphase, the phase of mitosis during which the two sister chromatids of each chromosome-now called daughter chromosomes, each containing one double-stranded DNA molecule-move to opposite ends of the spindle.

Discussion

Importation and exportation of solid material did return engineering results; however, in regards to the voluntary human/object that interacts with the flooring the results are out of context. Therefore, the engineering solutions are ignored and not presented here. Considering the flow of *electrical energy*, and the respective functions, the Design Repository returned several engineering component solutions. Solutions of electrical wire, electrical switch, circuit board and battery are commonplace, but offer a wide range of functionality. Although a biological solution may not have practical uses for *importing, regulating, transferring, transmitting* and *exporting electrical energy*, the concept generator results are interesting and informative. Essentially, what an engineer can gain from these biological results is that natural systems do utilize electrical energy for communication, sensing, regulating metabolism, photosynthesis and many other processes.

Results that are more intriguing are the biological phenomena relevant to the function of *detect*. Of the twelve returned solutions, hair cells, electroreceptors, echolocation, carotiod and aortic stretch receptors, membrane receptor proteins, graded action potentials and DNA offer the greatest potential to inspire a detection mechanism. The hair cell operates like a cantilever and would detect a presence when disturbed, such as being stepped upon. Electroreception and echolocation could be used like radar and even detect the presence of an object when it is just above the flooring. Detecting a certain material could adapt the idea of carotiod and aortic stretch receptors, by monitoring deformation in a material. Membrane receptor proteins and graded action potentials alter the flow of ions, thus indicating a difference in the environment. Chemical modifications within the flooring material, as suggested by DNA, could also signal the presence of a solid object. Perhaps the natural phenomena that most readily allow analogy discovery and inspiration are the hair cells and carotiod and aortic stretch receptors. These two solutions offer natural tactile responses that could be exploited to achieve the customer requirements.

Functional results for *indicating* a *status signal* are a mixed set of analog and digital methods. Engineered solutions suggest lights, LCD screen, tube, or an analog gauge, and the biological solutions suggest a change in color or shape, swelling, molecule production and expelling of pheromones. Again, the biological results may be more informative than useful; however, in the event that the resultant engineered solutions are not useful the designer has the opportunity to be inspired by the biological solutions.

Through the incorporation of the engineering-to-biology thesaurus in this computational technique, a 33% increase of relevant biological solutions for the function of *detect* resulted when compared to prior organized search tool results in [46]. It should also be noted that the 21 non-relevant results in [46] were ignored with this technique. It can be concluded that the computational concept generation technique was successful at extracting specific biological systems that perform the functions/flows of the conceptual functional model.

Conclusions

Concept generation and synthesis is perhaps the most exciting, important, and challenging step of engineering design. The research proposed in this paper makes fundamental contributions to engineering design through the creation of a computational concept generation technique that identifies biological solutions based on engineering function. The key contributions of this research extend beyond computational design practices and biological information retrieval. The tool proposed here represents a first step toward enabling widespread biologically-inspired concept generation, and subsequently, biologically-inspired design. Also, this research will enable engineers knowledgeable of customer needs driven design activities, but a limited biological background, to begin biomimetic design activities. Mimicking nature offers more than just the observable aspects that conjure up engineering solutions performing similar functions, but also less obvious strategic and sustainable aspects. It is these less obvious aspects that this research aims to facilitate as they hold the greatest potential for impact.

The computational concept generation technique assists with developing biomimetic designs by presenting the designer with short descriptions of biological systems that perform an engineering function of interest. From these descriptions the designer can make connections, similar to the process of Synectics [47], that link the biological solutions to engineering solutions, principles, components and materials. A key part of the connection making process is considering the biological systems from several viewpoints. Multiple viewpoints can spur novel and innovative ideas [47]. Integrating a strategic search method for indexing non-engineering information with an established computational concept generation method affords a computational foundation for accessing stored engineering information and, in this case, biological solutions for use with design activities. By placing the focus on function during concept generation, rather than form or component, the computational technique presented here has shown to successfully extract relevant biological solutions. Example results from the proposed technique were demonstrated with a smart flooring example.

Biological system solutions corresponding to the function/flow pairs of Figure 2 for designing a security or surveillance device were discussed. The connections between the biological solutions and engineered systems for *detect* were analyzed. Connection making points to the biological solutions of hair cells, electroreceptors, membrane receptor proteins, carotiod and aortic stretch receptors, echolocation, DNA, and graded action potentials as inspiration options to the function of *detect* for the smart flooring example. The other functions yielded informative results of how natural systems utilize electrical energy. The proposed computational concept generation algorithm provides targeted results and prompts designers to make connections, which result in creative solutions.

The biological domain provides many opportunities for identifying connections between what is found in the natural world and engineered systems. It is important to understand that the computational concept generation technique does not generate complete concepts; that is the task of the designer. However, the proposed technique does provide a systematic method for discovering biological inspiration based on function, so that it may be easier for the designer to make the necessary connections leading to biologically-inspired designs.

Future work for the proposed computational concept generation technique involves implementation of the algorithm and testing of the code. Further work would include the addition of hyperlinks to detailed biological information and the integration of images into the results of the computational concept generation technique presented in this paper. Visuals can stimulate designers in a different manner than text alone. Another avenue for this research is engineering education. The computational concept generation technique could be used to assist engineering students with discovering the connections between the biology and engineering domains.

Acknowledgements

This material is based in part upon work supported by the National Science Foundation under Grant CMMI-0800596.

References

1. Bar-Cohen, Y.: Biomimetics Biologically Inspired Technologies. CRC/Taylor & Francis, Boca Raton, FL (2006)
2. Brebbia, C.A., Sucharov, L.J., Pascolo, P.: Design and nature: Comparing design in nature with science and engineering. WIT, Southampton (2002)
3. Brebbia, C.A., Collins, M.W.: Design and nature II: Comparing design in nature with science and engineering. WIT, Southampton (2004)
4. Brebbia, C.A.: Design and nature III: Comparing design in nature with science and engineering. WIT, Southampton (2006)
5. Dym, C.L., Little, P.: Engineering design: A project-based introduction. John Wiley, New York (2004)
6. Otto, K.N., Wood, K.L.: Product Design: Techniques in Reverse Engineering and New Product Development. Prentice-Hall, Upper Saddle River (2001)

7. Pahl, G., Beitz, W.: Engineering Design: A Systematic Approach, 2nd edn. Springer, London (1984)
8. Ullman, D.G.: The Mechanical Design Process, 4th edn. McGraw-Hill, Inc., New York (2009)
9. Ulrich, K.T., Eppinger, S.D.: Product design and development. McGraw-Hill/Irwin, Boston (2004)
10. Voland, G.: Engineering By Design, 2nd edn. Pearson Prentice Hall, Upper Saddle River (2004)
11. Cross, N.: Engineering design methods: Strategies for product design. John Wiley & Sons, Chichester (2008)
12. Hyman, B.: Engineering design. Prentice-Hall, New Jersey (1998)
13. Gordon, W.J.J.: Synectics, the development of creative capacity. Harper, New York (1961)
14. Bohm, M., Vucovich, J., Stone, R.: Using a Design Repository to Drive Concept Generation. Journal of Computer and Information Science in Engineering 8(1), 14502-1-8 (2008)
15. Bryant Arnold, C.R., Stone, R.B., McAdams, D.A.: MEMIC: An Interactive Morphological Matrix Tool for Automated Concept Generation. Industrial Engineering Research Conference (2008)
16. Bryant, C., Bohm, M., McAdams, D., et al.: An Interactive Morphological Matrix Computational Design Tool: A Hybrid of Two Methods. In: ASME 2007 IDETC/CIE, Las Vegas, NV (2007)
17. Kurtoglu, T., Swantner, A., Campbell, M.I.: Automating the Conceptual Design Process: From Black-box to Component Selection. In: DCC 2008, Atlanta, Georgia, USA. Springer Science + Business Media B.V, Heidelberg (2008)
18. Hong-Zhong, H., Bo, R., Chen, W.: An integrated computational intelligence approach to product concept generation and evaluation. Mechanism and Machine Theory 41(5), 567–583 (2006)
19. Zu, Y., Xiao, R., Zhang, X.: Automated conceptual design of mechanisms using enumeration and functional reasoning. International Journal of Materials and Product Technology 34(3), 273–294 (2009)
20. Jin, Y., Li, W.: Design Concept Generation: A Hierarchical Coevolutionary Approach. Journal of Mech. Design 129(10), 1012–1022 (2007)
21. Design Engineering Lab (2009), `http://www.designengineeringlab.com` (Last accessed 2009)
22. Nagel, J.K.S., Stone, R.B., McAdams, D.A.: An Engineering-to-Biology Thesaurus for Engineering Design. In: 2010 ASME IDETC/CIE, Montreal, Quebec, Canada (2010)
23. Hirtz, J., Stone, R., McAdams, D., et al.: A Functional Basis for Engineering Design: Reconciling and Evolving Previous Efforts. Research in Engineering Design 13(2), 65–82 (2002)
24. Helms, M., Vattam, S.S., Goel, A.K.: Biologically Inspired Design: Products and Processes. Design Studies 30(5), 606–622 (2009)

25. Vincent, J.F.V., Bogatyreva, O.A., Bogatyrev, N.R., et al.: Biomimetics: its practice and theory. Journal of the Royal Society Interface 3, 471–482 (2006)
26. Nagel, R., Tinsley, A., Midha, P., et al.: Exploring the use of functional models in biomimetic design. Journal of Mech. Design 130(12), 11–23 (2008)
27. Wen, H.-I., Zhang, S.-J., Hapeshi, K., et al.: An Innovative Methodology of Product Design from Nature. Journal of Bionic Engineering 5(1), 75–84 (2008)
28. Linsey, J., Wood, K., Markman, A.: Modality and Representation in Analogy. AIEDAM 22(2), 85–100 (2008)
29. Mak, T.W., Shu, L.H.: Using descriptions of biological phenomena for idea generation. Research in Engineering Design 19(1), 21–28 (2008)
30. Bar-Cohen, Y.: Biomimetics - Using nature to inspire human innovation. Journal of Bioinspiration and Biomimetics 1, 1–12 (2006)
31. Lindemann, U., Gramann, J.: Engineering Design Using Biological Principles. In: International Design Conference - DESIGN 2004, Dubrovnik (2004)
32. Chakrabarti, A., Sarkar, P., Leelavathamma, B., et al.: A functional representation for aiding biomimetic and artificial inspiration of new ideas. AIEDAM 19, 113–132 (2005)
33. Srinivasan, V., Chakrabarti, A.: SAPPhIRE – An Approach to Analysis and Synthesis. In: International Conference on Engineering Design - ICED 2009, Stanford, USA (2009)
34. Wilson, J., Chang, P., Yim, S., et al.: Developing a Bio-inspired Design Repository Using Ontologies. In: 2009 ASME IDETC/CIE, California, USA (2009)
35. Chiu, I., Shu, L.H.: Using language as related stimuli for concept generation. AIEDAM 21(2), 103–121 (2007)
36. Chiu, I., Shu, L.H.: Biomimetic design through natural language analysis to facilitate cross-domain information retrieval. AIEDAM 21(1), 45–59 (2007)
37. Shu, L.H., Hansen, H.N., Gegeckaite, A., et al.: Case Study in Biomimetic Design: Handling and Assembly of Microparts. In: ASME 2006 IDETC/CIE, Philadelphia, PA (2006)
38. Wood, W.H., Yang, M.C., Cutkosky, M.R., et al.: Design Information Retrieval: Improving access to the informal side of design. In: ASME 1998 IDETC/CIE, Atlanta, GA (1998)
39. Bouchard, C., Omhover, J.-F., Mougenot, C., et al.: TRENDS: A Content-Based Information Retrieval System for Designers. In: DCC 2008, Atlanta, Georgia, USA. Springer Science + Business Media BV, Heidelberg (2008)
40. Cheong, H., Shu, L.H., Stone, R.B., et al.: Translating terms of the functional basis into biologically meaningful words. In: 2008 ASME IDETC/CIE, New York City, NY (2008)
41. Stone, R., Wood, K.: Development of a Functional Basis for Design. Journal of Mechanical Design 122(4), 359–370 (2000)

42. Bryant, C., Stone, R., McAdams, D., et al.: Concept Generation from the Functional Basis of Design. In: International Conference on Engineering Design, Melbourne, Australia (2005)
43. Lopez-Huertas, M.J.: Thesarus Structure Design: A Conceptual Approach for Improved Interaction. Journal of Documentation 53(2), 139–177 (1997)
44. Kurfman, M., Stone, R., Rajan, J., et al.: Experimental Studies Assessing the Repeatability of a Functional Modeling Derivation Method. Journal of Mechanical Design 125(4), 682–693 (2003)
45. Nagel, R.L., Stone, R., McAdams, D.: A Theory for the Development of Conceptual Functional Models for Automation of Manual Processes. In: 2007 ASME IDETC/CIE, Las Vegas, NV, USA (2007)
46. Stroble, J.K., Stone, R.B., McAdams, D.A., et al.: Automated Retrieval of Non-Engineering Domain Solutions to Engineering Problems. In: CIRP Design Conference 2009, Cranfield, Bedfordshire, UK (2009)
47. Prince, G.M.: The Practice of Creativity. Collier Books, New York (1970)

First Author Email Address

Alexiou, Katerina	k.alexiou@open.ac.uk
Alink, Thomas	alink@ipek.uni-karlsruhe.de
Bandini, Stefania	bandini@disco.unimib.it
Bitterman, Michael	M.S.Bittermann@tudelft.nl
Brown, Dave	dcb@cs.wpi.edu
Cascini, Gaetano	gaetano.cascini@polimi.it
Chen, Yong	mechenyong@sjtu.edu.cn
Conrardy, Céline	celine.conrardy@insa-strasbourg.fr
Dabbeeru, Madan	mmadan@iitk.ac.in
Elsen, Catherine	catherine.elsen@ulg.ac.be
Gadwal, Apeksha	apeksha.gadwal@gmail.com
Grabska, Ewa	uigrabsk@cyf-kr.edu.pl
Gyllenback, Katarina	katarina@sics.se
Hanna, Sean	s.hanna@ucl.ac.uk
Harding John	johnharding@fastmail.fm
Hasegawa, Hiroshi	h-hase@sic.shibaura-it.ac.jp
Heylighan, Ann	ann.heylighen@asro.kuleuven.be
Hoisi, Frank	frank.hoisl@pe.mw.tum.de
Indraprastha, Aswin	aswinindra@gmail.com
Jaafar, Rosidah	mnrj@leeds.ac.uk
Kan, Jeff	kan.jeff@gmail.com
Kannegngiesser, Udo	Udo.Kannengiesser@nicta.com.au
Koltsova, Anastasia	koltsova@arch.ethz.ch
Kumar, Mukund	kmukund87@gmail.com
Maher, Mary Lou	marylou.maher@gmail.com
Matthews, Peter	p.c.matthews@durham.ac.uk
Nagel, Jacquelyn	jkstroble@designengineeringlab.org
Nordin, Axel	nordinaxel@gmail.com
Oren, Michael	moren@iastate.edu
Paterso, Gareth	gareth.paterson@virgin.net
Piasecki, Michal	m.piasecki@ucl.ac.uk

Radhakrishnan, Pradeep	rkprad@yahoo.com
Schneider, Sven	schneider@caad.ar.tum.de
Sicilia, Alvaro	asicilia@salle.url.edu
Taura, Toshiharu	taura@kobe-u.ac.jp
Winkelman, Constance	winkelmann@psychologie.tu-dresden.de
Yilmaz, Seda	seda@umich.edu

Author Index

Acconci, Vito 171
Albers, Albert 405
Alexiou, Katerina 489
Alink, Thomas 405

Bandini, Stefania 171
Bittermann, Michael S. 505
Bjärnemo, Robert 701
Boman, Magnus 525
Bonomi, Andrea 171
Brown, David C. 157

Campbell, Matthew I. 465, 663
Cascini, Gaetano 77
Chau, Hau Hing 269
Chen, Lin 681
Chen, Yong 425
Conrardy, Céline 189

Dabbeeru, Madan Mohan 445
Daly, Shanna R. 3
Darses, Françoise 55
de Guio, Roland 189
de Pennington, Alan 269
Derix, Christian 305

Earl, Chris 251
Eckert, Claudia 405
Eckhardt, Claus Christian 701
Elsen, Catherine 55

Fantoni, Gualtiero 77
Fasiha, Mohd Yusof Nor 559
Fischer, Jan-Ruben 367
Frate, Luca Del 77

Gadwal, Apeksha 209
Gero, John S. 135, 621
Gilbert, Sam 489
Gilbert, Stephen B. 601
González, Mar 387
Gonzalez, Richard 3, 35
Grabska, Ewa 97
Gyllenbäck, Katarina Borg 525

Hacker, Winfried 545
Hanna, Sean 115, 347
Harding, John 305
Hasegawa, Hiroshi 229
Heylighen, Ann 23
Hoisl, Frank 643
Hopf, Andreas 701

Indraprastha, Aswin 285

Jaafar, Rosidah 269

Kan, Jeff W.T. 621
Kannengiesser, Udo 135
Koltsova, Anastasia 681

König, Reinhard 367
Kumar, Mukund 465

Leclercq, Pierre 55
Linsey, Julie 209
Liu, Ze-Lin 425

Madrazo, Leandro 387
Maher, Mary Lou 581
Matthews, Peter C. 327
McKay, Alison 269
Montagna, Francesca 77
Motte, Damien 701
Mukerjee, Amitabha 445
Murty, Paul 581

Nagai, Yukari 559
Nagel, Jacquelyn K.S. 721
Narang, Shipra 681
Nordin, Axel 701

Oren, Michael A. 601

Paterson, Gareth 251
Paulini, Mercedes 581
Piasecki, Michal 347

Radhakrishnan, Pradeep 663
Ruckpaul, Anne 405

Sato, Yusuke 229
Schmitt, Gerhard 681
Schneider, Sven 367
Schumacher, Patrik 681
Seifert, Colleen M. 3, 35
Shea, Kristina 643
Shinozaki, Michihiko 285
Sicilia, Álvaro 387
Ślusarczyk, Grażyna 97
Sonoda, Yuki 229
Stone, Robert B. 721
Sudo, Tomoyuki 681

Tang, Hsien-Hui 621
Taura, Toshiharu 559
Tsukamoto, Mika 229

Vizzari, Giuseppe 171

Winkelmann, Constance 545

Xie, You-Bai 425

Yamamoto, Eiko 559
Yilmaz, Seda 3, 35

Zamenopoulos, Theodore 489
Zuber, Bruno 189

Lightning Source UK Ltd.
Milton Keynes UK

174125UK00001B/1/P